Mechanics of Fluids

Third Edition, SI

SI Edition prepared by

K. K. Chaudhry

Institute of Technology & Management, Gurgaon
(Retd from Indian Institute of Technology, Delhi)

CENGAGE
Learning™

Australia • Brazil • Japan • Korea • Mexico • Singapore • Spain • United Kingdom • United States

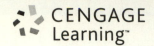

CENGAGE Learning™

Mechanics of Fluids, Third Edition, SI
Merle C. Potter and David C. Wiggert, with Miki Hondzo and Tom I-P. Shih
SI Edition prepared by K.K. Chaudhry

Director, Global Engineering Program: Chris Carson

Senior Developmental Editor: Hilda Gowans

Development Editor: Swati Meherishi

Editorial Assistant: Nancy Saundercook

Associate Marketing Manager: Lauren Betsos

Media Editor: Chris Valentine

Content Project Manager: Diane Bowdler

Typesetting and Proofreading Services, SI Edition:
   AnVi Composers

Production Service: RPK Editorial Services

Compositor: Better Graphics, Inc.

Compositor, SI Edition: Integra

Senior Art Director: Michelle Kunkler

Interior Design: Carmela Pereira

Interior Illustration: Rolin Graphics

Cover Designer: Andrew Adams

Cover Image: Dreamstime/Timurpix

Text and Image Permissions Researcher: Kristiina Paul

Senior First Print Buyer: Doug Wilke

Library of Congress Control Number: 2009929186

ISBN-13: 978-0-495-43857-1
ISBN-10: 0-495-43857-X

Cengage Learning
200 First Stamford Place, Suite 400
Stamford, CT 06902
USA

Cengage Learning is a leading provider of customized learning solutions with office locations around the globe, including Singapore, the United Kingdom, Australia, Mexico, Brazil, and Japan. Locate your local office at:
international.cengage.com/region.

Cengage Learning products are represented in Canada by Nelson Education Ltd.

For your course and learning solutions, visit **www.cengage.com/engineering.**

Purchase any of our products at your local college store or at our preferred online store **www.ichapters.com.**

Printed in the United States of America
1 2 3 4 5 6 7 13 12 11 10 09

To
our mentors:
C. S. Yih (deceased)
and
V. L. Streeter

Contents

Preface to the SI Edition

This edition of **Mechanics of Fluids** has been adapted to incorporate the International System of Units (*Le Système International d'Unités* or SI) throughout the book.

Le Système International d'Unités

The United States Customary System (USCS) of units uses FPS (foot/pound/second) units (also called English or Imperial units). SI units are primarily the units of the MKS (meter/kilogram/second) system. However, CGS (centimeter/gram/second) units are often accepted as SI units, especially in textbooks.

Using SI Units in this Book

In this book, we have used both MKS and CGS units. USCS units or FPS units used in the US Edition of the book have been converted to SI units throughout the text and problems. However, in case of data sourced from handbooks, government standards, and product manuals, it is not only extremely difficult to convert all values to SI, it also encroaches upon the intellectual property of the source. Also, some quantities such as the ASTM grain size number and Jominy distances are generally computed in FPS units and would lose their relevance if converted to SI. Some data in figures, tables, examples, and references, therefore, remains in FPS units. For readers unfamiliar with the relationship between the FPS and the SI systems, conversion tables have been provided inside the front and back covers of the book.

To solve problems that require the use of sourced data, the sourced values can be converted from FPS units to SI units just before they are to be used in a calculation. To obtain standardized quantities and manufacturers' data in SI units, the readers may contact the appropriate government agencies or authorities in their countries/regions.

Instructor Resources

A Printed Instructor's Solution Manual in SI units is available on request. An electronic version of the Instructor's Solutions Manual, and PowerPoint slides of the figures from the SI text are available through www.cengage.com/engineering.

The readers' feedback on this SI Edition will be highly appreciated and will help us improve subsequent editions.

The Publishers

Preface

The motivation to write a book is difficult to describe. Most often the authors suggest that the other texts on the subject have certain deficiencies that they will correct, such as an accurate description of entrance flows and flows around blunt objects, the difference between a one-dimensional flow and a uniform flow, the proper presentation of the control volume derivation, or a definition of laminar flow that makes sense. New authors, of course, introduce other deficiencies that future authors hope to correct! And life goes on. This is another fluids book that has been written in hopes of presenting an improved view of fluid mechanics so that the undergraduate can understand the physical concepts and follow the mathematics. This is not an easy task: Fluid mechanics is a subject that contains many difficult-to-understand phenomena. For example, how would you explain the hole scooped out in the snow by the wind on the upwind side of a tree during a snowstorm? Or the high concentration of smog contained in the Los Angeles area (it doesn't exist to the same level in New York)? Or the unexpected strong wind around the corner of a tall building in Chicago? Or the vibration and subsequent collapse of a large concrete-steel bridge due to the wind? We have attempted to present fluid mechanics so that the student can understand and analyze many of the important phenomena encountered by the engineer.

The mathematical level of this book is based on previous mathematics courses required in all engineering curricula. We use solutions to differential equations and vector algebra. Some use is made of vector calculus with the use of the gradient operator, but this is kept to a minimum since it tends to obscure the physics involved.

Most popular texts in fluid mechanics have not presented fluid flows as fields. That is, they have presented primarily those flows that can be approximated as one-dimensional flows and have treated other flows using experimental data. We must recognize that when a fluid flows around an object, such as a building or an abutment, its velocity possesses all three components which depend on all three space variables and often, time. If we present the equations that describe such a general flow, the equations are referred to as field equations and velocity and pressure fields become of interest. This is quite analogous to electrical and magnetic fields in electrical engineering. In order for the difficult problems of the future, such as large-scale environmental pollution, to be analyzed by engineers, it is imperative that we understand fluid fields. Thus in Chapter 5 we introduce the field equations and discuss several solutions for some relatively simple geometries. The more conventional manner of treating the flows individually is provided as an alternate route for those who wish this more standard approach. The field equations can then be included in a subsequent course.

Perhaps a listing of the additions made in this third edition would be of interest. We have:

- added a new chapter on computational fluid dynamics.
- presented all the examples and problems using SI units.
- added multiple-choice problems similar to those found on the Fundamentals of Engineering Exam.
- replaced the BASIC computer program for gradually varied flow in Chapter 10 with Microsoft® Excel spreadsheet solutions.
- replaced the BASIC computer code for pipe network analysis in Chapter 11 with EPANET output.
- added Mathcad® and MATLAB® solutions where applicable.
- improved many of the figures.
- highlighted important information in the margins.
- added chapter outlines and summaries.
- included a Nomenclature list after the Preface.
- added additional photos.
- improved the text material following suggestions by several reviewers.
- added additional explanatory material to the examples.
- added the names of scientists and engineers who have contributed to this subject.

The introductory material included in Chapters 1 through 9 has been selected carefully to introduce students to all fundamental areas of fluid mechanics. Not all of the material in each chapter need be covered in an introductory course. The instructor can fit the material to a selected course outline. Some sections at the end of each chapter may be omitted without loss of continuity in later chapters. In fact, Chapter 5 can be omitted in its entirety if it is decided to exclude field equations in the introductory course, a relatively common decision. That chapter can then be included in an intermediate fluid mechanics course. After the introductory material has been presented, there is sufficient material to present in one or two additional courses. This additional course or courses could include material that had been omitted in the introductory course and combinations of material from the more specialized Chapters 9 through 15. Much of the material is of interest to all engineers, although several of the chapters are of interest to only particular disciplines.

We have included examples worked out in detail to illustrate each important concept presented in the text material. Numerous home problems, many having multiple parts for better homework assignments, then provide the student with ample opportunity to gain experience solving problems of various levels of difficulty. Answers to selected home problems are presented just prior to the Index.

We have also included design-type problems in several of the chapters. After studying the material, reviewing the examples, and working several of the home problems, students should gain the needed capability to work many of the problems encountered in actual engineering situations. Of course, there are numerous classes of problems that are extremely difficult to solve, even for an experienced engineer. To solve these more difficult problems, the engineer must gain considerably more information than is included in this introductory text. There are, however, many problems that can be solved successfully using the material and concepts presented herein.

Many students take the FE/EIT exam at the end of their senior year, the first step in becoming a professional engineer in the USA. The problems in the FE/EIT exam are all four-part, multiple choice. Consequently, we have included this type of problem in the appropriate chapters and they are noted by use of the exam icon 📧 . The book is written emphasizing SI units. Multiple-choice problem have been presented using SI units since the FE/EIT exam uses SI units exclusively. Additional information on the FE/EIT exam can be obtained from a website at NCEES.com.

The authors are very much indebted to both their former professors and to their present colleagues. Professors C. S. Yih and V. L. Streeter of the University of Michigan demanded that each of us learn this subject well! Chapter 10 was written with inspiration from F. M. Henderson's book titled *Open Channel Flow* (1996), and D. Wood of the University of Kentucky encouraged us to incorporate comprehensive material on pipe network analysis in Chapter 11. Several illustrations in Chapter 11 relating to the water hammer phenomenon were provided by C. S. Martin of the Georgia Institute of Technology. R. D. Thorley provided some of the problems at the end of Chapter 12. Miki Hondzo of the University of Minnesota wrote Chapter 14 on Environmental Fluid Mechanics, and Tom Shih of Michigan State University wrote Chapter 15 on Computational Fluid Dynamics. Thanks to Richard Prevost for writing the MATLAB® solutions, and to Lori Hasse for typing the solutions manual. We would also like to thank our reviewers: Mohamed Alawady, Louisiana State University; John R. Biddle, California Polytechnical Institute—Pomona; Saeed Moaveni, Minnesota State University, Mankato; Julia Muccino, Arizona State University; Emmanuel U. Nzewi, North Carolina A & T State University; and Yiannis Ventikos, Swiss Federal Institute of Technology.

Merle C. Potter
David C. Wiggert

Nomenclature

A - area, constant

A_2, A_3 - profile type

a - acceleration, speed of a pressure wave

a - acceleration vector

a_x, a_y, a_z - acceleration components

B - constant, bulk modulus of elasticity, free surface width

b - channel bottom width

C - centroid, Chezy coefficient, Hazen-Williams coefficient, constant for curve fit, molar concentration

$\overline{C}$ - time average concentration

C - concentration fluctuation

C_1, C_3 - profile type

C_D - drag coefficient

C_d - discharge coefficient

C_f - skin friction coefficient

C_H - head coefficient

C_i - molar concentration

C_L - lift coefficient

C_P - pressure recovery factor, pressure coefficient

C_{NPSH} - net positive suction head coefficient

C_Q - flow rate coefficient

C_V - velocity coefficient

$C_{\dot{W}}$ - power coefficient

c - specific heat, speed of sound, chord length, celerity

c_1, c_2 - constants

c_f - local skin friction coefficient

c_p - constant pressure specific heat

c'_s - deviation from average concentration

c_v - constant volume specific heat

c.s. - control surface

c.v. - control volume

D - diameter, mass diffusion coefficient

D_t - turbulent diffusion coefficient

$\dfrac{D}{Dt}$ - substantial derivative

$\dfrac{d}{dx}$ - ordinary derivative

d - diameter

dx - differential distance

$d\theta$ - differential angle

E - energy, specific energy, modulus of elasticity, logitudinal heat dispersion coefficient

E_c - critical energy

EGL - energy grade line

Eu - Euler number

e - the exponential, specific energy, wall roughness height, pipe wall thickness

exp - the exponential e

F - force vector

F - force
F_B - buoyant force
F_H - horizontal force component
F_V - vertical force component
F_W - body force equal to the weight
f - friction factor, frequency
G - center of gravity
$\overline{GM}$ - metacentric height
$\mathbf{g}$ - gravity vector
g - gravity
H - enthalpy, height, total energy
H_2, H_3 - profile type
H_D - design head
H_P - pump head
H_T - turbine head
HGL - hydraulic grade line
h - distance, height, specific enthalpy
h_j - head loss across a hydraulic jump
I - second moment of an area
$\overline{I}$ - second moment about the centroidal axis
I_{xy} - product of inertia
$\hat{\mathbf{i}}$ - unit vector in the x-direction
$\mathbf{J}$ - mass flux vector
J - mass flux
$\hat{\mathbf{j}}$ - unit vector in the y-direction
$\hat{\mathbf{k}}$ - unit vector in the z-direction
K - thermal conductivity, flow coefficient, dispersion coefficient
K_c - contraction coefficient
K_e - expansion coefficient
K_y - tranverse dispersion coefficient
K_{uv} - correlation coefficient
k - ratio of specific heats
k_s - settling rate coefficient
L - length
L_E - entrance length
L_e - equivalent length
L_t - transverse length scale
L_x - distance between effluent injection and equilibrium zone
ℓ - length
ℓ_m - mixing length
M - molar mass, Mach number, momentum function
$\dot{M}$ - total mass flux of dye
M - Mach number
M_1, M_2, M_3 - profile type
m - mass, side-wall slope, constant for curve fit
$\dot{m}$ - mass flux
$\dot{m}_r$ - relative mass flux
m_a - added mass
m_1, m_2 - side-wall slopes
$\dot{mom}$ - momentum flux
N - general extensive property, an integer, number of jets, stability frequency
$NPSH$ - net positive suction head
n - normal direction number of moles, power-law exponent, Manning number
$\hat{\mathbf{n}}$ - unit normal vector
P - power, force, wetted perimeter
p - pressure
Q - flow rate (discharge), heat transfer
Q_D - design discharge
$\dot{Q}$ - rate of heat transfer
q - source strength, specific discharge, heat flux
R - radius, gas constant, hydraulic radius, resistance coefficient, radius of curvature
$\overline{R}_i$ - modified resistant coefficient
Re - Reynolds number

Re_{crit} - critical Reynolds number
R_u - universal gas constant
R_x, R_y - force components
r - radius, coordinate variable, rate of generation
$\bar{r}$ - time average rate of generation
$\mathbf{r}$ - position vector
S - specific gravity, entropy, distance, slope of chennel, slope of EGL, thermal energy source
S_1, S_2, S_3 - profile type
S_c - critical slope
St - Strouhal number
$\mathbf{S}$ - position vector
S_0 - slope of channel bottom
s - specific entropy, streamline coordinate
$\hat{s}$ - unit vector tangent to streamline
sys - system
T - temperature, torque, tension
t - time, tangential direction
t_{ad} - advection transport time
t_{diff} - diffusion time scale
U - average velocity
U_∞ - free-stream velocity away from a body
u - x-component velocity, circumferential blade speed
u' - velocity perturbation
u'_s - deviation from average velocity
$\tilde{u}$ - specific internal energy
$\bar{u}$ - time average velocity
u_τ - shear velocity
V - velocity
V_c - critical velocity
V_{ss} - steady-state velocity
$\mathbf{V}$ - velocity vector
$\overline{V}$ - spatial average velocity
Ψ - volume
V_B - blade velocity
V_n - normal component of velocity
V_r - relative speed
V_t - tangential velocity
v - velocity, y-component velocity
v' - velocity perturbation
$v_r, v_z, v_\theta, v_\phi$ - velocity components
W - work, weight, change in hydraulic grade line
$\dot{W}$ - work rate (power)
$\dot{W}_f$ - actual power
W_e - Weber number
$\dot{W}_S$ - shaft work (power)
w - z-component velocity, velocity of a hydraulic bore
X_T - distance where transition begins
x - coordinate variable
x_m - origin of moving reference frame
$\tilde{x}$ - distance relative to a moving reference frame
$\bar{x}$ - x-coordinate of centroid
Y - upstream water height above tope of wier
y - coordinate variable, flow energy head
y_p - distance to center of pressure
$\bar{y}$ - y-coordinate of centroid
y_c - critical depth
z - coordinate variable

α - angle, angle of attack, lapse rate, thermal diffusivity, kinetic-energy correction factor, blade angle
α_x - an empirical constant
β - angle, momentum correction factor, fixed jet angle, blade angle, diameter ratio
Δ - a small increment
∇ - gradient operator
∇^2 - Laplacian

δ - boundary layer thickness

$\delta(x)$ - Dirac-delta function

δ_d - displacement thickness

δ_ν - viscous wall layer thickness

δR - overall uncertainty interval

δK - precision

ε - a small volume

$\varepsilon_{xx}, \varepsilon_{xy}, \varepsilon_{xz}$ - rate-of-strain components

ϕ - angle, coordinate variable, velocity potential function, speed factor

Γ - circulation, vortex strength

γ - specific weight

η - a general intensive property, eddy viscosity, efficiency, a position variable

η_P - pump efficiency

η_T - turbine efficiency

λ - mean free path, a constant, wave length

μ - viscosity, doublet strength

ν - kinematic viscosity

π - a pi term

θ - angle, momentum thickness, laser beam angle

ρ - density

Ω - angular velocity

Ω_P - specific speed of a pump

Ω_T - specific speed of a turbine

$\boldsymbol{\Omega}$ - angular velocity vector

σ - surface tension, cavitation number, circumferential stress, standard deviation

σ^2 - variance

σ_t^2 - temporal variance

σ_x^2 - spatial variance

$\sigma_{xx}, \sigma_{yy}, \sigma_{zz}$ - normal stress components

τ - stress vector

$\bar{\tau}$ - time average stress

τ_{xy}, τ_{xz} - shear stress components

ω - angular velocity, vorticity

$\boldsymbol{\omega}$ - vorticity vector

ψ - stream function

$\dfrac{\partial}{\partial x}$ - partial derivative

Mechanics of Fluids

Left: Contemporary windmills are used to generate electricity at many locations in the United States. They are located in areas where consistent prevailing winds exist. (IRC/Shutterstock) *Top right*: Hurricane Bonnie, Atlantic Ocean about 800 km from Bermuda. At this stage in its development, the storm has a well-developed center, or "eye," where air currents are relatively calm. Vortex-like motion occurs away from the eye. (U.S. National Aeronautics and Space Administration) *Bottom right*: The Space Shuttle Discovery leaves the Kennedy Space Center on October 29, 1998. In 6 seconds the vehicle cleared the launch tower with a speed of 160 km/h, and in about two minutes it was 250 km down range from the Space Center, 47 km above the ocean with a speed of 6150 km/h. The wings and rudder on the tail are necessary for successful re-entry as it enters the earth's atmosphere upon completion of its mission. (U.S. National Aeronautics and Space Administration)

1

Basic Considerations

Outline

Chapter Objectives

The objectives of this chapter are to:
- ▲ Introduce many of the quantities encountered in fluid mechanics including their dimensions and units.
- ▲ Identify the liquids to be considered in this text.
- ▲ Introduce the fluid properties of interest.
- ▲ Present the thermodynamic laws and associated quantities.

1.1 INTRODUCTION

A proper understanding of the mechanics of fluids is extremely important in many areas of engineering. In biomechanics the flow of blood and cerebral fluid are of particular interest; in meteorology and ocean engineering an understanding of the motions of air movements and ocean currents requires a knowledge of the mechanics of fluids; chemical engineers must understand fluid mechanics to design the many different kinds of chemical-processing equipment; aeronautical engineers use their knowledge of fluids to maximize lift and minimize drag on aircraft and to design fan-jet engines; mechanical engineers design pumps, turbines, internal combustion engines, air compressors, air-conditioning equipment, pollution-control equipment, and power plants using a proper understanding of fluid mechanics; and civil engineers must also utilize the results obtained from a study of the mechanics of fluids to understand the transport of river sediment and erosion, the pollution of the air and water, and to design piping systems, sewage treatment plants, irrigation channels, flood control systems, dams, and domed athletic stadiums.

It is not possible to present the mechanics of fluids in such a way that all of the foregoing subjects can be treated specifically; it is possible, however, to present the fundamentals of the mechanics of fluids so that engineers are able to understand the role that the fluid plays in a particular application. This role may involve the proper sizing of a pump (the horsepower and flow rate) or the calculation of a force acting on a structure.

> **KEY CONCEPT** *We will present the fundamentals of fluids so that engineers are able to understand the role that fluid plays in particular applications.*

In this book we present the general equations, both integral and differential, that result from the conservation of mass principle, Newton's second law, and the first law of thermodynamics. From these a number of particular situations will be considered that are of special interest. After studying this book the engineer should be able to apply the basic principles of the mechanics of fluids to new and different situations.

In this chapter topics are presented that are directly or indirectly relevant to all subsequent chapters. We include a macroscopic description of fluids, fluid properties, physical laws dominating fluid mechanics, and a summary of units and dimensions of important physical quantities. Before we can discuss quantities of interest, we must present the units and dimensions that will be used in our study of fluid mechanics.

1.2 DIMENSIONS, UNITS, AND PHYSICAL QUANTITIES

Before we begin the more detailed studies of the mechanics of fluids, let us discuss the dimensions and units that will be used throughout the book. Physical quantities require quantitative descriptions when solving an engineering problem. Density is one such physical quantity. It is a measure of the mass contained in a unit volume. Density does not, however, represent a fundamental dimension. There are nine quantities that are considered to be fundamental dimensions: length, mass, time, temperature, amount of a substance, electric current, luminous intensity, plane angle, and solid angle. The dimensions of all other quantities can be expressed in terms of the fundamental dimensions. For example, the quantity "force" can be related to the fundamental dimensions of mass, length, and time. To do this, we use Newton's second law, named after Sir Isaac Newton (1642–1727), expressed in simplified form in one direction as

$$F = ma \qquad (1.2.1)$$

Using brackets to denote "the dimension of," this is written dimensionally as

$$[F] = [m][a]$$
$$F = \frac{ML}{T^2} \qquad (1.2.2)$$

where F, M, L, and T are the dimensions of force mass, length, and time, respectively. If force had been selected as a fundamental dimension rather than mass, a common alternative, mass would have dimensions of

$$[m] = \frac{[F]}{[a]}$$
$$M = \frac{FT^2}{L} \qquad (1.2.3)$$

where F is the dimension[1] of force.

There are also systems of dimensions in which both mass and force are selected as fundamental dimensions. In such systems conversion factors, such as a gravitational constant, are required; we do not consider these types of systems in this book, so they will not be discussed.

To give the dimensions of a quantity a numerical value, a set of units must be selected. Different systems of units are prevalent in different parts of the world. However the International System or Système International (SI) is preferred internationally. Hence, we have used SI units in this book.

KEY CONCEPT *SI units are preferred and are used internationally.*

The fundamental dimensions and their units are presented in Table 1.1; some derived units appropriate to fluid mechanics are given in Table 1.2. Other units that are acceptable are the hectare (ha), which is 10 000 m², used for large areas; the metric ton (t), which is 1000 kg, used for large masses; and the liter (L), which is 0.001 m³. Also, density is occasionally expressed as grams per liter (g/L).

In chemical calculations the mole is often a more convenient unit than the kilogram. In some cases it is also useful in fluid mechanics. For gases the kilogram-mole (kg-mol) is the quantity that fills the same volume as 32 kilograms of oxygen at the same temperature and pressure. The mass (in kilograms) of a gas filling that volume is equal to the molecular weight of the gas; for example, the mass of 1 kg-mol of nitrogen is 28 kilograms.

---

[1]Unfortunately, the quantity force F and the dimension of force $[F]$ use the same symbol.

TABLE 1.1 Fundamental Dimensions and Their Units

Quantity	Dimensions	SI units	
Length ℓ	L	meter	m
Mass m	M	kilogram	kg
Time t	T	second	s
Electric current i		ampere	A
Temperature T	Θ	kelvin	K
Amount of substance	M	kg-mole	kg-mol
Luminous intensity		candela	cd
Plane angle		radian	rad
Solid angle		steradian	sr

When expressing a quantity with a numerical value and a unit, prefixes have been defined so that the numerical value may be between 0.1 and 1000. These prefixes are presented in Table 1.3. Using scientific notation, however, we use powers of 10 rather than prefixes (e.g., 2×10^6 N rather than 2 MN). If larger numbers are written the comma is not used; twenty thousand would be written as 20 000 with a space and no comma.

TABLE 1.2 Derived Units

Quantity	Dimensions	SI units
Area A	L^2	m^2
Volume V	L^3	m^3
		L (liter)
Velocity V	L/T	m/s
Acceleration a	L/T^2	m/s^2
Angular velocity ω	T^{-1}	s^{-1}
Force F	ML/T^2	kg $\cdot$ m/s^2
		N (newton)
Density ρ	M/L^3	kg/m^3
Specific weight γ	M/L^2T^2	N/m^3
Frequency f	T^{-1}	s^{-1}
Pressure p	M/LT^2	N/m^2
		Pa (pascal)
Stress τ	M/LT^2	N/m^2
		Pa (pascal)
Surface tension σ	M/T^2	N/m
Work W	ML^2/T^2	N$\cdot$m
		J (joule)
Energy E	ML^2/T^2	N$\cdot$m
		J $\cdot$ (joule)
Heat rate $\dot{Q}$	ML^2/T^3	J/s
Torque T	ML^2/T^2	N $\cdot$ m
Power P	ML^2/T^3	J/s
		W (watt)
Viscosity μ	M/LT	N $\cdot$ s/m^2
Mass flux $\dot{m}$	M/T	kg/s
Flow rate Q	L^3/T	m^3/s
Specific heat c	$L^2/T^2\Theta$	J/kg $\cdot$ K
Conductivity K	$ML/T^3\Theta$	W/m $\cdot$ K

TABLE 1.3   SI Prefixes

Multiplication factor	Prefix	Symbol
10^{12}	tera	T
10^{9}	giga	G
10^{6}	mega	M
10^{3}	kilo	k
10^{-2}	centi[a]	c
10^{-3}	milli	m
10^{-6}	micro	μ
10^{-9}	nano	n
10^{-12}	pico	p

[a]Permissible if used alone as cm, cm^2, or cm^3.

Newton's second law relates a net force acting on a rigid body to its mass and acceleration. This is expressed as

$$\Sigma \mathbf{F} = m\mathbf{a} \tag{1.2.4}$$

Consequently, the force needed to accelerate a mass of 1 kilogram at 1 meter per second squared in the direction of the net force is 1 newton. This allows us to relate the units by

$$N = kg \cdot m/s^2 \tag{1.2.5}$$

KEY CONCEPT *The relationship* $N = kg \cdot m/s^2$ *is often used in the conversion of units.*

which are included in Table 1.2. These relationships between units are often used in the conversion of units. In the SI system, weight is always expressed in newtons, never in kilograms. To relate weight to mass, we use

$$W = mg \tag{1.2.6}$$

where g is the local gravity. The standard value for gravity is 9.80665 m/s^2 and it varies from a minimum of 9.77 m/s^2 at the top of Mt. Everest to a maximum of 9.83 m/s^2 in the deepest ocean trench. A nominal value of 9.81 m/s^2 will be used unless otherwise stated.

Finally, a note on significant figures. In engineering calculations we often do not have confidence in a calculation beyond three significant digits since the information given in the problem statement is often not known to more than three significant digits; in fact, viscosity and other fluid properties may not be known to even three significant digits. The diameter of a pipe may be stated as 2 cm; this would, in general, not be as precise as 2.000 cm would imply. If information used in the solution of a problem is known to only two significant digits, it is incorrect to express a result to more than two significant digits. In the examples and problems we will assume that all information given is known to three significant digits, and the results will be expressed accordingly. If the numeral 1 begins a number, it is not counted in the number of significant digits, i.e., the number 1.210 has three significant digits.

Example 1.1

A mass of 100 kg is acted on by a 400-N force acting vertically upward and a 600-N force acting upward at a 45° angle. Calculate the vertical component of the acceleration. The local acceleration of gravity is 9.81 m/s².

Solution

The first step in solving a problem involving forces is to draw a free-body diagram with all forces acting on it, as shown in Fig. E1.1.

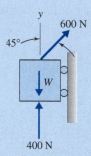

FIGURE E1.1

Next, apply Newton's second law (Eq. 1.2.4). It relates the net force acting on a mass to the acceleration and is expressed as

$$\Sigma F_y = ma_y$$

Using the appropriate components in the y-direction, we have

$$400 + 600 \sin 45° - 100 \times 9.81 = 100a_y$$
$$a_y = -1.567 \text{ m/s}^2$$

Note: We have used only three significant digits in the answer since the information given in the problem is assumed known to three significant digits. (The number 1.567 has three significant digits. A leading "1" is not counted as a significant digit.)

1.3 CONTINUUM VIEW OF GASES AND LIQUIDS

Substances referred to as fluids may be **liquids** or **gases**. In our study of the mechanics of fluids we restrict the liquids that are studied. Before we state the restriction, we must define a shearing stress. A force ΔF that acts on an area ΔA can be decomposed into a normal component ΔF_n and a tangential component ΔF_t, as shown in Fig. 1.1. The force divided by the area upon which it acts is called a *stress*. The force vector divided by the area is a **stress vector**, the normal component of force divided by the area is a **normal stress**, and the tangential force divided by the area is a **shear stress**. In this discussion we are interested in the shear stress τ. Mathematically, it is defined as

$$\tau = \lim_{\Delta A \to 0} \frac{\Delta F_t}{\Delta A} \qquad (1.3.1)$$

Our restricted family of fluids may now be identified; the fluids considered in this book are *those liquids and gases that move under the action of a shear stress, no matter how small that shear stress may be*. This means that even a very small shear stress results in motion in the fluid. Gases obviously fall within this category of fluids, as do water and tar. Some substances, such as plastics and catsup, may resist small shear stresses without moving; a study of these substances is included in the subject of *rheology* and is not included in this book.

It is worthwhile to consider the microscopic behavior of fluids in more detail. Consider the molecules of a gas in a container. These molecules are not stationary but move about in space with very high velocities. They collide with each other and strike the walls of the container in which they are confined, giving rise to the pressure exerted by the gas. If the volume of the container is increased while the temperature is maintained constant, the number of molecules impacting on a given area is decreased and as a result the pressure decreases. If the temperature of a gas in a given volume increases (i.e., the velocities of the molecules increase), the pressure increases due to increased molecular activity.

Molecular forces in liquids are relatively high, as can be inferred from the following example. The pressure necessary to compress 20 grams of water vapor at 20°C into 20 cm³, assuming that no molecular forces exist, can be shown by the ideal gas law to be approximately 1340 times the atmospheric pressure. Of course, this pressure is not required because 20 g of water occupies 20 cm³. It follows that the cohesive forces in the liquid phase must be very large.

Despite the high molecular attractive forces in a liquid, some of the molecules at the surface escape into the space above. If the liquid is contained, an equilibrium is established between outgoing and incoming molecules. The presence of molecules above the liquid surface leads to a so-called vapor pressure. This pressure increases with temperature. For water at 20°C this pressure is 0.02 times the atmospheric pressure.

Liquid: *A state of matter in which the molecules are relatively free to change their positions with respect to each other but restricted by cohesive forces so as to maintain a relatively fixed volume.[2]*

Gas: *A state of matter in which the molecules are practically unrestricted by cohesive forces. A gas has neither definite shape nor volume.*

Stress vector: *The force vector divided by the area.*

Normal stress: *The normal component of force divided by the area.*

Shear stress: *The tangential force divided by the area.*

> **KEY CONCEPT**   *Fluids considered in this text are those that move under the action of a shear stress.*

FIGURE 1.1   Normal and tangential components of a force.

[2]*Handbook of Chemistry and Physics*, 40th ed. CRC Press, Boca Raton, Fla.

Continuum: *Continuous distribution of a liquid or gas throughout a region of interest.*

In our study of the mechanics of fluids it is convenient to assume that both gases and liquids are continuously distributed throughout a region of interest, that is, the fluid is treated as a **continuum**. The primary property used to determine if the continuum assumption is appropriate is the *density* ρ, defined by

$$\rho = \lim_{\Delta v \to 0} \frac{\Delta m}{\Delta V} \tag{1.3.2}$$

Standard atmospheric conditions: *A pressure of 101.3 kPa and temperature of 15°C.*

where Δm is the incremental mass contained in the incremental volume ΔV. The density for air at **standard atmospheric conditions**, that is, at a pressure of 101.3 kPa and a temperature of 15°C, is 1.23 kg/m^3. For water, the nominal value of density is 1000 kg/m^3.

Physically, we cannot let $\Delta V \to 0$ since, as ΔV gets extremely small, the mass contained in ΔV would vary discontinuously depending on the number of molecules in ΔV; this is shown graphically in Fig. 1.2. Actually, the zero in the definition of density should be replaced by some small volume ε, below which the continuum assumption fails. For most engineering applications, the small volume ε shown in Fig. 1.2 is extremely small. For example, there are 2.7×10^{16} molecules contained in a cubic millimeter of air at standard conditions; hence, ε is much smaller than a cubic millimeter. An appropriate way to determine if the continuum model is acceptable is to compare a characteristic length l (e.g., the diameter of a rocket) of the device or object of interest with the **mean free path** λ, the average distance a molecule travels before it collides with another molecule; if $l \gg \lambda$, the continuum model is acceptable. The mean free path is derived in molecular theory. It is

KEY CONCEPT *To determine if the continuum model is acceptable, compare a length l with the mean free path.*

Mean free path: *The average distance a molecule travels before it collides with another molecule.*

$$\lambda = 0.225 \frac{m}{\rho d^2} \tag{1.3.3}$$

where m is the mass (kg) of a molecule, ρ the density (kg/m^3) and d the diameter (m) of a molecule. For air $m = 4.8 \times 10^{-26}$ kg and $d = 3.7 \times 10^{-10}$ m. At standard atmospheric conditions the mean free path is approximately 6×10^{-6} cm,

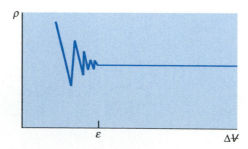

FIGURE 1.2 Density at a point in a continuum.

at an elevation of 100 km it is 10 cm, and at 160 km it is 5000 cm. Obviously, at higher elevations the continuum assumption is not acceptable and the theory of rarefied gas dynamics (or free molecular flow) must be utilized. Satellites are able to orbit the earth if the primary dimension of the satellite is of the same order of magnitude as the mean free path.

With the continuum assumption, fluid properties can be assumed to apply uniformly at all points in the region at any particular instant in time. For example, the density ρ can be defined at all points in the fluid; it may vary from point to point and from instant to instant; that is, in Cartesian coordinates ρ is a continuous function of x, y, z, and t, written as $\rho(x, y, z, t)$.

1.4 PRESSURE AND TEMPERATURE SCALES

In fluid mechanics pressure results from a normal compressive force acting on an area. The *pressure p* is defined as (see Fig. 1.3)

$$p = \lim_{\Delta A \to 0} \frac{\Delta F_n}{\Delta A} \qquad (1.4.1)$$

where ΔF_n is the incremental normal compressive force acting on the incremental area ΔA. The metric units to be used on pressure are newtons per square meter (N/m^2) or pascal (Pa). Since the pascal is a very small unit of pressure, it is more conventional to express pressure in units of kilopascal (kPa). For example, standard atmospheric pressure at sea level is 101.3 kPa.

Both pressure and temperature are physical quantities that can be measured using different scales. There exist absolute scales for pressure and temperature and there are scales that measure these quantities relative to selected reference points. In many thermodynamic relationships (see Section 1.7) absolute scales must be used for pressure and temperature. Figures 1.4 and 1.5 summarize the commonly used scales. Note that atmospheric pressure is often expressed as millimeters of mercury or meters of water, as shown in Fig. 1.4; such a column of fluid creates the pressure at the bottom of the column.

The **absolute pressure** reaches zero when an ideal vacuum is achieved, that is, when no molecules are left in a space; consequently, a negative absolute pressure is an impossibility. A second scale is defined by measuring pressures relative

KEY CONCEPT   *In many relationships, absolute scales must be used for pressure and temperature.*

Absolute pressure: *Scale measuring pressures, where zero is reached when an ideal vacuum is achieved.*

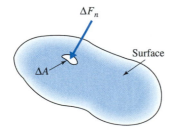

FIGURE 1.3   Definition of pressure.

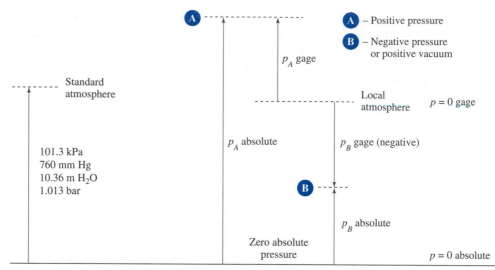

FIGURE 1.4   Gage pressure and absolute pressure.

Gage pressure: *A scale measuring pressures relative to the local atmospheric pressure.*

to the local atmospheric pressure. This pressure is called **gage pressure**. A conversion from gage pressure to absolute pressure can be carried out using

$$p_{\text{absolute}} = p_{\text{atmospheric}} + p_{\text{gage}} \qquad (1.4.2)$$

KEY CONCEPT
Whenever the absolute pressure is less than the atmospheric pressure, it may be called a vacuum.

Vacuum: *Whenever the absolute pressure is less than the atmospheric pressure.*

Note that the atmospheric pressure in Eq. 1.4.2 is the local atmospheric pressure, which may change with time, particularly when a weather "front" moves through. However, if the local atmospheric pressure is not given, we use the value given for a particular elevation, as given in Table B.3 of Appendix B, and assume zero elevation if the elevation is also unknown. The gage pressure is negative whenever the absolute pressure is less than atmospheric pressure; it may then be called a **vacuum**. In this book the word "absolute" will generally follow the pressure value if the pressure is given as an absolute pressure (e.g., $p = 50$ kPa absolute). If it were stated as $p = 50$ kPa, the pressure would be taken as a gage pressure, except that atmospheric pressure is always an absolute pressure. Most often in fluid mechanics gage pressure is used.

	°C	K	°F	°R
Steam point	100°	373	212°	672°
Ice point	0°	273	32°	492°
Zero absolute temperature				

FIGURE 1.5   Temperature scales.

Two temperature scales are commonly used, the Celsius (C) and Fahrenheit (F) scales. Both scales are based on the ice point and steam point of water at an atmospheric pressure of 101.3 kPa (14.7 psi). Figure 1.5 shows that the ice and steam point are 0 and 100°C on the Celsius scale and 32 and 212°F on the Fahrenheit scale. There are two corresponding absolute temperature scales. The absolute scale corresponding to the Celsius scale is the kelvin (K) scale. The relation between these scales is

$$K = °C + 273.15 \qquad (1.4.3)$$

The absolute scale corresponding to the Fahrenheit scale is the Rankine scale (°R). The relation between these scales is

$$°R = °F + 459.67 \qquad (1.4.4)$$

However, in this book, we will use only the celsius and the kelvin scales, in keeping with the International System of units. Note that in the SI system we do not write 100°K but simply 100 K, which is read "100 kelvins," similar to other units.

Reference will often be made to "standard atmospheric conditions" or "standard temperature and pressure." This refers to sea-level conditions at 40° latitude, which are taken to be 101.3 kPa for pressure and 15°C for temperature.

Example 1.2

A pressure gage attached to a rigid tank measures a vacuum of 42 kPa inside the tank shown in Fig. E1.2, which is situated at a site in Colorado where the elevation is 2000 m. Determine the absolute pressure inside the tank.

air   —42 kPa

FIGURE E1.2

Solution

To determine the absolute pressure, the atmospheric pressure must be known. If the elevation were not given, we would assume the standard atmospheric pressure of 101 kPa. However, with the elevation given, the atmospheric pressure is found from Table B.3 in Appendix B to be 79.5 kPa. Thus

$$p = -42 + 79.5 = 37.5 \text{ kPa absolute}$$

Note: A vacuum is always a negative gage pressure.

1.5 FLUID PROPERTIES

In this section we present several of the more common fluid properties. If density variation or heat transfer is significant, several additional properties, not presented here, become important.

1.5.1   Density and Specific Weight

Specific weight: *Weight per unit volume ($\gamma = \rho g$).*

Fluid density was defined in Eq. 1.3.2 as mass per unit volume. A fluid property directly related to density is the **specific weight** γ or weight per unit volume. It is defined by

$$\gamma = \rho g \tag{1.5.1}$$

where g is the local gravity. The units of specific weight are N/m^3. For water we use the nominal value of 9800 N/m^3.

Specific gravity: *The ratio of the density of a substance to that of water.*

The **specific gravity** S is often used to determine the specific weight or density of a fluid (usually a liquid). It is defined as the ratio of the density of a substance to that of water at a reference temperature of 4°C:

$$S = \frac{\rho}{\rho_{\text{water}}} = \frac{\gamma}{\gamma_{\text{water}}} \tag{1.5.2}$$

KEY CONCEPT *Specific gravity is often used to determine the density of a fluid.*

For example, the specific gravity of mercury is 13.6, a dimensionless number; that is, the mass of mercury is 13.6 times that of water for the same volume. The density, specific weight, and specific gravity of air and water at standard conditions are given in Table 1.4.

The density and specific weight of water do vary slightly with temperature; the approximate relationships are

$$\rho_{\text{H}_2\text{O}} = 1000 - \frac{(T-4)^2}{180}$$

$$\gamma_{\text{H}_2\text{O}} = 9800 - \frac{(T-4)^2}{18} \tag{1.5.3}$$

TABLE 1.4   Density, Specific Weight, and Specific Gravity of Air and Water at Standard Conditions

	Density ρ kg/m^3	Specific weight γ N/m^3	Specific gravity S
Air	1.23	12.1	0.00123
Water	1000	9810	1

For mercury the specific gravity relates to temperature by

$$S_{Hg} = 13.6 - 0.0024T \qquad (1.5.4)$$

Temperature in the three equations above is measured in degrees Celsius. For temperatures under 50°C, using the nominal values stated earlier for water and mercury, the error is less than 1%, certainly within engineering limits for most design problems. Note that the density of water at 0°C is less than that at 4°C; consequently, the lighter water at 0°C rises to the top of a lake where freezing occurs. For most other liquids the density at freezing is greater than the density just above freezing.

1.5.2   Viscosity

Viscosity can be thought of as the internal stickiness of a fluid. It is one of the properties that influences the power needed to move an airfoil through the atmosphere. It accounts for the energy losses associated with the transport of fluids in ducts, channels, and pipes. Further, viscosity plays a primary role in the generation of turbulence. Needless to say, viscosity is an extremely important fluid property in our study of fluid flows.

The rate of deformation of a fluid is directly linked to the viscosity of the fluid. For a given stress, a highly viscous fluid deforms at a slower rate than a fluid with a low viscosity. Consider the flow of Fig. 1.6 in which the fluid particles move in the x-direction at different speeds, so that particle velocities u vary with the y-coordinate. Two particle positions are shown at different times; observe how the particles move relative to one another. For such a simple flow field, in which $u = u(y)$, we can define the **viscosity** μ of the fluid by the relationship

$$\tau = \mu \frac{du}{dy} \qquad (1.5.5)$$

where τ is the shear stress of Eq. 1.3.1 and u is the velocity in the x-direction. The units of τ are N/m^2 or Pa, and of μ are N·s/m^2. The quantity du/dy is a velocity gradient and can be interpreted as a **strain rate**. Stress velocity-gradient relationships for more complicated flow situations are presented in Chapter 5.

KEY CONCEPT   *Viscosity plays a primary role in the generation of turbulence.*

Viscosity: *The internal stickiness of a fluid.*

Strain rate: *The rate at which a fluid element deforms.*

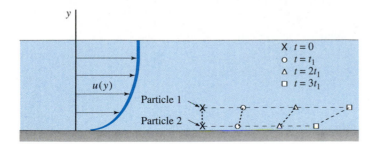

FIGURE 1.6   Relative movement of two fluid particles in the presence of shear stresses.

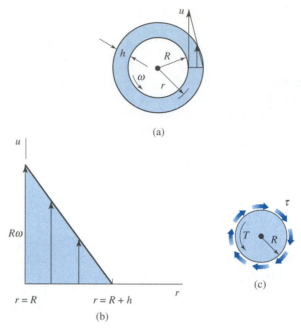

FIGURE 1.7   Fluid being sheared between cylinders with a small gap: (a) rotating inner cylinder; (b) velocity distribution; (c) the inner cylinder. The outer cylinder is fixed and the inner cylinder is rotating.

The concept of viscosity and velocity gradients can also be illustrated by considering a fluid within the small gap between two concentric cylinders, as shown in Fig. 1.7. A torque is necessary to rotate the inner cylinder at constant rotational speed ω while the outer cylinder remains stationary. This resistance to the rotation of the cylinder is due to viscosity. The only stress that exists to resist the applied torque for this simple flow is a shear stress, which is observed to depend directly on the velocity gradient; that is,

$$\tau = \mu \left| \frac{du}{dr} \right| \tag{1.5.6}$$

where du/dr is the velocity gradient and u is the tangential velocity component, which depends only on r. For a small gap ($h \ll R$), this gradient can be approximated by assuming a linear velocity distribution[3] in the gap. Thus

$$\left| \frac{du}{dr} \right| = \frac{\omega R}{h} \tag{1.5.7}$$

where h is the gap width. We can thus relate the applied torque T to the viscosity and other parameters by the equation

$$T = \text{stress} \times \text{area} \times \text{moment arm}$$
$$= \tau \times 2\pi RL \times R$$
$$= \mu \frac{\omega R}{h} \times 2\pi RL \times R = \frac{2\pi R^3 \omega L \mu}{h} \tag{1.5.8}$$

[3]If the gap is not small relative to R, the velocity distribution will not be linear (see Section 7.5). The distribution will also not be linear for relatively small values of ω.

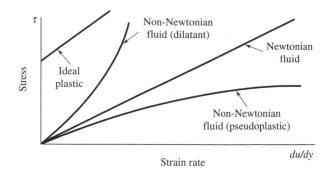

FIGURE 1.8   Newtonian and non-Newtonian fluids.

where we have neglected the shearing stress acting on the ends of the cylinder; L represents the length of the rotating cylinder. Note that the torque depends directly on the viscosity, thus the cylinders could be used as a *viscometer*, a device that measures the viscosity of a fluid.

If the shear stress of a fluid is directly proportional to the velocity gradient, as was assumed in Eqs. 1.5.5 and 1.5.6, the fluid is said to be a **Newtonian fluid**. Fortunately, many common fluids, such as air, water, and oil, are Newtonian. *Non-Newtonian fluids*, with shear stress versus strain rate relationships as shown in Fig. 1.8, often have a complex molecular composition.

Dilatants (quicksand, slurries) become more resistant to motion as the strain rate increases, and *pseudoplastics* become less resistant to motion with increased strain rate. *Ideal plastics* (or *Bingham fluids*) require a minimum shear stress to cause motion. Clay suspensions and toothpaste are examples that also require a minimum shear to cause motion, but they do not have a linear stress-strain rate relationship.

An important effect of viscosity is to cause the fluid to adhere to the surface; this is known as the **no-slip condition**. This was assumed in the example of Fig. 1.7. The velocity of the fluid at the rotating cylinder was taken to be ωR and the velocity of the fluid at the stationary cylinder was set equal to zero, as shown in Fig. 1.7b.

The viscosity is very dependent on temperature in liquids in which cohesive forces play a dominant role; note that the viscosity of liquids decreases with increased temperature, as shown in Fig. B.1 in Appendix B. The curves are often approximated by the equation

$$\mu = A e^{Bt} \tag{1.5.9}$$

known as *Andrade's equation*; the constants A and B would be determined from measured data. For a gas it is molecular collisions that provide the internal stresses, so that as the temperature increases, resulting in increased molecular activity, the viscosity increases. This can be observed in the bottom curve for a gas of Fig. B.1 in Appendix B. Note, however, that the percentage change of viscosity in a liquid is much greater than in a gas for the same temperature difference. Also, one can show that cohesive forces and molecular activity are quite insensitive to pressure, so that $\mu = \mu(T)$ only for both liquids and gases.

Newtonian fluid: *where the shear stress of a fluid is directly proportional to the velocity gradient.*

KEY CONCEPT   *Viscosity causes fluid to adhere to a surface.*

No-slip condition: *Condition where viscosity causes fluid to adhere to the surface.*

Since the viscosity is often divided by the density in the derivation of equations, it has become useful and customary to define *kinematic viscosity* to be

$$v = \frac{\mu}{\rho} \tag{1.5.10}$$

where the units of v are m^2/s. Note that for a gas, the kinematic viscosity will also depend on the pressure since the density is pressure sensitive.

Example 1.3

A viscometer is constructed with two 30-cm-long concentric cylinders, one 20.0 cm in diameter and the other 20.2 cm in diameter. A torque of 0.13 N·m is required to rotate the inner cylinder at 400 rpm (revolutions per minute). Calculate the viscosity.

Solution

The applied torque is just balanced by a resisting torque due to the shear stresses (see Fig. 1.7c). This is expressed by the small gap equation, Eq. 1.5.8.

The radius is $R = d/2 = 10$ cm; the gap $h = (d_2 - d_1)/2 = 0.1$ cm; the rotational speed, expressed as rad/s, is $\omega = 400 \times 2\pi/60 = 41.89$ rad/s.

Equation 1.5.8 provides:

$$\mu = \frac{Th}{2\pi R^3 \omega L}$$

$$= \frac{0.13(0.001)}{2\pi(0.1)^3(41.89)(0.3)} = 0.001646 \text{ N·s/m}^2$$

Note: All lengths are in meters so that the desired units on μ are obtained. The units can be checked by substitution:

$$[\mu] = \frac{\text{N·m·m}}{\text{m}^3(\text{rad/s})\text{m}} = \frac{\text{N·s}}{\text{m}^2}$$

1.5.3 Compressibility

In the preceding section we discussed the deformation of fluids that results from shear stresses. In this section we discuss the deformation that results from pressure changes. All fluids compress if the pressure increases, resulting in an increase in density. A common way to describe the compressibility of a fluid is by the following definition of the **bulk modulus of elasticity** B:

Bulk modulus of elasticity:
The ratio of change in pressure to relative change in density.

$$B = \lim_{\Delta V \to 0} \left[-\frac{\Delta p}{\Delta V/V} \right]_T = \lim_{\Delta \rho \to 0} \frac{\Delta p}{\Delta \rho/\rho} \Big|_T$$

$$= -V \frac{\partial p}{\partial V} \Big|_T = \rho \frac{\partial p}{\partial \rho} \Big|_T \tag{1.5.11}$$

In words, the bulk modulus, also called the *coefficient of compressibility*, is defined as the ratio of the change in pressure (Δp) to relative change in density ($\Delta \rho / \rho$) while the temperature remains constant. The bulk modulus obviously has the same units as pressure.

The bulk modulus for water at standard conditions is approximately 2100 MPa, or 21 000 times the atmospheric pressure. For air at standard conditions, B is equal to 1 atm. In general, B for a gas is equal to the pressure of the gas. To cause a 1% change in the density of water a pressure of 21 MPa (210 atm) is required. This is an extremely large pressure needed to cause such a small change; thus liquids are often assumed to be incompressible. For gases, if significant changes in density occur, say 4%, they should be considered as compressible; for small density changes under 3% they may also be treated as incompressible. This occurs for atmospheric airspeeds under about 100 m/s, which includes many airflows of engineering interest; air flow around automobiles, landing and take-off of aircraft, and air flow in and around buildings.

Small density changes in liquids can be very significant when large pressure changes are present. For example, they account for "water hammer," which can be heard shortly after the sudden closing of a valve in a pipeline; when the valve is closed an internal pressure wave propagates down the pipe, producing a hammering sound due to pipe motion when the wave reflects from the closed valve or pipe elbows.

The bulk modulus can also be used to calculate the speed of sound in a liquid; it is given by

$$c = \sqrt{\left.\frac{\Delta p}{\Delta \rho}\right|_T} = \sqrt{\frac{B}{\rho}} \qquad (1.5.12)$$

This yields approximately 1450 m/s for the speed of sound in water at standard conditions.

1.5.4 Surface Tension

Surface tension is a property that results from the attractive forces between molecules. As such, it manifests itself only in liquids at an interface, usually a liquid-gas interface. The forces between molecules in the bulk of a liquid are equal in all directions, and as a result, no net force is exerted on the molecules. However, at an interface the molecules exert a force that has a resultant in the interface layer. This force holds a drop of water suspended on a rod and limits the size of the drop that may be held. It also causes the small drops from a sprayer or atomizer to assume spherical shapes. It may also play a significant role when two immiscible liquids (e.g., oil and water) are in contact with each other.

Surface tension: A property resulting from the attractive forces between molecules.

Surface tension has units of force per unit length N/m. The force due to surface tension results from a length multiplied by the surface tension; the length to use is the length of fluid in contact with a solid, or the circumference in the case of a bubble. A surface tension effect can be illustrated by considering the free-body diagrams of half a droplet and half a bubble as shown in Fig. 1.9. The droplet has one surface and the bubble is composed of a thin film of liquid with

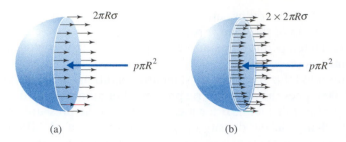

FIGURE 1.9 Internal forces in (a) a droplet and (b) a bubble.

surface and an outside surface. The pressure inside the droplet and bubble can now be calculated.

The pressure force $p\pi R^2$ in the droplet balances the surface tension force around the circumference. Hence

$$p\pi R^2 = 2\pi R\sigma$$

$$\therefore p = \frac{2\sigma}{R} \tag{1.5.13}$$

Similarly, the pressure force in the bubble is balanced by the surface tension forces on the two circumferences. Therefore,

$$p\pi R^2 = 2(2\pi R\sigma)$$

$$\therefore p = \frac{4\sigma}{R} \tag{1.5.14}$$

From Eqs. 1.5.13 and 1.5.14 we can conclude that the internal pressure in a bubble is twice as large as that in a droplet of the same size.

Figure 1.10 shows the rise of a liquid in a clean glass capillary tube due to surface tension. The liquid makes a contact angle β with the glass tube. Experiments have shown that this angle for water and most liquids in a clean glass tube is zero. There are also cases for which this angle is greater than 90° (e.g., mercury); such liquids have a capillary drop. If h is the capillary rise, D the diameter, ρ the density, and σ the surface tension, h can be determined from equating the vertical component of the surface tension force to the weight of the liquid column:

$$\sigma\pi D \cos\beta = \gamma\frac{\pi D^2}{4}h \tag{1.5.15}$$

or

$$h = \frac{4\sigma\cos\beta}{\gamma D} \tag{1.5.16}$$

Surface tension may influence engineering problems when, for example, laboratory modeling of waves is conducted at a scale that surface tension forces are of the same order of magnitude as gravitational forces.

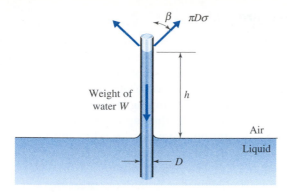

FIGURE 1.10   Rise in a capillary tube.

Example 1.4

A 2-mm-diameter clean glass tube is inserted in water at 15°C (Fig. E1.4). Determine the height that the water will climb up the tube. The water makes a contact angle of 0° with the clean glass.

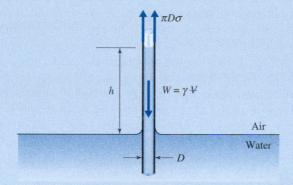

FIGURE E1.4

Solution

A free-body diagram of the water shows that the upward surface-tension force is equal and opposite to the weight. Writing the surface-tension force as surface tension times distance, we have

$$\sigma \pi D = \gamma \frac{\pi D^2}{4} h$$

or

$$h = \frac{4\sigma}{\gamma D} = \frac{4(0.0741)}{9800(0.002)} = 0.01512 \text{ m} \quad \text{or} \quad 15.12 \text{ mm}$$

The numerical values for σ and ρ were obtained from Table B.1 in Appendix B. Note that the nominal value used for the specific weight of water is $\gamma = \rho g = 9800$ N/m³. Here is would more accurately be $\gamma = 999.1 \times 9.81 = 9801$ N/m³, but 9800 is sufficiently accurate for engineering calculations.

FIGURE 1.11 Cooking food in boiling water takes a longer amount of time at a high altitude. It would take longer to boil these eggs in Denver than in New York City. (Courtesy Thomas Firak/FoodPix/Jupiterimages)

Vapor pressure: *The pressure resulting from molecules in a gaseous state.*

KEY CONCEPT *Cavitation can be very damaging.*

Boiling: *The point where vapor pressure is equal to the atmospheric pressure.*

Cavitation: *Bubbles form in a liquid when the local pressure falls below the vapor pressure of the liquid.*

1.5.5 Vapor Pressure

When a small quantity of liquid is placed in a closed container, a certain fraction of the liquid will vaporize. Vaporization will terminate when equilibrium is reached between the liquid and gaseous states of the substance in the container—in other words, when the number of molecules escaping from the water surface is equal to the number of incoming molecules. The pressure resulting from molecules in the gaseous state is the **vapor pressure**.

The vapor pressure is different from one liquid to another. For example, the vapor pressure of water at standard conditions (15°C, 101.3 kPa) is 1.70 kPa absolute and for ammonia it is 33.8 kPa absolute.

The vapor pressure is highly dependent on pressure and temperature; it increases significantly when the temperature increases. For example, the vapor pressure of water increases to 101.3 kPa if the temperature reaches 100°C. Water vapor pressures for other temperatures at atmospheric pressure are given in Appendix B.

It is, of course, no coincidence that the water vapor pressure at 100°C is equal to the standard atmospheric pressure. At that temperature the water is **boiling**; that is, the liquid state of the water can no longer be sustained because the attractive forces are not sufficient to contain the molecules in a liquid phase. In general, a transition from the liquid state to the gaseous state occurs if the local absolute pressure is less than the vapor pressure of the liquid. At high elevations where the atmospheric pressure is relatively low, resulting in boiling occurring at temperatures less than 100°C, see Fig. 1.11.

In liquid flows, conditions can be created that lead to a pressure below the vapor pressure of the liquid. When this happens, bubbles are formed locally. This phenomenon, called **cavitation**, can be very damaging when these bubbles are transported by the flow to higher-pressure regions. What happens is that the bubbles collapse upon entering the higher-pressure region, and this collapse produces local pressure spikes which have the potential of damaging a pipe wall or a ship's propeller. Cavitation on a propeller is shown in Fig. 1.12. Additional information on cavitation is included in Section 8.3.4.

Example 1.5

Calculate the vacuum necessary to cause cavitation in a water flow at a temperature of 80°C in Colorado where the elevation is 2500 m.

Solution
The vapor pressure of water at 80°C is given in Table B.1. It is 47.3 kPa absolute. The atmospheric pressure is found by interpolation using Table B.3 to be $79.48 - (79.48 - 61.64)500/2000 = 75.02$. The required pressure is then

$$p = 47.3 - 75.02 = -27.72 \text{ kPa} \quad \text{or} \quad 27.72 \text{ kPa vacuum}$$

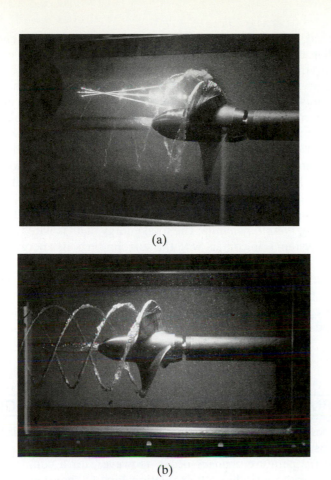

FIGURE 1.12   Photographs of cavitating propeller inside MIT's water tunnel.
(Courtesy of Prof. S. A. Kinnas, Ocean Engineering Group, University of Texas — Austin.)

1.6 CONSERVATION LAWS

From experience it has been found that fundamental laws exist that appear exact; that is, if experiments are conducted with the utmost precision and care, deviations from these laws are very small and in fact, the deviations would be even smaller if improved experimental techniques were employed. Three such laws form the basis for our study of fluid mechanics. The first is the **conservation of mass**, which states that matter is indestructible. Even though Einstein's theory of relativity postulates that under certain conditions, matter is convertible into energy and leads to the statement that the extraordinary quantities of radiation from the sun are associated with a conversion of 3.3×10^{14} kg of matter per day into energy, the destructibility of matter under typical engineering conditions is not measurable and does not violate the conservation of mass principle.

Conservation of mass:
Matter is indestructible.

System: *A fixed quantity of matter.*

For the second and third laws it is necessary to introduce the concept of a system. A **system** is defined as a fixed quantity of matter upon which attention is focused. Everything external to the system is separated by the system boundaries. These boundaries may be fixed or movable, real or imaginary. With this definition we can now present our second fundamental law, the **conservation of momentum**: The momentum of a system remains constant if no external forces are acting on the system. A more specific law based on this principle is *Newton's second law*: The sum of all external forces acting on a system is equal to the time rate of change of linear momentum of the system. A parallel law exists for the moment of momentum: The rate of change of angular momentum is equal to the sum of all torques acting on the system.

Conservation of momentum: *The momentum of a system remains constant if no external forces act on the system.*

The third fundamental law is the **conservation of energy**, which is also known as the *first law of thermodynamics*: The total energy of an isolated system remains constant. If a system is in contact with the surroundings, its energy increases only if the energy of the surroundings experiences a corresponding decrease. It is noted that the total energy consists of potential, kinetic, and internal energy, the latter being the energy content due to the temperature of the system. Other forms of energy are not considered in fluid mechanics. The first law of thermodynamics and other thermodynamic relationships are presented in the following section.

Conservation of energy: *The total energy of an isolated system remains constant. Also known as the* first law of thermodynamics.

1.7 THERMODYNAMIC PROPERTIES AND RELATIONSHIPS

For incompressible fluids, the three laws mentioned in the preceding section suffice. This is usually true for liquids but also for gases if relatively small pressure, density, and temperature changes occur. However, for a compressible fluid, it may be necessary to introduce other relationships, so that density, temperature, and pressure changes are properly taken into account. An example is the prediction of changes in density, pressure, and temperature when compressed gas is released from a container.

Extensive property: *A property that depends on the system's mass.*

Thermodynamic properties, quantities that define the state of a system, either depend on the system's mass or are independent of the mass. The former is called an **extensive property** and the latter is called an **intensive property**. An intensive property can be obtained by dividing the extensive property by the mass of the system. Temperature and pressure are intensive properties; momentum and energy are extensive properties.

Intensive property: *A property that is independent of the system's mass.*

1.7.1   Properties of an Ideal Gas

The behavior of gases in most engineering applications can be described by the ideal-gas law, also called the perfect-gas law. When the temperature is relatively low and/or the pressure relatively high, caution should be exercised and real-gas laws should be applied. For air with temperatures higher than $-50°C$ the ideal-gas law approximates the behavior of air to an acceptable degree provided that the pressure is not extremely high.

The *ideal-gas law* is given by

$$p = \rho R T \tag{1.7.1}$$

where p is the absolute pressure, ρ the density, T the absolute temperature, and R the gas constant. The gas constant is related to the universal gas constant R_u by the relationship

$$R = \frac{R_u}{M} \tag{1.7.2}$$

where M is the molar mass. Values of M and R are tabulated in Table B.4 in Appendix B. The value of R_u is

$$R_u = 8.314 \text{ kJ/kgmol} \cdot \text{K} \tag{1.7.3}$$

For air $M = 28.97$ kg/kgmol, so that for air $R = 0.287$ kJ/kg $\cdot$ K, a value used extensively in calculations involving air.

Other forms that the ideal-gas law takes are

$$p V = m R T \tag{1.7.4}$$

and

$$p V = n R_u T \tag{1.7.5}$$

where n is the number of moles.

Example 1.6

A tank with a volume of 0.2 m³ contains 0.5 kg of nitrogen. The temperature is 20°C. What is the pressure?

Solution
Assume this is an ideal gas. Apply Eq. 1.7.1, (R can be found in Appendix B.4). Solving the equation, $p = \rho R T$, we obtain

$$p = \frac{0.5 \text{ kg}}{0.2 \text{ m}^3} \times 0.2968 \, \frac{\text{kJ}}{\text{kg} \cdot \text{K}} \, (273 + 20) \text{ K} = 218 \text{ kPa absolute}$$

Note: The resulting units are kJ/m³ = kN·m/m³ = kN/m² = kPa. The ideal-gas law requires that pressure and temperature be in absolute units.

1.7.2 First Law of Thermodynamics

In the study of incompressible fluids, the first law of thermodynamics is particularly important. The first law of thermodynamics states that when a system, which is a fixed quantity of fluid, changes from state 1 to state 2, its energy content changes from E_1 to E_2 by energy exchange with its surroundings. The energy exchange is in the form of heat transfer or work. If we define heat transfer to the system as positive and work done by the system as positive,[4] the first law of thermodynamics can be expressed as

KEY CONCEPT *Energy exchange with surroundings is heat transfer or work.*

$$Q_{1\text{-}2} - W_{1\text{-}2} = E_2 - E_1 \qquad (1.7.6)$$

where $Q_{1\text{-}2}$ is the amount of heat transfer to the system and $W_{1\text{-}2}$ is the amount of work done by the system. The energy E represents the total energy, which consists of kinetic energy $(mV^2/2)$, potential energy (mgz), and internal energy $(m\tilde{u})$, where $\tilde{u}$ is the internal energy per unit mass; hence

$$E = m\left(\frac{V^2}{2} + gz + \tilde{u}\right) \qquad (1.7.7)$$

Note that $V^2/2$, gz, and $\tilde{u}$ are all intensive properties and E is an extensive property.

For an isolated system, one that is thermodynamically disconnected from the surroundings (i.e., $Q_{1\text{-}2} = W_{1\text{-}2} = 0$), Eq. 1.7.6 becomes

$$E_1 = E_2 \qquad (1.7.8)$$

This equation represents the conservation of energy.

KEY CONCEPT *Work results from a force moving through a distance.*

The work term in Eq. 1.7.6 results from a force F moving through a distance as it acts on the system's boundary; if the force is due to pressure, it is given by

$$W_{1\text{-}2} = \int_{l_1}^{l_2} F\, dl$$

$$= \int_{l_1}^{l_2} pA\, dl = \int_{V_1}^{V_2} p\, dV \qquad (1.7.9)$$

where $A\, dl = dV$. An example that demonstrates an application of the first law of thermodynamics follows.

---

[4]In some presentations the work done on the system is positive, so that Eq. 1.7.6 would appear as $Q + W = \Delta E$. Either choice is acceptable.

Example 1.7

A cart with a mass of 30 kg is pushed up a ramp with an initial force of 400 N (Fig. E1.7). The force decreases according to

$$F = 20(20 - l) \text{ N}$$

If the cart starts from rest at $l = 0$, determine its velocity after it has traveled 6 m along the ramp. Neglect friction.

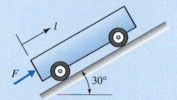

FIGURE E1.7

Solution

The energy equation (Eq. 1.7.6) allows us to relate the quantities of interest. Neglecting heat transfer we have

$$-W_{1\text{-}2} = E_2 - E_1$$

Recognizing that the force is doing work on the system, the work is negative. Hence the energy equation becomes

$$-\left[-\int_0^6 20(20 - l)\, dl\right] = m\left(\frac{V_2^2}{2} + gz_2\right) - m\left(\frac{V_1^{2\;0}}{2} + gz_1^{\,0}\right)$$

Taking the datum as $z_1 = 0$, we have $z_2 = 6 \sin 30° = 3$ m Thus

$$400 \times 6 - 20 \times \frac{6^2}{2} = 30\left(\frac{V_2^2}{2} + 9.81 \times 3\right)$$

$$\therefore V_2 = 8.78 \text{ m/s}$$

Note: We have assumed no internal energy change and no heat transfer.

1.7.3   Other Thermodynamic Quantities

In compressible fluids it is sometimes useful to define thermodynamic quantities that are combinations of other thermodynamic quantities. One such combination is the sum $(m\tilde{u} + p\Psi)$, which can be considered a system property; it is encountered in numerous thermodynamic processes. This property is defined as **enthalpy** H:

Enthalpy: *A property created to aid in thermodynamic calculations.*

$$H = m\tilde{u} + p\Psi \qquad\qquad (1.7.10)$$

The corresponding intensive property (H/m) is

$$h = \tilde{u} + \frac{p}{\rho} \qquad (1.7.11)$$

Other useful thermodynamic quantities are the constant-pressure specific heat c_p and the constant-volume specific heat c_v; they are used to calculate the enthalpy and the internal energy changes in an ideal gas as follows:

$$\Delta h = \int c_p \, dT \qquad (1.7.12)$$

and

$$\Delta \tilde{u} = \int c_v \, dT \qquad (1.7.13)$$

For many situations we can assume constant specific heats in the foregoing relationships. Specific heats for common gases are listed in Table B.4. For an ideal gas c_p is related to c_v by using Eq. 1.7.11 in differential form:

$$c_p = c_v + R \qquad (1.7.14)$$

The **ratio of specific heats** k is often of use for an ideal gas; it is expressed as

$$k = \frac{c_p}{c_v} \qquad (1.7.15)$$

For liquids and solids we use $\Delta u = c \, \Delta T$ where c is the specific heat of the fluid. For water $c \cong 4.18$ kJ/kg·°C.

A process in which pressure, temperature, and other properties are essentially constant at any instant throughout the system is called a **quasi-equilibrium** or *quasi-static* process. An example of such a process is the compression and expansion in an internal combustion engine. If, in addition, no heat is transferred ($Q_{1\text{-}2} = 0$), the process is called an *adiabatic*, quasi-equilibrium process or an *isentropic*[5] process. For such an isentropic[5] process the following relationships may be used:

$$\frac{p_1}{p_2} = \left(\frac{\rho_1}{\rho_2}\right)^k \qquad \frac{T_1}{T_2} = \left(\frac{p_1}{p_2}\right)^{(k-1)/k} \qquad \frac{T_1}{T_2} = \left(\frac{\rho_1}{\rho_2}\right)^{k-1} \qquad (1.7.16)$$

[5]An isentropic process occurs when the entropy is constant. We will not define or calculate entropy here; it is discussed in Section 9.1.

For a small pressure wave traveling in a gas at relatively low frequency, the wave speed is given by an isentropic process so that

$$c = \sqrt{\left.\frac{dp}{d\rho}\right|_s} = \sqrt{kRT} \tag{1.7.17}$$

If the frequency is relatively high, entropy is not constant and we use

$$c = \sqrt{\left.\frac{dp}{d\rho}\right|_T} = \sqrt{RT} \tag{1.7.18}$$

Example 1.8

A cylinder fitted with a piston has an initial volume of 0.5 m^3. It contains 2.0 kg of air at 400 kPa absolute. Heat is transferred to the air while the pressure remains constant until the temperature is 300°C. Calculate the heat transfer and the work done. Assume constant specific heats.

Solution

Using the first law, Eq. 1.7.9, and the definition of enthalpy, we see that

$$Q_{1\text{-}2} = p_2 V_2 - p_1 V_1 + m\tilde{u}_2 - m\tilde{u}_1$$
$$= m\tilde{u}_2 + p_2 V_2 - (m\tilde{u}_1 + p_1 V_1)$$
$$= H_2 - H_1 = m(h_2 - h_1) = mc_p(T_2 - T_1)$$

where Eq. 1.7.12 is used assuming c_p to be constant. The initial temperature is

$$T_1 = \frac{p_1 V_1}{mR} = \frac{400 \text{ kN/m}^2 \times 0.5 \text{ m}^3}{2.0 \text{ kg} \times 0.287 \text{ kJ/kg·K}} = 348.4 \text{ K}$$

(Use kJ = kN·m to check the units.) Thus the heat transfer is (c_p is found in Table B.4)

$$Q_{1\text{-}2} = 2.0 \times 1.0[(300 + 273) - 348.4] = 449 \text{ kJ}$$

The final volume is found using the ideal-gas law:

$$V_2 = \frac{mRT_2}{p_2} = \frac{2 \times 0.287 \times 573}{400} = 0.822 \text{ m}^3$$

The work done for the constant-pressure process is, using Eq. 1.7.9 with p = const,

$$W_{1\text{-}2} = p(V_2 - V_1)$$
$$= 400(0.822 - 0.5) = 129 \text{ kJ}$$

Example 1.9

The temperature on a cold winter day in the mountains of Wyoming is $-30°C$ at an elevation of 3000 m. Calculate the density of the air assuming the same pressure as in the normal atmosphere; also find the speed of sound.

Solution

From Table B.3 we find the atmospheric pressure at an elevation of 3000 m to be 70.56 kPa by interpolation. The absolute temperature is found to be

$$T = -30 + 273.15 = 243.15 \text{ K}$$

Using the ideal-gas law, the density is calculated as

$$\rho = \frac{p}{RT}$$

$$= \frac{70.56 \times 10^3 \text{ N/m}^2}{287 \text{ N} \cdot \text{m} / \text{kg} \cdot \text{K} \times 243.15 \text{ K}} = 1.0111 \text{ kg/m}^3$$

The speed of sound, using Eq. 1.7.17, is determined to be

$$c = \sqrt{kRT}$$

$$= \sqrt{1.4 \times 287 \times 243.15} = 312.5 \text{ m/s}$$

Note: The gas constant in the foregoing equations has units of $N \cdot m / kg \cdot K$ so that the appropriate units result. Express $kg = N \cdot s^2/m$ (from $m = F/a$) to observe that this is true.

1.8 SUMMARY

To relate units we often use Newton's second law which allows us to write

$$N = kg \cdot m/s^2 \tag{1.8.1}$$

When making engineering calculations, an answer should have the same number of significant digits as the least accurate number used in the calculations. Most fluid properties are known to at most four significant digits. Hence, answers should be expressed to at most four significant digits, and often to only three significant digits.

In fluid mechanics pressure is expressed as gage pressure unless stated otherwise. This is unlike thermodynamics, in which pressure is assumed to be absolute. If absolute pressure is needed, add 101 kPa if the atmospheric pressure is not given in the problem statement.

The density, or specific weight, of a fluid is known if the specific gravity is known:

$$\rho_x = S_x \rho_{\text{water}} \qquad \gamma_x = S_x \gamma_{\text{water}} \qquad\qquad (1.8.2)$$

The shear stress due to viscous effects in a simple flow where $u = u(y)$ is given by

$$\tau = \mu \frac{du}{dy} \qquad\qquad (1.8.3)$$

This stress can be used to calculate the torque needed to rotate a shaft in a bearing.

Many air flows, and other gases too, are assumed to be incompressible at low speeds, speeds under about 100 m/s for atmospheric air.

The three fundamental laws used in our study of fluid mechanics are the conservation of mass, Newton's second law, and the first law of thermodynamics. These will take on various forms depending on the problem being studied. Much of our study of fluid mechanics will be expressing these laws in mathematical forms so that quantities of interest can be calculated.

PROBLEMS

1.1 State the three basic laws that are used in the study of the mechanics of fluids. State at least one global (integral) quantity that occurs in each. State at least one quantity that may be defined at a point that occurs in each.

Dimensions, Units, and Physical Quantities

1.2 Verify the dimensions given in Table 1.2 for the following quantities:
   (a) Density
   (b) Pressure
   (c) Power
   (d) Energy
   (e) Mass
   (f) Flow rate

1.3 Express the dimensions of the following quantities using the *F-L-T* system:
   (a) Density
   (b) Pressure
   (c) Power
   (d) Energy
   (e) Mass flux
   (f) Flow rate

1.4 If force, length, and time are selected as the three fundamental dimensions, the units of mass in the SI system could be written as:
   A. FT^2/L
   B. FL/T^2
   C. $N \cdot s^2/m$
   D. $N \cdot m/s^2$

1.5 Select the dimensions of viscosity using the *F-L-T* system:
   A. FT^2/L
   B. FT/L^2
   C. $N \cdot s/m^2$
   D. $N \cdot s^2/m$

1.6 Recognizing that all terms in an equation must have the same dimensions, determine the dimensions on the constants in the following equations:
   (a) $d = 4.9\, t^2$ where d is distance and t is time.
   (b) $F = 9.8\, m$ where F is a force and m is mass.
   (c) $Q = 80AR^{2/3} S_0^{1/2}$ where A is area, R is a radius, S_0 is a slope and Q is a flow rate with dimensions of L^3/T.

1.7 Determine the units on each of the constants in the following equations, recognizing that all terms in an equation have the same dimensions:
(a) $d = 4.9\,t^2$ where d is in meters and t is in seconds.
(b) $F = 9.8\,m$ where F is in newtons and m is in kilograms.
(c) $Q = 80AR^{2/3}S_0^{1/2}$ where A is in meters squared, R is in meters, S_0 is the slope, and Q has units of meters cubed per second.

1.8 State the SI units of each of the following:
(a) Pressure
(b) Energy
(c) Power
(d) Viscosity
(e) Heat flux
(f) Specific heat

1.9 Determine the units on c, k and $f(t)$ in
$$m\frac{d^2y}{dt^2} + c\frac{dy}{dt} + ky = f(t)$$
if m is in kilograms, y is in meters, and t is in seconds.

1.10 Write the following with the use of prefixes:
(a) 2.5×10^5 N
(b) 5.72×10^{11} Pa
(c) 4.2×10^{-8} Pa
(d) 1.76×10^{-5} m³
(e) 1.2×10^{-4} m²
(f) 7.6×10^{-8} m³

1.11 Write the following with the use of powers; do not use a prefix:
(a) 125 MN
(b) 32.1 µs
(c) 0.67 GPa
(d) 0.0056 mm³
(e) 520 cm²
(f) 7.8 km³

1.12 The quantity 2.36×10^{-8} Pa can be written as:
A. 23.6 nPa
B. 236 µPa
C. 236×10^{-3} mPa
D. 236 nPa

1.13 Rewrite Eq. 1.8.3 as a dimensional formula using SI units.

1.14 Using the table of conversions on the inside front cover, express each of the following in the SI units of Table 1.2:
(a) 20 cm/hr
(b) 2000 rpm
(c) 500 hp
(d) 10^5 cm³/min
(e) 2000 kN/cm²
(f) 500 g/min
(g) 500 g/L
(h) 500 kWh

1.15 What net force is needed to accelerate a 10-kg mass at the rate of 40 m/s²:
(a) Horizontally?
(b) Vertically upward?
(c) On an upward slope of 30°?

1.16 A body that weighs 300 N on earth would weigh how much on the moon where $g \cong 1.65$ m/s²
A. 5030 N
B. 250 N
C. 40.77 N
D. 6.2 N

1.17 A particular body weighs 300 N on earth. Calculate its weight on the moon, where $g \approx 1.65$ m/s².

1.18 A 4200-N force acts on a 250-cm area at an angle of 30° to the normal. The shear stress acting on the area is:
A. 84 Pa
B. 84 mPa
C. 84 kPa
D. 84 MPa

1.19 Calculate the mean free path in the atmosphere using Eq. 1.3.3 and Table B.3 in the appendix at an elevation of:
(a) 30 000 m
(b) 50 000 m
(c) 80 000 m

Pressure and Temperature

1.20 A gage pressure of 52.3 kPa is read on a gage. Find the absolute pressure if the elevation is:
(a) At sea level
(b) 1000 m
(c) 5000 m
(d) 10 000 m
(e) 30 000 m

1.21 A vacuum of 31 kPa is measured in an airflow at sea level. Find the absolute pressure in:
(a) kPa
(b) mm Hg
(c) m of H_2O

1.22 For a constant-temperature atmosphere, the pressure as a function of elevation is given by $p(z) = p_0 e^{-gz/RT}$, where g is gravity, $R = 287$ J/kg · K, and T is the absolute temperature. Use this equation and estimate the pressure at 4000 m assuming that $p_0 = 101$ kPa and $T = 15°C$. What is the error?

1.23 Estimate the pressure and temperature at an elevation of 6880 m using Table B.3. Employ:
(a) A linear interpolation: $f \approx f_0 + n(f_1 - f_0)$.
(b) A parabolic interpolation: $f \approx f_0 + n(f_1 - f_0) + (n/2)(n-1)(f_2 - 2f_1 + f_0)$.

1.24 Estimate the temperature in °C at 9500 m, an elevation at which many commercial airplanes fly. Use Table B.3.

1.25 The temperature at 11 000 m in the standard atmosphere, using a parabolic interpolation of the entries in Table B.3, is nearest:

A.	$-62.4°C$	**B.**	$-53.6°C$
C.	$-32.8°C$	**D.**	$-17.3°C$

1.26 An applied force of 26.5 MN is distributed uniformly over a 152-cm^2 area; however, it acts at an angle of 42° with respect to a normal vector (see Fig. P1.26). If it produces a compressive stress, calculate the resulting pressure.

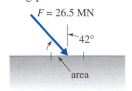

FIGURE P1.26

1.27 The force on an area of 0.2 cm^2 is due to a pressure of 120 kPa and a shear stress of 20 Pa, as shown in Fig. P1.27. Calculate the magnitude of the force acting on the area and the angle of the force with respect to a normal coordinate.

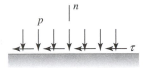

FIGURE P1.27

Density and Specific Weight

1.28 Calculate the density and specific weight of water if 0.5 kg occupies 500 cm^3.

1.29 Use Eq. 1.5.3 to determine the density and specific gravity of water at 70°C. What is the error in the calculation for density? Use Table B.1.

1.30 The specific gravity of mercury is usually taken as 13.6. What is the percent error in using a value of 13.6 at 50°C?

1.31 The specific weight of an unknown liquid is 12 400 N/m^3. What mass of the liquid is contained in a volume of 500 cm^3? Use:

(a) The standard value of gravity.
(b) The minimum value of gravity on the earth.
(c) The maximum value of gravity on the earth.

1.32 A liquid with a specific gravity of 1.2 fills a volume. If the mass in the volume is 150 kg, what is the magnitude of the volume?

1.33 Using an equation, estimate the density of water at 80°C:

A.	980 kg/m^3	**B.**	972 kg/m^3
C.	972 kg/m^3	**D.**	968 kg/m^3

Viscosity

1.34 A velocity distribution in a 5-cm diameter pipe is measured to be $u(r) = 10(1 - r^2/r_0^2)$ m/s, where r_0 is the radius of the pipe. Calculate the shear stress at the wall if water at 25°C is flowing.

1.35 For two 0.2-m-long rotating concentric cylinders, the velocity distribution is given by $u(r) = 0.4/r - 1000r$ m/s. If the diameters of the cylinders are 2 cm and 4 cm, respectively, calculate the fluid viscosity if the torque on the inner cylinder is measured to be 0.0026 N·m.

1.36 A 1.2-m long 25-mm diameter shaft rotates inside an equally long cylinder that is 26 mm in diameter. Calculate the torque required to rotate the inner shaft at 2000 rpm if SAE-30 oil at 20°C fills the gap. Also, calculate the horsepower required. Assume symmetric motion.

1.37 A 60-cm-wide belt moves as shown in Fig. P1.37. Calculate the horsepower requirement assuming a linear velocity profile in the 10°C water.

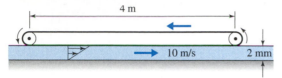

4 m

10 m/s   2 mm

FIGURE P1.37

1.38 A 150 mm-diameter horizontal disk rotates a distance of 2 mm above a solid surface. Water at 15°C fills the gap. Estimate the torque required to rotate the disk at 400 rpm.

1.39 The velocity distribution in a 1.0-cm-diameter pipe is given by $u(r) = 16(1 - r^2/r_0^2)$ m/s, where r_0 is the pipe radius. Calculate the shearing stress at the centerline, at $r = 0.25$ cm, and at the wall if water at 20°C is flowing.

1.40 The velocity distribution in a 4-cm-diameter pipe transporting 20°C water is given by $u(r) = 10(1 - 2500r^2)$ m/s. The shearing stress at the wall is nearest:

A.	1.0 Pa	**B.**	0.1 Pa
C.	0.01 Pa	**D.**	0.001 Pa

1.41 Calculate the torque needed to rotate the cone shown in Fig. P1.41 at 2000 rpm if SAE-30 oil at 40°C fills the gap. Assume a linear velocity profile.

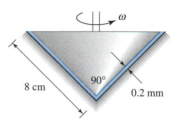

ω

8 cm   90°

0.2 mm

FIGURE P1.41

1.42 A free-body diagram of the liquid between a moving belt and a fixed wall shows that the shear stress in the liquid is constant. If the temperature varies according to $T(y) = K/y$, where y is measured from the wall (the temperature at the wall is very large), what would be the shape of the velocity profile if the viscosity varies according to Andrade's equation $\mu = Ae^{B/T}$?

1.43 The viscosity of water at 20°C is 0.001 N·s/m² and at 80°C it is 3.57×10^{-4} N·s/m². Using Andrade's equation $\mu = Ae^{B/T}$ estimate the viscosity of water at 40°C. Determine the percent error.

Compressibility

1.44 Show that $d\rho/\rho = -d\mathcal{V}/\mathcal{V}$, as was assumed in Eq. 1.5.11.

1.45 What is the volume change of 2 m³ of water at 20°C due to an applied pressure of 10 MPa?

1.46 Two engineers wish to estimate the distance across a lake. One pounds two rocks together under water on one side of the lake and the other submerges his head and hears a small sound 0.62 s later, as indicated by a very accurate stopwatch. What is the distance between the two engineers?

1.47 A pressure is applied to 20 L of water. The volume is observed to decrease to 18.7 L. Calculate the applied pressure.

1.48 Calculate the speed of propagation of a small-amplitude wave through water at:
(a) 5°C
(b) 40°C
(c) 90°C

1.49 The change in volume of a liquid with temperature is given by $\Delta \mathcal{V} = \alpha_T \mathcal{V} \, \Delta T$, where α_T is the **coefficient of thermal expansion**. For water at 40°C, $\alpha_T = 3.8 \times 10^{-4}$ K^{-1}. What is the volume change of 1 m³ of 40°C water if $\Delta T = -20°C$? What pressure change would be needed to cause that same volume change?

Surface Tension

1.50 Calculate the pressure in the small 10-μm-diameter droplets that are formed by spray machines. Assume the properties to be the same as water at 15°C. Calculate the pressure for bubbles of the same size.

1.51 A small 1.5-mm-diameter bubble is formed by a stream of 15°C water. Estimate the pressure inside the bubble.

1.52 Determine the height that 20°C water would climb in a vertical 0.02-cm-diameter tube if it attaches to the wall with an angle β of 30° to the vertical.

1.53 The distance 20°C water would climb in a long 10-μm-diameter, clean glass tube is nearest:
   A. 50 cm
   B. 100 cm
   C. 200 cm
   D. 300 cm

1.54 Mercury makes an angle of 130° (β in Fig. 1.10) when in contact with clean glass. What distance will mercury depress in a vertical 20-mm-diameter glass tube? Use $\sigma = 0.4$ N/m.

1.55 Find and expression for the rise of liquid between two parallel plates a distance t apart. Use a contact angle β and surface tension σ.

1.56 Write an expression for the maximum diameter d of a needle of length L that can float in a liquid with surface tension σ. The density of the needle is ρ.

1.57 Would a 7-cm-long 4-mm-diameter steel needle be able to float in 15°C water? Use $\rho_{\text{steel}} = 7850$ kg/m³.

1.58 Find an expression for the maximum vertical force F needed to lift a thin wire ring of diameter D slowly from a liquid with surface tension σ.

1.59 Two flat plates are positioned as shown in Fig. P1.59 with a small angle α in an open container with a small amount of liquid. The plates are vertical and the liquid rises between the plates. Find an expression for the location $h(x)$ of the surface of the liquid assuming that $\beta = 0$.

FIGURE P1.59

Vapor Pressure

1.60 Water is transported through the pipe of Fig. p1.60 such that a vacuum of 80 kPa exists at a particular location. What is the maximum possible temperature of the water? Use $p_{\text{atm}} = 92$ kPa.

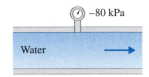

FIGURE P1.60

1.61 A group of explorers desired their elevation. An engineer boiled water and measured the temperature to be 82°C. They found a fluid mechanics book in a backpack and the engineer told the group their elevation! What elevation should the engineer have quoted?

1.62 A tank half-filled with 40°C water is to be evacuated. What is the minimum pressure that can be expected in the space above the water?

1.63 Water is forced through a contraction causing low pressure. The water is observed to boil at a pressure of −75 kPa. If atmospheric pressure is 100 kPa, what is the temperature of the water?

1.64 Oil is transported through a pipeline by a series of pumps that can produce a pressure of 10 MPa in the oil leaving each pump. The losses in the pipeline cause a pressure drop of 600 kPa each kilometer. What is the maximum possible spacing of the pumps?

1.65 Which of the following is an intensive property?
   A. kinetic energy   **B.** enthalpy
   C. density   **D.** momentum

Ideal Gas

1.66   Determine the density and the specific gravity of air at standard conditions (i.e., 15°C and 101.3 kPa absolute).

1.67   Calculate the density of air inside a house and outside a house using 20°C inside and −25°C outside. Use an atmospheric pressure of 85 kPa. Do you think there would be a movement of air from the inside to the outside (infiltration), even without a wind? Explain.

1.68   A 0.4 m³ air tank is pressurized to 5 MPa. When the temperature reaches −15°C, calculate the density and the air mass.

1.69   The mass of propane contained in a 4-m³ tank maintained at 800 kPa and 10°C is nearest:
  A.   100 kg   **B.**   80 kg
  C.   60 kg   **D.**   20 kg

1.70   Estimate the weight of air contained in a classroom that measures 10 m × 20 m × 4 m. Assume reasonable values for the variables.

1.71   A car tire is pressurized to 250 kPa in Michigan when the temperature is −25°C. The car is driven to Arizona, where the temperature on the highway, and in the tire, reaches 65°C. Estimate the maximum pressure in the tire.

1.72   The mass of all the air in the atmosphere contained above a 1-m² area is to be contained in a spherical volume. Estimate the diameter of the sphere if the air is at standard conditions.

First Law

1.73   A body falls from rest. Determine its velocity after 3 m and 6 m, using the energy equation.

1.74   Determine the final velocity of the 15-kg mass of Fig. P1.74 moving horizontally if it starts at 10 m/s and moves a distance of 10 m while the following net force acts in the direction of motion (where s is the distance in the direction of motion):
  (a)   200 N   **(b)** $20s$ N
  (c) $200 \cos s\pi/20$ N

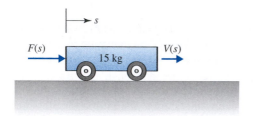

FIGURE P1.74

1.75   The 10-kg mass shown in Fig. P1.75 is traveling at 40 m/s and strikes a plunger that is attached to a piston. The piston compresses 0.2 kg of air contained in a cylinder. If the mass is brought to rest, calculate the maximum rise of temperature in the air. What effects could lead to a lower temperature rise?

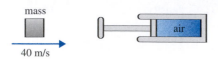

FIGURE P1.75

1.76   A 1500-kg automobile traveling at 100 km/h is suddenly grabbed by a hook and all of its kinetic energy is dissipated in a hydraulic absorber containing 2000 cm³ of water. Calculate the maximum temperature rise in the water.

1.77   A fuel mass of 0.2 kg contains 40 MJ/kg of energy. Calculate the temperature rise of 100 kg of water if complete combustion occurs and the water, which surrounds the fuel, is completely insulated from the surroundings.

1.78   Five 40-cm³ ice cubes completely melt in 2 liters of warm water (it takes 320 kJ to melt a kilogram of ice). The temperature drop in the water is nearest:
  A.   10°C   **B.**   8°C
  C.   6°C   **D.**   4°C

1.79   2 kg of air is compressed in a cylinder–piston arrangement while the temperature remains constant at 20°C. If the initial pressure is 200 kPa absolute, calculate the work needed to compress the air such that the absolute pressure doubles. Also calculate the heat transfer.

1.80   Determine the heat transfer needed to double the absolute pressure in a fixed 2-m³ volume containing air at 200 kPa absolute if the initial temperature is:
  (a)   20°C   **(b)**   100°C
  (c)   200°C

1.81   Heat is transferred to 2 kg of air in a cylinder so that the temperature doubles while the pressure remains constant. What work is required if the initial temperature is:
  (a)   60°C?   **(b)**   150°C?   **(c)**   200°C?

Isentropic Flow

1.82 Air flows from a tank maintained at 5 MPa absolute and 20°C. It exits a hole and reaches a pressure of 500 kPa absolute. Assuming an adiabatic, quasi-equilibrium process, calculate the exiting temperature.

1.83 An airstream flows with no heat transfer such that the temperature changes from 20°C to 150°C. If the initial pressure is measured to be 150 kPa, estimate the maximum final pressure.

1.84 Air is compressed in an insulated cylinder from 20°C to 200°C. If the initial pressure is 100 kPa absolute, what is the maximum final pressure? What work is required?

Speed of Sound

1.85 Calculate the speed of sound at 20°C in:
   (a) Air
   (b) Carbon dioxide
   (c) Nitrogen
   (d) Hydrogen
   (e) Steam

1.86 The speed of sound of a dog whistle in the atmosphere at a location where the temperature is 50°C is nearest:
   A. 396 m/s   **B.** 360 m/s
   C. 332 m/s   **D.** 304 m/s

1.87 Compare the speed of sound in the atmosphere at an elevation of 10 000 m with that at sea level by calculating a percentage decrease.

1.88 A lumberman, off in the distance, is chopping with an axe. An observer, using her digital stopwatch, measures a time of 8.32 s from the instant the axe strikes the tree until the sound is heard. How far is the observer from the lumberman if:
   (a) $T = -20°C$?
   (b) $T = 20°C$?
   (c) $T = 45°C$?

The Morrow Point Dam is an example of an arch-type dam. The curved wall enables large hydrostatic loads to exist on the upstream face while minimizing the required thickness of the structure. (U.S. Bureau of Reclamation)

2

Fluid Statics

Outline

Chapter Objectives

The objectives of this chapter are to:
- ▲ Establish the variation of pressure in a fluid at rest.
- ▲ Learn how to use manometers to measure pressure.
- ▲ Calculate forces on plane and curved surfaces including buoyant forces.
- ▲ Determine the stability of submerged and floating objects.
- ▲ Calculate pressure and forces in accelerating and rotating containers.
- ▲ Present numerous examples and problems that demonstrate how pressures and forces are calculated in fluids at rest.

2.1 INTRODUCTION

Fluid statics: *The study of fluids in which there is no relative motion between fluid particles.*

KEY CONCEPT   *The only stress that exists where there is no motion is pressure.*

Fluid statics is the study of fluids in which there is no relative motion between fluid particles. If there is no relative motion, no shearing stresses exist, since velocity gradients, such as *du/dy*, are required for shearing stresses to be present. The only stress that exists is a normal stress, the pressure, so it is the pressure that is of primary interest in fluid statics.

Three situations, depicted in Fig. 2.1, involving fluid statics will be investigated. These include fluids at rest, such as water pushing against a dam, fluids contained in devices that undergo linear acceleration, and fluids contained in rotating cylinders. In each of these three situations the fluid is in static equilibrium with respect to a reference frame attached to the boundary surrounding the fluid. In addition to the examples shown for fluids at rest, we consider instruments called manometers, and investigate the forces of buoyancy. Finally, the stability of floating bodies such as ships will also be presented.

2.2 PRESSURE AT A POINT

We have defined pressure as being the infinitesimal normal compressive force divided by the infinitesimal area over which it acts. This defines the pressure at a point. One might question whether the pressure at a given point varies as the normal to the area changes direction. To show that this is not the case, even for fluids in motion with no shear, consider the wedge-shaped element of unit depth (in the *z*-direction) shown in Fig. 2.2. Assume that a pressure *p* acts on the hypotenuse and that a different pressure acts on each of the other areas, as shown. Since the forces on the two end faces are in the *z*-direction, we have not included them on the element. Now, let us apply Newton's second law to the element, for both the *x*- and *y*-directions:

$$\Sigma F_x = ma_x: \qquad p_x \Delta y - p \Delta s \sin \theta = \rho \frac{\Delta x\, \Delta y}{2} a_x$$

$$\Sigma F_y = ma_y: \quad p_y \Delta x - \rho g \frac{\Delta x \Delta y}{2} - p \Delta s \cos \theta = \rho \frac{\Delta x \Delta y}{2} a_y$$

$$(2.2.1)$$

where we have used $\Delta V = \Delta x\, \Delta y/2$ (we could include Δz in each term to account

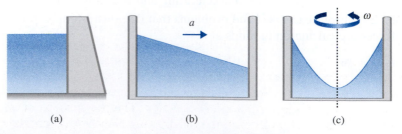

(a)   (b)   (c)

FIGURE 2.1 Examples included in fluid statics: (a) liquids at rest; (b) linear acceleration; (c) angular rotation.

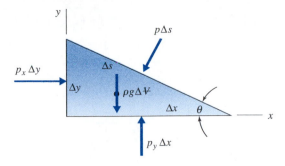

FIGURE 2.2   Pressure at a point in a fluid.

for the depth). The pressures shown are due to the surrounding fluid and are the average pressure on the areas. Substituting

$$\Delta s \sin \theta = \Delta y \qquad \Delta s \cos \theta = \Delta x \qquad (2.2.2)$$

we see that Eqs. 2.2.1 take the form

$$p_x - p = \frac{\rho a_x \, \Delta x}{2}$$

$$p_y - p = \frac{\rho(a_y + g) \, \Delta y}{2} \qquad (2.2.3)$$

Note that in the limit as the element shrinks to a point, $\Delta x \rightarrow 0$ and $\Delta y \rightarrow 0$. Hence the right-hand sides in the equations above go to zero, even for fluids in motion, providing us with the result that, at a point,

$$p_x = p_y = p \qquad (2.2.4)$$

Since θ is arbitrary, this relationship holds for all angles at a point. We could have analyzed an element in the xz-plane and concluded that $p_x = p_z = p$. Thus we conclude that the pressure in a fluid is constant at a point; that is, pressure is a scalar function. It acts equally in all directions at a given point for both a static fluid and a fluid that is in motion in the absence of shear stress.

KEY CONCEPT   *Pressure in a fluid acts equally in all directions at a given point.*

2.3 PRESSURE VARIATION

A general equation is derived to predict the pressure variation of fluids at rest or fluids undergoing an acceleration while the relative position of fluid elements to one another remains the same (this eliminates shear stress). To determine the pressure variation in such fluids, consider the infinitesimal element displayed in Fig. 2.3, where the z-axis is in the vertical direction. The pressure variation from one point to another will be determined by applying Newton's second law; that is, the sum of the forces acting on the fluid element is equal to the mass times the acceleration of the element.

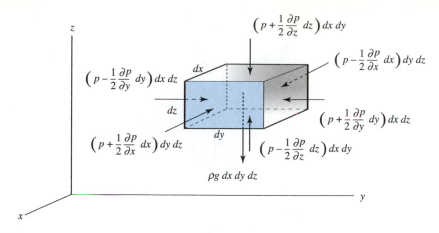

FIGURE 2.3 Forces acting on an infinitesimal element that is at rest in the *xyz*-reference frame. The reference frame may be accelerating or rotating.

If we assume that a pressure *p* exists at the center of this element, the pressures at each of the sides can be expressed by using the chain rule from calculus with $p(x, y, z)$:

$$dp = \frac{\partial p}{\partial x} dx + \frac{\partial p}{\partial y} dy + \frac{\partial p}{\partial z} dz \qquad (2.3.1)$$

If we move from the center to a face a distance $(dx/2)$ away, we see that the pressure is

$$p\left(x + \frac{dx}{2}, y, z\right) = p(x, y, z) + \frac{\partial p}{\partial x} \frac{dx}{2} \qquad (2.3.2)$$

The pressures at all faces are expressed in this manner, as shown in Fig. 2.3. Newton's second law is written in vector form for a constant-mass system as

$$\Sigma \mathbf{F} = m\mathbf{a} \qquad (2.3.3)$$

This results in the three component equations, assuming *z* to be vertical and using the mass as $\rho \, dx \, dy \, dz$,

$$-\frac{\partial p}{\partial x} dx \, dy \, dz = \rho a_x \, dx \, dy \, dz$$

$$-\frac{\partial p}{\partial y} dx \, dy \, dz = \rho a_y \, dx \, dy \, dz \qquad (2.3.4)$$

$$-\frac{\partial p}{\partial z} dx \, dy \, dz = \rho(a_z + g) \, dx \, dy \, dz$$

where a_x, a_y, and a_z are the components of the acceleration of the element. Division by the element's volume $dx\ dy\ dz$ yields

$$\frac{\partial p}{\partial x} = -\rho a_x$$

$$\frac{\partial p}{\partial y} = -\rho a_y \qquad (2.3.5)$$

$$\frac{\partial p}{\partial z} = -\rho(a_z + g)$$

The pressure differential in any direction can now be determined from Eq. 2.3.1 as

$$dp = -\rho a_x dx - \rho a_y dy - \rho(a_z + g)dz \qquad (2.3.6)$$

where z is always vertical. Pressure differences between specified points can be found by integrating Eq. 2.3.6. This equation is useful in a variety of problems, as will be demonstrated in the remaining sections of this chapter.

2.4 FLUIDS AT REST

A fluid at rest does not undergo any acceleration. Therefore, set $a_x = a_y = a_z = 0$ and Eq. 2.3.6 reduces to

$$dp = -\rho g\ dz \qquad (2.4.1)$$

or

$$\frac{dp}{dz} = -\gamma \qquad (2.4.2)$$

This equation implies that there is no pressure variation in the x- and y-directions, that is, in the horizontal plane. The pressure varies in the z-direction only. Also note that dp is negative if dz is positive; that is, the pressure decreases as we move up and increases as we move down, a rather obvious result.

2.4.1 Pressures in Liquids at Rest

If the density can be assumed constant, Eq. 2.4.2 is integrated to yield

$$\Delta p = -\gamma \Delta z \quad \text{or} \quad p + \gamma z = \text{constant} \quad \text{or} \quad \frac{p}{\gamma} + z = \text{constant} \qquad (2.4.3)$$

so that pressure increases with depth. Note that z is positive in the upward direction. The quantity $(p/\gamma + z)$ is often referred to as the *piezometric head*. If the point of interest were a distance h below a **free surface** (a surface separating a gas from a liquid), as shown in Fig. 2.4, Eq. 2.4.3 would result in

Free surface: *A surface separating a gas from a liquid.*

KEY CONCEPT *The equation $p = \gamma h$ is used to convert pressure to a height of liquid.*

$$p = \gamma h \tag{2.4.4}$$

where $p = 0$ at $h = 0$. This equation will be quite useful in converting pressure to an equivalent height of liquid. For example, atmospheric pressure is often expressed as millimeters of mercury; that is, the atmospheric pressure is equal to the pressure at a certain depth in a mercury column, and by knowing the specific weight of mercury, we can then determine that depth using Eq. 2.4.4.

2.4.2 Pressures in the Atmosphere

Standard atmosphere: *Position at 40° latitude where calculations are standardized.*

For the atmosphere where the density depends on height [i.e., $\rho = \rho(z)$], we must integrate Eq. 2.4.1 along a vertical path. The atmosphere is divided into four layers: the *troposphere* (nearest Earth), the *stratosphere*, the *ionosphere*, and the *exosphere*. Because conditions change with time and latitude in the atmosphere with the layers being thicker at the equator and thinner at the poles, we base calculations on the **standard atmosphere**, which is at 40° latitude. In the standard atmosphere the temperature in the troposphere varies linearly with elevation, $T(z) = T_0 - \alpha z$, where the *lapse rate* $\alpha = 0.0065$ K/m and T_0 is 288 K. In the part of the stratosphere between 11 and 20 km the temperature is constant at −56.5°C. (In the lower part of this constant-temperature region is where commercial aircraft usually fly.) The temperature then increases again and reaches a maximum near 50 km; it then decreases to the edge of the ionosphere. Figure 2.5 shows how atmospheric pressure varies with altitude on three mountains. The standard atmosphere is sketched in Fig. 2.6. Because the density of the air in the ionosphere and the exosphere is so low, it is possible for satellites to orbit the earth in either of these layers.

To determine the pressure variation of the troposphere, we can use the ideal-gas law $p = \rho RT$ and Eq. 2.4.1; there results

$$dp = -\frac{pg}{RT}\,dz$$

or

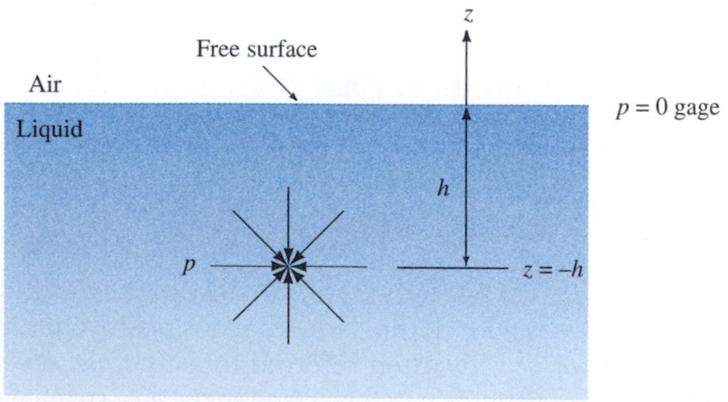

FIGURE 2.4 Pressure below a free surface.

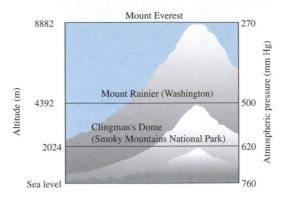

FIGURE 2.5   Atmospheric Pressure and Altitude   A column of air from the outer atmosphere to a given point on Earth contains gases that exert a force equal to 10.13 N on each square cm. This pressure is 1 atm or 760 mm Hg. At a higher altitude the pressure is less because the mass of the column of air from the outer atmosphere to that point is less. Examples of pressure on three mountains are given on the right.

$$\frac{dp}{p} = -\frac{g}{RT}\,dz \qquad (2.4.5)$$

This can be integrated, between sea level and an elevation z:

$$\int_{p_{atm}}^{p} \frac{dp}{p} = -\frac{g}{R}\int_{0}^{z} \frac{dz}{T_0 - \alpha z} \qquad (2.4.6)$$

Upon integration this gives

$$\ln \frac{p}{p_{atm}} = \frac{g}{\alpha R}\ln \frac{T_0 - \alpha z}{T_0} \qquad (2.4.7)$$

which can be put in the form

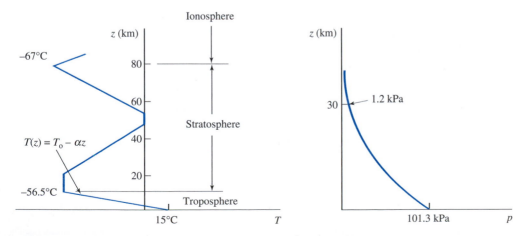

FIGURE 2.6   Standard atmosphere.

$$p = p_{atm}\left(\frac{T_0 - \alpha z}{T_0}\right)^{g/\alpha R} \tag{2.4.8}$$

If we use standard atmospheric conditions in Eq. 2.4.8, we find that $p/p_{atm} = 0.999$ at $z = 10$ m. Consequently, we ignore changes in pressure in a gas such as air unless z is relatively large.

In the stratosphere, where the temperature is constant, Eq. 2.4.5 is integrated again as follows:

$$\int_{p_s}^{p} \frac{dp}{p} = -\frac{g}{RT_s}\int_{z_s}^{z} dz \tag{2.4.9}$$

$$\ln\frac{p}{p_s} = -\frac{g}{RT_s}(z - z_s) \tag{2.4.10}$$

or

$$p = p_s \exp\left[\frac{g}{RT_s}(z_s - z)\right] \tag{2.4.11}$$

The subscript s denotes conditions at the troposphere–stratosphere interface. Properties of the standard atmosphere up to 80 km are listed in Appendix B.3.

Example 2.1

The atmospheric pressure is given as 680 mm Hg at a mountain location. Convert this to kilopascals and meters of water. Also, calculate the pressure decrease due to 500 m elevation increase, starting at 2000 m elevation, assuming constant density.

Solution
Use Eq. 2.4.4 and find, using $S_{Hg} = 13.6$ with Eq. 1.5.2,

$$p = \gamma_{Hg}h$$
$$= (9810 \times 13.6) \times 0.680 = 90\,700 \text{ Pa} \quad \text{or} \quad 90.7 \text{ kPa}$$

To convert this to meters of water, we have

$$h = \frac{p}{\gamma_{H_2O}}$$
$$= \frac{90\,700}{9810} = 9.25 \text{ m of water}$$

To find the pressure decrease, we use Eq. 2.4.3 and find the density in Table B.3:

$$\Delta p = -\gamma\Delta z = -\rho g\Delta z$$
$$= -1.007 \times 9.81 \times 500 = -4940 \text{ Pa}$$

Note: Since gravity is known to three significant digits, we express the answer to three significant digits.

Example 2.2

Assume an isothermal atmosphere and approximate the pressure at 10 000 m. Calculate the percent error when compared with the values using Eq. 2.4.8 and from Appendix B.3. Use a temperature of 256 K, the temperature at 5000 m.

Solution

Integrate Eq. 2.4.5 assuming that T is constant, as follows:

$$\int_{101}^{p} \frac{dp}{p} = -\frac{g}{RT} \int_{0}^{z} dz$$

$$\ln \frac{p}{101} = -\frac{gz}{RT} \quad \text{or} \quad p = 101e^{-gz/RT}$$

Substituting $z = 10\ 000$ m and $T = 256$ K, there results

$$p = 101e^{-9.81 \times 10\ 000/(287 \times 256)}$$

$$= 26.57 \text{ kPa}$$

Using Eq. 2.4.8 we have

$$p = p_{\text{atm}}\left(\frac{T_0 - \alpha z}{T_0}\right)^{g/\alpha R}$$

$$= 101\left(\frac{288 - 0.0065 \times 10\ 000}{288}\right)^{9.81/0.0065 \times 287} = 26.3 \text{ kPa}$$

The actual pressure at 10 000 m is found from Table B.3 to be 26.50 kPa. Hence the percent errors are

$$\% \text{ error} = \left(\frac{26.57 - 26.3}{26.3}\right) \times 100 = 1.03\%$$

$$\% \text{ error} = \left(\frac{26.57 - 26.50}{26.50}\right) \times 100 = 0.26\%$$

Because the error is so small, we often assume the atmosphere to be isothermal. *Note:* When evaluating gz/RT we use $R = 287$ J/kg·K, not 0.287 kJ/kg·K. To observe that gz/R is dimensionless, use N = kg·m/s² so that

$$\left[\frac{gz}{RT}\right] = \frac{(\text{m/s}^2)\text{m}}{(\text{J/kg·K})\text{K}} = \frac{\text{m}^2/\text{s}^2}{\text{N·m/kg}} = \frac{\text{m}^2/\text{s}^2}{(\text{kg·m}^2/\text{s}^2)/\text{kg}} = \frac{\text{m}^2/\text{s}}{\text{m}^2/\text{s}}$$

2.4.3 Manometers

Manometers are instruments that use columns of liquids to measure pressures. Three such instruments, shown in Fig. 2.7, are discussed to illustrate their use. Part (a) displays a U-tube manometer, used to measure relatively small pressures. In this case the pressure in the pipe can be determined by defining a point 1 at the center of the pipe and a point 2 at the surface of the right column. Then, using Eq. 2.4.3,

$$p_1 + \gamma z_1 = p_2 + \gamma z_2$$

where the datum from which z_1 and z_2 are measured is located at any desired position, such as through point 1. Since $p_2 = 0$ (gage pressure is selected; if absolute pressure is desired, we would select $p_2 = p_{atm}$) and $z_2 - z_1 = h$,

$$p_1 = \gamma h \tag{2.4.12}$$

Figure 2.7b shows a manometer used to measure relatively large pressures since we can select γ_2 to be quite large; for example, we could select γ_2 to be that of mercury so that $\gamma_2 = 13.6\ \gamma_{water}$. The pressure can be determined by introducing the points indicated. This is necessary because Eq. 2.4.3 applies throughout one fluid; γ must be constant. The value of γ changes abruptly at point 2. The pressure at point 2 and at point 2' is the same since the points are at the same elevation in the same fluid. Hence

$$p_2 = p_2{}'$$
$$p_1 + \gamma_1 h = p_3 + \gamma_2 H \tag{2.4.13}$$

Setting $p_3 = 0$ (gage pressure is used) results in

$$p_1 = -\gamma_1 h + \gamma_2 H \tag{2.4.14}$$

Figure 2.7c shows a micromanometer that is used to measure very small pressure changes. Introducing five points as indicated, we can write

$$p_1 + \gamma_1(z_1 - z_2) + \gamma_2(z_2 - z_3) = p_5 + \gamma_2(z_5 - z_4) + \gamma_3(z_4 - z_3) \tag{2.4.15}$$

Observing that $z_2 - z_3 + h = H + z_5 - z_4$ and setting $p_5 = 0$ lead to

$$
\begin{aligned}
p_1 &= \gamma_1(z_2 - z_1) + \gamma_2(h - H) + \gamma_3 H \\
&= \gamma_1(z_2 - z_1) + \gamma_2 h + (\gamma_3 - \gamma_2)H
\end{aligned}
\tag{2.4.16}
$$

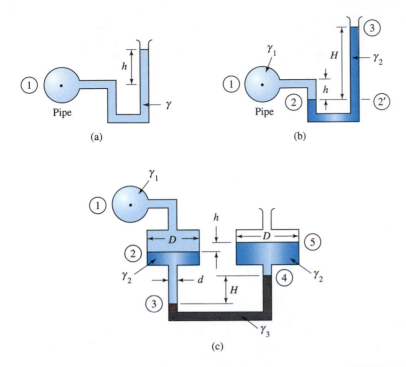

FIGURE 2.7   Manometers: (a) U-tube manometer (small pressures); (b) U-tube manometer (large pressures); (c) micromanometer (very small pressure changes).

Note that in all of the equations above for all three manometers, we have identified all interfaces with a point. This is always necessary when analyzing a manometer. The micromanometer is capable of measuring small pressure changes because a small pressure change in p_1 results in a relatively large deflection H. The change in H due to a change in p_1 can be determined using Eq. 2.4.16. Suppose that p_1 increases by Δp_1 and, as a result, z_2 decreases by Δz; then h and H also change. Using the fact that a decrease in z_2 is accompanied by an increase in z_5 leads to an increase in h of $2\Delta z$ and, similarly, assuming that the volumes are conserved, it can be shown that H increases by $2\Delta z D^2/d^2$. Hence a pressure change Δp_1 can be evaluated from changes in deflections as follows:

$$\Delta p_1 = \gamma_1(-\Delta z) + \gamma_2(2\Delta z) + \frac{(\gamma_3 - \gamma_2)2\Delta z D^2}{d^2} \qquad (2.4.17)$$

The rate of change in H with p_1 is

$$\frac{\Delta H}{\Delta p_1} = \frac{2\Delta z D^2/d^2}{\Delta p_1} \qquad (2.4.18)$$

Using Eq. 2.4.17 we have

$$\frac{\Delta H}{\Delta p_1} = \frac{2D^2/d^2}{-\gamma_1 + 2\gamma_2 + 2(\gamma_3 - \gamma_2)D^2/d^2} \tag{2.4.19}$$

An example of this type of manometer is given in Example 2.4.

Example 2.3

Water and oil flow in horizontal pipelines. A double U-tube manometer is connected between the pipelines, as shown in Fig. E2.3. Calculate the pressure difference between the water pipe and the oil pipe.

Solution

We first identify the relevant points as shown in the figure. Begin at point ① and add pressure when the elevation decreases and subtract pressure when the elevation increases until point ⑤ is reached:

$$p_1 + \gamma(z_1 - z_2) - \gamma S_1(z_3 - z_2) - \gamma S_{air}(z_4 - z_3) + \gamma S_2(z_4 - z_5) = p_5$$

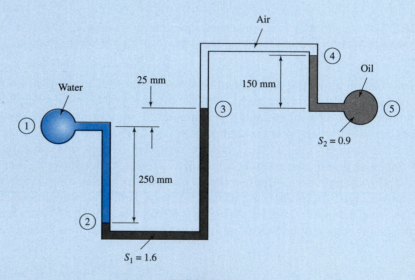

FIGURE E2.3

where $\gamma = 9810 \text{ N/m}^3$, $S_1 = 1.6$, $S_2 = 0.9$, and $S_{air} \approx 0$. Thus

$$p_1 - p_5 = 9810 \left(-\frac{250}{1000} + 1.6 \times \frac{275}{1000} + 0 \times \frac{150}{1000} - 0.9 \times \frac{150}{1000} \right)$$

$$= 539.6 \text{ Pa} \quad \text{or} \quad 0.5396 \text{ kPa}$$

Note that by neglecting the weight of the air, the pressure at point 3 is equal to the pressure at point 4.

Example 2.4

For a given condition the liquid levels in Fig. 2.7c are $z_1 = 0.95$ m, $z_2 = 0.70$ m, $z_3 = 0.52$ m, $z_4 = 0.65$ m, and $z_5 = 0.72$ m. Further, $\gamma_1 = 9810$ N/m^3, $\gamma_2 = 11\,500$ N/m^3, and $\gamma_3 = 14\,000$ N/m^3. The diameters are $D = 0.2$ m and $d = 0.01$ m. (a) Calculate the pressure p_1 in the pipe, (b) calculate the change in H if p_1 increases by 100 Pa, and (c) calculate the change in h of the manometer of Fig. 2.7a if $h = 0.5$ m of water and $\Delta p = 100$ Pa.

Solution

(a) Referring to Fig. 2.7c, we have

$$h = 0.72 - 0.70 = 0.02 \text{ m}$$

$$H = 0.65 - 0.52 = 0.13 \text{ m}$$

Substituting the given values into Eq. (2.4.16) leads to

$$p_1 = \gamma_1(z_2 - z_1) + \gamma_2 h + (\gamma_3 - \gamma_2)H$$
$$= 9810(0.70 - 0.95) + 11\,500(0.02) + (14\,000 - 11\,500)(0.13)$$
$$= -1898 \text{ Pa}$$

(b) If the pressure p_1 is increased by 100 Pa to $p_1 = -1798$ Pa, the change in H is, using Eq. 2.4.19,

$$\Delta H = \Delta p_1 \frac{2D^2/d^2}{-\gamma_1 + 2\gamma_2 + 2(\gamma_3 - \gamma_2)D^2/d^2}$$

$$\Delta H = 100 \frac{2(20^2)}{-9810 + 2(11\,500) + 2(14\,000 - 11\,500) \times 20^2} = 0.0397 \text{ m}$$

Thus H increases by 3.97 cm as a result of increasing the pressure by 100 Pa.

(c) For the manometer in Fig. 2.7a, the pressure p_1 is given by $p = \gamma h$. Assume that initially $h = 0.50$ m. Thus the pressure initially is

$$p_1 = 9810 \times 0.50 = 4905 \text{ Pa}$$

Now if p_1 is increased by 100 Pa, h can be found:

$$p_1 = \gamma h$$

$$h = \frac{p_1}{\gamma} = \frac{5005}{9810} = 0.510 \text{ m}. \quad \therefore \Delta h = 0.510 - 0.5 = 0.01 \text{ m}$$

Thus an increase of 100 Pa increases h by 1 cm in the manometer shown in part (a), 25% of the change in the micromanometer.

2.4.4 Forces on Plane Areas

In the design of devices and objects that are submerged, such as dams, flow obstructions, surfaces on ships, and holding tanks, it is necessary to calculate the magnitudes and locations of forces that act on both plane and curved surfaces. In this section we consider only plane surfaces, such as the plane surface of general shape shown in Fig. 2.8. Note that a side view is given as well as a view showing

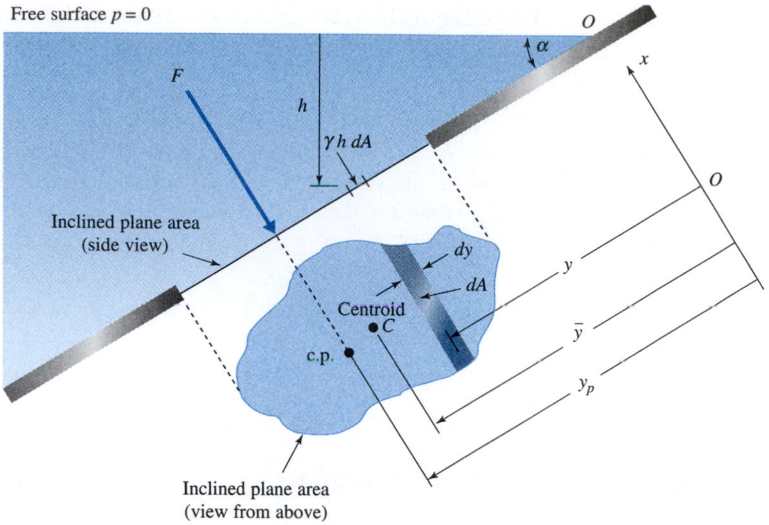

FIGURE 2.8 Force on an inclined plane area.

the shape of the plane. The total force of the liquid on the plane surface is found by integrating the pressure over the area, that is,

$$F = \int_A p \, dA \tag{2.4.20}$$

where we usually use gage pressure. Atmospheric pressure cancels out since it acts on both sides of the area. The x and y coordinates are in the plane of the plane surface, as shown. Assuming that $p = 0$ at $h = 0$, we know that

$$
\begin{aligned}
p &= \gamma h \\
&= \gamma y \sin \alpha
\end{aligned}
\tag{2.4.21}
$$

where h is measured vertically down from the free surface to the elemental area dA and y is measured from point O on the free surface. The force may then be expressed as

$$
\begin{aligned}
F &= \int_A \gamma h \, dA \\
&= \gamma \sin \alpha \int_A y \, dA
\end{aligned}
\tag{2.4.22}
$$

The distance to a centroid is defined as

$$\bar{y} = \frac{1}{A} \int_A y \, dA \tag{2.4.23}$$

The expression for the force then becomes

$$
\begin{aligned}
F &= \gamma \bar{y} A \sin \alpha \\
&= \gamma \bar{h} A = p_c A
\end{aligned}
\tag{2.4.24}
$$

where $\bar{h}$ is the vertical distance from the free surface to the centroid of the area and p_c is the pressure at the centroid. Thus we see that the magnitude of the force on a plane surface is the pressure at the centroid multiplied by the area. The force does not, in general, act at the centroid.

To find the location of the resultant force F, we note that the sum of the moments of all the infinitesimal pressure forces acting on the area A must equal the moment of the resultant force. Let the force F act at the point (x_p, y_p), the **center of pressure** (c.p.). The value of y_p can be obtained by equating moments about the x-axis:

$$y_p F = \int_A yp \, dA$$

$$= \gamma \sin \alpha \int_A y^2 \, dA = \gamma I_x \sin \alpha \tag{2.4.25}$$

where the second moment of the area about the x-axis is

$$I_x = \int_A y^2 \, dA \tag{2.4.26}$$

The second moment of an area is related to the second moment of an area $\bar{I}$ about the centroidal axis by the parallel-axis-transfer theorem,

$$I_x = \bar{I} + A\bar{y}^2 \tag{2.4.27}$$

Substitute Eqs. 2.4.24 and 2.4.27 into Eq. 2.4.25, and obtain

$$y_p = \frac{\gamma(\bar{I} + A\bar{y}^2)\sin \alpha}{\gamma \bar{y} A \sin \alpha}$$

$$= y + \frac{\bar{I}}{A\bar{y}} \tag{2.4.28}$$

where $\bar{y}$ is measured parallel to the plane area along the y-axis.

Centroids and moments for several areas are presented in Appendix C. Using the expression above, we can show that the force on a rectangular gate, with the top edge even with the liquid surface as shown in Fig. 2.9, acts two-thirds of the way down. This is also obvious considering the triangular pressure distribution acting on the gate. Note that Eq. 2.4.28 shows that y_p is always greater

KEY CONCEPT   *The force on a plane surface is the pressure at the centroid multiplied by the area.*

Center of pressure: *The point where the resultant force acts.*

KEY CONCEPT   *The force on a rectangular gate, with the top edge even with the liquid surface, acts two-thirds of the way down.*

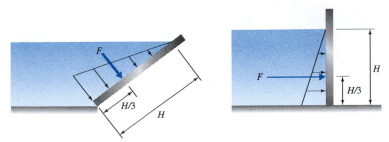

FIGURE 2.9   Force on a plane area with top edge in a free surface.

than $\bar{y}$; that is, the resultant force of the liquid on a plane surface always acts below the centroid of the area, except on a horizontal area for which $\bar{y} = \infty$; then the center of pressure and the centroid coincide.

Similarly, to locate the x-coordinate x_p of the c.p., we write

$$x_p F = \int_A xp \, dA$$

$$= \gamma \sin \alpha \int_A xy \, dA = \gamma I_{xy} \sin \alpha \qquad (2.4.29)$$

where the product of inertia of the area A is

$$I_{xy} = \int_A xy \, dA \qquad (2.4.30)$$

Using the transfer theorem for the product of inertia,

$$I_{xy} = \bar{I}_{xy} + A\bar{x}\bar{y} \qquad (2.4.31)$$

Equation 2.4.29 becomes

$$x_p = \bar{x} + \frac{\bar{I}_{xy}}{A\bar{y}}$$

$$(2.4.32)$$

We now have expressions for the coordinates locating the center of pressure.

Finally, we should note that the force F in Fig. 2.8 is the result of a *pressure prism* acting on the area. For the rectangular area shown in Fig. 2.10, the pressure increases, as shown by the pressure distribution in Fig. 2.10b. If we form the integral $\int p \, dA$, we obtain the volume of the pressure prism, which equals the force F acting on the area, shown in Fig. 2.10c. The force acts through the centroid of the volume. For the rectangular area shown in Fig. 2.10a, the volume could be divided into two volumes: a rectangular volume with centroid at its center, and a triangular volume with centroid one third the distance from the appropriate base. The location of the force is then found by locating the centroid of the composite volume.

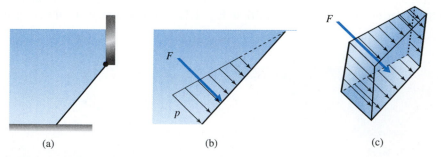

(a)   (b)   (c)

FIGURE 2.10   Pressure prism: (a) rectangular area; (b) pressure distribution on the area; (c) pressure prism.

Example 2.5

A plane area of 80 cm $\times$ 80 cm acts as a window on a submersible in the Great Lakes. If it is on a 45° angle with the horizontal, what force applied normal to the window at the bottom edge is needed to just open the window, if it is hinged at the top edge when the top edge is 10 m below the surface? The pressure inside the submersible is assumed to be atmospheric.

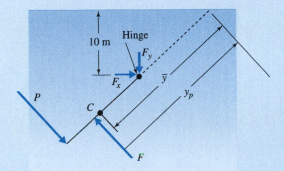

FIGURE E2.5

Solution

First, a sketch of the window would be very helpful as in Fig. E25. The force of the water acting on the window is

$$F = \gamma \bar{h} A$$
$$= 9810(10 + 0.4 \times \sin 45°)(0.8 \times 0.8) = 64\,560 \text{ N}$$

The distance $\bar{y}$ is

$$\bar{y} = \frac{\bar{h}}{\sin 45°} = \frac{10 + 0.4 \times \sin 45°}{\sin 45°} = 14.542 \text{ m}$$

so that

$$y_p = \bar{y} + \frac{\bar{I}}{A\bar{y}}$$
$$= 14.542 + \frac{0.8 \times 0.8^3/12}{(0.8 \times 0.8) \times 14.542} = 14.546 \text{ m}$$

Taking moments about the hinge provides the needed force P to open the window:

$$0.8P = (y_p - \bar{y} + 0.4)F$$

$$\therefore \quad P = \frac{14.546 - 14.542 + 0.4}{0.8} 64\,560 = 32\,610 \text{ N}$$

Alternatively, we could have sketched the pressure prism, composed of a rectangular volume and a triangular volume. Moments about the top hinge would provide the desired force.

Example 2.6

Find the location of the resultant force F of the water on the triangular gate and the force P necessary to hold the gate in the position shown in Fig. E2.6a.

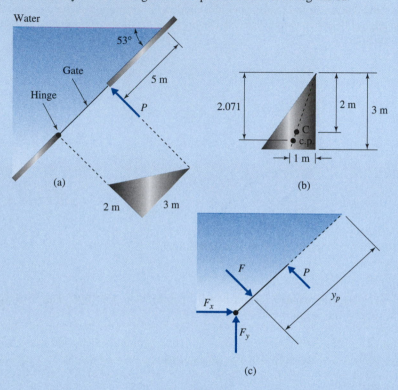

(a)

(b)

(c)

FIGURE E2.6

Solution

First we draw a free-body diagram of the gate, including all the forces acting on the gate (Fig. E2.6c). We have neglected the weight of the gate. The centroid of the gate is shown in Fig. E2.6b. The y-coordinate of the location of the resultant F can be found using Eq. 2.4.28 as follows:

$$\bar{y} = 2 + 5 = 7$$

$$y_p = \bar{y} + \frac{\bar{I}}{A\bar{y}}$$

$$= 7 + \frac{2 \times 3^3/36}{3 \times 7} = 7.071 \text{ m}$$

To find x_p we could use Eq. 2.4.32. Rather than that, we recognize that the resultant force must act on a line connecting the vertex and the midpoint of the opposite side since each infinitesimal force acts on this line (the moment of the resultant must equal the moment of its components). Thus using similar triangles we have

$$\frac{x_p}{1} = \frac{2.071}{3}$$

$$\therefore \quad x_p = 0.690 \text{ m}$$

The coordinates x_p and y_p locate where the force due to the water acts on the gate.

If we take moments about the hinge, assumed to be frictionless, we can determine the force P necessary to hold the gate in the position shown:

$$\Sigma \, M_{\text{hinge}} = 0$$

$$\therefore \quad 3 \times P = (3 - 2.071)F$$

$$= 0.929 \times \gamma \bar{h} A$$

$$= 0.929 \times 9810 \times (7 \sin 53°) \times 3$$

where $\bar{h}$ is the vertical distance from the centroid to the free surface. Hence

$$P = 50\,900 \text{ N} \quad \text{or} \quad 50.9 \text{ kN}$$

2.4.5   Forces on Curved Surfaces

We do not use a direct method of integration to find the force due to the hydro-static pressure on a curved surface. Rather, a free-body diagram that contains the curved surface and the liquids directly above or below the curved surface is iden-tified. Such a free-body diagram contains only plane surfaces upon which unknown fluid forces act; these unknown forces can be found as in the preceding section.

As an example, let us determine the force of the curved gate on the stop, shown in Fig. 2.11a. The free-body diagram, which includes the gate and some of the water contained directly above the gate, is shown in Fig. 2.11b; the forces F_x and F_y are the horizontal and vertical components, respectively, of the force act-ing on the hinge; F_1 and F_2 are due to the surrounding water and are the resul-tant forces of the pressure distributions shown; the body force F_W is due to the weight of the water shown. By summing moments about an axis passing through the hinge, we can determine the force P acting on the stop.

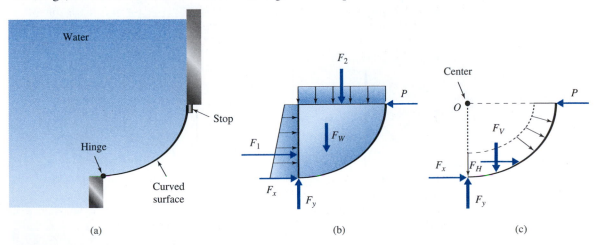

FIGURE 2.11   Forces acting on a curved surface: (a) curved surface; (b) free-body diagram of water and gate; (c) free-body diagram of gate only.

If the curved surface is a quarter circle, the problem can be greatly simplified. This is observed by considering a free-body diagram of the gate only (see Fig. 2.11c). The horizontal force F_H acting on the gate is equal to F_1 of Fig. 2.11b, and the component F_V is equal to the combined force $F_2 + F_W$ of Figure 2.11b. Now, F_H and F_V are due to the differential pressure forces acting on the circular arc; each differential pressure force acts through the center of the circular arc. Hence the resultant force $\mathbf{F}_H + \mathbf{F}_V$ (this is a vector addition) must act through the center. Consequently, we can locate the components F_H and F_V at the center of the quarter circle, resulting in a much simpler problem. Example 2.7 will illustrate.

KEY CONCEPT *The resultant force $F_H + F_V$ must act through the center of the circular arc.*

If the pressure on the free surface is p_0, we can simply add a depth of liquid necessary to provide p_0 at the location of the free surface, and then work the resulting problem, with a fictitious free surface located the appropriate distance above the original free surface. Or, the pressure force $p_0 A$ is added to the force F_2 of Fig. 2.11b.

Example 2.7

Calculate the force P necessary to hold the 4-m-wide gate in the position shown in Fig. E2.7a. Neglect the weight of the gate.

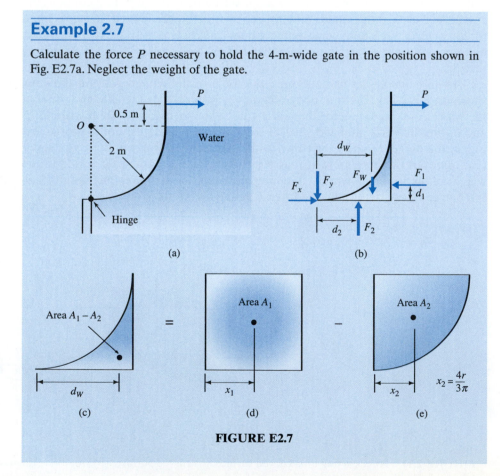

FIGURE E2.7

Solution

The first step is to draw a free-body diagram of the gate. One choice is to select the gate and the water directly below the gate as shown in Fig. E2.7b. To calculate P, we must determine F_1, F_2, F_W, d_1, d_2, and d_W; then moments about the hinge will allow us to find P. The force components are given by

$$F_1 = \gamma \bar{h}_1 A_1$$
$$= 9810 \times 1 \times 8 = 78\ 480 \text{ N}$$
$$F_2 = \gamma \bar{h}_2 A_2$$
$$= 9810 \times 2 \times 8 = 156\ 960 \text{ N}$$
$$F_W = \gamma \mathcal{V}_{\text{water}}$$
$$= 9810 \times 4\left(4 - \frac{\pi \times 2^2}{4}\right) = 33\ 700 \text{ N}$$

The distance d_W is the distance to the centroid of the volume. It can be determined by considering the area as the difference of a square and a quarter circle as shown in Fig. E2.7c–e. Moments of areas yield

$$d_W(A_1 - A_2) = x_1 A_1 - x_2 A_2$$
$$d_W = \frac{x_1 A_1 - x_2 A_2}{A_1 - A_2}$$
$$= \frac{1 \times 4 - (4 \times 2/3\pi) \times \pi}{4 - \pi} = 1.553 \text{ m}$$

The distance $d_2 = 1$ m. Because F_1 is due to a triangular pressure distribution (see Fig. 2.9), d_1 is given by

$$d_1 = \frac{1}{3}(2) = 0.667 \text{ m}$$

Summing moments about the frictionless hinge gives

$$2.5P = d_1 F_1 + d_2 F_2 - d_W F_W$$
$$P = \frac{0.667 \times 78.5 + 1 \times 157.0 - 1.553 \times 33.7}{2.5} = 62.8 \text{ kN}$$

Rather than the somewhat tedious procedure above, we could observe that all the infinitesimal forces that make up the resultant force $(\mathbf{F}_H + \mathbf{F}_V)$ acting on the circular arc pass through the center O, as noted in Fig. 2.11c. Since each infinitesimal force passes through the center, the resultant force must also pass through the center. Hence we could have located the resultant force $(\mathbf{F}_H + \mathbf{F}_V)$ at point O. If F_V and F_H were located at O, F_V would pass through the hinge, producing no moment about the hinge. Then, realizing that $F_H = F_1$ and summing moments about the hinge gives

$$2.5P = 2F_H$$

Therefore,

$$P = 2 \times \frac{78.48}{2.5} = 62.8 \text{ kN}$$

This was obviously much simpler. All we needed to do was calculate F_H and then sum moments!

Example 2.8

Find the force P needed to hold the gate in the position shown in Fig. E2.8a if P acts 3 m from the y-axis. The parabolic gate is 150 cm wide.

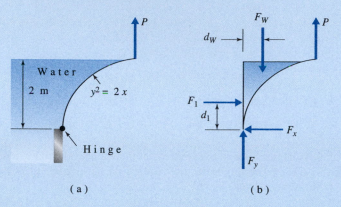

(a)   (b)

FIGURE E2.8

Solution

A free-body diagram of the gate and the water directly above the gate is shown in Fig. E2.8b. The forces are found to be

$$F_1 = \gamma \bar{h} A$$
$$= 9810 \times 1 \times (2 \times 1.5) = 29\,430 \text{ N}$$
$$F_W = \gamma \mathcal{V}$$
$$= 9810 \int_0^2 1.5x \, dy = 14\,715 \int_0^2 \frac{y^2}{2} \, dy = 14\,715 \frac{2^3}{6} = 19\,620 \text{ N}$$

The distance d_1 is $\frac{1}{3}(2) = 0.667$ m since the top edge is in the free surface. The distance d_W through the centroid is found using a horizontal strip:

$$d_W = \frac{\int_0^2 x(x/2) \, dy}{\int_0^2 x \, dy} = \frac{\frac{1}{8}\int_0^2 y^4 \, dy}{\frac{1}{2}\int_0^2 y^2 \, dy} = \frac{1}{4}\frac{2^5/5}{2^3/3} = 0.6 \text{ m}$$

Sum moments about the hinge and find P as follows:

$$3P = d_1 F_1 + d_W F_W$$
$$= 0.667 \times 29\,430 + 0.6 \times 19\,620 \therefore P = 10470 \text{ N}$$

2.4.6 Buoyancy

The law of buoyancy, known as Archimedes' principle, dates back some 2200 years to the Greek philosopher Archimedes. Legend has it that Hiero, king of Syracuse, suspected that his new gold crown may have been constructed of materials other than pure gold, so he asked Archimedes to test it. Archimedes probably made a lump of pure gold that weighed the same as the crown. The lump was discovered to weigh more in water than the crown weighed in water, thereby proving to Archimedes that the crown was not pure gold. The fake material possessed a larger volume to have the same weight as gold, hence it displaced more water. *Archimedes' principle* is: There is a buoyancy force on an object equal to the weight of displaced liquid.

To prove the law of buoyancy, consider the submerged body shown in Fig. 2.12a. In part (b) a cylindrical free-body diagram is shown that includes the submerged body with weight W and liquid having a weight F_W; the cross-sectional area A is the maximum cross-sectional area of the body. From the diagram we see that the resultant vertical force acting on the free-body diagram due to the water only (do not include W) is equal to

$$\Sigma F = F_2 - F_1 - F_W \tag{2.4.33}$$

This resultant force is by definition the *buoyant force* F_B. It can be expressed as

$$F_B = \gamma(h_2 A - h_1 A - V_W) \tag{2.4.34}$$

> **KEY CONCEPT**
> *Archimedes' principle states the buoyancy force on an object equals the weight of displaced liquid.*

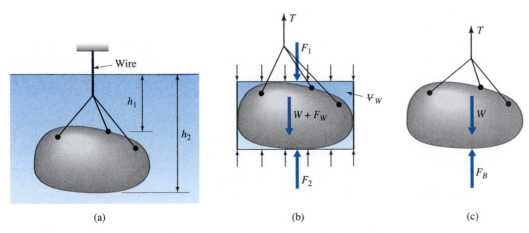

|(a)|(b)|(c)|

FIGURE 2.12   Forces on a submerged body: (1) submerged body; (b) free-body diagram; (c) free body showing F_B.

where V_W is the liquid volume included in the free-body diagram. Recognizing that the volume of the submerged body is

$$V_B = (h_2 - h_1)A - V_W \qquad (2.4.35)$$

we see from Eq. 2.4.34 that

$$F_B = \gamma V_{\text{displaced liquid}} \qquad (2.4.36)$$

thereby proving the law of buoyancy.

The force necessary to hold the submerged body in place (see Fig. 2.12c) is equal to

$$T = W - F_B \qquad (2.4.37)$$

where W is the weight of the submerged body.

KEY CONCEPT *The buoyant force acts through the centroid of the displaced liquid volume.*

For a floating object, as in Fig. 2.13, the buoyant force is

$$F_B = \gamma V_{\text{displaced liquid}} \qquad (2.4.38)$$

Obviously, $T = 0$, so that Eq. 2.4.36 gives

$$F_B = W \qquad (2.4.39)$$

where W is the weight of the floating object.

From the foregoing analysis it is apparent that the buoyant force F_B acts through the centroid of the displaced liquid volume. For the floating object, the weight of the object acts through its center of gravity, so the center of gravity of the object must lie on the same vertical line as the centroid of the liquid volume.

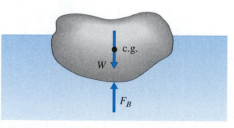

FIGURE 2.13 Forces on a floating object.

A **hydrometer**, an instrument used to measure the specific gravity of liquids, operates on the principle of buoyancy. A sketch is shown in Fig. 2.14. The upper part, the stem, has a constant diameter. When placed in pure water the specific gravity is marked to read 1.0. The force balance is

$$W = \gamma_{water} \forall \qquad (2.4.40)$$

where W is the weight of the hydrometer and $\forall$ is the submerged volume below the $S = 1.0$ line. In an unknown liquid of specific weight γ_x, a force balance would be

$$W = \gamma_x(\forall - A\Delta h) \qquad (2.4.41)$$

where A is the cross-sectional area of the stem. Equating these two expressions gives

$$\Delta h = \frac{\forall}{A}\left(1 - \frac{1}{S_x}\right) \qquad (2.4.42)$$

where $S_x = \gamma_x/\gamma_{water}$. For a given hydrometer, $\forall$ and A are fixed so that the quantity Δh is dependent only on the specific gravity S_x. Thus the stem can be calibrated to read S_x directly. Hydrometers are used to measure the amount of antifreeze in the radiator of an automobile, or the charge in a battery since the density of the fluid changes as H_2SO_4 is consumed or produced.

Hydrometer: *An instrument used to measure the specific gravity of liquids.*

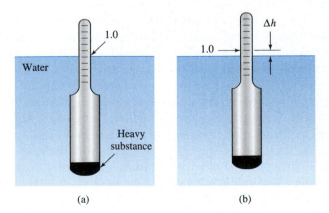

Water

Heavy substance

1.0

Δh

1.0

(a)

(b)

FIGURE 2.14 Hydrometer: (a) in water; (b) in an unknown liquid.

Example 2.9

The specific weight and the specific gravity of a body of unknown composition are desired. Its weight in air is found to be 1000 N and in water it weighs 750 N.

Solution

The volume is found from a force balance when submerged as follows (see Fig. 2.12c):

$$T = W - F_B$$
$$750 = 1000 - 9810 V \quad \therefore V = 0.02548 \text{ m}^3$$

The specific weight is then

$$\gamma = \frac{W}{V} = \frac{1000}{0.02548} = 39246 \text{ N/m}^3$$

The specific gravity is found to be

$$S = \frac{\gamma}{\gamma_{\text{water}}} = \frac{39246}{9810} = 4.00$$

2.4.7 Stability

The notion of stability can be demonstrated by considering the vertical stability of a floating object. If the object is raised a small distance, the buoyant force decreases and the object's weight returns the object to its original position. Conversely, if a floating object is lowered slightly, the buoyant force increases and the larger buoyant force returns the object to its original position. Thus a floating object has vertical stability since a small departure from equilibrium results in a restoring force.

KEY CONCEPT *A floating object has vertical stability.*

Consider now the rotational stability of a submerged body, shown in Fig. 2.15. In part (a) the center of gravity, G, of the body is above the centroid C (also referred to as the **center of buoyancy**) of the displaced volume and a small angular rotation results in a moment that will continue to increase the rotation; hence the body is unstable and overturning would result. If the center of gravity is below the centroid, as in part (c), a small angular rotation provides a restoring moment and the body is stable. Part (b) shows neutral stability for a body in which the center of gravity and the centroid coincide, a situation that is encountered whenever the density is constant throughout the submerged body.

Center of buoyancy: *Centroid of a floating body.*

Next, consider the rotational stability of a floating body. If the center of gravity is below the centroid, the body is always stable, as with the submerged body of Fig. 2.15c. The body may be stable, though, even if the center of gravity is above the centroid, as sketched in Fig. 2.16a. When the body rotates the centroid of the volume of displaced liquid moves to the new location C', shown in part (b). If the centroid C' moves sufficiently far, a restoring moment develops and the body is stable, as shown. This is determined by the *metacentric height* $\overline{GM}$ defined as the distance from G to the point of intersection of the buoyant force before rotation

KEY CONCEPT *If GM is positive, the body is stable.*

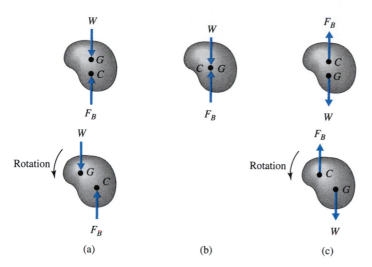

FIGURE 2.15 Stability of a submerged body: (a) unstable; (b) neutral; (c) stable.

with the buoyant force after rotation. If $\overline{GM}$ is positive, as shown, the body is stable; if $\overline{GM}$ is negative (M lies below G), the body is unstable.

To determine a quantitative relationship for the distance $\overline{GM}$ refer to the sketch of Fig. 2.17, which shows the uniform cross section. Let us find an expression for $\overline{x}$, the x-coordinate of the centroid of the displaced volume. It can be found by considering the volume to be the original volume plus the added wedge with cross-sectional area DOE minus the subtracted wedge with cross-sectional area AOB; to locate the centroid of a composite volume, we take moments as follows:

$$\overline{x}\,V = \overline{x}_0 V_0 + \overline{x}_1 V_1 - \overline{x}_2 V_2 \tag{2.4.43}$$

where V_0 is the original volume below the water line, V_1 is the area DOE times the length, V_2 is the area AOB times the length; the cross section is assumed to be uniform so that the length l is constant for the body. The quantity $\overline{x}_0$, the x-coordinate of point C, is zero. The remaining two terms can best be represented by integrals so that

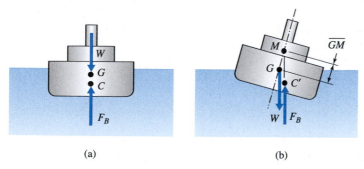

FIGURE 2.16 Stability of a floating body: (a) equilibrium position; (b) rotated position.

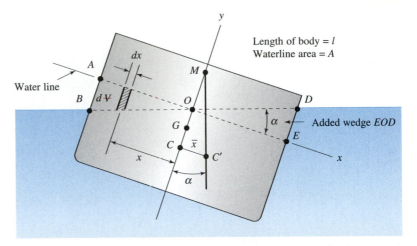

FIGURE 2.17 Uniform cross section of a floating body.

$$\bar{x}\,V = \int_{V_1} x\,dV - \int_{V_2} x\,dV \tag{2.4.44}$$

Then $dV = x \tan \alpha\, dA$ in volume 1 and $dV = -x \tan \alpha\, dA$ in volume 2, where $dA = l\,dx$, l being the constant length of the body. The equation above becomes

$$\bar{x}\,V = \tan \alpha \int_{A_1} x^2\,dA + \tan \alpha \int_{A_2} x^2\,dA$$

$$= \tan \alpha \int_A x^2\,dA$$

$$= \tan \alpha\, I_O \tag{2.4.45}$$

where I_O is the second moment (moment of inertia) of the waterline area about an axis passing through the origin O. The waterline area would be the length $\overline{AE}$ times the length l of the body if l were of constant length. Using $\bar{x} = \overline{CM} \tan \alpha$, we can write

$$\overline{CM}\,V = I_O \tag{2.4.46}$$

or, with $\overline{CG} + \overline{GM} = \overline{CM}$, we have

$$\overline{GM} = \frac{I_O}{V} - \overline{CG} \tag{2.4.47}$$

For a given body orientation, if $\overline{GM}$ is positive, the body is stable. Even though this relationship (2.4.47) is derived for a floating body with uniform cross section, it is applicable for floating bodies in general. We will apply it to a floating cylinder in the following example.

Example 2.10

A 0.25-m-diameter cylinder is 0.25 m long and composed of material with specific weight 8000 N/m³. Will it float in water with the ends horizontal?

Solution

With the ends horizontal, I_O will be the second moment of the circular cross section,

$$I_O = \frac{\pi d^4}{64} = \frac{\pi \times 0.25^4}{64} = 0.000192 \text{ m}^4$$

The displaced volume will be

$$V = \frac{W}{\gamma_{\text{water}}} = \frac{8000 \times \pi/4 \times 0.25^2 \times 0.25}{9810} = 0.0100 \text{ m}^3$$

The depth the cylinder sinks in the water is

$$depth = \frac{V}{A} = \frac{0.01}{\pi \times 0.25^2/4} = 0.204 \text{ m}$$

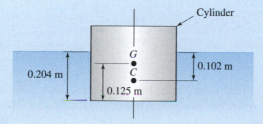

FIGURE E2.10

Hence, the distance $\overline{CG}$, as shown in Fig. E2.10, is

$$\overline{CG} = 0.125 - \frac{0.204}{2} = 0.023 \text{ m}$$

Finally,

$$\overline{GM} = \frac{0.000192}{0.01} - 0.023 = -0.004 \text{ m}$$

This is a negative value showing that the cylinder will not float with ends horizontal. It would undoubtedly float on its side.

2.5 LINEARLY ACCELERATING CONTAINERS

In this section the fluid will be at rest relative to a reference frame that is linearly accelerating with a horizontal component a_x and a vertical component a_z. Then Eq. 2.3.6 simplifies to

$$dp = -\rho a_x \, dx - \rho(g + a_z) \, dz \qquad (2.5.1)$$

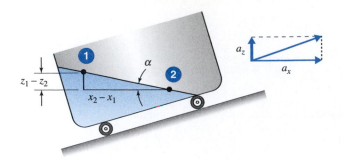

FIGURE 2.18 Linearly accelerating tank.

Integrating between two arbitrary points 1 and 2 results in

$$p_2 - p_1 = -\rho a_x(x_2 - x_1) - \rho(g + a_z)(z_2 - z_1) \qquad (2.5.2)$$

If points 1 and 2 lie on a constant-pressure line, such as the free surface in Fig. 2.18, $p_2 - p_1 = 0$ and we have

$$\frac{z_1 - z_2}{x_2 - x_1} = \tan \alpha = \frac{a_x}{g + a_z} \qquad (2.5.3)$$

where α is the angle that the constant-pressure line makes with the horizontal.

In the solution of problems involving liquids, we must often utilize the conservation of mass and equate the volumes before and after the acceleration is applied. After the acceleration is initially applied, sloshing may occur. Our analysis will assume that sloshing is not present; either sufficient time passes to dampen out time-dependent motions, or the acceleration is applied in such a way that such motions are minimal.

Example 2.11

The tank shown in Fig. E2.11a is accelerated to the right. Calculate the acceleration a_x needed to cause the free surface, shown in Fig. E2.11b, to touch point A. Also, find p_B and the total force acting on the bottom of the tank if the width is 1 m.

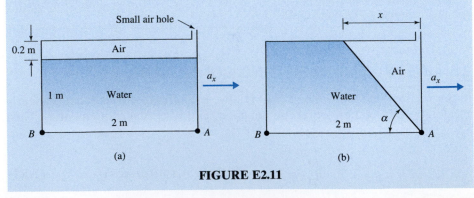

FIGURE E2.11

Solution

The angle the free surface takes is found by equating the air volume (actually, areas since the width is constant) before and after since no water spills out,

$$0.2 \times 2 = \tfrac{1}{2}(1.2x)$$

$$x = 0.667 \text{ m}$$

The quantity $\tan \alpha$ is now known. It is

$$\tan \alpha = \frac{1.2}{0.667} = 1.8$$

Using Eq. 2.5.3, we find a_x to be, letting $a_z = 0$,

$$a_x = g \tan \alpha$$

$$= 9.81 \times 1.8 = 17.66 \text{ m/s}^2$$

We can find the pressure at B by noting the pressure dependence on x. At A, the pressure is zero. Hence, Eq. 2.5.2 yields

$$p_B - \cancelto{0}{p_A} = -\rho a_x(x_B - x_A)$$

$$p_B = -1000 \times 17.66(-2)$$

$$= 35\,300 \text{ Pa} \quad \text{or} \quad 35.3 \text{ kPa}$$

To find the total force acting on the bottom of the tank, we realize that the pressure distribution is decreasing linearly from $p = 35.3$ kPa at B to $p = 0$ kPa at A. Hence, we can use the average pressure over the bottom of the tank:

$$F = \frac{p_B + p_A}{2} \times \text{area}$$

$$= \frac{35\,300 + 0}{2} \times 2 \times 1 = 35\,300 \text{ N}$$

2.6 ROTATING CONTAINERS

In this section we consider the situation of a liquid contained in a rotating container, such as that shown in Fig. 2.19. After a relatively short time the liquid reaches static equilibrium with respect to the container and the rotating rz-reference frame. The horizontal rotation will not alter the pressure distribution in the vertical direction. There will be no variation of pressure with respect to the θ-coordinate. Applying Newton's second law ($\Sigma F_r = ma_r$) in the r-direction to the element shown, using $\sin d\theta/2 \cong d\theta/2$, yields

$$-\frac{\partial p}{\partial r} dr\, rd\theta\, dz - prd\theta\, dz - p\, dr\, d\theta\, dz - \frac{\partial p}{\partial r}(dr)^2 d\theta\, dz$$

$$+ 2\frac{d\theta}{2p} dr\, dz + prd\theta\, dz = -\rho\, rd\theta\, dr\, dz\, r\omega^2 \qquad (2.6.1)$$

KEY CONCEPT Horizontal rotation will not alter the pressure distribution in the vertical direction.

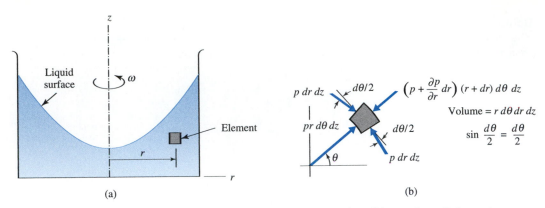

FIGURE 2.19 Rotating container: (a) liquid cross section; (b) top view of element.

where the acceleration is $r\omega^2$ toward the center of rotation. Simplify and divide by the volume $rd\theta\, dr\, dz$; then

$$\frac{\partial p}{\partial r} = \rho r \omega^2 \tag{2.6.2}$$

where we have neglected the higher-order term that contains the differential dr. The pressure differential then becomes

$$dp = \frac{\partial p}{\partial r}\, dr + \frac{\partial p}{\partial z}\, dz$$
$$= \rho r \omega^2\, dr - \rho g\, dz \tag{2.6.3}$$

where we have used the static pressure variation given by Eq. 2.3.5 with $a_z = 0$. We can now integrate between any two points (r_1, z_1) and (r_2, z_2) to obtain

$$p_2 - p_1 = \frac{\rho \omega^2}{2}(r_2^2 - r_1^2) - \rho g(z_2 - z_1) \tag{2.6.4}$$

If the two points are on a constant-pressure surface, such as the free surface, locating point 1 on the z-axis so that $r_1 = 0$, there results

$$\frac{\omega^2 r_2^2}{2} = g(z_2 - z_1) \tag{2.6.5}$$

KEY CONCEPT *The free surface is a paraboloid of revolution.*

which is the equation of a parabola. Hence the free surface is a paraboloid of revolution. The equations above can now, with the conservation of mass, be used to solve problems of interest.

Example 2.12

The cylinder shown in Fig. E2.12 is rotated about its centerline. Calculate the rotational speed that is necessary for the water to just touch the origin O and the pressures at A and B.

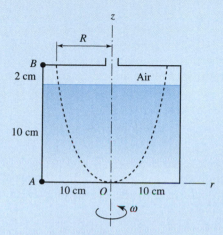

FIGURE E2.12

Solution

Since no water spills from the container, the air volume remains constant, that is,

$$\pi \times 10^2 \times 2 = \tfrac{1}{2}\pi R^2 \times 12$$

where we have used the fact that the volume of a paraboloid of revolution is one-half that of a circular cylinder with the same height and radius. This gives the value

$$R = 5.77 \text{ cm}$$

Using Eq. 2.6.5 with $r_2 = R$, we have

$$\frac{\omega^2 \times 0.0577^2}{2} = 9.81 \times 0.12$$

$$\omega = 26.6 \text{ rad/s}$$

To find the pressure at point A, we simply calculate the pressure difference between A and O. Using Eq. 2.6.4 with $r_2 = r_A = 0.1$ m, $r_1 = r_0 = 0$, and $p_1 = p_0 = 0$, there results

$$p_A = \frac{\rho\omega^2}{2}(r_A^2 - r_0^2) = \frac{1000 \times 26.6^2}{2} \times 0.1^2 = 3540 \text{ Pa} \quad \text{or} \quad 3.54 \text{ kPa}$$

The pressure at B can be found by applying Eq. 2.6.4 to points A and B. This equation simplifies to

$$p_B - p_A = -\rho g(z_B - z_A)$$

Hence

$$p_B = 3540 - 1000 \times 9.81 \times 0.12 = 2360 \text{ Pa} \quad \text{or} \quad 2.36 \text{ kPa}$$

2.7 SUMMARY

The pressure variation in the vertical z-direction in a constant density fluid is found using

$$\Delta p = -\gamma \Delta z \qquad (2.7.1)$$

This is used to interpret manometers and to establish the force on a plane as

$$F = \gamma \bar{h} A \qquad (2.7.2)$$

where $\bar{h}$ is the vertical distance to the centroid of the area. The force is located a distance from the free surface to the center of pressure parallel to the area given by

$$y_p = \bar{y} + \frac{\bar{I}}{A\bar{y}} \qquad (2.7.3)$$

where $\bar{I}$ is about the centroidal axis. Forces on curved surfaces are found using the above relationships and the weight of liquid contained above the surface.

Pressures and forces in linearly accelerating containers are determined using the angle α of a constant-pressure line:

$$\tan \alpha = \frac{a_x}{g + a_z} \qquad (2.7.4)$$

Quite often the acceleration a_z in the vertical direction is zero.

In a container rotating with angular velocity ω, a constant-pressure surface is described by

$$\tfrac{1}{2} \omega^2 r_2^2 = g(z_2 - z_1) \qquad (2.7.5)$$

where point 1 is on the axis of rotation and point 2 is anywhere on the constant-pressure surface.

PROBLEMS

Pressure

2.1 Assume the element of Fig. 2.2 to be in the yz-plane with unit depth in the x-direction. Find a result similar to that of Eq. 2.2.4. Assume gravity to be in the z-direction.

2.2 Calculate the pressure at a depth of 10 m in a liquid with specific gravity of:
(a) 1.0
(b) 0.8
(c) 13.6
(d) 1.59
(e) 0.68

2.3 What depth is necessary in a liquid to produce a pressure of 250 kPa if the specific gravity is:
(a) 1.0?
(b) 0.8?
(c) 13.6?
(d) 1.59?
(e) 0.68?

2.4 A meteorologist states that the barometric pressure is 724 mm of mercury. Convert this pressure to kilopascals.
A. 98.6 kPa
B. 97.2 kPa
C. 96.6 kPa
D. 95.6 kPa

2.5 A pressure of 140 kPa is measured at a depth of 6 m. Calculate the specific gravity and the density of the liquid if $p = 0$ on the surface.

2.6 How many meters of water are equivalent to:
(a) 760 mm Hg?
(b) 75 cm Hg?
(c) 10 mm Hg?

2.7 Determine the pressure at the bottom of an open tank if it contains layers of:
(a) 20 cm of water and 2 cm of mercury
(b) 52 mm of water and 26 mm of carbon tetrachloride
(c) 3 m of oil, 2 m of water, and 10 cm of mercury

2.8 Assuming the density of air to be constant at 1.23 kg/m³, calculate the pressure change from the top of a mountain to its base if the elevation change is 3000 m.

2.9 The pressure in the foothills of the Rockies near Boulder, Colorado is 84 kPa. The pressure, assum-ing a constant density of 1.00 kg/m³, at the top of a nearby 4000-m-high mountain is closest to:
A. 60 kPa
B. 55 kPa
C. 50 kPa
D. 45 kPa

2.10 Assume that the air pressure is 100 kPa absolute at the top of a 3-m-high wall. Assuming constant density, estimate the difference in pressure at the base of the wall if outside the wall the temperature is $-20°C$ and inside the wall it is $20°C$. This pressure difference induces an infiltration even though no wind is present.

2.11 The specific gravity of a liquid varies linearly from 1.0 at the surface to 1.1 at a depth of 10 m. Calculate the pressure at $h = 10$ m.

2.12 If the gradient of $p(x, y, z)$ in rectangular coordinates is $\nabla p = \dfrac{\partial p}{\partial x}\,\hat{\mathbf{i}} + \dfrac{\partial p}{\partial y}\,\hat{\mathbf{j}} + \dfrac{\partial p}{\partial z}\,\hat{\mathbf{k}}$, write the simplest expression for ∇p using Eq. 2.3.5, recognizing that $\mathbf{a} = a_x\hat{\mathbf{i}} + a_y\hat{\mathbf{j}} + a_z\hat{\mathbf{k}}$.

2.13 Use Eq. 2.4.8 to determine the pressure at the top of a building 300 m tall. Then assume that the density is constant at the $z = 0$ value and estimate p at 300 m; also calculate the percent error in this second calculation. Use standard conditions at $z = 0$. Comment as to the advisability of an engineer assuming the atmosphere to be incompressible over heights of up to 300 m.

2.14 Calculate the pressure change of air over a height of 20 m assuming standard conditions and using Eq. 2.4.8. Comment as to the advisability of an engineer to ignore completely changes in pressure up to heights of 20 m, or so, in a gas such as air.

2.15 Assume the bulk modulus to be constant and find an expression for the pressure as a function of depth h in the ocean. Use this expression and estimate the pressure assuming $\rho_0 = 1000$ kg/m³. Now, assume a constant density of 1000 kg/m³ and calculate the pressure and the percent error, assuming the estimate in the first calculation to be correct. Use depths of **(a)** 450 m, **(b)** 1500 m, and **(c)** 4500 m.

2.16 Estimate the pressure at 10 000 m assuming an isothermal atmosphere with temperature:
(a) 0°C
(b) 15°C
(c) −15°C

2.17 The temperature in the atmosphere is approximated by $T(z) = 15 - 0.0065z$°C for elevations less than 11 000 m. Calculate the pressure at elevations of:
(a) 3000 m   (b) 6000 m
(c) 9000 m   (d) 11 000 m

2.18 Determine the elevation where $p = 7$ Pa assuming an isothermal atmosphere with $T = -20$°C.

Manometers

2.19 Calculate the pressure in a pipe transporting air if a U-tube manometer measures 25 cm Hg. Note that the weight of air in the manometer is negligible.

2.20 If the pressure of air in a pipe is 450 kPa, what will a U-tube manometer with mercury read? Use $h = 1.5$ cm in Fig. 2.7b.

(a) Neglect the weight of the air column.
(b) Include the weight of the air column, assuming that $T_{air} = 20$°C, and calculate the percent error of part (a).

2.21 A U-tube manometer is attached to a pipe transporting a liquid. It is known that the pressure in the pipe at the location of the manometer is 2.4 kPa. Select the liquid from Table B.5 that is most likely being transported if the manometer indicates the following height of liquid above the pipe:

(a) 36.0 cm   (b) 27.2 cm
(c) 24.5 cm   (d) 15.4 cm

2.22 It is known that the pressure at the nose of an airplane traveling at relatively low speed is related to its speed by $p = \frac{1}{2}\rho V^2$, where ρ is the density of air. Determine the speed of an airplane traveling near the earth's surface if a U-tube manometer, that measures the nose pressure, reads:

(a) 6 cm of water
(b) 10 cm of water

2.23 Estimate the pressure in the water pipe shown in Fig. P2.23. the manometor is open to the atmosphere.
A. 10 kPa   B. 9 kPa
C. 8 kPa   D. 7 kPa

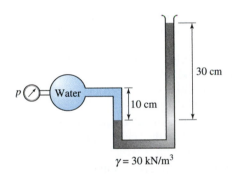

$\gamma = 30$ kN/m³

FIGURE P2.23

2.24 Oil with $S = 0.86$ is being transported in a pipe. Calculate the pressure if a U-tube manometer reads 240 mm Hg. The oil in the manometer is depressed 125 mm below the pipe centerline.

2.25 Several liquids are layered inside a tank with pressurized air at the top. If the air pressure is 3.2 kPa, calculate the pressure at the bottom of the tank if the layers include 20 cm of SAE 10 oil, 10 cm of water, 15 cm of glycerin, and 18 cm of carbon tetrachloride.

2.26 For the setup shown in Fig. P2.26, calculate the manometer reading H.

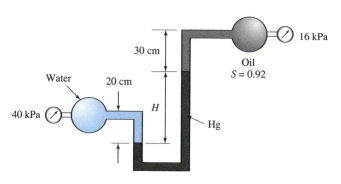

FIGURE P2.26

2.28 What is the pressure in the water pipe shown in Fig. P2.28?

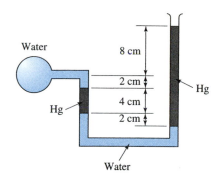

FIGURE P2.28

2.27 For the setup shown in Fig. P2.27, find the difference in pressure between the oil pipe and the water pipe.

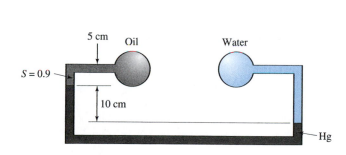

FIGURE P2.27

2.29 Determine the pressure difference between the water pipe and the oil pipe shown in Fig. P2.29.

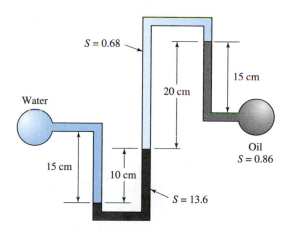

FIGURE P2.29

2.30 What is the pressure in the oil pipe shown in Fig. P2.30 if the pressure in the water pipe is 15 kPa?

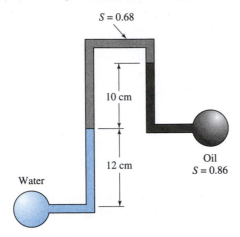

$S = 0.68$

10 cm

12 cm

Oil
$S = 0.86$

Water

FIGURE P2.30

2.31 For the tank shown in Fig. P2.31, determine the reading of the pressure gage if:
(a) $H = 2$ m, $h = 10$ cm.
(b) $H = 0.8$ m, $h = 20$ cm.

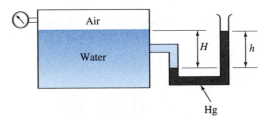

Air

Water

H

h

Hg

FIGURE P2.31

2.32 For the tank shown in Fig. P2.32, if $H = 16$ cm, what will the pressure gage read?

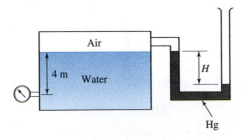

Air

4 m   Water

H

Hg

FIGURE P2.32

2.33 If the pressure in the air of Problem 2.32 is increased by 10 kPa, the magnitude of H will be nearest:
A.   8.5 cm
B.   10.5 cm
C.   16 cm
D.   24.5 cm

2.34 The pressure in the water pipe of Fig. 2.7b is 8.2 kPa, with $h = 25$ cm and $S_2 = 1.59$. Find the pressure in the water pipe if the H reading increases by 27.3 cm.

2.35 Find the pressure in the water pipe shown in Fig. P2.35.

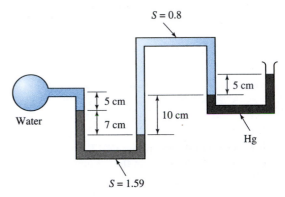

$S = 0.8$

5 cm

Water

5 cm

7 cm

10 cm

Hg

$S = 1.59$

FIGURE P2.35

2.36 In Fig. P2.36, with the manometer top open the mercury level is 20 cm below the air pipe; there is no pressure in the air pipe. The manometer top is then sealed. Calculate the manometer reading H for a pressure of 200 kPa in the air pipe. Assume an isothermal process for the air in the sealed tube.

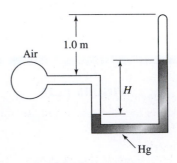

Air

1.0 m

H

Hg

FIGURE P2.36

2.37 Referring to Fig. 2.6c, determine the manometer reading H for the following conditions:
$z_i = (22, 16, 10, z_4, 17)$ cm, $p_1 = 4$ kPa,
$\gamma_i = (9800, 15\,600, 133\,400)$ N/m^3,
$d = 5$ mm, $D = 100$ mm.

2.38 Calculate the percentage increase in the manometer reading if the pressure p_1 is increased by 10% in problem 2.37.

2.39 The pressure in the water pipe of Problem 2.30 is increased to 15.5 kPa, while the pressure in the oil pipe remains constant. What will be the new manometer reading?

2.40 Determine the new manometer reading h if the air pressure is increased by 10% in:
(a) Problem 2.31a
(b) Problem 2.31b

Forces on Plane Areas

2.41 Calculate the force acting on a 30-cm-diameter porthole of a ship if the center of the porthole is 10 m below water level.

2.42 A swimming pool is filled with 2 m of water. Its bottom is square and measures 4 m on a side. Two opposite sides are vertical; one end is at 45° and the other makes an angle of 60° with the horizontal. Calculate the force of the water on:
(a) The bottom
(b) A vertical side
(c) The 45° end
(d) The 60° end

2.43 An enclosed rectangular concrete vault with outside dimensions 2 m × 1 m × 1.5 m and wall thickness 10 cm is buried with the top surface even with the ground surface. Will the vault tend to rise out of the ground if the ground becomes completely saturated with water? Use $S_{concrete} = 2.4$.

2.44 Gasoline fills a tank 4 m in diameter and 6 m long. Calculate the force that the gasoline exerts on one end of the tank. Assume that the tank is not pressurized and the ends are vertical.

2.45 The sides of a triangular area measure 2 m, 3 m, and 3 m, respectively. Calculate the force of water on one side of the area if the 2-m side is horizontal and 10 m below the surface and the triangle is:
(a) Vertical.
(b) Horizontal.
(c) On a 60° angle sloped upward.

2.46 The triangular gate shown in Fig. P2.46 has its 2-m side parallel and 10 m below the water surface. Calculate the magnitude and location of the force acting on the gate if it is:
(a) Vertical   **(b)** Horizontal
(c) On a 45° angle sloped upward

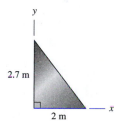

FIGURE P2.46

2.47 The rectangular gate shown in Fig. P2.47 is 3 m wide. The force P needed to hold the gate in the position shown is most nearly:
A. 24.5 kN   **B.** 32.7 kN
C. 98 kN   **D.** 147 kN

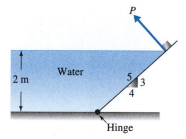

FIGURE P2.47

2.48   The top of each gate shown in Fig. P2.48 lies 4 m below the water surface. Find the location and magnitude of the force acting on one side assuming a vertical orientation.

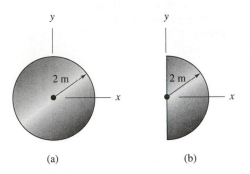

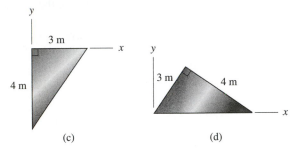

(c)   (d)

FIGURE P2.48

2.49   A 2-m-wide 3-m-high vertical rectangular gate has its top edge 2 m below the water level. It is hinged along its lower edge. What force, acting on the top edge, is necessary to hold the gate shut?

2.50   Determine the force P needed to hold the 4-m-wide gate in the position shown in Fig. P2.50.

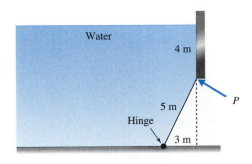

FIGURE P2.50

2.51   Calculate the force P necessary to hold the 4-m-wide gate in the position shown in Fig. P2.51 if:
   (a) $H = 6$ m   **(b)** $H = 8$ m
   (c) $H = 10$ m

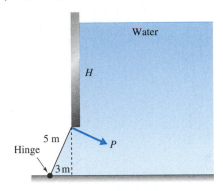

FIGURE P2.51

2.52   Use Eq. 2.4.28 and show that the force F in Fig. 2.8 acts one-third the way up on a vertical rectangular area and also on a sloped rectangular area. Assume the sloped gate to be on an angle α with the horizontal.

2.53   Find the force P needed to hold the 3-m-wide rectangular gate as shown in Fig. P2.53 if:
   (a) $l = 2$ m   **(b)** $l = 4$ m   **(c)** $l = 5$ m

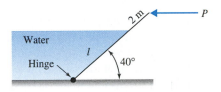

FIGURE P2.53

2.54   A trapezoidal channel, with cross section shown in Fig. P2.54, is gated at one end. What is the minimum force P needed to hold the vertical gate closed if it is hinged at the bottom? The gate has the same dimensions as the channel and the force P acts at the water surface.

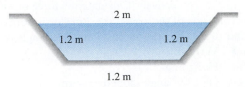

FIGURE P2.54

2.55 A vertical gate at the end of a channel (Fig. P2.55) opens when the water above the hinge produces a moment greater than the moment of the water below the hinge. What height h of water is needed to open the gate if:
(a) $H = 0.9$ m
(b) $H = 1.2$ m
(c) $H = 1.5$ m

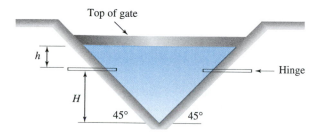

FIGURE P2.55

2.56 At what height H will the rigid gate, hinged at a central point as shown in Fig. P2.56, open if h is:
(a) 0.6 m?
(b) 0.8 m?
(c) 1.0 m?

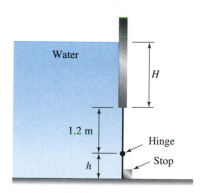

FIGURE P2.56

2.57 The rigid gate hinged at a central point as shown in Fig. P2.56 opens when $H = 5$ m. How far is the hinge above the water bottom?
A. 1.08 m
B. 1.10 m
C. 1.12 m
D. 1.14 m

2.58 For the gate shown in Fig. P2.58, calculate the height H that will result in the gate opening automatically if (neglect the weight of the gate):

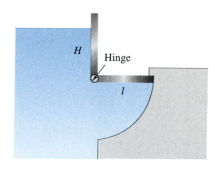

FIGURE P2.58

(a) $l = 2$ m
(b) $l = 1$ m

2.59 The pressure distribution over the base of a concrete ($S = 2.4$) dam varies linearly, as shown in Fig. P2.59, producing an *uplift*. Will the dam topple over (sum moments of all forces about the bottom, right-hand corner)? Use:
(a) $H = 45$ m
(b) $H = 60$ m
(c) $H = 75$ m

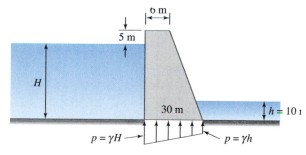

FIGURE P2.59

2.60 Assume a linear pressure distribution over the base of the concrete ($S = 2.4$) dam shown in Fig. P2.60. Will the dam topple over (sum moments about the bottom, right-hand corner)? Use:
- **(a)** $H = 15$ m
- **(b)** $H = 20$ m
- **(c)** $H = 25$ m

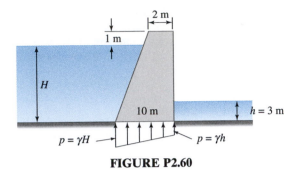

FIGURE P2.60

Forces on Curved Surfaces

2.61 In Example 2.7 assume that the water is above the gate rather than below the gate. Water above the gate will produce the same pressure distribution on the gate and hence the same forces (except the forces will be in opposite directions). Consequently, the force P will be numerically the same (it will act to the left). With water above the gate, draw a free-body diagram and calculate P. Compare with the details of the first method of Example 2.7.

2.62 Find the force P needed to hold the 10-m-long cylindrical object in position as shown in Fig. P2.62.

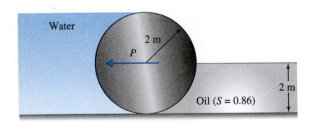

FIGURE P2.62

2.63 Find the force P needed to just open the gate shown in Fig. P2.63 if:
$H = 6$ m, $R = 2$ m, and the gate is 4 m wide.

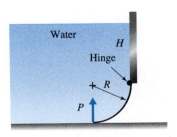

FIGURE P2.63

2.64 A force $P = 300$ kN is need to just open the gate of Fig. P2.63 with $R = 1.2$ m and $H = 4$ m. How wide is the gate?
- **A.** 2.98 m
- **B.** 3.67 m
- **C.** 4.32 m
- **D.** 5.16 m

2.65 What P is needed to hold the 4-m-wide gate shown in Fig. P2.65 closed?

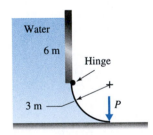

FIGURE P2.65

2.66 Find the force P required to hold the gate in the position shown in Fig. P2.66. The gate is 5 m wide.

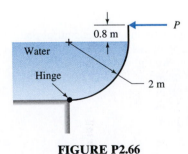

FIGURE P2.66

2.67 The 3-m-wide circular gate shown in Fig. P2.67 weighs 400 N with center of gravity 0.9 m to the left of the hinge. Estimate the force P needed to open the gate.

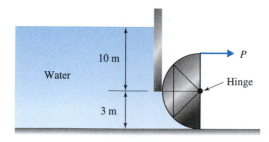

FIGURE P2.67

2.68 The quarter-circle cylindrical gate (Fig. P2.68; $S = 0.2$) is in equilibrium, as shown. Calculate the value of γ_x in SI units

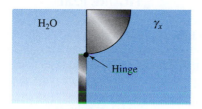

FIGURE P2.68

2.69 A log is in equilibrium, as shown in Fig. P2.69. Calculate the force pushing it against the dam and the specific gravity of the log if:
Its length is 6 m and $R = 0.6$ m.

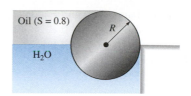

FIGURE P2.69

2.70 Find the force on the weld shown in Fig. P2.70 if:
(a) Air fills the hemisphere
(b) Oil fills the hemisphere

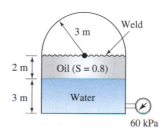

FIGURE P2.70

2.71 Find the force P if the parabolic gate shown in Fig. P2.71 is:
2 m wide and $H = 2$ m

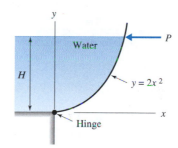

FIGURE P2.71

Buoyancy

2.72 The rectangular barge of Fig. P2.72 is known to be 15 m long. A load having a mass of 900 kg is added to the barge causing it to sink 10 mm. How wide is the barge?
A. 6 m
B. 9.2 m
C. 7.5 m
D. 0.62 m

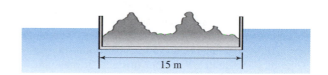

FIGURE P2.72

2.73 The 3-m-wide barge shown in Fig. P2.72 weighs 20 kN empty. It is proposed that it carry a 250-kN load. Predict the draft in:
(a) Fresh water
(b) Salt water ($S = 1.03$)

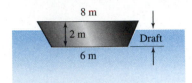

8 m

2 m

Draft

6 m

FIGURE P2.73

2.74 An object weighs 100 N in the air and 25 N when submerged in water. Calculate its volume and specific weight.

2.75 A car ferry is essentially rectangular with dimensions 8 m wide and 100 m long. If 60 cars, with an average weight per car of 15 kN, are loaded on the ferry, how much farther will it sink into the water?

2.76 A 30-m-long vessel, with cross section shown in Fig. P2.76, is to carry a load of 6000 kN. How far will the water level be from the top of the vessel if its mass is 100 000 kg?

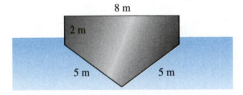

8 m

2 m

5 m   5 m

FIGURE P2.76

2.77 A body, with a volume of 2 m³, weighs 40 kN. Determine its weight when submerged in a liquid with $S = 1.59$.

2.78 A hot-air balloon carries a load of 1000 N, including its own weight. If it is 10 m in diameter, estimate the average temperature of the air inside if the air outside is at 20°C.

2.79 A large blimp is proposed to travel near the Earth's surface. If the blimp resembles a large cylinder 1500 m long with a diameter of 300 m, estimate the payload if its own weight is 10% of the payload. How many 800-N people could it carry? The blimp is filled with helium and standard conditions prevail. (This vehicle will not cause sea sickness and sunsets are spectacular!)

2.80 An object is constructed of a material lighter than water. It weighs 50 N in air and a force of 10 N is required to hold it under water. What is its density, specific weight, and specific gravity?

2.81 The plug and empty cylinder shown in Fig. P2.81 weigh 6000 N. Calculate the height h needed to lift the plug if the radius R of the 3-m-long cylinder is:
(a) 30 cm   **(b)** 40 cm   **(c)** 50 cm

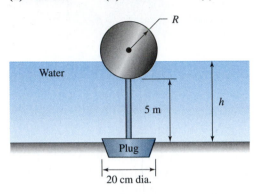

R

Water

5 m

h

Plug

20 cm dia.

FIGURE P2.81

2.82 The hydrometer shown in Fig. P2.82 with no mercury has a mass of 0.01 kg. It is designed to float at the midpoint of the 12-cm-long stem in pure water.
(a) Calculate the mass of mercury needed.
(b) What is the specific gravity of the liquid if the hydrometer is just submerged?
(c) What is the specific gravity of the liquid if the stem of the hydrometer is completely exposed?

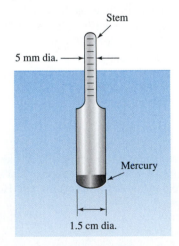

Stem

5 mm dia.

Mercury

1.5 cm dia.

FIGURE P2.82

2.83 The hydrometer of Problem 2.82 is weighted so that in fresh water the stem is just submerged.
  (a) What is the maximum specific gravity that can be read?
  (b) What mass of mercury is required?

Stability

2.84 A 25-cm-diameter cylinder is composed of material with specific gravity 0.8. Will it float in water with the ends horizontal if its length is:
  (a) 30 cm? **(b)** 25 cm? **(c)** 20 cm?

2.85 Over what range of specific weights will a circular cylinder with uniform specific weight γ_x float in water with ends horizontal if its height equals its diameter?

2.86 Over what range of specific weights will a homogeneous cube float with sides horizontal and vertical?

2.87 For the object shown in Fig. P2.87, calculate S_A for neutral stability when submerged.

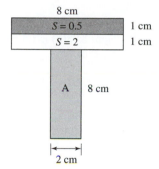

FIGURE P2.87

2.88 Orient the object shown in Fig. P2.88 for rotational stability when submerged if $t = 2$ cm

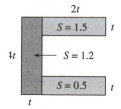

FIGURE P2.88

2.89 The barge shown in Fig. P2.89 is loaded such that the center of gravity of the barge and the load is at the waterline. Is the barge stable?

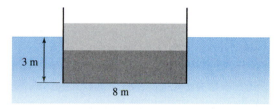

FIGURE P2.89

2.90 Is the barge shown in Fig. P2.90 stable? The center of gravity of the barge and load is located as shown.

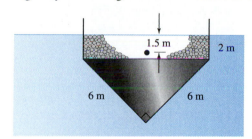

FIGURE P2.90

Linearly Accelerating Containers

2.91 The tank, with an initial pressure of $p = 20$ kPa, is accelerated as shown in Fig. P.2.91 at the rate of 5 m/s^2. The force on the 4-cm-diameter plug is nearest:

   A.   30 N   **B.**   50 N

   C.   130 N   **D.**   420 N

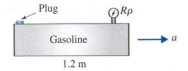

FIGURE P2.91

2.92 The tank shown in Fig. P2.92 is completely filled with water and accelerated. Calculate the maximum pressure in the tank if:

   (a) $a_x = 20$ m/s^2, $a_z = 0$, $L = 2$ m

   (b) $a_x = 0$, $a_z = 20$ m/s^2, $L = 2$ m

FIGURE P2.92

2.93 The tank shown in Fig. P2.93 is accelerated to the right at 10 m/s^2. Find:

   (a) p_A **(b)** p_B **(c)** p_C

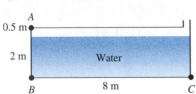

FIGURE P2.93

2.94 The tank of Problem 2.93 is accelerated so that $p_B = 60$ kPa. Find a_x assuming that:

   (a) $a_z = 0$

   (b) $a_z = 10$ m/s^2

   (c) $a_z = 5$ m/s^2

2.95 The tank shown in Fig. P2.95 is filled with water and accelerated. Find the pressure at A if:

   (a) $a = 20$ m/s^2, $L = 1$ m

   (b) $a = 10$ m/s^2, $L = 1.5$ m

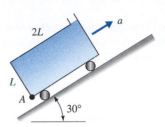

FIGURE P2.95

2.96 The tank of Problem 2.93 is 4 m wide. Find the force acting on:

   (a)   The end AB

   (b)   The bottom

   (c)   The top

2.97 The tank of Problem 2.95(a) is 1.5 m wide. Calculate the force on:

   (a)   The bottom

   (b)   The top

   (c)   The left end

2.98 For the U-tube shown in Fig. P2.98, determine the pressure at points A, B, and C if:

   (a) $a_x = 0$, $a_z = 10$ m/s^2, $L = 60$ cm

   (b) $a_x = 20$ m/s^2, $a_z = 0$, $L = 60$ cm

   (c) $a_x = 20$ m/s^2, $a_z = 10$ m/s^2, $L = 60$ cm

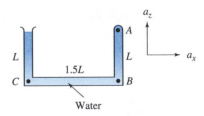

FIGURE P2.98

Rotating Containers

2.99 The U-tube of Problem 2.98 is rotated about the left leg at 50 rpm. Find p_A, p_B, and p_C if:
  (a) $L = 60$ cm   **(b)** $L = 40$ cm

2.100 The U-tube of Problem 2.98 is rotated about the right leg at 10 rad/s. Find the pressures at points A, B, and C if:
  (a) $L = 60$ cm   **(b)** $L = 40$ cm

2.101 The U-tube of Problem 2.98 is rotated about the center of the horizontal leg so that the pressure at the center is zero. Calculate ω if:
  (a) $L = 60$ cm   **(b)** $L = 40$ cm

2.102 For the cylinder shown in Fig. P2.102, determine the pressure at point A for a rotational speed of:
  (a) 5 rad/s   **(b)** 7 rad/s
  (c) 10 rad/s   **(d)** 20 rad/s

2.103 The hole in the cylinder of Problem 2.102 is closed and the air pressurized to 25 kPa. Find the pressure at point A if the rotational speed is:
  (a) 5 rad/s
  (b) 7 rad/s
  (c) 10 rad/s
  (d) 20 rad/s

2.104 Find the force on the bottom of the cylinder of:
  (a) Problem 2.102a
  (b) Problem 2.102b
  (c) Problem 2.102c
  (d) Problem 2.102d

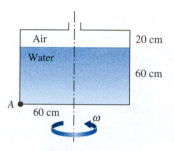

FIGURE P2.102

White water rafting is a popular sport in North America. It embodies the thrill of experiencing rapidly moving water in a craft that demands quick thinking and manipulative skills. (ArtmannWitte / Shutterstock)

3

Introduction to Fluids in Motion

Outline

Chapter Objectives

The objectives of this chapter are to:

▲ Mathematically describe the motion of a fluid.

▲ Express the acceleration of a fluid particle given the velocity components.

▲ Describe the deformation of a fluid particle.

▲ Classify various fluid flows. Is a flow viscous, is it turbulent, is it incompressible, is it a uniform flow?

▲ Derive the Bernoulli equation and identify its restrictions.

▲ Present several examples and numerous problems that demonstrate how fluid flows are described, how flows are classified, and how Bernoulli's equation is used to estimate flow variables.

3.1 INTRODUCTION

This chapter serves as an introduction to all the following chapters that deal with fluid motions. Fluid motions manifest themselves in many different ways. Some can be described very easily, while others require a thorough understanding of physical laws. In engineering applications, it is important to describe the fluid motions as simply as can be justified. This usually depends on the required accuracy. Often, accuracies of ± 10% are acceptable, although in some applications higher accuracies have to be achieved. The general equations of motion are very difficult to solve; consequently, it is the engineer's responsibility to know which simplifying assumptions can be made. This, of course, requires experience and, more important, an understanding of the physics involved.

Some common assumptions used to simplify a flow situation are related to fluid properties. For example, under certain conditions, the viscosity can affect the flow significantly; in others, viscous effects can be neglected greatly simplifying the equations without significantly altering the predictions. It is well known that the compressibility of a gas in motion should be taken into account if the velocities are very high. But compressibility effects do not have to be taken into account to predict wind forces on buildings or to predict any other physical quantity that is a direct effect of wind. Wind speeds are simply not high enough. Numerous examples could be cited. After our study of fluid motions, the appropriate assumptions used should become more obvious.

> **KEY CONCEPT** *Under certain conditions, viscous effects can be neglected.*

This chapter has three sections. In the first section we introduce the reader to some important general approaches used to analyze fluid mechanics problems. In the second section we give a brief overview of different types of flow, such as compressible and incompressible flows, and viscous and inviscid flows. Detailed discussions of each of these flow types follow in later chapters. The third section introduces the reader to the commonly used Bernoulli equation, an equation that establishes how pressures and velocities vary in a flow field. The use of this equation, however, requires many simplifying assumptions, and its application is, therefore, limited.

3.2 DESCRIPTION OF FLUID MOTION

The analysis of complex fluid flow problems is often aided by the visualization of flow patterns, which permit the development of a better intuitive understanding and help in formulating the mathematical problem. The flow in a washing machine is a good example. An easier, yet difficult problem is the flow in the vicinity of where a wing attaches to a fuselage, or where a bridge support interacts with the water at the bottom of a river. In Section 3.2.1 we discuss the description of physical quantities as a function of space and time coordinates. The second topic in this section introduces the different flow lines that are useful in our objective of describing a fluid flow. Finally, the mathematical description of motion is presented.

3.2.1 Lagrangian and Eulerian Descriptions of Motion

In the description of a flow field, it is convenient to think of individual particles each of which is considered to be a small mass of fluid, consisting of a large number of molecules, that occupies a small volume $\Delta \mathcal{V}$ that moves with the flow. If the fluid is incompressible, the volume does not change in magnitude but may deform. If the fluid is compressible, as the volume deforms, it also changes its magnitude. In both cases the particles are considered to move through a flow field as an entity.

In the study of particle mechanics, where attention is focused on individual particles, motion is observed as a function of time. The position, velocity, and acceleration of each particle are listed as $\mathbf{s}(x_0, y_0, z_0, t)$, $\mathbf{V}(x_0, y_0, z_0, t)$, and $\mathbf{a}(x_0, y_0, z_0, t)$ and quantities of interest can be calculated. The point (x_0, y_0, z_0) locates the starting point — the name — of each particle. This is the **Lagrangian** description, named after Joseph L. Lagrange (1736–1813), of motion that is used in a course on dynamics. In the Lagrangian description many particles can be followed and their influence on one another noted. This becomes, however, a difficult task as the number of particles becomes extremely large, as in a fluid flow.

Lagrangian: Description of motion where individual particles are observed as a function of time.

An alternative to following each fluid particle separately is to identify points in space and then observe the velocity of particles passing each point; we can observe the rate of change of velocity as the particles pass each point, that is, $\partial \mathbf{V} / \partial x$, $\partial \mathbf{V} / \partial y$, and $\partial \mathbf{V} / \partial z$, and we can observe if the velocity is changing with time at each particular point, that is, $\partial \mathbf{V} / \partial t$. In this **Eulerian** description, named after Leonhard Euler (1707–1783), of motion, the flow properties, such as velocity, are functions of both space and time. In rectangular, Cartesian coordinates the velocity is expressed as $\mathbf{V} = \mathbf{V}(x, y, z, t)$. The region of flow that is being considered is called a **flow field**.

Eulerian: Description of motion where the flow properties are functions of both space and time.

Flow field: The region of flow of interest.

An example may clarify these two ways of describing motion. An engineering firm is hired to make recommendations that would improve the traffic flow in a large city. The engineering firm has two alternatives: Hire college students to travel in automobiles throughout the city recording the appropriate observations (the Lagrangian approach), or hire college students to stand at the intersections and record the required information (the Eulerian approach). A correct interpretation of each set of data would lead to the same set of recommendations, that is, the same solution. In this example it may not be obvious which approach would be preferred; in an introductory course in fluids, however, the Eulerian description is used exclusively since the physical laws using the Eulerian description are easier to apply to actual situations. Yet, there are examples where a Lagrangian description is needed, such as drifting buoys used to study ocean currents.

If the quantities of interest do not depend on time, that is, $\mathbf{V} = \mathbf{V}(x, y, z)$, the flow is said to be a **steady flow**. Most of the flows of interest in this introductory textbook are steady flows. For a steady flow all flow quantities at a particular point are independent of time, that is,

Steady flow: Where quantities do not depend on time.

$$\frac{\partial \mathbf{V}}{\partial t} = 0 \qquad \frac{\partial p}{\partial t} = 0 \qquad \frac{\partial \rho}{\partial t} = 0 \qquad (3.2.1)$$

to list a few. It is implied that x, y, and z are held fixed in the above. Note that the properties of a fluid particle do, in general, vary with time; the velocity and pressure vary with time as a fluid particle progresses along its path in a flow, even in a steady flow. In a steady flow, however, properties do not vary with time at a fixed point.

3.2.2 Pathlines, Streaklines, and Streamlines

Pathline: A history of the particle's locations.

Three different lines help us in describing a flow field. A **pathline** is the locus of points traversed by a given particle as it travels in a field of flow; the pathline provides us with a "history" of the particle's locations. A photograph of a pathline would require a time exposure of an illuminated particle. A photograph showing pathlines of particles below a water surface with waves is given in Fig. 3.1.

Streakline: An instantaneous line.

A **streakline** is defined as an instantaneous line whose points are occupied by all particles originating from some specified point in the flow field. Streaklines tell us where the particles are "right now." A photograph of a streakline would be a snapshot of the set of illuminated particles that passed a certain point. Figure 3.2 shows streaklines produced by the continuous release of a small-diameter stream of smoke as it moves around a cylinder.

Streamline: The velocity vector is tangent to the streamline.

A **streamline** is a line in the flow possessing the following property: the velocity vector of each particle occupying a point on the streamline is tangent to the streamline. This is shown graphically in Fig. 3.3. An equation which expresses that the velocity vector is tangent to a streamline is

$$\mathbf{V} \times d\mathbf{r} = 0 \tag{3.2.2}$$

since $\mathbf{V}$ and $d\mathbf{r}$ are in the same direction, as shown in the figure; recall that the cross product of two vectors in the same direction is zero. This equation will be used in future chapters as the mathematical expression of a streamline. A photograph of a streamline cannot be made directly. For a general unsteady flow the streamlines can be inferred from photographs of short pathlines of a large number of particles.

FIGURE 3.1 Pathlines underneath a wave in a tank of water.
(Photograph by Wallet and Ruellan. Courtesy of M. C. Vasseur.)

FIGURE 3.2   Streaklines in the unsteady flow around a cylinder. (Photograph by Sadatashi Taneda.)

A **streamtube** is a tube whose walls are streamlines. Since the velocity is tangent to a streamline, no fluid can cross the walls of a streamtube. The streamtube is of particular interest in fluid mechanics. A pipe is a streamtube since its walls are streamlines; an open channel is a streamtube since no fluid crosses the walls of the channel. We often sketch a streamtube with a small cross section in the interior of a flow for demonstration purposes.

In a steady flow, pathlines, streaklines, and streamlines are all coincident. All particles passing a given point will continue to trace out the same path since nothing changes with time; hence the pathlines and streaklines coincide. In addition, the velocity vector of a particle at a given point will be tangent to the line that the particle is moving along; thus the line is also a streamline. Since the flows that we observe in laboratories are invariably steady flows, we call the lines that we observe streamlines even though they may actually be streaklines, or for the case of time exposures, pathlines.

Streamtube: *A tube whose walls are streamlines.*

KEY CONCEPT   *In a steady flow, pathlines, streaklines, and streamlines are all coincident.*

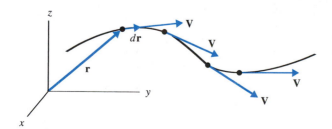

FIGURE 3.3   Streamline in a flow field.

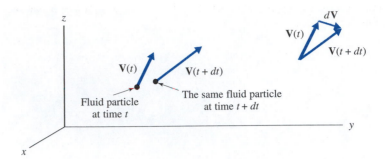

FIGURE 3.4 Velocity of a fluid particle.

3.2.3 Acceleration

The acceleration of a fluid particle is found by considering a particular particle shown in Fig. 3.4. Its velocity changes from $\mathbf{V}(t)$ at time t to $\mathbf{V}(t + dt)$ at time $t + dt$. The acceleration is, by definition,

$$\mathbf{a} = \frac{d\mathbf{V}}{dt} \tag{3.2.3}$$

where $d\mathbf{V}$ is shown in Fig. 3.4. The velocity vector $\mathbf{V}$ is given in component form as

$$\mathbf{V} = u\hat{\mathbf{i}} + v\hat{\mathbf{j}} + w\hat{\mathbf{k}} \tag{3.2.4}$$

where (u, v, w) are the velocity components in the x-, y-, and z- directions, respectively, and $\hat{\mathbf{i}}, \hat{\mathbf{j}}$ and $\hat{\mathbf{k}}$ are the unit vectors. The quantity $d\mathbf{V}$ is, using the chain rule from calculus with $\mathbf{V} = \mathbf{V}(x, y, z, t)$,

$$d\mathbf{V} = \frac{\partial \mathbf{V}}{\partial x}\, dx + \frac{\partial \mathbf{V}}{\partial y}\, dy + \frac{\partial \mathbf{V}}{\partial z}\, dz + \frac{\partial \mathbf{V}}{\partial t}\, dt \tag{3.2.5}$$

This gives the acceleration as

$$\mathbf{a} = \frac{\partial \mathbf{V}}{\partial x}\frac{dx}{dt} + \frac{\partial \mathbf{V}}{\partial y}\frac{dy}{dt} + \frac{\partial \mathbf{V}}{\partial z}\frac{dz}{dt} + \frac{\partial \mathbf{V}}{\partial t} \tag{3.2.6}$$

Since we have followed a particular particle as in Fig. 3.4, we recognize that

$$\frac{dx}{dt} = u \qquad \frac{dy}{dt} = v \qquad \frac{dz}{dt} = w \tag{3.2.7}$$

The acceleration is then expressed as

$$\mathbf{a} = u\frac{\partial \mathbf{V}}{\partial x} + v\frac{\partial \mathbf{V}}{\partial y} + w\frac{\partial \mathbf{V}}{\partial z} + \frac{\partial \mathbf{V}}{\partial t} \tag{3.2.8}$$

The scalar component equations of the above vector equation for rectangular coordinates are written as

$$
\begin{aligned}
a_x &= \frac{\partial u}{\partial t} + u\frac{\partial u}{\partial x} + v\frac{\partial u}{\partial y} + w\frac{\partial u}{\partial z} \\[6pt]
a_y &= \frac{\partial v}{\partial t} + u\frac{\partial v}{\partial x} + v\frac{\partial v}{\partial y} + w\frac{\partial v}{\partial z} \\[6pt]
a_z &= \frac{\partial w}{\partial t} + u\frac{\partial w}{\partial x} + v\frac{\partial w}{\partial y} + w\frac{\partial w}{\partial z}
\end{aligned}
\tag{3.2.9}
$$

We often return to Eq. 3.2.3 and write Eq. 3.2.8 in a simplified form as

$$
\mathbf{a} = \frac{D\mathbf{V}}{Dt} \tag{3.2.10}
$$

where, in rectangular coordinates,

$$
\frac{D}{Dt} = u\frac{\partial}{\partial x} + v\frac{\partial}{\partial y} + w\frac{\partial}{\partial z} + \frac{\partial}{\partial t} \tag{3.2.11}
$$

This derivative is called the **substantial derivative**, or **material derivative**. It is given a special name and special symbol (D/Dt instead of d/dt) because we followed a particular fluid particle, that is, we followed the substance (or material). It represents the relationship between a Lagrangian derivation in which a quantity depends on time t and a Eulerian derivation in which a quantity depends on position (x, y, z) and time t. The substantial derivative can be used with other dependent variables; for example, DT/Dt would represent the rate of change of the temperature of a fluid particle as we followed the particle along.

The substantial derivative and acceleration components in cylindrical and spherical coordinates are presented in Table 3.1 on page 96.

The time-derivative term on the right side of Eqs. 3.2.8 and 3.2.9 for the acceleration is called the **local acceleration** and the remaining terms on the right side in each equation form the **convective acceleration**. Hence the acceleration of a fluid particle is the sum of the local acceleration and convective acceleration. In a pipe, local acceleration results if, for example, a valve is being opened or closed; and convective acceleration occurs in the vicinity of a change in the pipe geometry, such as a pipe contraction or an elbow. In both cases fluid particles change speed, but for very different reasons.

We must note that the foregoing expressions for acceleration give the acceleration relative to an observer in the observer's reference frame only. In certain

Substantial or material derivative: *The derivative* D/Dt.

Local acceleration: *The time-derivative term* $\partial V/\partial t$ *for acceleration.*

Convective acceleration: *All other terms other than the local acceleration term.*

KEY CONCEPT *Convective acceleration occurs in the vicinity of a change in the geometry.*

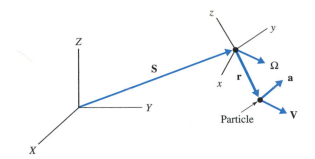

FIGURE 3.5   Motion relative to a noninertial reference frame.

situations the observer's reference frame may be accelerating; then the acceleration of a particle relative to a fixed reference frame may be needed. It is given by

$$\mathbf{A} = \mathbf{a} \; + \; \underbrace{\frac{d^2\mathbf{S}}{dt^2}}_{\substack{\text{acceleration of} \\ \text{reference frame}}} \; + \; \underbrace{2\mathbf{\Omega} \times \mathbf{V}}_{\substack{\text{Coriolis} \\ \text{acceleration}}} \; + \; \underbrace{\mathbf{\Omega} \times (\mathbf{\Omega} \times \mathbf{r})}_{\substack{\text{normal} \\ \text{acceleration}}} + \; \underbrace{\frac{d\mathbf{\Omega}}{dt} \times \mathbf{r}}_{\substack{\text{angular} \\ \text{acceleration}}} \qquad (3.2.12)$$

where $\mathbf{a}$ is given by Eq. 3.2.8, $d^2\mathbf{S}/dt^2$ is the acceleration of the observer's reference frame, $\mathbf{V}$ and $\mathbf{r}$ are the velocity and position vectors of the particle, respectively, in the observer's reference frame, and $\mathbf{\Omega}$ is the angular velocity of the
observer's reference frame (see Fig. 3.5). Note that all vectors are written using
the unit vectors of the XYZ-reference frame. For most engineering applications,
reference frames attached to the earth yield $\mathbf{A} = \mathbf{a}$, since the other terms in
Eq. 3.2.12 are often negligible with respect to $\mathbf{a}$. We may decide, however, to
attach the xyz-reference frame to an accelerating device (a rocket), or to a rotating device (a sprinkler arm); then certain terms of Eq. 3.2.12 must be included
along with $\mathbf{a}$ of Eq. 3.2.8.

   If the acceleration of all fluid particles is given by $\mathbf{A} = \mathbf{a}$ in a selected reference frame, that is an *inertial* reference frame. If $\mathbf{A} \neq \mathbf{a}$, it is a *noninertial* reference frame. A reference frame that moves with constant velocity without rotat-
ing is an inertial reference frame. When analyzing flow about, for example, an
airfoil, we attach the reference frame to the airfoil so that a steady flow is
observed in that reference frame.

3.2.4   Angular Velocity and Vorticity

A fluid flow may be thought of as the motion of a collection of fluid particles. As
a particle travels along it may rotate or deform. The rotation and deformation of
the fluid particles are of particular interest in our study of fluid mechanics. There
are certain flows, or regions of a flow, in which the fluid particles do not rotate;
such flows are of special importance, particularly in flows around objects, and are
referred to as **irrotational flows**. Flow outside a thin boundary layer on airfoils,
outside the separated flow region around autos and other moving vehicles, in the

Irrotational flows: *Flows
where the fluid particles do not
rotate.*

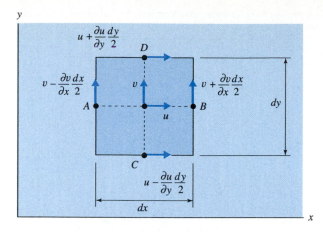

FIGURE 3.6   Fluid particle occupying an infinitesimal parallelepiped at a particular instant.

flow around submerged objects, and many other flows are examples of irrotational flows. Irrotational flows are extremely important.

Let us consider a small fluid particle that occupies an infinitesimal volume that has the xy-face as shown in Fig. 3.6. The **angular velocity** about the z-axis, Ω_z, is the average of the angular velocity of line segment AB and line segment CD. The two angular velocities, counterclockwise being positive, are

Angular velocity: The average velocity of two perpendicular line segments of a fluid particle.

$$\Omega_{AB} = \frac{v_B - v_A}{dx}$$

$$= \left[v + \frac{\partial v}{\partial x} \frac{dx}{2} - \left(v - \frac{\partial v}{\partial x} \frac{dx}{2} \right) \right] \Big/ dx = \frac{\partial v}{\partial x} \tag{3.2.13}$$

$$\Omega_{CD} = -\frac{u_D - u_C}{dy}$$

$$= -\left[u + \frac{\partial u}{\partial y} \frac{dy}{2} - \left(u - \frac{\partial u}{\partial y} \frac{dy}{2} \right) \right] \Big/ dy = -\frac{\partial u}{\partial y} \tag{3.2.14}$$

Consequently, the angular velocity Ω_z of the fluid particle is

$$\Omega_z = \tfrac{1}{2} \left(\Omega_{AB} + \Omega_{CD} \right)$$

$$= \frac{1}{2} \left(\frac{\partial v}{\partial x} - \frac{\partial u}{\partial y} \right) \tag{3.2.15}$$

If we had considered the xz-face, we would have found the angular velocity about the y-axis to be

$$\Omega_y = \frac{1}{2} \left(\frac{\partial u}{\partial z} - \frac{\partial w}{\partial x} \right) \tag{3.2.16}$$

TABLE 3.1 The Substantial Derivative, Acceleration, and Vorticity in Cartesian, Cylindrical and Spherical Coordinates

Substantive Derivative

Cartesian

$$\frac{D}{Dt} = u\frac{\partial}{\partial x} + v\frac{\partial}{\partial y} + w\frac{\partial}{\partial z} + \frac{\partial}{\partial t}$$

Cylindrical

$$\frac{D}{Dt} = v_r\frac{\partial}{\partial r} + \frac{v_\theta}{r}\frac{\partial}{\partial \theta} + v_z\frac{\partial}{\partial z} + \frac{\partial}{\partial t}$$

Spherical

$$\frac{D}{Dt} = v_r\frac{\partial}{\partial r} + \frac{v_\theta}{r}\frac{\partial}{\partial \theta} + \frac{v_\phi}{r\sin\theta}\frac{\partial}{\partial \phi} + \frac{\partial}{\partial t}$$

Acceleration

Cartesian

$$a_x = \frac{\partial u}{\partial t} + u\frac{\partial u}{\partial x} + v\frac{\partial u}{\partial y} + w\frac{\partial u}{\partial z}$$

$$a_y = \frac{\partial v}{\partial t} + u\frac{\partial v}{\partial x} + v\frac{\partial v}{\partial y} + w\frac{\partial v}{\partial z}$$

$$a_z = \frac{\partial w}{\partial t} + u\frac{\partial w}{\partial x} + v\frac{\partial w}{\partial y} + w\frac{\partial w}{\partial z}$$

Cylindrical

$$a_r = \frac{\partial v_r}{\partial t} + v_r\frac{\partial v_r}{\partial r} + \frac{v_\theta}{r}\frac{\partial v_r}{\partial \theta} + v_z\frac{\partial v_r}{\partial z} - \frac{v_\theta^2}{r}$$

$$a_\theta = \frac{\partial v_\theta}{\partial t} + v_r\frac{\partial v_\theta}{\partial r} + \frac{v_\theta}{r}\frac{\partial v_\theta}{\partial \theta} + v_z\frac{\partial v_\theta}{\partial z} + \frac{v_r v_\theta}{r}$$

$$a_z = \frac{\partial v_z}{\partial t} + v_r\frac{\partial v_z}{\partial r} + \frac{v_\theta}{r}\frac{\partial v_z}{\partial \theta} + v_z\frac{\partial v_z}{\partial z}$$

Spherical

$$a_r = \frac{\partial v_r}{\partial t} + v_r\frac{\partial v_r}{\partial r} + \frac{v_\theta}{r}\frac{\partial v_r}{\partial \theta} + \frac{v_\phi}{r\sin\theta}\frac{\partial v_r}{\partial \phi} - \frac{v_\phi^2 + v_\theta^2}{r}$$

$$a_\theta = \frac{\partial v_\theta}{\partial t} + v_r\frac{\partial v_\theta}{\partial t} + \frac{v_\theta}{r}\frac{\partial v_\theta}{\partial \theta} + \frac{v_\phi}{r\sin\phi}\frac{\partial v_\theta}{\partial \phi} + \frac{v_r v_\theta - v_\phi^2\cot\theta}{r}$$

$$a_\phi = \frac{\partial v_\phi}{\partial t} + v_r\frac{\partial v_\phi}{\partial r} + \frac{v_\theta}{r}\frac{\partial v_\phi}{\partial \theta} + \frac{v_\phi}{r\sin\theta}\frac{\partial v_\phi}{\partial \phi} + \frac{v_r v_\phi + v_\theta v_\phi\cot\theta}{r}$$

Vorticity

Cartesian

$$\omega_x = \frac{\partial w}{\partial y} - \frac{\partial v}{\partial z} \qquad \omega_y = \frac{\partial u}{\partial z} - \frac{\partial w}{\partial x} \qquad \omega_z = \frac{\partial v}{\partial x} - \frac{\partial u}{\partial y}$$

Cylindrical

$$\omega_r = \frac{1}{r}\left(\frac{\partial v_z}{\partial \theta}\right) - \frac{\partial v_\theta}{\partial z} \qquad \omega_\theta = \frac{\partial v_r}{\partial z} - \frac{\partial v_z}{\partial r} \qquad \omega_z = \frac{1}{r}\left(\frac{\partial(rv_\theta)}{\partial r} - \frac{\partial v_r}{\partial \theta}\right)$$

Spherical

$$\omega_r = \frac{1}{r\sin\theta}\left[\frac{\partial}{\partial \theta}(v_\phi\sin\theta) - \frac{\partial v_\theta}{\partial \phi}\right] \qquad \omega_\phi = \frac{1}{r}\left[\frac{\partial}{\partial r}(rv_\theta) - \frac{\partial v_r}{\partial \theta}\right]$$

$$\omega_\theta = \frac{1}{r}\left[\frac{1}{\sin\theta}\frac{\partial v_r}{\partial \phi} - \frac{\partial}{\partial r}(rv_\phi)\right]$$

and the *yz*-face would provide us with the angular velocity about the *x*-axis:

$$\Omega_x = \frac{1}{2}\left(\frac{\partial w}{\partial y} - \frac{\partial v}{\partial z}\right) \tag{3.2.17}$$

These are the three components of the angular velocity vector. A cork placed in a water flow in a wide channel (the *xy*-plane) would rotate with the angular velocity given by Eq. 3.2.15.

It is common to define the **vorticity ω** to be twice the angular velocity; its three components are then

Vorticity: *Twice the angular velocity.*

$$\omega_x = \frac{\partial w}{\partial y} - \frac{\partial v}{\partial z} \qquad \omega_y = \frac{\partial u}{\partial z} - \frac{\partial w}{\partial x} \qquad \omega_z = \frac{\partial v}{\partial x} - \frac{\partial u}{\partial y} \tag{3.2.18}$$

The vorticity components in cylindrical and spherical coordinates are included in Table 3.1 above.

An irrotational flow possesses no vorticity; the cork mentioned above would not rotate in an irrotational flow. We consider this special flow in Section 8.5.

The deformation of the particle of Fig. 3.6 is the rate of change of the angle that line segment AB makes with line segment CD. If AB is rotating with an angular velocity different from that of CD, the particle is deforming. The deformation is represented by the **rate-of-strain tensor**; its component ϵ_{xy} in the xy-plane is given by

Rate-of-strain tensor: *The rate at which deformation occurs.*

$$\epsilon_{xy} = \tfrac{1}{2}(\Omega_{AB} - \Omega_{CD})$$

$$= \frac{1}{2}\left(\frac{\partial v}{\partial x} + \frac{\partial u}{\partial y}\right) \tag{3.2.19}$$

For the xz-plane and the yz-plane we have

$$\epsilon_{xz} = \frac{1}{2}\left(\frac{\partial w}{\partial x} + \frac{\partial u}{\partial z}\right) \qquad \epsilon_{yz} = \frac{1}{2}\left(\frac{\partial w}{\partial y} + \frac{\partial v}{\partial z}\right) \tag{3.2.20}$$

Observe that $\epsilon_{xy} = \epsilon_{yx}$, $\epsilon_{xz} = \epsilon_{zx}$, and $\epsilon_{yz} = \epsilon_{zy}$. The rate-of-strain tensor is symmetric.

The fluid particle could also deform by being stretched or compressed in a particular direction. For example, if point B of Fig. 3.6 is moving faster than point A, the particle would be stretching in the x-direction. This normal rate of strain is measured by

$$\epsilon_{xx} = \frac{u_B - u_A}{dx}$$

$$= \left[u + \frac{\partial u}{\partial x}\frac{dx}{2} - \left(u - \frac{\partial u}{\partial x}\frac{dx}{2}\right)\right]\bigg/dx = \frac{\partial u}{\partial x} \tag{3.2.21}$$

Similarly, in the y- and z-directions we would find that

$$\epsilon_{yy} = \frac{\partial v}{\partial y} \qquad \epsilon_{zz} = \frac{\partial w}{\partial z} \tag{3.2.22}$$

The rate-of-strain tensor can be displayed as

$$\epsilon_{ij} = \begin{pmatrix} \epsilon_{xx} & \epsilon_{xy} & \epsilon_{xz} \\ \epsilon_{yx} & \epsilon_{yy} & \epsilon_{yz} \\ \epsilon_{zx} & \epsilon_{zy} & \epsilon_{zz} \end{pmatrix} \tag{3.2.23}$$

where the subscripts i and j take on numerical values 1, 2, or 3. Then ϵ_{12} represents ϵ_{xy} in row 1 column 2.

We will see in Chapter 5 that the normal and shear stress components in a flow are related to the foregoing rate-of-strain components. In fact, in the one-dimensional flow of Fig. 1.6, the shear stress was related to $\partial u/\partial y$ with Eq. 1.5.5; note that $\partial u/\partial y$ is twice the rate-of-strain component given by Eq. 3.2.19 with $v = 0$.

Example 3.1

The velocity field is given by $\mathbf{V} = 2x\hat{\mathbf{i}} - yt\hat{\mathbf{j}}$ m/s, where x and y are in meters and t is in seconds. Find the equation of the streamline passing through $(2, -1)$ and a unit vector normal to the streamline at $(2, -1)$ at $t = 4$ s.

Solution

The velocity vector is tangent to a streamline so that $\mathbf{V} \times d\mathbf{r} = 0$ (the cross product of two parallel vectors is zero). For the given velocity vector we have, at $t = 4$ s,

$$(2x\hat{\mathbf{i}} - 4y\hat{\mathbf{j}}) \times (dx\hat{\mathbf{i}} + dy\hat{\mathbf{j}}) = (2x\,dy + 4y\,dx)\hat{\mathbf{k}} = 0$$

where we have used $\hat{\mathbf{i}} \times \hat{\mathbf{j}} = \hat{\mathbf{k}}$, $\hat{\mathbf{j}} \times \hat{\mathbf{i}} = -\hat{\mathbf{k}}$, and $\hat{\mathbf{i}} \times \hat{\mathbf{i}} = 0$. Consequently,

$$2x\,dy = -4y\,dx \qquad \text{or} \qquad \frac{dy}{y} = -2\frac{dx}{x}$$

Integrate both sides:

$$\ln y = -2\ln x + \ln C$$

where we used $\ln C$ for convenience. This is written as

$$\ln y = \ln x^{-2} + \ln C = \ln(Cx^{-2})$$

Hence

$$x^2 y = C$$

At $(2, -1)$ $C = -4$, so that the streamline passing through $(2, -1)$ has the equation

$$x^2 y = -4$$

A normal vector is perpendicular to the streamline, hence the velocity vector, so that using $\hat{\mathbf{n}} = n_x\hat{\mathbf{i}} + n_y\hat{\mathbf{j}}$ we have at $(2, -1)$ and $t = 4$ s

$$\mathbf{V} \cdot \hat{\mathbf{n}} = (4\hat{\mathbf{i}} + 4\hat{\mathbf{j}}) \cdot (n_x\hat{\mathbf{i}} + n_y\hat{\mathbf{j}}) = 0$$

Using $\hat{\mathbf{i}} \cdot \hat{\mathbf{i}} = 1$ and $\hat{\mathbf{i}} \cdot \hat{\mathbf{j}} = 0$, this becomes

$$4n_x + 4n_y = 0 \qquad \therefore n_x = -n_y$$

Then, because $\hat{\mathbf{n}}$ is a unit vector, $n_x^2 + n_y^2 = 1$ and we find that

$$n_x^2 = 1 - n_x^2 \qquad \therefore n_x = \frac{\sqrt{2}}{2}$$

The unit vector normal to the streamline is written as

$$\hat{\mathbf{n}} = \frac{\sqrt{2}}{2}(\hat{\mathbf{i}} - \hat{\mathbf{j}})$$

Example 3.2

A velocity field in a particular flow is given by $\mathbf{V} = 20y^2\hat{\mathbf{i}} - 20xy\hat{\mathbf{j}}$ m/s. Calculate the acceleration, the angular velocity, the vorticity vector, and any nonzero rate-of-strain components at the point $(1, -1, 2)$.

Solution

We could use Eq. 3.2.9 and find each component of the acceleration, or we could use Eq. 3.2.8 and find a vector expression. Using Eq. 3.2.8, we have

$$\mathbf{a} = u\frac{\partial \mathbf{V}}{\partial x} + v\frac{\partial \mathbf{V}}{\partial y} + w\overset{0}{\cancel{\frac{\partial \mathbf{V}}{\partial z}}} + \overset{0}{\cancel{\frac{\partial \mathbf{V}}{\partial t}}}$$

$$= 20y^2(-20y\hat{\mathbf{j}}) - 20xy(40y\hat{\mathbf{i}} - 20x\hat{\mathbf{j}})$$

$$= -800xy^2\hat{\mathbf{i}} - 400(y^3 - x^2y)\hat{\mathbf{j}}$$

where we have used $u = 20y^2$ and $v = -20xy$, as given by the velocity vector. All particles at the point $(1, -1, 2)$ have the acceleration

$$\mathbf{a} = -800\hat{\mathbf{i}} \text{ m/s}^2$$

The angular velocity has two zero components:

$$\Omega_x = \frac{1}{2}\left(\overset{0}{\cancel{\frac{\partial w}{\partial y}}} - \overset{0}{\cancel{\frac{\partial v}{\partial z}}}\right) = 0, \quad \Omega_y = \frac{1}{2}\left(\overset{0}{\cancel{\frac{\partial u}{\partial z}}} - \overset{0}{\cancel{\frac{\partial w}{\partial x}}}\right) = 0$$

The non-zero z-component is, at the point $(1, -1, 2)$,

$$\Omega_z = \frac{1}{2}\left(\frac{\partial v}{\partial x} - \frac{\partial u}{\partial y}\right)$$

$$= \frac{1}{2}(-20y - 40y) = 30 \text{ rad/s}$$

The vorticity vector is twice the angular velocity vector:

$$\boldsymbol{\omega} = 2\Omega_z\hat{\mathbf{k}} = 60\hat{\mathbf{k}} \text{ rad/s}$$

The nonzero rate-of-strain components are

$$\epsilon_{xy} = \frac{1}{2}\left(\frac{\partial v}{\partial x} + \frac{\partial u}{\partial y}\right)$$

$$= \frac{1}{2}(-20y + 40y) = -10 \text{ rad/s}$$

$$\epsilon_{yy} = \frac{\partial v}{\partial y}$$

$$= -20x = -20 \text{ rad/s}$$

All other rate-of-strain components are zero.

3.3 CLASSIFICATION OF FLUID FLOWS

In this section we provide an overview of some of the aspects of fluid mechanics that are considered in more depth in subsequent chapters and sections. Although most of the notions presented here are redefined and discussed in detail later, it will be helpful at this point to introduce the general classification of fluid flows.

3.3.1 One-, Two-, and Three-Dimensional Flows

Three-dimensional flow: *The velocity vector depends on three space variables.*

Stagnation point: *The point where the fluid comes to rest.*

In the Eulerian description of motion the velocity vector, in general, depends on three space variables and time, that is, $\mathbf{V} = \mathbf{V}(x, y, z, t)$. Such a flow is a **three-dimensional flow**, because the velocity vector depends on three space coordinates. The solutions to problems in such a flow are very difficult and are beyond the scope of an introductory course. Even if the flow could be assumed to be steady [i.e., $\mathbf{V} = \mathbf{V}(x, y, z)$], it would remain a three-dimensional flow. A particular flow is shown in Fig. 3.7. This flow is normal to a plane surface; the fluid decelerates and comes to rest at the **stagnation point**. The velocity components, u, v, and w depend on x, y and z; that is, $u = u(x, y, z)$, $v = v(x, y, z)$ and $w = w(x, y, z)$.

Two-dimensional flow: *The velocity vector is dependent on only two space variables.*

Plane flow: *The velocity vector depends on the two coordinates x and y.*

Often a three-dimensional flow can be approximated as a two-dimensional flow. For example, the flow over a wide dam is three-dimensional because of the end conditions, but the flow in the central portion away from the ends can be treated as two-dimensional. In general, a **two-dimensional flow** is a flow in which the velocity vector depends on only two space variables. An example is a **plane flow**, in which the velocity vector depends on two spatial coordinates, x and y, but not z [i.e., $\mathbf{V} = \mathbf{V}(x, y)$].

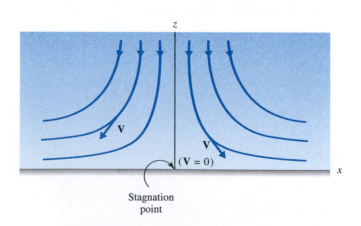

Stagnation point

FIGURE 3.7 A stagnation point flow.

A **one-dimensional flow** is a flow in which the velocity vector depends on only one space variable. Such flows occur in long, straight pipes or between parallel plates, as shown in Fig. 3.8. The velocity in the pipe varies only with r i.e., $u = u(r)$. The velocity between parallel plates varies only with the coordinate y i.e., $u = u(y)$. Even if the flow is unsteady so that $u = u(y, t)$, as would be the situation during startup, the flow is one-dimensional.

The flows shown in Fig. 3.8 may also be referred to as **developed flows**; that is, the velocity profiles do not vary with respect to the space coordinate in the direction of flow. This demands that the region of interest be a substantial distance from an entrance or a sudden change in geometry.

There are many engineering problems in fluid mechanics in which a flow field is simplified to a **uniform flow**: the velocity, and other fluid properties, are constant over the area as in Fig. 3.9. This simplification is made when the velocity is essentially constant over the area, a rather common occurrence. Examples of such flows are relatively high speed flow in a pipe section, and flow in a stream. The average velocity may change from one section to another; the flow conditions depend only on the space variable in the flow direction. For large conduits, however, it may be necessary to consider hydrostatic variation in the pressure normal to the streamlines.

One-dimensional flow: The velocity vector depends on only one space variable.

Developed flows: The velocity profiles do not vary with respect to the space coordinates in the direction of flow.

Uniform flow: The fluid properties are constant over the area.

3.3.2 Viscous and Inviscid Flows

A fluid flow may be broadly classified as either a viscous flow or an inviscid flow. An **inviscid flow** is one in which viscous effects do not significantly influence the flow and are thus neglected. In a **viscous flow** the effects of viscosity are important and cannot be ignored.

To model an inviscid flow analytically, we can simply let the viscosity be zero; this will obviously make all viscous effects zero. It is more difficult to create an inviscid flow experimentally, because all fluids of interest (such as water and air)

Inviscid flow: Viscous effects do not significantly influence the flow.

Viscous flow: The effects of viscosity are significant.

(a)   (b)

FIGURE 3.8   One-dimensional flow: (a) flow in a pipe; (b) flow between parallel plates.

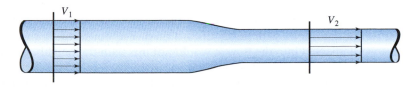

FIGURE 3.9   Uniform velocity profiles.

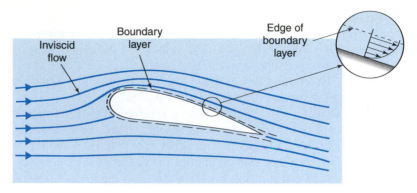

FIGURE 3.10 Flow around an airfoil.

have viscosity. The question then becomes: Are there flows of interest in which the viscous effects are negligibly small? The answer is "yes, if the shear stresses in the flow are small and act over such small areas that they do not significantly affect the flow field." This statement is very general, of course, and it will take considerable analysis to justify the inviscid flow assumption.

Based on experience, it has been found that the primary class of flows, which can be modeled as inviscid flows, is **external flows**, that is, flows which exist exterior to a body. Inviscid flows are of primary importance in flows around streamlined bodies, such as flow around an airfoil or a hydrofoil. Any viscous effects that may exist are confined to a thin layer, called a **boundary layer**, that is attached to the boundary, such as that shown in Fig. 3.10; the velocity in a boundary layer is always zero at a fixed wall, a result of viscosity. For many flow situations, boundary layers are so thin that they can simply be ignored when studying the gross features of a flow around a streamlined body. For example, the inviscid flow solution provides an excellent prediction to the flow around the airfoil, except inside the boundary layer and possibly near the trailing edge. Inviscid flow is also encountered in contractions inside piping systems and in short regions of internal flows where viscous effects are negligible.

Viscous flows include the broad class of internal flows, such as flows in pipes and conduits and in open channels. In such flows viscous effects cause substantial "losses" and account for the huge amounts of energy that must be used to transport oil and gas in pipelines. The no-slip condition resulting in zero velocity at the wall, and the resulting shear stresses, lead directly to these losses.

3.3.3   Laminar and Turbulent Flows

A viscous flow can be classified as either a laminar flow or a turbulent flow. In a **laminar flow** the fluid flows with no significant mixing of neighboring fluid particles. If dye were injected into the flow, it would not mix with the neighboring fluid except by molecular activity; it would retain its identity for a relatively long period of time. Viscous shear stresses always influence a laminar flow. The flow may be highly time dependent, due to the erratic motion of a piston as shown by the output of a velocity probe in Fig. 3.11a, or it may be steady, as shown in Fig. 3.11b.

External flows: *Flows which exist exterior to a body.*

Boundary layer: *A thin layer attached to the boundary in which viscous effects are concentrated.*

KEY CONCEPT   *Inviscid flow provides an excellent prediction to the flow around an airfoil.*

Laminar flow: *A flow with no significant mixing of particles but with significant viscous shear stresses.*

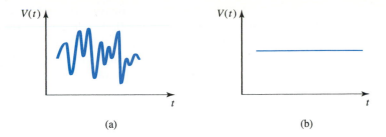

FIGURE 3.11   Velocity as a function of time in a laminar flow: (a) unsteady flow; (b) steady flow.

In a **turbulent flow** fluid motions vary irregularly so that quantities such as velocity and pressure show a random variation with time and space coordinates. The physical quantities are often described by statistical averages. In this sense we can define a "steady" turbulent flow: a flow in which the time-average physical quantities do not change in time. Figure 3.12 shows instantaneous velocity measurements in an unsteady and a steady turbulent flow. A dye injected into a turbulent flow would mix immediately by the action of the randomly moving fluid particles; it would quickly lose its identity in this diffusion process.

A laminar flow and a turbulent flow can be observed by performing a simple experiment with a water faucet. Turn the faucet on so the water flows out very slowly as a silent stream. This is laminar flow. Open the faucet slowly and observe the flow becoming turbulent. Note that a turbulent flow develops with a relatively small flow rate.

The reason why a flow can be laminar or turbulent has to do with what happens to a small flow disturbance, a perturbation to the velocity components. A flow disturbance can either increase or decrease in size. If a flow disturbance in a laminar flow increases (i.e., the flow is unstable), the flow may become turbulent; if the disturbance decreases, the flow remains laminar. In certain situations the flow may develop into a different laminar flow, as is the case between concentrically rotating cylinders shown in Fig. 3.13. At low rotational speed the flow would be in simple circles. But at a sufficiently high speed the flow becomes unstable and the vertices suddenly appear; it is a much more complex laminar flow called Taylor-Conette flow.

Turbulent flow: *Flow varies irregularly so that flow quantities show random variation.*

KEY CONCEPT *A dye injected into a turbulent flow would mix immediately.*

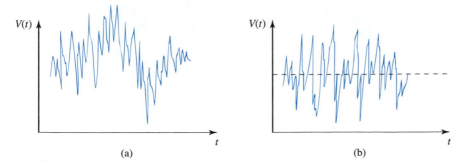

FIGURE 3.12   Velocity as a function of time in a turbulent flow: (a) unsteady flow; (b) "steady" flow.

The flow regime depends on three physical parameters describing the flow conditions. The first parameter is a length scale of the flow field, such as the thickness of a boundary layer or the diameter of a pipe. If this length scale is sufficiently large, a flow disturbance may increase and the flow may be turbulent. The second parameter is a velocity scale such as a spatial average of the velocity; for a large enough velocity the flow may be turbulent. The third parameter is the kinematic viscosity; for a small enough viscosity the flow may be turbulent.

The three parameters can be combined into a single parameter that can serve as a tool to predict the flow regime. This quantity is the **Reynolds number**, named after Osborne Reynolds (1842–1912), a dimensionless parameter, defined as

$$\mathrm{Re} = \frac{VL}{\nu} \tag{3.3.1}$$

where L and V are a characteristic length and velocity, respectively, and ν is the kinematic viscosity; for example, in a pipe flow L could be the pipe diameter and V could be the average velocity. If the Reynolds number is relatively small, the flow is laminar as shown in Figs. 3.13 and 3.14; if it is large, the flow is turbulent. This is more precisely stated by defining a **critical Reynolds number**, $\mathrm{Re_{crit}}$, so that the flow is laminar if $\mathrm{Re} < \mathrm{Re_{crit}}$. For example, in a flow inside a rough-walled pipe it is found that $\mathrm{Re_{crit}} \approx 2000$. This is the minimum critical Reynolds number and is used for most engineering applications. If the pipe wall is extremely smooth and free of vibration, the critical Reynolds number can be

FIGURE 3.13 Laminar flow between rotating cylinders. A secondary flow occurs as regularly spaced toroidal vortices. ("Steady supercritical Taylor vortex flow," by Burkhalter and Koschmieder. From *Journal of Fluid Mechanics* vol. 58, pp. 547-560 (1973). Reprinted with the permission of Cambridge University Press.)

Reynolds number:
Parameter combining a length of scale, a velocity scale, and the kinematic viscosity into

$$\mathrm{Re} = \frac{VL}{\nu}$$

Critical Reynolds number:
The number above which a primary laminar flow ceases to exist.

FIGURE 3.14   Streamlines around a semicircular arc. At this Reynolds number of 0.031 the centers of the pair of eddies in the cavity are separated by 0.52 diameter, in good agreement with a solution of the differential equations. Aluminum powder dispersed in glycerine is illuminated by a slit of light. (Courtesy of The Parabolic Press, Stanford, California. Reprinted with permission.)

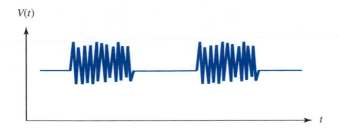

FIGURE 3.15   Velocity versus time signal from a velocity probe in an intermittent flow.

increased as the fluctuation level in the flow is decreased; values in excess of 40 000 have been measured. The critical Reynolds number is different for every geometry, e.g, it is 1500 for flow between parallel plates using the average velocity and the distance between the plates.

The flow can also be intermittently turbulent and laminar; this is called an *intermittent flow*. This phenomenon can occur when the Reynolds number is close to Re_{crit}. Figure 3.15 shows the output of a velocity probe for such a flow.

In a boundary layer that exists on a flat plate, due to a constant-velocity fluid stream, as shown in Fig. 3.16, the length scale changes with distance from the upstream edge. A Reynolds number is calculated using the length x as the characteristic length. For a certain x_T, Re becomes Re_{crit} and the flow undergoes transition from laminar to turbulent. For a smooth rigid plate in a uniform flow with a low free-stream fluctuation level, values as high as $\text{Re}_{crit} = 10^6$ have been observed. In most engineering applications we assume a rough wall, or high free-stream fluctuation level, with an associated critical Reynolds number of approximately 3×10^5.

It is not appropriate to refer to an inviscid flow as laminar or as turbulent. The inviscid flow of Fig. 3.10 is often called the **free stream**. The free stream can be irrotational or it can possess vorticity; most often it is irrotational.

KEY CONCEPT   *In most applications, we assume a critical Reynolds number of 3×10^5 in flow on a flat plate.*

Free stream: *The inviscid flow outside the boundary layer in an external flow.*

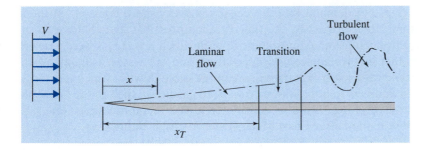

FIGURE 3.16   Boundary layer flow on a flat plate.

Example 3.3

The 2-cm-diameter pipe of Fig. E3.3 is used to transport water at 20°C. What is the maximum average velocity that may exist in the pipe for which laminar flow is guaranteed?

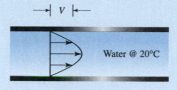

Water @ 20°C

FIGURE E3.3

Solution

The kinematic viscosity is found in Appendix B to be $\nu = 10^{-6} \text{ m}^2/\text{s}$. Using a Reynolds number of 2000 so that a laminar flow is guaranteed, we find that

$$V = \frac{2000\nu}{D}$$

$$= \frac{2000 \times 10^{-6}}{0.02} = 0.1 \text{ m/s}$$

This average velocity is quite small. Velocities this small are not usually encountered in actual situations; hence laminar flow is seldom of engineering interest except for specialized topics such as lubrication. Most internal flows are turbulent flows and thus the study of turbulence gains much attention.

3.3.4   Incompressible and Compressible Flows

The last major classification of fluid flows to be considered in this chapter separates flows into incompressible and compressible flows. An **incompressible flow** exists if the density of each fluid particle remains relatively constant as it moves through the flow field, that is,

Incompressible flow: *The density of each fluid particle remains constant.*

$$\frac{D\rho}{Dt} = 0 \tag{3.3.2}$$

This does not demand that the density is everywhere constant. If the density is constant, then obviously, the flow is incompressible, but that would be a more restrictive condition. Atmospheric flow, in which $\rho = \rho(z)$, where z is vertical, and flows that involve adjacent layers of fresh and salt water, as happens when rivers enter the ocean, are examples of incompressible flows in which the density varies.

KEY CONCEPT *Constant density is more restrictive than incompressibility.*

In addition to liquid flows, low-speed gas flows, such as the atmospheric flow referred to above, are also considered to be incompressible flows. The **Mach number**, named after Ernst Mach (1838–1916), is defined as

Mach number: *A perameter in a gas flow defined as*
$$M = \frac{V}{c}.$$

$$\mathrm{M} = \frac{V}{c} \tag{3.3.3}$$

where V is the gas speed and the wave speed $c = \sqrt{kRT}$. Equation 3.3.3 is useful in deciding whether a particular gas flow can be studied as an incompressible flow. If M < 0.3, density variations are at most 3% and the flow is assumed to be incompressible; for standard air this corresponds to a velocity below about 100 m/s. If M > 0.3, the density variations influence the flow and compressibility effects should be accounted for; such flows are **compressible flows** and are considered in Chapter 9.

Incompressible gas flows include atmospheric flows, the aerodynamics of landing and takeoff of commercial aircraft, heating and air-conditioning airflows, flow around automobiles and through radiators, and the flow of air around buildings, to name a few. Compressible flows include the aerodynamics of high-speed aircraft, airflow through jet engines, steam flow through the turbine in a power plant, airflow in a compressor, and the flow of the air-gas mixture in an internal combustion engine.

3.4 THE BERNOULLI EQUATION

In this section we present an equation that is probably used more often in fluid flow applications than any other equation. It is also often misused; it is thus important to understand its limitations. Its limitations are a result of several assumptions made in the derivation. One of the assumptions is that viscous effects are neglected. In other words, in view of Eq. 1.5.5, shear stresses introduced by velocity gradients are not taken into consideration. These stresses are often very small compared with pressure differences in the flow field. Locally, these stresses have little effect on the flow field and the assumption is justified. However, over long distances or in regions of high-velocity gradients, these stresses may affect the flow conditions so that viscous effects must be included.

The derivation of this important equation, the Bernoulli equation, starts with the application of Newton's second law to a fluid particle. Let us use an infinitesimal cylindrical particle positioned as shown in Fig. 3.17, with length ds and cross-sectional area dA. The forces acting on the particle are the pressure forces and the weight, as shown. Summing forces in the direction of motion, the s-direction, there results

$$p\, dA - \left(p + \frac{\partial p}{\partial s} ds\right) dA - \rho g\, ds\, dA \cos\theta = \rho\, ds\, dA\, a_s \qquad (3.4.1)$$

where a_s is the acceleration of the particle in the s-direction. It is given by[1]

$$a_s = V\frac{\partial V}{\partial s} + \frac{\partial V}{\partial t} \qquad (3.4.2)$$

where $\partial V/\partial t = 0$ since we will assume steady flow. Also, we see that

$$dh = ds \cos\theta = \frac{\partial h}{\partial s} ds \qquad (3.4.3)$$

so that

$$\cos\theta = \frac{\partial h}{\partial s} \qquad (3.4.4)$$

---

[1] This can be verified by considering Eq. 3.2.9a, assuming that $v = w = 0$. Think of the x-direction being tangent to the streamline at the instant shown, so that $u = V$.

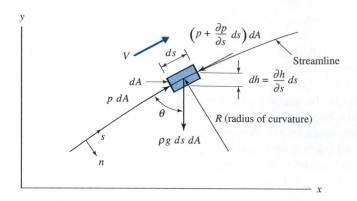

FIGURE 3.17   Particle moving along a streamline.

Then, after dividing by $ds\, dA$, and using the above equations for a_s and $\cos\theta$ Eq. 3.4.1 takes the form

$$-\frac{\partial p}{\partial s} - \rho g\, \frac{\partial h}{\partial s} = \rho V\, \frac{\partial V}{\partial s} \tag{3.4.5}$$

Now, we assume constant density and note that $V\,\partial V/\partial s = \partial(V^2/2)/\partial s$; then we can write Eq. 3.4.5 as

$$\frac{\partial}{\partial s}\left(\frac{V^2}{2} + \frac{p}{\rho} + gh\right) = 0 \tag{3.4.6}$$

This is satisfied if, along the streamline,

$$\frac{V^2}{2} + \frac{p}{\rho} + gh = \text{const} \tag{3.4.7}$$

where the constant may have a different value on a different streamline. Between two points on the same streamline,

$$\frac{V_1^2}{2} + \frac{p_1}{\rho} + gh_1 = \frac{V_2^2}{2} + \frac{p_2}{\rho} + gh_2 \tag{3.4.8}$$

This is the well-known *Bernoulli equation*, named after Daniel Bernoulli (1700–1782). Note the assumptions:

> Inviscid flow (no shear stresses)
> Steady flow ($\partial V/\partial t = 0$)
> Along a streamline ($a_s = V\,\partial V/\partial s$)
> Constant density ($\partial \rho/\partial s = 0$)
> Inertial reference frame ($\mathbf{A} = \mathbf{a}$ in Eq. 3.2.15)

If Eq. 3.4.8 is divided by g, this equation becomes

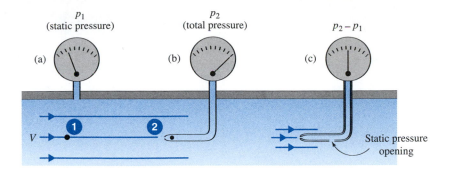

FIGURE 3.18   Pressure probes: (a) piezometer; (b) pitot probe; (c) pitot-static probe.

$$\frac{V_1^2}{2g} + \frac{p_1}{\gamma} + h_1 = \frac{V_2^2}{2g} + \frac{p_2}{\gamma} + h_2 \qquad (3.4.9)$$

KEY CONCEPT   *The total pressure is $p + \rho V^2/2$.*

The sum of the two terms $(p/\gamma + h)$ is called the *piezometric head* and the sum of the three terms the *total head*. The pressure p is often referred to as **static pressure** and the sum of the two terms

Static pressure: *The pressure p, usually expressed as gage pressure.*

$$p + \rho \frac{V^2}{2} = p_T \qquad (3.4.10)$$

is called the *total pressure p_T* or **stagnation pressure**, the pressure at a stagnation point (see Fig. 3.7) in the flow.

Stagnation pressure: *The pressure that exists at a stagnation point.*

The static pressure in a pipe can be measured simply by installing a so-called **piezometer**, shown[2] in Fig. 3.18a. A device, known as a **pitot probe**, sketched in Fig. 3.18b, is used to measure the total pressure in a fluid flow. Point 2 just inside the pitot tube is a stagnation point; the velocity there is zero. The difference between the readouts can be used to determine the velocity at point 1. A **pitot-static probe** is also used to measure the difference between total and static pressure with one probe (Fig. 3.18c). The velocity at point 1 (using the readings of the piezometer and pitot probes, or the reading from the pitot-static probe) can be determined by applying the Bernoulli equation between points 1 and 2:

Piezometer: *Gauge designed to measure static pressure.*

Pitot probe: *Gauge designed to measure total pressure.*

Pitot-static probe: *Gauge designed to measure the difference between total and static pressure.*

$$\frac{V_1^2}{2g} + \frac{p_1}{\gamma} = \frac{p_2}{\gamma} \qquad (3.4.11)$$

where we have assumed point 2 to be a stagnation point so that $V_2 = 0$. This gives

KEY CONCEPT   *Never confuse Bernoulli's equations with the energy equation.*

$$V_1 = \sqrt{\frac{2}{\rho}(p_2 - p_1)} \qquad (3.4.12)$$

We will find many uses for Bernoulli's equation in our study of fluids. We must be careful, however, never to use it in an unsteady flow or if viscous effects are significant (the primary reasons for making Bernoulli's equation inapplicable). We must also never confuse Bernoulli's equation with the energy equation; they are independent equations as Example 3.6 illustrates.

[2] When drilling the hole in the wall necessary for the piezometer, burrs are often formed on the inner surface. It is important that such burrs be removed since they may cause errors as high as 30% in the pressure reading.

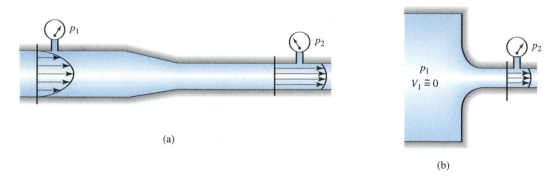

(a)

(b)

FIGURE 3.19 Internal inviscid flows: (a) flow through a contraction; (b) flow from a plenum.

The Bernoulli equation can be used to determine how high the water from a fireman's hose will reach, to find the pressure on the surface of a low-speed airfoil,[3] and to find the wind force on the window in a house. These examples are all external flows, flows around objects submerged in the fluid.

Another class of problems where inviscid flow can be assumed and where the Bernoulli equation finds frequent application involves internal flows over relatively short distances, for example, flow through a contraction, as shown in Fig. 3.19a, or flow from a plenum, as shown in Fig. 3.19b. For a given velocity profile entering the short contraction, the pressure drop ($p_1 - p_2$) and the velocity profile at section 2 can be approximated assuming an inviscid flow. Viscous effects are typically very small and require substantial distances and areas over which to operate in order to become significant; so in situations such as those shown in Fig. 3.19, viscous effects can often be neglected.

Inviscid flow does not always give a good approximation to the actual flow that exists around a body. Consider the inviscid flow around the sphere, shown in Fig. 3.20. A stagnation point where $V = 0$ exists at both the front and the back of the sphere. Bernoulli's equation predicts a maximum pressure at the stagnation points A and C because the velocity is zero at such points. A maximum velocity, and thus a minimum pressure, would exist at point B. In the inviscid flow of part (a) the fluid flowing from B to C must flow from the low-pressure region near B

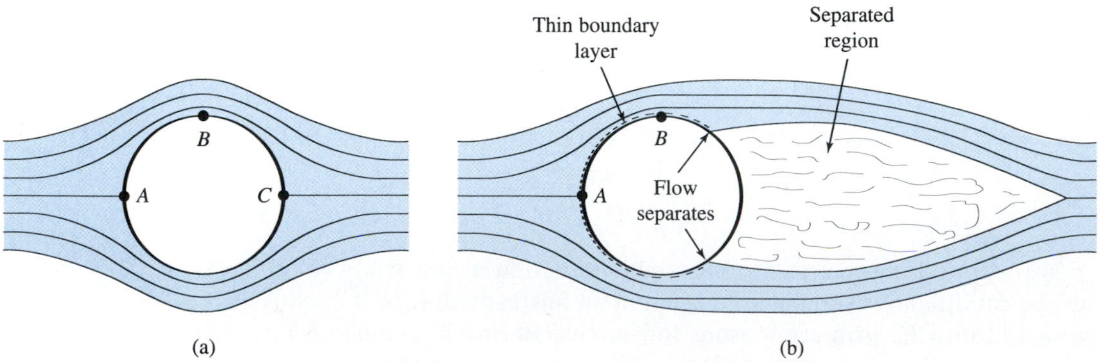

KEY CONCEPT *Viscous effects require substantial areas in order to be significant.*

Thin boundary layer

Separated region

B

A

C

B

A Flow separates

(a)

(b)

FIGURE 3.20 Flow around a sphere: (a) inviscid flow; (b) actual flow.

[3] To consider the flow around an aircraft as a steady flow, we simply make the aircraft stationary and move the air, as is done in model studies using a wind tunnel. The pressures and forces remain unchanged.

to the high-pressure region near C. In the actual flow there exists a thin boundary layer in which the velocity goes to zero at the surface of the sphere. This slow-moving fluid near the boundary does not have sufficient momentum to move into the higher-pressure region near C; the result is that the fluid *separates* from the boundary — the boundary streamline leaves the boundary — creating a **separated region**, a region of recirculating flow, as shown in the actual flow sketched in part (b). The pressure does not increase but remains relatively low over the rear part of the sphere. The high pressure that exists near the front stagnation point is never recovered on the rear of the sphere, resulting in a relatively large drag force in the direction of flow. A similar situation occurs in the flow around an automobile.

The flow on the front of the sphere is well approximated by an inviscid flow; but it is obvious that the flow over the rear of the sphere deviates radically from an inviscid flow. Viscous effects in the boundary layer have led to a separated flow, a phenomenon that is often undesirable. For example, separated flow on an airfoil is called **stall** and must never occur, except on the wings of special stunt planes. On the blades of a turbine separated flows lead to substantially reduced efficiency. The air deflector on the roof of the cab of a semitruck reduces the separated region, thereby reducing drag and fuel consumption.

If viscous effects are negligible in a steady liquid flow, we can use Bernoulli's equation to locate points of possible *cavitation*. This condition occurs when the local pressure becomes equal to the vapor pressure of the liquid. It is to be avoided, if at all possible, because of damage to solid surfaces or because the vaporized liquid may cause devices to not operate effectively. Figure 3.21 shows cavitating flow just downstream of a contraction in a pipe. At the point where cavitation occurs, small vapor bubbles are generated and these bubbles collapse when they enter a higher-pressure region. The collapse is accompanied by very large local pressures that last for only a small fraction of a second. These pressure spikes may reach a wall, where they can, after repeated applications, result in significant damage.

An important observation regarding pressure changes in a fluid needs to be made regarding entrances and exits to a pipe or conduit. Consider a flow from a reservoir through a pipe, as sketched in Fig. 3.22. At the entrance the streamlines are curved and the pressure is not constant across section 1, so the pressure at section 1 cannot be assumed to be uniform. At the exit, however, the streamlines

Separated region: *A region of recirculating flow due to the fluid separating from the boundary.*

KEY CONCEPT *Pressure remains relatively low over the rear part of a sphere.*

KEY CONCEPT *The flow on the front of a sphere is approximated by an inviscid flow.*

Stall: *Separated flow on an airfoil.*

KEY CONCEPT *Cavitation occurs when the local pressure equals the vapor pressure.*

KEY CONCEPT *At a pipe entrance, the streamlines are curved and the pressure is not constant across the inlet area.*

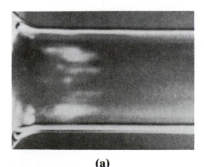

(a)   (b)

FIGURE 3.21 Cavitation in a venturi nozzle, with water flowing at a velocity of 15 m/s: (a) incandescent lamp, exposure time $\frac{1}{30}$ s; (b) strobe light exposure time 5 μs. (Photograph courtesy of the Japan Society of Mechanical Engineers and Pergamon Press.)

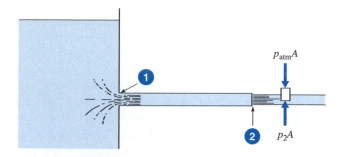

FIGURE 3.22 Exit flow into the atmosphere.

are straight, so that no acceleration exists normal to the streamlines; hence the pressure forces acting on the ends of the small cylindrical control volume must be equal. We write this as $p_2 = p_{atm}$ or $p_2 = 0$ gage pressure.

In Fig. 3.17 we summed forces on the fluid element along the streamline and derived Bernoulli's equation. We can gain additional insight into the pressure field if we sum forces normal to the streamline. Let us consider the fluid particle to be a parallel-piped with thickness dn in the n-direction and area dA_s on the side with length ds. Applying Newton's second law in the n-direction provides

$$p dA_s - \left(p + \frac{\partial p}{\partial n} \, dn \right) dA_s = \rho dA_s dn \, \frac{V^2}{R} \tag{3.4.13}$$

where we have neglected the weight since we are not intending to integrate over significant distances. We have assumed the acceleration in the normal direction to be V^2/R, where R is the radius of curvature in this plane flow (in a three-dimensional flow there would be a principal radius of curvature and a binormal radius of curvature). Equation (3.4.13) reduces to

$$-\frac{\partial p}{\partial n} = \rho \, \frac{V^2}{R} \tag{3.4.14}$$

From this equation we can qualitatively describe how the pressure changes normal to a streamline (Bernoulli's equation predicts the pressure changes along a streamline). If we replace $\partial p/\partial n$ with $\Delta p/\Delta n$, the incremental pressure change Δp over the short distance Δn normal to the streamline is given by

KEY CONCEPT *The pressure decreases in the n-direction.*

$$\Delta p = -\rho \, \frac{V^2}{R} \, \Delta n \tag{3.4.15}$$

This says that the pressure decreases in the n-direction; this decrease is directly proportional to ρ and V^2 and inversely proportional to R. Consequently, a tornado, with $p = 0$ outside the tornado, will have a very low pressure at its center where R is relatively small and V is quite large.

In Fig. 3.22 the pressure would be relatively low at the corner of section 1 and relatively high at the center of section 1. Such qualitative descriptions can be quite helpful in understanding fluid flow behavior.

Example 3.4

The wind reaches a speed of 105 kmph in a storm. Calculate the force acting on the 1 m × 2 m window of Fig. E3.4 facing the storm. The window is in a high-rise building, so the wind speed is not reduced due to ground effects. Use $\rho = 1.22$ kg/m^3.

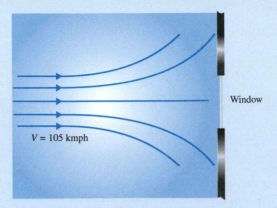

FIGURE E3.4

Solution

The window facing the storm will be in a stagnation region where the wind speed is brought to zero. Working with gage pressures, the pressure p upstream in the wind is zero. The velocity V must have units of m/sec. It is

$$V = 105 \, \frac{\text{km}}{\text{hr}} \times \frac{1 \, \text{hr}}{3600 \, \text{sec}} \times \frac{1000 \, \text{m}}{1 \, \text{km}} = 29.17 \, \text{m/s}$$

Bernoulli's equation can be used in this situation since we can neglect viscous effects, and steady flow occurs along a streamline at constant density (air is incompressible at speeds below about 300 mph). We calculate the pressure on the window selecting state 1 in the free stream and state 2 on the window, as follows:

$$\frac{V_1^2}{2g} + \frac{\cancel{p_1}^{\,0}}{\cancel{\gamma}} + h_1 = \frac{\cancel{V_2^2}^{\,0}}{\cancel{2g}} + \frac{p_2}{\gamma} + h_2$$

$$\therefore p_2 = \frac{\rho V_1^2}{2}$$

$$= \frac{1.22 \times (29.17)^2}{2} = 519 \, \text{N/m}^2$$

where we have used $\gamma = \rho g, h_2 = h_1, p_1 = 0,$ and $V_2 = 0$. Multiply by the area and find the force to be

$$F = pA$$

$$= 519 \times 1 \times 2 = 1038 \, \text{N}$$

We recommend that you verify the units of N/m^2 on the pressure calculation above. To do this, use $F = ma$ which provides kg = N · s^2/m.

Example 3.5

The static pressure head in an air pipe (Fig. E3.5) is measured with a piezometer as 16 mm of water. A pitot probe at the same location indicates 24 mm of water. Calculate the velocity of the 20°C air. Also, calculate the Mach number and comment as to the compressibility of the flow.

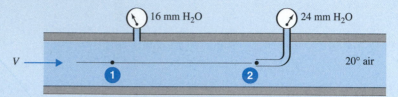

FIGURE E3.5

Solution

Bernoulli's equation is applied between two points on the streamline that terminates at the stagnation point of the pitot probe. Point 1 is upstream and p_2 is the total pressure at point 2; then, with no elevation change,

$$\frac{V_1^2}{2g} + \frac{p_1}{\gamma} = \frac{p_T}{\gamma}$$

The pressure measured with the piezometer is $p_1 = \gamma h = 9810 \times 0.016 = 157$ Pa. We use the ideal gas law to calculate the density:

$$\rho = \frac{p}{RT}$$

$$= \frac{157 + 101\ 000}{287 \times (273 + 20)} = 1.203 \text{ kg/m}^3$$

where standard atmospheric pressure, which is 101 000 Pa (if no elevation is given, assume standard conditions), is added since absolute pressure is needed in the preceding equation. The units are checked by using Pa = N/m^2 and J = N · m. The velocity is then

$$V_1 = \sqrt{\frac{2}{\rho}(p_T - p_1)}$$

$$= \sqrt{\frac{2(0.024 - 0.016) \times 9810}{1.203}} = 11.42 \text{ m/s}$$

where the units can be verified by using kg = N · s^2/m. To find the Mach number, we must calculate the speed of sound. It is

$$c = \sqrt{kRT}$$

$$= \sqrt{1.4 \times 287 \times 293} = 343 \text{ m/s}$$

The Mach number is then

$$M = \frac{V}{c} = \frac{11.44}{343} = 0.0334$$

Obviously, the flow can be assumed to be incompressible since M < 0.3. The velocity would have to be much higher before compressibility would be significant.

Example 3.6

Bernoulli's equation, in the form of Eq. 3.4.8, looks very much like the energy equation developed in thermodynamics for a control volume. Discuss the differences between the two equations.

Solution

From thermodynamics we recall that the steady-flow energy equation for a control volume with one inlet and one outlet takes the form

$$\dot{Q} - \dot{W}_s = \dot{m}\left(\frac{V_2^2}{2} + \frac{p_2}{\rho_2} + \tilde{u}_2 + gz_2\right) - \dot{m}\left(\frac{V_1^2}{2} + \frac{p_1}{\rho_1} + \tilde{u}_1 + gz_1\right)$$

Eq. 3.4.8 becomes, after dividing through by g,

$$\frac{V_2^2}{2g} + \frac{p_2}{\gamma} + z_2 = \frac{V_1^2}{2g} + \frac{p_1}{\gamma} + z_1$$

where we have made the following assumptions:

No heat transfer ($\dot{Q} = 0$)
No shaft work ($\dot{W}_s = 0$)
No temperature change ($\tilde{u}_2 = \tilde{u}_1$, i.e., no losses due to shear stresses)
Uniform velocity profiles at the two sections
Steady flow
Constant density ($\gamma_2 = \gamma_1$)

Even though several of these assumptions are the same as those made in the derivation of the Bernoulli equation (steady flow, constant density, and no shear stress), we must not confuse the two equations; the Bernoulli equation is derived from Newton's second law and is valid along a streamline, whereas the energy equation is derived from the first law of thermodynamics and is valid between two sections in a fluid flow. The energy equation can be used across a pump to determine the horsepower required to provide a particular pressure rise; the Bernoulli equation can be used along a stagnation streamline to determine the pressure at a stagnation point, a point where the velocity is zero. The equations are quite different, and just because the energy equation degenerates to the Bernoulli equation for particular situations, the two should not be used out of context.

Example 3.7

Explain why a burr on the upstream side of the piezometer opening of Fig. E3.7 will result in a low reading of the pressure.

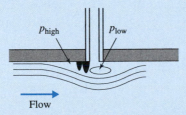

p_{high} p_{low}

Flow

FIGURE E3.7   *(continued)*

Solution

A burr on the upstream side of the piezometer opening would result in a flow in the vicinity of the burr somewhat like that shown in Fig. E3.7. A streamline pattern would develop so that a relatively high pressure would occur on the upstream side of the burr and a relatively low pressure on the downstream side at the opening of the piezometer tube. Consequently, since the center of the curvature of the streamline is in the vicinity of the opening, a low reading of the pressure will be recorded.

3.5 SUMMARY

The Eulerian description of motion was used to express the acceleration as

$$\mathbf{a} = u\frac{\partial \mathbf{V}}{\partial x} + v\frac{\partial \mathbf{V}}{\partial y} + w\frac{\partial \mathbf{V}}{\partial z} + \frac{\partial \mathbf{V}}{\partial t} \tag{3.5.1}$$

Fluid motion may result in fluid particles being rotated and/or deformed. For a flow in the xy-plane a particle would rotate with angular velocity

$$\Omega_z = \frac{1}{2}\left(\frac{\partial v}{\partial x} - \frac{\partial u}{\partial y}\right) \tag{3.5.2}$$

and deform with

$$\epsilon_{xx} = \frac{\partial u}{\partial x}, \qquad \epsilon_{yy} = \frac{\partial v}{\partial y}, \qquad \epsilon_{xy} = \frac{1}{2}\left(\frac{\partial v}{\partial x} + \frac{\partial u}{\partial y}\right) \tag{3.5.3}$$

Fluid flows are classified as steady or unsteady; viscous or inviscid; laminar, turbulent, or free stream; and incompressible or compressible. Any of these can be uniform, one-, two-, and three-dimensional flows. Experience and practice are needed to appropriately classify a particular flow of interest. It is only for the simpler flows (e.g., a steady, laminar, incompressible, uniform flow) that we hope to obtain a relatively simple solution.

Finally, the famous Bernoulli equation

$$\frac{V_1^2}{2g} + \frac{p_1}{\gamma} + gh_1 = \frac{V_2^2}{2g} + \frac{p_2}{\gamma} + gh_2 \tag{3.5.4}$$

was presented for a steady, inviscid, constant-density flow along a streamline in an inertial reference frame. The estimate of the pressure change normal to a streamline,

$$\Delta p = -\rho\frac{V^2}{R}\Delta n \tag{3.5.5}$$

was also obtained.

PROBLEMS

Flow Fields

3.1   A fire is started and the smoke from the chimney goes straight up; no wind is present. After a few minutes a wind arises but the smoke continues to rise slowly. Sketch the streakline of the smoke, the pathline of the first particles leaving the chimney, and a few streamlines, assuming that the wind blows parallel to the ground in a constant direction.

3.2   A researcher has a large number of small flotation devices each equipped with a battery and light bulb. Explain how she would determine the pathlines and streaklines near the surface of a stream with some unknown currents that vary with time.

3.3   A little boy chases his dad around the yard with the water hose of Fig. P3.3. Sketch a pathline and a streakline if the boy is running perpendicular to the water jet.

FIGURE P3.3

3.4   The hot-air balloon of Fig. P3.4 travels with the wind. The wind velocity vector is $\mathbf{V} = 6\hat{\mathbf{i}} + 10\hat{\mathbf{j}}$ m/s for the first hour and then it is $10\hat{\mathbf{i}} + 5\hat{\mathbf{j}}$ m/s for 2 hr. On xy-coordinates, sketch the pathline of the balloon and streamlines at $t = 2$ hrs. If a number of hot-air balloons started from the same location, sketch the streakline formed by the balloons at $t = 3$ hr. The balloons start at the origin.

FIGURE P3.4

3.5   A velocity field is given by $\mathbf{V} = (2t + 2)\hat{\mathbf{i}} + 2t\hat{\mathbf{j}}$ m/s. Sketch the pathlines of two particles up to $t = 5$ s, one that originates from the origin at $t = 0$, and another that originates from the origin at $t = 2$ s. Also, sketch the streamlines at $t = 5$ s.

3.6   Using rectangular coordinates, express the z-component of Eq. 3.2.2.

3.7   The traffic situation on Mackinac Island, Michigan, where no automobiles are allowed (bikes are permitted), is to be studied. Comment on how such a study could be performed using a Lagrangian approach and an Eulerian approach.

3.8   Determine the speed of a fluid particle at the origin and at the point $(1, -2, 0)$ for each of the following velocity fields when $t = 2$ s. All distances are in meters and t is in seconds.
(a) $\mathbf{V} = (x + 2)\hat{\mathbf{i}} + xt\hat{\mathbf{j}} - z\hat{\mathbf{k}}$ m/s
(b) $\mathbf{V} = xy\hat{\mathbf{i}} - 2y^2\hat{\mathbf{j}} + tyz\hat{\mathbf{k}}$ m/s
(c) $\mathbf{V} = x^2t\hat{\mathbf{i}} - (xz + 2t)\hat{\mathbf{j}} + xyt\hat{\mathbf{k}}$ m/s

3.9   Determine the unit vector normal to the streamline at a point where $\mathbf{V} = 3\hat{\mathbf{i}} - 4\hat{\mathbf{j}}$ in a plane flow.
A. $0.6\hat{\mathbf{i}} + 0.8\hat{\mathbf{j}}$ B. $-0.6\hat{\mathbf{i}} + 0.8\hat{\mathbf{j}}$
C. $0.8\hat{\mathbf{i}} - 0.6\hat{\mathbf{j}}$ D. $0.8\hat{\mathbf{i}} + 0.6\hat{\mathbf{j}}$

3.10   Calculate the angle that the velocity vector makes with the x-axis; and a unit vector normal to the streamline at $(1, -2)$ for the following velocity fields when $t = 2$ s. All distances are in meters and t is in seconds.
(a) $\mathbf{V} = (x + 2)\hat{\mathbf{i}} + xt\hat{\mathbf{j}}$ m/s
(b) $\mathbf{V} = xy\hat{\mathbf{i}} - 2y^2\hat{\mathbf{j}}$ m/s
(c) $\mathbf{V} = (x^2 + 4)\hat{\mathbf{i}} - y^2t\hat{\mathbf{j}}$ m/s

3.11   Find the equation of the streamline that passes through $(1, -2)$ at $t = 2$ s for the flow of:
(a)   Problem 3.10a   (b)   Problem 3.10b
(c)   Problem 3.10c

3.12   A velocity field is given by $\mathbf{V} = 2xy\hat{\mathbf{i}} - y^2\hat{\mathbf{j}}$ m/s. The magnitude of the acceleration at $(-1$ m, 2 m$)$ is nearest:
A. 11.21 m/s^2 B. 14.69 m/s^2
C. 17.89 m/s^2 D. 1.2 m/s^2

3.13   Find the acceleration vector field for a fluid flow that possesses the following velocity field where x, y, z are in meters. Evaluate the acceleration at $(2, -1, 3)$ at $t = 2$ s.
(a) $\mathbf{V} = 20 (1 - y^2)\hat{\mathbf{i}}$ m/s
(b) $\mathbf{V} = 2x\hat{\mathbf{i}} + 2y\hat{\mathbf{j}}$ m/s
(c) $\mathbf{V} = x^2 t\hat{\mathbf{i}} + 2xyt\hat{\mathbf{j}} + 2yzt\hat{\mathbf{k}}$ m/s
(d) $\mathbf{V} = x\hat{\mathbf{i}} - 2xyz\hat{\mathbf{j}} + tz\hat{\mathbf{k}}$ m/s

3.14 Find the angular velocity vector for the following flow fields. Evaluate the angular velocity at $(2, -1, 3)$ at $t = 2$ s.
   (a) Problem 3.13a
   (b) Problem 3.13b
   (c) Problem 3.13c
   (d) Problem 3.13d

3.15 Find the vorticity vector for the following flow fields. Evaluate the vorticity at $(2, -1, 3)$ at $t = 2$ s.
   (a) Problem 3.13a
   (b) Problem 3.13b
   (c) Problem 3.13c
   (d) Problem 3.13d

3.16 Determine the components of the rate-of-strain tensor for the velocity field of the following at $(2, -1, 3)$ at $t = 2$ s.
   (a) Problem 3.13a
   (b) Problem 3.13b
   (c) Problem 3.13c
   (d) Problem 3.13d

3.17 The velocity components in cylindrical coordinates are given by

$$v_r = \left(10 - \frac{40}{r^2}\right) \cos \theta \text{ m/s}, \quad v_\theta = -\left(10 + \frac{40}{r^2}\right) \sin \theta \text{ m/s}$$

   (a) Calculate the acceleration of a fluid particle occupying the point $(4 \text{ m}, 180°)$.
   (b) Calculate the vorticity component at $(4 \text{ m}, 180°)$.

3.18 The velocity components in spherical coordinates are given by

$$v_r = \left(10 - \frac{80}{r^3}\right) \cos \theta \text{ m/s}, \quad v_\theta = -\left(10 + \frac{80}{r^3}\right) \sin \theta \text{ m/s}$$

   (a) Calculate the acceleration of a fluid particle occupying the point $(4 \text{ m}, 180°)$.
   (b) Calculate the vorticity component at $(4 \text{ m}, 180°)$.

3.19 An unsteady flow occurs between parallel plates such that $u = u(y, t)$, $v = 0$, and $w = 0$. Write an expression for the acceleration. What is the acceleration if the flow is steady, that is, $u = u(y)$, $v = 0$, and $w = 0$?

3.20 Consider a symmetrical steady flow in a pipe with the axial and radial velocity components designated $u(r, x)$ and $v(r, x)$, respectively. Write the equations for the two acceleration components a_r and a_x. Use equations from Table 3.1. See Fig. P3.20 for the coordinates.

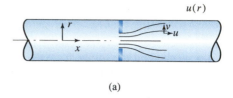

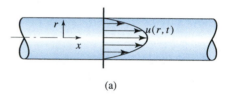

(a)

FIGURE P3.20

3.21 The velocity in the 2-cm-diameter pipe of Fig. P3.21 has only one nonzero velocity component given by $u(r, t) = 2(1 - r^2/r_0^2)(1 - e^{-t/10})$ m/s, where r_0 is the radius of the pipe and t is in seconds. Calculate the maximum velocity and the maximum acceleration:
   (a) Along the centerline of the pipe.
   (b) Along a streamline at $r = 0.5$ cm.
   (c) Along a streamline just next to the pipe wall.
   [*Hint:* Let $v_z = u(r, t)$, $v_r = 0$, and $v_\theta = 0$ in the appropriate equations in Table 3.1.]

(a)

FIGURE P3.21

3.22 The temperature changes periodically in a flow according to $T(y, t) = 20(1 - y^2) \cos \pi t/100°C$. If the velocity is given by $u = 2(1 - y^2)$ m/s, determine the rate of change of the temperature of a fluid particle located at $y = 0$ if $t = 20$ s.

3.23 The density of air in the atmosphere varies according to $\rho(z) = 1.23e^{-10^{-4}z}$ kg/m³. Air flowing over the mountain of Fig. P3.23 has the velocity vector $\mathbf{V} = 20\hat{\mathbf{i}} + 10\hat{\mathbf{k}}$ m/s at a location of interest where $z = 3000$ m. Find the rate at which a particle's density is changing at that location.

3000 m

FIGURE P3.23

3.24 The density variation with elevation is given by $\rho(z) = 1000\,(1 - z/4)$ kg/m³. At a location where $\mathbf{V} = 10\hat{\mathbf{i}} + 10\hat{\mathbf{k}}$ m/s, find $D\rho/Dt$.

3.25 Salt is slowly added to water in a pipe so that $\partial\rho/\partial x = 0.01$ kg/m⁴. Determine $D\rho/Dt$ if the velocity is uniform at 4 m/s.

3.26 The velocity shown in Fig. P3.26 is given by $V(x) = 10/(4 - x)^2$ m/s. The acceleration at $x = 2$ m is nearest:

A. 52.5 m/s² B. 42.5 m/s²
C. 25 m/s² D. 6.25 m/s²

FIGURE P3.26

3.27 Express the substantial derivative in terms of the gradient ∇ and the velocity vector $\mathbf{V}$. Recall from calculus that in rectangular coordinates

$$\nabla = \frac{\partial}{\partial x}\hat{\mathbf{i}} + \frac{\partial}{\partial y}\hat{\mathbf{j}} + \frac{\partial}{\partial z}\hat{\mathbf{k}}.$$

3.28 We can write Eqs. 3.2.9 in a simplified vector form. The gradient is a vector operator expressed in rectangular coordinates as $\nabla = \dfrac{\partial}{\partial x}\hat{\mathbf{i}} + \dfrac{\partial}{\partial y}\hat{\mathbf{j}} + \dfrac{\partial}{\partial z}\hat{\mathbf{k}}$.

Write the dot product of the velocity vector $\mathbf{V}$ and the gradient ∇ and then write Eqs. 3.2.9 as one vector equation expressing the acceleration $\mathbf{a}$ as the sum of the local acceleration and the convective acceleration.

3.29 For the flow shown in Fig. P3.29, relative to a fixed reference frame, find the acceleration of a fluid particle at:
(a) Point A
(b) Point B

The water at B makes an angle of 30° with respect to the ground and the sprinkler arm is horizontal.

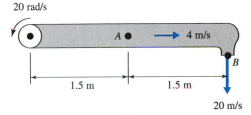

FIGURE P3.29

3.30 A river is flowing due south at 5 m/s at a latitude of 45°. Calculate the acceleration of a particle floating with the river relative to a fixed reference frame. The radius of the Earth is 6000 km.

Classification of Fluid Flows

3.31 Consider each of the following flows and state whether it could be approximated as a one-, two-, or three-dimensional flow or as a uniform flow:
(a) Flow from a vertical pipe striking a horizontal wall
(b) Flow in the waves of the ocean near a beach
(c) Flow near the entrance of a pipe
(d) Flow around a rocket with a blunt nose
(e) Flow around an automobile
(f) Flow in an irrigation channel
(g) Flow through an artery
(h) Flow through a vein

3.32 Which of the flows in Problem 3.31 could be assumed to be a steady flow? Which must be modeled as an unsteady flow?

3.33 Which flow in Problem 3.31 could best be modeled as a plane flow?

3.34 Select the flows in Problem 3.31 that would possess a stagnation point. Sketch each of the selected flows indicating the location of the stagnation point.

3.35 Which of the flows in Problem 3.31 could be modeled as a developed flow?

3.36 State whether each of the flows in Problem 3.31 could be considered as primarily an inviscid flow or a viscous flow.

3.37 Select the flows in Problem 3.31 that are external flows. Does each of the external flows possess a stagnation point?

3.38 Sketch the flow around a razor blade positioned parallel to the flow showing the boundary layers.

3.39 The flow in the section of a conduit shown in Fig. P3.26 is a:
   A.   developed flow
   B.   uniform flow
   C.   one-dimensional flow
   D.   two-dimensional flow

3.40 The 32°C water exiting the 1.5-cm-diameter faucet of Fig. P3.40 has an average velocity of 2 m/s. Would you expect the flow to be laminar or turbulent?

FIGURE P3.40

3.41 The Red Cedar River flows placidly through Michigan State University's campus. In a certain section the depth is 0.8 m and the average velocity is 0.2 m/s. Is the flow laminar or turbulent?

3.42 Air at 40°C flows in a rectangular 30 cm × 6 cm heating duct at an average velocity of 4 m/s. Is the flow laminar or turbulent?

3.43 The sphere of diameter D of Fig. P3.43 is traveling with a velocity V of 1.2 m/s in 20°C atmospheric air. If Re = VD/ν is less than 4×10^4, the boundary layer around the front of the sphere is completely laminar. Determine if the boundary layer is completely laminar on a sphere of diameter:
   (a)   1 cm   (b)   1 m

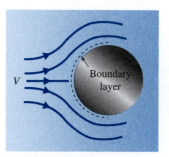

FIGURE P3.43

3.44 The airfoil on a commercial airliner is approximated as a flat plate sketched in Fig. P3.44. How long would you expect the laminar portion of the boundary layer to be if it is flying:

   (a)   At an altitude of 10 000 m and a speed of 900 km/h?

   (b)   At an altitude of 30,000 ft and a speed of 600 mph?

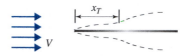

FIGURE P3.44

3.45 A leaf keeps cool by transpiration, a process in which water flows from the leaf to the atmosphere. An experimenter wonders if the boundary layer on a leaf influences the transpiration, so an "experimental" leaf is set up in the laboratory and air is blown over it at 6 m/s. Comment as to whether the boundary layer is expected to be laminar or turbulent.

3.46 For the following situations state whether a compressible flow is required or if the flow can be approximated with an incompressible flow:

   (a)   An aircraft flying at 100 m/s at an elevation of 8000 m

   (b)   A golf ball traveling at 80 m/s

   (c)   Flow around an object being studied in a high-temperature wind tunnel if the temperature is 100°C and the air velocity is 100 m/s

3.47 Write out Eq. 3.3.2 using Eq. 3.2.11. For a steady, plane flow what relationship must exist for an incompressible flow in which the density is allowed to vary?

3.48 If $\rho = \rho_0 (1 + cz)$ models the density variation in a channel (there is heavy salt water at the bottom and fresh water at the top) in which $u(y, z)$ is the only velocity component, is the flow incompressible?

Bernoulli's Equation

3.49 The velocity of an airplane is measured with a pitot tube. If the pitot tube measures 800 mm of water, estimate the speed of the airplane. Use $\rho_{air} = 1.23$ kg/m³.

A.	125 m/s	**B.**	113 m/s
C.	80 m/s	**D.**	36 m/s

3.50 A pitot tube is used to measure the velocity of a small aircraft flying at 1000 m. Calculate its velocity if the pitot tube measures:

  (a) 2 kPa **(b)** 6 kPa **(c)** 600 Pa

3.51 Approximate the force acting on the 15-cm-diameter headlight shown in Fig. P3.51 of an automobile traveling at 120 kph.

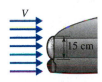

FIGURE P3.51

3.52 A vacuum cleaner is capable of creating a vacuum of 2 kPa just inside the hose of Fig. P3.52. What maximum average velocity would be expected in the hose?

FIGURE P3.52

3.53 A pitot tube measures 600 mm of water in a pipe transporting water. A static pressure probe at the same location measures 200 mm of water. The velocity of water in the pipe is nearest:

A.	1.10 m/s	**B.**	1.98 m/s
C.	2.8 m/s	**D.**	3.43 m/s

3.54 A manometer, utilizing a pitot probe, measures 10 mm of mercury. If the velocity is desired in a pipe transporting water to which the manometer is attached, what additional information in the following list is needed?

I. The temperature of the water	**A.** I and II
II. The pressure in the pipe	**B.** II and III
III. The density of the mercury	**C.** III and IV
IV. The diameter of the pipe	**D.** III and IV

3.55 The inviscid, incompressible flow in the vicinity of a stagnation point (Fig. P3.55) is approximated by $u = -10x, v = 10y$. If the pressure at the origin is p_0, find an expression for the pressure neglecting gravity effects:

  (a) Along the negative x-axis
  (b) Along the positive y-axis

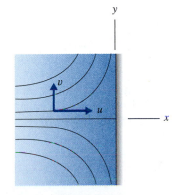

FIGURE P3.55

3.56 The inviscid, incompressible flow field exterior to the cylinder shown in Fig. P3.56 is given by

$$v_r = U_\infty\left(1 - \frac{r_c^2}{r^2}\right)\cos\theta \text{ and } v_\theta = U_\infty\left(1 + \frac{r_c^2}{r^2}\right)$$

$\sin\theta$. If the pressure at $r = \infty$ is zero (i.e., $p_\infty = 0$), find an expression for the pressure neglecting gravity effects:

  (a) Along the negative x-axis
  (b) At the stagnation point
  (c) On the surface of the cylinder
  (d) On the surface of the cylinder at $\theta = 90°$

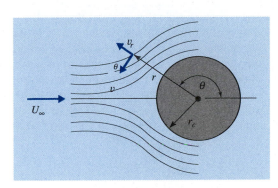

FIGURE P3.56

3.57 The inviscid, incompressible flow field exterior to a sphere (see Fig. P3.56) is given by $v_r =$

$$U_\infty\left(1 - \frac{r_c^3}{r^3}\right)\cos\theta \text{ and } v_\theta = U_\infty\left(1 + \frac{r_c^3}{r^3}\right)\sin\theta.$$

If the pressure at $r = \infty$ is zero, find an expression for the pressure neglecting gravity effects:
(a) Along the negative x-axis
(b) At the stagnation point
(c) On the surface of the sphere
(d) On the surface of the sphere at $\theta = 90°$

3.58 The velocity along the negative x-axis in the inviscid, incompressible flow field exterior to the body shown in Fig. P3.58 is given by $u(x) = U_\infty + q/2\pi x$. If the pressure at $x = -\infty$ is zero, find an expression for the pressure neglecting gravity effects:
(a) Along the negative x-axis if $U_\infty = 10$ m/s and $q = 20\pi$ m²/s
(b) At the stagnation point if $U_\infty = 10$ m/s and $q = 20\pi$ m²/s

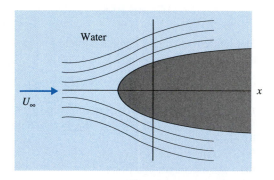

FIGURE P3.58

3.59 The incompressible flow of water through the short contraction of Fig. 3.19a is assumed to be inviscid. If a pressure drop of 20 kPa is measured, estimate the velocity at the wall at section 2 just downstream of the contraction. (Actually, a boundary layer would develop, and the velocity calculated at the wall would be the velocity at the edge of the boundary layer; see the inset of Fig. 3.10.)

3.60 Air flows from a relatively large plenum in a furnace out a relatively small rectangular duct. If the pressure in the plenum measures 60 Pa and in the duct 10.2 Pa, estimate the velocity of the 40°C air in the duct.

3.61 What is the velocity of the water in the pipe if the manometer shown in Fig. P3.61 reads:
(a) 4 cm?
(b) 10 cm?

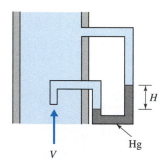

FIGURE P3.61

3.62 A manometer, positioned inside a cylinder as shown in Fig. P3.62, reads 4 cm of water. Estimate U_∞ assuming an inviscid flow. Refer to the velocity field in Problem 3.56.

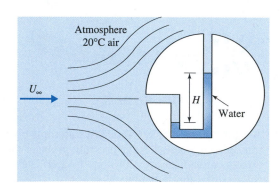

FIGURE P3.62

3.63 A manometer, positioned as shown in Figure P3.62 inside a sphere, reads 4 cm of water. Estimate U_∞ assuming an inviscid flow. Refer to the velocity field in Problem 3.57.

3.64 Air at 20°C is drawn into the hose of a vacuum cleaner through a head that is relatively free of obstructions (the flow can be assumed to be inviscid). Estimate the velocity in the hose if the vacuum in the hose measures:

(a) 2 cm of water
(b) 8 cm of water

3.65 A wind tunnel is designed to draw in air from the atmosphere and produce a velocity of 100 m/s in the test section. The fan is located downstream of the test section. What pressure is to be expected in the test section if the atmospheric temperature and pressure are:

(a) −20°C, 90 kPa?
(b) 0°C, 95 kPa?
(c) 20°C, 92 kPa?
(d) 40°C, 100 kPa?

3.66 A water hose is pressurized to 800 kPa with a nozzle in the off position. If the nozzle is opened a small amount, as shown in Fig. P3.66, estimate the exiting velocity of the water. Assume the velocity inside the hose to be negligible.

A. 40 m/s
B. 30 m/s
C. 20 m/s
D. 10 m/s

FIGURE P3.66

3.67 The pump shown in Fig. P3.67 creates a flow such that $V = 14$ m/s. Predict the pressure at the gage shown assuming an inviscid flow in the entrance and a uniform flow at the gage. Use a streamline starting at:

(a) Point A.
(b) Point B.

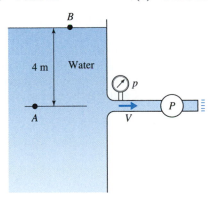

FIGURE P3.67

3.68 For the flow shown in Fig. P3.68, estimate the pressure p_1 and velocity V_1 if $V_2 = 20$ m/s and:

(a) $H = 1$ cm
(b) $H = 5$ cm
(c) $H = 10$ cm

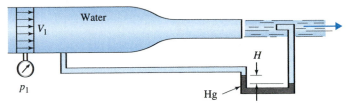

FIGURE P3.68

3.69 A fireman reduces the exit area on a nozzle so that the velocity inside the hose is quite small relative to the exiting velocity. What is the maximum exiting velocity and what is the maximum height the water can reach if the pressure inside the hose is:

(a) 700 kPa?
(b) 1400 kPa?

3.70 The velocity downstream of a sluice gate is assumed to be uniform (Fig. P3.70). Express V in terms of H and h for this inviscid flow. Use a streamline:

(a) Along the top
(b) Along the bottom

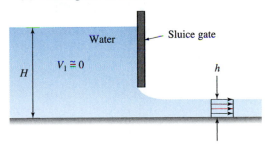

FIGURE P3.70

3.71 To what maximum velocity can water be accelerated before it reaches the turbine blades of a hydroturbine if it enters with relatively low velocity at:

(a) 600 kPa?
(b) 300 kPa?

3.72 Water exists in a city's water system at a pressure of 500 kPa at a particular location. The water pipe must traverse over a hill. How high could the hill be, above that location, for the system to possibly supply water to the other side of the hill?

3.73 Fluid flows between the radial disks shown in Fig. P3.73. Estimate the pressure in the 2-cm-diameter pipe if the fluid exits to the atmosphere. Neglect viscous effects. The fluid is:

   (a) water   **(b)** benzene
   (c) gasoline   **(d)** air

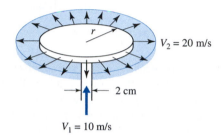

$V_2 = 20$ m/s

2 cm

$V_1 = 10$ m/s

FIGURE P3.73

3.74 Estimate the pressure at $r = 10$ cm if the velocity there is 8 m/s in Problem 3.73d.

3.75 Benzene flows out through the disks shown in Fig. P3.75. If $V_2 = 30$ m/s, the pressure p_1, is nearest:

   A. 150 kPa   **B.** 200 kPa
   C. 250 kPa   **D.** 300 kPa

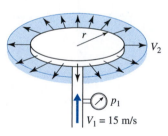

V_2

p_1

$V_1 = 15$ m/s

FIGURE P3.75

3.76 It is proposed that air blown through a tube attached to a metal disk be used to pick up envelopes, as shown in Fig. P3.76. Would such a setup actually pick up an envelope? Explain. Assume an inviscid flow with the air slowing down as it moves radially outward.

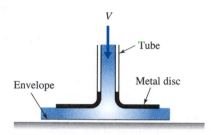

V

Tube

Envelope

Metal disc

FIGURE P3.76

3.77 From which of the following objects would you expect the flow to separate and form a substantial separated region?

   (a) A golf ball
   (b) A telephone wire
   (c) A wind machine blade
   (d) A 2-mm-diameter wire in a low-speed wind tunnel
   (e) An automobile
   (f) An aircraft

(*Note:* Separation occurs whenever the Reynolds number exceeds a value around 20 over a blunt object.)

3.78 Explain, with the use of a sketch, why a burr on the downstream side of a piezometer opening in a pipe (see Fig. E3.7) will result in a reading of the pressure that is too high.

3.79 An incompressible, inviscid flow of water enters a bend with a uniform velocity of $V_1 = 10$ m/s (Fig. P3.79). Estimate the pressure difference between points A and B if the average radius of curvature at the bend is 5 cm. Sketch an anticipated velocity profile along AB. Assume that $p_A < p_1$ and $p_B > p_1$.

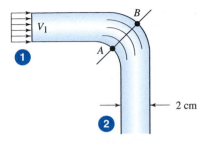

FIGURE P3.79

3.80 Viscosity causes a fluid to stick to a surface. If the fluid in Problem 3.79 sticks to the surface, explain why, in a viscous flow, a secondary flow is caused by the bend. Sketch such a flow on a circular cross section at section 2.

3.81 In Fig. P3.81, assuming an inviscid flow, insert one of these signs between the pressure: $>$, $<$, $\cong$.

p_A	p_B
p_C	p_D
p_B	p_D

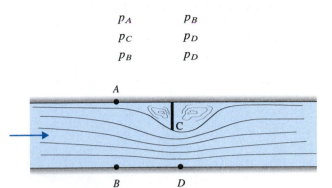

FIGURE P3.81

Charles Regatta. With each stroke, the work performed by the crew is transferred to the hull to overcome drag forces. (John Kropewnicki/Shutterstock)

4

The Integral Forms of the Fundamental Laws

Outline

Chapter Objectives

The objectives of this chapter are to:
▲ Derive an equation that will allow us to convert the three basic laws that are stated for a system to a form that is applicable to a control volume.
▲ Apply the conservation of mass to control volumes of interest.
▲ Analyze the work-rate term of the energy equation.
▲ Apply the energy equation to numerous engineering situations.
▲ Apply Newton's second law to control volumes of interest.
▲ Apply the moment-of-momentum equation to rotating devices.

▲ Present numerous examples of the basic laws applied to control volumes so that students will be able to correctly solve fluid flow problem involving many of the control volumes of interest to engineers.

▲ Express the basic laws in their most general control-volume form so that complex problems encountered in engineering applications can be properly analyzed and hopefully solved.

4.1 INTRODUCTION

Quantities of interest to engineers can often be expressed in terms of integrals. For example, volume flow rate is the integral of the velocity over an area; heat transfer is the integral of the heat flux over an area; force is the integral of a stress over an area; mass is the integral of the density over a volume; and kinetic energy is the integral of $V^2/2$ over each mass element in a volume. There are, of course, many other integral quantities. To determine an integral quantity, the integrand must be known, or information must be available so that a good approximation to the integrand can be made. If the integrand is not known or cannot be approximated with any degree of certainty, appropriate differential equations (see Chapter 5) must be solved yielding the needed integrand; the integration is then performed giving the engineer the desired integral quantity.

> **KEY CONCEPT** *To determine an integral quantity the integrand must be known, or information must be available so that a good approximation to the integrand can be made.*

In this chapter we present the integral quantities of interest, develop equations that relate the integral quantities, and work a number of problems for which the integrands are given or can be approximated. This includes a surprisingly large variety of problems. There are, however, many integral quantities that cannot be determined since the integrands are unknown. These would include the lift and drag on an airfoil, the torque on the blades of a wind machine, and the kinetic energy in the wake of a submarine. To determine such integrands, it would be necessary to solve the appropriate differential equations, a task that is often quite difficult; some relatively simple situations are considered in subsequent chapters.

Also, there are many quantities of interest that are not integral in nature. Included would be the point of separation of the flow around a body, the concentration of a pollutant in a stream at a certain location, the pressure distribution on the side of a building, and the wave-shore interaction along a lake. To study subjects such as these, it is necessary to consider the differential equations that describe the flow situation. Most of the topics mentioned are relegated to specialized graduate courses; however, some topics that require the solution of the more easily solvable differential equations are included in this book.

4.2 THE THREE BASIC LAWS

The integral quantities of primary interest in fluid mechanics are contained in three basic laws: conservation of mass, first law of thermodynamics, and Newton's second law. These basic laws are expressed using a Lagrangian description in terms of a **system**, a fixed collection of material particles. For example, if we consider flow through a pipe, we could identify a fixed quantity of fluid at time t as the system (Fig. 4.1); this system would then move due to velocity to a down-

> **System:** *A fixed collection of material particles.*

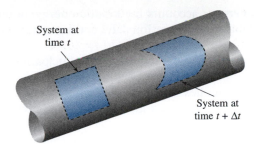

System at
time t

System at
time $t + \Delta t$

FIGURE 4.1 Example of a system in fluid mechanics.

stream location at time $t + \Delta t$. Any of the three basic laws could be applied to this system. This is, however, not an easy task. First let us state the basic laws in their general form.

Conservation of Mass: The law stating that mass must be conserved is:

> The mass of a system remains constant.

The mass of a fluid particle is $\rho \, d V$, where $d V$ is the volume occupied by the particle and ρ is its density. Knowing that the density can change from point to point in the system, the conservation of mass can be expressed in integral form as

$$\frac{D}{Dt} \int_{\text{sys}} \rho \, d V = 0 \qquad (4.2.1)$$

where D/Dt is used since we are following a specified group of material particles, a system.

First Law of Thermodynamics: The law that relates heat transfer, work, and energy change is the first law of thermodynamics; it states:

> The rate of heat transfer to a system minus the rate at which the system does work equals the rate at which the energy of the system is changing.

Recognizing that both density and specific energy may change from point to point in the system, it may be expressed as

$$\dot{Q} - \dot{W} = \frac{D}{Dt} \int_{\text{sys}} e\rho \, d V \qquad (4.2.2)$$

Specific energy: *Accounts for kinetic energy, potential energy, and internal energy per unit mass.*

where the **specific energy** e accounts for kinetic energy, potential energy, and internal energy per unit mass. Equation 4.2.2 is often referred to as the energy equation.

Other forms of energy — chemical, electrical, nuclear — are not included in an elementary course in fluid mechanics. In its basic form stated here, the first law of thermodynamics applies only to a system, a collection of fluid particles; therefore, D/Dt is used. We study $\dot{Q}$ and $\dot{W}$ in Section 4.5, where we consider the energy equation in detail.

Newton's Second Law: Newton's second law, also called the momentum equation, states:

> The resultant force acting on a system equals the rate at which the momentum of the system is changing.

The momentum of a fluid particle of mass is a vector quantity given by $\mathbf{V}\rho \, d\Psi$; consequently, Newton's second law may be expressed in an inertial reference frame as

$$\Sigma \, \mathbf{F} = \frac{D}{Dt} \int_{\text{sys}} \mathbf{V}\rho \, d\Psi \qquad (4.2.3)$$

recognizing that both density and velocity may change from point to point in the system. This equation reduces to $\Sigma \, \mathbf{F} = m\mathbf{a}$ if $\mathbf{V}$ and ρ are constant throughout the entire system; ρ is often a constant, but in fluid mechanics the velocity vector invariably changes from point to point. Again, D/Dt is used to provide the rate-of-change since Newton's second law is applied to a system.

Moment-of-Momentum Equation: The moment-of-momentum equation results from Newton's second law; it states:

> The resultant moment acting on a system equals the rate of change of the angular momentum of the system.

In equation form this becomes, relative to an inertial reference frame,

$$\Sigma \, \mathbf{M} = \frac{D}{Dt} \int_{\text{sys}} \mathbf{r} \times \mathbf{V}\rho \, d\Psi \qquad (4.2.4)$$

where $\mathbf{r} \times \mathbf{V}\rho \, d\Psi$ represents the angular momentum of a fluid particle with mass $\rho \, d\Psi$. The vector $\mathbf{r}$ locates the volume element $d\Psi$ and is measured from the origin of the coordinate axes, the point relative to which the resultant moment is measured.

Note that in each of the basic laws the integral quantity is an extensive property of the system (see Section 1.7). We will use the symbol N_{sys} to denote this extensive property; for example, N_{sys} could be the mass, the momentum, or the energy of the system. The left-hand side of Eq. 4.2.1 and the right-hand sides of Eqs. 4.2.2, 4.2.3, and 4.2.4 may all be expressed as

$$\frac{DN_{sys}}{Dt} \qquad (4.2.5)$$

where N_{sys} represents an integral quantity, either a scalar quantity or a vector quantity.

It is also useful to introduce the variable η for the intensive property, the property of the system per unit mass. The relation between N_{sys} and η is given by

$$N_{sys} = \int_{sys} \eta \rho \, d\Psi \qquad (4.2.6)$$

As an example, the extensive property of Newton's second law is the momentum

$$\mathbf{momentum}_{system} = \int_{sys} \mathbf{V} \, \rho \, d\Psi \qquad (4.2.7)$$

which is a vector quantity. The corresponding intensive property would be the velocity vector $\mathbf{V}$. Note that the density and velocity, which may vary from point to point within the system, may also be functions of time, as in unsteady flow.

Our interest is most often focused on a device, or a region of space, into which fluid enters and/or from which fluid leaves; we identify this region as a **control volume**. An example of a fixed control volume is shown in Fig. 4.2a. A control volume need not be fixed; it could deform as in a piston-cylinder during exhaust or in a balloon as it deflates. We will, however, consider only fixed control volumes in this book. This will not limit us in most problem situations.

Control volume: *A region of space, into which fluid enters and/or from which fluid leaves.*

The difference between a control volume and a system is illustrated in Fig. 4.2b. The figure indicates that the system occupies the control volume at time t and has partially moved out of it at time $t + \Delta t$. Since it is often more convenient to focus on a control volume (e.g., a pump) rather than on a system,

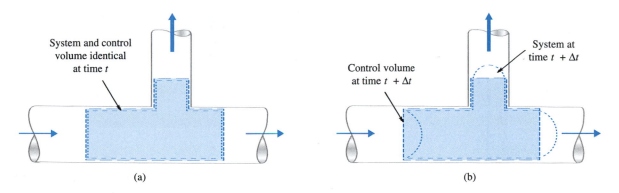

FIGURE 4.2 Example of a fixed control volume and a system: (a) time t; (b) time $t + \Delta t$.

the first order of business is to find a transformation that will allow us to express the substantial derivative of a system (a Lagrangian description) in terms of quantities associated with a control volume (an Eulerian description) so that the basic laws can be applied directly to a control volume. This will be done in general and then applied to the specific laws.

4.3 SYSTEM-TO-CONTROL-VOLUME TRANSFORMATION

We are interested in the time rate of change of the extensive property N_{sys} as we follow the system along, that is, DN_{sys}/Dt, and we would like to express this in terms of quantities that pertain to the control volume. In this section we present the derivation of the transformation.

The derivation involves fluxes of the extensive property in and out of the control volume. A flux is a measure of the rate at which an extensive property crosses an area; for example, a mass flux is the rate at which mass crosses an area. It is useful to introduce vector notation to describe these fluxes. Consider an area element dA of the **control surface**, the surface area that completely encloses the control volume. The property flux across an elemental area dA (see Fig. 4.3) may be expressed by

Control surface: *The surface area that completely encloses the control volume.*

$$\text{flux across } dA = \eta\rho\hat{\mathbf{n}}\cdot\mathbf{V}\, dA \qquad (4.3.1)$$

where $\hat{\mathbf{n}}$, a unit vector normal to the area element dA, always points out of the control volume, and η represents the intensive property associated with N_{sys}. Note that this expression yields a negative value if it concerns a property influx. Only the normal component $\hat{\mathbf{n}}\cdot\mathbf{V}$ of the velocity vector contributes to this flux term. If there is no normal component of velocity on a particular area, such as the wall of a pipe, no flux occurs across that area. A positive $\hat{\mathbf{n}}\cdot\mathbf{V}$ indicates a flux out of the volume; a negative $\hat{\mathbf{n}}\cdot\mathbf{V}$, that is, $\mathbf{V}$ has a component in the opposite direction of $\hat{\mathbf{n}}$, indicates a flux into the volume. We must always use $\hat{\mathbf{n}}$ pointing out of the volume. The velocity vector $\mathbf{V}$ may be at some angle to the unit vector

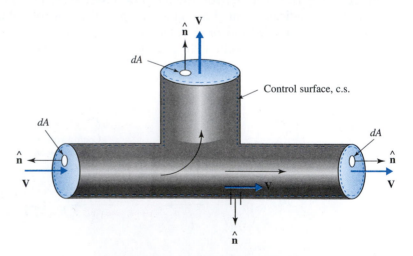

FIGURE 4.3   Illustration showing the flux of an extensive property.

$\hat{\mathbf{n}}$; the dot product $\hat{\mathbf{n}} \cdot \mathbf{V}$ accounts for the appropriate component of $\mathbf{V}$ that produces a flux through the area.

The net property flux out of the control surface is then obtained by integrating over the entire control surface:

$$\text{net flux of property} = \int_{\text{c.s.}} \eta \rho \hat{\mathbf{n}} \cdot \mathbf{V} \, dA \tag{4.3.2}$$

If the net flux is positive, the flux out is larger than the flux in.

Let us return now to the derivative DN_{sys}/Dt. The definition of a derivative allows us to write

$$\frac{DN_{\text{sys}}}{Dt} = \lim_{\Delta t \to 0} \frac{N_{\text{sys}}(t + \Delta t) - N_{\text{sys}}(t)}{\Delta t} \tag{4.3.3}$$

The system is shown in Fig. 4.4 at times t and $t + \Delta t$. Assume that the system occupies the full control volume at time t; if we were considering a device, such as a pump, the particles of the system would just fill the device at time t. Since the device, the control volume shown in Fig. 4.4, is assumed to be fixed in space, the system will move through the device. Equation 4.3.3 can then be written

$$\frac{DN_{\text{sys}}}{Dt} = \lim_{\Delta t \to 0} \frac{N_3(t + \Delta t) + N_2(t + \Delta t) - N_2(t) - N_1(t)}{\Delta t}$$

$$= \lim_{\Delta t \to 0} \frac{N_2(t + \Delta t) + N_1(t + \Delta t) - N_2(t) - N_1(t)}{\Delta t}$$

$$+ \lim_{\Delta t \to 0} \frac{N_3(t + \Delta t) - N_1(t + \Delta t)}{\Delta t} \tag{4.3.4}$$

where, in this second expression, we have simply added and subtracted $N_1(t + \Delta t)$ in the numerator. In the equations above, the numerical subscript denotes the region; for example, $N_2(t)$ signifies the extensive property in region 2 at time t. Now, we observe that the first limit on the right-hand side refers to the control volume, so we can write

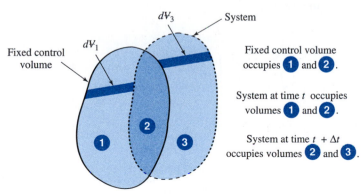

FIGURE 4.4   System and fixed control volume.

$$\frac{DN_{sys}}{Dt} = \lim_{\Delta t \to 0} \frac{N_{c.v.}(t + \Delta t) - N_{c.v.}(t)}{\Delta t} + \lim_{\Delta t \to 0} \frac{N_3(t + \Delta t) - N_1(t + \Delta t)}{\Delta t} \qquad (4.3.5)$$

The first ratio on the right-hand side is $dN_{c.v.}/dt$, where we use an ordinary derivative since we are not following specific fluid particles. Thus there results

$$\frac{DN_{sys}}{Dt} = \frac{dN_{c.v.}}{dt} + \lim_{\Delta t \to 0} \frac{N_3(t + \Delta t) - N_1(t + \Delta t)}{\Delta t} \qquad (4.3.6)$$

Now, we must find expressions for the extensive quantities $N_3(t + \Delta t)$ and $N_1(t + \Delta t)$. They, of course, depend on the mass contained in the volume elements shown in Fig. 4.4 and enlarged in Fig. 4.5. Note that the unit vector $\hat{\mathbf{n}}$ always points out of the volume, and hence to obtain a positive differential volume a negative sign is required for region 1. Also, note that the cosine of the angle between the velocity vector and the normal vector is required,[1] thus the presence of the dot product. Referring to Fig. 4.5, we have

$$N_3(t + \Delta t) = \int_{A_3} \eta \rho \hat{\mathbf{n}} \cdot \mathbf{V} \, \Delta t \, dA_3$$

$$\qquad (4.3.7)$$

$$N_1(t + \Delta t) = -\int_{A_1} \eta \rho \hat{\mathbf{n}} \cdot \mathbf{V} \, \Delta t \, dA_1$$

Recognizing that A_3 plus A_1 completely surrounds the control volume, we combine the two integrals into one integral. That is,

$$N_3(t + \Delta t) - N_1(t + \Delta t) = \int_{c.s.} \eta \rho \hat{\mathbf{n}} \cdot \mathbf{V} \, \Delta t \, dA \qquad (4.3.8)$$

where the control surface, denoted by c.s., is an area that completely surrounds the control volume. Substituting Eq. 4.3.8 back into Eq. 4.3.6 yields the desired result, the system-to-control-volume transformation, or equivalently, the **Reynolds transport theorem**:

Reynolds transport theorem:
The system-to-control-volume
transformation.

$$\frac{DN_{sys}}{Dt} = \frac{d}{dt} \int_{c.v.} \eta \rho \, dV + \int_{c.s.} \eta \rho \hat{\mathbf{n}} \cdot \mathbf{V} \, dA \qquad (4.3.9)$$

---

[1]To obtain the volume of a box, we multiply the height by the area of the base, provided that the box is upright. If it is completely collapsed, its volume is zero. Hence for some intermediate position the volume is the height times the area of the base times the cosine of the appropriate angle.

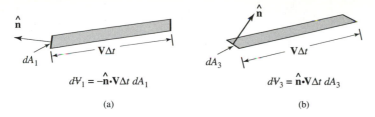

FIGURE 4.5   Differential volume elements.

This is a Lagrangian-to-Eulerian transformation of the rate of change of an extensive integral quantity.

The first integral represents the rate of change of the extensive property in the control volume. The second integral represents the flux of the extensive property across the control surface; it is nonzero only where fluid crosses the control surface. We study this flux term in considerable detail in the following sections. Thus we can now express the basic laws in terms of a fixed volume in space. We will do this in subsequent sections for each of the basic laws.

We can move the time derivative of the control volume term inside the integral since, for a fixed control volume, the limits on the volume integral are independent of time; we then write

$$\frac{DN_{sys}}{Dt} = \int_{c.v.} \frac{\partial}{\partial t}(\rho\eta)\, d\mathcal{V} + \int_{c.s.} \eta\rho\hat{\mathbf{n}}\cdot\mathbf{V}\, dA \tag{4.3.10}$$

KEY CONCEPT   *The time derivative of the control volume term can be moved inside the integral for a fixed control volume.*

In this form we have used $\partial/\partial t$ since ρ and η are, in general, dependent on the position variables.

4.3.1   Simplifications of the System-to-Control-Volume Transformation

Many flows of interest are steady flows, so that $\partial(\eta\rho)/\partial t = 0$. Our system-to-control-volume transformation then takes the form

$$\frac{DN_{sys}}{Dt} = \int_{c.s.} \eta\rho\hat{\mathbf{n}}\cdot\mathbf{V}\, dA \tag{4.3.11}$$

Furthermore, there is often only one area A_1 across which fluid enters the control volume and one area A_2 across which fluid leaves the control volume; assuming that the velocity vector is normal to the area (see Fig. 4.6), we can write $\hat{\mathbf{n}}\cdot\mathbf{V}_1 = -V_1$ over area A_1 and $\hat{\mathbf{n}}\cdot\mathbf{V}_2 = V_2$ over area A_2. Then Eq. 4.3.11 becomes

$$\frac{DN_{sys}}{Dt} = \int_{A_2} \eta_2\rho_2 V_2\, dA - \int_{A_1} \eta_1\rho_1 V_1\, dA \tag{4.3.12}$$

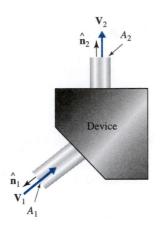

FIGURE 4.6 Flow into and from a device.

Finally, there are many situations that are modeled acceptably by assuming uniform properties over each plane area (see Fig. 3.9); then the equation simplifies to

$$\frac{DN_{sys}}{Dt} = \eta_2\rho_2 V_2 A_2 - \eta_1\rho_1 V_1 A_1 \tag{4.3.13}$$

We will find that the system-to-control-volume transformation in this simplified form is most often used in the application of the basic laws to problems of interest in an introductory course in fluid mechanics. Some applications will, however, be included that will illustrate nonuniform distributions and unsteady flows.

If we generalize Eq. 4.3.13 to include several areas across which the fluid flows, we could write

$$\frac{DN_{sys}}{Dt} = \sum_{i=1}^{N} \eta_i\rho_i \mathbf{V}_i \cdot \hat{\mathbf{n}}_i A_i \tag{4.3.14}$$

where N is the number of areas. The dot product $\hat{\mathbf{n}} \cdot \mathbf{V}$ would provide us with the appropriate sign at each area; for an inlet area, $\hat{\mathbf{n}} \cdot \mathbf{V}$ introduces a negative sign, and for an exit area, $\hat{\mathbf{n}} \cdot \mathbf{V}$ introduces a positive sign.

For an unsteady flow in which flow properties are assumed to be uniform throughout the control volume, the system-to-control-volume equation takes the form

$$\frac{DN_{sys}}{Dt} = V_{c.v.}\frac{d(\eta\rho)}{dt} + \eta_2\rho_2 V_2 A_2 - \eta_1\rho_1 V_1 A_1 \tag{4.3.15}$$

for one inlet and one outlet with uniform properties.

4.4 CONSERVATION OF MASS

A system is a given collection of fluid particles; hence its mass remains fixed:

$$\frac{Dm_{sys}}{Dt} = \frac{D}{Dt} \int_{sys} \rho \, d\forall = 0 \qquad (4.4.1)$$

In Eq. 4.2.6, N_{sys} represents the mass of the system, so we simply let $\eta = 1$. Thus the conservation of mass, referring to Eq. 4.3.9, becomes

$$0 = \frac{d}{dt} \int_{c.v.} \rho \, d\forall + \int_{c.s.} \rho \hat{\mathbf{n}} \cdot \mathbf{V} \, dA \qquad (4.4.2)$$

or, if we prefer,

$$0 = \int_{c.v.} \frac{\partial \rho}{\partial t} \, d\forall + \int_{c.s.} \rho \hat{\mathbf{n}} \cdot \mathbf{V} \, dA \qquad (4.4.3)$$

If the flow is steady, there results

$$\int_{c.s.} \rho \hat{\mathbf{n}} \cdot \mathbf{V} \, dA = 0 \qquad (4.4.4)$$

which, for a uniform flow with one entrance and one exit, takes the form

$$\rho_2 A_2 V_2 = \rho_1 A_1 V_1 \qquad (4.4.5)$$

where for an inlet we have used $\hat{\mathbf{n}}_1 \cdot \mathbf{V}_1 = -V_1$ and for an exit $\hat{\mathbf{n}} \cdot \mathbf{V}_2 = V_2$. Recall $\hat{\mathbf{n}}$ always points out of the control volume.

KEY CONCEPT $\hat{\mathbf{n}}$ always points out of the control volume.

If the density is constant in the control volume, the derivative $\partial \rho / \partial t = 0$ even if the flow is unsteady. The continuity equation (4.4.3) then reduces to

$$A_1 V_1 = A_2 V_2 \qquad (4.4.6)$$

This form of the continuity equation is used quite often, particularly with liquids and low-speed gas flows.

At this point we wish to discuss again the use of uniform velocity profiles (see also Section 3.3.1). Suppose that the velocity profiles at the entrance and the exit are not uniform, such as sketched in Fig. 4.7. Furthermore, suppose that the density is uniform over each area. Then the continuity equation takes the form

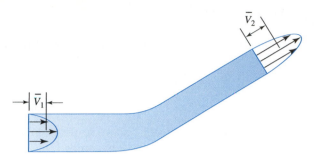

FIGURE 4.7 Nonuniform velocity profiles.

$$\rho_1 \int_{A_1} V_1 \, dA = \rho_2 \int_{A_2} V_2 \, dA \tag{4.4.7}$$

or, letting an overbar denote an average, we can write

$$\rho_1 \overline{V}_1 A_1 = \rho_2 \overline{V}_2 A_2 \tag{4.4.8}$$

where $\overline{V}_1$ and $\overline{V}_2$ are the *average velocities* over the areas at sections 1 and 2, respectively. In examples and problems the overbar is often omitted. It should be kept in mind, however, that actual velocity profiles are usually not uniform; Eqs. 4.4.5 and 4.4.6 are used with the velocities representing average velocities.

Any one of the above equations (4.4.2) through (4.4.8) is referred to as the *continuity equation.*

Before presenting some examples applying the continuity equation, two fluxes are defined that will be useful in specifying the quantity of flow. The *mass flux ṁ*, or the mass rate of flow, is

$$\dot{m} = \int_A \rho V_n \, dA \tag{4.4.9}$$

and has units of kg/s; V_n is the normal component of velocity. The *flow rate Q*, or the volume rate of flow, is

$$Q = \int_A V_n \, dA \tag{4.4.10}$$

and has units of m³/s or sometimes L/s. The mass flux is usually used in specifying the quantity of flow for a compressible flow and the flow rate for an incompressible flow. We often refer to the flow rate as **discharge.**

In terms of average velocity, we have

Discharge: *Another term for flow rate.*

$$Q = A\overline{V} \tag{4.4.11}$$

$$\dot{m} = \rho A \overline{V} \tag{4.4.12}$$

where for the mass flux we assume a uniform density profile; we also assume that the velocity is normal to the area.

The following examples are solved by first selecting a control volume. If you study the examples carefully, you will notice that often there is only one proper choice for the control volume. We must position the inlet and exit areas at locations where the integrands are either known or where they can be approximated; also, the quantity being sought is often included at an inlet or exit area. In a few cases there may be more freedom in the selection of the control volume (Example 4.5).

This first example represents the primary use of the continuity equation. It allows us to calculate the velocity at one section if it is known at another section.

Example 4.1

Water flows at a uniform velocity of 3 m/s into a nozzle that reduces the diameter from 10 cm to 2 cm (Fig. E4.1). Calculate the water's velocity leaving the nozzle and the flow rate.

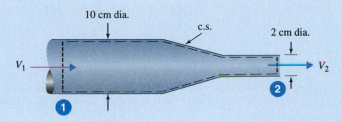

FIGURE E4.1

Solution

The control volume is selected to be the inside of the nozzle as shown. Flow enters the control volume at section 1 and leaves at section 2. The simplified continuity equation (4.4.6) is used since the density is assumed constant and the velocity profiles are uniform:

$$A_1 V_1 = A_2 V_2$$

$$\therefore V_2 = V_1 \frac{A_1}{A_2} = 3 \, \frac{\pi \times 0.1^2/4}{\pi \times 0.02^2/4} = 75 \text{ m/s}$$

The flow rate, or discharge, is found to be

$$Q = V_1 A_1$$

$$= 3 \times \pi \times 0.1^2/4 = 0.0236 \text{ m}^3/\text{s}$$

Example 4.2

Water flows in and out of a device as shown in Fig. E4.2a. Calculate the rate of change of the mass of water (dm/dt) in the device.

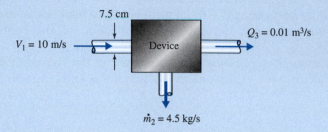

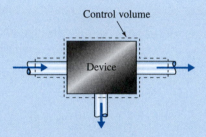

FIGURE E4.2

Solution

The control volume selected is shown in Fig. E4.2b. For the control surface surrounding the device, the continuity equation (4.4.2), with three surfaces across which water flows, takes the following form:

$$0 = \frac{d}{dt} \int_{c.v.} \rho \, d\mathcal{V} + \int_{c.s.} \rho \hat{\mathbf{n}} \cdot \mathbf{V} \, dA$$

$$= \frac{dm}{dt} - \rho_1 A_1 V_1 + \rho_2 A_2 V_2 + \rho_3 A_3 V_3$$

where we have assumed the density to be constant over the volume and we have used $\mathbf{V}_1 \cdot \mathbf{n} = -V_1$ since $\hat{\mathbf{n}}_1$ points out of the volume, opposite to the direction of $\mathbf{V}_1$. The last three terms come from the area integral. In terms of the quantities given, the above can be expressed as

$$0 = \frac{dm}{dt} - \rho_1 A_1 V_1 + \dot{m}_2 + \rho_3 Q_3$$

$$= \frac{dm}{dt} - 1000 \times \pi \times \frac{3.75^2}{10^4} \times 10 + 4.5 + 1000 \times 0.01$$

This is solved to yield

$$\frac{dm}{dt} = 29.68 \text{ kg/s}$$

Hence the mass is increasing at the rate of 29.68 kg/s. To accomplish this, the device could have a spongelike material that absorbs water.

Example 4.3

Uniform flow approaches a cylinder as shown in Fig. E4.3a. The symmetrical velocity distribution at the location shown downstream in the wake of the cylinder is approximated by

$$u(y) = 1.25 + \frac{y^2}{4} \quad -1 < y < 1$$

where $u(y)$ is in m/s and y is in meters. Determine the mass flux across the surface AB per meter of depth. Use $\rho = 1.23$ kg/m^3.

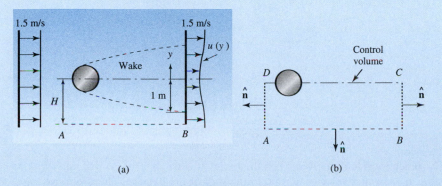

(a)   (b)

FIGURE E4.3

Solution

Select $ABCD$ as the control volume (Fig. E4.3b). Outside the wake (a region of retarded flow) the velocity is constant at 1.5 m/s. Hence the velocity normal to plane AD is 1.5 m/s. Obviously, no mass flux crosses the surface CD because of symmetry. Assuming a steady flow, the continuity equation (4.4.3) becomes

$$0 = \int_{c.s.} \rho \mathbf{V} \cdot \hat{\mathbf{n}} \, dA$$

Mass flux occurs across three surfaces: AB, BC, and AD. Thus the equation above takes the form

$$0 = \int_{A_{AB}} \rho \mathbf{V} \cdot \hat{\mathbf{n}} \, dA + \int_{A_{BC}} \rho \mathbf{V} \cdot \hat{\mathbf{n}} \, dA + \int_{A_{AD}} \rho \mathbf{V} \cdot \hat{\mathbf{n}} \, dA$$

$$= \dot{m}_{AB} + \int_0^H \rho u(y) \, 1 \times dy - \rho H \times 1 \times 1.5$$

where the negative sign for surface AD results from the fact that the unit vector points out of the volume to the left while the velocity vector points to the right. Recall that a negative sign in the steady-flow continuity equation is always associated with an influx and a positive sign with an outflux. Now, we integrate out to 1 m instead of H, since the mass that enters on the left beyond 1 m simply leaves on the right with no net gain or loss. So, letting $H = 1$ m, we have

$$0 = \dot{m}_{AB} + \int_0^1 1.23\left(1.25 + \frac{y^2}{4}\right) dy - 1.23 \times 1 \times 1.5$$

Perform the integration and there results

$$\dot{m}_{AB} = 0.205 \text{ kg/s per meter}$$

Example 4.4

A balloon is being inflated with a water supply of 0.6 m³/s (Fig. E4.4). Find the rate of growth of the radius at the instant when $R = 0.5$ m.

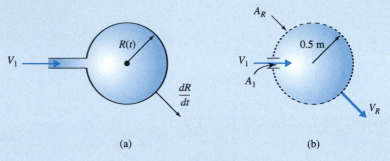

(a)   (b)

FIGURE E4.4

Solution

The objective is to find dR/dt when the radius $R = 0.5$ m. This growth rate dR/dt is the same as the water velocity normal to the wall of the balloon. Therefore, we select as our fixed control volume a sphere with a constant radius of 0.5 m so that we can calculate the velocity of the water at the surface at the instant shown moving radially out at $R = 0.5$ m. The continuity equation is written as

$$0 = \int_{c.v.} \underset{0}{\underbrace{\frac{\partial \rho}{\partial t}}}\, d\mathcal{V} + \int_{c.s.} \rho \mathbf{V} \cdot \hat{\mathbf{n}}\, dA$$

The first term is zero because the density of water inside the control volume does not change in time. Further, the water crosses two areas: the inlet area A_1 with a velocity V_1 and the remainder of the sphere surface A_R with a velocity V_R. We will assume that $A_1 \ll A_R$. The continuity equation then takes the form

$$0 = -\rho A_1 V_1 + \rho A_R V_R$$

Since the flow rate into the volume is $A_1 V_1 = 0.6$ m³/s and $A_R \simeq 4\pi R^2$ assuming that A_1 is quite small, we can solve for V_R. At $R = 0.5$ m

$$V_R = \frac{A_1 V_1}{4\pi R^2} = \frac{0.6}{4\pi \times 0.5^2} = 0.191 \text{ m/s}$$

$$\therefore \frac{dR}{dt} = 0.191 \text{ m/s}$$

We have used a fixed control volume and allowed the moving surface of the balloon to pass through it at the instant considered. With this approach it is possible to model situations in which surfaces, such as a piston, are allowed to move.

Example 4.5

This example shows that there may be more than one good choice for a control volume. We want to determine the rate at which the water level rises in an open container if the water coming in through a 0.10-m^2 pipe has a velocity of 0.5 m/s and the flow rate going out is 0.2 m^3/s (Fig. E4.5a). The container has a circular cross section with a diameter of 0.5 m.

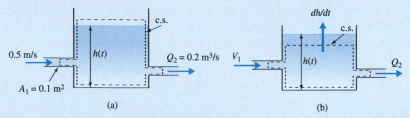

(a)   (b)

FIGURE E4.5

Solution

First we select a control volume that extends above the water surface as shown in Fig. E4.5a. Apply the continuity equation (Eq. 4.3.2):

$$\frac{d}{dt}\int_{c.v.} \rho d\mathcal{V} + \rho(-V_1)A_1 + \rho V_2 A_2 = 0$$

in which the first term describes the rate of change of mass in the control volume. Hence, neglecting the airmass above the water, we have

$$\frac{d(\rho h \pi D^2/4)}{dt} - \rho V_1 A_1 + \rho Q_2 = 0$$

Divide by the constant ρ,

$$\frac{\pi D^2}{4}\frac{dh}{dt} - V_1 A_1 + Q_2 = 0$$

The rate at which the water level is rising is then

$$\frac{dh}{dt} = \frac{V_1 A_1 - Q_2}{\pi D^2/4}$$

Thus

$$\frac{dh}{dt} = \frac{0.5 \times 0.1 - 0.2}{\pi \times 0.5^2/4} = -0.764 \text{ m/s}$$

The negative sign indicates that the water level is actually decreasing.

Let's solve this problem again but with another choice for the control volume, one with its top surface below the water level (Fig. E4.5b). The velocity at the top surface is then equal to the rate at which the surface rises. The flow condition inside the control volume is steady. Hence we can apply Eq. 4.3.4. There are three areas across which fluid flows. On the third area, the velocity is dh/dt; hence the continuity equation takes the form

$$\rho(-V_1)A_1 + \rho Q_2 + \rho \frac{dh}{dt}\frac{\pi}{4}D^2 = 0$$

so that

$$\frac{dh}{dt} = \frac{V_1 A_1 - Q_2}{\pi D^2/4}$$

This is the same result as given above.

4.5 ENERGY EQUATION

Many problems involving fluid motion demand that the first law of thermodynamics, often referred to as the *energy equation*, be used to relate quantities of interest. If the heat transferred to a device (a boiler or compressor), or the work done by a device (a pump or turbine), is desired, the energy equation is obviously needed. It is also used to relate pressures and velocities when Bernoulli's equation is not applicable; this is the case whenever viscous effects cannot be neglected, such as flow through a piping system or in an open channel. Let us express the energy equation in control volume form. For a system it is

$$\dot{Q} - \dot{W} = \frac{D}{Dt} \int_{\text{sys}} e\rho \, d\mathcal{V} \tag{4.5.1}$$

where the specific energy e includes specific kinetic energy $V^2/2$, specific potential energy gz, and specific internal energy $\tilde{u}$; that is,

$$e = \frac{V^2}{2} + gz + \tilde{u} \tag{4.5.2}$$

We will not include other forms of energy, such as energy due to magnetic or electric field-flow field interactions or those due to chemical reactions. In terms of a control volume, Eq. 4.5.1 becomes

$$\dot{Q} - \dot{W} = \frac{d}{dt} \int_{\text{c.v.}} e\rho \, d\mathcal{V} + \int_{\text{c.s.}} \rho e \mathbf{V} \cdot \hat{\mathbf{n}} \, dA \tag{4.5.3}$$

This can be put in simplified forms for certain restricted flows, but first let us discuss the rate-of-heat transfer term $\dot{Q}$ and the work-rate term $\dot{W}$.

The term $\dot{Q}$ represents the rate-of-energy transfer across the control surface due to a temperature difference. (Do not confuse this term with the flow rate Q.) The rate-of-heat transfer term is either given or results from using Eq. 4.5.3. The calculation of $\dot{Q}$ is the objective of a course in heat transfer and is quite difficult, in general, to calculate. The work-rate term is discussed in detail in the following section.

KEY CONCEPT $\dot{Q}$ represents the rate-of-energy transfer across the control surface due to a temperature difference.

4.5.1 Work-Rate Term

The work-rate term results from work being done by the system. Or, since we consider the instant that the system occupies the control volume, we can also state that the work-rate term results from work being done by the control volume. Work is due to a force moving through a distance while it acts on the control volume. The rate of doing work $\dot{W}$ or power, is given by the dot product of a force $\mathbf{F}$ with its velocity:

KEY CONCEPT *The rate of doing work is given by the dot product of a force* $\mathbf{F}$ *with its velocity.*

$$\dot{W} = -\mathbf{F} \cdot \mathbf{V}_I \tag{4.5.4}$$

where $\mathbf{V}_I$ is the velocity measured with respect to a fixed reference frame. The negative sign results because we have selected the convention that work done on the control volume is negative.

If the force results from a variable stress acting over the control surface, we must integrate,

$$\dot{W} = -\int_{\text{c.s.}} \boldsymbol{\tau} \cdot \mathbf{V}_I \, dA \qquad (4.5.5)$$

where $\boldsymbol{\tau}$ is the stress vector acting on the elemental area dA, the differential force being represented by $d\mathbf{F} = \boldsymbol{\tau}\, dA$, as shown in Fig. 4.8.

For a moving control volume, such as a car, we have to evaluate the velocity with respect to a fixed reference frame. For example, let us consider an automobile traveling at constant speed (see Example 4.10). If we want to apply the energy equation, we could make the car the control volume. In that case, the velocity in Eq. 4.5.4 would be measured relative to a fixed reference and not relative to the car. If the velocity relative to the car were used, the drag force would have a zero velocity which would result in no work done; but we know that at high speed, energy from the gasoline goes primarily to overcome the drag. Thus the stationary reference frame is needed.

In general, for moving control volumes the velocity vector $\mathbf{V}_I$ is related to a relative velocity $\mathbf{V}$, observed in a reference frame attached to the control volume by

$$\mathbf{V}_I = \mathbf{V} + \dot{\mathbf{S}} + \boldsymbol{\Omega} \times \mathbf{r} \qquad (4.5.6)$$

where $\dot{\mathbf{S}}$ is the velocity of the control volume (see Fig. 3.5). We can now write the work rate as

$$\dot{W} = -\int \boldsymbol{\tau} \cdot \mathbf{V} \, dA + \dot{W}_I \qquad (4.5.7)$$

where the "inertial work-rate" term is given by

$$\dot{W}_I = -\int_{\text{c.s.}} \boldsymbol{\tau} \cdot (\dot{\mathbf{S}} + \boldsymbol{\Omega} \times \mathbf{r}) \, dA \qquad (4.5.8)$$

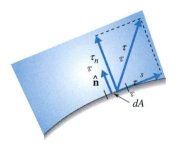

FIGURE 4.8   Stress vector acting on the control surface.

Next, express the stress vector as the sum of a normal component and a shear component, that is,

$$\boldsymbol{\tau} = -p\hat{\mathbf{n}} + \boldsymbol{\tau}_s \tag{4.5.9}$$

where the pressure p is assumed to be positive in a compressive state. Then

$$\dot{W} = \int_{\text{c.s.}} p\hat{\mathbf{n}}\cdot\mathbf{V}\,dA - \int_{\text{c.s.}} \boldsymbol{\tau}_s\cdot\mathbf{V}\,dA + \dot{W}_I \tag{4.5.10}$$

KEY CONCEPT *Shaft work* $\dot{W}_S$ *is transmitted by a rotating shaft that is cut by the control surface.*

We will allow the shear stress term to consist of two parts. One part accounts for work called *shaft work* $\dot{W}_S$ transmitted by a rotating shaft that is cut by the control surface; this term is important when we deal with flows in pumps and turbines. The other part will be denoted *shear work* $\dot{W}_{\text{shear}}$ and results from moving boundaries; this term is required if the control surface itself moves relative to the control volume as occurs with a moving belt.

Hence the work-rate term becomes

$$\dot{W} = \int_{\text{c.s.}} p\hat{\mathbf{n}}\cdot\mathbf{V}\,dA + \dot{W}_S + \dot{W}_{\text{shear}} + \dot{W}_I \tag{4.5.11}$$

The terms are summarized as follows:

$\int p\hat{\mathbf{n}}\cdot\mathbf{V}\,dA$ Work rate resulting from the force due to pressure moving at the control surface. It is often referred to as **flow work**.

Flow work: *Work rate resulting from the force due to pressure moving at the control surface.*

$\dot{W}_S$ Work rate resulting from rotating shafts such as that of a pump or turbine, or the equivalent electric power.

$\dot{W}_{\text{shear}}$ Work rate due to the shear acting on a moving boundary such as a moving belt.

$\dot{W}_I$ Work rate that occurs when the control volume moves relative to a fixed reference frame.

We should note that the work-rate terms $\dot{W}_{\text{shear}}$ and $\dot{W}_I$ are seldom encountered in problems in an introductory course and are often omitted from textbooks. They are included here for completeness.

4.5.2 General Energy Equation

When the work-rate term of Eq. 4.5.11 is substituted into Eq. 4.5.3, we obtain the energy equation in the form

$$\dot{Q} - \dot{W}_S - \dot{W}_{\text{shear}} - \dot{W}_I = \frac{d}{dt}\int_{\text{c.v.}} e\rho\,d\Psi + \int_{\text{c.s.}} \left(e + \frac{p}{\rho}\right)\rho\hat{\mathbf{n}}\cdot\mathbf{V}\,dA \tag{4.5.12}$$

Note that the work-rate term needed to move the pressure force has been moved to the right-hand side and is treated like an energy flux term.

Substitution of Eq. 4.5.2 results in

$$\dot{Q} - \dot{W}_S - \dot{W}_{\text{shear}} - \dot{W}_I = \frac{d}{dt} \int_{\text{c.v.}} \left(\frac{V_I^2}{2} + gz + \tilde{u} \right) \rho \, dV$$

$$+ \int_{\text{c.s.}} \left(\frac{V_I^2}{2} + gz\tilde{u} + \frac{p}{\rho} \right) \rho \mathbf{V} \cdot \hat{\mathbf{n}} \, dA \qquad (4.5.13)$$

This general form of the energy equation is useful in analyzing fluid flow problems that may include time-dependent effects and nonuniform profiles. Before we simplify the equation for steady flow and uniform profiles, let us introduce the notion of "losses."

In many fluid flows, useful forms of energy (kinetic energy and potential energy) and flow work are converted into unusable energy forms (internal energy or heat transfer). If we assume that the temperature of the control volume remains unchanged, the internal energy does not change and the losses are balanced by heat transfer across the control surface. This heat transfer can be the result of convection, radiation, or conduction at the control surfaces. The theory of heat transfer is aimed at the detailed description of these effects. However, in an introductory fluid mechanics course the sum of these effects is lumped together and denoted $\dot{Q}$. Thus we define **losses** as the sum of all the terms representing unusable forms of energy:

Losses: *The sum of all the terms representing unusable forms of energy.*

$$\text{losses} = -\dot{Q} + \frac{d}{dt} \int_{\text{c.v.}} \tilde{u}\rho dV + \int_{\text{c.s.}} \tilde{u}\rho \, \mathbf{V} \cdot \hat{\mathbf{n}} \, dA \qquad (4.5.14)$$

We can now rewrite the energy equation as

$$-\dot{W}_S - \dot{W}_{\text{shear}} - \dot{W}_I = \frac{d}{dt} \int_{\text{c.v.}} \left(\frac{V_I^2}{2} + gz \right) \rho \, dV$$

$$+ \int_{\text{c.s.}} \left(\frac{V_I^2}{2} + gz + \frac{p}{\rho} \right) \rho \, \mathbf{V} \cdot \hat{\mathbf{n}} \, dA + \text{losses} \qquad (4.5.15)$$

Losses are due to two primary effects:

KEY CONCEPT *Losses are due primarily to internal friction and separated flows.*

1. Viscosity causes internal friction that results in increased internal energy (temperature increase) or heat transfer.
2. Changes in geometry result in separated flows that require useful energy to maintain the resulting secondary motions in which viscous dissipation occurs.

In a conduit, the losses due to viscous effects are distributed over the entire length, whereas the loss due to a geometry change (a valve, an elbow, an enlargement) is concentrated in the vicinity of the geometry change.

It turns out that the analytical calculation of losses is rather difficult, particularly when the flow is turbulent. In general, the prediction of losses is based on empirical formulas. Such formulas will be given in subsequent chapters. In this chapter we discuss losses qualitatively and, in examples and problems, losses will be given. For a pump or a turbine the losses are expressed in terms of the efficiency. For example, if the efficiency of a pump is 80%, the losses would be 20% of the energy input to the pump.

It may be that the objective in a particular fluid flow is to change the internal energy of the fluid, such as in the steam generator (boiler) of a power plant, by the transfer of heat; then the definition of losses above must be altered so that the loss term includes only the dissipative effects of the viscosity of the fluid. Generally, for problems of interest in fluid mechanics, Eq. 4.5.15 is acceptable.

4.5.3   Steady Uniform Flow

Consider a steady-flow situation in which there is one entrance and one exit across which uniform profiles can be assumed. Also, assume that $\dot{W}_{\text{shear}} = \dot{W}_I = 0$ with $V_I = V$. For such a flow the term $(V^2/2 + gz + p/\rho)$ in Eq. 4.5.15 is constant across the cross section because V is constant (we assume a uniform velocity profile) and the sum of $p/\rho + gz$ is constant if the streamlines at each section are parallel. The energy equation (Eq. 4.5.15) then simplifies to

$$-\dot{W}_S = \rho_2 V_2 A_2 \left(\frac{V_2^2}{2} + \frac{p_2}{\rho_2} + gz_2 \right) - \rho_1 V_1 A_1 \left(\frac{V_1^2}{2} + \frac{p_1}{\rho_1} + gz_1 \right) + \text{losses} \quad (4.5.16)$$

where the subscripts 1 and 2 refer to the entrance and exit, respectively. The mass flux is given by $\dot{m} = \rho_1 A_1 V_1 = \rho_2 A_2 V_2$. After dividing by $\dot{m}g$ we have

$$-\frac{\dot{W}_S}{\dot{m}g} = \frac{V_2^2 - V_1^2}{2g} + \frac{p_2}{\gamma_2} - \frac{p_1}{\gamma_1} + z_2 - z_1 + h_L \quad (4.5.17)$$

where we have introduced the *head loss* h_L, defined to be

$$h_L = \frac{\tilde{u}_2 - \tilde{u}_1}{g} - \frac{\dot{Q}}{\dot{m}g} \quad (4.5.18)$$

It is often written in terms of a *loss coefficient K* as

$$h_L = K \frac{V^2}{2g} \quad (4.5.19)$$

where V may be either V_1 or V_2; if it is not obvious, it will be specified. Loss coefficients are discussed in some detail in Chapter 7 and are tabulated in Table 7.2.

The head loss is referred to as a "head" since it has dimensions of length. We may also refer to $V^2/2g$ as the *velocity head* and p/γ as the *pressure head* since those terms also have dimensions of length. Also recall from Chapter 3 that $p/\gamma + z$ is called the *piezometric head*. Further, the sum of the piezometric head and the velocity head is called the *total head*.

The energy equation, in the form of Eq. 4.5.17, is useful in many applications and is, perhaps, the most often used form of the energy equation. If the losses are negligible and if there is no shaft work, we note that the energy equation takes the form

$$\frac{V_2^2}{2g} + \frac{p_2}{\gamma_2} + z_2 = \frac{V_1^2}{2g} + \frac{p_1}{\gamma_1} + z_1 \tag{4.5.20}$$

Observe that the energy equation has been reduced to a form identical with Bernoulli's equation when $\gamma_2 = \gamma_1$ (a constant density flow). We must remember, however, that Bernoulli's equation is a momentum equation applicable along a streamline and the equation above is an energy equation applied between two sections of a flow. It is not surprising that both should predict identical results from the conditions stated because the velocity head is constant over a cross section and the sum of pressure head and elevation remains constant over a cross section.

The energy equation (4.5.17) may be applied to any steady, uniform flow with one entrance and one exit. The control volume is usually selected such that the entrance and exit sections have a uniform total head. For example, it may be applied to water flow through a long pipeline; the total head at the entrance and exit may then be evaluated conveniently at the center of the pipe entrance and exit. The energy equation may be applied to the flow passing a gate (Fig. 4.9). An appropriate control volume is shown. The total head at the entrance and exit can be evaluated at any point at the entrance and exit, respectively. However, a convenient choice would be the points at the water surface. Thus the energy equation becomes

$$\frac{V_1^2}{2g} + \overset{0}{\cancel{\frac{p_1}{\gamma}}} + h_1 = \frac{V_2^2}{2g} + \overset{0}{\cancel{\frac{p_2}{\gamma}}} + h_2 + h_L \tag{4.5.21}$$

where the shaft work has been set to zero. If we had picked the centroids of the entrance and exit, as shown in Fig. 4.9, we would have obtained

$$\frac{V_1^2}{2g} + \frac{p_1}{\gamma} + \frac{h_1}{2} = \frac{V_2^2}{2g} + \frac{p_2}{\gamma} + \frac{h_2}{2} + h_L \tag{4.5.22}$$

We see that this result is the same as in Eq. 4.5.21 if we substitute $p_1 = \gamma h_1/2$ and $p_2 = \gamma h_2/2$. For completeness, it is noted that the losses between 1 and 2 in

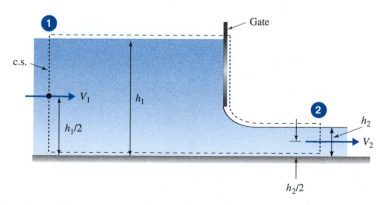

FIGURE 4.9   Application of the energy equation to a gate in an open channel.

Fig. 4.9 could be neglected because internal viscous effects occur only over a relatively short distance and no significant secondary flows are generated.

The energy equation (4.5.15) can be applied to any control volume. For example, consider steady, uniform incompressible flow through a T-section in a pipe (Fig. 4.10) in which there is one entrance and two exits. The energy equation can be applied to each of two control volumes, one for the mass flux that exits section 2 and the other for the mass flux that exits section 3:

$$
\frac{V_1^2}{2g} + \frac{p_1}{\gamma} + z_1 = \frac{V_2^2}{2g} + \frac{p_2}{\gamma} + z_2 + h_{L_{1-2}}
$$
$$
\frac{V_1^2}{2g} + \frac{p_1}{\gamma} + z_1 = \frac{V_3^2}{2g} + \frac{p_3}{\gamma} + z_3 + h_{L_{1-3}}
$$
(4.5.23)

The loss terms in Eq. 4.5.23 include the losses between the inlet and the respective exits. If the losses are negligible, the energy equation reduces to a form similar to the Bernoulli equation being applied along a streamline going from 1 to 2 or a streamline going from 1 to 3.

A final note to this section regards nomenclature for pumps and turbines in a flow system. It is often conventional to call the energy term ($\dot{W}_S/\dot{m}g$) associated with a pump the *pump head H_P*, and the term ($\dot{W}_S/\dot{m}g$) associated with a turbine the *turbine head H_T*. Then the energy equation, for an incompressible flow, takes the form

$$
H_P + \frac{V_1^2}{2g} + \frac{p_1}{\gamma} + z_1 = H_T + \frac{V_2^2}{2g} + \frac{p_2}{\gamma} + z_2 + h_L
$$
(4.5.24)

In this form we have equated the energy at the inlet plus added energy to the energy at the exit plus extracted energy (energy per unit weight, of course). If any of the quantities is zero (e.g., there is no pump), the appropriate term is simply omitted. The terms H_P and H_T above represent the energy that is transferred to and from the fluid, respectively. If the energy delivered by the turbine or required by the pump is desired, the efficiency of each device must be used.

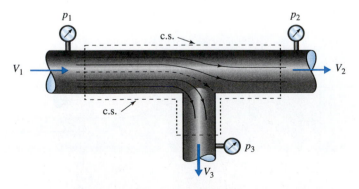

FIGURE 4.10   Application of the energy equation to a T-section.

The power generated by the turbine with an efficiency of η_T is simply

$$\dot{W}_T = \dot{m}gH_T\eta_T = \gamma Q H_T\eta_T \tag{4.5.25}$$

The power requirement by a pump with an efficiency of η_P would be

$$\dot{W}_P = \frac{\dot{m}gH_P}{\eta_P} = \frac{\gamma Q H_P}{\eta_P} \tag{4.5.26}$$

We will calculate power in watts or horsepower. Recall that one horsepower is equivalent to 746 W.

4.5.4 Steady Nonuniform Flow

If the assumption of uniform velocity profiles is not acceptable for a problem of interest, as is sometimes the situation, we have to consider the control-surface integral in Eq. 4.5.15 with the proper expression for the velocity distribution. In practice, a velocity distribution can be accounted for by introducing the *kinetic-energy correction factor* α, defined by

$$\alpha = \frac{\int V^3\, dA}{\overline{V}^3 A} \tag{4.5.27}$$

where $\overleftrightarrow{V}$ is the average velocity over the area A, given by Eq. 4.4.11. Then the term that accounts for the flux of kinetic energy in Eq. 4.5.15 is

$$\frac{1}{2}\rho \int_A V^3\, dA = \frac{1}{2}\alpha\rho \overline{V}^3 A \tag{4.5.28}$$

where we have used $\mathbf{V}\cdot\hat{\mathbf{n}} = V$ and $V_I = V$. Using this factor, we can account for nonuniform velocity distributions by modifying Eq. 4.5.24 to read

$$H_P + \alpha_1\frac{\overline{V}_1^2}{2g} + \frac{p_1}{\gamma} + z_1 = H_T + \alpha_2\frac{\overline{V}_2^2}{2g} + \frac{p_2}{\gamma} + z_2 + h_L \tag{4.5.29}$$

where $\overline{V}_1$ and $\overline{V}_2$ are the average velocities at sections 1 and 2, respectively. For a flow with a parabolic profile in a pipe we can calculate $\alpha = 2.0$ (see Example 4.9). For most internal turbulent flows, however, the profile is nearly uniform with $\alpha \simeq 1.05$. Hence we simply let $\alpha = 1$ since it is so close to unity; this will always be done unless otherwise stated, since most of the internal flows that we encounter are, in fact, turbulent flows.

KEY CONCEPT *For most internal turbulent flows, we let $\alpha = 1$.*

Example 4.6

The pump of Fig. E4.6 is to increase the pressure of 0.2 m³/s of water from 200 kPa to 600 kPa. If the pump is 85% efficient, how much electrical power will the pump require? The exit area is 20 cm above the inlet area. Assume inlet and exit areas are equal.

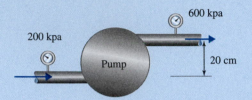

200 kpa

600 kpa

Pump

20 cm

FIGURE E4.6

Solution

Equation (4.5.24) across the pump provides

$$H_P = \frac{p_2 - p_1}{\gamma} + z_2 - z_1$$

$$= \frac{600\,000 - 200\,000}{9810} + 0.2 = 41.0 \text{ m}$$

where $V_2 = V_1$ since the inlet and exit areas are equal, and any losses are accounted for with the efficiency of Eq. 4.5.26. That equation provides the power:

$$\dot{W}_P = \frac{\gamma Q H_P}{\eta_P}$$

$$= \frac{9810 \times 0.2 \times 41.0}{0.85} = 94\,600 \text{ W} \quad \text{or} \quad 94.6 \text{ kW}$$

Note: The units work out if length is in meters ($H_P = 41$ m), weight is in newtons ($\gamma = 9810$ N/m³), and time is in seconds ($Q = 0.2$ m³/s). Check the units to make sure they provide the answer in watts, i.e., J/s.

Example 4.7

Water flows from a reservoir through a 0.8-m-diameter pipeline to a turbine-generator unit and exits to a river that is 30 m below the reservoir surface. If the flow rate is 2.8 m³/s, and the turbine-generator efficiency is 88%, calculate the power output. Assume the loss coefficient in the pipeline (including the exit) to be $K = 2$.

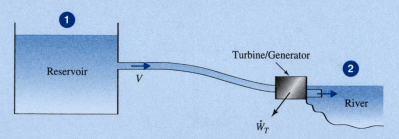

FIGURE E4.7

Solution

Referring to Fig. E4.7, we select the control volume to extend from section 1 to section 2 on the reservoir and river surfaces, where we know the velocities, pressures, and elevations; we consider the water surface of the left reservoir to be the entrance and the water surface of the river to be the exit. The velocity in the pipe is

$$V = \frac{Q}{A} = \frac{2.8}{\pi \times 0.8^2 / 4} = 5.57 \text{ m/s}$$

Now, consider the energy equation. We will use gage pressures so that $p_1 = p_2 = 0$; the datum is placed through the lower section 2 so that $z_2 = 0$; the velocities V_1 and V_2 on the reservoir surfaces are negligibly small; K is assumed to be based on the 0.8-m-diameter pipe velocity. The energy equation (4.5.24) then becomes

$$\cancel{H_P}^{0} + \cancel{\frac{V_1^2}{2g}}^{0} + \cancel{\frac{p_1}{\gamma}}^{0} + z_1 = H_T + \cancel{\frac{V_2^2}{2g}}^{0} + \cancel{\frac{p_2}{\gamma}}^{0} + \cancel{z_2}^{0} + K\frac{V^2}{2g}$$

$$30 = H_T + 2\,\frac{5.57^2}{2 \times 9.81}$$

$$\therefore H_T = 26.84 \text{ m}$$

From this the power output is found using Eq. 4.5.25 to be

$$\dot{W}_T = Q\gamma H_T \eta_T$$
$$= 2.8 \times 9810 \times 26.84 \times 0.88 = 648772 \text{ N·m/s} \quad \text{or} \quad 869.7 \text{ hp}$$

In this example we have used gage pressure; the potential-energy datum was assumed to be placed through section 2, V_1 and V_2 were assumed to be insignificantly small, and K was assumed to be based on the 0.8-m-diameter pipe velocity.

Example 4.8

The venturi meter shown reduces the pipe diameter from 10 cm to a minimum of 5 cm (Fig. E4.8). Calculate the flow rate and the mass flux assuming ideal conditions.

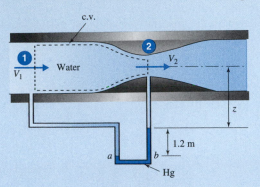

FIGURE E4.8

Solution

The control volume is selected as shown such that the entrance and exit correspond to the sections where the pressure information of the manometer can be applied. The manometer's reading is interpreted as follows:

$$p_a = p_b$$

$$p_1 + \gamma(z + 1.2) = p_2 + \gamma z + 13.6\gamma \times 1.2$$

where z is the distance from the pipe centerline to the top of the mercury column. The manometer then gives

$$\frac{p_1 - p_2}{\gamma} = 15.12 \text{ m}$$

Continuity allows us to relate V_2 to V_1 by

$$V_1 A_1 = V_2 A_2$$

$$\therefore V_2 = \frac{A_1}{A_2} V_1 = 4V_1$$

The energy equation (4.5.17) assuming ideal conditions (no losses and uniform flow) with $h_L = \dot{W}_S = 0$ takes the form

$$0 = \frac{V_2^2 - V_1^2}{2g} + \frac{p_2 - p_1}{\gamma} + (z_2 - z_1)^0$$

$$= \frac{16V_1^2 - V_1^2}{2g} - 15.12$$

$$\therefore V_1 = 4.45 \text{ m/s}$$

The flow rate is

$$Q = A_1 V_1 = \pi \times 0.05^2 \times 4.45 = 0.0350 \text{ m}^3/\text{s}$$

The mass flux is

$$\dot{m} = \rho Q = 1000 \times 0.035 = 35.0 \text{ kg/s}$$

Example 4.9

The velocity distribution for a certain flow in a pipe is $V(r) = V_{max}(1 - r^2/r_0^2)$, where r_0 is the pipe radius (Fig. E4.9). Determine the kinetic-energy correction factor.

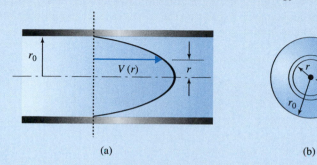

(a)   (b)

FIGURE E4.9

Solution

To find the kinetic-energy correction factor α, we must know the average velocity. It is (combine Eqs. 4.4.10 and 4.4.11)

$$\overline{V} = \frac{\int V \, dA}{A}$$

$$= \frac{1}{\pi r_0^2} \int_0^{r_0} V_{max}\left(1 - \frac{r^2}{r_0^2}\right) 2\pi r \, dr = \frac{2\pi V_{max}}{\pi r_0^2} \int_0^{r_0} \left(r - \frac{r^3}{r_0^2}\right) dr$$

$$= \frac{2 V_{max}}{r_0^2}\left(\frac{r_0^2}{2} - \frac{r_0^4}{4 r_0^2}\right) = \frac{1}{2} V_{max}$$

Using Eq. 4.5.27, there results

$$\alpha = \frac{\int V^3 \, dA}{\overline{V}^3 A}$$

$$= \frac{\int_0^{r_0} V_{max}^3 (1 - r^2/r_0^2)^3 \, 2\pi r \, dr}{\left(\frac{1}{2} V_{max}\right)^3 \pi r_0^2} = \frac{16}{r_0^2} \int_0^{r_0}\left(1 - \frac{3r^2}{r_0^2} + \frac{3r^4}{r_0^4} - \frac{r^6}{r_0^6}\right) r \, dr$$

$$= \frac{16}{r_0^2}\left(\frac{r_0^2}{2} - \frac{3r_0^2}{4} + \frac{3r_0^2}{6} - \frac{r_0^2}{8}\right) = 2$$

Consequently, the kinetic energy flux associated with a parabolic velocity distribution across a circular area is given by

$$\int \rho \mathbf{V} \cdot \hat{\mathbf{n}} \frac{V^2}{2} \, dA = 2 \times \frac{\dot{m}\overline{V}^2}{2}$$

Parabolic velocity distributions are encountered in laminar flows in pipes and between parallel plates, downstream of inlets and geometry changes (valves, elbows, etc.). The Reynolds number must be quite small, usually less than about 2000.

Example 4.10

The drag force on an automobile (Fig. E4.10) is approximated by the expression $0.15\,\rho V_\infty^2 A$, where A is the projected cross-sectional area and V_∞ is the automobile's speed. If $A = 1.2\ \text{m}^2$, calculate the efficiency η of the engine if the rate of fuel consumption $\dot{f}$ (the gas mileage) is 15 km/L and the automobile travels at 90 km/h. Assume that the fuel releases 44 000 kJ/kg during combustion. Neglect the energy lost due to the exhaust gases and coolant and assume that the only resistance to motion is the drag force. Use $\rho_{\text{air}} = 1.12\ \text{kg/m}^3$ and $\rho_{\text{fuel}} = 0.68\ \text{kg/L}$.

FIGURE E4.10

Solution

If the car is taken as the moving control volume (note that the control volume is fixed), as shown, we can simplify the energy equation (Eq. 4.5.3 in combination with 4.5.11) to

$$\dot{Q} - \dot{W}_I = 0$$

since all other terms are negligible; there is no velocity crossing the control volume, so $\mathbf{V} \cdot \hat{\mathbf{n}} = 0$ (neglect the energy of the exhaust gases); there is no shear or shaft work; the energy of the c.v. remains constant. The energy input $\dot{Q}$ which accomplishes useful work is η times the energy released during combustion, that is,

$$\dot{Q} = \dot{m}_f \times 44\,000\eta \ \text{kJ/s}$$

where $\dot{m}_f$ is the mass flux of the fuel. The mass flux of fuel is determined knowing the rate of fuel consumption $\dot{f}$ and the density of fuel as 0.68 kg/L, as follows:

$$\dot{f} = \frac{\text{distance}}{\text{volume}} = \frac{V_\infty \times \text{time}}{Q \times \text{time}} = \frac{V_\infty}{\dot{m}_f/\rho_f} = \frac{\rho_f V_\infty}{\dot{m}_f}$$

with $V_\infty = 90\,000/3600 = 25$ m/s, we have, using $\dot{f} = 15 \times 1000$ m/L,

$$15 \times 1000 = \frac{0.68 \times 25}{\dot{m}_f}$$

$$\therefore \dot{m}_f = 0.001133 \ \text{kg/s}$$

The inertial work-rate term is

$$\dot{W}_I = V_\infty \times \text{drag}$$
$$= 0.15\rho V_\infty^3 A = 0.15 \times 1.12 \times 25^3 \times 1.2 = 3150 \ \text{J/s}$$

Equating $\dot{Q} = \dot{W}_I$, we have

$$44\,000\eta \times 0.001133 = 3.15$$

$$\therefore \eta = 0.0632 \quad \text{or} \quad 6.32\%$$

This is obviously a very low percentage, perhaps surprisingly low to the reader. Very little power (3.15 kJ/s = 4.22 hp) is actually needed to propel the automobile at 90 km/h. The relatively large engine, needed primarily for acceleration, is quite inefficient when simply propelling the automobile. Note the importance of using a stationary reference frame. The reference frame attached to the automobile is an inertial reference frame since it is moving at constant velocity. Yet the energy equation demands a stationary reference frame so that the energy required by the drag force can properly be included.

4.6 MOMENTUM EQUATION

4.6.1 General Momentum Equation

Newton's second law, often called the *momentum equation*, states that the resultant force acting on a system equals the rate of change of momentum of the system when measured in an inertial reference frame; that is,

$$\Sigma \, \mathbf{F} = \frac{D}{Dt} \int_{\text{sys}} \rho \mathbf{V} \, d\Psi \qquad (4.6.1)$$

KEY CONCEPT *The momentum equation is used primarily to determine the forces induced by the flow.*

Using Eq. 4.3.9, with η replaced by $\mathbf{V}$, this is written for a control volume as

$$\Sigma \, \mathbf{F} = \frac{d}{dt} \int_{\text{c.v.}} \rho \mathbf{V} \, d\Psi + \int_{\text{c.s.}} \rho \mathbf{V}(\mathbf{V} \cdot \hat{\mathbf{n}}) \, dA \qquad (4.6.2)$$

where $\mathbf{V} \cdot \hat{\mathbf{n}}$ is simply a scalar for each differential area dA. The control surface integral on the right represents the net momentum flux across the control surface of the fluid entering and/or leaving the control volume.

When applying Newton's second law the quantity $\Sigma \, \mathbf{F}$ represents all forces acting on the control volume. The forces include the surface forces resulting from the surroundings acting on the control surface and body forces that result from gravity and magnetic fields. The momentum equation is often used to determine the forces induced by the flow. For example, the equation allows us to calculate the force on the support of an elbow in a pipeline or the force on a submerged body in a free-surface flow.

When we apply the momentum equation the surrounding fluid and sometimes the entire conduit or container is separated from the control volume. For example, in the horizontal nozzle of Fig. 4.11a, the nozzle and the fluid in the nozzle are isolated. Thus care must be taken to include the pressure forces shown and the force $\mathbf{F}_{\text{joint}}$. It is convenient to use gage pressures so that the pressure acting

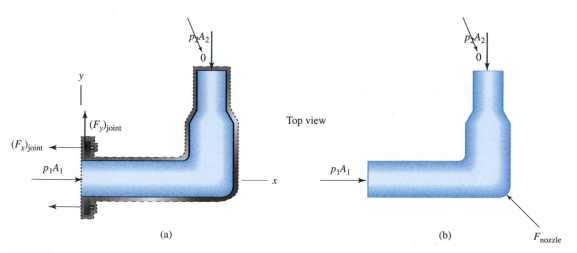

(a)   (b)

FIGURE 4.11   Forces acting on the control volume of a horizontal nozzle: (a) control volume includes nozzle and fluid in nozzle; (b) control volume includes fluid in nozzle only. We have neglected body forces.

on the exterior of the pipe is then zero. Alternatively, we could have selected a control volume that includes only the fluid in the nozzle (Fig. 4.11b). In that case we have to consider the pressure forces at the entrance and exit and the resultant pressure force $\mathbf{F}_{\text{nozzle}}$ of the interior wall of the nozzle on the fluid. Of course, the force $\mathbf{F}_{\text{joint}}$ and $\mathbf{F}_{\text{nozzle}}$ are equal in magnitude, as is obvious from a free body of the nozzle excluding the fluid. If the problem is to determine the force exerted by the flow on the nozzle (Fig. 4.11b), we have to reverse the direction of the calculated force $\mathbf{F}_{\text{nozzle}}$. Examples at the end of this section illustrate this.

4.6.2   Steady Uniform Flow

Equation 4.6.2 can be simplified considerably if a device has entrances and exits across which the flow may be assumed to be uniform and if the flow is steady. Then there results

$$\Sigma\,\mathbf{F} = \sum_{i=1}^{N} \rho_i A_i \mathbf{V}_i (\mathbf{V}_i \cdot \hat{\mathbf{n}}) \tag{4.6.3}$$

where N is the number of flow exit/entrance areas.

At an entrance $\mathbf{V} \cdot \hat{\mathbf{n}} = -V$ since the unit vector points out of the volume and at the exit $\mathbf{V} \cdot \hat{\mathbf{n}} = V$. If there is only one entrance and one exit, as in Fig. 4.11, the momentum equation becomes

$$\Sigma\,\mathbf{F} = \rho_2 A_2 V_2 \mathbf{V}_2 - \rho_1 A_1 V_1 \mathbf{V}_1 \tag{4.6.4}$$

Using continuity,

$$\dot{m} = \rho_1 A_1 V_1 = \rho_2 A_2 V_2 \tag{4.6.5}$$

the momentum equation takes the simplified form

$$\Sigma\,\mathbf{F} = \dot{m}\,(\mathbf{V}_2 - \mathbf{V}_1) \tag{4.6.6}$$

Note that the momentum equation is a vector equation which represents three scalar equations.

$$\Sigma \, \mathbf{F}_x = \dot{m}(V_{2x} - V_{1x})$$

$$\Sigma \, \mathbf{F}_y = \dot{m}(V_{2y} - V_{1y}) \qquad (4.5.7)$$

$$\Sigma \, \mathbf{F}_z = \dot{m}(V_{2z} - V_{1z})$$

> **KEY CONCEPT** *The momentum equation is a vector equation which represents three scalar equations.*

If we consider the nozzle of Fig. 4.11a and we want to determine the x-component of the force of the joint on the nozzle, $(\mathbf{V}_1)_x = V_1$ and $(\mathbf{V}_2)_x = 0$ so that the momentum equation for the x-direction becomes

$$\Sigma \, F_x = -(F_x)_{\text{joint}} + p_1 A_1 = -\dot{m} \, V_1 \qquad (4.6.8a)$$

Similarly, we could write the y-component equation and find an expression for $(F_y)_{\text{joint}}$.

An example of a free-surface flow in a rectangular channel is shown in Fig. 4.12. If we want to determine the force of the gate on the flow, the following expression can be derived from the momentum equation:

$$\Sigma \, F_x = -F_{\text{gate}} + F_1 - F_2 = \dot{m}(V_2 - V_1) \qquad (4.6.8b)$$

where F_1 and F_2 are pressure forces (see Fig. 4.12).

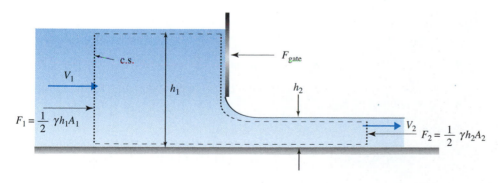

FIGURE 4.12   Force of the flow on a gate in a free-surface flow.

Example 4.11

Water flows through a horizontal pipe bend and exits into the atmosphere (Fig. E4.11a). The flow rate is 10 L/s. Calculate the force in each of the rods holding the pipe bend in position. Neglect body forces and viscous effects and shear force in the rods.

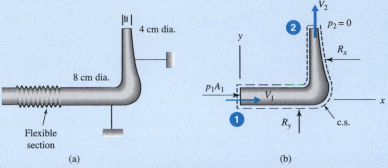

FIGURE E4.11

Solution

We have selected a control volume that surrounds the bend, as shown in Fig. E4.11b. Since the rods have been cut, the forces that the rods exert on the control volume are included. The pressure force at the entrance of the control volume is also shown. The flexible section is capable of resisting the interior pressure but it transmits no axial force or moment. The body force (weight of the control volume) does not act in the x- or y-direction but normal to it. Therefore, no other forces are shown. The average velocities are found to be

$$V_1 = \frac{Q}{A_1} = \frac{(10)/1000}{\pi \times (8/100)^2/4} = 1.99 \text{ m/s}; \quad V_2 = \frac{Q}{A_2} = \frac{(10)/1000}{\pi \times (4/100)^2/4} = 7.96 \text{ m/s}$$

Before we can calculate the forces R_x and R_y we need to find the pressures p_1 and p_2. The pressure p_2 is zero because the flow exits into the atmosphere. The pressure at section 1 can be determined using the energy equation or the Bernoulli equation. Neglecting losses between sections 1 and 2, the energy equation gives

$$\frac{V_1^2}{2g} + \frac{p_1}{\gamma} = \frac{V_2^2}{2g} + \frac{p_2^{\ 0}}{\gamma}$$

$$\therefore p_1 = \frac{\gamma}{2g}(V_2^2 - V_1^2) = \frac{9810}{2 \times 9.81}(7.96^2 - 1.99^2) = 29.7 \times 10^3 \text{ N/m}^2$$

Now we can apply the momentum equation (4.6.6) in the x-direction to find R_x and in the y-direction to find R_y:

x-direction:
$$p_1A_1 - R_x = \dot{m}(V_{2x}^{\ 0} - V_{1x})$$

$$\left(29.7 \times 10^3\right) \times \frac{\pi}{4}\left(\frac{8}{100}\right)^2 - R_x = 1000 \times (10/1000) \times (-1.99)$$

$$\therefore R_x = 169.2 \text{ N}$$

y-direction:
$$R_y = \dot{m}(V_{2y} - V_{1y}^{\ 0})$$
$$= 1000 \times (10/1000) \times 7.96 = 79.6 \text{ N}$$

Note that we have assumed uniform profiles and steady flow and used $\dot{m} = \rho Q$. These are the usual assumptions if information is not given otherwise.

Example 4.12

When the velocity of a flow in an open rectangular channel of width w is relatively large, it is possible for the flow to "jump" from a depth y_1 to a depth y_2 over a relatively short distance, as shown in Fig. E4.12; this is referred to as a *hydraulic jump*. Express y_2 in terms of y_1 and V_1; assume a horizontal uniform flow.

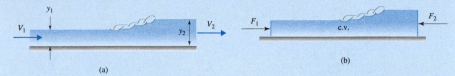

(a)

(b)

FIGURE E4.12

Solution

A control volume is selected as shown with inlet and exit areas upstream and downstream of the "jump" sufficiently far that the streamlines are parallel to the wall with hydrostatic pressure distributions. Neglecting the drag that is present on the walls (if the distance between sections is relatively small, the drag force should be negligible), the momentum equation can be manipulated as follows:

$$\Sigma F_x = \dot{m}(V_{2x} - V_{1x})$$

$$F_1 - F_2 = \rho A_1 V_1(V_2 - V_1)$$

$$\gamma \frac{y_1}{2}(y_1 w) - \gamma \frac{y_2}{2}(y_2 w) = \rho y_1 w V_1\left(V_1 \frac{y_1}{y_2} - V_1\right)$$

where we have expressed F_1 and F_2 using Eq. 2.4.24, and continuity in the form of Eq. 4.4.6, so that

$$V_2 = \frac{y_1}{y_2} V_1$$

The above momentum equation can be simplified to

$$\frac{\gamma}{2}(y_1^2 - y_2^2) = \rho y_1 V_1^2 \frac{y_1 - y_2}{y_2}$$

or

$$\frac{g}{2}(y_1 - y_2)(y_1 + y_2) = \frac{y_1}{y_2} V_1^2 (y_1 - y_2)$$

The factor $(y_1 - y_2)$ is divided out and y_2 is found assuming y_1 and V_1 are known as follows:

$$\frac{g}{2}(y_1 + y_2) = \frac{y_1}{y_2} V_1^2$$

$$y_2^2 + y_1 y_2 - \frac{2}{g} y_1 V_1^2 = 0$$

$$\therefore y_2 = \frac{1}{2}\left(-y_1 + \sqrt{y_1^2 + \frac{8}{g} y_1 V_1^2}\right)$$

where the quadratic formula has been used. The energy equation could now be used to provide an expression for the losses in the hydraulic jump.

Example 4.13

Consider the symmetrical flow of air around the cylinder. The control volume, excluding the cylinder, is shown in Fig. E4.13. The velocity distribution downstream of the cylinder is approximated with the parabola, as shown. Determine the drag force per meter of length acting on the cylinder. Use $\rho = 1.23$ kg/m³.

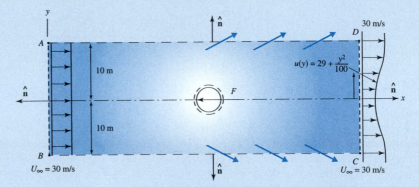

FIGURE E4.13

Solution

First, we must recognize that not all of the mass flux entering through AB exits through CD; consequently, some mass flux must exit AD and BC, as shown. The momentum equation (4.6.2) for the steady flow, applied to the control volume $ABCD$, takes the form

$$-F = \int_{c.s.} \rho V_x \mathbf{V} \cdot \hat{\mathbf{n}} \, dA = \int_{A_{CD}} \rho u \mathbf{V} \cdot \hat{\mathbf{n}} \, dA + \int_{A_{AD}} \rho u \mathbf{V} \cdot \hat{\mathbf{n}} \, dA + \int_{A_{BC}} \rho u \mathbf{V} \cdot \hat{\mathbf{n}} \, dA$$

$$+ \int_{A_{AB}} \rho u \mathbf{V} \cdot \hat{\mathbf{n}} \, dA$$

$$= \int_{A_{CD}} \rho u^2 \, dA + U_\infty \dot{m}_{AD} + U_\infty \dot{m}_{BC} - \int_{A_{AB}} \rho u^2 \, dA$$

$$= 2 \int_0^{10} 1.23 \left(29 + \frac{y^2}{100} \right)^2 dy + 2 \times 30 \dot{m}_{AD} - 1.23 \times 30^2 \times 20$$

where $\dot{m}_{BC} = \dot{m}_{AD}$ is the mass flux crossing BC and AD with the x-component velocity equal to 30 m/s. We have used Eq. 4.4.9 for $\dot{m}_{AD}$ and $\dot{m}_{BC}$ recognizing that $\mathbf{V} \cdot \hat{\mathbf{n}} = V_n$ which would be the small y-component velocity. We now use continuity to find $\dot{m}_{AD}$:

$$0 = \int \rho \hat{\mathbf{n}} \cdot \mathbf{V} \, dA = \int_{A_{AD}} \rho \hat{\mathbf{n}} \cdot \mathbf{V} \, dA + \int_{A_{BC}} \rho \hat{\mathbf{n}} \cdot \mathbf{V} \, dA + \int_{A_{AB}} \rho \hat{\mathbf{n}} \cdot \mathbf{V} \, dA + \int_{A_{CD}} \rho \hat{\mathbf{n}} \cdot \mathbf{V} \, dA$$

$$= \dot{m}_{AD} + \dot{m}_{BC} + 2 \int_0^{10} \rho u(y) \, dy - \rho \times 20 \times 30$$

$$= 2 \dot{m}_{AD} + 2 \int_0^{10} 1.23 \times \left(29 + \frac{y^2}{100} \right) dy - 1.23 \times 20 \times 30$$

$$\therefore \dot{m}_{AD} = 8.2 \text{ kg/s per meter of length}$$

Evaluating the terms in the momentum equation above gives us

$$F = -21\,170 - 492 + 22\,140$$

$$= 478 \text{ N/m}$$

Example 4.14

Find an expression for the head loss in a sudden expansion in a pipe in terms of V_1 and the area ratio (Fig. E4.14a). Assume uniform velocity profiles and assume that the pressure at the sudden enlargement is p_1.

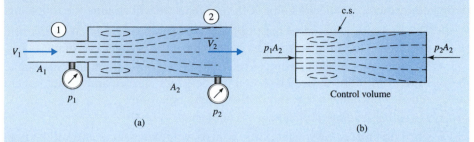

(a)

(b)

FIGURE E4.14

Solution

Figure E4.14a shows a sudden expansion with the diameter changing from d_1 to d_2. The pressure at the sudden enlargement is closest to p_1 since the streamlines are approximately parallel as shown (there is no pressure variation normal to parallel streamlines); they take some distance to again fill the pipe. Hence the force acting on the left end of the control volume shown in Fig. E4.14b is $p_1 A_2$. Newton's second law applied to the control volume yields, assuming uniform profiles,

$$\Sigma F_x = \dot{m}(V_2 - V_1)$$

$$(p_1 - p_2)A_2 = \rho A_2 V_2 (V_2 - V_1)$$

$$\therefore \frac{p_1 - p_2}{\rho} = V_2(V_2 - V_1)$$

The energy equation (4.5.17) provides

$$0 = \frac{V_2^2 - V_1^2}{2g} + \frac{p_2 - p_1}{\gamma} + z_2 \cancel{}^{0} z_1 + h_L$$

$$\therefore h_L = \frac{p_1 - p_2}{\gamma} - \frac{V_2^2 - V_1^2}{2g}$$

$$= \frac{V_2(V_2 - V_1)}{g} - \frac{(V_2 + V_1)(V_2 - V_1)}{2g} = \frac{(V_1 - V_2)^2}{2g}$$

To express this in terms of only V_1, we can use continuity and relate

$$V_2 = \frac{A_1}{A_2} V_1$$

Then the expression above for the head loss becomes

$$h_L = \left(1 - \frac{A_1}{A_2}\right)^2 \frac{V_1^2}{2g}$$

4.6.3   Momentum Equation Applied to Deflectors

The application of the momentum equation to deflectors forms an integral part of the analysis of many turbomachines, such as turbines, pumps, and compressors. In this section we illustrate the steps in such an analysis. It will be separated into two parts: fluid jets deflected by stationary deflectors and fluid jets deflected by moving deflectors. For both problems we will assume the following:

- The pressure external to the fluid jets is everywhere constant so that the pressure in the fluid as it moves over a deflector remains constant.
- The frictional resistance due to the fluid-deflector interaction is negligible so that the relative speed between the deflector surface and the jet stream remains unchanged, a result of Bernoulli's equation.
- Lateral spreading of a plane jet is neglected.
- The body force, the weight of the control volume, is small and will be neglected.

Stationary Deflector.   Let us first consider the stationary deflector, illustrated in Fig. 4.13. Bernoulli's equation allows us to conclude that the magnitudes of the velocity vectors are equal (i.e., $V_2 = V_1$), since the pressure is assumed to be constant external to the fluid jet and elevation changes are negligible (see Eq. 3.4.9). Assuming steady, uniform flow the momentum equation takes the form of Eq. 4.6.6, which for the x- and y-directions becomes

$$-R_x = \dot{m}(V_2 \cos \alpha - V_1) = \dot{m}V_1(\cos \alpha - 1)$$

$$R_y = \dot{m}V_2 \sin \alpha = \dot{m}V_1 \sin \alpha \tag{4.6.9}$$

For given jet conditions the reactive force components can be calculated.

Moving Deflectors.   The situation involving a moving deflector depends on whether a single deflector is moving (a water scoop used to slow a high-speed train) or whether a series of deflectors is moving (the vanes on a turbine). Let us first consider a single deflector shown in Fig. 4.14 to be moving in the positive x-direction with the speed V_B. In a reference frame attached to the stationary nozzle, from which the fluid jet issues, the flow is unsteady; that is, at a particular point in space, the flow situation varies with time.[2] A steady flow is observed,

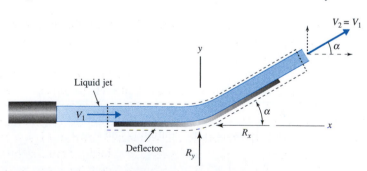

FIGURE 4.13   Stationary deflector.

<hr>

[2]This is a rather subtle point. To determine whether a flow is steady, we observe the flow at a given point in space. If a flow property changes with time at that point, the flow is unsteady. In this situation, if we focus our attention on a particular point just before the blade, such as point A in Fig. 4.14, first there is no flow, then the blade and the jet pass through the point; then there is again no flow. This is an unsteady flow.

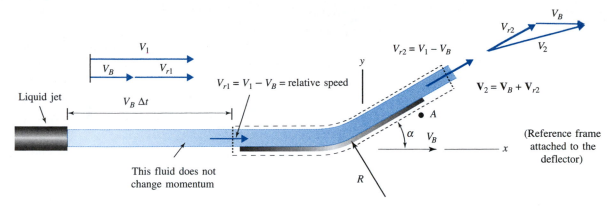

FIGURE 4.14 Moving deflector.

however, from a reference frame attached to the deflector. From this inertial reference frame, moving with the constant velocity V_B, we observe the relative speed V_{r1} entering the control volume to be $V_1 - V_B$, as shown. It is this relative speed that remains constant as the fluid flows relative to the deflector; it does not change since the pressure does not change. Hence, from this moving frame, the momentum equation takes the forms

$$-R_x = \dot{m}_r(V_1 - V_B)(\cos \alpha - 1)$$
$$R_y = \dot{m}_r(V_1 - V_B) \sin \alpha \qquad\qquad (4.6.10)$$

where $\dot{m}_r$ represents only that part of the mass flux exiting the fixed jet that has its momentum changed. Since the deflector moves away from the fixed jet some of the fluid that exits the fixed jet never experiences a momentum change; this fluid is represented by the distance $V_B\Delta t$, shown in Fig. 4.14. Hence

$$\dot{m}_r = \rho A(V_1 - V_B) \qquad\qquad (4.6.11)$$

where the relative speed $(V_1 - V_B)$ is used in the calculation; the mass flux $\rho A V_B$ is subtracted from the exiting mass flux $\rho A V_1$ to provide the mass flux $\dot{m}_r$ that experiences a momentum change.

For a series of vanes (a cascade) the jets may be oriented to the side, as shown in Fig. 4.15. The actual force on a particular vane would be zero until the jet strikes the vane; then the force would increase to a maximum and decrease to zero as the vane leaves the jet. We will idealize the situation as follows: Assume that, on the average, the jet is deflected by the vanes as shown in Figs. 4.15 and 4.16a as viewed from a stationary reference frame; the fluid jet enters the vanes with an angle β_1 and exits with an angle β_2. What is desired, however, is that the relative velocity enter the vanes tangent to the leading edge of the vanes, that is, V_{r1} in Fig. 4.16b is at the angle α_1. The relative speed then remains constant as the fluid travels over the vane with the exiting relative velocity V_{r2} leaving with the vane angle α_2. The relative and absolute velocities are related with the velocity equations which are displayed by the velocity polygons of Fig. 4.16b and c.

KEY CONCEPT *The relative speed remains constant as the fluid travels over a moving vane.*

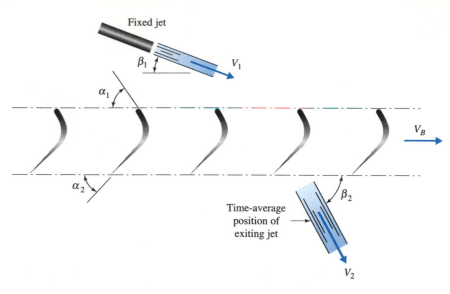

FIGURE 4.15 Fluid striking a series of vanes.

Assuming that all of the mass exiting the fixed jet has its momentum changed, we can write the momentum equation as

$$-R_x = \dot{m} \, (V_{2x} - V_{1x}) \tag{4.6.12}$$

Example 4.17 will illustrate the details.

Interest is usually focused on the x-component of force since it is this component that is related to the power output (or requirement). The power would be found by multiplying the x-component force by the blade speed for each jet; this takes the form

$$\dot{W} = NR_x V_B \tag{4.6.13}$$

where N represents the number of jets. The y-component force does not move in the y-direction so it does no work.

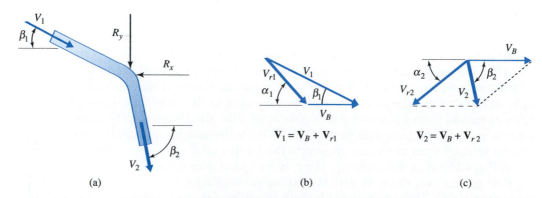

V₁ = V_B + V_r1 $\mathbf{V}_1 = \mathbf{V}_B + \mathbf{V}_{r1}$

$\mathbf{V}_2 = \mathbf{V}_B + \mathbf{V}_{r2}$

(a)   (b)   (c)

FIGURE 4.16 Detail of the flow situation involving a series of vanes: (a) average position of jet; (b) entrance velocity polygon; (c) exit velocity polygon.

Example 4.15

A deflector turns a sheet of water through an angle of 30° as shown in Fig. E4.15. What force is necessary to hold the deflector in place if $\dot{m} = 32$ kg/s?

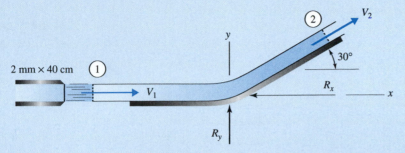

FIGURE E4.15

Solution

The control volume we have selected includes the deflector and the water adjacent to it. The only force that is acting on the control volume is due to a support needed to hold the deflector. This force has been decomposed into R_x and R_y.

The velocity V_1 is found to be

$$V_1 = \frac{\dot{m}}{\rho A_1}$$

$$= \frac{32}{1000 \times 0.002 \times 0.4} = 40 \text{ m/s}$$

Bernoulli's equation (3.4.8) shows that if the pressure does not change, then the magnitude of the velocity does not change, provided that there is no significant change in elevation and that viscous effects are negligible; thus we can conclude that $V_2 = V_1$ since $p_2 = p_1$. Next, the momentum equation is applied in the x-direction to find R_x and then in the y-direction for R_y:

x-direction:
$$-R_x = \dot{m}\,(V_{2x} - V_{1x})$$

$$= 32(40 \cos 30° - 40)$$

$$\therefore R_x = 172 \text{ N}$$

y-direction:
$$R_y = \dot{m}(V_{2y} - \overset{0}{\cancel{V_{1y}}})$$

$$= 32(40 \sin 30°) = 640 \text{ N}$$

Example 4.16

The deflector shown in Fig. E4.16 moves to the right at 30 m/s while the nozzle remains stationary. Determine (a) the force components needed to support the deflector, (b) $\mathbf{V}_2$ as observed from a fixed observer, and (c) the power generated by the vane. The jet velocity is 80 m/s.

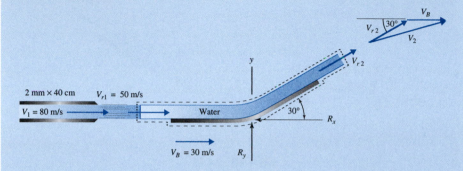

FIGURE E4.16

Solution

(a) To solve the problem of a moving deflector, we observe the flow from a reference frame attached to the deflector. In this moving reference frame the flow is steady and Bernoulli's equation with $p_1 = p_2$ can then be used to show that $V_{r1} = V_{r2} = 50$ m/s, the velocity of the sheet of water as observed from the deflector. Note that we cannot apply Bernoulli's equation in a fixed reference frame since the flow would not be steady. Applying the momentum equation to the moving control volume, which is indicated again by the dashed line, we obtain the following:

$$x\text{-direction:} \qquad -R_x = \dot{m}_r[(V_{r2})_x - (V_{r1})_x]$$
$$= 1000 \times 0.002 \times 0.4 \times 50(50 \cos 30° - 50)$$
$$\therefore R_x = 268 \text{ N}$$

$$y\text{-direction:} \qquad R_y = \dot{m}_r[(V_{r2})_y - (\overset{0}{\cancel{V_{r1}}})_y]$$
$$= 1000 \times 0.002 \times 0.4 \times 50(50 \sin 30°) = 1000 \text{ N}$$

When calculating $\dot{m}_r$ we must use only that water which has its momentum changed; hence the velocity used is 50 m/s.

(b) Observed from a fixed observer the velocity $\mathbf{V}_2$ of the fluid after the deflection is $\mathbf{V}_2 = \mathbf{V}_{r2} + \mathbf{V}_B$, where $\mathbf{V}_{r2}$ is directed tangential to the deflector at the exit and has a magnitude equal to V_{r1} (see the velocity diagram above). Thus

$$(V_2)_x = V_{r2} \cos 30° + V_B$$
$$= 50 \times 0.866 + 30 = 73.3 \text{ m/s}$$
$$(V_2)_y = V_{r2} \sin 30°$$
$$= 50 \times 0.5 = 25 \text{ m/s}$$

Finally,

$$\mathbf{V}_2 = 73.3\,\hat{\mathbf{i}} + 25\,\hat{\mathbf{j}} \text{ m/s}$$

(c) The power generated by the moving vane is equal to the velocity of the vane times the force the vane exerts in the direction of the motion. Therefore,

$$\dot{W} = V_B \times R_x = 30 \times 268 = 8040 \text{ W}$$

Example 4.17

High-speed air jets strike the blades of a turbine rotor tangentially while the 1.5-m-diameter rotor rotates at 140 rad/s (Fig. E4.17a). There are 10 such 4-cm-diameter jets. Calculate the maximum power output. The air density is 2.4 kg/m^3.

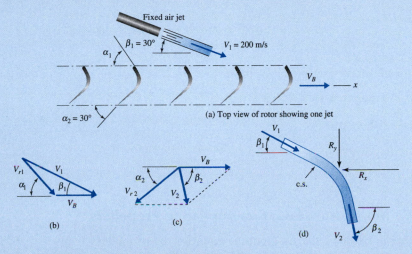

(a) Top view of rotor showing one jet

(b)   (c)   (d)

FIGURE E4.17

Solution

The blade angle α_1 is set by demanding that the air jet enter the blades tangentially, as observed from the moving blade; that is, the relative velocity vector $\mathbf{V}_r$ must make the angle α_1 with respect to the blade velocity $\mathbf{V}_B$. This is shown in Fig. E4.17b. The relative entrance velocity is $\mathbf{V}_{r1}$ (Fig. E4.17b) and the relative exit velocity is $\mathbf{V}_{r2}$ (Fig. E4.17c). Both velocity polygons are presented by the vector equation

$$\mathbf{V} = \mathbf{V}_r + \mathbf{V}_B$$

which states that the absolute velocity equals the relative velocity plus the blade velocity. From the polygon at the entrance we have

$$V_1 \sin \beta_1 = V_{r1} \sin \alpha_1$$

$$V_1 \cos \beta_1 = V_{r1} \cos \alpha_1 + V_B$$

$$\therefore 200 \sin 30° = V_{r1} \sin \alpha_1$$

$$200 \cos 30° = V_{r1} \cos \alpha_1 + 0.75 \times 140$$

where V_B is the radius multiplied by the angular velocity. A simultaneous solution yields

$$V_{r1} = 121 \text{ m/s} \qquad \alpha_1 = 55.7°$$

The friction between the air and the blade is quite small and can be neglected when calculating the maximum output. This allows us to assume $V_{r2} = V_{r1}$. From the exiting velocity polygon we can write

$$V_B - V_{r2} \cos \alpha_2 = V_2 \cos \beta_2$$

$$V_{r2} \sin \alpha_2 = V_2 \sin \beta_2$$

$$\therefore 0.75 \times 140 - 121 \cos 30° = V_2 \cos \beta_2$$

$$121 \sin 30° = V_2 \sin \beta_2 \qquad \textit{(continued)}$$

A simultaneous solution results in

$$V_2 = 60.5 \text{ m/s} \qquad \beta_2 = 89.8°$$

The momentum equation applied to the control volume, shown in Fig. E4.17d, gives

$$-R_x = \dot{m}(V_{2x} - V_{1x})$$
$$= 2.4 \times \pi \times 0.02^2 \times 200(60.5 \cos 89.8° - 200 \cos 30°)$$
$$\therefore R_x = 104.3 \text{ N}$$

There are 10 jets, each producing the force above. The maximum power output is then

$$\text{power} = 10 \times R_x \times V_B$$
$$= 10 \times 104.3 \times (0.75 \times 140) = 109\ 600 \text{ W} \quad \text{or} \quad 109.6 \text{ kW}$$

4.6.4 Momentum Equation Applied to Propellers

The application of the momentum equation to propellers is also of sufficient interest that this section will be devoted to illustrating the procedure. Consider the propeller of Fig. 4.17 with the streamlines shown forming the surface of a control volume in which the fluid enters with a uniform velocity V_1 and exits with a uniform velocity V_2. This flow situation can be seen to be identical to that of a propeller moving with velocity V_1 in a stagnant fluid by adding V_1 to the left in Fig. 4.17. The momentum equation, applied to the large control volume shown, gives

$$F = \dot{m}(V_2 - V_1) \tag{4.6.14}$$

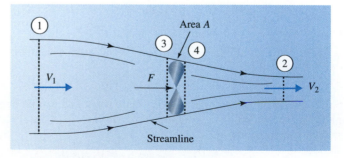

FIGURE 4.17 Propeller in a fluid flow.

This control volume is not sufficient, however, since the areas A_1 and A_2 are unknown. We know the flow area A of the propellar. So a control volume is drawn close to the propeller such that $V_3 \cong V_4$ and $A_3 \cong A_4 = A$. The momentum equation (4.6.6) in the x-direction gives

$$F + p_3 A - p_4 A = 0 \qquad (4.6.15)$$

or

$$F = (p_4 - p_3)A \qquad (4.6.16)$$

Now, since viscous effects would be quite small in this flow situation the energy equation up to the propeller and then downstream from the propeller is used to obtain

$$\frac{V_1^2 - V_3^2}{2} + \frac{p_1 - p_3}{\rho} = 0 \qquad \frac{V_4^2 - V_2^2}{2} + \frac{p_4 - p_2}{\rho} = 0 \qquad (4.6.17)$$

Adding these equations together, recognizing that $p_1 = p_2 = p_{atm}$, we have

$$(V_2^2 - V_1^2)\frac{\rho}{2} = p_4 - p_3 \qquad (4.6.18)$$

Inserting this and Eq. 4.6.16 into Eq. 4.6.14 results in

$$V_3 = \frac{1}{2}(V_2 + V_1) \qquad (4.6.19)$$

KEY CONCEPT *The velocity of the fluid moving through the propeller is the average of the upstream and downstream velocities.*

where we have used $\dot{m} = \rho A V_3$ since the propeller area is the only area known. This result shows that the velocity of the fluid moving through the propeller is the average of the upstream and downstream velocities.

The input power needed to produce this effect is found by applying the energy equation between sections 1 and 2, where the pressures are atmospheric; neglecting losses, Eq. 4.5.17 takes the form

$$\dot{W}_{fluid} = \frac{V_2^2 - V_1^2}{2}\dot{m} \qquad (4.6.20)$$

where $\dot{W}_{\text{fluid}}$ is the energy input between the two sections. The moving propeller requires power given by

$$\dot{W}_{\text{prop}} = F \times V_1$$
$$= \dot{m}V_1(V_2 - V_1) \tag{4.6.21}$$

The theoretical propeller efficiency is then

$$\eta_P = \frac{\dot{W}_{\text{prop}}}{\dot{W}_{\text{fluid}}} = \frac{V_1}{V_3} \tag{4.6.22}$$

KEY CONCEPT   *In a wind machine, the downstream velocity is reduced and the diameter is increased.*

In contrast to the propeller, a wind machine extracts energy from the airflow; the downstream velocity is reduced and the diameter is increased.

4.6.5   Steady Nonuniform Flow

If we cannot assume uniform velocity profiles, we can let

$$\int_A V^2 \, dA = \beta \overline{V}^2 A \tag{4.6.23}$$

where we have introduced the *momentum-correction factor* β, expressed explicitly as

$$\beta = \frac{\int V^2 \, dA}{\overline{V}^2 A} \tag{4.6.24}$$

The momentum equation (4.6.6), for a steady flow with one inlet and one exit, can then be written as

$$\Sigma \mathbf{F} = \dot{m}(\beta_2 \mathbf{V}_2 - \beta_1 \mathbf{V}_1) \tag{4.6.25}$$

For a laminar flow with a parabolic profile in a circular pipe, $\beta = \frac{4}{3}$. If a profile is given, however, the integral is usually simply integrated and Eq. 4.6.2 is used.

Example 4.18

Calculate the momentum correction factor for a parabolic profile (a) between parallel plates and (b) in a circular pipe. The parabolic profiles are shown in Fig. E4.18.

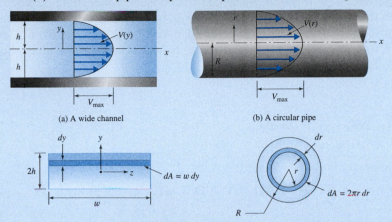

(a) A wide channel   (b) A circular pipe

FIGURE E4.18

Solution

(a) A parabolic profile between parallel plates can be expressed as

$$V(y) = V_{max}\left(1 - \frac{y^2}{h^2}\right)$$

where y is measured from the centerline, the velocity is zero at the walls where $y = \pm h$, and V_{max} is the centerline velocity at $y = 0$. First, let us find the average velocity. It is

$$\overline{V} = \frac{1}{A}\int V\, dA$$

$$= \frac{1}{hw}\int_0^h V_{max}\left(1 - \frac{y^2}{h^2}\right)w\, dy = \frac{V_{max}}{h}\left(h - \frac{1}{3}h\right) = \frac{2}{3}V_{max}$$

where we have integrated over the top half of the cross-section.

Then

$$\beta = \frac{\int V^2\, dA}{\overline{V}^2 A} = \frac{2}{\frac{4}{9}V_{max}^2 \times 2hw}\int_0^h V_{max}^2\left(1 - \frac{y^2}{h^2}\right)^2 w\, dy = \frac{6}{5}$$

(b) For a circular pipe a parabolic profile can be written as

$$V(r) = V_{max}\left(1 - \frac{r^2}{R^2}\right)$$

where R is the pipe radius and $V = 0$ at $r = R$. The average velocity is found to be

$$\overline{V} = \frac{1}{A}\int V\, dA = \frac{1}{\pi R^2}\int_0^R V_{max}\left(1 - \frac{r^2}{R^2}\right)2\pi r\, dr = \frac{1}{2}V_{max}$$

The momentum correction factor is then

$$\beta = \frac{\int V^2\, dA}{\overline{V}^2 A} = \frac{1}{\frac{1}{4}V_{max}^2 \pi R^2}\int_0^R V_{max}^2\left(1 - \frac{r^2}{R^2}\right)^2 2\pi r\, dr = \frac{4}{3}$$

The correction factors above can be used to express the momentum flux across a cross-sectional area as $\beta\rho A\overline{V}^2$.

4.6.6 Noninertial Reference Frames

In certain situations it may be necessary to choose a noninertial reference frame in which the velocity is measured. This would be the case if we were to study the flow through a dishwasher arm, around a turbine blade, or from a rocket. Relative to a noninertial reference frame, Newton's second law takes the form (refer to Eq. 3.2.15)

$$\Sigma \mathbf{F} = \frac{D}{Dt} \int_{\text{sys}} \rho \mathbf{V} \, d\mathcal{V}$$

$$+ \int_{\text{sys}} \left[\frac{d^2\mathbf{S}}{dt^2} + 2\mathbf{\Omega} \times \mathbf{V} + \mathbf{\Omega} \times (\mathbf{\Omega} \times \mathbf{r}) + \frac{d\mathbf{\Omega}}{dt} \times \mathbf{r} \right] \rho \, d\mathcal{V} \quad (4.6.26)$$

where $\mathbf{V}$ is the velocity relative to the noninertial frame and where the acceleration $\mathbf{a}$ of each particle in the system is already accounted for in the first integral. Equation 4.6.26 is often written as

$$\Sigma \mathbf{F} - \mathbf{F}_I = \frac{D}{Dt} \int_{\text{sys}} \rho \mathbf{V} \, d\mathcal{V}$$

$$= \frac{d}{dt} \int_{\text{c.v.}} \rho \mathbf{V} \, d\mathcal{V} + \int_{\text{c.s.}} \rho \mathbf{V} (\mathbf{V} \cdot \hat{\mathbf{n}}) \, dA \quad (4.6.27)$$

where $\mathbf{F}_I$ is called the "inertial body force," given by

$$\mathbf{F}_I = \int_{\text{sys}} \left[\frac{d^2\mathbf{S}}{dt^2} + 2\mathbf{\Omega} \times \mathbf{V} + \mathbf{\Omega} \times (\mathbf{\Omega} \times \mathbf{r}) + \frac{d\mathbf{\Omega}}{dt} \times \mathbf{r} \right] \rho \, d\mathcal{V} \quad (4.6.28)$$

Since the system and control volume are identical at time t the system integration can be replaced with a control volume integration in the integral of Eq. 4.6.28. Example 4.19 will illustrate the use of a noninertial reference frame.

Example 4.19

The rocket shown in Fig. E4.19, with an initial mass of 150 kg, burns fuel at the rate of 10 kg/s with a constant exhaust velocity of 700 m/s. What is the initial acceleration of the rocket and the velocity after 4 s? Neglect the drag on the rocket.

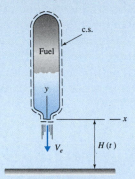

FIGURE E4.19

Solution

The control volume is sketched and includes the entire rocket. The reference frame attached to the rocket is accelerating upward at d^2H/dt^2. Newton's second law is written as, using z upward,

$$\Sigma F_z - (F_I)_z = \frac{d}{dt} \int_{c.v.} \rho V_z \, dV + \int_{c.s.} \rho V_z \mathbf{V} \cdot \hat{\mathbf{n}} \, dA$$

$$\therefore -W - \frac{d^2H}{dt^2} M_{c.v.} = \rho_e(-V_e)V_e A_e$$

where

$$\frac{d}{dt} \int_{c.v.} \rho V_z dV \approx 0$$

since V_z is the velocity of each mass element $\rho \, dV$ relative to the reference frame attached to the control volume; the only vertical force is the weight W; and $M_{c.v.}$ is the mass of the control volume. From continuity we see that

$$M_{c.v.} = 150 - \dot{m}t = 150 - 10t$$

$$\therefore W = (150 - 10t) \times 9.81$$

The momentum equation becomes

$$-(150 - 10t) \times 9.81 - \frac{d^2H}{dt^2}(150 - 10t) = -\dot{m}_e V_e = -10 \times 700 = -7000$$

This is written as

$$\frac{d^2H}{dt^2} = \frac{700}{15 - t} - 9.81$$

The initial acceleration is found by letting $t = 0$:

$$\frac{d^2H}{dt^2}\bigg|_{t=0} = \frac{700}{15} - 9.81 = 36.9 \text{ m/s}^2$$

Integrate the expression for d^2H/dt^2 and obtain

$$\frac{dH}{dt} = -700 \ln (15 - t) - 9.81t + C$$

The constant $C = 700 \ln 15$ since $dH/dt = 0$ at $t = 0$. Thus at $t = 4$ s the velocity is

$$\frac{dH}{dt} = 700 \ln \frac{15}{11} - 9.81 \times 4 = 178 \text{ m/s}$$

4.7 MOMENT-OF-MOMENTUM EQUATION

In the preceding section we determined the magnitude of force components in a variety of flow situations. To determine the line of action of a given force component, it is often necessary to apply the moment-of-momentum equation. Also, in analyzing the flow situation in devices that have rotating components the moment-of-momentum equation is needed to relate the rotational speed to the other flow parameters. Since it may be advisable to attach the reference frame to the rotating component, we will write the general equation with the inertial forces included. It is (see Eq. 4.2.4)

$$\Sigma \mathbf{M} - \mathbf{M}_I = \frac{D}{Dt} \int_{\text{sys}} \mathbf{r} \times \mathbf{V} \, \rho \, d\mathcal{V} \tag{4.7.1}$$

where

$$\mathbf{M}_I = \int \mathbf{r} \times \left[\frac{d^2\mathbf{S}}{dt^2} + 2\boldsymbol{\Omega} \times \mathbf{V} + \boldsymbol{\Omega} \times (\boldsymbol{\Omega} \times \mathbf{r}) + \frac{d\boldsymbol{\Omega}}{dt} \times \mathbf{r} \right] \rho \, d\mathcal{V} \tag{4.7.2}$$

KEY CONCEPT *The inertial moment* $\mathbf{M}_I$ *accounts for the fact that a noninertial reference frame was selected.*

This inertial moment $\mathbf{M}_I$ accounts for the fact that a noninertial reference frame was selected; it is simply the moment of $\mathbf{F}_I$ (see Eq. 4.6.28). Applying the system-to-control volume transformation, the moment-of-momentum equation for a control volume becomes

$$\Sigma \mathbf{M} - \mathbf{M}_I = \frac{d}{dt} \int_{\text{c.v.}} \mathbf{r} \times \mathbf{V} \, \rho \, d\mathcal{V} + \int_{\text{c.s.}} \mathbf{r} \times \mathbf{V}(\mathbf{V} \cdot \hat{\mathbf{n}}) \, \rho \, dA \tag{4.7.3}$$

Examples will illustrate the application of this equation.

Example 4.20

A sprinkler has four 50-cm-long arms with nozzles at right angles with the arms and 45° with the ground (Fig. E4.20). If the total flow rate is 0.01 m³/s and a nozzle exit diameter is 12 mm, find the rotational speed of the sprinkler. Neglect friction.

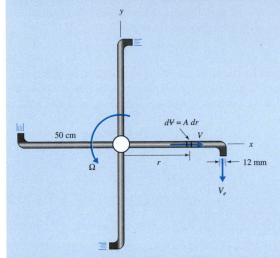

FIGURE E4.20

Solution

The velocity exiting a nozzle as shown is

$$V_e = \frac{Q}{A}$$

$$= \frac{0.01/4}{\pi \times 0.006^2} = 22.1 \text{ m/s}$$

where the factor 4 accounts for the four exit areas. Attach the reference frame to the rotating arms as shown. Then, recognizing that $\mathbf{r} \times [\mathbf{\Omega} \times (\mathbf{\Omega} \times \mathbf{r})] = 0$ and assuming a stationary sprinkler so that $d^2\mathbf{S}/dt^2 = 0$ and constant angular velocity so that $d\mathbf{\Omega}/dt = 0$, we have

$$\mathbf{M}_I = \int_{c.v.} \mathbf{r} \times (2\mathbf{\Omega} \times \mathbf{V})\rho \, d\mathcal{V}$$

$$= 4 \int_0^{0.5} r\hat{\mathbf{i}} \times (2\Omega\hat{\mathbf{k}} \times V\hat{\mathbf{i}})\rho A \, dr$$

$$= 8\rho AV\Omega\hat{\mathbf{k}} \int_0^{0.5} r \, dr = \rho AV\Omega\hat{\mathbf{k}}$$

where the factor 4 again accounts for the four arms (each arm would provide the unit vector $\hat{\mathbf{k}}$). Since there are no external moments applied to the sprinkler about the vertical z-axis, $\Sigma M_z = 0$. For the steady flow Eq. 4.7.3 provides

$$(\Sigma \mathbf{M})_z^{\;0} - (\mathbf{M}_I)_z = \int_{c.s.} (\mathbf{r} \times \mathbf{V})_z \, \mathbf{V} \cdot \hat{\mathbf{n}} \, \rho \, dA$$

$$-\rho AV\Omega = 4 \int_{A_{\text{exit}}} [0.5\hat{\mathbf{i}} \times (0.707V_e\hat{\mathbf{k}} - 0.707V_e\hat{\mathbf{j}})]_z \, V_e\rho \, dA$$

$$-VA\Omega = -4 \times 0.5 \times 0.707V_e^2 A_e$$

$$\therefore \Omega = 4 \times 0.5 \times 0.707 \times 22.1 = 31.25 \text{ rad/s}$$

where we have used $AV = A_e V_e$ from continuity considerations. Note that we have neglected the small mass in the short nozzles on the ends of the arms.

Example 4.21

The nozzles of Example 4.20 make an angle of $0°$ with the ground and $90°$ with the arms. The water is suddenly turned on at $t = 0$ with the sprinkler motionless. Determine the resulting $\Omega(t)$ if the arm diameter is 24 mm. Neglect friction.

Solution

The reference frame is again attached to the rotating arms, as sketched in Example 4.20. Referring to the control volume integral of Eq. 4.7.3, we observe that $\mathbf{r} \times \mathbf{V} = 0$ since $\mathbf{r}$ is in the same direction as $\mathbf{V}$ along an arm. Thus Eq. 4.7.3, along with Eq. 4.7.2, takes the form

$$\Sigma \overset{0}{\cancel{\mathbf{M}}} - 4\int_0^{0.5} r\hat{\imath} \times \left[2\Omega\hat{\mathbf{k}} \times V\hat{\imath} + \Omega\hat{\mathbf{k}} \times (\Omega\hat{\mathbf{k}} \times r\hat{\imath}) + \frac{d\Omega}{dt}\hat{\mathbf{k}} \times r\hat{\imath}\right]\rho A\, dr$$

$$= \frac{d}{dt}\int_{\text{c.v.}} r\hat{\imath} \times V\hat{\imath}\,\rho\, d\mathcal{V} + 4\int_{A_{\text{exit}}} 0.5\hat{\imath} \times V_e(-\hat{\mathbf{j}})V_e\rho\, dA$$

Perform the vector operations and divide by 4ρ,

$$-2AV\Omega\int_0^{0.5} r\, dr - \frac{d\Omega}{dt}A\int_0^{0.5} r^2\, dr = -0.5V_e^2 A_e$$

The required integration, using $AV = A_eV_e = 0.01 \text{ m}^3/\text{s}$ and $V_e = 22.1$ m/s gives

$$\frac{d\Omega}{dt} + 132.6\Omega = 5862$$

This linear, first-order differential equation is solved by adding the homogeneous solution (suppress the right-hand side) to the particular solution to obtain

$$\Omega(t) = Ce^{-132.6t} + 44.2$$

Using the initial condition $\Omega(0) = 0$, we find that $C = -44.2$. Then

$$\Omega(t) = 44.2(1 - e^{-132.6t}) \text{ rad/s}$$

Observe that as time becomes large, the angular velocity is limited to 44.2 rad/s. If friction were included, this value would be reduced. If 44.2 is multiplied by 0.707 to account for the $45°$ angle, we obtain the value of Example 4.20.

4.8 SUMMARY

In this chapter we have presented the control-volume formulation of the fundamental laws. This formulation is useful when the integrands (the velocities and
pressure) are known or can be approximated with an acceptable degree of accuracy. If this is not the case, the differential equations of Chapter 5 must be solved
(numerically as in Chapter 15 or analytically as in Chapter 7), or experimental
methods must be used to obtain the desired information; much of the remainder
of this book is devoted to this task. After unknown velocities and pressures are
determined, we often return to the control-volume formulation and calculate
integral quantities of interest. Examples would include the lift and drag on an airfoil, the torque on a row of turbine blades, and the oscillating force on a suspension cable of a bridge.

   As we have observed in the examples and problems of this chapter, the task
of applying the control-volume equations is highly dependent on the appropriate
selection of the boundaries of the control volume. Such boundaries are selected
at locations where either information is known or where the unknowns appear.
Experience is often required in the selection of a control volume, as it is in the
selection of a free-body diagram in dynamics and solid mechanics. The student
has undoubtedly gained some of this experience while working through the sections of this chapter. Table 4.1 presents the various forms of the fundamental
laws to aid the user in selecting an appropriate form for a particular problem.

TABLE 4.1 Integral Forms of the Fundamental Laws

Continuity	Energy	Momentum

General Form

$$0 = \frac{d}{dt}\int_{c.v.} \rho\, dV + \int_{c.s.} \rho\mathbf{V}\cdot\hat{\mathbf{n}}\, dA$$

$$-\Sigma\dot{W} = \frac{d}{dt}\int_{c.v.}\left(\frac{V^2}{2}+gz\right)\rho\, dV$$

$$\Sigma\mathbf{F} = \frac{d}{dt}\int_{c.v.}\rho\mathbf{V}\, dV + \int_{c.s.}\rho\mathbf{V}(\mathbf{V}\cdot\hat{\mathbf{n}})\, dA$$

$$+\int_{c.s.}\left(\frac{V^2}{2}+\frac{p}{\rho}+gz\right)\rho\mathbf{V}\cdot\hat{\mathbf{n}}\, dA + \text{losses}$$

Steady Flow

$$0 = \int_{c.s.}\rho\mathbf{V}\cdot\hat{\mathbf{n}}\, dA$$

$$-\Sigma\dot{W} = \int_{c.s.}\left(\frac{V^2}{2}+\frac{p}{\rho}+gz\right)\rho\mathbf{V}\cdot\hat{\mathbf{n}}\, dA + \text{losses}$$

$$\Sigma\mathbf{F} = \int_{c.s.}\rho\mathbf{V}(\mathbf{V}\cdot\hat{\mathbf{n}})\, dA$$

Steady Nonuniform Flow[a]

$$\dot{m} = \rho_1 A_1 \overline{V}_1 = \rho_2 A_2 \overline{V}_2$$

$$\frac{-\Sigma\dot{W}}{\dot{m}g} = \alpha_2\frac{\overline{V}_2^2}{2g}+\frac{p_2}{\gamma_2}+z_2-\alpha_1\frac{\overline{V}_1^2}{2g}-\frac{p_1}{\gamma_1}-z_1+h_L$$

$$\Sigma F_x = \dot{m}(\beta_2\overline{V}_{2x}-\beta_1\overline{V}_{1x})$$

$$\Sigma F_y = \dot{m}(\beta_2\overline{V}_{2y}-\beta_1\overline{V}_{1y})$$

Steady Uniform Form[a]

$$\dot{m} = \rho_1 A_1 V_1 = \rho_2 A_2 V_2$$

$$-\frac{\Sigma\dot{W}}{\dot{m}g} = \frac{V_2^2}{2g}+\frac{p_2}{\gamma_2}+z_2-\frac{V_1^2}{2g}-\frac{p_1}{\gamma_1}-z_1+h_L$$

$$\Sigma\mathbf{F} = \dot{m}(\mathbf{V}_2-\mathbf{V}_1)$$

Steady Uniform Incompressible Flow[a]

$$Q = A_1 V_1 = A_2 V_2$$

$$-\frac{\Sigma\dot{W}}{\dot{m}g} = \frac{V_2^2}{2g}+\frac{p_2}{\gamma}+z_2-\frac{V_1^2}{2g}-\frac{p_1}{\gamma}-z_1+h_L$$

$$\Sigma\mathbf{F} = \dot{m}(\mathbf{V}_2-\mathbf{V}_1)$$

or

$$H_P + \frac{V_1^2}{2g}+\frac{p_1}{\gamma}+z_1 = H_T + \frac{V_2^2}{2g}+\frac{p_2}{\gamma}+z_2+h_L$$

$\dot{m}$ = mass flux	α = kinetic energy correction factor	h_L = head loss
Q = flow rate	$= \dfrac{\int V^3 dA}{\overline{V}^3 A}$	$\Sigma\dot{W} = \dot{W}_S + \dot{W}_{\text{shear}} + \dot{W}_I$
$\overline{V}$ = average velocity	β = momentum correction factor	H_P = pump head = $-\dot{W}_P/\dot{m}g$
$= \dfrac{\int V dA}{A}$	$= \dfrac{\int V^2 dA}{\overline{V}^2 A}$	H_T = turbine head $= \dot{W}_T/\dot{m}g$

[a]The control volume has one entrance (section 1) and one exit (section 2).

PROBLEMS

Basic Laws

4.1 **(a)** State the necessary conditions for momentum of a system to remain constant.
   (b) State the necessary conditions for energy of a system to remain constant.
   (c) Show the detailed steps and state the assumptions that allow Eq. 4.2.3 to be reduced to $\Sigma \mathbf{F} = m\mathbf{a}$.

4.2 Make a list of five extensive properties that are of interest in fluid mechanics. Also, list their associated intensive properties. In addition, list five other intensive properties.

4.3 Select the extensive property from the following:
   A. Temperature   **B.** Volume
   C. Pressure   **D.** Density

4.4 A control volume is identified as the volume interior to a balloon. At an instant the system is also identified as the air inside the balloon. Air escapes during a short time increment Δt. Sketch the system and the control volume at t and at $t + \Delta t$.

4.5 At an instant in time the control volume and the system occupy the volume inside the pump shown in Fig. P4.5 and a few diameters of the pipe at the inlet side. Sketch the system and the control volume at times t and $t + \Delta t$.

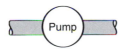

FIGURE P4.5

4.6 Indicate which fundamental equation would be most useful in determining the following quantity:
   (a) The horsepower output of a pump
   (b) The mass flux from closing baffles
   (c) The drag force on an airfoil
   (d) The head loss in a pipeline
   (e) The rotational speed of a wind machine

4.7 Sketch the unit vector $\hat{\mathbf{n}}$ and the velocity vector $\mathbf{V}$ on each of the areas listed:
   (a) The exit area of a fireman's nozzle
   (b) The inlet area of a pump
   (c) The wall area of a pipe
   (d) The porous bottom area of a river into which a small amount of water flows
   (e) The cylindrical exit area of a rotating impeller

4.8 Fluid flows through the enlargement shown in Fig. P4.8 with a velocity distribution $v_1(r)$ at the inlet and $v_2(r)$ at the exit. Sketch a control volume showing $\mathbf{V}$ and $\hat{\mathbf{n}}$ at selected locations on the control volume. Include locations on the lateral sides as well as the ends.

FIGURE P4.8

4.9 Sketch the unit vector $\hat{\mathbf{n}}$ and the velocity vector $\mathbf{V}$ at several positions on a rectangular box surrounding the airfoil shown in Fig. P4.9.

FIGURE P4.9

System-to-Control-Volume Transformation

4.10 Assume that $V_1 = V_2 = V_3 = 10$ m/s for the control volume shown in Fig. P4.10. Write $\hat{\mathbf{n}}_1$, $\hat{\mathbf{n}}_2$, and $\hat{\mathbf{n}}_3$ in terms of $\hat{\mathbf{i}}$, $\hat{\mathbf{j}}$, and $\hat{\mathbf{k}}$, and calculate the normal component of the velocity vector on each of the three plane areas. The volume is of uniform depth in the z-direction.

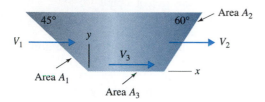

FIGURE P4.10

4.11 Write an expression for the flux of a property across each of the three areas of the control volume of Problem 4.10 if η and ρ are both constant over the entire control volume. Let the cross sectional area (normal to the xy-plane) be A. Use $V_1 = V_2 = V_3 = 10$ m/s.

4.12 Show that $(\mathbf{B} \cdot \hat{\mathbf{n}})A$ is the volume of the parallelepiped of depth 12 cm (Fig. P4.12). Note that $\hat{\mathbf{n}}$ is normal to the area A.

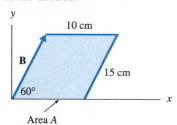

FIGURE P4.12

4.13 We recognize that

$$\frac{d}{dt} \int_{c.v.} \rho\eta \, dV = \int_{c.v.} \frac{\partial}{\partial t} \rho\eta \, dV$$

What condition allows this equivalence? Why is an ordinary derivative used on the left and a partial derivative used on the right?

4.14 A spray can for mosquitos is activated at $t = 0$. Select the chemical inside the can as the system and sketch the system at $t = \Delta t$. Select a control volume and sketch the control volume at $t = \Delta t$.

4.15 The air inside the lungs at the end of an inhaled breath is identified as the system at $t = 0$. Select a control volume and sketch both the system and control volume at $t = \Delta t$ if air is exhaled through the nose only.

4.16 The control volume selected to analyze flow around an airfoil is the rectangular box as sketched in the figure of Problem 4.9. The system occupies the box at t. Sketch the system at $t + \Delta t$.

Conservation of Mass

4.17 Show that Eq. 4.4.5 results from Eq. 4.4.4 by assuming one inlet and one exit and uniform flow (constant properties).

4.18 An incompressible fluid enters a volume filled with an absorbing material with a mass flux $\dot{m}$ and leaves the volume with a flow rate Q. Determine an expression for the rate of change of mass in the volume.

4.19 A liquid of density ρ flows into a volume filled with a sponge at a flow rate Q_1. It exits one area with a mass flux $\dot{m}_2$ and a second area A_3 with an average velocity V_3 as shown in Fig. P4.19. Write an expression for dm_{sponge}/dt, the rate of change of mass in the sponge.

FIGURE P4.19

4.20 Air flows through an 8-cm-diameter pipe with an average velocity of 70 m/s with a temperature of 20°C and a pressure of 200 kPa. The mass flux is nearest:

A. 3.7 kg/s B. 2.37 kg/s
C. 1.26 kg/s D. 0.84 kg/s

4.21 Water is flowing in a 6-cm-diameter pipe at 20 m/s. If the pipe enlarges to a diameter of 12 cm, calculate the reduced velocity. Also, calculate the mass flux and the flow rate.

4.22 Water flows in the 5-cm-diameter pipe shown in Fig. P4.22 with an average velocity of 10 m/s. It turns a 90° angle and flows radially between two parallel plates. What is the velocity at a radius of 60 cm? What are the mass flux and the discharge?

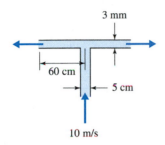

3 mm

60 cm

5 cm

10 m/s

FIGURE P4.22

4.23 A pipe transports 200 kg/s of water. The pipe tees into a 5-cm-diameter pipe and a 7-cm-diameter pipe (Fig. P4.23). If the average velocity in the smaller-diameter pipe is 25 m/s, calculate the flow rate in the larger pipe.

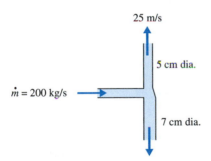

25 m/s

5 cm dia.

$\dot{m}$ = 200 kg/s

7 cm dia.

FIGURE P4.23

4.24 Air at 15°C and 275 kPa flows in a 10-cm-diameter pipe with a mass flux of 3 kg/s. The pipe undergoes a conversion to a 5 cm by 8 cm rectangular duct in which $T = 65°C$ and $p = 50$ kPa. Calculate the velocity in each section.

4.25 Air at 120°C and 500 kPa absolute is flowing in a pipe at 600 m/s and suddenly undergoes an abrupt change to 249°C and 1246 kPa absolute at a location where the diameter is 10 cm. Calculate the velocity after the sudden change (a shock wave) illustrated in Fig P4.25. Also, calculate the mass flux and the flow rates before and after the abrupt change.

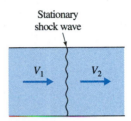

Stationary
shock wave

V_1 V_2

FIGURE P4.25

4.26 A laser velocimeter is used to measure velocities of 40 m/s and 120 m/s before and after an abrupt change in the diameter of a pipe from 10 cm to 6 cm, respectively. The pressure in the air before and after the change is measured to be 200 kPa and 120 kPa, respectively. If the temperature before the change is 20°C, what is the temperature after the change?

4.27 Water flows with a velocity of 3 m/s at a depth of 1.5 m in a trapezoidal channel with a 2-m base and sides sloping at 45°. It empties into a circular pipe and flows at 2 m/s. What is the diameter if:
(a) The pipe flows full?
(b) The pipe flows half full?
(c) The water in the pipe flows at a depth of one-half the radius?

4.28 Water flows with a velocity of 3 m/s in an 80-cm-diameter sewer pipe at a depth of 30 cm. The flow rate is nearest:
A. 516 L/s
B. 721 L/s
C. 938 L/s
D. 1262 L/s

4.29 Water flows in a 8-cm-diameter pipe with the profiles shown in Fig. P4.29. Find the average velocity, the mass flux, and the flow rate.

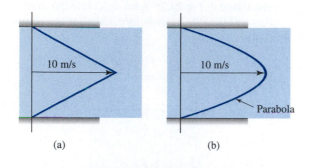

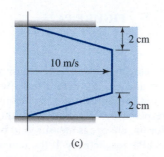

(c)

FIGURE P4.29

4.30 The profiles above are assumed to exist in a wide rectangular channel, 8 cm high and 80 cm wide. Find the average velocity, the mass flux, and the flow rate.

4.31 A constant-density fluid flows as shown in Fig. P4.31. Find the equation of the parabola if the conduit is:
(a) A pipe with $d = 2$ cm and $V = 2$ m/s
(b) A wide rectangular channel with $d = 2$ cm and $V = 2$ m/s

FIGURE P4.31

4.32 In Example 4.2 let $\dot{m}_2$ be unknown and V_1 and Q_3 be as shown in the figure. Calculate $\dot{m}_2$ so that dm/dt of the device is zero.

4.33 A parabolic profile exists in a 10-mm-diameter pipe. The pipe contracts to a 5-mm-diameter pipe in which the velocity profile is essentially uniform at 2 m/s. Write the equation for the parabola. Assume an incompressible flow.

4.34 As air flows as shown in Fig. P4.34 over a flat plate, the velocity is reduced to zero at the wall. If $u(y) = 10(20y - 100y^2)$ m/s, find the mass flux $\dot{m}$ through a surface parallel to the plate and 0.2 m above the plate. The plate is 2 m wide and $\rho = 1.23$ kg/m^3.

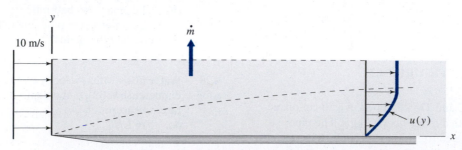

FIGURE 4.34

4.35 A streamline is 5 cm above the plate shown in Fig. P4.34 at the leading edge. How far from the plate is that same streamline at the location of the profile $u(y) = 10(20y - 100y^2)$?

4.36 Stratified salt water flows at a depth of 10 cm in a channel with a velocity distribution $0.5(6y - 9y^2)$ m/s, where y is measured in meters. If the density varies linearly from 2200 kg/m³ at the bottom to clear water at the top, find $\dot{m}$. Also, show that $\dot{m} \neq \bar{\rho}\,\bar{V}A$. The channel is 12 cm wide.

4.37 Water flows as shown in Fig. P4.37. Calculate V_2.

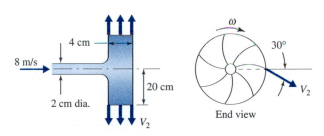

FIGURE P4.37

4.38 Rain is falling vertically downward with an average velocity of 5.0 m/s onto a 9000-m² parking lot. All of the water flows from the lot in an open rectangular ditch with an average velocity of 1.5 m/s. Estimate the depth of flow in the 1.5-m-wide ditch if two thousand 3-mm-diameter drops of water are contained in each cubic meter of rain.

4.39 Air at 250 kPa gage and 15°C is being forced into a tire, which has a volume of 0.5 m³, at a velocity of 60 m/s through a 6-mm-diameter opening. Determine the rate of change of density in the tire.

4.40 In Fig. P4.40, if the mass of the control volume is not changing, find $\bar{V}_3$.

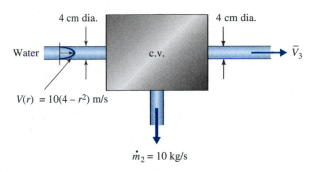

FIGURE P4.40

4.41 The average velocity $\bar{V}_3 = 10$ m/s in Problem 4.40. Find the rate at which the mass of the control volume is changing.

4.42 Find the velocity of the gas-fuel interface shown in Fig. P4.42. Use $R_{gas} = 0.28$ kJ/kg·K and $d_e = 30$ cm.

FIGURE P4.42

4.43 Conditioned air is delivered to a large conference room through four inlets, each transferring 2500 m³/hr. If the air is returned to the conditioner through a single 0.6 m × 1.2 m rectangular duct, estimate the average velocity in the duct. Make any needed assumptions.

4.44 A rectangular scoop, 80 cm deep, collects air as shown in Fig. P4.44, and delivers it through a 30-cm-diameter pipe. Estimate the average velocity of air in the pipe if $u(y) = 20\,y^{1/5}$ m/s, where y is in meters.

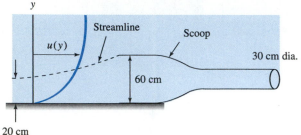

FIGURE P4.44

4.45 The jet pump operates by inducing a flow due to the high velocity in the 5-cm-diameter pipe as shown in Fig. P4.45. The velocity in the small pipe is $200[1 - (r/R)^2]$. Estimate the average velocity at the exit.

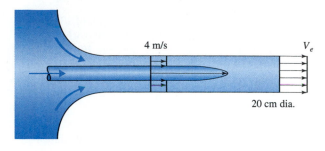

FIGURE P4.45

4.46 The experimental setup shown in Fig. P4.46 is used to provide liquid for the tissue. Derive an expression for the storage rate of liquid in the tissue in terms of the relevant information.

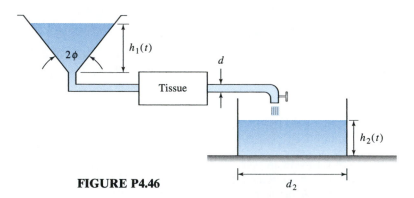

FIGURE P4.46

4.47 The water is 4 m deep behind a sluice gate in a rectangular channel that is suddenly opened (Fig. P4.47). Find the initial dh/dt if $V_2 = 8$ m/s and $V_1 = 0.2$ m/s. The upstream channel length is 100 m.

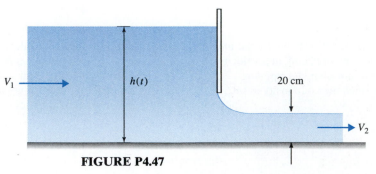

FIGURE P4.47

4.48 Water enters a kidney through a tube at 10 mL/min. It exits through a 6-mm-diameter tube at 20 mm/s. What is the rate of change of mass of water in the kidney?

4.49 The solid fuel in a rocket is burning at the rate of $400\,e^{-t/100}$ cm³/s (Fig. P4.49). If the density of the fuel is 900 kg/m³, estimate the exit velocity V_e when $t = 10$ s assuming the density of the exiting gases to be 0.2 kg/m³.

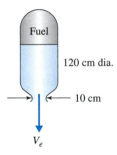

Fuel

120 cm dia.

← 10 cm

V_e

FIGURE P4.49

4.50 In Fig. P4.50, find the rate of change of $h(t)$ if water is the fluid at all locations:
(a) $V_1 = 10$ m/s, $\dot{m}_2 = 10$ kg/s, $Q_3 = 600$ L/min

(b) $V_1 = 0$, $\dot{m}_2 = 20$ kg/s, $Q_3 = 10$ L/s
(c) $V_1 = 5$ m/s, $\dot{m}_2 = 10$ kg/s, $Q_3 = 1000$ L/min

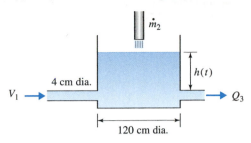

$\dot{m}_2$

4 cm dia.

$h(t)$

$V_1 \rightarrow$ $\rightarrow Q_3$

120 cm dia.

FIGURE P4.50

4.51 The velocity of the water entering the volume shown in Fig. P4.51 is $V(t) = 10e^{-t/10}$ m/s. Assuming $h(0) = 0$, find $h(t)$ if the volume is:
(a) A cone.
(b) A 10-m-long trough.

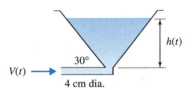

$h(t)$

30°

$V(t) \rightarrow$

4 cm dia.

FIGURE P4.51

Energy Equation

4.52 In Fig. P4.52, determine the work rate by the air at the instant shown if $V_{\text{piston}} = 10$ m/s, the torque $T = 20$ N·m, the velocity gradient at the belt surface is $100\ \text{s}^{-1}$, and the pressure acting on the piston is 400 Pa. The belt is 80 cm × 50 cm and the 40-cm-high piston is 50 cm deep (into the paper).

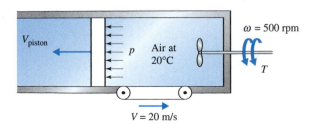

V_{piston}

p Air at 20°C

$\omega = 500$ rpm

T

$V = 20$ m/s

FIGURE P4.52

4.53 Assuming that the internal energy of natural gas depends on temperature only, what happens to the losses as natural gas is pumped from Texas to Michigan? The temperature remains essentially constant. Refer to Eq. 4.5.14.

4.54 An insulated water pump requires 500 W while pumping 0.02 m³/s with an efficiency of 80%. What is the temperature rise of the water from inlet to exit of the pump assuming equal inlet and exit areas? The specific heat of water is 4.18 kJ/kg°C.

4.55 What is the energy requirement of an 85% efficient pump that transports 40 L/s of water if the pressure increases from 200 kPa to 1200 kPa?

    A.   4.8 kW

    B.   14.2 kW

    C.   34.0 kW

    D.   47.1 kW

4.56 A water pump requires 5 hp to create a pump head of 20 m. If its efficiency is 87%, what is the flow rate of water?

4.57 An 89% efficient hydro turbine operates with a turbine head of 40 m. What is the turbine output if the mass flux is:

    (a) 200 kg/s?

    (b) 90 000 kg/min?

    (c) 8×10^6 kg/h?

4.58 The desired output from a set of 89% efficient turbines on a river is 10 MW. If the maximum turbine head attainable is 50 m, determine the average velocity at a location where the river is 60 m wide and 3 m deep.

4.59 Water flows in an open rectangular channel at a depth of 1 m with a velocity of 4 m/s. The bottom of the channel drops over a short length a distance of 1 m. Calculate the two possible depths of flow after the drop. Neglect all losses.

4.60 If the head loss in Problem 4.59 across the channel drop is 0.2 m, determine the two possible depths of flow.

4.61 Find the velocity V_1 of the water in the vertical pipe shown in Fig. P4.61. Assume no losses.

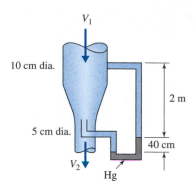

FIGURE P4.61

4.62 If the head-loss coefficient (based on V_2) between sections 1 and 2 of Problem 4.61 is 0.05, determine V_1 of the water.

4.63 A high-speed water jet is used to cut a material. If the velocity issuing from the 2-mm-diameter jet is 120 m/s, the maximum pressure on the material at the point of impact is nearest:

    A. 7200 kPa   **B.** 3600 kPa

    C. 735 kPa   **D.** 452 kPa

4.64 The flow rate of water in a 5-cm-diameter horizontal pipeline at a pressure of 400 kPa is 8 L/s. If the pipeline increases to 8 cm diameter, calculate the increased pressure if the loss coefficient (based on V_1) is 0.37.

4.65 Water flows at the rate of 600 L/min in a 4-cm-diameter horizontal pipeline with a pressure of 690 kPa. If the pressure after an enlargement to 6 cm diameter is measured to be 700 kPa, calculate the head loss across the enlargement.

4.66 Calculate the pressure p_1 shown in Fig. P4.66 needed to maintain a flow rate of 0.08 m³/s of water in a 6-cm-diameter horizontal pipe leading to a nozzle if a loss coefficient based on V_1 is 0.2 between the pressure gage and the exit.

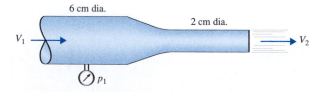

FIGURE P4.66

4.67 In Fig. P4.67, neglect all losses and predict the value of H and p if:
    (a) $h = 15$ cm   **(b)** $h = 20$ cm

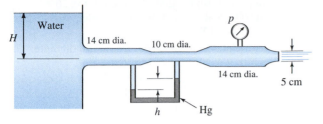

FIGURE P4.67

4.68 Water flows from the rectangular outlet shown in Fig. P4.68. Estimate the flow rate per unit width for each if $h = 80$ cm, $H = 2$ m. Neglect all losses.

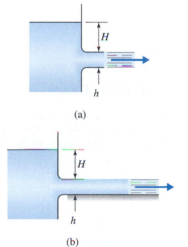

FIGURE P4.68

4.69 The overall loss coefficient for the pipe shown in Fig. P4.69 is 5; up to A it is 0.8, from A to B it is 1.2, from B to C it is 0.8, from C to D it is 2.2. Estimate the flow rate and the pressures at A, B, C, and D. The elevations are shown.

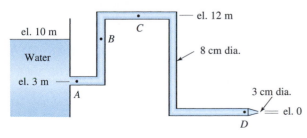

FIGURE P4.69

4.70 Water exits from a pressurized reservoir as shown in Fig. P4.70. Calculate the flow rate if on section A we:
    (a) Attach a nozzle with exit diameter 5 cm
    (b) Attach a diffuser with exit diameter 18 cm
    (c) Leave as an open pipe as shown
Neglect losses for all cases.

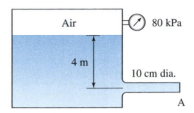

FIGURE P4.70

4.71 Rework Problem 4.70 assuming that $K_{pipe} = 1.5$, $K_{nozzle} = 0.04$ (based on V_1), and $K_{diffuser} = 0.8$ (based on V_1).

4.72 Estimate V_1. Assume the air to be incompressible with $\rho = 1.2$ kg/m^3.
    A. 62 m/s   **B.** 40 m/s
    C. 18 m/s   **D.** 10 m/s

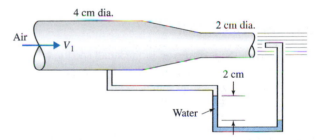

FIGURE P4.72

4.73 Relate the flow rate of the water through the venturi meter shown in Fig. P4.73 to the diameter and the manometer reading. Assume no losses.

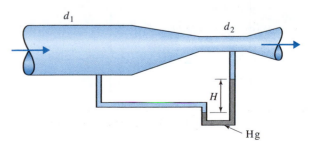

FIGURE P4.73

4.74 In the venturi meter of Problem 4.73, calculate the flow rate if:
(a) $H = 20$ cm, $d_1 = 2d_2 = 16$ cm
(b) $H = 40$ cm, $d_1 = 3d_2 = 24$ cm

4.75 The pressure drop across a valve, through which 40 L/s of water flows is measured to be 100 kPa. Estimate the loss coefficient if the nominal diameter of the valve is 8 cm.
A. 0.79   B. 3.2
C. 8.7   D. 31

4.76 In Fig. P4.76, determine the maximum possible height H if cavitation is to be avoided. Let: $d = 10$ cm and $T = 20°C$

Neglect all losses and assume $P_{atm} = 100$ kPa

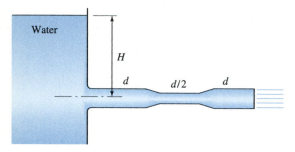

FIGURE P4.76

4.77 Cavitation is observed in the small section of pipe of Problem 4.76 when $H = 65$ cm. Estimate the temperature of the water. Neglect all losses and assume $p_{atm} = 100$ kPa. Use:
(a) $d = 10$ cm   (b) $d = 12$ cm

4.78 In Fig. P4.78, what is the minimum possible depth H if cavitation is to be avoided? Assume a vapor pressure of 6 kPa absolute and an overall loss coefficient of 8 based on V_2 and including the exit loss. Neglect losses up to the enlargement.

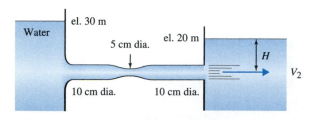

FIGURE P4.78

4.79 A contraction occurs in a 10-cm-diameter pipe to 6 cm followed by an enlargement back to 10 cm. The pressure upstream is measured to be 200 kPa when cavitation is first observed in the 20°C water. Calculate the flow rate. Neglect losses. Use $p_{atm} = 100$ kPa.

4.80 In Fig. P4.80, calculate the maximum diameter D such that cavitation will be avoided if:
(a) $d = 20$ cm, $H = 5$ m, and $T_{water} = 20°C$
(b) $d = 8$ in., $H = 15$ ft, and $T_{water} = 70°F$

Neglect all losses and use $p_{atm} = 100$ kPa (14.7 psi).

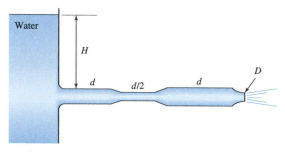

FIGURE P4.80

4.81 The overall loss coefficient in the siphon shown in Fig. P4.81 is 4; up to section A it is 1.5. At what height H will the siphon cease to function?

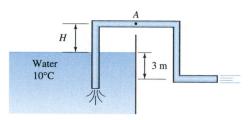

FIGURE P4.81

4.82 The pump shown in Fig. P4.82 is 85% efficient. If the pressure rise is 825 kPa, calculate the required energy input in horsepower.

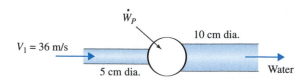

FIGURE P4.82

4.83 The pump shown in Fig. P4.82 is powered by a 20-kW motor. If the pump is 82% efficient, determine the pressure rise.

4.84 An 89%-efficient pump is inserted in a 4-cm-diameter line transporting 40 L/s of water. A pressure rise of 400 kPa is desired. The power required by the pump is nearest:
- **A.** 12 kW
- **B.** 16 kW
- **C.** 18 kW
- **D.** 22 kW

4.85 A turbine, which is 87% efficient, accepts 2 m³/s of water from a 50-cm-diameter pipe. The pressure drop is 600 kPa and the exit velocity is small. What is the turbine output?

4.86 A turbine receives 12.75 m³/s of water from a 1.8-m-diameter pipe at a pressure of 825 kPa and delivers 10 000 kW. The pressure in the 2.4-m-diameter exit pipe is 125 kPa. Calculate the turbine efficiency.

4.87 Air enters a compressor with negligible velocity at 85 kPa absolute and 20°C. It leaves with a velocity of 200 m/s at 600 kPa absolute. For a mass flux of 5 kg/s, calculate the exit temperature if the power required is 1500 kW and:

(a) No heat transfer occurs

(b) The heat transfer rate is 60 kW

4.88 Air enters a compressor at standard conditions with negligible velocity. At the 2.5-cm-diameter exit the pressure, temperature, and velocity are 400 kPa, 150°C, and 180 m/s, respectively. If the heat transfer is 20 kJ/kg of air, find the power required by the compressor.

4.89 A small river with a flow rate of 15 m³/s feeds the reservoir shown in Fig. P4.89. Calculate the energy that is available continuously if the turbine is 80% efficient. The loss coefficient for the overall piping system is $K = 4.5$.

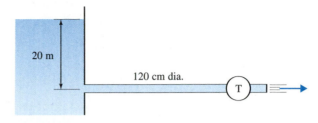

FIGURE P4.89

4.90 A hydroturbine generates power by transporting 0.2 m³/s of water from a dam. The water surface is 10 m above the turbine outlet. the overall loss coefficient for the 24-cm-diameter connecting pipe is 3.2. The maximum turbine output is nearest:
- **A.** 42 kW
- **B.** 21 kW
- **C.** 18 kW
- **D.** 13 kW

4.91 For the system shown in Fig. P4.91 the average velocity in the pipe is 10 m/s. Up to point A, $K = 1.5$, from B to C, $K = 6.2$, and the pump is 80% efficient. If $p_C = 200$ kPa find p_A and p_B and the power required by the pump.

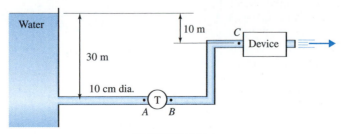

FIGURE P4.91

4.92 Determine the power output of the turbine shown in Fig. P4.92 for a water flow rate of 500 L/s. The turbine is 90% efficient.

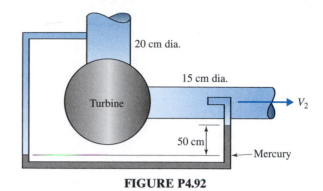

FIGURE P4.92

4.93 Find the power requirement of the 85%-efficient pump shown in Fig. P4.93 if the loss coefficient up to A is 3.2, and from B to C, $K = 1.5$. Neglect the losses through the exit nozzle. Also, calculate p_A and p_B.

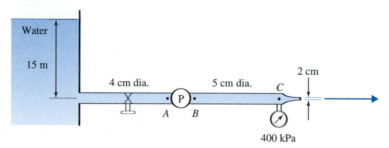

FIGURE P4.93

4.94 A 75%-efficient pump delivers 0.1 m³/s of water from a reservoir to a device at an elevation of 50 m above the reservoir. The pressure at the 8-cm-diameter entrance to the device is 180 kPa. If the piping loss coefficient is 5.6, the necessary power input to the pump is nearest:

A.	263 kW	**B.**	203 kW
C.	121 kW	**D.**	91.3 kW

4.95 Water leaves a reservoir with a head of 10 m and flows through a horizontal 10-cm-diameter pipe, then exits to the atmosphere. At the end of the pipe a short length of smaller-diameter pipe is added. The loss coefficient, including the reduction, is 2.2, and after the reduction the losses are negligible. Calculate the minimum diameter of the smaller pipe if the flow rate is 0.02 m³/s.

4.96 Neglect losses and find the depth of water on the raised section of the rectangular channel shown in Fig. P4.96. Assume uniform velocity profiles.

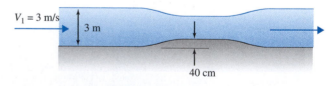

FIGURE P4.96

4.97 Determine the rate of kinetic energy loss, in watts, due to the cylinder shown in Fig. P4.97. Assume a plane flow with $\rho = 1.23$ kg/m^3 and make the calculation per meter of cylinder length.

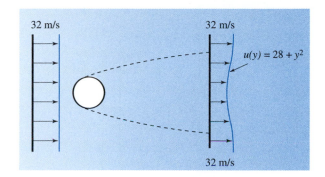

32 m/s   32 m/s

$u(y) = 28 + y^2$

32 m/s

FIGURE P4.97

4.98 Calculate the head loss between the two sections shown in Fig. P4.98. Assume:
(a) A pipe with $d = 1.2$ cm
(b) A rectangular duct 1.2 cm $\times$ 8 cm

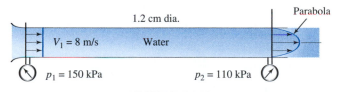

1.2 cm dia.   Parabola

$V_1 = 8$ m/s   Water

$p_1 = 150$ kPa $p_2 = 110$ kPa

FIGURE P4.98

4.99 Calculate the kinetic-energy correction factor for the velocity profile at the downstream location in Problem 4.97.

4.100 Determine the kinetic-energy correction factor if:
(a) $u(r) = 10(1 - r^2/R^2)$ in a 2-cm-diameter pipe.
(b) $u(y) = 10 (1 - y^2/h^2)$ in a 2-cm-high channel.

4.101 A turbulent velocity profile in a pipe is often written as $u(r) = u_{max}(1 - r/R)^{1/n}$, where n ranges between 5 and 9, 7 being the most common value. Calculate an expression for the kinetic energy passing a section of pipe and the kinetic-energy correction factor if:
(a) $n = 5$
(b) $n = 7$
(c) $n = 9$

4.102 A jet aircraft is flying with a velocity V_∞. Use the energy equation to relate the fuel consumption $\dot{m}_f$ to other flow variables such as exhaust gas velocity V_2 and temperature T_2, inlet velocity V_1 and temperature T_1, drag F_D acting on the aircraft, inlet air mass flux $\dot{m}$, and the fuel heating value q_f (kJ/kg).

4.103 An automobile moves at 100 km/hr with a drag force of 1340 N. The gas consumption is noted to be 5 km/L. If the engine efficiency is 15%, determine the energy released per kilogram of fuel. The fuel density is 680 kg/m^3.

4.104 A 180-m-long 2-cm-diameter siphon provides water at 20°C from a reservoir to a field for irrigation. It exits 35 cm below the reservoir surface. The velocity distribution is assumed to be $u(r) = 2V(1 - r^2/r_0^2)$, where V is the average velocity. Determine the flow rate if the head loss is given by $32\nu LV/(D^2 g)$, where L is the siphon's length, D its diameter, and ν the kinematic viscosity of the water.

4.105 The manufacturer's pump curve for the pump in the flow system shown in Fig. P4.105a is provided in Fig. P4.105b. Estimate the flow rate. The overall loss coefficient is:

(a) $K = 5$ **(b)** $K = 20$

The solution involves a trial-and-error procedure, or the energy equation can be written as $H_p = H_p(Q)$ and plotted on the pump curve.

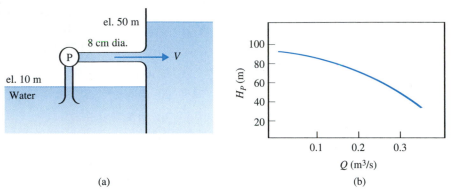

(a)   (b)

FIGURE P4.105

4.106 A water pump has one inlet and two outlets as shown in Fig. P4.106, all at the same elevation. What pump power is required if the pump is 85% efficient? Neglect pipe losses.

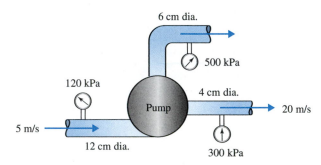

FIGURE P4.106

Momentum Equation

4.107 A strong wind blows directly against a window on a building. The force on the window can be approximated using:
   A. Bernoulli's equation
   B. The continuity equation
   C. The momentum equation
   D. All of the above

4.108 Water flows in a pipe of diameter d at a pressure p. It flows out a nozzle of diameter $d/2$ to the atmosphere. Calculate the force of the water on the nozzle if:
   (a) $d = 6$ cm, $p = 200$ kPa
   (b) $d = 6$ cm, $p = 400$ kPa
   (c) $d = 12$ cm, $p = 200$ kPa

4.109 A nozzle and hose are attached to the ladder of a fire truck. What force is needed to hold a nozzle supplied by a 9-cm-diameter hose with a pressure of 2000 kPa? The nozzle outlet diameter is 3 cm.

4.110 Water flows in a 10-cm-diameter pipe at a pressure of 400 kPa out a straight nozzle. Calculate the force of the water on the nozzle if the outlet diameter is:
   (a) 8 cm   **(c)** 4 cm
   (b) 6 cm   **(d)** 2 cm

4.111 A nozzle with an exit diameter of 4 cm is attached to a 10-cm-diameter pipe transporting 0.1 m³/s of water. The force needed to hold the nozzle on the pipe is nearest:
   A. 6.7 kN   **B.** 12.2 kN
   C. 17.5 kN   **D.** 24.2 kN

4.112 Find the horizontal force of the water on the horizontal bend shown in Fig. P4.112.

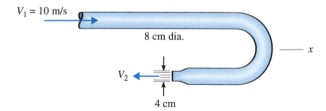

FIGURE P4.112

4.113 Find the horizontal force components of the water on the horizontal bend shown in Fig. P4.113 if p_1 is:
   (a) 200 kPa
   (b) 400 kPa
   (c) 800 kPa

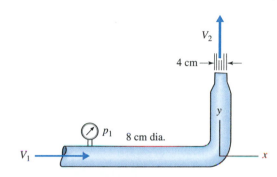

FIGURE P4.113

4.114 What is the net force needed to hold the orifice plate shown in Fig. P4.114 onto the pipe?

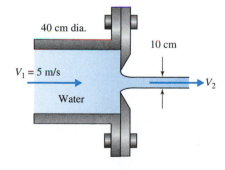

FIGURE P4.114

4.115 Assuming uniform velocity profiles, find F needed to hold the plug in the pipe shown in Fig. P4.115. Neglect viscous effects.

FIGURE P4.115

4.116 Neglect viscous effects, assume uniform velocity profiles, and find the horizontal force component acting on the obstruction shown in Fig. P4.116.

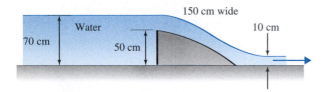

FIGURE P4.116

4.117 Assuming hydrostatic pressure distributions, uniform velocity profiles, and negligible viscous effects, find the horizontal force needed to hold the sluice gate in the position shown in Fig. P4.117.

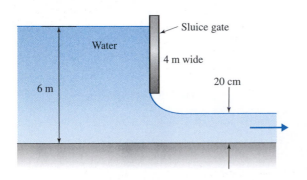

FIGURE P4.117

4.118 A sudden jump (a hydraulic jump) occurs in a rectangular channel as shown in Fig. P4.118. Find y_2 and V_2 if:
   (a) $V_1 = 8$ m/s, $y_1 = 60$ cm
   (b) $V_1 = 12$ m/s, $y_1 = 40$ cm

FIGURE P4.118

4.119 A hydraulic jump, as shown in Fig. P4.118, occurs so that $V_2 = \frac{1}{4}V_1$. Find V_1 and y_2 if: $y_1 = 80$ cm

4.120 For a flow rate of 9 m³/s, find V_2 and y_2 for the hydraulic jump shown in Fig. P4.120. The channel is 3 m wide. Neglect losses up to the jump.

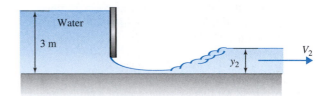

FIGURE P4.120

4.121 The velocity is measured to be 3 m/s downstream of a hydraulic jump where the depth is 2 m. Calculate the velocity and depth prior to the jump.

4.122 For the system shown in Fig. P4.122, estimate the down stream pressure p_2 if $p_1 = 60$ kPa and $V_1 = 20$ m/s. Neglect losses. (*Note:* The pressure immediately after the pipe expansion is p_1.)

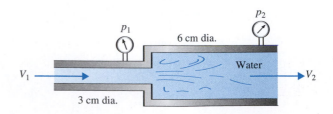

FIGURE P4.122

4.123 Water flows at 15 m/s in a 10-cm-diameter stem of a horizontal T-section that branches into 5-cm-diameter pipes. Find the force of the water on the T-section if the branches exit to the atmosphere. Neglect viscous effects.

4.124 Find the *x*- and *y*-force components on the horizontal T-section shown in Fig. P4.124. Neglect viscous effects.

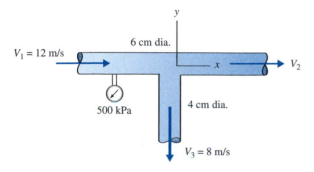

FIGURE P4.124

4.125 A horizontal 10-cm-diameter jet of water with $\dot{m} = 300$ kg/s strikes a vertical plate. Calculate:
(a) The force needed to hold the plate stationary
(b) The force needed to move the plate away from the jet at 10 m/s
(c) The force needed to move the plate into the jet at 10 m/s

4.126 A horizontal 10-cm-diameter jet of water strikes a vertical plate. Determine the velocity issuing from the jet if a force of 1000 N is needed to:
(a) Hold the plate stationary
(b) Move the plate away from the jet at 10 m/s
(c) Move the plate into the jet at 10 m/s

4.127 Determine the mass flux issuing from the jet shown in Fig. P4.127 if a force of 700 N is needed to:
(a) Hold the cone stationary
(b) Move the cone away from the jet at 8 m/s
(c) Move the cone into the jet at 8 m/s

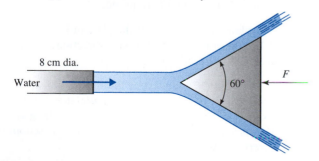

FIGURE P4.127

4.128 A 1 cm × 20 cm sheet of water is deflected as shown in Fig. P4.128. The magnitude of the total force acting on the stationary deflector is nearest:
  A. 6830 N   **B.** 5000 N
  C. 4330 N   **D.** 2500 N

FIGURE P4.128

4.129 Calculate the force components of the water acting on the deflecting blade shown in Fig. P4.129 if:
(a) The blade is stationary
(b) The blade moves to the right at 20 m/s
(c) The blade moves to the left at 20 m/s

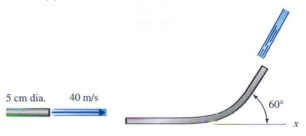

FIGURE P4.129

4.130 The blade of Problem 4.129 is one of a series of blades that are attached to a 50-cm-radius rotor that has a rotational speed of 30 rad/s. If there are 10 such water jets, find the power output.

4.131 Determine the force components of superheated steam acting on the blade shown in Fig. P4.131 if:
(a) The blade is stationary
(b) The blade moves to the right at 100 m/s
(c) The blade moves to the left at 100 m/s

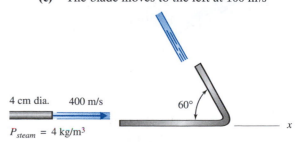

FIGURE P4.131

4.132 The blade of Problem 4.131 is one of a series of blades that are attached to a 1.2-m-radius rotor that rotates at 150 rad/s. Calculate the power output if there are 15 such steam jets.

4.133 The water impacts one of the turbine blades as shown in Fig. P4.133. For a blade speed of 20 m/s, the maximum power output for a single jet is nearest:

  A. 18 kW **B.** 154 kW
  C. 206 kW **D.** 309 kW

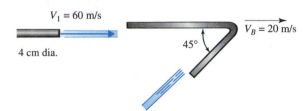

FIGURE P4.133

4.134 Superheated steam jets strike the turbine blades shown in Fig. P4.134. Find the power output of the turbine if there are 15 jets and α_1 is:

  (a) 45° **(b)** 60° **(c)** 90°

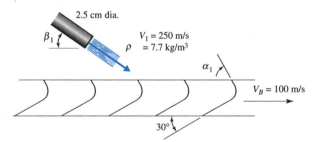

FIGURE P4.134

4.135 Twelve high-speed water jets strike the blades as shown in Fig. P4.135. Find the power output and the blade angles if V_B is:

  (a) 20 m/s **(b)** 40 m/s **(c)** 50 m/s

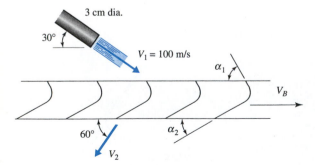

FIGURE P4.135

4.136 Fifteen water jets strike the blades of a turbine as shown in Fig. P4.136. Calculate the power output and the blade angles if β_2 is:

  (a) 60°
  (b) 70°
  (c) 80°

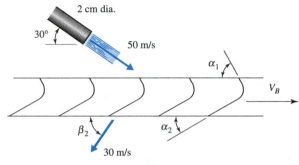

FIGURE P4.136

4.137 Water flows from the rectangular jet as shown in Fig. P4.137. Find the force F and the mass fluxes $\dot{m}_2$ and $\dot{m}_3$ if $b = 20$, $h = 40$ cm, $V_1 = 40$ m/s

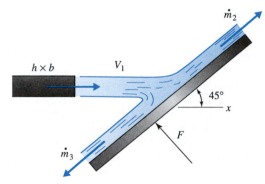

FIGURE P4.137

4.138 The plate of Problem 4.137a moves to the left at 20 m/s. Find the power required.

4.139 Calculate the velocity that the plate of Problem 4.137a must move with (in the x-direction) to produce the maximum power output.

4.140 A large vehicle is slowed by lowering a 2-m-wide scoop into a reservoir of water. Estimate the force exerted on the scoop if the vehicle is travelling at 60 m/s and it scoops off 5 cm of water? The scoop diverts the water through 180°.

(a) 720 kN   (b) 360 kN
(c) 12 kN   (d) 7.2 kN

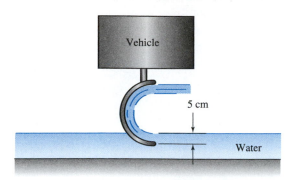

FIGURE P4.140

4.141 A large vehicle with a mass of 100 000 kg is slowed by inserting a 180° deflector in a trough of water. If the 60-cm-wide deflector scoops off 10 cm of water, calculate the initial deceleration if the vehicle is traveling at 120 km/h. Also, find the time necessary to reach a speed of 60 km/h.

4.142 A 2.5-m-wide snowplow travels at 50 km/h plowing snow at a depth of 0.8 m. The snow leaves the blade normal to the direction of motion of the snowplow. What power does the plowing operation require if the density of the snow is 90 kg/m³?

4.143 A vehicle with a mass of 5000 kg is traveling at 900 km/h. It is decelerated by lowering a 20-cm-wide scoop into water a depth of 6 cm (Fig. P4.143). If the water is deflected through 180°, calculate the distance the vehicle must travel for the speed to be reduced to 100 km/h.

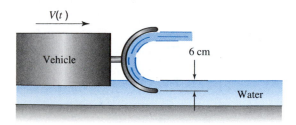

FIGURE P4.143

4.144 A 300-kg vehicle is required to have an initial acceleration of 2 m/s². It is proposed that a 6-cm-diameter water jet strike a vane built into the rear of the vehicle that will divert the water through an angle of 180°. What jet velocity is necessary? What speed will be achieved after 2 s?

4.145 The swamp boat shown in Fig. P4.145 is propelled at 50 km/h by a 2-m-diameter propeller that requires a 20-kW engine. Calculate the thrust on the boat, the flow rate of air through the propeller, and the propeller efficiency.

FIGURE P4.145

4.146 An aircraft is propelled by a 2.2-m-diameter propeller at a speed of 200 km/h. The air velocity downstream of the propeller is 320 km/h relative to the aircraft. Determine the pressure difference across the propeller blades and the required power. Use $\rho = 1.2$ kg/m³.

4.147 The 50-cm-diameter propeller on a boat moves at 30 kmph by causing a velocity of 60 kmph relative to the boat. Calculate the required power and the mass flux of water through the propeller.

4.148 A jet boat takes in 0.2 m³/s of water and discharges it at a velocity of 20 m/s relative to the boat. If the boat travels at 10 m/s, calculate the thrust produced and the power required.

4.149 Calculate the change in the momentum flux of the water flowing through the plane contraction shown in Fig. P4.149 if the flow rate is 0.2 m³/s. The slope of the two profiles is the same. The upstream profile is created by a plate containing slots of various widths.

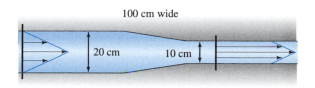

FIGURE P4.149

4.150 Determine the momentum correction factor for the following profile shown in Problem 4.149:
   (a) The entering profile
   (b) The exiting profile

4.151 Water at 15°C flows in a 4-cm-diameter horizontal pipe and experiences a pressure drop of 200 Pa over a 10-m length of pipe. Calculate the velocity gradient at the wall. Recall that $\tau = \mu \,|\, du/dr \,|$.

4.152 Find the drag force on the walls between the two sections of the horizontal pipe shown in Fig. P4.152.

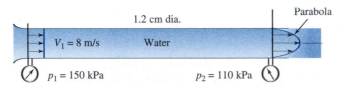

FIGURE P4.152

4.153 The velocity distribution downstream of a 10-m-long circular cylinder is as shown in Fig. P4.153. Determine the force of the air on the cylinder. Use $\rho = 1.23 \text{ kg/m}^3$.

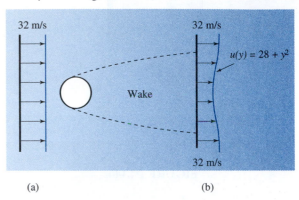

FIGURE P4.153

4.154 Calculate the drag force acting on the flat 2-m-wide plate shown in Fig. P4.154. Outside the viscous region the velocity is uniform. Select (use $\rho = 1.23 \text{ kg/m}^3$):

(a) A rectangular control volume extending outside the viscous region (mass flux crosses the top).

(b) A control volume with the upper boundary a streamline (no mass flux crosses a streamline).

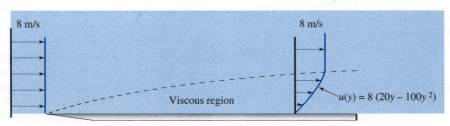

FIGURE P4.154

Momentum and Energy

4.155 Determine the power lost in the hydraulic jump of:
   (a) Problem 4.118a.   **(b)** Problem 4.120.

4.156 Calculate the loss coefficient for the expansion of Problem 4.122. Base the coefficient on the velocity V_1.

4.157 Find a relationship between the acceleration of the cylindrical cart and the variables shown in Fig. P4.157. Neglect friction. The initial mass of the cart and water is m_0.

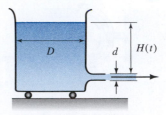

FIGURE P4.157

4.158 Set up the equations necessary to determine $H(t)$ for the air/water rocket shown in Fig. P.4.158.

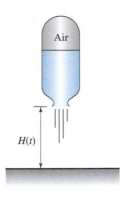

FIGURE P4.158

Moment of Momentum

4.159 A four-armed water sprinkler has nozzles at right angles to the 30-cm-long arms and at 45° angles with the ground. If the outlet diameters are 8 mm and 4 kg/s of water exits the four nozzles, find the rotational speed.

4.160 A four-armed rotor has 1-cm-diameter nozzles that exit water at 60 m/s relative to the arm. The nozzles are at right angles to the 25-cm-long arms and are parallel to the ground. If the rotational speed is 30 rad/sec, find the power output. The arms are 4 cm in diameter.

4.161 Water flows out the 6-mm slots as shown in Fig. P4.161. Calculate Ω if 20 kg/s is delivered by the two arms.

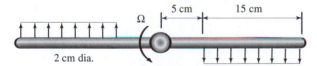

FIGURE P4.161

4.162 A 1-kW motor, drives the rotor shown in Fig. P4.162 at 500 rad/s. Determine the flow rate neglecting all losses. Use $\rho = 1.23$ kg/m³.

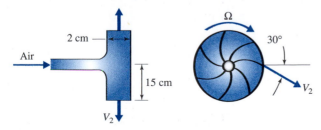

FIGURE P4.162

4.163 Find an expression for $\Omega(t)$ if the sprinkler of Problem 4.159 is suddenly turned on at $t = 0$. Assume the arms to be 2 cm in diameter.

4.164 Air enters the centrifugal-type air pump of a leaf blower through the shaded area shown in Fig. P4.164. The 10-cm-diameter 1.2-m-long tube has an attached nozzle with a 30-cm² exit area. The exit velocity is 240 km/h.
   (a) Calculate the discharge.
   (b) If the overall loss coefficient is 1.2, estimate the pump head.
   (c) What power must the pump supply to the air?
   (d) If the pump is 65% efficient, what is the required horsepower of the gasoline engine?
   (e) Estimate the pressure at the tube entrance (just downstream of the pump).
   (f) If the 10-kg blower hangs from a strap, what force must be applied at the handle located 30 cm above the nozzle? The center of gravity is 70 cm above and 120 cm to the left of the exit.

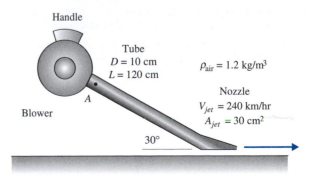

FIGURE P4.164

A commercial jetliner taking off. Current improvements in aircraft efficiency and speed are due in part to solving the differential equations of motion of air as particles move relative to the jet at subsonic and transonic speeds. (Mayskyphoto-Shutterstock)

5

The Differential Forms of the Fundamental Laws

Outline

Chapter Objectives

The objectives of this chapter are to derive the differential equations and state the boundary and initial conditions needed to solve for the velocity and pressure fields in a fluid flow. The partial differential equations include:

▲ The continuity equation
▲ The momentum equations for inviscid flows (Euler's equations)
▲ The momentum equations for viscous flows (Navier–Stokes equations)
▲ The vorticity equations
▲ The energy equation

Numerous examples and problems will illustrate several relatively simple applications of the differential equations along with how to simplify the equations depending on the flow situation.

5.1 INTRODUCTION

The material in this chapter can be omitted in an introductory course. The subsequent chapters in this book have been designed to allow two possible routes: The general differential equations presented in this chapter may be used, or equations unique to a particular geometry may be derived without referring to these general equations.

In Chapter 4 the basic laws were expressed in terms of a fixed control volume, a finite volume in space. This is often described as the global approach to the mechanics of fluids. To find a solution using the control volume it was necessary either to assume an approximation to the integrands (primarily the velocity and pressure distributions) or expressions for the integrands were given. Suppose that we wish to find an integral quantity, such as the flow rate under a dam or the lift on an airfoil, and we cannot make a reasonable assumption for the velocity or pressure distribution. It is then necessary to determine the distributions that enter the integrands before the integral quantity can be found. This is done by solving the partial differential equations that express the basic laws.

> **KEY CONCEPT** *It is necessary to determine the distributions that enter the integrands before the integral quantity can be found.*

Not only do solutions of the differential forms of the basic laws allow us to determine integral quantities of interest, they often contain information in themselves. For example, we may wish to know the exact location of the minimum pressure on a body, or the region of separated flow from a surface may be of interest. So we often solve the differential equations to answer a specific question raised about a special flow.

There are two primary methods used in deriving the differential forms of the fundamental laws. One method involves the application of Gauss's theorem, which allows the area integrals of the basic equations of Chapter 4 to be transformed to volume integrals; the integrands are then collected under one integral, which can be set equal to zero. The integration is valid over any arbitrary control volume, and thus the integrand itself can be set equal to zero, providing us with the differential form of the basic law. The other approach, the one used in this book, is to identify an infinitesimal element in space and apply the basic laws directly to that element. Both methods result in the differential forms of the basic laws; the first method, however, demands the use of vector and tensor calculus, mathematics that is usually considered unnecessary in a first course in fluids. We introduce some vector calculus but it will not be used at the operational level demanded by Gauss's theorem.

The conservation of mass, applied to an infinitesimal element, leads to the differential continuity equation; it relates the density and velocity fields.[1] Newton's second law (a vector relationship) results in three partial differential equations known as the Navier-Stokes equations; they relate the velocity, pressure, and density fields and introduce the viscosity and the gravity vector in a fluid flow. The first law of thermodynamics provides us with the differential energy equation, which relates the temperature field to the velocity, density, and pressure fields and introduces the specific heat and thermal conductivity. Most problems considered in an introductory course are for isothermal, incompress-

---

[1]When a dependent variable depends on more than one independent variable, it is referred to as a field, that is, $\mathbf{V}(x, y)$ is a velocity field and $p(x, y, z, t)$ is a pressure field. The partial differential equations describing the field quantities are often called *field equations*.

ible flows in which the temperature field does not play a role; for such flows the three Navier-Stokes equations and the continuity equation provide four partial differential equations that relate the three velocity components and the pressure. Thus the energy equation would not be needed. We will, however, derive the differential energy equation for use in a limited number of situations.

Partial differential equations require conditions that specify certain values for the dependent variables at particular values of the independent variables. If the independent variable is time, the conditions are called **initial conditions**; if the independent variable is a space coordinate, the conditions are **boundary conditions**. The total problem is referred to as an *initial-value problem* or a *boundary-value problem*.

In fluid mechanics the boundary conditions result from:

- The no-slip conditions for a viscous flow. The viscosity causes the fluid to stick to the wall and thus the velocity of the fluid at the wall assumes the velocity of the wall. Usually the velocity of the wall is zero.

- The normal component of velocity in an inviscid flow, a flow in which viscous effects are negligible. Near a non-porous wall the velocity vector must be tangent to the wall, demanding that the normal component be zero.

- The pressure in a flow involving a free surface. For problems with a free surface, such as a flow with a liquid–gas interface, the pressure is known on the interface. This would be the situation involving wave motion, or in a separated flow involving cavitation.

- The temperature of the boundary or the temperature gradient at the boundary. If the temperature of the boundary is held constant, the temperature of the fluid next to the boundary will equal the boundary temperature. If the boundary is insulated, the temperature gradient will be zero at the boundary.

In an unsteady flow the initial condition demands that the three velocity components be specified at all points in the flow at a particular instant in time, usually taken as $t = 0$. This information would be very difficult to obtain in many situations, e.g., the atmosphere; to obtain the three velocity components everywhere in the atmosphere at an instant is obviously an improbable task.

Differential equations take very different forms depending on the coordinate system chosen. We derive the equations using rectangular coordinates and then express the equations in vector form. Forms using both cylindrical and spherical coordinates are presented in Table 5.1 at the end of the chapter.

Initial conditions: *Conditions that depend on time.*

Boundary conditions: *Conditions that depend on a space coordinate.*

KEY CONCEPT *Viscosity causes the velocity of the fluid at the wall to assume the velocity of the wall.*

5.2 DIFFERENTIAL CONTINUITY EQUATION

Let us begin our quest for the partial differential equations that model the detailed motion of a fluid by applying the conservation of mass to a small volume in a fluid flow. Consider the mass flux through each face of the fixed infinitesimal control volume shown in Fig. 5.1. We set the net flux of mass entering the element equal to the rate of change of the mass of the element; that is,

$$\dot{m}_{\text{in}} - \dot{m}_{\text{out}} = \frac{\partial}{\partial t} m_{\text{element}} \tag{5.2.1}$$

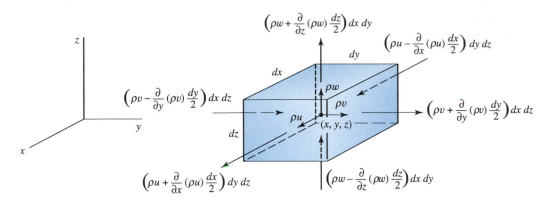

FIGURE 5.1 Infinitesimal control volume using rectangular coordinates.

To perform this mass balance we identify ρu, ρv, and ρw at the center of the element and then treat each of these quantities as a single variable. See the analysis of pressure associated with Fig. 2.3. Refer to Fig. 5.1 showing the mass flux through each of the six faces; Eq. 5.2.1 takes the form

$$
\left[\rho u - \frac{\partial(\rho u)}{\partial x}\frac{dx}{2}\right] dy\, dz - \left[\rho u + \frac{\partial(\rho u)}{\partial x}\frac{dx}{2}\right] dy\, dz
$$

$$
+ \left[\rho v - \frac{\partial(\rho v)}{\partial y}\frac{dy}{2}\right] dx\, dz - \left[\rho v + \frac{\partial(\rho v)}{\partial y}\frac{dy}{2}\right] dx\, dz
$$

$$
+ \left[\rho w - \frac{\partial(\rho w)}{\partial z}\frac{dz}{2}\right] dx\, dy - \left[\rho w + \frac{\partial(\rho w)}{\partial z}\frac{dz}{2}\right] dx\, dy
$$

$$
= \frac{\partial}{\partial t}(\rho\, dx\, dy\, dz) \tag{5.2.2}
$$

Subtracting the appropriate terms and dividing by $dx\, dy\, dz$ yields

$$
\frac{\partial}{\partial x}(\rho u) + \frac{\partial}{\partial y}(\rho v) + \frac{\partial}{\partial z}(\rho w) = -\frac{\partial \rho}{\partial t} \tag{5.2.3}
$$

Since density is considered a variable, we differentiate the products and put Eq. 5.2.3 in the form

$$
\frac{\partial \rho}{\partial t} + u\frac{\partial \rho}{\partial x} + v\frac{\partial \rho}{\partial y} + w\frac{\partial \rho}{\partial z} + \rho\left(\frac{\partial u}{\partial x} + \frac{\partial v}{\partial y} + \frac{\partial w}{\partial z}\right) = 0 \tag{5.2.4}
$$

or, in terms of the substantial derivative (see Eq. 3.2.11),

$$
\frac{D\rho}{Dt} + \rho\left(\frac{\partial u}{\partial x} + \frac{\partial v}{\partial y} + \frac{\partial w}{\partial z}\right) = 0 \tag{5.2.5}
$$

This is the most general form of the **differential continuity equation** expressed in rectangular coordinates.

Differential continuity equation: *The differential equation that results from the conservation of mass.*

We can introduce the *gradient operator*, called "del," which, in rectangular coordinates, is

$$\nabla = \frac{\partial}{\partial x}\,\hat{\mathbf{i}} + \frac{\partial}{\partial y}\,\hat{\mathbf{j}} + \frac{\partial}{\partial z}\,\hat{\mathbf{k}} \qquad (5.2.6)$$

The continuity equation can then be written in the form

$$\frac{D\rho}{Dt} + \rho\,\nabla\cdot\mathbf{V} = 0 \qquad (5.2.7)$$

where $\mathbf{V} = u\hat{\mathbf{i}} + v\hat{\mathbf{j}} + w\hat{\mathbf{k}}$ and $\nabla\cdot\mathbf{V}$ is called the *divergence* of the velocity. This form of the continuity equation does not refer to any particular coordinate system. It is the form used to express the continuity equation using various coordinate systems.

For the case of *incompressible flow*, a flow in which the density of a fluid particle does not change as it travels along, we see that

$$\frac{D\rho}{Dt} = \frac{\partial\rho}{\partial t} + u\frac{\partial\rho}{\partial x} + v\frac{\partial\rho}{\partial y} + w\frac{\partial\rho}{\partial z} = 0 \qquad (5.2.8)$$

Note that this is less restrictive than the assumption of constant density, which would require each term in Eq. 5.2.8 to be zero. Incompressible flows that have density gradients are sometimes referred to as *stratified flows* or *nonhomogeneous flows*; atmospheric and oceanic flows are examples of such flows. Using Eq. 5.2.5, the continuity equation, for an incompressible flow, takes the form

KEY CONCEPT *For an incompressible flow,* $\nabla\cdot\mathbf{V} = 0$.

$$\frac{\partial u}{\partial x} + \frac{\partial v}{\partial y} + \frac{\partial w}{\partial z} = 0 \qquad (5.2.9)$$

or, in vector form,

$$\nabla\cdot\mathbf{V} = 0 \qquad (5.2.10)$$

The divergence of the velocity vector is zero for an incompressible flow.

In cylindrical and spherical coordinates the continuity equation for an incompressible flow is presented in Table 5.1. The expression for D/Dt in cylindrical and spherical coordinates can also be found in Table 5.1.

Example 5.1

The x-component velocity is given by $u(x, y) = Ay^2$ in an incompressible plane flow. Determine $v(x, y)$ if $v(x, 0) = 0$, as would be the case in flow between parallel plates.

Solution

The differential continuity equation for an incompressible, plane flow is

$$\frac{\partial u}{\partial x} + \frac{\partial v}{\partial y} = 0$$

since in a plane flow the two velocity components depend only on x and y. Using the given $u(x, y)$ we find that

$$\frac{\partial v}{\partial y} = -\frac{\partial u}{\partial x} = -\frac{\partial}{\partial x}(Ay^2) = 0$$

Since this is a partial differential equation, its solution is

$$v(x, y) = f(x)$$

But $v(x, 0) = 0$ requiring that $f(x) = 0$. Consequently,

$$v(x, y) = 0$$

is the y-component velocity demanded by the conservation of mass. In order for $v(x, y)$ to be non-zero, $u(x, y)$ would have to vary with x or $v(x, 0)$ would have to be non-zero.

Example 5.2

Air flows in a pipe and the velocity at three neighboring points A, B, and C, 100 mm apart, is measured to be 91, 95, and 97 m/s, respectively, shown in Fig. E5.2. The temperature and pressure are 10°C and 350 kPa, respectively, at point B. Approximate $d\rho/dx$ at that point, assuming steady, uniform flow.

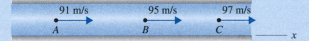

FIGURE E5.2

Solution

The continuity equation (5.2.4) for this steady $\left(\frac{\partial}{\partial t} = 0\right)$, uniform $\left(\frac{\partial}{\partial y} = \frac{\partial}{\partial z} = 0\right)$ flow reduces to

$$u\frac{d\rho}{dx} + \rho\frac{du}{dx} = 0$$

We used ordinary derivatives since u and ρ depend only on x. The velocity derivative is approximated by

$$\frac{du}{dx} \simeq \frac{\Delta u}{\Delta x} = \frac{97 - 91}{100/1000} = 60.0 \; \frac{m/s}{m}$$

[2]Using a forward difference $\Delta u/\Delta x = (97 - 95)/0.05 = 40$; a backward difference provides $\Delta u/\Delta x = (95 - 91)/0.05 = 80$. These two approximations are less accurate than the central difference used in the example. A sketch of a general curve $u(x)$ could graphically show this by displaying three points at $x - \Delta x$, x, and $x + \Delta x$ and sketching the slopes.

where the more accurate central difference has been used.[2] The density is

$$\rho = \frac{p}{RT} = \frac{350 \times 10^3}{287 \times (10 + 273.15)} = 4.307 \text{ kg / m}^3$$

where absolute pressure and temperature are used. The density derivative is then approximated to be

$$\frac{d\rho}{dx} = -\frac{\rho}{u}\frac{du}{dx}$$

$$= -\frac{4.307}{95} \times 60 = -2.72 \text{ kg / m}^4$$

Example 5.3

The x-component of velocity at points A, B, C, and D, which are 10 mm apart, is measured to be 5.76, 6.72, 7.61, and 8.47 m/s, respectively, in the plane steady, symmetrical, incompressible flow shown in Fig. E5.3 in which $w = 0$. Approximate the x-component acceleration at C and the y-component of velocity 6 mm above B.

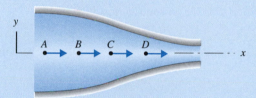

FIGURE E5.3

Solution
The desired acceleration component is found from Eq. 3.2.9 to be

$$a_x = \cancel{\frac{\partial u}{\partial t}}^{0} + u\frac{\partial u}{\partial x} + \cancel{v\frac{\partial u}{\partial y}}^{0} + \cancel{w\frac{\partial u}{\partial z}}^{0}$$

$$\cong u\frac{\Delta u}{\Delta x} = 7.61\frac{8.47 - 6.72}{0.02} = 666 \text{ m/s}^2$$

where we have assumed a symmetrical flow so that v along the centerline is zero. We have used central differences to approximate $\partial u/\partial x$ at point C, as done in Example 5.2 (see footnote 2).

The y-component of velocity 6 mm above B is found using the continuity equation (5.2.9) as follows:

$$\frac{\partial v}{\partial y} = -\frac{\partial u}{\partial x}$$

$$\frac{\Delta v}{\Delta y} \cong -\frac{\Delta u}{\Delta x} = -\frac{7.61 - 5.76}{0.02} = -92.5$$

$$\therefore \Delta v = -92.5\, \Delta y = -92.5 \times 0.006 = -0.555 \text{ m/s}$$

We know that $v = 0$ at B; hence at the desired location, with $\Delta v = v - v_B$, there results

$$v = -0.555 \text{ m/s}$$

Example 5.4

The continuity equation can be used to change the form of an expression. Write the expression $\rho\, D\tilde{u}/Dt + p\, \boldsymbol{\nabla} \cdot \mathbf{V}$ in terms of enthalpy h rather than internal energy $\tilde{u}$. Recall that $h = \tilde{u} + p/\rho$ (see Eq. 1.7.11).

Solution
Using the definition of enthalpy, we can write

$$\frac{D\tilde{u}}{Dt} = \frac{Dh}{Dt} - \frac{1}{\rho}\frac{Dp}{Dt} + \frac{p}{\rho^2}\frac{D\rho}{Dt}$$

where we used

$$\frac{D}{Dt}\left(\frac{p}{\rho}\right) = \frac{1}{\rho}\frac{Dp}{Dt} - \frac{p}{p^2}\frac{Dp}{Dt}$$

The desired expression is then

$$\rho\frac{D\tilde{u}}{Dt} + p\,\boldsymbol{\nabla}\cdot\mathbf{V} = \rho\frac{Dh}{Dt} - \frac{Dp}{Dt} + \frac{p}{\rho}\frac{D\rho}{Dt} + p\,\boldsymbol{\nabla}\cdot\mathbf{V}$$

The continuity equation (5.2.7) is introduced resulting in

$$\rho\frac{D\tilde{u}}{Dt} + p\,\boldsymbol{\nabla}\cdot\mathbf{V} = \rho\frac{Dh}{Dt} - \frac{Dp}{Dt} + \frac{p}{\rho}\frac{Dp}{Dt} + p\left(-\frac{1}{\rho}\frac{D\rho}{Dt}\right)$$

$$= \rho\frac{Dh}{Dt} - \frac{Dp}{Dt}$$

and enthalpy has been introduced.

5.3 DIFFERENTIAL MOMENTUM EQUATION

5.3.1 General Formulation

Suppose that we do not know the velocity field or the pressure field in an incompressible[3] flow of interest and we wish to solve differential equations to provide us with that information. The differential continuity equation is one differential equation to help us toward this end; however, it has three unknowns, the three velocity components. The differential momentum equation is a vector equation and thus provides us with three scalar equations. These component equations will aid us in our attempt to determine the velocity and pressure fields. There is a difficulty in deriving these equations, however, since we must use the stress components to determine the forces required in the momentum equation. Let us identify these stress components.

There are nine stress components that act at a particular point in a fluid flow. They are the nine components of the stress tensor τ_{ij}. We will not study the properties of a stress tensor in detail in this study of fluid mechanics since we do not have to maximize or minimize the stress (as would be required in a solid mechan-

---

[3]An incompressible flow, when referred to in a general discussion such as in this section, will generally refer to a constant-density flow. This is true in most fluid mechanics literature, including textbooks on the subject.

ics course); we must, however, use the nine stress components in our derivations, then relate the stress components to the velocity and pressure fields with the appropriate equations. The stress components that act at a point are displayed on two- and three-dimensional rectangular elements in Fig. 5.2. These elements are considered to be an exaggerated point, a cubical point; the stress components act in the positive direction on a positive face (a normal vector points in the positive coordinate direction) and in the negative direction on a negative face (a normal vector points in the negative coordinate direction). The first subscript on a stress component denotes the face upon which the component acts, and the second subscript denotes the direction in which it acts; the component τ_{xy} acts in the positive y-direction on a positive x-face and in the negative y-direction on a negative x-face, as displayed in Fig. 5.2a. A stress component that acts perpendicular to a face is referred to as a **normal stress**; the components σ_{xx}, σ_{yy}, and σ_{zz} are normal stresses. A stress component that acts tangential to a face is called a **shear stress**; the components τ_{xy}, τ_{yx}, τ_{xz}, τ_{zx}, τ_{yz}, and τ_{zy} are the shear stress components.

Normal stress: *A stress component that acts perpendicular to an area.*

Shear stress: *A stress component that acts tangential to an area.*

There are nine stress components that act at a particular point in a fluid. To derive the differential momentum equation, consider the forces acting on the infinitesimal fluid particle shown in Fig. 5.3. Only the forces acting on positive faces are shown. The stress components are assumed to be function of x, y, z, and t, and hence the values of the stress components change from face to face since the location of each face is slightly different. The body force is shown acting in an arbitrary direction.

Newton's second law applied to a fluid particle, for the x-component direction, is $\Sigma F_x = ma_x$. For the particle shown, this takes the form

$$\left(\sigma_{xx} + \frac{\partial \sigma_{xx}}{\partial x}\frac{dx}{2}\right) dy\, dz + \left(\tau_{yx} + \frac{\partial \tau_{yx}}{\partial y}\frac{dy}{2}\right) dx\, dz + \left(\tau_{zx} + \frac{\partial \tau_{zx}}{\partial z}\frac{dz}{2}\right) dx\, dy$$

$$-\left(\sigma_{xx} - \frac{\partial \sigma_{xx}}{\partial x}\frac{dx}{2}\right) dy\, dz - \left(\tau_{yx} - \frac{\partial \tau_{yx}}{\partial y}\frac{dy}{2}\right) dx\, dz$$

$$-\left(\tau_{zx} - \frac{\partial \tau_{zx}}{\partial z}\frac{dz}{2}\right) dx\, dy + \rho g_x\, dx\, dy\, dz = \rho\, dx\, dy\, dz\, \frac{Du}{Dt} \qquad (5.3.1)$$

FIGURE 5.2 Stress components in rectangular coordinates: (a) two-dimensional stress components; (b) three-dimensional stress components.

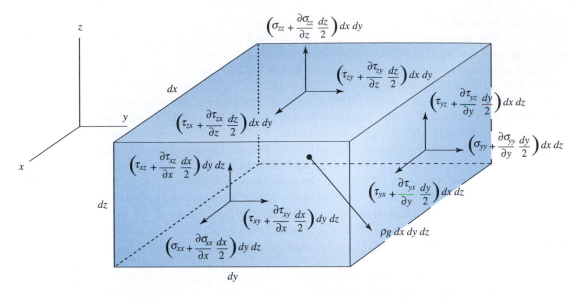

FIGURE 5.3 Forces acting on an infinitesimal fluid particle.

where the component of the gravity vector **g** acting in the x-direction is g_x, and Du/Dt is the x-component acceleration of the fluid particle (see Eq. 3.2.9). After we divide by the volume $dx\, dy\, dz$, the equation above simplifies to

$$\rho \frac{Du}{Dt} = \frac{\sigma_{xx}}{\partial x} + \frac{\partial \tau_{yx}}{\partial y} + \frac{\partial \tau_{zx}}{\partial z} + \rho g_x \tag{5.3.2}$$

Similarly, for the y- and z-directions we would have

$$\rho \frac{Dv}{Dt} = \frac{\partial \tau_{xy}}{\partial x} + \frac{\sigma_{yy}}{\partial y} + \frac{\partial \tau_{zy}}{\partial z} + \rho g_y$$

$$\rho \frac{Dw}{Dt} = \frac{\partial \tau_{xz}}{\partial x} + \frac{\partial \tau_{yz}}{\partial y} + \frac{\sigma_{zz}}{\partial z} + \rho g_z \tag{5.3.3}$$

We can show, by taking moments about axes passing through the center of the infinitesimal element, that

$$\tau_{yx} = \tau_{xy} \qquad \tau_{yz} = \tau_{zy} \qquad \tau_{xz} = \tau_{zx} \tag{5.3.4}$$

That is, the stress tensor is symmetric; so there are actually six independent stress components.

The stress tensor may be displayed in the usual way as

$$\tau_{ij} = \begin{pmatrix} \sigma_{xx} & \tau_{xy} & \tau_{xz} \\ \tau_{yx} & \sigma_{yy} & \tau_{yz} \\ \tau_{zx} & \tau_{zy} & \sigma_{zz} \end{pmatrix} \tag{5.3.5}$$

The subscripts i and j take on numerical values 1, 2, or 3. Then τ_{12} represents the element τ_{xy} in the first row, second column.

5.3.2 Euler's Equations

Good approximations to the components of the stress tensor for many flows, especially for flow away from a boundary (flow around an airfoil) or in regions of sudden change (flow through a contraction) are displayed by the array

$$\tau_{ij} = \begin{pmatrix} -p & 0 & 0 \\ 0 & -p & 0 \\ 0 & 0 & -p \end{pmatrix} \qquad (5.3.6)$$

For such flows, we have assumed the shear stress components that result from viscous effects to be negligibly small and the normal stress components to be equal to the negative of the pressure; this is precisely what we did in Fig. 3.16 when deriving Bernoulli's equation. If these stress components are introduced back into Eqs. 5.3.2 and 5.3.3 there results, for this frictionless flow,

KEY CONCEPT *We often assume shear stress components to be negligibly small.*

$$\rho \frac{Du}{Dt} = -\frac{\partial p}{\partial x} + \rho g_x$$

$$\rho \frac{Dv}{Dt} = -\frac{\partial p}{\partial y} + \rho g_y \qquad (5.3.7)$$

$$\rho \frac{Dw}{Dt} = -\frac{\partial p}{\partial z} + \rho g_z$$

Let us assume that the z-axis is vertical so that $g_x = g_y = 0$ and $g_z = -g$. The scalar equations above can then be written as the vector equation

$$\rho \frac{D}{Dt} (u\hat{\mathbf{i}} + v\hat{\mathbf{j}} + w\hat{\mathbf{k}}) = -\left(\frac{\partial p}{\partial x} \hat{\mathbf{i}} + \frac{\partial p}{\partial y} \hat{\mathbf{j}} + \frac{\partial p}{\partial z} \hat{\mathbf{k}} \right) - \rho g\hat{\mathbf{k}} \qquad (5.3.8)$$

In vector form, we have the well-known **Euler's equation**

$$\rho \frac{D\mathbf{V}}{Dt} = -\nabla p - \rho g\hat{\mathbf{k}} \qquad (5.3.9)$$

If we assume a constant-density, steady flow, Eq. 5.3.9 can be integrated along a streamline to yield Bernoulli's equation, a result that does not surprise us since the same assumptions were imposed when deriving Bernoulli's equation in Chapter 3; this will be illustrated in Example 5.6.

Euler's equation: *The three differential equations that result from applying Newton's second law and neglecting viscous effects.*

With the differential momentum equations in the form of Eqs. 5.3.7, we have added three additional equations to the continuity equation to give four equations and four unknowns, u, v, w, and p. With the appropriate boundary and initial conditions, a solution, yielding the velocity and pressure fields for this inviscid, incompressible flow, would be possible.

Example 5.5

A velocity field is proposed to be

$$u = \frac{10y}{x^2 + y^2} \qquad v = -\frac{10x}{x^2 + y^2} \qquad w = 0$$

(a) Is this a possible incompressible flow? (b) If so, find the pressure gradient ∇p assuming a frictionless air flow with the z-axis vertical. Use $\rho = 1.23$ kg/m^3.

Solution

(a) The continuity equation (5.2.9) is used to determine if the velocity field is possible. For this incompressible flow we have

$$\frac{\partial u}{\partial x} + \frac{\partial v}{\partial y} + \overset{0}{\cancel{\frac{\partial w}{\partial z}}} = 0$$

Substituting in the velocity components, we have

$$\frac{\partial}{\partial x}\left(\frac{10y}{x^2 + y^2}\right) + \frac{\partial}{\partial y}\left(-\frac{10x}{x^2 + y^2}\right) = \frac{-10y(2x)}{(x^2 + y^2)^2} - \frac{-10x(2y)}{(x^2 + y^2)^2} = \frac{1}{(x^2 + y^2)^2}[-20xy + 20xy] = 0$$

The quantity in brackets is obviously zero; hence the velocity field given is a possible incompressible flow.

(b) The pressure gradient is found using Euler's equation. In component form we have the following:

$$\rho \frac{Du}{Dt} = -\frac{\partial p}{\partial x} + \overset{0}{\cancel{\rho g_x}}$$

$$\therefore \frac{\partial p}{\partial x} = -\rho\left[u\frac{\partial u}{\partial x} + v\frac{\partial u}{\partial y} + w\overset{0}{\cancel{\frac{\partial u}{\partial z}}} + \overset{0}{\cancel{\frac{\partial u}{\partial t}}}\right]$$

$$= -1.23\left[\frac{10y}{x^2 + y^2}\frac{-20xy}{(x^2 + y^2)^2} + \frac{-10x}{x^2 + y^2}\frac{(x^2 + y^2)10 - 10y(2y)}{(x^2 + y^2)^2}\right]$$

$$= \frac{123x}{(x^2 + y^2)^2}$$

$$\rho \frac{Dv}{Dt} = -\frac{\partial p}{\partial y} + \overset{0}{\cancel{\rho g_y}}$$

$$\therefore \frac{\partial p}{\partial y} = -\rho\left[u\frac{\partial v}{\partial x} + v\frac{\partial v}{\partial y} + w\overset{0}{\cancel{\frac{\partial v}{\partial z}}} + \overset{0}{\cancel{\frac{\partial v}{\partial t}}}\right]$$

$$= -1.23\left[\frac{10y}{x^2 + y^2}\frac{(x^2 + y^2)(-10) + 10x(2x)}{(x^2 + y^2)^2} + \frac{-10x}{x^2 + y^2}\frac{20xy}{(x^2 + y^2)^2}\right]$$

$$= \frac{123y}{(x^2 + y^2)^2}$$

$$\rho \overset{0}{\cancel{\frac{Dw}{Dt}}} = -\frac{\partial p}{\partial z} + \rho g_z$$

$$\therefore \frac{\partial p}{\partial z} = \rho g_z = 1.23 \times (-9.81) = -12.07 \text{ Pa/m}$$

Thus $\qquad \nabla p = \frac{\partial p}{\partial x}\hat{\mathbf{i}} + \frac{\partial p}{\partial y}\hat{\mathbf{j}} + \frac{\partial p}{\partial z}\hat{\mathbf{k}} = \frac{123}{(x^2 + y^2)^2}(x\hat{\mathbf{i}} + y\hat{\mathbf{j}}) - 12.07\hat{\mathbf{k}} \text{ Pa/m}$

Example 5.6

Assume a steady, constant-density flow and integrate Euler's equation along a streamline in a plane flow.

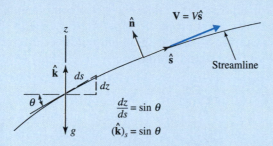

FIGURE E5.6

Solution

First, let us express the substantial derivative in streamline coordinates. Since the velocity vector is tangent to the streamline, we can write

$$\mathbf{V} = V\hat{\mathbf{s}}$$

where $\hat{\mathbf{s}}$ is the unit vector tangent to the streamline and V is the magnitude of the velocity, as shown in Fig. E5.6. The substantial derivative is then, for this plane flow,

$$\frac{D\mathbf{V}}{Dt} = \frac{\partial \mathbf{V}}{\partial t} + V\frac{\partial (V\hat{\mathbf{s}})}{\partial s} + \cancelto{0}{(V)_n}\frac{\partial \mathbf{V}}{\partial n} = \frac{\partial \mathbf{V}}{\partial t} + V\frac{\partial V}{\partial s}\hat{\mathbf{s}} + V^2\frac{\partial \hat{\mathbf{s}}}{\partial s}$$

The quantity $\partial\hat{\mathbf{s}}/\partial s$ results from the change of the unit vector $\hat{\mathbf{s}}$; the unit vector cannot change magnitude (it must always have a magnitude of 1), it can only change direction. Hence the derivative $\partial\hat{\mathbf{s}}/\partial s$ is in a direction normal to the streamline and does not enter the streamwise component equation. For a steady flow $\partial\mathbf{V}/\partial t = 0$. Consequently, in the streamwise direction, Euler's equation (5.3.9) takes the form

$$\rho V\frac{\partial V}{\partial s} = -\frac{\partial p}{\partial s} - \rho g\frac{\partial z}{\partial s}$$

recognizing that the component of $\hat{\mathbf{k}}$ along the streamline can be expressed as $(\hat{\mathbf{k}})_s = \partial z/\partial s$ (see the sketch above). Note that we use partial derivatives in this equation since velocity and pressure also vary with the normal coordinate.

The equation above can be written, assuming constant density so that $\partial\rho/\partial s = 0$, as

$$\frac{\partial}{\partial s}\left(\rho\frac{V^2}{2} + p + \rho g z\right) = 0$$

Integrating along the streamline results in

$$\rho\frac{V^2}{2} + p + \rho g z = \text{const.}$$

or

$$\frac{V^2}{2} + \frac{p}{\rho} + g z = \text{const.}$$

This is, of course, Bernoulli's equation. We have integrated along a streamline assuming constant density, steady flow, negligible viscous effects, and an inertial reference frame, so it is to be expected that Bernoulli's equation will emerge.

5.3.3 Navier–Stokes Equations

Newtonian fluids: *Fluids that possess a linear relationship between stress and the velocity gradients.*

Isotropic fluid: *A fluid whose properties are independent of direction at a given position.*

Many fluids exhibit a linear relationship between the stress components and the velocity gradients. Such fluids are called **Newtonian fluids** and include common fluids such as water, oil, and air. If in addition to linearity, we require that the fluid be **isotropic**,[4] it is possible to relate the stress components and the velocity gradients using only two fluid properties, the *viscosity* μ and the *second coefficient of viscosity* λ. The stress–velocity gradient relations, often referred to as the *constitutive equations*,[5] are stated as follows:

$$\sigma_{xx} = -p + 2\mu\frac{\partial u}{\partial x} + \lambda\,\nabla\cdot\mathbf{V} \qquad \tau_{xy} = \mu\left(\frac{\partial u}{\partial y} + \frac{\partial v}{\partial x}\right)$$

$$\sigma_{yy} = -p + 2\mu\frac{\partial v}{\partial y} + \lambda\,\nabla\cdot\mathbf{V} \qquad \tau_{xz} = \mu\left(\frac{\partial u}{\partial z} + \frac{\partial w}{\partial x}\right) \qquad (5.3.10)$$

$$\sigma_{zz} = -p + 2\mu\frac{\partial w}{\partial z} + \lambda\,\nabla\cdot\mathbf{V} \qquad \tau_{yz} = \mu\left(\frac{\partial v}{\partial z} + \frac{\partial w}{\partial y}\right)$$

For most gases, and for monatomic gases exactly, the second coefficient of viscosity is related to the viscosity by

$$\lambda = -\frac{2}{3}\mu \qquad (5.3.11)$$

a condition that is known as *Stokes's hypothesis*. With this relationship the negative average of the three normal stresses is equal to the pressure, that is,

$$-\frac{1}{3}(\sigma_{xx} + \sigma_{yy} + \sigma_{zz}) = p \qquad (5.3.12)$$

Using Eqs. 5.3.10, this can be shown to always be true for a liquid in which $\nabla\cdot\mathbf{V} = 0$, and with Stokes' hypothesis it is also true for a gas.

If we substitute the constitutive equations into the differential momentum equations (5.3.2) and (5.3.3), there results, using Stokes' hypothesis,

$$\rho\frac{Du}{Dt} = -\frac{\partial p}{\partial x} + \rho g_x + \mu\left(\frac{\partial^2 u}{\partial x^2} + \frac{\partial^2 u}{\partial y^2} + \frac{\partial^2 u}{\partial z^2}\right) + \frac{\mu}{3}\frac{\partial}{\partial x}\left(\frac{\partial u}{\partial x} + \frac{\partial v}{\partial y} + \frac{\partial w}{\partial z}\right)$$

$$\rho\frac{Dv}{Dt} = -\frac{\partial p}{\partial y} + \rho g_y + \mu\left(\frac{\partial^2 v}{\partial x^2} + \frac{\partial^2 v}{\partial y^2} + \frac{\partial^2 v}{\partial z^2}\right) + \frac{\mu}{3}\frac{\partial}{\partial y}\left(\frac{\partial u}{\partial x} + \frac{\partial v}{\partial y} + \frac{\partial w}{\partial z}\right) \qquad (5.3.13)$$

$$\rho\frac{Dw}{Dt} = -\frac{\partial p}{\partial z} + \rho g_z + \mu\left(\frac{\partial^2 w}{\partial x^2} + \frac{\partial^2 w}{\partial y^2} + \frac{\partial^2 w}{\partial z^2}\right) + \frac{\mu}{3}\frac{\partial}{\partial z}\left(\frac{\partial u}{\partial x} + \frac{\partial v}{\partial y} + \frac{\partial w}{\partial z}\right)$$

Homogeneous fluid: *A fluid whose properties are independent of position.*

where we have assumed a **homogeneous fluid**, that is, fluid properties (e.g., the viscosity) are independent of position.

For an incompressible flow the continuity equation allows the equations above to be reduced to

---

[4]The condition of isotropy exists if the fluid properties are independent of direction. Polymers are examples of anisotropic fluids.

[5]Details of the development of the constitutive equations can be found in any textbook on the subject of continuum mechanics.

$$\rho \frac{Du}{Dt} = -\frac{\partial p}{\partial x} + \rho g_x + \mu \left(\frac{\partial^2 u}{\partial x^2} + \frac{\partial^2 u}{\partial y^2} + \frac{\partial^2 u}{\partial z^2} \right)$$

$$\rho \frac{Dv}{Dt} = -\frac{\partial p}{\partial y} + \rho g_y + \mu \left(\frac{\partial^2 v}{\partial x^2} + \frac{\partial^2 v}{\partial y^2} + \frac{\partial^2 v}{\partial z^2} \right) \qquad (5.3.14)$$

$$\rho \frac{Dw}{Dt} = -\frac{\partial p}{\partial z} + \rho g_z + \mu \left(\frac{\partial^2 w}{\partial x^2} + \frac{\partial^2 w}{\partial y^2} + \frac{\partial^2 w}{\partial z^2} \right)$$

These are the **Navier–Stokes equations**, named after Louis M. H. Navier (1785–1836) and George Stokes (1819–1903); with these three differential equations and the differential continuity equation we have four equations and four unknowns, u, v, w, and p. The viscosity and density are fluid properties that are assumed to be known. With the appropriate boundary and initial conditions the equations can hopefully be solved. Several relatively simple geometries allow for analytical solutions; some of the solutions are presented in Chapter 7. Numerical solutions have also been determined for many flows of interest; computational methods are presented in Chapter 15. Because the equations are nonlinear partial differential equations (the acceleration terms cause the equations to be nonlinear as observed in Eqs. 3.2.9), we cannot be assured that the solution we find will actually be realized in the laboratory; that is, the solutions are not unique. For example, a laminar flow and a turbulent flow may have the identical initial and boundary conditions, yet the two flows (the two solutions) are very different.

Navier–Stokes equations: The three differential equations that result from applying Newton's second law.

The Navier–Stokes equations have not been solved for a turbulent flow. All turbulent flows are unsteady and three-dimensional and hence the time-derivative terms must be retained. This requires an initial condition on all dependent variables; i.e., u, v, w, and p must be known at all points in the flow field at $t = 0$. Such information would be extremely difficult, if not impossible, to obtain. To avoid this situation, time-averaged quantities are introduced for turbulent flows. This subject will be studied in a later chapter.

We can express the Navier–Stokes equations in vector form by multiplying equations 5.3.14 by $\hat{\mathbf{i}}$, $\hat{\mathbf{j}}$, and $\hat{\mathbf{k}}$, respectively, and adding. We recognize that

$$\frac{Du}{Dt}\hat{\mathbf{i}} + \frac{Dv}{Dt}\hat{\mathbf{j}} + \frac{Dw}{Dt}\hat{\mathbf{k}} = \frac{D\mathbf{V}}{Dt}$$

$$\frac{\partial p}{\partial x}\hat{\mathbf{i}} + \frac{\partial p}{\partial y}\hat{\mathbf{j}} + \frac{\partial p}{\partial z}\hat{\mathbf{k}} = \nabla p \qquad (5.3.15)$$

$$\nabla^2 u\hat{\mathbf{i}} + \nabla^2 v\hat{\mathbf{j}} + \nabla^2 w\hat{\mathbf{k}} = \nabla^2 \mathbf{V}$$

where we have used the Laplacian

$$\nabla^2 = \frac{\partial^2}{\partial x^2} + \frac{\partial^2}{\partial y^2} + \frac{\partial^2}{\partial z^2} \qquad (5.3.16)$$

Combining the above, the Navier–Stokes equations (5.3.14) take the vector form

$$\rho \frac{D\mathbf{V}}{Dt} = -\nabla p + \rho \mathbf{g} + \mu \nabla^2 \mathbf{V} \qquad (5.3.17)$$

Using this vector form we can express the Navier–Stokes equations using other coordinate systems. The equations are listed for cylindrical and spherical coordinates in Table 5.1.

Example 5.7

Simplify the x-component Navier–Stokes equation for steady flow in a horizontal, rectangular channel assuming all streamlines parallel to the walls. Let the x-direction be in the direction of flow (Fig. E5.7).

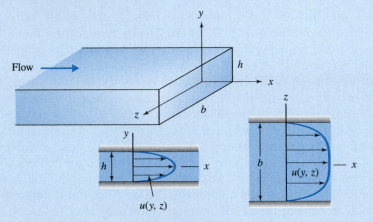

FIGURE E5.7

Solution

If the streamlines are parallel to the walls, only the x-component of velocity will be nonzero. Letting $v = w = 0$ the continuity equation (5.2.9) for an incompressible flow becomes

$$\frac{\partial u}{\partial x} = 0$$

showing that $u = u(y, z)$. The acceleration is then

$$\frac{Du}{Dt} = \overset{0}{\cancel{\frac{\partial u}{\partial t}}} + u\overset{0}{\cancel{\frac{\partial u}{\partial x}}} + \overset{0}{\cancel{v}}\,\frac{\partial u}{\partial y} + \overset{0}{\cancel{w}}\,\frac{\partial u}{\partial z} = 0$$

The x-component momentum equation then simplifies to

$$0 = -\frac{\partial p}{\partial x} + \overset{0}{\cancel{\rho g_x}} + \mu\left(\overset{0}{\cancel{\frac{\partial^2 u}{\partial x^2}}} + \frac{\partial^2 u}{\partial y^2} + \frac{\partial^2 u}{\partial z^2}\right)$$

or

$$\frac{\partial p}{\partial x} = \mu\left(\frac{\partial^2 u}{\partial y^2} + \frac{\partial^2 u}{\partial z^2}\right)$$

With the appropriate boundary conditions (the no-slip conditions), a solution to the foregoing equation could be sought. It would provide the velocity profiles sketched in Fig. E5.7.

5.3.4 Vorticity Equations

There are certain fluid flow phenomena that cannot be explained or understood without reference to the vorticity equations, equations that are derived from the Navier–Stokes equations (Example 5.8 provides such a phenomenon). In addition to providing insight into such phenomena, the vorticity equations do not contain the pressure or gravity terms found in the Navier–Stokes equations but contain terms involving the velocity only. Since boundary conditions most often involve only the velocity, the vorticity equations are often the equations of choice for numerical solutions.

To derive the vorticity equations, we take the curl of Eq. 5.3.17, the vector form of the Navier–Stokes equations. This is a difficult task, so we will not display all the steps here but simply outline the process. First, let us define the vorticity of Eqs. 3.2.21 in vector form using the del operator; using Eq. 5.2.6, we see that the three scalar equations 3.2.21 can be written as the single vector equation

$$\boldsymbol{\omega} = \nabla \times \mathbf{V} \tag{5.3.18}$$

where $\nabla \times \mathbf{V}$ is the curl of the velocity. The *curl* is the cross-product of the del operator and a vector function. Second, let us write the acceleration in vector form as

$$\mathbf{a} = \frac{D\mathbf{V}}{Dt} = \frac{\partial \mathbf{V}}{\partial t} + (\mathbf{V} \cdot \nabla)\,\mathbf{V} \tag{5.3.19}$$

where we have used Eqs. 3.2.8, 3.2.10, and 5.2.6. Finally, let us take the curl of the Navier–Stokes vector equation (5.3.17):

$$\nabla \times \left[\rho \frac{\partial \mathbf{V}}{\partial t} + \rho(\mathbf{V} \cdot \nabla)\mathbf{V} \right] = -\nabla \times \nabla p + \rho \nabla \times \mathbf{g} + \mu \nabla \times \nabla^2 \mathbf{V} \tag{5.3.20}$$

The curl of the gradient of a scalar function and the curl of a constant are both zero. Also, since we can interchange differentiation, we can write

$$\nabla \times \frac{\partial \mathbf{V}}{\partial t} = \frac{\partial}{\partial t} \nabla \times \mathbf{V} = \frac{\partial \boldsymbol{\omega}}{\partial t}$$

$$\nabla \times \nabla^2 \mathbf{V} = \nabla^2 (\nabla \times \mathbf{V}) = \nabla^2 \boldsymbol{\omega} \tag{5.3.21}$$

The difficult step, which we will leave as a homework problem, comes in showing that

$$\nabla \times [(\mathbf{V} \cdot \nabla)\mathbf{V}] = (\mathbf{V} \cdot \nabla)\boldsymbol{\omega} - (\boldsymbol{\omega} \cdot \nabla)\mathbf{V} \tag{5.3.22}$$

Vorticity equation: *Equation derived by taking the curl of the Navier–Stokes equation. The vorticity equation does not contain terms involving pressure or gravity.*

Equation 5.3.20 then becomes, assuming that ρ and μ are constants, the **vorticity equation**,

$$\frac{D\boldsymbol{\omega}}{Dt} = (\boldsymbol{\omega} \cdot \nabla)\,\mathbf{V} + \nu \nabla^2 \boldsymbol{\omega} \tag{5.3.23}$$

The vorticity equation can be written as three scalar equations. Using rectangular coordinates, the three vorticity equations are

$$\frac{D\omega_x}{Dt} = \omega_x \frac{\partial u}{\partial x} + \omega_y \frac{\partial u}{\partial y} + \omega_z \frac{\partial u}{\partial z} + \nu \nabla^2 \omega_x$$

$$\frac{D\omega_y}{Dt} = \omega_x \frac{\partial v}{\partial x} + \omega_y \frac{\partial v}{\partial y} + \omega_z \frac{\partial v}{\partial z} + \nu \nabla^2 \omega_y \tag{5.3.24}$$

$$\frac{D\omega_z}{Dt} = \omega_x \frac{\partial w}{\partial x} + \omega_y \frac{\partial w}{\partial y} + \omega_z \frac{\partial w}{\partial z} + \nu \nabla^2 \omega_z$$

KEY CONCEPT *The vorticity equations involve only the velocity and its derivatives.*

Vortex line: *A line to which the vorticity vector is tangent.*

Vortex tube: *A bundle of vortex lines.*

Since the vorticity is the curl of the velocity, note that all of the terms in the vorticity equations involve only the velocity and its derivatives. Consequently, the vorticity equations often become the equations of choice when solving problems requiring the differential equations of motion. In fact, we refer to vortex lines and vortex tubes as being similar to streamlines and streamtubes. A **vortex line** is a line to which the vorticity vector is tangent. A **vortex tube**, or simply a *vortex*, is a tube whose walls contain vortex lines. A vortex is shown in Fig. 5.4.

An interesting conclusion can be made by considering the vorticity equation 5.3.23. If an inviscid flow is everywhere irrotational (i.e., $\boldsymbol{\omega} = 0$ at all points in the flow), it must remain irrotational, since $D\boldsymbol{\omega}/Dt = 0$. This is referred to as the *persistence of irrotationality*. Also, if a uniform flow approaches an object, vorticity (rotation of fluid particles) is introduced to the flow only by the action of viscosity. Without viscous effects, vorticity cannot be created in an oncoming irrotational flow.

Plane flows are often of particular interest. If $w = 0$, $u = u(x, y)$, and $v = v(x, y)$ the only nonzero vorticity component is ω_z. For such a flow, the vorticity equation takes the simplified form

$$\frac{D\omega_z}{Dt} = \nu \nabla^2 \omega_z \tag{5.3.25}$$

KEY CONCEPT *Viscous effects are needed to cause vorticity changes in a plane flow.*

We observe that viscous effects are needed to cause vorticity changes in a plane flow.

The following example illustrates a flow phenomenon that can be explained quite easily with the use of the vorticity equation.

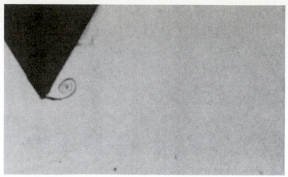

$t = 1.0$ s

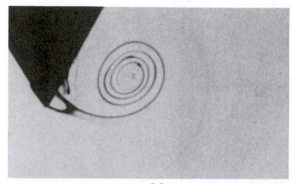

$t = 5.0$ s

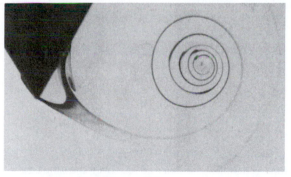

$t = 9.0$ s

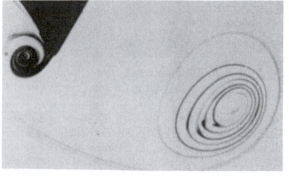

$t = 13.0$ s

FIGURE 5.4 Starting vortex on a wedge. A piston drives water normal to the axis of a wedge. Dye is injected into the water from small holes in the wedge surface. The characteristic Reynolds number is of order 1000. The piston stops at 12.5 s, producing a stopping vortex in the last photograph. (Courtesy of The Parabolic Press, Stanford, California. Reprinted with permission.)

Example 5.8

In a snowstorm, the snow is actually scooped out in front of a tree, or post, as shown in Fig. E5.8a. Explain this phenomenon by referring to the vorticity equations.

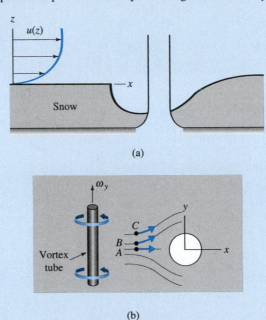

(a)

(b)

FIGURE E5.8

Solution

Let the velocity approaching the tree be in the x-direction with a velocity gradient $\partial u/\partial z$ near the ground. The vorticity components are then (refer to Eqs. 3.2.21)

$$\omega_x = 0 \qquad \omega_y = \frac{\partial u}{\partial z} \qquad \omega_z = 0$$

The vorticity equation (5.3.24) for ω_y, ignoring viscous effects over the short flow length, reduces to

$$\frac{D\omega_y}{Dt} = \omega_y \frac{\partial v}{\partial y}$$

Observe from Fig. E5.8b that in the vicinity of the tree $\partial v/\partial y$ is positive since $v_C > v_B > v_A$. ($\partial v/\partial y$ can be shown to be positive for negative y also.) Since ω_y and $\partial v/\partial y$ are both positive, $D\omega_y/Dt$ is positive and ω_y increases as the vortex tubes approach the tree. This increased vorticity creates a strong vortex in front of the tree resulting in the snow being scooped out as shown. This same phenomenon occurs in a sandstorm or in a water flow around a post in a riverbed.

5.4 DIFFERENTIAL ENERGY EQUATION

Most problems of interest in fluid mechanics do not involve temperature gradients; they involve flows in which the temperature is everywhere constant. For such flows it is not necessary to introduce the differential energy equation. There are situations, however, for both compressible and incompressible flows, in which temperature gradients are important, and for such flows the differential energy equation may be needed. We will derive the differential energy equation assuming negligible viscous effects, an assumption that significantly simplifies the derivation. Since the shear stresses that result from viscosity are quite small for many applications, this assumption may be acceptable. These shear stresses do, however, account for the high temperatures that burn up satellites on reentry to the atmosphere; if they are significant, they must be included in any analysis.

Consider the infinitesimal fluid element, shown in Fig. 5.5. The heat transfer rate $\dot{Q}$ through an area A is given by *Fourier's law of heat transfer*, named after Jean B. J. Fourier (1768–1830):

$$\dot{Q} = -KA \frac{\partial T}{\partial n} \qquad (5.4.1)$$

where n is the direction normal to the area, T is the temperature, and K is the *thermal conductivity*, assumed constant. The work rate done by a force is the magnitude of the force multiplied by the velocity in the direction of the force, that is,

$$\dot{W} = pAV \qquad (5.4.2)$$

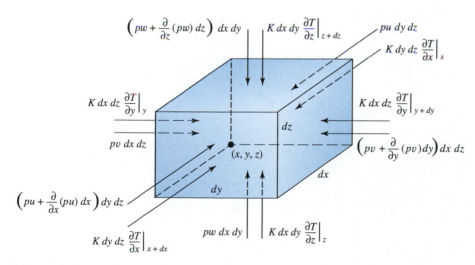

FIGURE 5.5   Rate of heat transfer and work rate on an infinitesimal fluid element.

where V is the velocity in the direction of the pressure force pA. The first law of thermodynamics (refer to Eq. 1.7.6) applied to a fluid particle is

$$\dot{Q} - \dot{W} = \frac{DE}{Dt} \tag{5.4.3}$$

where D/Dt is used since we are following a fluid particle at the instant shown. For the particle occupying the infinitesimal element of Fig. 5.5, the relationships above allow us to write

$$K \, dy \, dz \left(\frac{\partial T}{\partial x}\Big|_{x+dx} - \frac{\partial T}{\partial x}\Big|_x \right) - \frac{\partial}{\partial x}(pu) \, dx \, dy \, dz$$

$$+ K \, dx \, dz \left(\frac{\partial T}{\partial y}\Big|_{y+dy} - \frac{\partial T}{\partial y}\Big|_y \right) - \frac{\partial}{\partial y}(pv) \, dx \, dy \, dz$$

$$+ K \, dx \, dy \left(\frac{\partial T}{\partial z}\Big|_{z+dz} - \frac{\partial T}{\partial z}\Big|_z \right) - \frac{\partial}{\partial z}(pw) \, dx \, dy \, dz$$

$$= \rho \, dx \, dy \, dz \, \frac{D}{Dt}\left(\frac{u^2 + v^2 + w^2}{2} + gz + \tilde{u} \right) \tag{5.4.4}$$

where $\tilde{u}$ is the internal energy, E has included kinetic, potential, and internal energy, and the z-axis is assumed vertical. Also, since the mass of a fluid particle is constant $\rho \, dx \, dy \, dz$ is outside the D/Dt-operator. Divide both sides by $dx \, dy \, dz$. The result is

$$K\left(\frac{\partial^2 T}{\partial x^2} + \frac{\partial^2 T}{\partial y^2} + \frac{\partial^2 T}{\partial z^2} \right) - \frac{\partial}{\partial x}(\rho u) - \frac{\partial}{\partial y}(\rho v) - \frac{\partial}{\partial z}(\rho w)$$

$$= \rho \frac{D}{Dt}\left(\frac{u^2 + v^2 + w^2}{2} + gz + \tilde{u} \right) \tag{5.4.5}$$

This can be rearranged as follows:

$$K\left(\frac{\partial^2 T}{\partial x^2} + \frac{\partial^2 T}{\partial y^2} + \frac{\partial^2 T}{\partial z^2} \right) - p\left(\frac{\partial u}{\partial x} + \frac{\partial v}{\partial y} + \frac{\partial w}{\partial z} \right) - u \frac{\partial p}{\partial x} - v \frac{\partial p}{\partial y} - w \frac{\partial p}{\partial z}$$

$$= \rho u \frac{Du}{Dt} + \rho v \frac{Dv}{Dt} + \rho w \frac{Dw}{Dt} + \rho g \frac{Dz}{Dt} + \rho \frac{D\tilde{u}}{Dt} \tag{5.4.6}$$

The Euler's equations (5.3.7) are applicable for this inviscid flow; hence the last three terms on the left equal the first four terms on the right if we recognize that

$$\frac{Dz}{Dt} = \overset{0}{\cancel{\frac{\partial z}{\partial t}}} + u \overset{0}{\cancel{\frac{\partial z}{\partial x}}} + v \overset{0}{\cancel{\frac{\partial z}{\partial y}}} + w \frac{\partial z}{\partial z} = w \tag{5.4.7}$$

since x, y, z, t are all independent variables. The simplified energy equation then takes the form

$$\rho \frac{D\tilde{u}}{Dt} = K\left(\frac{\partial^2 T}{\partial x^2} + \frac{\partial^2 T}{\partial y^2} + \frac{\partial^2 T}{\partial z^2}\right) - p\left(\frac{\partial u}{\partial x} + \frac{\partial v}{\partial y} + \frac{\partial w}{\partial z}\right) \qquad (5.4.8)$$

In vector form this is expressed as

$$\rho \frac{D\tilde{u}}{Dt} = K\nabla^2 T - p\,\boldsymbol{\nabla}\cdot\mathbf{V} \qquad (5.4.9)$$

Before we simplify this equation for incompressible gas flow, let us write it in terms of enthalpy rather than internal energy. Using

$$\tilde{u} = h - \frac{p}{\rho} \qquad (5.4.10)$$

the energy equation becomes, using Eq. 5.2.7,

$$\rho \frac{Dh}{Dt} = K\nabla^2 T + \frac{Dp}{Dt} \qquad (5.4.11)$$

See Example 5.4 for the details of this conversion.

We have two special cases to be considered. First, for a liquid flow we can use $\boldsymbol{\nabla}\cdot\mathbf{V} = 0$ and with $\tilde{u} = c_p T$, c_p being the specific heat,[6] Eq. 5.4.9 simplifies to

$$\frac{DT}{Dt} = \alpha\nabla^2 T \qquad (5.4.12)$$

where we have introduced the *thermal diffusivity* α defined by

$$\alpha = \frac{K}{\rho c_p} \qquad (5.4.13)$$

---

[6]The specific heat for a liquid is often listed in tables as c_p. The specific heat at constant volume for a liquid is approximately equal to c_p. Hence we often simply drop the subscript and let $c_p = c$ for a liquid. It is assumed to be a constant, but it does depend on temperature. Here we will use c_p.

For an incompressible gas flow, an interesting result occurs. In Example 5.10 we will show that

$$\left| \frac{Dp}{Dt} \right| \ll |p \, \boldsymbol{\nabla} \cdot \mathbf{V}| \tag{5.4.14}$$

Thus, when comparing Eqs. 5.4.9 and 5.4.11, it is Eq. 5.4.11 that simplifies to

$$\rho c_p \frac{DT}{Dt} = K \nabla^2 T \tag{5.4.15}$$

for an incompressible gas flow if we make the ideal-gas assumption that

$$dh = c_p \, dT \tag{5.4.16}$$

If viscous effects are not negligible, the derivation would include the work input due to the shear stress components. This would add a term to the right-hand side of all of the differential energy equations above; this term is called the *dissipation function* Φ, which, in rectangular coordinates, is

$$\Phi = 2\mu \left[\left(\frac{\partial u}{\partial x} \right)^2 + \left(\frac{\partial v}{\partial y} \right)^2 + \left(\frac{\partial w}{\partial z} \right)^2 + \frac{1}{2} \left(\frac{\partial u}{\partial y} + \frac{\partial v}{\partial x} \right)^2 + \frac{1}{2} \left(\frac{\partial v}{\partial z} + \frac{\partial w}{\partial y} \right)^2 \right.$$

$$\left. + \frac{1}{2} \left(\frac{\partial u}{\partial z} + \frac{\partial w}{\partial x} \right)^2 \right] \tag{5.4.17}$$

With the addition of the energy equation, problems involving temperature variations in a flow can now be considered. Such problems are obviously present in compressible flows in which pressure and density are related to temperature by an equation of state. In incompressible gas flows (Mach number < 0.3) and liquid flows, temperature variations are often negligible so that the differential energy equation is not of interest. If, however, a temperature field does exist in a liquid flow or an incompressible gas flow (heat exchangers, atmospheric flows, lake inversions, lubrication flows, free convection flows), the energy equation provides an additional equation relating the quantities of interest. For liquid flows involving temperature gradients it is often necessary to assume that $\mu = \mu(T)$; in free convection flows we must assume that $\rho = \rho(T)$. In incompressible gas flows we can usually assume viscosity to be constant since the temperature variation is quite small.

Example 5.9

A constant-density liquid flows into a wide, rectangular horizontal channel, the walls of which are maintained at a higher temperature than the liquid, as shown in Fig. E5.9. Assume a variable μ, include viscous dissipation, and write the describing differential equations for a steady flow.

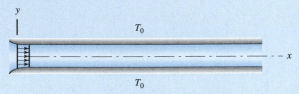

FIGURE E5.9

Solution

Let the x-axis coincide with centerline of the channel and the y-axis be vertical. The continuity equation would take the form

$$\nabla \cdot \mathbf{V} = \frac{\partial u}{\partial x} + \frac{\partial v}{\partial y} = 0$$

since $w = 0$ for the wide channel.

The flow will be primarily in the x-direction, but we must allow for variation of the y-component v. There will be no variation in the z-direction. The accelerations for this steady flow will be

$$\frac{Du}{Dt} = u \frac{\partial u}{\partial x} + v \frac{\partial u}{\partial y}$$

$$\frac{Dv}{Dt} = u \frac{\partial v}{\partial x} + v \frac{\partial v}{\partial y}$$

The stress terms of Eqs. 5.3.2 and 5.3.3 using Eqs. 5.3.10 with $\nabla \cdot \mathbf{V} = 0$, assuming a variable μ, become

$$\frac{\partial \sigma_{xx}}{\partial x} + \frac{\partial \tau_{xy}}{\partial y} = -\frac{\partial p}{\partial x} + \mu \left(\frac{\partial^2 u}{\partial x^2} + \frac{\partial^2 u}{\partial y^2} \right) + 2 \frac{\partial \mu}{\partial x} \frac{\partial u}{\partial x} + \frac{\partial \mu}{\partial y} \left(\frac{\partial u}{\partial y} + \frac{\partial v}{\partial x} \right)$$

$$\frac{\partial \tau_{xy}}{\partial x} + \frac{\partial \sigma_{yy}}{\partial y} = -\frac{\partial p}{\partial y} + \mu \left(\frac{\partial^2 v}{\partial x^2} + \frac{\partial^2 v}{\partial y^2} \right) + 2 \frac{\partial \mu}{\partial y} \frac{\partial v}{\partial y} + \frac{\partial \mu}{\partial x} \left(\frac{\partial u}{\partial y} + \frac{\partial v}{\partial x} \right)$$

The momentum equations are then

$$\rho \left(u \frac{\partial u}{\partial x} + v \frac{\partial u}{\partial y} \right) = -\frac{\partial p}{\partial x} + \mu \left(\frac{\partial^2 u}{\partial x^2} + \frac{\partial^2 u}{\partial y^2} \right) + 2 \frac{\partial \mu}{\partial x} \frac{\partial u}{\partial x} + \frac{\partial \mu}{\partial y} \left(\frac{\partial u}{\partial y} + \frac{\partial v}{\partial x} \right)$$

$$\rho \left(u \frac{\partial v}{\partial x} + v \frac{\partial v}{\partial y} \right) = -\frac{\partial p}{\partial y} + \mu \left(\frac{\partial^2 v}{\partial x^2} + \frac{\partial^2 v}{\partial y^2} \right) + 2 \frac{\partial \mu}{\partial y} \frac{\partial v}{\partial y} + \frac{\partial \mu}{\partial x} \left(\frac{\partial u}{\partial y} + \frac{\partial v}{\partial x} \right)$$

The energy equation simplifies to

$$u \frac{\partial T}{\partial x} + v \frac{\partial T}{\partial y} = \alpha \left(\frac{\partial^2 T}{\partial x^2} + \frac{\partial^2 T}{\partial y^2} \right) + \frac{2\mu}{c_p} \left[\left(\frac{\partial u}{\partial x} \right)^2 + \left(\frac{\partial v}{\partial y} \right)^2 + \frac{1}{2} \left(\frac{\partial u}{\partial y} + \frac{\partial v}{\partial x} \right)^2 \right]$$

where we have assumed K to be constant. The nonlinear, partial differential equations above, although quite formidable when attempting an analytic solution, could be solved numerically with the appropriate boundary conditions, and for a sufficiently low flow rate so that laminar flow exists (a turbulent flow is always unsteady and three-dimensional).

Example 5.10

Show that for an ideal gas, $|Dp/Dt| \ll |p\, \nabla \cdot \mathbf{V}|$ in a low-speed flow, thereby concluding that Eq. 5.4.15 is the appropriate equation.

Solution

Let us consider a steady, uniform flow in a pipe so that $|\mathbf{V}| = u$ and $Dp/Dt = u\, \partial p/\partial x$. Then the problem becomes: Show that

$$\left| u\, \frac{\partial p}{\partial x} \right| \ll \left| p\, \frac{\partial u}{\partial x} \right|$$

Viscous effects are small and would not change the conclusion, so we can ignore any possible viscous effects. Then Euler's equation (5.3.7) allows us to use

$$\frac{\partial p}{\partial x} = -\rho u\, \frac{\partial u}{\partial x}$$

Using the definition of the speed of sound (Eq. 1.7.17) and the equation of state, we see that

$$c = \sqrt{\frac{kp}{\rho}} \quad \text{or} \quad p = c^2 \frac{\rho}{k}$$

Thus

$$p\, \frac{\partial u}{\partial x} = \frac{c^2}{k}\, \rho\, \frac{\partial u}{\partial x}$$

Our problem can now be stated: Show that

$$\left| \rho u^2\, \frac{\partial u}{\partial x} \right| \ll \left| \frac{c^2}{k}\, \rho\, \frac{\partial u}{\partial x} \right|$$

Or, more simply, is it true that

$$u^2 \ll \frac{c^2}{k}?$$

This can be seen to be true since we have assumed for a low-speed gas flow that the speed of the gas is much less than the speed of sound (e.g., $u < 0.3c$ or M < 0.3). We know that k is of order unity ($k = 1.4$ for air), so it will not affect our conclusion that

$$\left| \frac{Dp}{Dt} \right| \ll |p\, \nabla \cdot \mathbf{V}|$$

5.5 SUMMARY

We have now completed our derivation of the partial differential equations that are used in describing flows of interest. Let's summarize the equations in vector form, for an incompressible flow:

Continuity: $$\nabla \cdot \mathbf{V} = 0 \tag{5.5.1}$$

Momentum: $$\rho \frac{D\mathbf{V}}{Dt} = -\nabla p - \rho \mathbf{g} + \mu \nabla^2 \mathbf{V} \tag{5.5.2}$$

Energy: $$\rho c_p \frac{DT}{Dt} = K\nabla^2 T + \Phi \quad \text{Liquids} \tag{5.5.3}$$

Energy: $$\rho c_p \frac{DT}{Dt} = K\nabla^2 T + \Phi \quad \text{Incompressible gases} \tag{5.5.4}$$

To express these equations in the forms above, we have assumed:

- A Newtonian fluid (a linear relationship between the stress components and the velocity gradients).
- An isotropic fluid (the fluid properties are independent of direction).
- A homogeneous fluid (the fluid properties μ, c_p, and K do not depend on position).
- An incompressible flow (the density of a particle is constant, that is, $D\rho/Dt = 0$; we do not demand that ρ = constant. For a gas flow we require that M < 0.3.)
- An inertial reference frame.

The vorticity equation is also of interest, It is

$$\frac{D\boldsymbol{\omega}}{Dt} = (\boldsymbol{\omega}\cdot\nabla)\mathbf{V} + p\,\nabla^2\boldsymbol{\omega} \tag{5.5.5}$$

Numerical methods often make use of this vorticity equation.

In the derivations of the differential equations in this chapter we have made no mention of laminar or turbulent flow. The equations are applicable to either class of flow. Some laminar flows in relatively simple geometries have been solved analytically and many others have been solved numerically. Turbulent flows, however, have not been solved even for the simplest geometry. A turbulent flow is always an unsteady flow and the presence of the time-derivative terms demands initial conditions; that is, at time $t = 0$ we must specify u, v, w and p at all points in the region of interest, information that is difficult to obtain even in a simple pipe flow.

KEY CONCEPT *The differential equations are applicable to laminar and turbulent flows.*

TABLE 5.1 Fundamental Laws for Incompressible Flows

Continuity

Cartesian

$$\frac{\partial u}{\partial x} + \frac{\partial v}{\partial y} + \frac{\partial w}{\partial z} = 0$$

Cylindrical

$$\frac{1}{r}\frac{\partial}{\partial r}(rv_r) + \frac{1}{r}\frac{\partial v_\theta}{\partial \theta} + \frac{\partial v_z}{\partial z} = 0$$

Spherical

$$\frac{1}{r^2}\frac{\partial}{\partial r}(r^2 v_r) + \frac{1}{r\sin\theta}\frac{\partial}{\partial\theta}(v_\theta \sin\theta) + \frac{1}{r\sin\theta}\frac{\partial v_\phi}{\partial\phi} = 0$$

Momentum

Cartesian

$$\frac{Du}{Dt} = -\frac{1}{\rho}\frac{\partial p}{\partial x} + g_x + \nu\nabla^2 u$$

$$\frac{Dv}{Dt} = -\frac{1}{\rho}\frac{\partial p}{\partial y} + g_y + \nu\nabla^2 v$$

$$\frac{Dw}{Dt} = -\frac{1}{\rho}\frac{\partial p}{\partial z} + g_z + \nu\nabla^2 w$$

$$\frac{D}{Dt} = \frac{\partial}{\partial t} + u\frac{\partial}{\partial x} + v\frac{\partial}{\partial y} + w\frac{\partial}{\partial z}$$

$$\nabla^2 = \frac{\partial^2}{\partial x^2} + \frac{\partial^2}{\partial y^2} + \frac{\partial^2}{\partial z^2}$$

Cylindrical

$$\frac{Dv_r}{Dt} - \frac{v_\theta^2}{r} = -\frac{1}{\rho}\frac{\partial p}{\partial r} + g_r + \nu\left(\nabla^2 v_r - \frac{v_r}{r^2} - \frac{2}{r^2}\frac{\partial v_\theta}{\partial\theta}\right)$$

$$\frac{Dv_\theta}{Dt} + \frac{v_r v_\theta}{r} = -\frac{1}{\rho r}\frac{\partial p}{\partial\theta} + g_\theta + \nu\left(\nabla^2 v_\theta + \frac{2}{r^2}\frac{\partial v_r}{\partial\theta} - \frac{v_\theta}{r^2}\right)$$

$$\frac{Dv_z}{Dt} = -\frac{1}{\rho}\frac{\partial p}{\partial z} + g_z + \nu\nabla^2 v_z$$

$$\frac{D}{Dt} = \frac{\partial}{\partial t} + v_r\frac{\partial}{\partial r} + \frac{v_\theta}{r}\frac{\partial}{\partial\theta} + v_z\frac{\partial}{\partial z}$$

$$\nabla^2 = \frac{\partial^2}{\partial r^2} + \frac{1}{r}\frac{\partial}{\partial r} + \frac{1}{r^2}\frac{\partial^2}{\partial\theta^2} + \frac{\partial^2}{\partial z^2}$$

Spherical

$$\frac{Dv_r}{Dt} - \frac{v_\theta^2 + v_\phi^2}{r}$$
$$= -\frac{1}{\rho}\frac{\partial p}{\partial r} + g_r + \nu\left(\nabla^2 v_r - \frac{2v_r}{r^2} - \frac{2}{r^2}\frac{\partial v_\phi}{\partial\phi}\right.$$
$$\left. - \frac{2v_\theta\cot\theta}{r^2} - \frac{2}{r^2\sin\theta}\frac{\partial v_\phi}{\partial\phi}\right)$$

$$\frac{Dv_\theta}{Dt} + \frac{v_r v_\theta - v_\phi^2\cot\theta}{r}$$
$$= -\frac{1}{\rho r}\frac{\partial p}{\partial\theta} + g_\theta + \nu\left(\nabla^2 v_\theta + \frac{2}{r^2}\frac{\partial v_r}{\partial\theta}\right.$$
$$\left. - \frac{v_\theta}{r^2\sin^2\theta} - \frac{2\cos\theta}{r^2\sin^2\theta}\frac{\partial v_\phi}{\partial\phi}\right)$$

$$\frac{Dv_\phi}{Dt} + \frac{v_\phi v_r + v_\theta v_\phi \cot\theta}{r}$$
$$= -\frac{1}{\rho r\sin\theta}\frac{\partial p}{\partial\phi} + g_\phi + \nu\left(\nabla^2 v_\phi - \frac{v_\phi}{r^2\sin^2\theta}\right.$$
$$\left. + \frac{2}{r^2\sin\theta}\frac{\partial v_r}{\partial\phi} + \frac{2\cos\theta}{r^2\sin^2\theta}\frac{\partial v_\theta}{\partial\phi}\right)$$

$$\frac{D}{Dt} = \frac{\partial}{\partial t} + v_r\frac{\partial}{\partial r} + \frac{v_\theta}{r}\frac{\partial}{\partial\theta} + \frac{v_\phi}{r\sin\theta}\frac{\partial}{\partial\phi}$$

$$\nabla^2 = \frac{1}{r^2}\frac{\partial}{\partial r}\left(r^2\frac{\partial}{\partial r}\right) + \frac{1}{r^2\sin\theta}\frac{\partial}{\partial\theta}\left(\sin\theta\frac{\partial}{\partial\theta}\right) + \frac{1}{r^2\sin^2\theta}\frac{\partial^2}{\partial\phi^2}$$

Energy

Cartesian

$$\rho\frac{Dh}{Dt} = K\nabla^2 T + 2\mu\left[\left(\frac{\partial u}{\partial x}\right)^2 + \left(\frac{\partial v}{\partial y}\right)^2 + \left(\frac{\partial w}{\partial z}\right)^2\right.$$
$$+ \frac{1}{2}\left(\frac{\partial u}{\partial y} + \frac{\partial v}{\partial x}\right)^2 + \frac{1}{2}\left(\frac{\partial v}{\partial z} + \frac{\partial w}{\partial y}\right)^2$$
$$\left. + \frac{1}{2}\left(\frac{\partial u}{\partial z} + \frac{\partial w}{\partial x}\right)^2\right]$$

Cylindrical

$$\rho\frac{Dh}{Dt} = K\nabla^2 T + 2\mu\left[\left(\frac{\partial v_r}{\partial r}\right)^2 + \left(\frac{1}{r}\frac{\partial v_\theta}{\partial\theta} + \frac{v_r}{r}\right)^2 + \left(\frac{\partial v_z}{\partial z}\right)^2\right.$$
$$+ \frac{1}{2}\left(\frac{1}{r}\frac{\partial v_z}{\partial\theta} + \frac{\partial v_\theta}{\partial z}\right)^2 + \frac{1}{2}\left(\frac{\partial v_r}{\partial z} + \frac{\partial v_z}{\partial r}\right)^2$$
$$\left. + \frac{1}{2}\left(\frac{1}{r}\frac{\partial v_r}{\partial\theta} + \frac{\partial v_\theta}{\partial r} - \frac{v_\theta}{r}\right)^2\right]$$

Spherical

$$\rho\frac{Dh}{Dt} = K\nabla^2 T + 2\mu\left[\left(\frac{\partial v_r}{\partial r}\right)^2 + \left(\frac{1}{r}\frac{\partial v_\theta}{\partial\theta} + \frac{v_r}{r}\right)^2\right.$$
$$+ \left(\frac{1}{r\sin\theta}\frac{\partial v_\phi}{\partial\phi} + \frac{v_r}{r} + \frac{v_\theta\cot\theta}{r}\right)^2$$
$$+ \mu\left[\left(\frac{1}{r\sin\theta}\frac{\partial v_r}{\partial\phi} + \frac{\sin\theta}{r}\frac{\partial}{\partial\theta}\left(\frac{v_\theta}{\sin\theta}\right)\right)^2\right.$$
$$+ \left(\frac{1}{r\sin\theta}\frac{\partial v_r}{\partial\phi} + r\frac{\partial}{\partial r}\left(\frac{v_\phi}{r}\right)\right)^2$$
$$\left. + \left(r\frac{\partial}{\partial r}\left(\frac{v_\theta}{r}\right) + \frac{1}{r}\frac{\partial v_r}{\partial\theta}\right)^2\right]$$

Stresses

Cartesian

$$\sigma_{xx} = -p + 2\mu\frac{\partial u}{\partial x} \qquad \tau_{xy} = \mu\left(\frac{\partial u}{\partial y} + \frac{\partial v}{\partial x}\right)$$

$$\sigma_{yy} = -p + 2\mu\frac{\partial v}{\partial y} \qquad \tau_{yz} = \mu\left(\frac{\partial v}{\partial z} + \frac{\partial w}{\partial y}\right)$$

$$\sigma_{zz} = -p + 2\mu\frac{\partial w}{\partial z} \qquad \tau_{xz} = \mu\left(\frac{\partial u}{\partial z} + \frac{\partial w}{\partial x}\right)$$

Cylindrical

$$\sigma_{rr} = -p + 2\mu\frac{\partial v_r}{\partial r} \qquad \tau_{r\theta} = \mu\left[r\frac{\partial}{\partial r}\left(\frac{v_\theta}{r}\right) + \frac{1}{r}\frac{\partial v_r}{\partial\theta}\right]$$

$$\sigma_{\theta\theta} = -p + 2\mu\left(\frac{1}{r}\frac{\partial v_\theta}{\partial\theta} + \frac{v_r}{r}\right) \qquad \tau_{\theta z} = \mu\left[\frac{\partial v_\theta}{\partial z} + \frac{1}{r}\frac{\partial v_z}{\partial\theta}\right]$$

$$\sigma_{zz} = -p + 2\mu\frac{\partial v_z}{\partial z} \qquad \tau_{rz} = \mu\left[\frac{\partial v_r}{\partial z} + \frac{\partial v_z}{\partial r}\right]$$

Spherical

$$\sigma_{rr} = -p + 2\mu\frac{\partial v_r}{\partial r}$$

$$\sigma_{\theta\theta} = -p + 2\mu\left(\frac{1}{r}\frac{\partial v_\theta}{\partial\theta} + \frac{v_r}{r}\right)$$

$$\sigma_{\phi\phi} = -p + 2\mu\left(\frac{1}{r\sin\theta}\frac{\partial v_\phi}{\partial\phi} + \frac{v_r}{r} + \frac{v_\theta\cot\theta}{r}\right)$$

$$\tau_{r\theta} = \mu\left[r\frac{\partial}{\partial r}\left(\frac{v_\theta}{r}\right) + \frac{1}{r}\frac{\partial v_r}{\partial\theta}\right]$$

$$\tau_{\theta\phi} = \mu\left[\frac{\sin\theta}{r}\frac{\partial}{\partial\theta}\left(\frac{v_\phi}{\sin\theta}\right) + \frac{1}{r\sin\theta}\frac{\partial v_\theta}{\partial\phi}\right]$$

$$\tau_{r\phi} = \mu\left[\frac{1}{r\sin\theta}\frac{\partial v_r}{\partial\phi} + r\frac{\partial}{\partial r}\left(\frac{v_\phi}{r}\right)\right]$$

PROBLEMS

Differential Continuity Equation

5.1 The divergence theorem (also knows as Gauss's theorem) states that

$$\int_A \mathbf{V} \cdot \hat{\mathbf{n}}\, dA = \int_V \nabla \cdot \mathbf{V}\, dV$$

where $\mathbf{V}$ represents any vector and A completely surrounds the volume V. Apply this theorem to the integral continuity equation 4.3.3 and derive the differential continuity equation 5.2.5.

5.2 Use the infinitesimal element shown in Fig. P5.2 and derive the differential continuity equation in cylindrical coordinates. The velocity vector is $\mathbf{V} = (v_r, v_\theta, v_z)$.

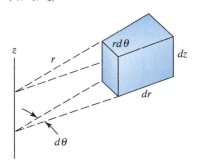

FIGURE P5.2

5.3 Use the infinitesimal elements shown in Fig. P5.3 and derive the differential continuity equation in spherical coordinates. The velocity vector is $\mathbf{V} = (v_r, v_\theta, v_\phi)$.

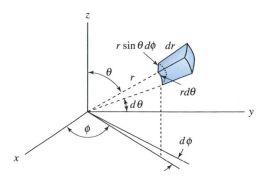

FIGURE P5.3

5.4 A uniform compressible flow occurs in a constant-diameter pipe. Write the simplified differential continuity equation for the steady flow.

5.5 An incompressible flow of air over the mountain range shown in Fig. P5.5 can be approximated by a plane, steady flow. If the z-axis is vertical and density is allowed to vary, write the differential equations that result from conservation of mass considerations.

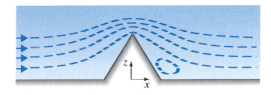

FIGURE P5.5

5.6 A stratified flow of salt water in which the density increases with depth occurs over an obstruction in the bottom of the channel of Fig. P5.6. Assuming a plane steady flow with the z-axis vertical, write the equations that result from the differential continuity equation.

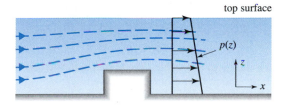

FIGURE P5.6

5.7 Show that for an isothermal compressible flow,

$$\frac{1}{p}\frac{Dp}{Dt} = -\nabla \cdot \mathbf{V}$$

5.8 An incompressible fluid is flowing radially into a sink (treated as a line or a point at the origin). Determine an expression for the radial velocity component if it is:

(a) A line sink   (b) A point sink

5.9 A compressible flow occurs such that

$$u = 200xy \qquad v = 200(x^2 + y^2) \qquad w = 0 \text{ m/s}$$

Find the rate at which the density is changing at the point (2 m, 1 m) where $\rho = 2.3 \text{ kg/m}^3$.

5.10 If, in an incompressible plane flow, the velocity component $u = \text{const}$, what can we say about the y-component of velocity? About the density?

5.11 In an incompressible flow we know that u and v are both nonzero but constant in magnitude. What can we infer about w from the differential continuity equation? About the density?

5.12 In an incompressible plane flow $u = Ax$. Find $v(x, y)$ if $v(x, 0) = 0$.

5.13 If the velocity component u is given by

$$u(x, y) = 10 + \frac{5x}{x^2 + y^2}$$

in an incompressible plane flow, determine $v(x, y)$. Let $v(x, 0) = 0$.

5.14 The θ-component of velocity is given by

$$v_\theta = -\left(10 + \frac{0.4}{r^2}\right) \cos \theta$$

Find the r-component of velocity for the incompressible plane flow if $v_r(0.2, \theta) = 0$.

5.15 In an incompressible plane flow

$$v_\theta = 20\left(1 + \frac{1}{r^2}\right) \sin \theta - \frac{40}{r}$$

Find $v_r(r, \theta)$ if $v_r(1, \theta) = 0$.

5.16 In an incompressible axisymmetric flow ($v_\phi = 0$) the velocity component v_θ is given by

$$v_\theta = -\left(10 + \frac{40}{r^3}\right) \sin \theta$$

Find $v_r(r, \theta)$ if $v_r(2, \theta) = 0$.

5.17 The velocity of air in a pipe is measured at points 5 cm apart to be 151, 162, and 175 m/s, respectively. At the middle point the temperature is 5°C and the pressure is 125 kPa. Find $d\rho/dx$ at the middle point of this steady, uniform flow.

5.18 The x-component velocity on the x-axis (Fig. P5.18) is proposed to be $u(x) = -20 (1 - e^{-x})$ m/s. Approximate the y-component velocity at the point

(2, 0.2) in this plane incompressible flow. The coordinates are in meters.

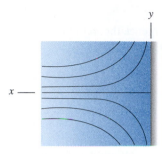

FIGURE P5.18

5.19 Assume the flow in Problem 5.18 is axisymmetric and replace y with r and x with z. Approximate the r-component velocity at (2, 0.2). The coordinates are in meters.

5.20 The velocity component along the x-axis (Fig. P5.20) is $u(x) = 10 - 40/x^2$ m/s. What is the radius of the cylinder? Approximate the y-component velocity at $(-3, 0.1)$ assuming an incompressible flow. The coordinates are in meters.

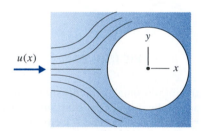

FIGURE P5.20

5.21 Assume the flow in Problem 5.20 represents flow around a sphere and let $v_r(r) = (40/r^2) - 10$ along the negative x-axis. What is the radius of the sphere? Approximate the θ-component of velocity at $(-3, 0.1)$ assuming an incompressible flow. The coordinates are in meters.

5.22 The x-component of the velocity vector is measured, at points A, B, and C 5 mm apart, as 11.3, 12.6, and 13.5 m/s, respectively, in the incompressible, steady plane flow shown in Fig. P5.22.

Estimate:

(a) The y-component of velocity 4 mm above point B.

(b) The acceleration at point B.

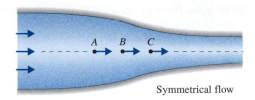

Symmetrical flow

FIGURE P5.22

Differential Momentum Equation

5.23 Sum forces on the element of Fig. 5.3 in the y-direction and show that Eq. 5.3.3a results.

5.24 Does the velocity field

$$u = \frac{10x}{x^2 + y^2} \qquad v = \frac{10y}{x^2 + y^2} \qquad w = 0$$

represent a possible incompressible flow? If so, find the pressure gradient ∇p assuming a frictionless flow with negligible body forces.

5.25 Does the velocity field

$$v_r = 10\left(1 - \frac{1}{r^2}\right)\cos\theta$$

$$v_\theta = -10\left(1 + \frac{1}{r^2}\right)\sin\theta$$

$$v_z = 0$$

represent a possible incompressible flow? If so, find the pressure gradient ∇p assuming a frictionless flow with negligible body forces.

5.26 Consider the velocity field

$$v_r = 10\left(1 - \frac{8}{r^3}\right)\cos\theta$$

$$v_\theta = -10\left(1 + \frac{4}{r^3}\right)\sin\theta$$

$$v_\phi = 0$$

Does it represent an incompressible flow? If so, find the pressure gradient ∇p assuming a frictionless flow, neglecting the body force.

5.27 For a steady plane flow shown in Fig. P5.27, find an expression for DV/Dt in terms of coordinates (s, n) tangential and normal to a streamline. Let R be the radius of curvature.

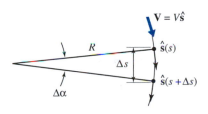

FIGURE P5.27

5.28 Write Euler's equation if the velocity is referred to a reference frame that is rotating with constant angular velocity.

5.29 A velocity field is given by $u = 10\,(y - 30y^2)$ m/s, $v = 0$, and $w = 0$. Display the stress components at $y = 2.5$ mm using $\mu = 5 \times 10^{-4}$ N-s/m^2 and $p = 200$ kPa. Find the ratio τ_{xy}/σ_{xx}.

5.30 The velocity field near a surface is approximated by $u = 10\,(2y/\delta - y^2/\delta^2)$, where $\delta = Cx^{4/5}$. If $\delta = 8$ m at $x = 1000$ m, find $v(x, y)$ assuming that $w = 0$ and $v(x, 0) = 0$. Also, display the stress components at $(1000, 0)$ using $\mu = 2 \times 10^{-5}$ N·s/m^2 and $p = 100$ kPa. Assume an incompressible flow.

5.31 Show that for a steady flow Du/Dt can be written as $(\mathbf{V}\cdot\nabla)\,u$, and that $D\mathbf{V}/Dt = (\mathbf{V}\cdot\nabla)\,\mathbf{V}$. Verify using rectangular coordinates.

5.32 Write the compressible flow differential momentum equations (5.3.13) as one equation in vector form.

5.33 Simplify the Navier–Stokes equations for incompressible steady flow between horizontal parallel plates assuming that $u = u(y)$; $w = 0$. Write all three equations.

5.34 Simplify the Navier–Stokes equations for incompressible steady flow in a horizontal pipe assuming that $v_z = v_z(r)$, $v_\theta = 0$. Write all three equations.

5.35 Fluid flows in the small gap between concentrically rotating spheres such that $v_\theta = v_\theta(r)$ and $v_\phi = 0$. Simplify the Navier–Stokes equations neglecting gravity for steady incompressible flow.

5.36 Substitute the constitutive equations (5.3.10) for incompressible flow into the differential momentum equations 5.3.2 and 5.3.3 and derive the Navier–Stokes equations 5.3.14.

5.37 In Eqs. 5.3.14 the viscosity is assumed to be constant. If the temperature is not constant, as in a liquid flow with temperature gradients, we must let $\mu = \mu(T)$ so that $\mu = \mu(x, y, z)$ since $T = T(x, y, z)$. Modify Eqs. 5.3.14 to account for variable viscosity.

5.38 A large flat plate oscillates beneath a liquid as shown in Fig. P5.38. Write the differential equation that describes the motion if the plane laminar flow moves only parallel to the plate. Assume that $\mu = $ const.

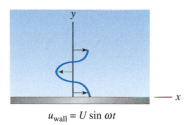

$u_{\text{wall}} = U \sin \omega t$

FIGURE P.538

5.39 For a gas flow in which Stokes's hypothesis is not applicable, the negative average of the three normal stresses, denoted $\bar{p}$, may be different from the pressure p. Find an expression for $(\bar{p} - p)$.

Vorticity

5.40 Show that the relationship (5.3.22) is indeed true by using rectangular coordinates.

5.41 A uniform flow exists on the flat plate of Fig. P5.41 oriented parallel to the flow. The plate has a very sharp leading edge. Identify the term that is responsible for the creation of vorticity.

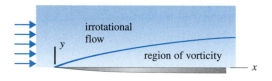

irrotational flow

region of vorticity

FIGURE P5.41

5.42 Neglect viscous effects and determine the velocity profile just downstream of the contraction at section 2 in the plane flow shown in Fig. P5.42.

Hint: In an inviscid flow, the fluid does not stick to the wall.

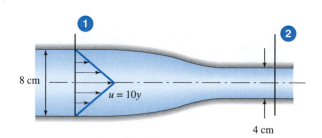

8 cm

$u = 10y$

4 cm

FIGURE P5.42

5.43 In Example 5.8, does ω_x remain zero as the vortex tube nears the tree? If not, explain why.

Differential Energy Equation

5.44 Derive the incompressible differential energy equation by applying Gauss's theorem (see Problem 5.1) to the integral energy equation 4.5.13 assuming that $\dot{W}_s = \dot{W}_{\text{shear}} = \dot{W}_l = 0$ and using

$$\dot{Q}_s = \int_{\text{c.s.}} K\nabla T \cdot \hat{\mathbf{n}} \, dA \text{ assuming no viscous effects.}$$

5.45 Verify that Eq. 5.4.5 follows from Eq. 5.4.4. Recall the definition of a second derivative.

5.46 Verify that Eq. 5.4.11 follows from Eq. 5.4.9.

5.47 Simplify the differential energy equation for a liquid flow in which the temperature gradients are quite large and the velocity components are very small, such as in a lake heated from above.

5.48 Explain which term in the differential energy equation accounts for the extremely high temperatures that exist on satellites during reentry.

5.49 The velocity distribution in a 2.0-cm-diameter pipe is given by $u(r) = 10\,(1 - 10\,000r^2)$ m/s. Find the magnitude of the dissipation function at the wall, at the centerline, and halfway between for air at 20°C.

5.50 Let the plate in Problem 5.38 be heated. Write the simplified differential energy equation and the simplified Navier–Stokes equation assuming:

(a) $\mu = \text{const.}$

(b) $\mu = \mu(T)$.

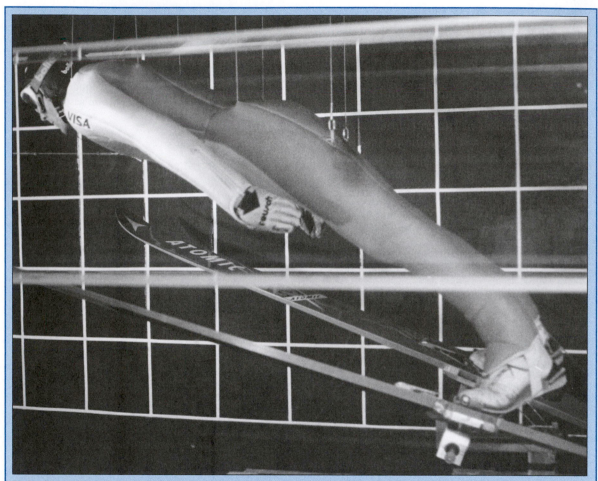

U.S. Olympian ski jumpers have used wind tunnels to improve their technique. The jumper is suspended by cables attached to his chest and thighs, enabling his lift and drag to be measured. The sport has recently been revolutionized by opening skis like a *V* into the wind. (Calspan Corporation, Buffalo, New York)

6

Dimensional Analysis and Similitude

Outline

Chapter Objectives

The objectives of this chapter are to:
▲ Establish the parameters necessary to guide experimental studies.
▲ Present the technique used to apply the results of model studies to prototypes for a variety of flow situations.
▲ Extract the flow parameters from the differential equations and boundary conditions used to guide computational studies.
▲ Provide examples and problems that illustrate how dimensionless flow parameters are used, how model studies allow us to predict quantities of interest on a prototype, and how normalized differential equations are utilized.

6.1 INTRODUCTION

There are many problems of interest in the field of fluid mechanics in the real world of design that cannot be solved using the differential and integral equations only. It is often necessary to resort to experimental methods to establish relationships between the variables of interest. Since experimental studies are usually quite expensive, it is necessary to keep the required experimentation to a minimum. This is done using a technique called *dimensional analysis*, which is based on the notion of **dimensional homogeneity**—that all terms in an equation must have the same dimensions. For example, if we write Bernoulli's equation in the form

$$\frac{V_1^2}{2g} + \frac{p_1}{\gamma} + z_1 = \frac{V_2^2}{2g} + \frac{p_2}{\gamma} + z_2 \tag{6.1.1}$$

we note that the dimension of each term is length. Furthermore, if we factored out z_1 from the left-hand side and z_2 from the right-hand side, we would have

$$\frac{V_1^2}{2gz_1} + \frac{p_1}{\gamma z_1} + 1 = \left(\frac{V_2^2}{2gz_2} + \frac{p_2}{\gamma z_2} + 1 \right) \frac{z_2}{z_1} \tag{6.1.2}$$

In this form of Bernoulli's equation the terms are all dimensionless and we have written the equation as a combination of dimensionless parameters, the basic idea in dimensional analysis that will be presented in the next section.

Often in experimental work we are required to perform experiments on objects that are quite large, too large to experiment with for a reasonable cost. This would include flows over weirs and dams; wave interactions with piers and breakwaters; flows around submarines and ships; subsonic and supersonic flows around aircraft; flows around stadiums and buildings, as shown in Fig. 6.1; flows through large pumps and turbines; and flows around automobiles and trucks. Such flows are usually studied in laboratories using models that are smaller than the prototype, the actual device. This substantially reduces the costs when compared with full-scale studies and allows for the study of various configurations or flow conditions.

There are also flows of interest that involve rather small dimensions, such as flow around a turbine blade, flow into a capillary tube, flow around a microorganism, flow through a small control valve, and flow around and inside a falling droplet. These flows would require that the model be larger than the prototype so that observations could be made with an acceptable degree of accuracy.

Similitude is the study of predicting prototype conditions from model observations. This will be presented following dimensional analysis. Similitude involves the use of the dimensionless parameters obtained in dimensional analysis.

There are two approaches that can be used in the study of dimensional analysis and similitude. We will present both approaches. First, we use the **Buckingham π-theorem**, which organizes the steps of ensuring dimensional homogeneity; it requires some knowledge of the phenomenon being studied in order that the appropriate quantities of interest are included. Second, we extract the dimensionless parameters that influence a particular flow situation from the differential equations and boundary conditions that are needed to describe the phenomenon being investigated.

FIGURE 6.1   Scale model of the large buildings in a city. Air flow around the buildings is studied. The roughening elements on the floor generate the desired wall turbulence. (Courtesy of Fluid Mechanics and Diffusion Laboratory, Colorado State University.)

6.2 DIMENSIONAL ANALYSIS

6.2.1 Motivation

In the study of phenomena involving fluid flows, either analytically or experimentally, there are invariably many flow and geometric parameters involved. In the interest of saving time and money the fewest possible combinations of parameters should be utilized. For example, consider the pressure drop across the slider valve of Fig. 6.2. We may suspect that the pressure drop depends on such parameters as pipe mean velocity V, the density ρ of the fluid, the fluid viscosity μ, the pipe diameter d, and the gap height h. This could be expressed as

$$\Delta p = f(V, \rho, \mu, d, h) \tag{6.2.1}$$

Now, if we attempt an experimental study of this problem, consider the strategy for finding the dependence of the pressure drop on the parameters involved. We could fix all parameters except the velocity and investigate the dependence of the pressure drop on the average velocity. Then the diameter could be changed and the experiment repeated. This would lead to the set of results shown in Fig. 6.3a. Following that set of experiments the gap height h could be changed, leading to the curves of Fig. 6.3b. Again, different fluids could be studied, leading to curves with ρ and μ changing values.

Consider next the notion that any equation that relates a certain set of variables, such as Eq. 6.2.1, can be written in terms of dimensionless parameters, as

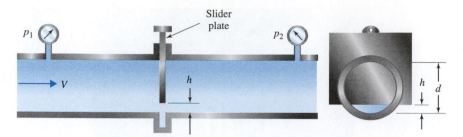

FIGURE 6.2 Flow around a slider valve.

was done with the Bernoulli equation (6.1.2). We can organize the variables of Eq. 6.2.1 into dimensionless parameters (the steps needed to do this will be presented in a subsequent section) as follows:

$$\frac{\Delta p}{\rho V^2} = f\left(\frac{V\rho d}{\mu}, \frac{h}{d}\right) \tag{6.2.2}$$

Obviously, this is a much simpler relationship. We could perform an experiment with a fixed h/d (say, $h/d = 0.1$) by varying $V\rho d/\mu$ (this is done by simply varying V), resulting in a curve as shown in Fig. 6.4. The quantity h is changed so that $h/d = 0.5$ and the testing is repeated. Finally, the entire experiment is presented in one figure, as in Fig. 6.4. This has greatly reduced the effort and the cost in determining the actual form of $f(V\rho d/\mu, h/d)$; we would use only one pipe and one valve, and we would use only one fluid.

It is not always clear, however, which parameters should be included in an equation such as 6.2.1. The selection of these parameters requires a detailed understanding of the physics involved. In the selection of the parameters that affect the pressure drop across the slider valve, it was assumed that density and viscosity are important parameters, while parameters such as pipe pressure and fluid compressibility are not. It should be kept in mind that the selection of the proper parameters is a first crucial step in the application of dimensional analysis.

6.2.2 Review of Dimensions

Before we present the dimensional analysis technique let's review the dimensions of the quantities of interest in an introductory course in fluid mechanics. All of

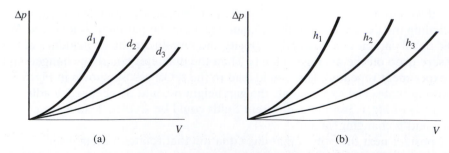

FIGURE 6.3 Pressure drop versus velocity curves: (a) ρ, μ, h fixed; (b) ρ, μ, d fixed.

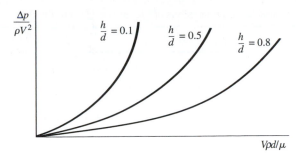

FIGURE 6.4   Dimensionless pressure drop versus dimensionless velocity.

the quantities have some combination of dimensions of length, time, mass, and force which are related by Newton's second law,

$$\Sigma \mathbf{F} = m\mathbf{a} \tag{6.2.3}$$

In terms of dimensions, it is written as

$$F = \frac{ML}{T^2} \tag{6.2.4}$$

where F, M, L, and T are the dimensions of force, mass, length, and time, respectively. Thus we see that it is sufficient to use only three basic dimensions. We will choose the M-L-T system because we can eliminate the force dimension with Eq. 6.2.4.

KEY CONCEPT *Through- out this chapter, we will use the M-L-T system.*

If we were considering more complicated flow situations such as those involving electromagnetic field interactions, or those involving temperature gradients, we would need to include the appropriate additional dimensions. However, in this book such phenomena will not be introduced, except for the compressible flow of an ideal gas; for that case an equation of state relates the thermal effects to the dimensions above. That is,

$$p = \rho RT \tag{6.2.5}$$

where T represents temperature. This allows us to write

$$[RT] = [p/\rho] = \frac{F}{L^2} \cdot \frac{L^3}{M} = \frac{ML/T^2}{L^2} \cdot \frac{L^3}{M} = \frac{L^2}{T^2} \tag{6.2.6}$$

where the brackets represent "the dimensions of." Notice that the equation of state does not introduce additional dimensions.

The quantities of interest in fluid mechanics are listed with their respective dimensions in Table 6.1. Reference to this table will simplify writing the dimensions of the quantities introduced in the problems.

6.2.3   Buckingham π-Theorem

In a given physical problem the dependent variable x_1 can be expressed in terms of the independent variables as

$$x_1 = f(x_2, x_3, x_4, \dots, x_n) \tag{6.2.7}$$

TABLE 6.1   Symbols and Dimensions of Quantities Used in Fluid Mechanics

Quantity	Symbol	Dimensions
Length	l	L
Time	t	T
Mass	m	M
Force	F	ML/T^2
Velocity	V	L/T
Acceleration	a	L/T^2
Frequency	ω	T^{-1}
Gravity	g	L/T^2
Area	A	L^2
Flow rate	Q	L^3/T
Mass flux	$\dot{m}$	M/T
Pressure	p	M/LT^2
Stress	τ	M/LT^2
Density	ρ	M/L^3
Specific weight	γ	M/L^2T^2
Viscosity	μ	M/LT
Kinematic viscosity	ν	L^2/T
Work	W	ML^2/T^2
Power, heat flux	$\dot{W}, \dot{Q}$	ML^2/T^3
Surface tension	σ	M/T^2
Bulk modulus	B	M/LT^2

where n represents the total number of variables. Referring to Eq. 6.2.1, Δp is the dependent variable and V, ρ, μ, d, and h are the independent variables. The *Buckingham π-theorem*, named after Edgar Buckingham (1867–1940), states that $(n - m)$ dimensionless groups of variables, called π-terms, where m is the number[1] of basic dimensions included in the variables, can be related by

$$\pi_1 = f_1(\pi_2, \pi_3, \ldots, \pi_{n-m}) \tag{6.2.8}$$

where π_1 includes the dependent variable and the remaining π-terms include only independent variables, as in Eq. 6.2.2.

Further, it is noted that a requirement for a successful application of dimensional analysis is that a dimension must occur at least twice or not at all. For example, the equation $\Delta p = f(V, l, d)$ is ill-stated since pressure involves the dimensions of force and V, l, and d do not contain such a dimension.

The procedure used in applying the π-theorem is summarized as follows:

KEY CONCEPT *A dimension must occur twice or not at all.*

1. Write the functional form of the dependent variable depending on the $(n - 1)$ independent variables. This step requires knowledge of the phenomenon being studied. All variables that effect the dependent variable must be included. These include geometric variables, fluid properties, and external effects that influence the variable being studied. Quantities that have no influence on the dependent variable must not be included. Also, do not include variables that depend on each other; e.g., both radius and area would not be included. The variables on the right-hand side of Eq. 6.2.7 should be independent.

---

[1] There are situations where m is less than the number of basic dimensions. Example 6.2 will illustrate.

2. Identify m **repeating variables**, variables that will be combined with each remaining variable to form the π-terms. The repeating variables selected from the independent variables must include all of the basic dimensions, but they must not form a π-term by themselves. An angle cannot be a repeating variable since it is dimensionless and forms a π-term itself.

3. Form the π-terms by combining the repeating variables with each of the remaining variables.

4. Write the functional form of the $(n - m)$ dimensionless π-terms.

Repeating variables: *Variables that will be combined with each remaining variable to form the π-terms.*

Step 3 can be accomplished by a relatively simple algebraic procedure; we will also illustrate a procedure in the examples that utilizes simple observation.

The algebraic procedure will now be illustrated with an example. Suppose that we desire to combine the variables surface tension σ, velocity V, density ρ, and length l into a π-term; this can be written as

$$\pi = \sigma^a V^b \rho^c l^d \tag{6.2.9}$$

The objective is to determine a, b, c, and d so that the grouping is dimensionless. In terms of dimensions, Eq. 6.2.9 is

$$M^0 L^0 T^0 = \left(\frac{M}{T^2}\right)^a \left(\frac{L}{T}\right)^b \left(\frac{M}{L^3}\right)^c L^d \tag{6.2.10}$$

Equating exponents on each of the basic dimensions:

$$
\begin{aligned}
M: & \quad 0 = a + c \\
L: & \quad 0 = b - 3c + d \\
T: & \quad 0 = -2a - b
\end{aligned}
\tag{6.2.11}
$$

The three algebraic equations are solved simultaneously to yield

$$a = -c \qquad b = 2c \qquad d = c \tag{6.2.12}$$

so that the π-term becomes

$$\pi = \left(\frac{\rho l V^2}{\sigma}\right)^c \tag{6.2.13}$$

A dimensionless parameter raised to any power remains dimensionless; consequently, we can select c to be any number other than zero. It is usually selected as $c = 1$, depending on the ratio desired. Selecting $c = 1$, the π-term is

KEY CONCEPT *A dimensionless parameter raised to any power remains dimensionless.*

$$\pi = \frac{\rho l V^2}{\sigma} \tag{6.2.14}$$

Actually, we could have selected $c = 1$ in Eq. 6.2.9 and proceeded with only three unknowns. Or if it were desired to have σ in the numerator to the first power, we could have set $a = 1$ and let b, c, and d be unknowns.

A final note: If only one π-term results, the functional form would state that the π-term must be a constant since the right-hand side of Eq. 6.2.8 would contain no additional π-terms. This would result in an expression that includes an arbitrary constant that could be determined through analysis or experimentation.

Example 6.1

The drag force F_D on a cylinder of diameter d and length l is to be studied. What functional form relates the dimensionless variables if a fluid with velocity V flows normal to the cylinder?

Solution

First, we must determine the variables that have some influence on the drag force. If we include variables that do not influence the drag force, we would have additional π-terms that experimentation would show to be unimportant; if we do not include a variable that does influence the drag force, experimentation would also reveal that problem. Experience is essential in choosing the correct variables; in this example we will include as influential variables the free stream velocity V, the viscosity μ, the density ρ of the fluid, in addition to the diameter d and the length l of the cylinder, resulting in $n = 6$ variables. This is written as

$$F_D = f(d, l, V, \mu, \rho)$$

The variables are observed to include $m = 3$ dimensions:

$$[F_D] = \frac{ML}{T^2} \qquad [V] = \frac{L}{T} \qquad [\mu] = \frac{M}{LT} \qquad [d] = L \qquad [l] = L \qquad [\rho] = \frac{M}{L^3}$$

Consequently, we can expect $n - m = 6 - 3 = 3$ π-terms.

We choose repeating variables with the simplest combinations of dimensions such that they do not form a π-term by themselves (we could not include d and l as repeating variables); the repeating variables are chosen to be d, V, and ρ. These three variables are combined with each of the remaining variables to form the π-terms. Rather than writing equations similar to Eq. 6.2.9 for the π-terms, let us form the π-terms by inspection. When the repeating variables are combined with F_D we observe that only F_D and ρ have the mass dimension; hence F_D must be divided by ρ. Only F_D and V have the time dimension; hence, F_D must be divided by V^2. Thus F_D divided by ρ has L^4 in the numerator; when divided by V^2 this results in L^2 remaining in the numerator. Hence we must have d^2 in the denominator resulting in

$$\pi_1 = \frac{F_D}{\rho V^2 d^2}$$

When d, V, and ρ are combined with l there obviously results

$$\pi_2 = \frac{l}{d}$$

The last π-term results from combining μ with d, V, and ρ. The mass dimension disappears if we divide μ by ρ. The time dimension disappears if we divide μ by V. This leaves one length dimension in the numerator; hence d is needed in the denominator resulting in

$$\pi_3 = \frac{\mu}{\rho V d}$$

The dimensionless, functional relationship relating the π-terms is

$$\pi_1 = f_1(\pi_2, \pi_3) \qquad \text{or} \qquad \frac{F_D}{\rho V^2 d^2} = f_1\left(\frac{l}{d}, \frac{\mu}{\rho V d}\right)$$

Rather than the original relationship of six variables we have reduced the problem to one involving three π-terms, a much simpler problem. To determine the particular form of the functional relationship above, we would actually have to solve the problem; experimentation would be needed if analytical or numerical methods were not available. This is often the case in fluid mechanics.

Note that we could have included several additional variables in our original list, such as gravity g, the angle θ that the velocity makes with the cylinder, and the roughness e of the cylinder surface. To not include variables that are significant, or to include variables that are not significant is a matter of experience. The novice must learn how to identify significant variables; however, even the experienced researcher is often at a loss to correlate certain phenomena; much experimentation is often needed to discover the appropriate parameters.

Example 6.2

The rise of liquid in a capillary tube is to be studied. It is anticipated that the rise h will depend on surface tension σ, tube diameter d, liquid specific weight γ, and angle β of attachment between the liquid and tube. Write the functional form of the dimensionless variables.

Solution

The expression relating the variables is

$$h = f(\sigma, d, \gamma, \beta)$$

The dimensions of the variables are

$$[h] = L \qquad [\gamma] = \frac{M}{L^2 T^2} \qquad [\beta] = 1 \text{ (dimensionless)} \qquad [\sigma] = \frac{M}{T^2} \qquad [d] = L$$

By observation we see that M/T^2 occurs as that combination in both σ and γ; hence M and T are not independent dimensions in this problem. There are only two independent groupings of basic dimensions, L and M/T^2. Thus $m = 2$ and we choose σ and d as the repeating variables. When combined with h, the first π-term is

$$\pi_1 = \frac{h}{d}$$

When σ and d are combined with γ, the second π-term is

$$\pi_2 = \frac{\gamma d^2}{\sigma}$$

Finally, since the angle β is dimensionless, it forms a π-term by itself; that is,

$$\pi_3 = \beta$$

The final functional form relating the π-terms is

$$\pi_1 = f_1(\pi_2, \pi_3) \quad \text{or} \quad \frac{h}{d} = f_1\left(\frac{\gamma d^2}{\sigma}, \beta\right)$$

Note: In this example we could not have chosen the angle β as a repeating variable since it already is a dimensionless π-term. Also, we could not have chosen three repeating variables since M and T were not independent.

Also, note that we may have thought that gravity should have been included in the problem. If it had been included above, it would not have appeared in any of the π-terms, indicating that it should not have been included. If density and gravity, rather than specific weight, had been included the relationship above would have resulted since $\gamma = \rho g$; this, by the way, would have avoided the necessity of observing that M/T^2 was a dimensional grouping.

A final note regarding the functional form of the π-terms: The relationship above could equally have been written as

$$\frac{h}{d} = f_1\left(\frac{\sigma}{\gamma d^2}, \beta\right)$$

Also, occasionally a different set of repeating variables could be selected. This simply expresses the final functional equation in a different but equivalent form. Actually, a second form can be shown to be a combination of the π-terms from an initial form.

6.2.4 Common Dimensionless Parameters

Consider a relatively general relationship between the pressure drop Δp, a characteristic length l, a characteristic velocity V, the density ρ, the viscosity μ, the gravity g, the surface tension σ, the speed of sound c, and an angular frequency ω, written as

$$\Delta p = f(l, V, \rho, \mu, g, c, \omega, \sigma) \tag{6.2.15}$$

The π-theorem applied to this problem, with l, V, and ρ as repeating variables, results in

$$\frac{\Delta p}{\rho V^2} = f_1\left(\frac{V\rho l}{\mu}, \frac{V^2}{lg}, \frac{V}{c}, \frac{l\omega}{V}, \frac{V^2\rho l}{\sigma}\right) \tag{6.2.16}$$

Each of the π-terms in this expression is a common dimensionless parameter that appears in numerous fluid flow situations. They are identified as follows:

$$\text{Euler number, Eu} = \frac{\Delta p}{\rho V^2}$$

$$\text{Reynolds number, Re} = \frac{V\rho l}{\mu}$$

$$\text{Froude number}^2, \text{Fr} = \frac{V}{\sqrt{lg}}$$

$$\text{Mach number, M} = \frac{V}{c} \tag{6.2.17}$$

$$\text{Strouhal number}^2, \text{St} = \frac{l\omega}{V}$$

$$\text{Weber number}^2, \text{We} = \frac{V^2 l\rho}{\sigma}$$

The physical significance of each parameter can be determined by observing that each dimensionless number can be written as the ratio of two forces. The forces are observed to be

---

[2] Froude, Strouhal and Weber number were named after William Froude (1810–1879), Vincenz Strouhal (1850–1922), and Moritz Weber (1871–1951), respectively.

$$F_p = \text{pressure force} = \Delta p A \sim \Delta p l^2$$

$$F_I = \text{inertial force} = mV \frac{dV}{ds} \sim \rho l^3 V \frac{V}{l} = \rho l^2 V^2$$

$$F_\mu = \text{viscous force} = \tau A = \mu \frac{du}{dy} A \sim \mu \frac{V}{l} l^2 = \mu l V$$

$$F_g = \text{gravity force} = mg \sim \rho l^3 g \qquad\qquad (6.2.18)$$

$$F_B = \text{compressibility force} = BA \sim \rho \frac{dp}{d\rho} l^2 = \rho c^2 l^2$$

$$F_\omega = \text{centrifugal force} = mr\omega^2 \sim \rho l^3 l \omega^2 = \rho l^4 \omega^2$$

$$F_\sigma = \text{surface tension force} = \sigma l$$

Thus we see that

$$\text{Eu} \propto \frac{\text{pressure force}}{\text{inertial force}}$$

$$\text{Re} \propto \frac{\text{inertial force}}{\text{viscous force}}$$

$$\text{Fr} \propto \frac{\text{inertial force}}{\text{gravity force}} \qquad\qquad (6.2.19)$$

$$\text{M} \propto \frac{\text{inertial force}}{\text{compressibility force}}$$

$$\text{St} \propto \frac{\text{centrifugal force}}{\text{inertial force}}$$

$$\text{We} \propto \frac{\text{inertial force}}{\text{surface tension force}}$$

Thinking of the dimensionless parameters in terms of the ratios of forces allows us to anticipate the significant parameters in a particular flow of interest. If viscous forces are important, such as in the pipe flow of Fig. 3.8 or the boundary layer flow of Fig. 3.10, we know that the Reynolds number is a significant dimensionless parameter. If surface tension forces are instrumental in affecting the flow, as in droplet formation or flow over a weir with a small head, we expect the Weber number to be important. Similar analysis can be applied to other fluid flow situations.

KEY CONCEPT *The ratios of forces allows us to anticipate the significant parameters in a flow.*

Obviously, all of the effects included in the general relationship (6.2.16) would not be of interest in any one situation. It would be very unlikely that both compressibility effects and surface tension effects would influence a flow simultaneously. In addition, there is often more than one length of importance, thereby introducing additional geometric, dimensionless ratios. We have, however, introduced the more common dimensionless flow parameters of interest in fluid mechanics. Table 6.2 summarizes this section.

6.3 SIMILITUDE

6.3.1 General Information

As stated in the introduction, *similitude* is the study of predicting prototype conditions from model observations. When an analytical or numerical solution is not practical, or when calculations are based on a simplified model so that uncertainty is introduced, it is usually advisable to perform tests on a model if testing is not practical on a full-scale prototype, be it too large or too small.

If it is decided that a model study is to be performed, it is necessary to develop the means whereby a quantity measured on the model can be used to predict the associated quantity on the prototype. We can develop such a means if we have **dynamic similarity** between model and prototype, that is, if the forces which act on corresponding masses in the model flow and the prototype flow are in the same ratio throughout the entire flow fields. Suppose that inertial forces, pressure forces, viscous forces, and gravity forces are present; then dynamic similarity requires that, at corresponding points in the flow fields,

Dynamic similarity: Forces which act on corresponding masses in the model flow and prototype flow are in the same ratio throughout the entire flows.

$$\frac{(F_I)_m}{(F_I)_p} = \frac{(F_p)_m}{(F_p)_p} = \frac{(F_\mu)_m}{(F_\mu)_p} = \frac{(F_g)_m}{(F_g)_p} = \text{const.} \tag{6.3.1}$$

These can be rearranged to read

$$\left(\frac{F_I}{F_p}\right)_m = \left(\frac{F_I}{F_p}\right)_p \qquad \left(\frac{F_I}{F_\mu}\right)_m = \left(\frac{F_I}{F_\mu}\right)_p \qquad \left(\frac{F_I}{F_g}\right)_m = \left(\frac{F_I}{F_g}\right)_p \tag{6.3.2}$$

which, in the preceding section, have been shown to be

$$\text{Eu}_m = \text{Eu}_p \qquad \text{Re}_m = \text{Re}_p \qquad \text{Fr}_m = \text{Fr}_p \tag{6.3.3}$$

TABLE 6.2 Common Dimensionless Parameters in Fluid Mechanics

Parameter	Expression	Flow situations where parameter is important
Euler number	$\dfrac{\Delta p}{\rho V^2}$	Flows in which pressure drop is significant: most flow situations
Reynolds number	$\dfrac{\rho l V}{\mu}$	Flows that are influenced by viscous effects: internal flows, boundary layer flows
Froude number	$\dfrac{V}{\sqrt{lg}}$	Flows that are influenced by gravity: primarily free surface flows
Mach number	$\dfrac{V}{c}$	Compressibility is important in these flows, usually if $V > 0.3\,c$
Strouhal number	$\dfrac{l\omega}{V}$	Flow with an unsteady component that repeats itself periodically
Weber number	$\rho\dfrac{V^2 l}{\sigma}$	Surface tension influences the flow; flow with an interface may be such a flow

If the forces above were the only ones present, we could write

$$F_I = f(F_p, F_\mu, F_g) \qquad (6.3.4)$$

Recognizing that there is only one basic dimension, namely force, dimensional analysis would allow us to write (see Eq. 6.2.8) the equation above in terms of force ratios or

$$Eu = f(Re, Fr) \qquad (6.3.5)$$

Hence we could conclude that if the Reynolds number and the Froude number are the same on the model and prototype, the Euler number must also be the same. Thus dynamic similarity between model and prototype is guaranteed by equating the Reynolds number and the Froude number of the model to those on the prototype, respectively. If compressibility forces were included here, the analysis above would result in the Mach number being included in Eq. 6.3.5.

We can write the inertial force ratio as

$$\frac{(F_I)_m}{(F_I)_p} = \frac{a_m m_m}{a_p m_p} = \text{const.} \qquad (6.3.6)$$

showing that the acceleration ratio between corresponding points on the model and prototype is a constant provided that the mass ratio of corresponding fluid elements is a constant. We can write the acceleration ratio as

$$\frac{a_m}{a_p} = \frac{V_m^2/l_m}{V_p^2/l_p} = \text{const.} \qquad (6.3.7)$$

showing that the velocity ratio between corresponding points is a constant providing the length ratio is a constant. The velocity ratio being a constant between all corresponding points in the flow fields is the statement of **kinematic similarity**. This would result in the streamline pattern around the model being the same as that around the prototype except for a scale factor. The length ratio being constant between all corresponding points in the flow fields is the demand of **geometric similarity** which results in the model having the same shape as the prototype. Hence, to ensure complete similarity between model and prototype, we demand that:

Kinematic similarity: *Condition where the velocity ratio is a constant between all corresponding points in the flows.*

KEY CONCEPT *The streamline pattern around the model is the same as that around the prototype.*

- Geometric similarity be satisfied
- The mass ratio of corresponding fluid elements be a constant
- The appropriate dimensionless parameters of Eq. 6.2.17 be equal

Assuming that complete similarity between model and prototype exists, we can now predict quantities of interest on a prototype from measurements on a model. If we measure a drag force F_D on a model and wish to predict the

Geometric similarity: *A condition where the model has the same shape as the prototype.*

corresponding drag on the prototype, we could equate the ratio of the drag forces to the ratio of the inertial forces (see Eq. 6.2.18) as

$$\frac{(F_D)_m}{(F_D)_p} = \frac{(F_I)_m}{(F_I)_p} = \frac{\rho_m V_m^2 l_m^2}{\rho_p V_p^2 l_p^2} \qquad (6.3.8)$$

If we measure the power input to a model and wish to predict the power requirement of the prototype, we would recognize that power is force times velocity and write

$$\frac{\dot{W}_m}{\dot{W}_p} = \frac{(F_I)_m V_m}{(F_I)_p V_p} = \frac{\rho_m V_m^2 l_m^2 V_m}{\rho_p V_p^2 l_p^2 V_p} \qquad (6.3.9)$$

Hence we can predict a prototype quantity if we select the model fluid (this provides ρ_m/ρ_p), the scale ratio (this provides l_m/l_p), and the appropriate dimensionless number from Table 6.2 (this provides V_m/V_p). Examples will illustrate.

6.3.2 Confined Flows

A confined flow is a flow that has no free surfaces (a liquid-gas surface) or interfaces (two different liquids forming an interface). It is confined to move within a specified region; such flows include external flows around objects, such as aircraft, buildings, and submarines, as well as internal flows in pipes and conduits.

KEY CONCEPT *Gravity does not influence the flow pattern in confined flows.*

Gravity does not influence the flow pattern in confined flows; that is, if gravity could be changed in magnitude, the flow pattern and associated flow quantities would not change. The dominant effect is that of viscosity in incompressible confined flows (all liquid flows and gas flows in which M < 0.3). Surface tension is obviously not a factor, as it would be in bubble formation, and for steady flows there would be no unsteady effects due to oscillations in the flow. The three relevant forces are pressure forces, inertial forces, and viscous forces. Therefore, in confined flows dynamic similarity is achieved if the ratios of viscous forces, inertial forces, and pressure forces between model and prototype are the same. This leads to the conclusion (see Eq. 6.3.5) that Eu = f(Re), so that it is only necessary to consider the Reynolds number as the dominant dimensionless parameter in a confined incompressible flow. If compressibility effects are significant, the Mach number would also be important.

KEY CONCEPT *The Reynolds number is the dominant dimensionless parameter in a confined incompressible flow.*

6.3.3 Free-Surface Flows

A free-surface flow is one in which part of the boundary involves a pressure boundary condition. This includes flows over weirs and dams, as shown in Fig. 6.5, flows in channels and spillways, flows involving two fluids separated by an interface, and flows around floating objects with waves and around submerged objects.

Example 6.3

A test is to be performed on a proposed design for a large pump that is to deliver $1.5\ \text{m}^3/\text{s}$ of water from a 40-cm-diameter impeller with a pressure rise of 400 kPa. A model with an 8-cm-diameter impeller is to be used. What flow rate should be used and what pressure rise is to be expected? The model fluid is water at the same temperature as the water in the prototype.

Solution

For similarity to exist in this confined incompressible flow problem, the Reynolds numbers must be equal, that is,

$$\text{Re}_m = \text{Re}_p$$

$$\frac{V_m d_m}{\nu_m} = \frac{V_p d_p}{\nu_p}$$

Recognizing that $\nu_m = \nu_p$ if the temperatures are equal, we see that

$$\frac{V_m}{V_p} = \frac{d_p}{d_m}$$

$$= \frac{0.4}{0.08} = 5$$

The ratio of flow rates is found recognizing that $Q = VA$:

$$\frac{Q_m}{Q_p} = \frac{V_m d_m^2}{V_p d_p^2}$$

$$= 5 \times \left(\frac{1}{5}\right)^2 = \frac{1}{5}$$

Thus we find that

$$Q_m = \frac{Q_p}{5} = \frac{1.5}{5} = 0.3\ \text{m}^3/\text{s}$$

The dimensionless pressure rise is found using the Euler number:

$$\left(\frac{\Delta p}{\rho V^2}\right)_m = \left(\frac{\Delta p}{\rho V^2}\right)_p$$

Hence the pressure rise for the model is

$$\Delta p_m = \Delta p_p \frac{\rho_m}{\rho_p} \frac{V_m^2}{V_p^2}$$

$$= 400 \times 1 \times 5^2 = 10\ 000\ \text{kPa}$$

Note that in this example we see that the velocity in the model is equal to the velocity in the prototype multiplied by the length ratio, and the pressure rise in the model is equal to the pressure rise in the prototype multiplied by the length ratio squared. If the length ratio were very large, it is obvious that to maintain Reynolds number equivalence would be quite difficult, indeed. This observation is discussed in more detail in Section 6.3.4.

with cavitation present. In all of these flows the location of the free surface is unknown and the velocity at the free surface is unknown; it is the pressure that must be the same[3] on either side of the interface. In free-surface flows, gravity controls both the location and the motion of the free surface. This introduces the Froude number because of the influence of the gravity forces. If we consider flows that do not exhibit periodic motions, which have negligible surface tension and compressibility effects, we can ignore the influence of St, M, and We. That leaves only the viscous effects to be considered. There are many free-surface flows in which viscous effects are significant. Consider, however, that in most model studies water is the only economical fluid to use; if the prototype fluid is also water, as it often is, we would find from the Froude numbers

$$\frac{V_m^2}{l_m g_m} = \frac{V_p^2}{l_p g_p} \qquad \therefore \frac{V_m}{V_p} = \left(\frac{l_m}{l_p}\right)^{1/2} \tag{6.3.10}$$

assuming that $g_m = g_p$. Matching the Reynolds numbers (using $\nu_m = \nu_p$):

$$\frac{V_m l_m}{\nu_m} = \frac{V_p l_p}{\nu_p} \qquad \therefore \frac{V_m}{V_p} = \frac{l_p}{l_m} \tag{6.3.11}$$

FIGURE 6.5   Model of the Bonneville lock and dam on the Columbia River. (Courtesy of the U.S. Army Corps of Engineers Waterways Experiment Station.)

[3] The surface tension, if significant, results in a pressure difference across the interface.

Thus we have a conflict. If we use the same fluid in the model study as in the prototype flow, we cannot satisfy both the Froude number criterion and the Reynolds number criterion. If we demand that both criteria be satisfied by using different fluids for model and prototype ($\nu_m \neq \nu_p$), we must select a model fluid with a viscosity $\nu_m = \nu_p(l_m/l_p)^{3/2}$ (this results from a matching of the Froude and Reynolds numbers). A fluid with this viscosity is probably either an impossibility or an impracticality. Hence, when modeling free-surface flows in which viscous effects are important, we equate the Froude numbers and include viscous effects by some other technique. For example, if we measured the total drag on the model of a ship, we would approximate the viscous drag (using some technique not included here) and subtract it from the total drag, leaving the drag due to wave resistance. The wave drag on the prototype would then be predicted using similitude, and the approximated viscous drag would be added to the wave drag, yielding the expected drag on the ship. For the better designs the viscous drag on ship hulls can be of the same order of magnitude as the wave drag.

Example 6.4

A 1:20 scale model of a surface vessel is used to test the influence of a proposed design on the wave drag. A wave drag of 30 N is measured at a model speed of 2.4 m/s. What speed does this correspond to on the prototype, and what wave drag is predicted for the prototype? Neglect viscous effects, and assume the same fluid for model and prototype.

Solution
The Froude number must be equated for both model and prototype. Thus

$$\text{Fr}_m = \text{Fr}_p \qquad \frac{V_m}{\sqrt{l_m g}} = \frac{V_p}{\sqrt{l_p g}}$$

This yields, recognizing that g does not vary significantly on the surface of the earth,

$$V_p = V_m \left(\frac{l_p}{l_m}\right)^{1/2} = 2.4\sqrt{20} = 10.7 \text{ m/s}$$

To find the wave drag on the prototype, we equate the drag ratio to the inertia force ratio:

$$\frac{(F_D)_m}{(F_D)_p} = \frac{\rho_m V_m^2 l_m^2}{\rho_p V_p^2 l_p^2}$$

This allows us to calculate the wave drag on the prototype as, using $\rho_p = \rho_m$,

$$(F_D)_p = (F_D)_m \frac{\rho_p V_p^2 l_p^2}{\rho_m V_m^2 l_m^2}$$

$$= 30 \times \frac{10.7^2}{2.4^2} \times 20^2 = 238\ 520 \text{ N} = 238.52 \text{ kN}$$

Note: We could have used the gravity force ratio rather than the inertial force ratio, but we could not have used the viscous force ratio since viscous forces were assumed negligible.

6.3.4 High-Reynolds-Number Flows

In a confined flow in which the Reynolds number is the dimensionless parameter that guarantees dynamic similarity, we note that if the same fluid is used in model and prototype, the velocity in the model study is $V_m = V_p l_p / l_m$; the velocity in the model is the velocity in the prototype multiplied by the scale factor. This often results in velocities that are prohibitively large in the model study. Also, the pressures encountered in the model study are large, as shown in Example 6.3, and the energy consumption is also very large. Because of these problems the Reynolds numbers may not be matched in studies involving large Reynolds numbers.

There is, however, some justification for not matching the Reynolds number in model studies. Consider a typical drag coefficient C_D versus Reynolds number curve, as shown in Fig. 6.6 (the complete curve is presented in Fig. 8.8). The drag coefficient is a dimensionless drag, defined as $C_D = \text{drag} / \frac{1}{2}\rho V^2 A$. At a sufficiently high Reynolds number, between 10^3 and 10^5, the flow is insensitive to changes in the Reynolds number; note that the drag coefficient is essentially constant and independent of Re. That implies that the flow field is similar at $\text{Re} = 10^3$ to that at $\text{Re} = 10^5$. Thus if $\text{Re}_p = 10^5$, it is only necessary that $10^3 < \text{Re}_m < 10^5$ for the viscous effects to have the same effect on model and prototype. This often allows another parameter of interest to be matched, such as the Froude number or the Mach number. There are, however, high-Reynolds-number flows in which compressibility effects and free-surface effects are negligible, so that neither the Froude number nor the Mach number are applicable. Examples include flow around automobiles, large smoke stacks, and dirigibles. For such flows we must only ensure that the Reynolds number be within the range where the drag coefficient is constant. We should note that for Reynolds numbers quite large ($\text{Re} > 5 \times 10^5$ in Fig. 6.6) the flow may also become Reynolds number independent. If that is the case, it is only necessary that Re_m be sufficiently large.

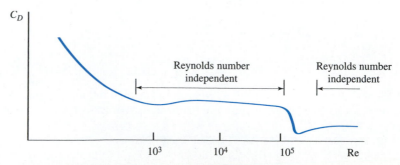

FIGURE 6.6   Drag coefficient versus Reynolds number.

Example 6.5

A 1:10 scale model of an automobile is used to measure the drag on a proposed design. It is to simulate a prototype speed of 90 km/h. What speed should be used in the wind tunnel if Reynolds numbers are equated? For this condition, what is the ratio of drag forces?

Solution

The same fluid exists on model and prototype, thus equating the Reynolds numbers results in

$$\frac{V_m l_m}{\nu_m} = \frac{V_p l_p}{\nu_p} \qquad \therefore \quad V_m = V_p \frac{l_p}{l_m}$$

$$= 90 \times 10 = 900 \text{ km/h}$$

This speed would, of course, introduce compressibility effects, effects that do not exist in the prototype. Hence the proposed model study would be inappropriate.
If we did use this velocity in the model, the drag force ratio would be

$$\frac{(F_D)_p}{(F_D)_m} = \frac{\rho_p V_p^2 l_p^2}{\rho_m V_m^2 l_m^2} \qquad \therefore \quad \frac{(F_D)_p}{(F_D)_m} = 1$$

Thus we see that the drag force on the model is the same as the drag force on the prototype if the same fluids are used when we equate Reynolds numbers.

Example 6.6

In Example 6.5, if the Reynolds numbers were equated, the velocity in the model study was observed to be in the compressible flow regime (i.e., M > 0.3 or V_m > 360 km/h). To conduct an acceptable model study, could we use a velocity of 90 km/h on a model with a characteristic length of 10 cm? Assume that the drag coefficient ($C_D = F_D / \frac{1}{2}\rho V^2 A$, where A is the projected area), is independent of Re for Re > 10^5. If so, what drag force on the prototype would correspond to a drag force of 1.2 N measured on the model?

Solution

The proposed model study in a wind tunnel is to be conducted with V_m = 90 km/h and l_m = 0.1 m. Using $\nu = 1.6 \times 10^{-5}$ m^2/s, the Reynolds number is

$$\text{Re}_m = \frac{V_m l_m}{\nu_m} = \frac{\dfrac{90 \times 1000}{3600} \times 0.1}{1.6 \times 10^{-5}} = 1.56 \times 10^5$$

This Reynolds number is greater than 10^5, so we will assume that similarity exists between model and prototype. The velocity of 90 km/h is sufficiently high.
The drag force on the prototype traveling at 90 km/h corresponding to 1.2 N on the model is found from

$$\frac{(F_D)_p}{(F_D)_m} = \frac{\rho_p V_p^2 l_p^2}{\rho_m V_m^2 l_m^2} \qquad \therefore (F_D)_p = (F_D)_m \frac{\rho_p}{\rho_m} \frac{V_p^2}{V_m^2} \frac{l_p^2}{l_m^2} = 1.2 \times 10^2 = 120 \text{ N}$$

Note that in this example we have assumed that the drag coefficient is independent of Re for Re > 10^5. If the drag coefficient continued to vary above Re = 10^5 (this would be evident from experimental data), the foregoing analysis would have to be modified accordingly.

6.3.5 Compressible Flows

For most compressible flow situations the Reynolds number is so large (refer to Fig. 6.6) that it is not a parameter of significance; the compressibility effects lead to the Mach number as the primary dimensionless parameter for model studies. Thus for a particular model study we require

$$M_m = M_p \qquad \text{or} \qquad \frac{V_m}{c_m} = \frac{V_p}{c_p} \qquad (6.3.12)$$

If the model study is carried out in a wind tunnel and the prototype fluid is air, we can assume that $c_m \simeq c_p$ if the temperature is the same in the two flows. For such a case, the velocity in the model study is equal to the velocity associated with the prototype. Of course, if the speeds of sound are different, the velocity ratio will be different from unity, accordingly.

Example 6.7

The pressure rise from free stream to the nose of a fusilage section of an aircraft is measured in a wind tunnel at 20°C to be 34 kPa with a wind-tunnel airspeed of 900 km/h. If the test is to simulate flight at an elevation of 12 km, what is the prototype velocity and the expected nose pressure rise?

Solution

To find the prototype velocity corresponding to a wind-tunnel airspeed of 900 km/h, we equate the Mach numbers

$$M_m = M_p \qquad \text{or} \qquad \frac{V_m}{\sqrt{kRT_m}} = \frac{V_p}{\sqrt{kRT_p}}$$

Thus

$$V_p = V_m \left(\frac{kRT_p}{kRT_m} \right)^{1/2} = 900 \left(\frac{216.7}{293} \right)^{1/2} = 774 \text{ km/h}$$

The pressure at the nose of the prototype fusilage is found using the Euler number as follows:

$$\frac{\Delta p_m}{\rho_m V_m^2} = \frac{\Delta p_p}{\rho_p V_p^2}$$

$$\therefore \quad \Delta p_p = \Delta p_m \frac{\rho_p}{\rho_m} \frac{V_p^2}{V_m^2}$$

$$= 34 \times 0.2546 \times \frac{774^2}{900^2} = 6.4 \text{ kPa}$$

The density ratio and temperature T_p were found in Appendix B.

6.3.6 Periodic Flows

In many flow situations there are regions of the flows in which periodic motions occur. Such flows include the periodic fluid motion (in Section 8.3.2 this is called vortex shedding) that takes place when a fluid flows past a cylindrical object such as a bridge, a TV tower, a cable, or a tall building; the flow past a wind machine; and the flow through turbomachinery. In flows such as these it is necessary to equate the Strouhal numbers, which can be written as

$$\frac{V_m}{\omega_m l_m} = \frac{V_p}{\omega_p l_p} \qquad (6.3.13)$$

in order that the periodic motion be properly modeled.

In addition to the Strouhal number, there may be additional dimensionless parameters that must be equated: in viscous flows, the Reynolds number; in free-surface flows, the Froude number; and in compressible flows, the Mach number.

Example 6.8

A large wind turbine, designed to operate at 50 km/h, is to be tested in a laboratory by constructing a 1:15 scale model. What airspeed should be used in the wind tunnel, what angular velocity should be used to simulate a prototype speed of 5 rpm, and what power output is expected from the model if the prototype output is designed to be 500 kW?

Solution

The speed in the wind tunnel can be any speed above that needed to provide a sufficiently large Reynolds number. Let us select the same speed with which the prototype is to operate, namely, 50 km/h, and calculate the minimum characteristic length that a Reynolds number of 10^5 would demand; this gives

$$\text{Re} = \frac{Vl}{\nu} \qquad 10^5 = \frac{(50 \times 1000/3600) \times l}{1.6 \times 10^{-5}} \qquad \therefore l = 0.12 \text{ m}$$

Obviously, in a reasonably large wind tunnel we can maintain a characteristic length (e.g., the blade length) that large.

The angular velocity is found by equating the Strouhal numbers. There results

$$\frac{V_m}{\omega_m l_m} = \frac{V_p}{\omega_p l_p} \qquad \therefore \omega_m = \omega_p \frac{\overset{1}{\cancel{V_m}}}{V_p} \frac{l_p}{l_m} = 5 \times 1 \times 15 = 75 \text{ rpm}$$

assuming that the wind velocities are equal.

The power is found by observing that power is force times velocity.

$$\frac{\dot{W}_m}{\dot{W}_p} = \frac{\rho_m V_m^3 l_m^2}{\rho_p V_p^3 l_p^2}$$

or

$$\dot{W}_m = \dot{W}_p \frac{\overset{1}{\cancel{\rho_m}}}{\cancel{\rho_p}} \frac{\overset{1}{\cancel{V_m^3}}}{\cancel{V_p^3}} \frac{l_m^2}{l_p^2} = 500 \times \left(\frac{1}{15}\right)^2 = 2.22 \text{ kW}$$

6.4 NORMALIZED DIFFERENTIAL EQUATIONS

In Chapter 5 we derived the partial differential equations used to describe any flow of interest that involves an isotropic, homogeneous, Newtonian fluid. Such a flow could be laminar or turbulent, steady or unsteady, compressible or incompressible, a confined flow or a free-surface flow, a flow with surface tension effects or one in which surface tension is negligible. Most often, when utilizing the differential equations, we express them in dimensionless, or normalized, form. This form of the equations provides us with information not contained in the dimensional form, information similar to that provided by dimensional analysis. Let us normalize the differential equations that would describe the motion of an incompressible, homogeneous flow. The energy equation will not be needed.

Before we normalize the equations, let's review them. In vector form the continuity equation and the Navier–Stokes equation are

$$\boldsymbol{\nabla \cdot V} = 0$$

$$\rho \frac{D\mathbf{V}}{Dt} = - \boldsymbol{\nabla} p - \rho g \boldsymbol{\nabla} h + \mu \nabla^2 \mathbf{V} \tag{6.4.1}$$

where we have assumed h to be vertical. Because the first two terms on the right are of the same form (the gradient of a scalar function) we can combine them as follows:

$$\boldsymbol{\nabla} p + \rho g \boldsymbol{\nabla} h = \boldsymbol{\nabla} p_k \tag{6.4.2}$$

where the *kinetic pressure* p_k is defined to be

$$p_k = p + \rho g h \tag{6.4.3}$$

It is the pressure that results from motion alone. As the fluid motion ceases, p_k goes to zero according to Fig. 6.4.1b. In terms of the kinetic pressure the Navier–Stokes equation becomes

$$\rho \frac{D\mathbf{V}}{Dt} = - \boldsymbol{\nabla} p_k + \mu \nabla^2 \mathbf{V} \tag{6.4.4}$$

If the pressure p is never used in a boundary condition, we can retain the kinetic pressure in the equations, thereby "hiding" the gravity effects. If the pressure enters a boundary condition, we must return to Eq. 6.4.3 and gravity becomes important.

To normalize the differential equations and boundary conditions, we must select characteristic quantities that best describe the problem of interest. For example, consider a flow that has an average velocity V and a primary dimension l. The dimensionless velocities and coordinate variables are

$$u^* = \frac{u}{V} \qquad v^* = \frac{v}{V} \qquad w^* = \frac{w}{V} \qquad x^* = \frac{x}{l} \qquad y^* = \frac{y}{l} \qquad z^* = \frac{z}{l} \qquad (6.4.5)$$

where asterisks are used to denote dimensionless quantities. The characteristic pressure will be twice the inviscid pressure rise between free stream and the stagnation point (i.e., ρV^2); the characteristic time, provided that no characteristic time such as a period of oscillation exists in the problem, will be the time it takes a fluid particle to travel the distance l at the velocity V (i.e., l/V). Hence

$$p^* = \frac{p}{\rho V^2} \qquad t^* = \frac{t}{l/V} \qquad (6.4.6)$$

Using the dimensionless quantities of Eq. 6.4.5, we see that

$$\mathbf{V}^* = u^*\hat{\mathbf{i}} + v^*\hat{\mathbf{j}} + w^*\hat{\mathbf{k}}$$

$$= \frac{u}{V}\hat{\mathbf{i}} + \frac{v}{V}\hat{\mathbf{j}} + \frac{w}{V}\hat{\mathbf{k}} = \frac{\mathbf{V}}{V}$$

$$\boldsymbol{\nabla}^* = \frac{\partial}{\partial x^*}\hat{\mathbf{i}} + \frac{\partial}{\partial y^*}\hat{\mathbf{j}} + \frac{\partial}{\partial z^*}\hat{\mathbf{k}} \qquad (6.4.7)$$

$$= l\frac{\partial}{\partial x}\hat{\mathbf{i}} + l\frac{\partial}{\partial y}\hat{\mathbf{j}} + l\frac{\partial}{\partial z}\hat{\mathbf{k}} = l\boldsymbol{\nabla}$$

Introducing the above into the Navier–Stokes equation and the continuity equation, we have

$$\rho \frac{V}{l/V}\frac{D^*\mathbf{V}^*}{Dt^*} = -\frac{\rho V^2}{l}\boldsymbol{\nabla}^* p_k^* + \mu \frac{V}{l^2}\nabla^{*2}\mathbf{V}^*$$

$$\frac{V}{l}\boldsymbol{\nabla}^* \cdot \mathbf{V}^* = 0 \qquad (6.4.8)$$

where D^*/Dt^* represents the dimensionless material derivative. Finally, in dimensionless form,

$$\frac{D^*\mathbf{V}^*}{Dt^*} = -\, \nabla^* p_k^* + \frac{\mu}{\rho Vl}\, \nabla^{*2}\mathbf{V}^*$$

$$\nabla^* \cdot \mathbf{V}^* = 0$$

(6.4.9)

Note that we have introduced the Reynolds number

$$\mathrm{Re} = \frac{\rho Vl}{\mu}$$

(6.4.10)

as a parameter in the normalized Navier–Stokes equation.

Consider now the boundary conditions. The no-slip condition on a fixed boundary $\mathbf{V}^* = 0$ introduces geometric parameters that are necessary to specify the geometry of the fixed boundary. These could include, for example, a roughness parameter e/l, where e is the average height of the roughness elements; a thickness parameter t/l, where t is the maximum thickness of a turbine blade; and a nose radius parameter r_0/l, where r_0 is the nose radius of an airfoil.

Another boundary condition would be the entering velocity distribution, for example, $u(y)/V$. This may introduce parameters having to do with the profile and the turbulent structure of the entering flow. Such features could be extremely significant for a particular flow.

Also, a portion of the boundary may be oscillating, like that of a rotating component. On such a boundary we would require that the velocity of the fluid be the same as the velocity of the rotating part, that is, $v = r\omega$. This would introduce the Strouhal number St,

$$v^* = \mathrm{St}\, r^*$$

(6.4.11)

where

$$\mathrm{St} = \frac{\omega l}{V}$$

(6.4.12)

Another boundary condition that introduces an additional parameter is that of a free surface. It is required that, neglecting surface tension effects, the pressure be constant across a free surface. On an air–water surface this requires that $p = 0$ on the free surface. Returning to Eq. 6.4.3, and normalizing, there results

$$\rho V^2 p_k^* = \rho V^2 p^* + \rho g l h^* \qquad \text{or} \qquad p_k^* = p^* + \frac{gl}{V^2}\, h^*$$

(6.4.13)

where we have introduced the Froude number

$$\text{Fr} = \frac{V}{\sqrt{lg}}$$

(6.4.14)

Finally, a boundary condition may involve surface tension, as in droplet formation problems. Such a boundary condition would involve the two radii of curvature leading to the normalized equation

$$\Delta p^* = \frac{\sigma}{l\rho V^2}\left(\frac{1}{r_1^*} + \frac{1}{r_2^*}\right)$$

(6.4.15)

where we have introduced the Weber number

$$\text{We} = \frac{V^2 l\rho}{\sigma}$$

(6.4.16)

Note that we have introduced all of the dimensionless parameters discussed in Section 6.3 with the exception of the Mach number. It could be introduced in a normalized energy equation, a task that is left to the home problems.

Obviously, if we can write the differential equations that describe a particular flow, those equations and the boundary conditions contain all the parameters of interest; hence the Buckingham π-theorem is not actually necessary for flow situations in which the equations and boundary conditions are known.

In addition, we can obtain the similitude conditions from the normalized equations and boundary conditions. The idea is to make the normalized differential equations and boundary conditions the same for both model and prototype. If they are the same they should possess the same solution.[4] The normalized boundary conditions being identical would demand that geometric similarity be satisfied. In order that the equations and boundary conditions be identical on both model and prototype, it is necessary that the dimensionless parameters be the same for both model and prototype. Thus, similitude conditions are also included in the normalized problem.

KEY CONCEPT *The differential equations and the boundary conditions contain all of the parameters of interest.*

KEY CONCEPT
Similitude conditions are also included in the normalized problem.

---

[4] Questions of uniqueness, which we will not consider, enter here. For example, it is possible to have identical conditions and equations, yet in one case a laminar flow could result and in the other it could be a turbulent flow or a second laminar flow different from the first; that is, different solutions would exist.

6.5 SUMMARY

Experimental studies are greatly simplified by reducing the number of variables that influence the phenomenon being studied. We demonstrated how the relationship

$$\Delta p = f(V, \rho, \mu, d, h) \tag{6.5.1}$$

could be written in the simplified form,

$$\frac{\Delta p}{\rho V^2} = f\left(\frac{V\rho d}{\mu}, \frac{h}{d}\right) \tag{6.5.2}$$

The most common flow parameters are, using l as the characteristic length,

$$\text{Re} = \frac{V\rho l}{\mu}, \quad \text{Fr} = \frac{V}{\sqrt{lg}}, \quad \text{M} = \frac{V}{c}, \quad \text{Eu} = \frac{\Delta p}{\rho V^2}, \quad \text{St} = \frac{l\omega}{V}, \quad \text{We} = \frac{V^2 l\rho}{\sigma} \tag{6.5.3}$$

The parameters, used to guide model studies, that are of primary significence in a particular flow are identified as follows:

Confined flows: $\qquad\qquad \text{Re} = \dfrac{V\rho l}{\mu}$ $\qquad$ (6.5.4)

Free-surface flows: $\qquad\qquad \text{Fr} = \dfrac{V}{\sqrt{lg}}$ $\qquad$ (6.5.5)

High-Reynolds number flows: $\text{Re} > (\text{Re})_{\text{minimum}}$ $\qquad$ (6.5.6)

Compressible flows: $\qquad\qquad \text{M} = \dfrac{V}{c}$ $\qquad$ (6.5.7)

Periodic flows: $\qquad\qquad \text{St} = \dfrac{V}{l\omega}$ $\qquad$ (6.5.8)

The Navier–Stokes equation is written using dimensionless variables as

$$\frac{D\mathbf{V}}{Dt} = -\nabla p_k + \frac{1}{\text{Re}} \nabla^2 \mathbf{V} \tag{6.5.9}$$

where p_k is the kinetic pressure. We know that all variables are dimensionless since the Reynolds number appears in the equation. We often write equations in dimensionless form without the asterisks for simplicity.

PROBLEMS

6.1 Write Bernoulli's equation in dimensionless form by dividing Eq. 6.1.1 by V_1^2 and multiplying by g. Express the equation in a form similar to that of Eq. 6.1.2.

6.2 If the F-L-T system of units were used, what would the dimensions be on each of the following?
(a) Mass flux   **(b)** Pressure
(c) Density   **(d)** Viscosity
(e) Work   **(f)** Power
(g) Surface tension

Dimensional Analysis

6.3 Combine power $\dot{W}$, diameter d, pressure drop Δp, and average velocity into a dimensionless group.

A. $\dfrac{\dot{W}}{d^2 V \Delta p}$

B. $\dfrac{\dot{W}}{V \Delta p}$

C. $\dfrac{\dot{W} d}{V \Delta p}$

D. $\dfrac{\dot{W}}{V d \Delta p}$

6.4 Verify that the dimensions of power are ML^2/T^3 as listed in Table 6.1.

6.5 It is proposed that the velocity of a flow depends on a diameter d, a length l, gravity g, rotational speed ω, and viscosity μ. Select the variable that would not influence the velocity.
A. μ
B. ω
C. g
D. both d and l

6.6 If the velocity V in a fluid flow depends on a dimension l, the fluid density ρ, and the viscosity μ, show that this implies that the Reynolds number $Vl\rho/\mu$ is constant.

6.7 If the velocity V in a fluid depends on the surface tension σ, the density ρ, and a diameter d, show that this implies that the Weber number $V^2 d\rho/\sigma$ is constant.

6.8 Assume that the velocity V of fall of an object depends on the height H through which it falls, the gravity g, and the mass m of the object. Find an expression for V.

6.9 Include the density ρ and viscosity μ of the surrounding fluid and repeat Problem 6.8. This would account for the resistance (drag) of the fluid.

6.10 Select l, V, and ρ as the repeating variables in Example 6.1 and find an expression for F_D. Show that this is an equivalent form to that of Example 6.1.

6.11 Select d, μ, and V as the repeating variables in Example 6.1 and find an expression for F_D. Show that this is equivalent to the expression of Example 6.1.

6.12 Include gravity g in the list of variables of Example 6.2 and determine the final expression that results for h.

6.13 Find an expression for the centrifugal force F_c if it depends on the mass m, the angular velocity ω, and the radius R of an impeller.

6.14 The normal stress σ in a beam depends on the bending moment M, the distance y from the neutral axis, and the moment of inertia I. Relate σ to the other variables if we know that σ varies linearly with y.

6.15 Find an expression for the average velocity in a smooth pipe if it depends on the viscosity, the diameter, and the pressure gradient $\partial p/\partial x$.

6.16 It is suggested that the velocity of the water flowing from a hole in the side of an open tank depends on the height H of the hole from the surface, the gravity, and the density of the water. What expression relates the variables?

6.17 Derive an expression for the velocity of liquid issuing from a hole in the side of an open tank if the velocity depends on the height H of the hole from the free surface, the fluid viscosity and density, gravity, and the hole diameter.

6.18 The pressure drop Δp in the pipe of Fig. P6.18 depends on the average velocity, the pipe diameter, the kinematic viscosity, the length L of pipe, the wall roughness height e, and the fluid density. Find an expression for Δp.

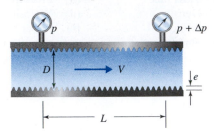

FIGURE P6.18

6.19 Select an appropriate set of variables that influence the drag force F_D on an airfoil (Fig. P6.19) and write the final form in terms of dimensionless parameters.

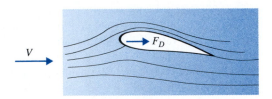

FIGURE P6.19

6.20 The flow rate Q in an open channel depends on the hydraulic radius R, the cross-sectional area A, the wall roughness height e, gravity g, and the slope S. Relate Q to the other variables using (a) the M-L-T system, and (b) the F-L-T system.

6.21 The velocity of propagation of ripples in a shallow liquid depends on the liquid depth h, gravity g, surface tension σ, and the liquid density ρ. Find an expression for the propagation velocity V. See Fig. P6.21.

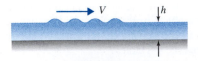

FIGURE P6.21

6.22 The drag force F_D on a sphere depends on the velocity V, the viscosity μ, the density ρ, the surface roughness height e, the freestream fluctuation intensity I (a dimensionless quantity), and the diameter D. Find an expression for F_D.

6.23 The drag force F_D on the smooth sphere of Fig. P6.23 falling in a liquid depends on the constant sphere speed V, the solid density ρ_s, the liquid density ρ and viscosity μ, the sphere diameter D, and gravity g. Find an expression for F_D using (a) the M-L-T system, and (b) the F-L-T system.

FIGURE P6.23

6.24 The drag force F_D on a golf ball depends on the velocity, the viscosity, the density, the diameter, the dimple depth, the dimple radius, and the concentration C of dimples measured by the number of dimples per unit area. What expression relates F_D to the other variables?

6.25 The frequency f with which vortices are shed from a cylinder depends on the viscosity, density, velocity, and diameter. Derive an expression for f.

6.26 The lift F_L on an airfoil depends on the aircraft speed V, the speed of sound c, the fluid density, the airfoil chord length l_c, the airfoil thickness t, and the angle of attack α. Find an expression for F_L.

6.27 Derive an expression for the torque T needed to rotate a disk of diameter d at the angular velocity ω in a fluid with density ρ and viscosity μ if the disk is a distance t from a wall. Also, find an expression for the power requirement.

6.28 The cables holding a suspension bridge (Fig. P6.28) are observed to undergo large vibrations under certain wind conditions. Select a set of variables that influence the periodic force acting on a cable and write a simplified relationship of dimensionless parameters.

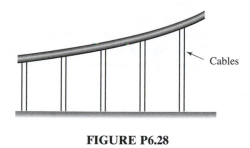

FIGURE P6.28

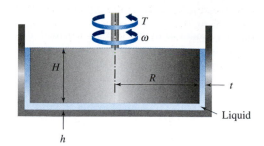

FIGURE P6.33

6.29 A vacuum cleaner creates a pressure drop Δp across its fan. Relate this pressure drop to the impeller diameter D and width h, its rotational speed ω, the air density ρ, and the inlet and outlet diameters d_i and d_o. Also, find an expression for the power requirement of the fan.

6.30 Derive an expression for the maximum torque needed to rotate an agitator if it depends on the frequency of oscillation, the angular velocity with which the agitator rotates during an oscillation, the diameter of the agitator, the nominal height of the paddles, the length of the paddles, the number of paddles, the depth of liquid, and the liquid density.

6.31 The flow rate of water over a weir depends on the head of water, the width of the weir, gravity, viscosity, density, and surface tension. Relate the flow rate to the other variables.

6.32 The size of droplets in the spray of a fruit sprayer depends on the air velocity, the spray jet velocity, the spray jet exit diameter, the liquid spray surface tension, density, and viscosity, and the air density. Relate the droplet diameter to the other variables.

6.33 Relate the torque T to the other variables shown in Fig. P6.33.

6.34 A viscometer is composed of an open tank in which the liquid level is held constant. A small-diameter tube empties into a calibrated volume. Find an expression for the viscosity using the relevant parameters.

6.35 Derive an expression for the torque T necessary to rotate the shaft of Fig. P6.35.

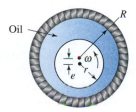

FIGURE P6.35

6.36 Derive an expression for the depth y_2 in the hydraulic jump of Fig. P6.36.

FIGURE P6.36

6.37 Derive an expression for the frequency with which a cylinder, suspended by a string in a liquid flow, oscillates.

Similitude

6.38 After a model study has been performed, it is necessary to predict quantities of interest that are to be expected on the prototype. Write expressions, in terms of density, velocity, and length, for the ratio of model to prototype of each of the following: flow rate Q, pressure drop Δp, pressure force F_p, shear stress τ_0, torque T, and heat transfer rate $\dot{Q}$.

6.39 What velocity should be selected in a wind tunnel where a 9:1 scale model of an automobile is to simulate a speed of 12 m/s? Neglect compressibility effects.

  A. 108 m/s   **B.** 12 m/s
  C. 4 m/s   **D.** 1.33 m/s

6.40 Flow around an underwater structural component is to be studied in a 20°C-wind tunnel with a 10:1 scale model. What speed should be selected in the wind tunnel to simulate a 10°C-water speed of 4 m/s?

  A. 4.61 m/s
  B. 40 m/s
  C. 31.6 m/s
  D. 461 m/s

6.41 A 1:7 scale model simulates the operation of a large turbine that is to generate 200 kW with a flow rate of 1.5 m³/s. What flow rate should be used in the model, and what power output is expected?
  (a) Water at the same temperature is used in both model and prototype.
  (b) The model water is at 25°C and the prototype water is at 10°C.

6.42 A 1:5 scale model of a large pump is used to test a proposed change. The prototype pump produces a pressure rise of 600 kPa at a mass flux of 800 kg/s. Determine the mass flux to be used in the model and the expected pressure rise.
  (a) Water at the same temperature is used in both model and prototype.
  (b) The water in the model study is at 30°C and the water in the prototype is at 15°C.

6.43 A force on a component of a 1:10 scale model of a large pump is measured to be 50 N. What force is expected on the prototype component if water is used for both model and prototype:
  (a) With the water at the same temperature?
  (b) With the prototype water at 10°C and the model water at 20°C?

6.44 A 1:10 scale model study of an automobile is proposed. A prototype speed of 100 km/h is desired. What wind-tunnel speed should be selected for the model study? Comment on the advisability of such a speed selection.

6.45 A 1:10 scale model study of a torpedo is proposed. Prototype speeds of 90 km/h are to be studied. Should a wind tunnel or a water facility be used?

6.46 It is proposed that a model study be carried out on a proposed low-speed airfoil that is to fly at low altitudes at a speed of 50 m/s. If a 1:10 scale model is to be constructed, what velocity should be used in a wind tunnel? Comment as to the advisability of such a test. Would it be more advisable to perform the test in a 20°C water channel? If a water channel study were undertaken, calculate the drag ratio between the model and prototype.

6.47 A model study of oil (SAE-10W) at 0°C flowing in a 80-cm-diameter pipe is performed by using water at 20°C. Pipe of what diameter should be used if the average velocities are to be the same? What pressure drop ratio is expected? $S_{oil} = 0.9$.

6.48 A 0.025-mm-long microorganism moves through water at the rate of 0.1 body length per second. Could a model study be performed for such a prototype in a water channel or a wind tunnel?

6.49 A 1:30 scale model of a ship is to be tested with complete similarity. What should be the viscosity of the model fluid? Is such a liquid possible? What conclusion can be stated?

6.50 What upstream velocity should be selected in a 16:1 scale model of a levee which has an upstream velocity of 2 m/s?

  A. 2 m/s   **B.** 1 m/s
  C. 0.5 m/s   **D.** 0.25 m/s

6.51 A force of 10 N is measured on a 25:1 scale model of a ship tested in a water channel. What force should be expected on the prototype ship? Neglect viscous effects.

  A. 156 kN   **B.** 62.5 kN
  C. 6250 N   **D.** 250 N

6.52 A 1:60 scale model of a ship is used in a water tank to simulate a ship speed of 10 m/s. What should be the model speed? If a towing force of 10 N is measured on the model, what force is expected on the prototype? Neglect viscous effects.

6.53   The flow rate over a weir is 2 m³/s of water. A 1:10 scale model of the weir is tested in a water channel.
   (a)   What flow rate should be used?
   (b)   If a force of 12 N is measured on the model, what force would be expected on the prototype?

6.54   A 1:10 scale model of a hydrofoil is tested in a water channel with a force of 4 N measured at a speed of 6 m/s. Determine the speed and force predicted for the prototype. Neglect viscous effects.

6.55   The propeller of a ship is to be studied with a 1:10 scale model.
   (a)   Assuming that the propeller operates close to the surface, select the propeller speed of the model if the prototype speed is 600 rpm.
   (b)   What torque would be expected if 1.2 N · m is measured on the model?

6.56   The pontoon of a seaplane is to be studied in a water channel that has a channel speed capability of 6 m/s. If the seaplane is to lift off at 100 m/s, what model scale should be selected?

6.57   A 1:30 scale model study of a submarine is proposed in an attempt to study the influence of a suggested shape modification. The prototype is 2 m in diameter and it is designed to travel at 15 m/s. The model is towed in a water tank at 2 m/s and a drag force of 2.15 N is measured. Does similitude exist for this test? If so, predict the power needed for the prototype.

6.58   The smoke emanating from the stacks on a ship has a tendency to make its way to the deck down the outside of the stacks causing discomfort to the passengers. This problem is studied with a 1:20 scale model of a 4-m-diameter stack. The ship is traveling at 10 m/s. What range of wind-tunnel airspeeds could be used in this study?

6.59   A model study of a dirigible (a large balloon that travels through the air) is to be performed. The 10-m-diameter dirigible is to travel at 20 m/s. If a 40-cm-diameter model is proposed for use in a wind tunnel, or a 10-cm-diameter model for use in a 20°C water channel, which should be selected? Suppose that the wind-tunnel model is used with a speed of 15 m/s and a drag force of 3.2 N is measured. What force would be expected on the water-channel model with a speed of 2.4 m/s in the water channel? What horsepower would be predicted to overcome drag on the prototype? Assume that the flow is Reynolds number independent for $Re > 10^5$.

6.60   A model study is to be performed of a 300-m-tall, 15-m-diameter smokestack of a power plant. It is known that the smokestack is submerged in a ground boundary layer that is 400-m-thick. Could the study be performed in a wind tunnel that produces a 1.25-m-thick boundary layer?

6.61   A 1:20 scale model of an aircraft is tested in a 23°C wind tunnel. A speed of 200 m/s is used in the model study. A drag force of 10 N is measured. What prototype speed and drag force does this simulate if the elevation is:
   (a)   Sea level?
   (b)   5000 m?
   (c)   10 000 m?

6.62   A 1:10 scale model of an airfoil is tested in a wind tunnel that uses outside air at 0°C. The test is to simulate an aircraft speed of 250 m/s at an elevation of 10 000 m. What wind-tunnel speed should be used? If a velocity of 290 m/s and a pressure of 80 kPa absolute are measured on the model at a particular location at a 5° angle of attack, what speed and pressure are expected on the prototype at the corresponding location, and what is the angle of attack?

6.63   A 1:10 scale model of a propeller on a ship is to be tested in a water channel. What should be the rotational speed of the model if the rotational speed of the propeller is 2000 rpm and:
   (a)   The Froude number governs the study?
   (b)   The Reynolds number governs the study?

6.64   A torque of 12 N·m is measured on a 1:10 scale model of a large wind machine with a wind tunnel speed of 60 m/s. This is to simulate a wind velocity of 15 m/s since viscous effects are considered negligible. What torque is expected on the prototype? If the model rotates at 500 rpm, what prototype angular velocity is simulated?

6.65   An underwater study of a porpoise is to be performed by using a 1:10 scale model. A porpoise swimming at 10 m/s and making one swimming motion each second is to be simulated. What speed could be used in the water channel, and for that speed, how many swimming motions per second should be used?

Normalized Differential Equations

6.66 Normalize the continuity equation

$$\frac{\partial \rho}{\partial t} + \frac{\partial}{\partial x}(\rho u) + \frac{\partial}{\partial y}(\rho v) = 0$$

using a characteristic velocity V, length l, density ρ_0, and time f^{-1}, where f is the frequency. What dimensionless parameter is introduced?

6.67 Normalize Euler's equation

$$\frac{\partial \mathbf{V}}{\partial t} + u\frac{\partial \mathbf{V}}{\partial x} + v\frac{\partial \mathbf{V}}{\partial y} + w\frac{\partial \mathbf{V}}{\partial z} = -\frac{\nabla p}{\rho}$$

using a characteristic velocity U, length l, pressure ρU^2, and time f^{-1}, where f is the frequency. Find any dimensionless parameters that are introduced.

6.68 Normalize Euler's equation

$$\rho\frac{D\mathbf{V}}{Dt} = -\nabla p - \rho g \nabla h$$

using a characteristic velocity U, length l, pressure ρU^2, and time l/U. Determine any dimensionless parameters introduced in the normalized equation.

6.69 A fluid is at rest between the large horizontal plates shown in Fig. P6.69. The top plate is suddenly given a velocity U. Show that the Navier–Stokes equation that describes the resulting motion simplifies to

$$\frac{\partial u}{\partial t} = \nu\frac{\partial^2 u}{\partial y^2}$$

Nondimensionalize this equation using characteristic velocity U, length h, and time (a) h/U and

(b) h^2/ν. Identify any dimensionless parameters that result.

FIGURE P6.69

6.70 A fluid flows in a horizontal pipe of diameter d (Fig. P6.70). The flow is suddenly increased to an average velocity V. Show that the appropriate Navier–Stokes equation, using the coordinates shown, simplifies to

$$\rho\frac{\partial u}{\partial t} = -\frac{\partial p}{\partial x} + \mu\left(\frac{\partial^2 u}{\partial r^2} + \frac{1}{r}\frac{\partial u}{\partial r}\right)$$

Normalize this equation using characteristic velocity V, length d, and time (a) d/V and (b) d^2/ν. Identify any dimensionless parameters that result.

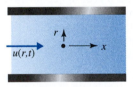

FIGURE P6.70

6.71 A highly viscous liquid such as honey is flowing down a vertical flat surface. Its thickness decreases as it progresses down the surface (Fig. P6.71). Show that the steady flow is described by

$$\rho u \frac{\partial u}{\partial x} = \mu \left(\frac{\partial^2 u}{\partial x^2} + \frac{\partial^2 u}{\partial y^2} \right) + g$$

where we have neglected the y-component of velocity. Normalize this equation using characteristic length h (measured at $x = 0$) and velocity V (the average velocity). Identify any dimensionless parameters that result.

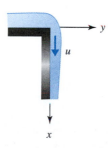

FIGURE P6.71

6.72 Nondimensionalize the energy equation

$$\rho c_p \left(u \frac{\partial T}{\partial x} + v \frac{\partial T}{\partial y} \right) = K \nabla^2 T$$

using a characteristic velocity U, length l, and temperature T_0. Express the dimensionless parameter that results in terms of the Prandtl number

$$\text{Pr} = \mu \frac{c_p}{K}$$

6.73 Nondimensionalize the compressible flow differential momentum equation

$$\rho \frac{D\mathbf{V}}{Dt} = -\nabla p + \mu \nabla^2 \mathbf{V} + \frac{\mu}{3} \nabla(\nabla \cdot \mathbf{V})$$

and the energy equation

$$\rho c_v \frac{DT}{Dt} = K \nabla^2 T - p \nabla \cdot \mathbf{V}$$

using characteristic quantities $\rho_0, p_0, T_0, U,$ and l. The characteristic time is l/U. The speed of sound is $c = \sqrt{kRT_0}$. Find any dimensionless parameters that result if $\text{Pr} = \mu c_p / K$.

The Trans Alaska Pipeline conveys crude oil over long distances and through a variety of terrain. Pumping stations are spaced along its length to overcome pressures lost due to viscous forces and elevation changes. (U.S. Bureau of Land Management)

7

Internal Flows

Outline

Chapter Objectives

The objectives of this chapter are to:

▲ Establish the length of the entrance region for both laminar and turbulent flows.

▲ Determine the laminar flow solution for pipes, parallel plates, and rotating cylinders.

▲ Present the quantities of interest for turbulent pipe flow. Losses are of particular interest.

▲ Calculate the flow rate with a centrifugal pump in a single piping system.

▲ Determine flow rates in open channels.

▲ Provide numerous examples and problems that demonstrate entrance and developed flow solutions for both laminar and turbulent flows, including losses due to wall friction and various devices. Open channel flows are also analyzed.

7.1 INTRODUCTION

In this chapter the effects of viscosity on an incompressible, internal flow will be studied. Such flows are of particular importance to engineers. Flow in a circular pipe is undoubtedly the most common internal fluid flow. It is encountered in the veins and arteries in a body, in a city's water system, in a farmer's irrigation system, in the piping systems transporting fluids in a factory, in the hydraulic lines of an aircraft, and in the ink jet of a computer's printer. Flows in noncircular ducts and open channels must also be included in our study. In Chapter 6 we discovered that the viscous effects in a flow resulted in the introduction of the Reynolds number,

$$\mathrm{Re} = \frac{V\rho l}{\mu} \tag{7.1.1}$$

The Reynolds number was observed to be a ratio of the inertial force to the viscous force. Hence, when this ratio becomes large, it is expected that the inertial forces may dominate the viscous forces. This is usually true when short, sudden geometric changes occur; for long reaches of pipe or open channels, this is not the situation. When the surface areas, such as the wall area of a pipe, are relatively large, viscous effects become quite important and must be included in our study.

KEY CONCEPT *When the surface areas are large, viscous effects become important.*

Internal flow between parallel plates, in a pipe, between rotating cylinders, and in an open channel will be considered in detail. For a sufficiently low Reynolds number (Re < 2000 in a pipe and Re < 1500 in a wide channel) a laminar flow results, and at sufficiently high Reynolds number a turbulent flow occurs. See Section 3.3.3 for a more detailed discussion. We consider laminar flow first and then turbulent flow.

7.2 ENTRANCE FLOW AND DEVELOPED FLOW

When considering internal flows we are interested primarily in developed flows in conduits. A **developed laminar flow** results when the velocity profile ceases to change in the flow direction. Let us first focus our attention on a laminar flow. In the *entrance region* of a laminar flow the velocity profile changes in the flow direction, as shown in Fig. 7.1. The idealized flow from a reservoir begins at the inlet as a uniform flow (in reality, there is a thin viscous layer on the wall, as shown); the viscous wall layer then grows over the *inviscid core length* L_i until the viscous stresses dominate the entire cross section; the profile then continues to change in the *profile development region* due to viscous effects until a developed flow is achieved. The inviscid core length is one-fourth to one-third of the *entrance length* L_E, depending on the conduit geometry, shape of the inlet velocity profile, and the Reynolds number.

Developed laminar flow: *A flow where the velocity profile ceases to change in the flow direction.*

KEY CONCEPT *The inviscid core length in a laminar flow is one-fourth to one-third of the entrance length.*

For a laminar flow in a circular pipe with a uniform profile at the inlet, the entrance length is given by

$$\frac{L_E}{D} = 0.065\mathrm{Re} \qquad \mathrm{Re} = \frac{VD}{\nu} \tag{7.2.1}$$

where the Reynolds number is based on the average velocity and the diameter. Laminar flow in a pipe has been observed at Reynolds numbers in excess of 40 000 for carefully controlled conditions in a smooth pipe. However, for engineering applications a value of about 2000 is the highest Reynolds number for which laminar flow is assured; this is due to vibrations of the pipe, fluctuations in the flow, or roughness elements on the pipe wall.

For a laminar flow in a high-aspect-ratio channel (the aspect ratio is the width divided by the distance between the top and bottom plates) with a uniform profile at the inlet the entrance length is

$$\frac{L_E}{h} = 0.04\text{Re} \qquad \text{Re} = \frac{Vh}{\nu} \qquad (7.2.2)$$

where the Reynolds number is based on the average velocity and the distance h between the plates. The inviscid core length is approximately one-third of the entrance length. A laminar flow cannot exist over Re = 7700; for engineering situations a value of 1500 is often used as the upper limit for laminar flow.

For a turbulent flow the situation is slightly different, as shown in Fig. 7.2 for flow in a pipe. A developed flow results when all characteristics of the flow cease to change in the flow direction; this includes details of the turbulence that will be introduced later in this chapter. The inviscid core exists followed by the velocity profile development region, which terminates at $x = L_d$. An additional length is needed, however, for the detailed structure of the turbulent flow to develop. The detailed structure is important in certain calculations such as accurate estimates of wall heat transfer. For large Reynolds number flow (Re $> 10^5$) in a pipe, where the turbulence fluctuations initiate near $x = 0$, tests have yielded

$$\frac{L_i}{D} \simeq 10 \qquad \frac{L_d}{D} \simeq 40 \qquad \frac{L_E}{D} \simeq 120 \qquad (7.2.3)$$

For turbulent flow with Re = 4000 the foregoing developmental lengths would be substantially higher, perhaps five times the values listed. This is true since, for low-Re turbulent flows, transition to a turbulent flow occurs in the profile development region; hence much of the entry region is laminar with relatively low wall shear. Experimental data are not available for low-Reynolds-number turbulent flow.

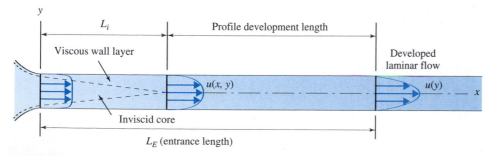

FIGURE 7.1   Entrance region of a laminar flow in a pipe or a wide rectangular channel.

In Fig. 7.3 the pressure variation is sketched. In the flow beyond a sufficiently large x it is noted that the pressure variation decreases linearly with x. If transition to turbulent flow occurs near the origin, the linear pressure variation begins near L_i and the pressure gradient [the slope of the $p(x)$ curve] in the inlet region is higher than in the developed flow region; if transition occurs near L_d, as it does for low Re, the linear variation begins at the end of the transition process and the pressure gradient in the inlet region is less than that of developed flow.

For a laminar flow, the pressure variation qualitatively resembles that associated with a large Reynolds number. The pressure gradient is higher than in the developed flow region due to the higher shear stress at the wall and the increasing momentum flux.

7.3 LAMINAR FLOW IN A PIPE

In this section we investigate incompressible, steady, developed laminar flow in a pipe, as sketched in Fig. 7.4 (see page 276). Two methods will be used: an elemental approach and a direct solution of the appropriate Navier–Stokes equation. Both develop the same equations so either can be used.

7.3.1 Elemental Approach

An elemental volume of the fluid is shown in Fig. 7.4 (see page 276). This volume can be considered an infinitesimal control volume into which and from which fluid flows, or it can be considered an infinitesimal fluid mass upon which forces are acting. If it is considered a control volume, we would apply the momentum equation (4.5.6); if it is a fluid mass, we would apply Newton's second law. Since the velocity profile does not change in the x-direction, the momentum flux in equals the momentum flux out and the resultant force is zero; since there is no acceleration of the mass element, the resultant force must also be zero. Consequently, a force balance in the x-direction yields

$$p\pi r^2 - (p + dp)\pi r^2 - \tau 2\pi r\, dx + \gamma\pi r^2\, dx \sin\theta = 0 \qquad (7.3.1)$$

which can be simplified to

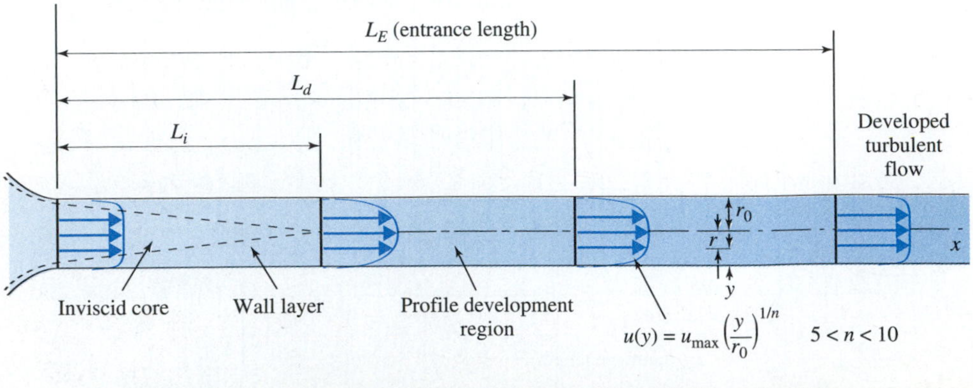

$$u(y) = u_{\max}\left(\frac{y}{r_0}\right)^{1/n} \qquad 5 < n < 10$$

FIGURE 7.2 Velocity profile development in a turbulent pipe flow.

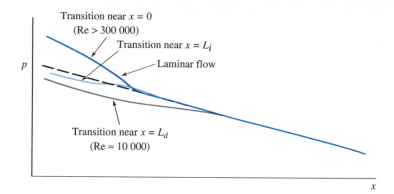

FIGURE 7.3   Pressure variation in a horizontal pipe flow for both laminar and turbulent flows. (From PhD Thesis of Dr. Jack Backus, Michigan State University.)

$$\tau = -\frac{r}{2}\frac{d}{dx}(p + \gamma h) \tag{7.3.2}$$

where we have used $\sin\theta = -dh/dx$, the vertical direction being denoted by h. Note that Eq. 7.3.2 can be applied to either a laminar or turbulent flow. The shear stress[1] in this laminar flow is related to the velocity gradient and the viscosity (see Eq. 1.5.5), giving

$$-\mu\frac{du}{dr} = -\frac{r}{2}\frac{d}{dx}(p + \gamma h) \tag{7.3.3}$$

Since $u = u(r)$ for this developed flow, we can conclude from the above that $d/dx(p + \gamma h)$ must be at most a constant; it cannot depend on x, and from fluid statics $(p + \gamma h)$ does not depend on x since there is no acceleration in the radial direction. Equation (7.3.3) can then be integrated to give the velocity distribution,

$$u(r) = \frac{r^2}{4\mu}\frac{d}{dx}(p + \gamma h) + A \tag{7.3.4}$$

where A is a constant of integration. Using $u = 0$ at $r = r_0$, we can evaluate A and find the velocity distribution to be

$$u(r) = \frac{1}{4\mu}\frac{d(p + \gamma h)}{dx}(r^2 - r_0^2) \tag{7.3.5}$$

a parabolic profile. It is often referred to as **Poiseuille flow**, named after Jean L. Poiseuille (1799–1869).

    The foregoing result can also be obtained by integrating the appropriate Navier–Stokes equation, as illustrated in the following section. If such an exercise is not of interest, go to Section 7.3.3.

Poiseuille flow: A flow with a parabolic profile in a pipe or between parallel plates.

---

[1]The shear stress is a positive quantity, as displayed in Fig. 7.4. The minus sign in $\tau = -\mu\, du/dr$ is necessary since du/dr is negative.

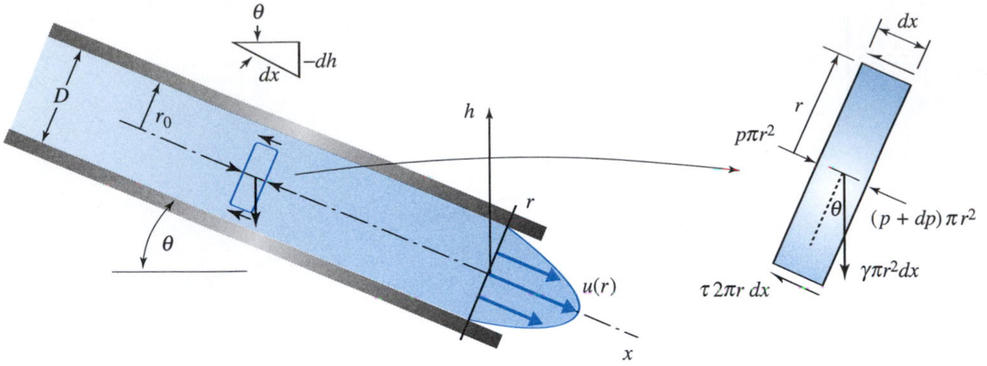

FIGURE 7.4   Developed flow in a circular pipe.

7.3.2   Solving the Navier–Stokes Equations

For developed flow in a circular pipe the streamlines are parallel to the wall with no swirl, so that $v_r = v_\theta = 0$ and $u = u(r)$ only. Referring to the momentum equations in cylindrical coordinates of Table 5.1 (note that $u = v_z$ and the z-coordinate has been replaced by x) the x-component Navier–Stokes equation is

$$\rho\left(\underbrace{\frac{\partial u}{\partial t}}_{\text{steady}} + \underbrace{v_r\frac{\partial u}{\partial r}}_{\substack{\text{II wall}\\ \text{Streamlines}}} + \underbrace{\frac{v_\theta}{r}\frac{\partial u}{\partial \theta}}_{\text{no swirl}} + \underbrace{u\frac{\partial u}{\partial x}}_{\substack{\text{developed}\\ \text{flow}}}\right)$$

$$= -\frac{\partial p}{\partial x} + \gamma \sin\theta + \mu\left(\frac{\partial^2 u}{\partial r^2} + \frac{1}{r}\frac{\partial u}{\partial r} + \underbrace{\frac{1}{r^2}\frac{\partial^2 u}{\partial \theta^2}}_{\substack{\text{symmetric}\\ \text{flow}}} + \underbrace{\frac{\partial^2 u}{\partial x^2}}_{\substack{\text{developed}\\ \text{flow}}}\right) \qquad (7.3.6)$$

Note that there is no acceleration (the left-hand side is zero) of the fluid particles as they move in the pipe. Equation (7.3.6) simplifies to, using $\sin\theta = -dh/dx$ (see Fig. 7.4),

$$\frac{1}{\mu}\frac{\partial}{\partial x}\left(p + \gamma h\right) = \frac{1}{r}\frac{\partial}{\partial r}\left(r\frac{\partial u}{\partial r}\right) \qquad (7.3.7)$$

where the first two terms in the parentheses on the right-hand side of Eq. 7.3.6 have been combined. We see that the left-hand side is at most a function of x and the right-hand side is at most a function of r. Since x and r can be varied independently, we must have

$$\frac{1}{r}\frac{d}{dr}\left(r\frac{du}{dr}\right) = \lambda \qquad (7.3.8)$$

KEY CONCEPT   *We use ordinary derivatives since u depends on only one variable.*

where λ is a constant[2] and we have used ordinary derivatives since u depends on only one variable r. Multiply both sides by r and integrate:

[2]If λ were a function of x, u would depend on x, which is not acceptable for this developed flow. If λ were a function of r, $p + \gamma h$ would depend on r, also unacceptable. Hence it is at most a constant.

$$r \frac{du}{dr} = \frac{\lambda}{2} r^2 + A \qquad (7.3.9)$$

Divide both sides by r and integrate:

$$u(r) = \frac{\lambda}{4} r^2 + A \ln r + B \qquad (7.3.10)$$

The velocity must remain finite at $r = 0$; hence $A = 0$. Also, at $r = r_0$, $u = 0$; thus B can be evaluated and we have

$$u(r) = \frac{\lambda}{4} (r^2 - r_0^2)$$

$$= \frac{1}{4\mu} \frac{d}{dx} (p + \gamma h)(r^2 - r_0^2) \qquad (7.3.11)$$

where we have used λ as the left-hand side of Eq. 7.3.7. This is the parabolic velocity distribution for flow in a pipe, often referred to as *Poiseuille flow*, named after Jean L. Poiseuille (1799–1869).

7.3.3   Pipe Flow Quantities

For steady, laminar, developed flow in a circular pipe, the velocity distribution has been shown to be

$$u(r) = \frac{1}{4\mu} \frac{d(p + \gamma h)}{dx} (r^2 - r_0^2) \qquad (7.3.12)$$

The average velocity V is found to be

$$V = \frac{Q}{A} = \frac{\displaystyle\int_0^{r_0} u(r) \, 2\pi r \, dr}{\pi r_0^2}$$

$$= \frac{2}{r_0^2} \int_0^{r_0} \frac{1}{4\mu} \frac{d(p + \gamma h)}{dx} (r^2 - r_0^2) r \, dr = -\frac{r_0^2}{8\mu} \frac{d(p + \gamma h)}{dx} \qquad (7.3.13)$$

Or, expressing the pressure drop Δp in terms of the average velocity, we have, for a horizontal[3] pipe,

$$\qquad (7.3.14)$$

$$\Delta p = \frac{8\mu V L}{r_0^2}$$

where we have used $\Delta p/L = -dp/dx$ since dp/dx is a constant for developed flow. Note that the pressure drop is a positive quantity, whereas the pressure gradient is negative.

KEY CONCEPT   *The pressure drop is a positive quantity, whereas the pressure gradient is negative.*

---

[3]For a pipe on an incline, simply replace p with $(p + \gamma h)$.

The maximum velocity at $r = 0$ from Eq. 7.3.12 is

$$u_{max} = -\frac{r_0^2}{4\mu} \frac{d(p + \gamma h)}{dx} \tag{7.3.15}$$

so that (see Eq. 7.3.13)

$$V = \tfrac{1}{2} u_{max} \tag{7.3.16}$$

The shearing stress is determined to be

$$\tau = -\mu \frac{du}{dr}$$

$$= -\frac{r}{2} \frac{d(p + \gamma h)}{dx} \tag{7.3.17}$$

Letting $\tau = \tau_0$ at $r = r_0$, we see that the pressure drop Δp over a length L of a horizontal section of pipe is

$$\Delta p = \frac{2\tau_0 L}{r_0} \tag{7.3.18}$$

Friction factor: *A quantity of substantial interest in pipe flow; a dimensionless wall shear.*

where we have again used $dp/dx = -\Delta p/L$.

If we introduce the **friction factor** f, a quantity of substantial interest in pipe flow, a dimensionless wall shear, defined by

$$f = \frac{\tau_0}{\frac{1}{8}\rho V^2} \tag{7.3.19}$$

we see that

$$\frac{\Delta p}{\gamma} = h_L = f \frac{L}{D} \frac{V^2}{2g} \tag{7.3.20}$$

where h_L is the head loss (see Eq. 4.5.17) with dimension of length. This equation is often referred to as the *Darcy–Weisbach equation*[4], named after Henri P. G. Darcy (1803–1858) and Julius Weisbach (1806–1871). Combining Eqs. 7.3.14 and 7.3.20, we find that

$$f = \frac{64}{Re} \tag{7.3.21}$$

for laminar flow in a pipe. Substituting this back into Eq. 7.3.20, we see that

$$h_L = \frac{32\mu L V}{\gamma D^2} \tag{7.3.22}$$

KEY CONCEPT *The head loss is directly proportional to the average velocity.*

The head loss is directly proportional to the average velocity (and hence the discharge also) to the first power, a result that generally is applied to developed, laminar flows in conduits, conduits of shape other than circular.

[4]This equation will be shown to be valid for both laminar and turbulent flows.

Example 7.1

A small-diameter horizontal tube is connected to a supply reservoir as shown in Fig. E7.1. If 6600 mm^3 is captured at the outlet in 10 s, estimate the viscosity of the water.

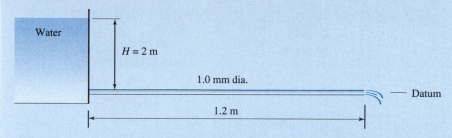

FIGURE E7.1

Solution

The tube is very small, so we expect viscous effects to limit the velocity to a small value. Using Bernoulli's equation from the surface to the entrance to the tube, and neglecting the velocity head, we have, letting 0 be a point on the surface,

$$\frac{\cancel{p_0}^{\,0}}{\gamma} + H = \frac{\cancel{V^2}^{\,0}}{2g} + \frac{p}{\gamma}$$

where we have used gage pressure with $p_0 = 0$. This becomes, assuming $V^2/2g \cong 0$,

$$p = \gamma H = 9800 \times 2 = 19\,600 \text{ Pa}$$

At the exit of the tube the pressure is zero; hence

$$\frac{\Delta p}{L} = \frac{19\,600}{1.2} = 16\,300 \text{ Pa/m}$$

The average velocity is found to be

$$V = \frac{Q}{A} = \frac{6600 \times 10^{-9}/10}{\pi \times 0.001^2/4} = 0.840 \text{ m/s}$$

Check to make sure the velocity head is negligible: $V^2/2g = 0.036$ m compared with $p/\gamma = 2$ m, so the assumption of negligible velocity head is valid. Our pressure calculation is acceptable. Using Eq. 7.3.14, we can find the viscosity of this assumed laminar flow to be

$$\mu = \frac{r_0^2}{8V} \frac{\Delta p}{L} = \frac{0.0005^2}{8 \times 0.84} (16\,300) = 6.06 \times 10^{-4} \text{ N·s/m}^2$$

We should check the Reynolds number to determine if our assumption of a laminar flow was acceptable. It is

$$\text{Re} = \frac{\rho V D}{\mu} = \frac{1000 \times 0.84 \times 0.001}{6.06 \times 10^{-4}} = 1390$$

This is obviously a laminar flow since Re < 2000, so the calculations are valid providing the entrance length is not too long. It is

$$L_E = 0.065 \text{ Re} \times D = 0.065 \times 1390 \times 0.001 = 0.09 \text{ m}$$

This is approximately 8% of the total length, a sufficiently small quantity; hence the calculations are assumed reliable.

Example 7.2

Derive an expression for the velocity distribution between horizontal, concentric pipes for a steady, incompressible developed flow (Fig. E7.2).

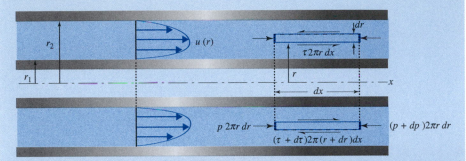

FIGURE E7.2

Solution

Let us use an elemental approach. The element is a hollow cylindrical shell as sketched in the figure. If we sum forces, we obtain

$$p2\pi r\, dr - (p + dp)2\pi r\, dr + \tau 2\pi r\, dx - (\tau + d\tau)2\pi(r + dr)\, dx = 0$$

Simplifying, there results, neglecting the term of differential magnitude,

$$\frac{dp}{dx} = -\frac{\tau}{r} - \frac{d\tau}{dr} - \underbrace{\frac{d\tau}{dr}}_{\text{neglect}}$$

Substituting $\tau = -\mu\, du/dr$ we have

$$\frac{dp}{dx} = \mu\left(\frac{1}{r}\frac{du}{dr} + \frac{d^2u}{dr^2}\right)$$

$$= \frac{\mu}{r}\frac{d}{dr}\left(r\frac{du}{dr}\right)$$

Multiply both sides by $r\,dr$ and divide by μ, then integrate:

$$r\frac{du}{dr} = \frac{1}{2\mu}\frac{dp}{dx}r^2 + A$$

Multiply both sides by dr/r and integrate again:

$$u(r) = \frac{1}{4\mu}\frac{dp}{dx}r^2 + A\ln r + B$$

where A and B are arbitrary constants. They are found by setting $u = 0$ at $r = r_1$ and at $r = r_2$; that is,

$$0 = \frac{1}{4\mu}\frac{dp}{dx}r_1^2 + A\ln r_1 + B$$

$$0 = \frac{1}{4\mu}\frac{dp}{dx}r_2^2 + A\ln r_2 + B$$

Solve for A and B:

$$A = \frac{1}{4\mu}\frac{dp}{dx}\frac{r_1^2 - r_2^2}{\ln(r_2/r_1)}$$

$$B = -A \ln r_2 - \frac{r_2^2}{4\mu}\frac{dp}{dx}$$

Thus

$$u(r) = \frac{1}{4\mu}\frac{dp}{dx}\left[r^2 - r_2^2 + \frac{r_2^2 - r_1^2}{\ln(r_1/r_2)}\ln(r/r_2)\right]$$

This is integrated to give the flow rate:

$$Q = \int_{r_1}^{r_2} u(r)2\pi r\, dr$$

$$= -\frac{\pi}{8\mu}\frac{dp}{dx}\left[r_2^4 - r_1^4 - \frac{(r_2^2 - r_1^2)^2}{\ln(r_2/r_1)}\right]$$

As $r_1 \to 0$ the velocity distribution approaches the parabolic distribution of pipe flow. As $r_1 \to r_2$ this distribution approaches that of parallel-plate flow. These two conclusions are not obvious and are presented as a problem at the end of this chapter.

If this problem were solved using the Navier–Stokes equation, we would simply use $u(r_1) = 0$ rather than $u(0) =$ finite. This change in the boundary condition would lead to the velocity distribution $u(r)$.

7.4 LAMINAR FLOW BETWEEN PARALLEL PLATES

Consider the incompressible, steady, developed flow of a fluid between parallel plates, with the upper plate moving with velocity U as shown in Fig. 7.5. We will derive the velocity distribution by two methods; either can be used.

7.4.1 Elemental Approach

Consider an elemental volume of unit depth in the z-direction, as sketched in Fig. 7.5. If we sum forces in the x-direction, we can write

$$p\, dy - (p + dp)\, dy - \tau\, dx + (\tau + d\tau)\, dx + \gamma\, dx\, dy \sin\theta = 0 \quad (7.4.1)$$

since there is no acceleration. The pressure is assumed to depend only on x; variation with y is assumed to be negligible since the dimension a is quite small for most applications. After dividing by $dx\, dy$, the above is simplified to

$$\frac{d\tau}{dy} = \frac{dp}{dx} - \gamma \sin\theta \quad (7.4.2)$$

Since this is one-dimensional flow, the shear stress is

$$\tau = \mu \frac{du}{dy} \quad (7.4.3)$$

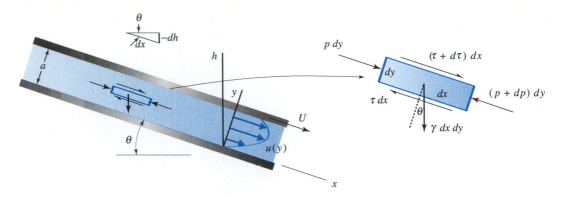

FIGURE 7.5   Developed flow between parallel plates.

Using this and $\sin \theta = -dh/dx$, there results

$$\mu \frac{d^2 u}{dy^2} = \frac{dp}{dx} + \gamma \frac{dh}{dx} = \frac{d}{dx}(p + \gamma h) \qquad (7.4.4)$$

Since $u = u(y)$ for this developed flow, the left-hand side is only a function of y; because the right-hand side is a function of x we conclude that it must be a constant. Hence, it can be integrated twice to yield (first, divide by μ)

$$u(y) = \frac{y^2}{2\mu} \frac{d}{dx}(p + \gamma h) + Ay + B \qquad (7.4.5)$$

where A and B are integration constants. If we require $u = 0$ at $y = 0$ and $u = U$ at $y = a$, we have

$$A = \frac{U}{a} - \frac{a}{2\mu} \frac{d}{dx}(p + \gamma h) \qquad B = 0 \qquad (7.4.6)$$

Thus the velocity distribution is the parabola

$$u(y) = \frac{1}{2\mu} \frac{d}{dx}(p + \gamma h)(y^2 - ay) + \frac{U}{a} y \qquad (7.4.7)$$

Couette flow: *A flow with a linear profile that results from the motion of the plate only.*

If the motion is due to the motion of the plate only (a linear profile), the flow is called a **Couette flow**, if the motion is due to the pressure gradient only, i.e., $U = 0$, it is Poiseuille flow.

Rather than summing forces on an element we can integrate the appropriate Navier–Stokes equation, as follows. If such an exercise is not of interest, go to Section 7.4.3.

7.4.2   Integrating the Navier–Stokes Equations

For developed flow between parallel plates the streamlines are parallel to the plates so that $u = u(y)$ only and $v = w = 0$. The Navier–Stokes equation for the x-direction is (see Eq. 5.3.14)

$$\rho\left(\underset{0}{\frac{\partial \cancel{u}}{\partial t}} + u\frac{\partial \cancel{u}}{\partial x} + \cancel{v}\frac{\partial u}{\partial y} + \cancel{w}\frac{\partial u}{\partial z}\right)$$

steady   developed flow

$$= -\frac{\partial p}{\partial x} + \mu\left(\frac{\partial^2 \cancel{u}}{\partial x^2} + \frac{\partial^2 u}{\partial y^2} + \frac{\partial^2 \cancel{u}}{\partial z^2}\right) + \gamma \sin\theta \qquad (7.4.8)$$

developed flow   wide channel

If the parallel plates are the top and bottom of a channel, the channel must be very wide; the analysis then applies to the midsection away from the side walls. Equation 7.4.8 reduces to, using $\sin\theta = -dh/dx$,

$$\frac{\partial^2 u}{\partial y^2} = \frac{1}{\mu}\frac{d}{dx}(p + \gamma h) \qquad (7.4.9)$$

The left-hand side is at most a function of y and the right-hand side is at most a function of x. Hence we must have

$$\frac{\partial^2 u}{\partial y^2} = \lambda \qquad (7.4.10)$$

where λ is a constant, since x and y are independent variables. This can be integrated twice to yield

$$u(y) = \frac{\lambda}{2}y^2 + Ay + B \qquad (7.4.11)$$

where A and B are arbitrary constants of integration. We demand that $u = 0$ at $y = 0$ and $u = U$ at $y = a$. This gives

$$A = \frac{U}{a} - \frac{\lambda a}{2} \qquad B = 0 \qquad (7.4.12)$$

Using these constants, we can write Eq. 7.4.11 as

$$u(y) = \frac{\lambda}{2}(y^2 - ay) + \frac{U}{a}y$$

$$= \frac{1}{2\mu}\frac{d}{dx}(p + \gamma h)(y^2 - ay) + \frac{U}{a}y \qquad (7.4.13)$$

where we have used λ as the right-hand side of Eq. 7.4.9. If the flow is due to the motion of the plate only (a linear profile), the flow is called a *Couette flow*, if the motion is due to the pressure gradient only, i.e., $U = 0$, it is Poiseuille flow.

7.4.3 Simplified Flow Situation

Using the results derived above, we can write the expression for the velocity distribution between fixed plates (let $U = 0$) as

$$u(y) = \frac{1}{2\mu} \frac{d(p + \gamma h)}{dx} (y^2 - ay) \tag{7.4.14}$$

Using this distribution, we can find the flow rate per unit width to be

$$Q = \int u \, dA$$

$$= \int_0^a \frac{1}{2\mu} \frac{d(p + \gamma h)}{dx} (y^2 - ay) \, dy = -\frac{a^3}{12\mu} \frac{d(p + \gamma h)}{dx} \tag{7.4.15}$$

The average velocity V is found to be

$$V = \frac{Q}{a \times 1}$$

$$= -\frac{a^2}{12\mu} \frac{d(p + \gamma h)}{dx} \tag{7.4.16}$$

This can be expressed as the pressure drop in terms of the average velocity; for a horizontal[5] channel we have

$$\Delta p = \frac{12\mu V L}{a^2} \tag{7.4.17}$$

where we have used $\Delta p/L = -dp/dx$ since dp/dx is constant for developed flow. Observe that the maximum velocity occurs at $y = a/2$ and from Eq. 7.4.14 is

$$u_{max} = -\frac{a^2}{8\mu} \frac{dp}{dx} \tag{7.4.18}$$

Thus the average velocity is related to the maximum velocity by

$$V = \tfrac{2}{3} u_{max} \tag{7.4.19}$$

---

[5]For plates on an incline, simply replace p with $(p + \gamma h)$.

The shearing stress is found to be

$$\tau = \mu \frac{du}{dy} = \frac{1}{2} \frac{dp}{dx} (2y - a)$$ (7.4.20)

At the wall where $y = 0$ there results

$$\tau_0 = -\frac{a}{2} \frac{dp}{dx}$$ (7.4.21)

The pressure drop Δp over a length L of horizontal channel is found to be

$$\Delta p = \frac{2\tau_0}{a} L$$ (7.4.22)

recognizing that $dp/dx = -\Delta p/L$. This can be expressed in a more convenient form as

$$\frac{\Delta p}{\gamma} = f \frac{L}{2a} \frac{V^2}{2g}$$ (7.4.23)

if we introduce the friction factor f, defined by

$$f = \frac{\tau_0}{\frac{1}{8}\rho V^2}$$ (7.4.24)

In terms of the head loss (see Eq. 4.5.17), Eq. 7.4.23 becomes

$$h_L = f \frac{L}{2a} \frac{V^2}{2g}$$ (7.4.25)

If we combine Eqs. 7.4.17, 7.4.21, and 7.4.24, we find that

$$f = \frac{8}{\rho V^2} \left(-\frac{a}{2} \frac{dp}{dx} \right) = \frac{8}{\rho V^2} \left(-\frac{a}{2} \right) \left(-\frac{12\mu V}{a^2} \right) = \frac{48\mu}{\rho a V} = \frac{48}{\text{Re}}$$ (7.4.26)

where the Reynolds number $\text{Re} = \rho a V/\mu$ has been introduced. If this is substituted into Eq. 7.4.25, we see that

$$h_L = \frac{12\mu L V}{ra^2}$$ (7.4.27)

The head loss is directly proportional to the average velocity, a conclusion that is true for laminar flows, in general.

We have calculated most of the quantities of interest for laminar flow of an incompressible, steady flow between parallel plates. This would, of course, be an acceptable approximation for a wide channel, one for which the width exceeds the height by at least a factor of 8. For channels with smaller aspect ratios, the edge effects become significant and must be accounted for by adding an additional shear stress to the sides of the element in Fig. 7.5, or by maintaining $\partial^2 u/\partial z^2$ in Eq. 7.4.8. The solution is beyond the scope of this book.

KEY CONCEPT   *A wide channel is one for which the width exceeds the height by at least a factor of 8.*

Example 7.3

Water at 15°C flows with a Reynolds number of 1500 between the 50-cm-wide, horizontal plates shown in Fig. E7.3. Calculate (a) the flow rate, (b) the wall shear stress, (c) the pressure drop over 3 m, and (d) the velocity at $y = 5$ mm.

FIGURE E7.3

Solution

Since the Reynolds number is 1500, the laminar flow equations are assumed applicable.

(a) Using the definition of the Reynolds number the average velocity is found as follows:

$$1500 = \frac{Va}{\nu}$$

$$\therefore V = \frac{1500\nu}{a} = \frac{1500 \times 1.141 \times 10^{-6}}{12.5/1000} = 0.137 \text{ m/s}$$

Thus

$$Q = VA = 0.137 \times (12.5/1000) \times (50/100) = 8.56 \times 10^{-4} \text{ m}^3/\text{s}$$

(b) Using Eq. 7.4.17, the pressure gradient is

$$\frac{\Delta p}{L} = \frac{12\mu V}{a^2} = \frac{12 \times 1.14 \times 10^{-3} \times 0.137}{(12.5/1000)^2} = 12 \text{ Pa/m}$$

The shearing stress at the wall is found, using Eq. 7.4.22, to be

$$\tau_0 = \frac{a}{2}\frac{\Delta p}{L} = \frac{12.5/1000}{2} \times 12 = 0.075 \text{ Pa}$$

(c) The pressure drop over 3 m is found to be

$$\Delta p = 12L = 12 \times 3 = 36 \text{ Pa}$$

(d) The velocity distribution of Eq. 7.4.14 is

$$u(y) = \frac{1}{2\mu}\frac{dp}{dx}(y^2 - ay)$$

$$= \frac{1}{2 \times 1.14 \times 10^{-3}}(12)\left(y^2 - \frac{12.5}{1000}y\right) = -5263\,(y^2 - 0.0125y)$$

where we have used $dp/dx = -\Delta p/L$. At $y = 5$ mm, the velocity is

$$u = -5263\left[\left(\frac{5}{1000}\right)^2 - 0.0125 \times \left(\frac{5}{1000}\right)\right] = 0.198 \text{ m/s}$$

We have used three significant digits since the fluid properties are known to three significant digits.

Example 7.4

Find an expression for the pressure gradient that results in a zero shear stress at the lower wall, where $y = 0$, of two parallel plates; also, sketch the velocity profiles for a top plate speed of U with various pressure gradients. Assume horizontal plates.

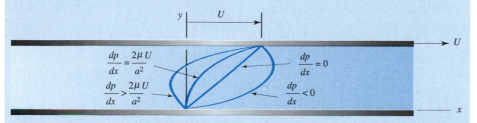

FIGURE E7.4

Solution

The velocity distribution for plates with the top plate moving with velocity U is given by Eq. 7.4.17. Letting $dh/dx = 0$, we have

$$u(y) = \frac{1}{2\mu}\frac{dp}{dx}(y^2 - ay) + \frac{U}{a}y$$

The shear stress is

$$\tau = \mu \frac{du}{dy}$$

$$= \frac{1}{2}\frac{dp}{dx}(2y - a) + \mu \frac{U}{a}$$

If $\tau = 0$ at $y = 0$, then $du/dy = 0$ at $y = 0$ and the pressure gradient is

$$\frac{dp}{dx} = \frac{2\mu U}{a^2}$$

If dp/dx is greater than this value, the slope du/dy at $y = 0$ is negative and thus the velocity u will be negative near $y = 0$. If $dp/dx = 0$, we observe that a linear velocity distribution results, namely,

$$u(y) = \frac{U}{a}y$$

If dp/dx is negative, $u(y)$ is greater at each y-location than the linear distribution since $(y^2 - ay)$ is a negative quantity for all y's of interest.

All of the results above can be qualitatively displayed on a sketch of $u(y)$ for several dp/dx as shown in Fig. E7.4.

7.5 LAMINAR FLOW BETWEEN ROTATING CYLINDERS

Fully developed, steady flow between concentric, rotating cylinders, as shown in Fig. 7.6, is another flow that yields a rather simple solution. It has particular application in the field of lubrication, where the fluid may be oil and the inner cylinder a rotating shaft. We will again use two approaches to find the velocity distribution. The laminar flow solution we will find will be valid up to a Reynolds number[6] of 1700, providing the angular velocity of the outer cylinder $\omega_2 = 0$, as is often the case. Above Re = 1700 a secondary laminar flow (a flow with a different velocity distribution) may develop and eventually a turbulent flow develops. In fact, numerous laminar flows (all different) have been observed for Re > 1700.

> **KEY CONCEPT**   *The laminar flow solution will be valid up to a Reynolds number of 1700.*

7.5.1   Elemental Approach

Body forces will be neglected in this derivation, or the cylinders will be assumed to be vertical. Since pressure does not vary with θ, an element in the form of a thin cylindrical shell will be used as shown in Fig. 7.6b. The resultant torque acting on this element is zero because it has no angular acceleration; this is expressed as

$$\tau 2\pi r L \times r - (\tau + d\tau)2\pi(r + dr)L \times (r + dr) = 0 \tag{7.5.1}$$

where L, the length of the cylinders, should be large with respect to the gap width $(r_2 - r_1)$ to avoid three-dimensional end effects. Equation 7.5.1 reduces to, neglecting the three higher-order terms which approach zero as $dr \to 0$,

$$2\tau + r \frac{d\tau}{dr} = 0 \tag{7.5.2}$$

The one-dimensional constitutive equation (see Table 5.1), and recognizing that $\tau = -\tau_{r\theta}$, provides the shear stress:

$$\tau = -\mu r \frac{d}{dr}\left(\frac{v_\theta}{r}\right) \tag{7.5.3}$$

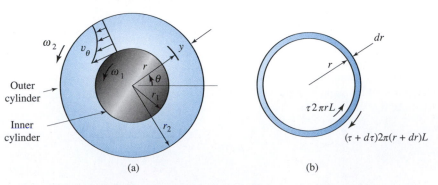

FIGURE 7.6   Flow between concentric cylinders: (a) basic flow variables; (b) element from between the cylinders.

[6]The Reynolds number is defined as Re = $\omega_1 r_1 \delta/\nu$, where $\delta = r_2 - r_1$.

There results

$$-2\mu r \frac{d}{dr}\left(\frac{v_\theta}{r}\right) - r\mu \frac{d}{dr}\left[r\frac{d}{dr}\left(\frac{v_\theta}{r}\right)\right] = 0 \tag{7.5.4}$$

Divide by μr, multiply by dr, and integrate to find

$$2\frac{v_\theta}{r} + r\frac{d}{dr}\left(\frac{v_\theta}{r}\right) = A \tag{7.5.5}$$

This can be rearranged as $\left[\text{perform the differentiation: } \dfrac{d}{dr}\left(\dfrac{v_\theta}{r}\right) = \dfrac{1}{r}\dfrac{dv_\theta}{dr} - \dfrac{v_\theta}{r^2}\right]$

$$\frac{dv_\theta}{dr} + \frac{v_\theta}{r} = A \tag{7.5.6}$$

or

$$\frac{1}{r}\frac{d}{dr}(rv_\theta) = A \tag{7.5.7}$$

Multiply by $r\,dr$ and integrate again to find that

$$v_\theta(r) = \frac{A}{2}r + \frac{B}{r} \tag{7.5.8}$$

The boundary conditions are $v_\theta = r_1\omega_1$ at $r = r_1$, and $v_\theta = r_2\omega_2$ at $r = r_2$. These conditions allow the constants to be evaluated as

$$A = 2\frac{\omega_2 r_2^2 - \omega_1 r_1^2}{r_2^2 - r_1^2} \qquad B = \frac{r_1^2 r_2^2(\omega_1 - \omega_2)}{r_2^2 - r_1^2} \tag{7.5.9}$$

We can obtain the same result by integrating the appropriate Navier–Stokes equation, or if that is not of interest, go directly to Section 7.5.3.

7.5.2 Solving the Navier–Stokes Equations

For steady, laminar flow between concentric cylinders we assume the streamlines to be circular so that $v_r = v_z = 0$, $v_\theta = v_\theta(r)$ only, and $\partial p/\partial\theta = 0$. The θ-component Navier–Stokes equation from Table 5.1 is

$$\left(\frac{\partial v_\theta}{\partial t} + v_r\frac{\partial v_\theta}{\partial r} + \frac{v_\theta}{r}\frac{\partial v_\theta}{\partial\theta} + v_z\frac{\partial v_\theta}{\partial z} + \frac{v_r v_\theta}{r}\right)$$
$$= -\frac{1}{r}\frac{\partial p}{\partial\theta} + \mu\left[\frac{\partial^2 v_\theta}{\partial r^2} + \frac{1}{r}\frac{\partial v_\theta}{\partial r} + \frac{1}{r^2}\frac{\partial^2 v_\theta}{\partial\theta^2} + \frac{\partial^2 v_\theta}{\partial z^2} + \frac{2}{r^2}\frac{\partial v_\theta}{\partial\theta} - \frac{v_\theta}{r^2}\right] \tag{7.5.10}$$

This reduces to

$$0 = \frac{\partial^2 v_\theta}{\partial r^2} + \frac{1}{r}\frac{\partial v_\theta}{\partial r} - \frac{v_\theta}{r^2} \qquad (7.5.11)$$

which can be written as

$$\frac{d^2 v_\theta}{dr^2} + \frac{d}{dr}\left(\frac{v_\theta}{r}\right) = 0 \qquad (7.5.12)$$

Integrating once, there results

$$\frac{dv_\theta}{dr} + \frac{v_\theta}{r} = A \qquad (7.5.13)$$

or

$$\frac{1}{r}\frac{d}{dr}(rv_\theta) = A \qquad (7.5.14)$$

A second integration yields

$$v_\theta = \frac{A}{2}r + \frac{B}{r} \qquad (7.5.15)$$

Applying the boundary conditions $v_\theta = r_1\omega_1$ at $r = r_1$ and $v_\theta = r_2\omega_2$ at $r = r_2$, we can find that

$$A = 2\frac{r_2^2\omega_2 - r_1^2\omega_1}{r_2^2 - r_1^2} \qquad B = \frac{r_1^2 r_2^2(\omega_1 - \omega_2)}{r_2^2 - r_1^2} \qquad (7.5.16)$$

7.5.3 Flow with the Outer Cylinder Fixed

For a number of situations such as a shaft rotating in a bearing the outer cylinder is fixed. Setting $\omega_2 = 0$, the velocity distribution is

$$v_\theta = \frac{r_1^2\omega_1}{r_2^2 - r_1^2}\left(\frac{r_2^2}{r} - r\right) \qquad (7.5.17)$$

The shearing stress τ_1 on the inner cylinder (see Table 5.1 and let $\tau = -\tau_{r\theta}$) is found to be

$$\tau_1 = -\left[\mu r \frac{d}{dr}\left(\frac{v_\theta}{r}\right)\right]_{r=r_1}$$

$$= \mu\frac{2}{r_1^2}\frac{r_1^2 r_2^2 \omega_1}{r_2^2 - r_1^2} = \frac{2\mu r_2^2 \omega_1}{r_2^2 - r_1^2} \qquad (7.5.18)$$

The torque T necessary to rotate the inner cylinder of length L is

$$T = \tau_1 A_1 r_1$$

$$= \frac{2\mu r_2^2 \omega_1}{r_2^2 - r_1^2} 2\pi r_1 L r_1 = \frac{4\pi\mu r_1^2 r_2^2 L \omega_1}{r_2^2 - r_1^2} \qquad (7.5.19)$$

The power $\dot{W}$ necessary to rotate the shaft is found by multiplying the torque by the rotational speed ω_1; it is

KEY CONCEPT *The power is found by multiplying the torque by the rotational speed.*

$$\dot{W} = T\omega_1$$

$$= \frac{4\pi\mu r_1^2 r_2^2 L \omega_1^2}{r_2^2 - r_1^2} \qquad (7.5.20)$$

This power is necessary to overcome the resistance of viscosity and results in an increase in the internal energy and thus the temperature of the fluid. The removal of this energy from the fluid often requires special heat exchangers.

Example 7.5

Estimate the viscosity of an oil contained in the annulus between two 25-cm-long cylinders, as shown in Fig. 7.6. The outer stationary cylinder is 8 cm in diameter. The 7.8-cm-diameter inner cylinder rotates at 3800 rpm when a torque of 0.12 N·m is applied. The specific gravity of the oil is 0.85. Neglect any torque due to the cylinder ends.

Solution

Assuming that the Reynolds number is less than 1700, Eq. 7.5.19 provides

$$\mu = \frac{T(r_2^2 - r_1^2)}{4\pi r_1^2 r_2^2 L \omega_1}$$

$$= \frac{0.12(0.04^2 - 0.039^2)}{4\pi \times 0.04^2 \times 0.039^2 \times 0.25 \times (3800 \times 2\pi/60)} = 0.00312 \text{ N·s/m}^2$$

Check the Reynolds number using $\nu = \mu/\rho$:

$$\text{Re} = \frac{\omega_1 r_1 \delta}{\nu} = \frac{(3800 \times 2\pi/60) \times 0.039 \times 0.002/2}{0.00312/(1000 \times 0.85)} = 845$$

This is less than 1700 so the calculation is assumed to be acceptable.

Example 7.6

Show that as the inner cylinder radius of Fig. E7.6 approaches the outer cylinder radius the velocity distribution approaches the linear distribution between parallel plates with one plate moving and a zero pressure gradient. This is Couette flow.

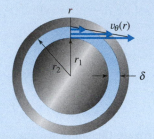

FIGURE E7.6

Solution

For this problem we will let $\omega_2 = 0$; the velocity distribution (7.5.17) is

$$v_\theta(r) = \frac{r_1^2 \omega_1}{r_2^2 - r_1^2} \left(\frac{r_2^2}{r} - r \right)$$

$$= \frac{r_1^2 \omega_1}{r_2^2 - r_1^2} \frac{r_2^2 - r^2}{r} = \frac{r_1^2 \omega_1}{r_2^2 - r_1^2} (r_2 - r) \frac{r_2 + r}{r}$$

Introduce the independent variable y, measured from the outer cylinder defined by $r + y = r_2$ (see Fig. 7.6); let $\delta = r_2 - r_1$. Then the above can be written as

$$v_\theta(r) = \frac{r_1^2 \omega_1 (r_2 - r)}{(r_2 - r_1)(r_2 + r_1)} \frac{r_2 + r}{r}$$

$$= \frac{r_1^2 \omega_1 y}{\delta(r_2 + r_1)} \frac{2r_2 - y}{r_2 - y}$$

As the inner radius approaches the outer radius we can write $r_1 \simeq r_2$. Letting $r_2 \simeq r_1 \simeq R$ we have $r_2 + r_1 \simeq 2R$ and

$$\frac{2R - y}{R - y} \simeq 2$$

since $y \ll R$. The velocity distribution then simplifies to

$$v_\theta(y) = \frac{R^2 \omega_1 y}{\delta \, 2R} \times 2 = \frac{R \omega_1}{\delta} y$$

This is a linear distribution and is a good approximation to the flow whenever $\delta \ll R$.

7.6 TURBULENT FLOW IN A PIPE

The study of developed turbulent flow in a circular pipe is of substantial interest in actual flows since most flows encountered in practical applications are turbulent flows in pipes. Even though for carefully controlled laboratory conditions, laminar flows have been observed up to Reynolds numbers of 40 000 in a pipe flow, turbulent flow is assumed to occur in a pipe under standard operating conditions whenever the Reynolds number

$$\mathrm{Re} = \frac{VD}{\nu}$$

exceeds 4000; between 2000 and 4000 the flow is assumed to randomly oscillate between being laminar and turbulent. Consider, for example, water at 20°C flowing in a rather small 5-mm-diameter pipe; the average velocity need only be 0.8 m/s for turbulent flow to occur. This is the situation whenever we drink water from a drinking fountain. For larger-diameter pipes the average velocity is sufficiently large so that a turbulent flow is produced in most engineering situations.

In a turbulent flow all three velocity components are nonzero. If we measure the components as a function of time, graphs similar to those shown in Fig. 7.7 would result for a flow in a pipe where u, v, w are in the x-, r-, and θ-directions, respectively. There is seldom any interest (to the engineer) in the details of the randomly fluctuating velocity components; hence we introduce the notion of a time-average quantity. The velocity components (u, v, w) are written as

$$u = \bar{u} + u' \qquad v = \bar{v} + v' \qquad w = \bar{w} + w' \qquad (7.6.1)$$

> **KEY CONCEPT** *Turbulent flow occurs whenever we drink water from a drinking fountain.*

where a bar over a quantity denotes a time average and a prime denotes the fluctuating part. Using the component u as an example, the *time average* is defined as,

$$\bar{u} = \frac{1}{T} \int_0^T u(t)\, dt \qquad (7.6.2)$$

> **KEY CONCEPT** *A bar over a quantity denotes a time average.*

where T is a time increment large enough[7] to eliminate all time dependence from $\bar{u}$. In a developed pipe flow $\bar{u}$ would be nonzero and $\bar{v} = \bar{w} = 0$, as is obvious from Fig. 7.7.

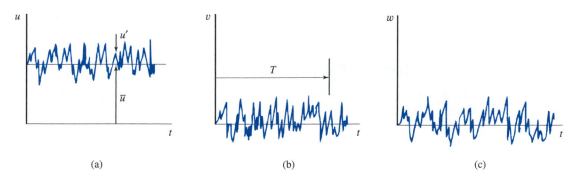

FIGURE 7.7 Velocity components in a turbulent pipe flow: (a) x-component velocity; (b) r-component velocity; (c) θ-component velocity.

[7]At a given location in a pipe this could be checked experimentally as follows: Start with a relatively large value of T and then decrease T for subsequent runs. If $\bar{u}$ remains unchanged as T decreases, T is sufficiently large. If T is too small, $\bar{u}$ will be different for each run.

Example 7.7

Show that $\overline{u'} = 0$ and $\overline{\dfrac{\partial u}{\partial y}} = \dfrac{\partial \overline{u}}{\partial y}$ for a turbulent flow.

Solution

To show that $\overline{u'} = 0$ we simply substitute the expression (7.6.1) for $u(t)$ into Eq. 7.6.2 and obtain

$$\overline{u} = \frac{1}{T} \int_0^T (\overline{u} + u')dt$$

$$= \frac{1}{T} \int_0^T \overline{u}\, dt + \frac{1}{T} \int_0^T u'\, dt$$

$$= \overline{u}\, \frac{1}{T} \int_0^T dt + \overline{u'}$$

$$= \overline{u} + \overline{u'}$$

Subtracting $\overline{u}$ from both sides results in

$$\overline{u'} = 0$$

Now, let us time average the derivative $\partial u / \partial y$. We have

$$\overline{\frac{\partial u}{\partial y}} = \frac{1}{T} \int_0^T \frac{\partial u}{\partial y}\, dt$$

$$= \frac{\partial}{\partial y} \left(\frac{1}{T} \int_0^T u\, dt \right) = \frac{\partial}{\partial y}\, \overline{u}$$

since T is a constant. Thus

$$\overline{\frac{\partial u}{\partial y}} = \frac{\partial \overline{u}}{\partial y}$$

7.6.1   Differential Equation

The differential equation that must be solved to provide us with the time-average velocity distribution is derived by using a particle approach. The Navier–Stokes equations could be time averaged, arriving at the same differential equation; that approach is included in the problems.

Consider the situation for a turbulent flow in a horizontal pipe, as shown in Fig. 7.8. We use velocity components u and v in the x- and y-directions, respectively. Fluid particles[8] move randomly throughout the flow. At an instant in time a fluid particle moves through the incremental area dA due to the velocity fluctuation v'; it enters a neighboring layer of fluid which is moving with a higher x-component velocity, thereby providing a retarding effect on the neighboring layer. A fluid particle that moves to a neighboring layer that is traveling with a lower x-component velocity would tend to accelerate the slower moving fluid. The x-component force that results due to the random motion of a fluid particle passing through the incremental area dA would be (see Eq. 4.6.6)

$$dF = -\rho v' dA \, u' \qquad (7.6.3a)$$

where u' is the negative change in x-component velocity due to the momentum exchange and $\rho v' \, dA$ is the mass flux through the area; the negative sign provides a positive dF. If we divide both sides by the area dA we obtain a "stress" that we call the *turbulent shear stress*. It is

$$\tau_{\text{turb}} = \frac{dF}{dA} = -\rho u' v' \qquad (7.6.3b)$$

where we know that $(u'v')$ is, on the average, a negative quantity since a positive v' produces a negative u'. This "shear stress" is actually a momentum exchange but since it has the same effect as a stress, we call it a shear stress.

The time-average turbulent shear stress, often called the *apparent shear stress*, which is of primary interest, is

$$\bar{\tau}_{\text{turb}} = -\rho \overline{u'v'} \qquad (7.6.4)$$

where Eq. 7.6.2 would be used for the time averages. Note that $\overline{u'w'}$ would be zero since a w'-component (in the θ-direction) would not move a fluid particle

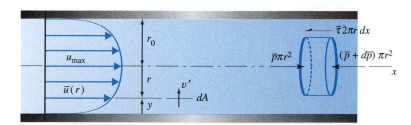

FIGURE 7.8   Turbulent flow in a horizontal pipe.

---

[8]A particle is a relatively small mass of the fluid, consisting of a large number of molecules, that tends to move as a unit.

into a layer of higher or lower x-component velocity. Also, $\overline{v'w'} = 0$ using the same logic.

The total shear stress at a particular location would be due to both the viscosity and the momentum exchange described above; that is,

$$\overline{\tau} = \overline{\tau}_{\text{lam}} + \overline{\tau}_{\text{turb}}$$

$$= \mu \frac{\partial \overline{u}}{\partial y} - \rho \overline{u'v'} \tag{7.6.5}$$

The shear stress can be related to the pressure gradient by summing forces on the horizontal cylindrical element shown in Fig. 7.8. There results

$$\overline{\tau} = -\frac{r}{2}\frac{d\overline{p}}{dx} = -\frac{r\Delta\overline{p}}{2L} \tag{7.6.6}$$

which shows that the shear stress distribution is linear for a turbulent flow as well as a laminar flow (see Eq. 7.3.17). The turbulent shear obviously goes to zero at the wall since the velocity perturbations are zero at the wall, and the total shear is zero at the centerline where $r = 0$ or $y = r_0$, as shown in Fig. 7.9. The viscous shear is nonzero only in a very thin viscous wall layer, of thickness δ_ν, near the wall, as shown in part (b). Note that the turbulent shear reaches a maximum near the wall in the viscous wall layer.

The differential equation that must be solved if the time-average velocity distribution is to be determined is found by combining the two preceding equations; it is

$$\frac{r}{2}\frac{d\overline{p}}{dx} = \mu \frac{d\overline{u}}{dr} + \rho \overline{u'v'} \tag{7.6.7}$$

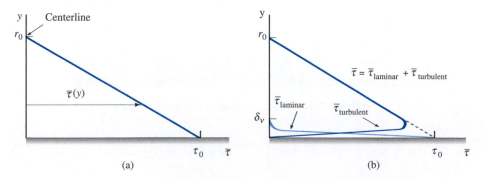

FIGURE 7.9 Shear stress distributions in a developed pipe flow.

where we have used $r + y = r_0$ so that $dy = -dr$. For developed flow we know that $d\bar{p}/dx =$ const.; hence if we know how $\overline{u'v'}$ varied with r, the differential equation could be solved. The quantity $\overline{u'v'}$ cannot be determined analytically, however, so the solution to Eq. 7.6.7 cannot be attempted until an empirical expression is found for $\overline{u'v'}$. Rather than finding an empirical expression for $\overline{u'v'}$ and then solving the differential equation for $\bar{u}$, we will simply present the empirical results obtained for the velocity profile $\bar{u}(y)$. However, before we do this in the next section we should introduce the eddy viscosity, the mixing length, and the correlation coefficient.

KEY CONCEPT   *The quantity $\overline{u'v'}$ cannot be determined analytically.*

Instead of using the quantity $\overline{u'v'}$ as the unknown in Eq. 7.6.7, we often introduce the *eddy viscosity* η, defined by the relationship

$$\overline{u'v'} = \eta \frac{d\bar{u}}{dy} \tag{7.6.8}$$

Note that it has the same dimensions as the kinematic viscosity. In terms of the eddy viscosity the differential equation becomes

$$\frac{r}{2}\frac{d\bar{p}}{dx} = \rho(\nu + \eta)\frac{d\bar{u}}{dr} \tag{7.6.9}$$

If we view the turbulent process as the random and chaotic mixing of particles of fluid, we may choose to introduce the *mixing length* l_m, the distance a particle travels before interacting with another particle. Based on reasoning related to momentum interchange we relate the eddy viscosity to the mixing length with

$$\eta = l_m^2 \left|\frac{d\bar{u}}{dy}\right| \tag{7.6.10}$$

The *correlation coefficient* K_{uv}, a normalized turbulent shear stress, often used when describing turbulent motions, has limits of ± 1 and is defined by

$$K_{uv} = \frac{\overline{u'v'}}{\sqrt{\overline{u'^2}}\,\sqrt{\overline{v'^2}}} \tag{7.6.11}$$

where the time-average quantities would be defined by Eq. 7.6.2

The quantities η, l_m, and K_{uv} are functions of r (or y) that simply replace the variable $\overline{u'v'}$. They do not simplify the differential equation 7.6.7; they allow the equation to take slightly different forms. Since we cannot derive an expression for η, l_m, or K_{uv} we cannot find $\bar{u}(r)$ using analytical methods. We must rely on experimental data. It follows in the next section.

KEY CONCEPT   *The quantities η, l_m and K_{uv} simply replace the variable $\overline{u'v'}$.*

Example 7.8

Note that in Fig. 7.9b there is a region near the wall where the turbulent shear is near its maximum and is relatively constant, as shown in Fig. E7.8, and the viscous shear is quite small. Assume that the mixing length is directly proportional to the distance from the wall. With this assumption show that $\bar{u}(y)$ is logarithmic in this region near the wall.

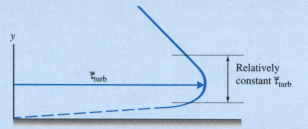

FIGURE E7.8

Solution

If the viscous shear is negligible, we have, combining Eqs. 7.6.5 and 7.6.8, and 7.6.10,

$$\bar{\tau}_{\text{turb}} = \rho\eta\frac{d\bar{u}}{dy} = \rho l_m^2\left(\frac{d\bar{u}}{dy}\right)^2$$

Now, if $\bar{\tau}_{\text{turb}} = \text{const.} = c_1$ and we assume that

$$l_m = c_2 y$$

there results

$$c_1 = \rho c_2^2 y^2\left(\frac{du}{dy}\right)^2$$

or

$$y\frac{d\bar{u}}{dy} = c_3$$

where $c_3 = \sqrt{c_1/\rho c_2^2}$. This is integrated to yield

$$\bar{u}(y) = c_3 \ln y + c_4$$

Hence, with the foregoing assumptions we see that a logarithmic profile is predicted for the region of constant turbulent shear near the wall. This is, in fact, observed from experimental data; so we conclude that the above assumptions are reasonable for a turbulent flow in a pipe.

7.6.2 Velocity Profile

The time-average velocity profile in a pipe is quite sensitive to the magnitude of the average wall roughness height e, as sketched in Fig. 7.10. All materials are "rough" when viewed with sufficient magnification, although glass and plastic are assumed to be smooth with $e = 0$. (Values for e are listed in Fig. 7.13.) As noted in the preceding section, the laminar shear is significant only near the wall in the

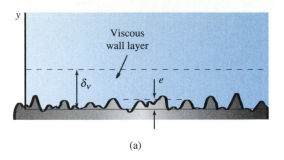
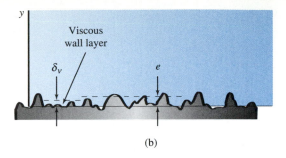

(a)   (b)

FIGURE 7.10   (a) A smooth wall and (b) a rough wall.

viscous wall layer with thickness δ_v. If the thickness δ_v is sufficiently large, it submerges the wall roughness elements so that they have negligible effect on the flow; it is as if the wall were smooth. Such a condition is often referred to as being *hydraulically smooth*. If the viscous wall layer is relatively thin, the roughness elements protrude out of this layer and the wall is rough. The *relative roughness e/D* and the Reynolds number can be used to determine if a pipe is smooth or rough. This will be observed from the friction factor data presented in the next section.

We do not solve for the velocity distribution in a developed turbulent flow since $\overline{u'v'}$ cannot be determined analytically. Thus we will present the empirical data for $\overline{u}(y)$ directly and not solve Eq. 7.6.7. The two most common expressions for turbulent flow are now presented.

The first method of empirically expressing the velocity distribution involves flows with smooth walls and flows with rough walls. If the flow has a smooth wall, as in Fig. 7.10a, we identify two regions of the flow, the *wall region* and the *outer region*. In the wall region the characteristic velocity and length are the **shear velocity** u_τ defined by[9] $u_\tau = \sqrt{\tau_0/\rho}$, and the **viscous length** ν/u_τ. The dimensionless velocity distribution in the wall region, for a smooth pipe, is

$$\frac{\overline{u}}{u_\tau} = \frac{u_\tau y}{\nu} \text{ (viscous wall layer)} \qquad 0 \le \frac{u_\tau y}{\nu} \lesssim 5 \qquad (7.6.12)$$

$$\frac{\overline{u}}{u_\tau} = 2.44 \ln \frac{u_\tau y}{\nu} + 4.9 \text{ (turbulent region)} \qquad 30 < \frac{u_\tau y}{\nu}, \frac{y}{r_0} < 0.15 \quad (7.6.13)$$

In the interval $5 < u_\tau y/\nu < 30$, the *buffer zone,* the experimental data do not fit either of the curves above but merge the two curves as shown in Fig. 7.11a. The viscous wall layer has thickness δ_v and it is in the viscous layer that turbulence is

KEY CONCEPT   *If the thickness δ_v is sufficiently large, it is as if the wall were smooth. The pipe is hydraulically smooth.*

Shear velocity: *The quantity $\sqrt{\tau_0/\rho}$ is the shear velocity u_τ.*

Viscous length: *The length ν/u_τ is a characteristic length in a turbulent flow.*

KEY CONCEPT   *It is in the viscous wall layer that turbulence is thought to be initiated.*

[9]The shear velocity u_τ is a fictitious velocity and has no relationship to an actual velocity in the flow. It is a quantity with dimensions of velocity that allows the experimental data to be presented in dimensionless form as *universal profiles* (profiles that are valid for all turbulent pipe flows). The viscous length ν/u_τ is also a fictitious length.

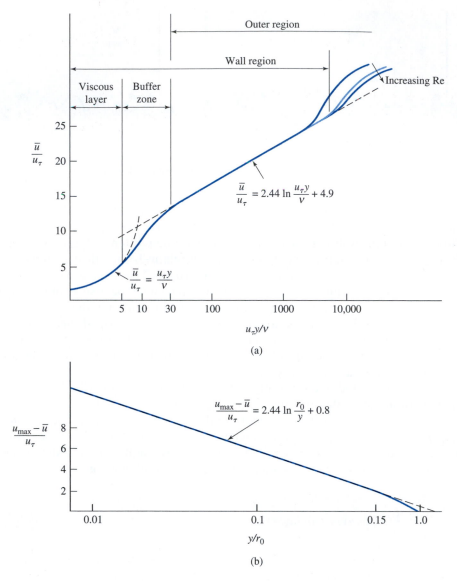

FIGURE 7.11 Empirical relations for turbulent flow in a smooth pipe: (a) wall region; (b) outer region. (Based on data from J. Laufer, The Structure of Turbulence in Fully Developed Pipe Flow, *NACA Report 1174*, 1954.)

thought to be initiated; this layer possesses a time-average, linear velocity distribution, but instantaneously the layer is very time dependent. The outer edge of the wall region is quite dependent on the Reynolds number, as shown; for low Re it may be located near $u_\tau y/\nu = 3000$.

For rough pipes the viscous wall layer does not play an important role since turbulence initiates from the protruding wall elements, so only a logarithmic profile is necessary in the wall region. The characteristic length is the average roughness height e; the dimensionless velocity profile for the rough pipe is

$$\frac{\bar{u}}{u_\tau} = 2.44 \ln \frac{y}{e} + 8.5 \qquad \frac{y}{r_0} < 0.15 \qquad \text{(wall region)} \qquad (7.6.14)$$

where the constants 8.5 and 2.44 allow a good fit with experimental data.

In the outer region, sketched in Fig. 7.11b, the characteristic length is r_0; the velocity defect $(u_{max} - \bar{u})$ is normalized with u_τ, and the empirical relation, for both smooth and rough pipes, is

$$\frac{u_{max} - \bar{u}}{u_\tau} = 2.44 \ln \frac{r_0}{y} + 0.8 \quad \text{(outer region)} \qquad \frac{y}{r_0} \leq 0.15 \qquad (7.6.15)$$

An additional empirical equation is needed to complete the profile for $0.15 < y/r_0 \leq 1$.

The wall region and the outer region overlap as shown in Fig. 7.11a. In this overlap region we can combine the equations above to obtain an expression for the maximum velocity; for a smooth pipe it is

$$\frac{u_{max}}{u_\tau} = 2.44 \ln \frac{u_\tau r_0}{\nu} + 5.7 \qquad \text{(smooth pipes)} \qquad (7.6.16)$$

and for a rough pipe we find that

$$\frac{u_{max}}{u_\tau} = 2.44 \ln \frac{r_0}{e} + 9.3 \qquad \text{(rough pipes)} \qquad (7.6.17)$$

Although we do not often desire the actual time-average velocity at a specified radial position in a pipe, the distributions above are of occasional use and are presented for completeness. Note, however, that before u_{max} can be found we must know u_τ; before u_τ can be found we must know τ_0. To find τ_0 we use the pressure gradient,

$$\tau_0 = -\frac{r_0}{2} \frac{dp}{dx} \qquad (7.6.18)$$

or the friction factor with Eq. 7.3.19. If neither dp/dx nor f is known, we may use the power-law form of the profile, described in the following paragraph, to approximate f.

An alternative, and simpler form that adequately describes the turbulent flow velocity distribution in a pipe is the *power-law profile*, namely,

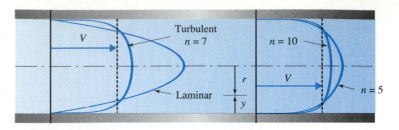

FIGURE 7.12   Turbulent velocity profile.

$$\frac{\bar{u}}{u_{max}} = \left(\frac{y}{r_0}\right)^{1/n}$$ (7.6.19)

where y is measured from the pipe wall and n is an integer between 5 and 10. Using this distribution the average velocity is found to be

$$V = \frac{\displaystyle\int_0^{r_0} \bar{u}(r)2\pi r\, dr}{\pi r_0^2} = \frac{2n^2}{(n+1)(2n+1)} u_{max}$$ (7.6.20)

This distribution is compared with a laminar profile in Fig. 7.12.

The value of n in the exponent is related to the friction factor f by the empirical expression

$$n = \frac{1}{\sqrt{f}}$$ (7.6.21)

The constant n varies from 5 to 10 depending on the Reynolds number and the pipe wall roughness e/D. For smooth pipes the exponent n is related to the Reynolds number as shown in Table 7.1.

The power-law profile cannot be used to obtain the slope at the wall since it will always yield $(d\bar{u}/dy_{wall} = \infty$ for all n. Thus it cannot be used to predict the wall shear stress. The wall shear is found by combining Eqs. 7.6.21 and 7.3.19. In addition, it gives a positive slope $d\bar{u}/dy$ at the centerline of the pipe, where the slope should be zero, so it is not valid near the centerline.

It should be noted that the kinetic-energy-correction factor α (see Eq. 4.5.29) in pipes is 1.11, 1.06, and 1.03 for $n = 5$, 7, and 10, respectively. Because it is close to unity for $n > 7$, it is often set equal to unity in the energy equation when working problems involving turbulent flow.

KEY CONCEPT   *The power-law profile cannot be used to obtain the slope at the wall or the slope at the centerline.*

TABLE 7.1   Exponent n for Smooth Pipes

$Re = VD/\nu$	4×10^3	10^5	10^6	$>2 \times 10^6$
n	6	7	9	10

Example 7.9

Water at 20°C flows in a 10-cm-diameter pipe at an average velocity of 1.6 m/s. If the roughness elements are 0.046 mm high, would the wall be rough or smooth? Refer to Fig. 7.10.

Solution

To determine if the wall is rough or smooth, we must compare the viscous wall layer thickness with the height of the roughness elements. So, let's find the viscous wall layer thickness. From Fig. 7.11 the viscous layer thickness is determined by letting $u_\tau y/\nu = 5$, where $y = \delta_\nu$. First, we must find u_τ. The Reynolds number is

$$\mathrm{Re} = \frac{VD}{\nu}$$

$$= \frac{1.6 \times 0.1}{10^{-6}} = 1.6 \times 10^5$$

From Table 7.1 $n \simeq 7.5$, so that, from Eq. 7.6.21,

$$f = \frac{1}{n^2}$$

$$= \frac{1}{7.5^2} = 0.018$$

The wall shear is calculated from Eq. 7.3.19:

$$\tau_0 = \tfrac{1}{8}\,\rho V^2 f$$

$$= \tfrac{1}{8} \times 1000 \times 1.6^2 \times 0.018 = 5.8 \text{ Pa}$$

The friction velocity is

$$u_\tau = \sqrt{\frac{\tau_0}{\rho}}$$

$$= \sqrt{\frac{5.8}{1000}} = 0.076 \text{ m/s}$$

This allows us to calculate the viscous wall layer thickness:

$$\delta_\nu = \frac{5\nu}{u_\tau}$$

$$= \frac{5 \times 10^{-6}}{0.076} = 6.6 \times 10^{-5} \text{ m} \quad \text{or} \quad 0.066 \text{ mm}$$

Since the roughness elements are only 0.046 mm high, they are submerged in the viscous wall layer. Consequently, the wall is smooth (see Fig. 7.10a). If the pipe were made of cast iron with $e = 0.26$ mm, the wall would be rough.

Note that the viscous wall layer, even at this relatively low velocity, is about 0.1% of the radius. The viscous wall layer is usually extremely thin.

Example 7.10

The 4-cm-diameter smooth, horizontal pipe of Fig. E7.10 transports 0.004 m³/s of water at 20°C. Using the power-law profile, approximate (a) the friction factor, (b) the maximum velocity, (c) the radial position where $u = V$, (d) the wall shear, (e) the pressure drop over a 10-m length, and (f) the maximum velocity using Eq. 7.6.16.

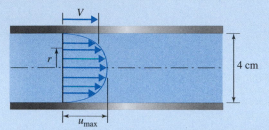

FIGURE E7.10

Solution

(a) The average velocity is calculated to be

$$V = \frac{Q}{A} = \frac{0.004}{\pi \times 0.02^2} = 3.18 \text{ m/s}$$

The Reynolds number is

$$\text{Re} = \frac{VD}{\nu} = \frac{3.18 \times 0.04}{10^{-6}} = 1.27 \times 10^5$$

From Table 7.1 we see that $n \cong 7.5$ and from Eq. 7.6.21,

$$f = \frac{1}{n^2}$$

$$= \frac{1}{7.5^2} = 0.018$$

(b) The maximum velocity is found using Eq. 7.6.20 to be

$$u_{max} = \frac{(n + 1)(2n + 1)}{2n^2} V$$

$$= \frac{8.5 \times 16}{2 \times 7.5^2} \times 3.18 = 3.84 \text{ m/s}$$

(c) The distance from the wall where $u = V = 3.18$ m/s is found using Eq. 7.6.19 as follows:

$$\frac{\bar{u}}{u_{max}} = \left(\frac{y}{r_0}\right)^{1/7.5}$$

$$\therefore y = r_0 \left(\frac{u}{u_{max}}\right)^{7.5}$$

$$= 2\left(\frac{3.18}{3.84}\right)^{7.5} = 0.49 \text{ cm}$$

The radial position is thus

$$r = r_0 - y$$

$$= 2 - 0.49 = 1.51 \text{ cm}$$

(d) The wall shear is found using Eq. 7.3.19 and is

$$\tau_0 = \tfrac{1}{8}\rho V^2 f$$

$$= \tfrac{1}{8} \times 1000 \times 3.18^2 \times 0.018 = 23 \text{ Pa}$$

(e) The pressure drop is calculated using Eq. 7.6.18 with $\Delta p/L = -dp/dx$ to be

$$\Delta p = \frac{2\tau_0 L}{r_0}$$

$$= \frac{2 \times 23 \times 10}{0.02} = 23\,000 \text{ Pa} \quad \text{or} \quad 23 \text{ kPa}$$

(f) To use Eq. 7.6.16 we must know the shear velocity. It is

$$u_\tau = \sqrt{\frac{\tau_0}{\rho}}$$

$$= \sqrt{\frac{23}{1000}} = 0.152 \text{ m/s}$$

Then we find u_{max} to be

$$u_{max} = 0.152\left(2.44 \ln \frac{0.152 \times 0.02}{10^{-6}} + 5.7\right) = 3.84 \text{ m/s}$$

the same as that given by the power-law formula in part (b). This answer is considered to be more accurate if it differs from that of Eq. 7.6.20. Note that the experimental data do not allow for accuracy in excess of three significant digits, and often to only two significant digits.

7.6.3 Losses in Developed Pipe Flow

Perhaps the most calculated quantity in pipe flow is the head loss. If the head loss is known in a developed flow, the pressure change can be calculated; for developed flow in a pipe the energy equation (4.5.17) provides us with

$$h_L = \frac{\Delta(p + \gamma h)}{\gamma} \tag{7.6.22}$$

The head loss that results from the wall shear in a developed flow is related to the friction factor (see Eq. 7.3.20) by the Darcy–Weisbach equation, namely,

$$h_L = f\frac{L}{D}\frac{V^2}{2g} \tag{7.6.23}$$

KEY CONCEPT *If the friction factor is known, we can find the head loss and the pressure drop.*

Consequently, if the friction factor is known, we can find the head loss and then the pressure drop.

The friction factor f depends on the various quantities that affect the flow, written as

$$f = f(\rho, \mu, V, D, e) \tag{7.6.24}$$

where the average wall roughness height e accounts for the influence of the wall roughness elements. A dimensional analysis following the steps of Section 6.2 provides us with

$$f = f\left(\frac{\rho VD}{\mu}, \frac{e}{D}\right) \tag{7.6.25}$$

where e/D is the relative roughness.

Experimental data that relate the friction factor to the Reynolds number have been obtained for fully developed pipe flow over a wide range of wall roughnesses. The results of these data are presented in Fig. 7.13, which is commonly referred to as the *Moody diagram*, named after Lewis F. Moody (1880–1953). There are several features of the Moody diagram that should be noted.

- For a given wall roughness, measured by the relative roughness e/D, there is a sufficiently large value of Re above which the friction factor is constant, thereby defining the *completely turbulent regime*. The average roughness element size e is substantially greater than the viscous wall layer thickness δ_ν, so that viscous effects are not significant; the resistance to the flow is produced primarily by the drag of the roughness elements that protrude into the flow.

- For the smaller relative roughness e/D values it is observed that, as Re decreases, the friction factor increases in the *transition zone* and eventually becomes the same as that of a smooth pipe. The roughness elements become submerged in the viscous wall layer so that they produce little effect on the main flow.

- For Reynolds numbers less than 2000, the friction factor of laminar flow is shown. The *critical zone* couples the turbulent flow to the laminar flow and may represent an oscillatory flow that alternately exists between turbulent and laminar flow.

- The e values in this diagram are for new pipes. With age a pipe will corrode and become fouled, changing both the roughness and the pipe diameter, with a resulting increase in the friction factor. Such factors should be included in design considerations; they will not be reviewed here.

The following empirical equations represent the Moody diagram for Re > 4000:

Smooth pipe flow:
$$\frac{1}{\sqrt{f}} = 0.86 \ln \mathrm{Re}\,\sqrt{f} - 0.8 \tag{7.6.26}$$

Completely turbulent zone:
$$\frac{1}{\sqrt{f}} = -0.86 \ln \frac{e}{3.7D} \tag{7.6.27}$$

Transition zone:
$$\frac{1}{\sqrt{f}} = -0.86 \ln \left(\frac{e}{3.7D} + \frac{2.51}{\mathrm{Re}\,\sqrt{f}} \right) \tag{7.6.28}$$

The transition zone equation (7.6.28) that couples the smooth pipe equation to the completely turbulent regime equation is known as the **Colebrook equation**. Note that Eq. 7.6.26 is the Colebrook equation with $e = 0$, and Eq. 7.6.27 is the Colebrook equation with Re = ∞.

Colebrook equation: *The equation that couples the smooth pipe equation to the completely turbulent regime equation.*

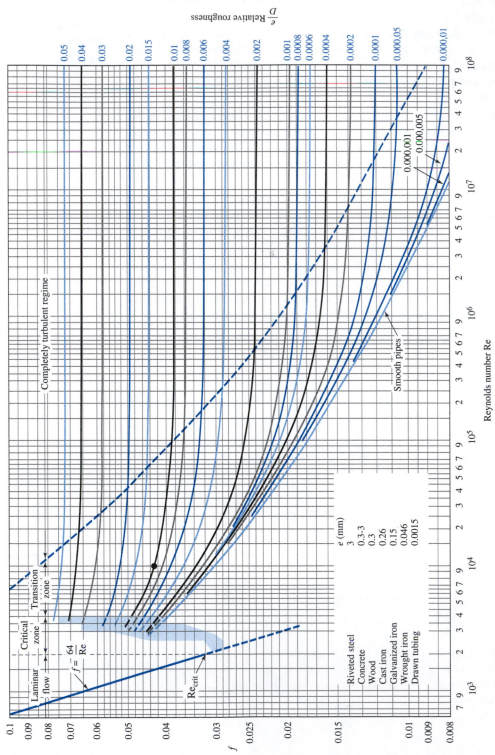

	e (mm)
Riveted steel	3
Concrete	0.3-3
Wood	0.3
Cast iron	0.26
Galvanized iron	0.15
Wrought iron	0.046
Drawn tubing	0.0015

FIGURE 7.13 Moody diagram. (From L. F. Moody, Trans. ASME, Vol. 66, 1944. Reproduced with permission of ASME.) (Note: If $e/D = 0.006$ and $Re = 10^4$, the dot locates $f = 0.043$.)

Three categories of problems can be identified for developed turbulent flow in a pipe of length L:

Category	Known	Unknown
1	Q, D, e, ν	h_L
2	D, e, ν, h_L	Q
3	Q, e, ν, h_L	D

A category 1 problem is straightforward and requires no iteration procedure when using the Moody diagram. Category 2 and 3 problems are more like problems encountered in engineering design situations and require an iterative trial-and-error process when using the Moody diagram. Each of these types will be illustrated with an example.

An alternative to using the Moody diagram that avoids any trial-and-error process is made possible by empirically derived formulas. Perhaps the best of such formulas were presented by Swamee and Jain (1976) for pipe flow; an explicit expression that provides an approximate value for the unknown in each category above is as follows:

$$h_L = 1.07 \frac{Q^2 L}{gD^5} \left\{ \ln\left[\frac{e}{3.7D} + 4.62 \left(\frac{\nu D}{Q}\right)^{0.9} \right] \right\}^{-2} \qquad \begin{array}{l} 10^{-6} < e/D < 10^{-2} \\ 3000 < \mathrm{Re} < 3 \times 10^8 \end{array} \qquad (7.6.29)$$

$$Q = -0.965 \left(\frac{gD^5 h_L}{L}\right)^{0.5} \ln\left[\frac{e}{3.7D} + \left(\frac{3.17\nu^2 L}{gD^3 h_L}\right)^{0.5} \right] \qquad \mathrm{Re} > 2000 \qquad (7.6.30)$$

$$D = 0.66 \left[e^{1.25} \left(\frac{LQ^2}{gh_L}\right)^{4.75} + \nu Q^{9.4} \left(\frac{L}{gh_L}\right)^{5.2} \right]^{0.04} \qquad \begin{array}{l} 10^{-6} < e/D < 10^{-2} \\ 5000 < \mathrm{Re} < 3 \times 10^8 \end{array} \qquad (7.6.31)$$

Equation 7.6.30 is as accurate as the Moody diagram, and Eqs. 7.6.29 and 7.6.31 are accurate to within approximately 2% of the Moody diagram. These tolerances are acceptable for engineering calculations. It is important to realize that the Moody diagram is based on experimental data that likely is accurate to within no more than 5%. Hence the foregoing three formulas of Swamee and Jain, which can easily be input on a programmable hand-held calculator, are often used by design engineers. The following examples also illustrate the use of these approximate formulas.

Example 7.11

Water at 20°C is transported for 500 m in a 4 cm-diameter wrought iron horizontal pipe with a flow rate of 3 L/s. Calculate the pressure drop over the 500-m length of pipe, using (a) the Moody diagram and (b) the alternate method.

Solution

(a) The average velocity is

$$V = \frac{Q}{A} = \frac{3 \times 10^{-3}}{\pi \times (2/100)^2} = 2.38 \text{ m/s}$$

The Reynolds number is

$$Re = \frac{VD}{\nu} = \frac{2.38 \times (4/100)}{1.007 \times 10^{-6}} = 9.52 \times 10^4$$

Obtaining e from Fig. 7.13, we have, using $D = 4/100$

$$\frac{e}{D} = \frac{0.046 \times 10^{-3}}{0.04} = 0.00115$$

The friction factor is read from the Moody diagram to be

$$f = 0.023$$

The head loss is calculated as

$$h_L = f \frac{L}{D} \frac{V^2}{2g}$$

$$= 0.023 \; \frac{500}{(4/100)} \frac{2.38^2}{2 \times 9.81} = 83 \text{ m}$$

This answer is given to two significant numbers since the friction factor is known to at most two significant numbers. The pressure drop is found by Eq. 7.6.22 to be

$$\Delta p = \gamma h_L$$

$$= 9810 \times 83 = 814230 \text{ N/m}^2 \quad \text{or} \quad 814.23 \text{ kPa}$$

(b) The alternate method for this Category 1 problem uses Eq. 7.6.29, with $D = 4/100 = 0.04$ m:

$$h_L = 1.07 \frac{\left(3 \times 10^{-3}\right)^2 \times 500}{9.81 \times (0.04)^5} \left\{ \ln \left[\frac{0.00115}{3.7} + 4.62 \left(\frac{1.007 \times 10^{-6} \times 0.04}{\left(3 \times 10^{-3}\right)} \right)^{0.9} \right] \right\}^{-2}$$

$$= 1.07 \times 4480 \times 0.01732 = 83 \text{ m}$$

This much simpler method provides the same value as that found using the Moody diagram.

Example 7.12

A pressure drop of 700 kPa is measured over a 300-m length of horizontal, 10-cm-diameter wrought iron pipe that transports oil ($S = 0.9$, $\nu = 10^{-5}$ m²/s). Calculate the flow rate using (a) the Moody diagram, and (b) the Alternate method.

Solution

(a) The relative roughness is

$$\frac{e}{D} = \frac{0.046}{100} = 0.00046$$

Assuming that the flow is completely turbulent (Re is not needed), the Moody diagram gives

$$f = 0.0165$$

The head loss is found to be

$$h_L = \frac{\Delta p}{\gamma_{oil}} = \frac{700\,000}{9800 \times 0.9} = 79.4 \text{ m}$$

The velocity is calculated from Eq. 7.6.23 to be

$$V = \left(\frac{2gDh_L}{fL}\right)^{1/2} = \left(\frac{2 \times 9.8 \times 0.1 \times 79.4}{0.0165 \times 300}\right)^{1/2} = 5.61 \text{ m/s}$$

This provides us with a Reynolds number of

$$\text{Re} = \frac{VD}{\nu} = \frac{5.61 \times 0.1}{10^{-5}} = 5.61 \times 10^4$$

Using this Reynolds number and $e/D = 0.00046$, the Moody diagram gives the friction factor as

$$f = 0.023$$

This corrects the original value for f. The velocity is recalculated to be

$$V = \left(\frac{2 \times 9.8 \times 0.1 \times 79.4}{0.023 \times 300}\right)^{1/2} = 4.75 \text{ m/s}$$

The Reynolds number is then

$$\text{Re} = \frac{4.75 \times 0.1}{10^{-5}} = 4.75 \times 10^4$$

From the Moody diagram $f = 0.023$ appears to be satisfactory. Thus the flow rate is

$$Q = VA = 4.75 \times \pi \times 0.05^2 = 0.037 \text{ m}^3/\text{s}$$

Only two significant numbers are given since f is known to at most two significant numbers.

   (b) The alternative method for this Category 2 problem uses the explicit relationship (7.6.30). We can directly calculate Q to be

$$Q = -0.965\left(\frac{9.8 \times 0.1^5 \times 79.4}{300}\right)^{0.5} \ln\left[\frac{0.00046}{3.7} + \left(\frac{3.17 \times 10^{-10} \times 300}{9.8 \times 0.1^3 \times 79.4}\right)^{0.5}\right]$$

$$= -0.965 \times 5.096 \times 10^{-3} \times (-7.655) = 0.038 \text{ m}^3/\text{s}$$

This much simpler method produces in value essentially the same as that obtained using the Moody diagram.

Example 7.13

Drawn tubing of what diameter should be selected to transport 0.002 m^3/s of 20°C water over a 400-m length so that the head loss does not exceed 30 m? (a) Use the Moody diagram and (b) the alternative method.

Solution

(a) In this problem we do not know D. Thus, a trial-and-error solution is anticipated. The average velocity is related to D by

$$V = \frac{Q}{A} = \frac{0.002}{\pi D^2/4} = \frac{0.00255}{D^2}$$

The friction factor and D are related as follows:

$$h_L = f \frac{L}{D} \frac{V^2}{2g}$$

$$30 = f \frac{400}{D} \frac{(0.00255/D^2)^2}{2 \times 9.8}$$

$$\therefore D^5 = 4.42 \times 10^{-6} f$$

The Reynolds number is

$$\text{Re} = \frac{VD}{\nu} = \frac{0.00255D}{D^2 \times 10^{-6}} = \frac{2550}{D}$$

Now, let us simply guess a value for f and check with the relations above and the Moody diagram. The first guess is $f = 0.03$ and the correction is listed in the following table. Note: the second guess is the value for f found from the calculations of the first guess.

f	D(m)	Re	e/D	f(Fig. 7.13)
0.03	0.0421	6.06×10^4	0.000036	0.02
0.02	0.0388	6.57×10^4	0.000039	0.02

The value of $f = 0.02$ is acceptable, yielding a diameter of 3.88 cm. Since this diameter would undoubtedly not be standard, a diameter of

$$D = 4 \text{ cm}$$

would be the tube size selected. This tube would have a head loss less than the limit of $h_L = 30$ m imposed in the problem statement. Any larger-diameter tube would also satisfy this criterion but would be more costly, so should not be selected.

　　(b) The alternative method for this Category 3 problem uses the explicit relationship (7.6.31). We can directly calculate D to be

$$D = 0.66\left[(1.5 \times 10^{-6})^{1.25}\left(\frac{400 \times 0.002^2}{9.81 \times 30}\right)^{4.75} + 10^{-6} \times 0.002^{9.4}\left(\frac{400}{9.81 \times 30}\right)^{5.2}\right]^{0.04}$$

$$= 0.66[5.163 \times 10^{-33} + 2.102 \times 10^{-31}]^{0.04} = 0.039 \text{ m}$$

Hence $D = 4$ cm would be the tube size selected. This is the same tube size as that selected using the Moody diagram.

7.6.4   Losses in Noncircular Conduits

Good approximations can be made for the head loss in conduits with noncircular cross sections by using the *hydraulic radius R*, defined by

$$R = \frac{A}{P} \tag{7.6.32}$$

where A is the cross-sectional area and P is the **wetted perimeter**, that perimeter where the fluid is in contact with the solid boundary. For a circular pipe flowing full the hydraulic radius is $R = r_0/2$. Hence we simply replace the radius r_0 with $2R$ and use the Moody diagram with

Wetted perimeter: *That perimeter where the fluid is in contact with the solid boundary.*

$$\text{Re} = \frac{4VR}{\nu} \qquad \frac{\text{relative}}{\text{roughness}} = \frac{e}{4R}, \tag{7.6.33}$$

The head loss becomes

$$h_L = f \frac{L}{4R} \frac{V^2}{2g} \tag{7.6.34}$$

To use this hydraulic radius technique the cross section should be fairly "open," such as a rectangle with aspect ratio less than 4:1, an equilateral triangle, or an oval. For other shapes, such as an annulus, the error would be significant.

Example 7.14

Air at standard conditions is to be transported through 500 m of a smooth, horizontal, 30 cm × 20 cm rectangular duct at a flow rate of 0.24 m³/s. Calculate the pressure drop.

Solution
The hydraulic radius is

$$R = \frac{A}{P} = \frac{0.3 \times 0.2}{(0.3 + 0.2) \times 2} = 0.06 \text{ m}$$

The average velocity is

$$V = \frac{Q}{A} = \frac{0.24}{0.3 \times 0.2} = 4.0 \text{ m/s}$$

This gives a Reynolds number of

$$\text{Re} = \frac{4VR}{\nu} = \frac{4 \times 4 \times 0.06}{1.5 \times 10^{-5}} = 6.4 \times 10^4$$

Using the smooth pipe curve of the Moody diagram, there results

$$f = 0.0196$$

Hence,

$$h_L = f \frac{L}{4R} \frac{V^2}{2g} = 0.0196 \frac{500}{4 \times 0.06} \frac{4^2}{2 \times 9.8} = 33.3 \text{ m}$$

The pressure drop is

$$\Delta p = \rho g h_L = 1.23 \times 9.8 \times 33.3 = 402 \text{ Pa}$$

7.6.5 Minor Losses in Pipe Flow

We now know how to calculate the losses due to a developed flow in a pipe. Pipe systems do, however, include valves, elbows, enlargements, contractions, inlets, outlets, bends, and other fittings that cause additional losses, referred to as *minor losses*, even though such losses can exceed the frictional losses of Eq. 7.6.23. Each of these devices causes a change in the magnitude and/or the direction of the velocity vectors and hence results in a loss. In general, if the flow is gradually accelerated by a device, the losses are very small; relatively large losses are associated with sudden enlargements or contractions because of the separated regions that result (a separated flow occurs when the primary flow separates from the wall).

A minor loss is expressed in terms of a loss coefficient K, defined by

$$h_L = K \frac{V^2}{2g} \tag{7.6.35}$$

Values of K have been determined experimentally for the various fittings and geometry changes of interest in piping systems. One exception is the sudden expansion from area A_1 to area A_2, for which the loss can be calculated; this was done in Example 4.14, where we found that

$$h_L = \left(1 - \frac{A_1}{A_2}\right)^2 \frac{V_1^2}{2g} \tag{7.6.36}$$

Thus, for the sudden expansion

$$K = \left(1 - \frac{A_1}{A_2}\right)^2 \tag{7.6.37}$$

If A_2 is extremely large (e.g., a pipe exiting into a reservoir), $K = 1.0$, an obvious result since the entire kinetic energy is lost.

A pipe fitting that has a relatively large loss coefficient with no change in cross-sectional area is the pipe bend, or the elbow. This results primarily from the secondary flow caused by the fluid flowing from the high-pressure region to the low-pressure region (see Eq. 3.4.15), as shown in Fig. 7.14; this secondary flow is eventually dissipated after the fluid leaves the long sweep bend or elbow. In addition, a separated region occurs at the sharp corner in a standard elbow. Energy is needed to maintain a secondary flow and the flow in the separated region. This wasted energy is measured in terms of the loss coefficient.

The loss coefficients for various geometries are presented in Table 7.2 and Fig. 7.15. A globe valve may be used to control the flow rate by introducing large losses by partially closing the valve. The other types of valves should not be used to control the flow; damage could result.

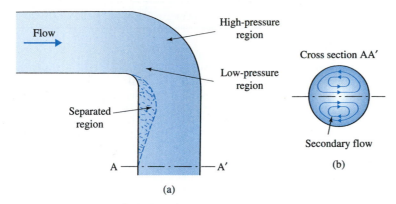

FIGURE 7.14   Flow in an elbow.

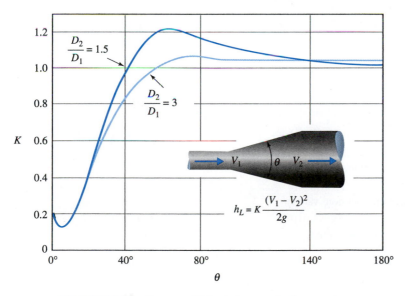

FIGURE 7.15   Loss coefficients in a conical expansion.
(From A. H. Gibson, *Engineering,* Vol. 93, 1912.)

TABLE 7.2 Nominal Loss Coefficients K (Turbulent Flow)[a]

Type of fitting		Screwed			Flanged		
Diameter		2.5 cm	5 cm	10 cm	5 cm	10 cm	20 cm
Globe valve	(fully open)	8.2	6.9	5.7	8.5	6.0	5.8
	(half open)	20	17	14	21	15	14
	(one-quarter open)	57	48	40	60	42	41
Angle valve (fully open)		4.7	2.0	1.0	2.4	2.0	2.0
Swing check valve (fully open)		2.9	2.1	2.0	2.0	2.0	2.0
Gate valve (fully open)		0.24	0.16	0.11	0.35	0.16	0.07
Return bend		1.5	.95	.64	0.35	0.30	0.25
Tee (branch)		1.8	1.4	1.1	0.80	0.64	0.58
Tee (line)		0.9	0.9	0.9	0.19	0.14	0.10
Standard elbow		1.5	0.95	0.64	0.39	0.30	0.26
Long sweep elbow		0.72	0.41	0.23	0.30	0.19	0.15
45° elbow		0.32	0.30	0.29			

Square-edged entrance		0.5
Reentrant entrance		0.8
Well-rounded entrance		0.03
Pipe exit		1.0
	Area ratio	
Sudden contraction[b]	2:1	0.25
	5:1	0.41
	10:1	0.46
	Area ratio A/A_0	
Orifice plate	1.5:1	0.85
	2:1	3.4
	4:1	29
	$\geq$ 6:1	$2.78 \left(\dfrac{A}{A_0} - 0.6\right)^2$
Sudden enlargement[c]		$\left(1 - \dfrac{A_1}{A_2}\right)^2$
90° miter bend (without vanes)		1.1
(with vanes)		0.2
General contraction	(30° included angle)	0.02
	(70° included angle)	0.07

[a]Values for other geometries can be found in *Technical Paper 410*, The Crane Company, 1957.
[b]Based on exit velocity V_2.
[c]Based on entrance velocity V_1.

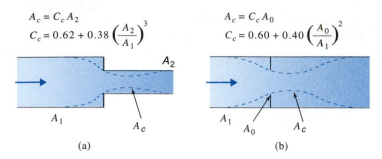

$$A_c = C_c A_2$$

$$C_c = 0.62 + 0.38 \left(\frac{A_2}{A_1}\right)^3$$

$$A_c = C_c A_0$$

$$C_c = 0.60 + 0.40 \left(\frac{A_0}{A_1}\right)^2$$

(a)   (b)

FIGURE 7.16 Vena contractas in contractions and orifices: (a) sudden contraction; (b) concentric orifice.

Loss coefficients for sudden contractions and orifice plates can be approximated by neglecting the losses in the converging flow up to the vena contracta and calculating the losses in the diverging flow using the loss coefficient for a sudden expansion. Figure 7.16 provides the information necessary to establish the area of the **vena contracta**, the minimum area; this minimum area results when the converging streamlines begin to expand to fill the downstream area.

Vena contracta: *The minimum area in a sudden contraction.*

It is often the practice to express a loss coefficient as an *equivalent length L_e* of pipe. This is done by equating Eq. 7.6.35 to Eq. 7.6.23:

$$K \frac{V^2}{2g} = f \frac{L_e}{D} \frac{V^2}{2g} \qquad (7.6.38)$$

giving the relationship

$$L_e = K \frac{D}{f} \qquad (7.6.39)$$

Hence the square-edged entrance of a 20-cm-diameter pipe with a friction factor of $f = 0.02$ could be replaced by an equivalent pipe length of $L_e = 5$ m.

Finally, a comment should be made concerning the magnitude of the minor losses. In piping systems involving intermediate lengths (i.e., 100 diameters) of pipe between minor losses, the minor losses may be of the same order of magnitude as the frictional losses; for relatively short lengths the minor losses may be substantially greater than the frictional losses; and for long lengths (e.g., 1000 diameters) of pipe, the minor losses are usually neglected.

KEY CONCEPT *For long lengths of pipe, the minor losses are usually neglected.*

Example 7.15

If the flow rate through a 10-cm-diameter wrought iron pipe (Fig. E7.15) is 0.04 m³/s, find the difference in elevation H of the two reservoirs.

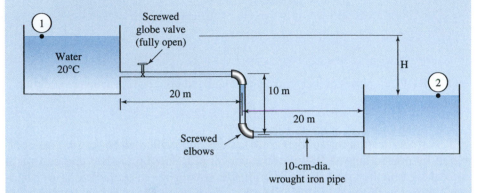

FIGURE E7.15

Solution

The energy equation written for a control volume that contains the two reservoir surfaces (see Eq. 4.5.17), where $V_1 = V_2 = 0$ and $p_1 = p_2 = 0$, is

$$0 = z_2 - z_1 + h_L$$

Thus, letting $z_1 - z_2 = H$, we have

$$H = (K_{\text{entrance}} + K_{\text{valve}} + 2K_{\text{elbow}} + K_{\text{exit}}) \frac{V^2}{2g} + f \frac{L}{D} \frac{V^2}{2g}$$

The average velocity, Reynolds number, and relative roughness are

$$V = \frac{Q}{A} = \frac{0.04}{\pi \times 0.05^2} = 5.09 \text{ m/s}$$

$$\text{Re} = \frac{VD}{\nu} = \frac{5.09 \times 0.1}{10^{-6}} = 5.09 \times 10^5$$

$$\frac{e}{D} = \frac{0.046}{100} = 0.00046$$

From the Moody diagram we find that

$$f = 0.0173$$

Using the loss coefficients from Table 7.2 for an entrance, a globe valve, screwed 10-cm-diameter standard elbows, and an exit there results

$$H = (0.5 + 5.7 + 2 \times 0.64 + 1.0) \frac{5.09^2}{2 \times 9.8} + 0.0173 \frac{50}{0.1} \frac{5.09^2}{2 \times 9.8}$$

$$= 11.2 + 11.4 = 22.6 \text{ m}$$

Note: The minor losses are about equal to the frictional losses as expected, since there are five minor loss elements in 500 diameters of pipe length.

Example 7.16

Approximate the loss coefficient for the sudden contraction $A_1/A_2 = 2$ by neglecting the losses in the contracting portion up to the vena contracta and assuming that all the losses occur in the expansion from the vena contracta to A_2 (see Fig. 7.16). Compare with that given in Table 7.2.

Solution

The head loss from the vena contracta to area A_2 is (see Table 7.2, sudden enlargement)

$$h_L = \left(1 - \frac{A_c}{A_2}\right)^2 \frac{V_c^2}{2g}$$

Continuity allows us to write

$$V_c = \frac{A_2}{A_c} V_2$$

Thus, the head loss based on V_2 is

$$h_L = \left(1 - \frac{A_c}{A_2}\right)^2 \left(\frac{A_2}{A_c}\right)^2 \frac{V_2^2}{2g}$$

so the loss coefficient of Eq. 7.6.35 is

$$K = \left(1 - \frac{A_c}{A_2}\right)^2 \left(\frac{A_2}{A_c}\right)^2$$

Using the expression of C_c given in Fig. 7.16, we have

$$\frac{A_c}{A_2} = C_c = 0.62 + 0.38 \left(\frac{1}{2}\right)^3 = 0.67$$

Finally,

$$K = (1 - 0.67)^2 \frac{1}{0.67^2} = 0.24$$

This compares favorably with the value of 0.25 given in Table 7.2.

7.6.6   Hydraulic and Energy Grade Lines

When the energy equation is written in the form of Eq. 4.5.17, that is,

$$-\frac{\dot{W}_s}{\dot{m}g} = \frac{V_2^2 - V_1^2}{2g} + \frac{p_2 - p_1}{\gamma} + z_2 - z_1 + h_L \qquad (7.6.40)$$

the terms have dimensions of length. This has led to the conventional use of the hydraulic grade line and the energy grade line. The **hydraulic grade line** (HGL), the dashed line in Fig. 7.17, in a piping system is formed by the locus of points located a distance p/γ above the center of the pipe, or $p/\gamma + z$ above a preselected datum; the liquid in a piezometer tube would rise to the HGL. The **energy grade line** (EGL), the solid line in Fig. 7.17, is formed by the locus of points a distance

Hydraulic grade line (HGL): *In a piping system, the HGL is located a distance p/γ above the center of the pipe.*

Energy grade line (EGL): *In a piping system, the EGL is located a distance $V^2/2g$ above the HGL.*

$V^2/2g$ above the HGL, or the distance $V^2/2g + p/\gamma + z$ above the datum; the liquid in a pitot tube would rise to the EGL. The following points are noted relating to the HGL and the EGL:

- As the velocity goes to zero, the HGL and the EGL approach each other. Thus, in a reservoir, they are identical and lie on the surface (see Fig. 7.17).
- The EGL and, consequently, the HGL slope downward in the direction of the flow due to the head loss in the pipe. The greater the loss per unit length, the greater the slope. As the average velocity in the pipe increases, the loss per unit length increases.
- A sudden change occurs in the HGL and the EGL whenever a loss occurs due to a sudden geometry change as represented by the valve or the sudden enlargement of Fig. 7.17.
- A jump occurs in the HGL and the EGL whenever useful energy is added to the fluid as occurs with a pump, and a drop occurs if useful energy is extracted from the flow as occurs with a turbine.
- At points where the HGL passes through the centerline of the pipe, the pressure is zero. If the pipe lies above the HGL, there is a vacuum in the pipe, a condition that is often avoided, if possible, in the design of piping systems; an exception would be in the design of a siphon.

KEY CONCEPT *If the pipe lies above the HGL, there is a vacuum in the pipe.*

The concepts of energy grade line and hydraulic grade line may also be applied to open-channel flows. The HGL coincides with the free surface and the EGL is a distance $V^2/2g$ above the free surface. Uniform flows in open channels will be discussed in the following section, and Chapter 10 is devoted to nonuniform open-channel flow.

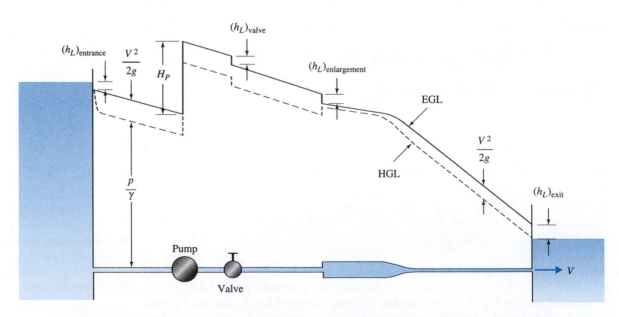

FIGURE 7.17 Hydraulic grade line (HGL) and energy grade line (EGL) for a piping system.

Example 7.17

Water at 20°C flows between two reservoirs at the rate of 0.06 m³/s as shown in Fig. E7.17. Sketch the HGL and the EGL. What is the minimum diameter D_B allowed to avoid the occurrence of cavitation?

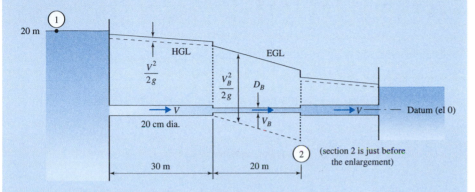

FIGURE E7.17

Solution

The EGL and the HGL are sketched on the figure, including sudden changes at the entrance, contraction, enlargement, and the exit. Note the large velocity head (the difference between the EGL and the HGL) in the smaller pipe because of the high velocity. The velocity, Reynolds number, and relative roughness in the 20-cm-diameter pipe are calculated to be

$$V = \frac{Q}{A} = \frac{0.06}{\pi \times 0.20^2/4} = 1.91 \text{ m/s}$$

$$\text{Re} = \frac{VD}{\nu} = \frac{1.91 \times 0.2}{10^{-6}} = 3.8 \times 10^5$$

$$\frac{e}{D} = \frac{0.26}{200} = 0.0013$$

Thus $f = 0.022$ from Fig. 7.13. The velocity, Reynolds number, and relative roughness in the smaller pipe are

$$V_B = \frac{0.06}{\pi D_B^2/4} = \frac{0.0764}{D_B^2}$$

$$\text{Re} = \frac{0.0764 \times D_B}{D_B^2 \times 10^{-6}} = \frac{76\ 400}{D_B}$$

$$\frac{e}{D_B} = \frac{0.00026}{D_B}$$

The minimum possible diameter is established by recognizing that the water vapor pressure (2450 Pa absolute) at 20°C is the minimum allowable pressure. Since the

(continued)

distance between the pipe and the HGL is an indication of the pressure in the pipe, we can conclude that the minimum pressure will occur at section 2. Hence the energy equation applied between section 1, the reservoir surface, and section 2 gives

$$\cancel{\frac{V_1^2}{2g}}^{\,0} + \frac{p_1}{\gamma} + z_1 = \frac{V_B^2}{2g} + \frac{p_2}{\gamma} + \cancel{z_2}^{\,0} + K_{ent}\frac{V_A^2}{2g} + K_{cont}\frac{V_B^2}{2g} + f_A\frac{L_A}{D_A}\frac{V^2}{2g} + f_B\frac{L_B}{D_B}\frac{V_B^2}{2g}$$

where the subscript A refers to the 20-cm-diameter pipe. This simplifies to

$$\frac{101\,000}{9810} + 20 = \frac{(0.0764/D_B^2)^2}{2\times 9.81}\left(1 + 0.25 + f_B\frac{20}{D_B}\right) + \frac{2450}{9810}$$

$$+ \left(0.5 + 0.022\frac{30}{0.2}\right)\frac{1.91^2}{2\times 9.81}$$

$$98\,600 = \frac{1.25}{D_B^4} + f_B\frac{20}{D_B^5}$$

where we have used $K_{ent} = 0.5$ and assumed that $K_{cont} = 0.25$. This requires a trial-and-error solution. The following illustrates the procedure.

Let $D_B = 0.1$ m. Then $e/D = 0.0026$ and Re $= 7.6 \times 10^5$. Therefore, $f = 0.026$:

$$98\,600 \stackrel{?}{=} 12\,500 + 52\,000$$

Let $D_B = 0.09$ m. Then $e/D = 0.0029$ and Re $= 8.4 \times 10^5$. Therefore, $f = 0.027$:

$$98\,600 \stackrel{?}{=} 19\,000 + 91\,000$$

We see that 0.1 m is too large and 0.09 m is too small. In fact, the value of 0.09 m is only slightly too small. Consequently, to be safe we must select the next larger pipe size, of 0.1 m diameter. If there were a pipe size of 9.5 cm diameter, that could be selected. Assuming that that size is not available, we select

$$D_B = 10 \text{ cm}$$

Note that the assumption of a 2:1 area ratio for the contraction is too small. It is actually 4:1. This would give $K_{cont} \simeq 0.4$. After a quick check we conclude that this value does not significantly influence the result.

7.6.7 Simple Pipe System with a Pump

The problems we have considered thus far in this section have not involved a pump. If a centrifugal pump is included in the piping system and the flow rate is specified, the solution is straightforward using the techniques we have already developed. If, on the other hand, the flow rate is not specified, as is often the case, a trial-and-error solution involving the centrifugal pump results since the head produced by a centrifugal pump and its efficiency η_P (see Eq. 4.5.26) depend on the discharge, as shown by the pump characteristic curves, the solid curves in Fig. 7.18. Companies furnish such characteristic curves for each centrifugal pump manufactured. Such a curve provides one equation relating the flow rate Q and pump head H_P. The other equation is provided by the energy equation, which can typically be written as (see the energy equation in the following example)

KEY CONCEPT *If the flow rate is specified, the solution is straightforward.*

$$H_P = c_1 + c_2 Q^2 \qquad (7.6.41)$$

This is the **system demand curve**. This along with the characteristic curve must be solved simultaneously to yield the desired flow rate. To determine the power requirement of the pump, the efficiency η_P must be used.

System demand curve: *The energy equation relating the pump head to the unknown flow note.*

Note that for the piping system, the required pump energy head H_P, demanded by the energy equation, increases with Q and from the pump characteristic curve we see that H_P decreases with Q; hence the two curves will intersect at a point, called the *operating point* of the system. An example will illustrate the solution technique.

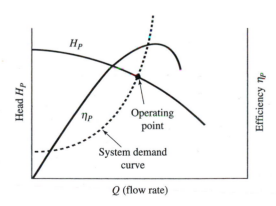

FIGURE 7.18 Pump characteristic curves and the system demand curve.

Example 7.18

Estimate the flow rate in the simple piping system of Fig. E7.18a if the pump characteristic curves are as shown in Fig. E7.18b. Also, find the pump power requirement.

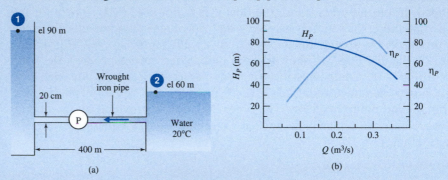

FIGURE E7.18

Solution

We will assume that the Reynolds number is sufficiently large that the flow is completely turbulent. So, using $e/D = 0.046/200 = 0.00023$, the friction factor from the Moody diagram is

$$f = 0.014$$

The energy equation (see Eq. 7.6.40), with $H_P = -\dot{W}_S/\dot{m}g$, applied between the two surfaces, yields

$$H_P = \frac{V_2^2 - \cancelto{0}{V_1^2}}{2g} + z_2 - z_1 + \frac{p_2 - \cancelto{0}{p_1}}{\gamma} + h_L$$

or

$$H_P = 90 - 60 + \left(K_{\text{entrance}} + K_{\text{exit}} + f\frac{L}{D}\right)\frac{V^2}{2g}$$

$$= 30 + \left(0.5 + 1.0 + 0.014\frac{400}{0.2}\right)\frac{Q^2}{2 \times 9.8 \times [\pi \times (0.1)^2]^2}$$

$$= 30 + 1520Q^2$$

This equation, the system demand curve, and the characteristic curve $H_P(Q)$ of the pump are now solved simultaneously by trial and error. Actually, the curve could be plotted on the same graph as the characteristic curve and the point of intersection, the operating point, would provide Q. Try $Q = 0.2$ m³/s: $(H_P)_{\text{energy}} = 91$ m, $(H_P)_{\text{char}} \cong 75$ m. Try $Q = 0.15$ m³/s: $(H_P)_{\text{energy}} = 64$ m, $(H_P)_{\text{char}} \cong 75$ m. Try $Q = 0.17$ m³/s: $(H_P)_{\text{energy}} = 74$ m, $(H_P)_{\text{char}} \cong 76$ m. This is our solution. We have

$$Q = 0.17 \text{ m}^3/\text{s}$$

Check the Reynolds number: $\text{Re} = DQ/A\nu = 0.2 \times 0.17/\pi \times 0.1^2 \times 10^{-6} = 1.08 \times 10^6$. This is sufficiently large, but marginally so.

The power requirement of the pump is given by Eq. 4.5.26:

$$\dot{W}_P = \frac{Q\gamma H_P}{\eta_P}$$

$$= \frac{0.17 \times 9800 \times 75}{0.65} = 198\,000 \text{ W} \quad \text{or} \quad 198 \text{ kW}$$

where the efficiency $\eta_P = 0.63$ is found from the characteristic curve at $Q = 0.17$ m³/s.
Note: Since $L/D > 1000$, minor losses due to the entrance and exit could have been neglected.

7.7 UNIFORM TURBULENT FLOW IN OPEN CHANNELS

The last internal flow situation we consider in this chapter is that of steady uniform (constant depth) flow in an open channel, shown in Fig. 7.19. We could already treat this flow by using the Darcy–Weisbach relation presented in Section 7.6.3. In fact, that technique predicts better results than does the more common method we present here. Both methods will be compared in two examples. However, unless otherwise stated, uniform flow in open, rough channels is commonly analyzed using the following less-complicated method.

If the energy equation is applied between two sections of the channel sketched in Fig. 7.19, we obtain

$$0 = \frac{V_2^2 - V_1^2}{2g} + \frac{p_2 - p_1}{\gamma} + z_2 - z_1 + h_L \qquad (7.7.1)$$

which shows that the head loss is

$$
\begin{aligned}
h_L &= z_1 - z_2 \\
&= L \sin \theta = LS
\end{aligned}
\qquad (7.7.2)
$$

where L is the length of the channel between the two sections and S is the slope of the channel, which is assumed small, so that $\sin \theta \simeq S$. (Do not confuse S with specific gravity.)

The Darcy–Weisbach equation (7.6.31), with $h_L = LS$ from Eq. 7.7.2, takes the form

$$LS = f\frac{L}{4R}\frac{V^2}{2g} \qquad \therefore RS = \frac{f}{8g}V^2 \qquad (7.7.3)$$

where R is the hydraulic radius. Since open channels are usually quite large with large Reynolds numbers, the friction factor is invariably constant (the flow is completely turbulent). Thus the equation above is written as

$$V = C\sqrt{RS} \qquad (7.7.4)$$

> **KEY CONCEPT** *Since open channels are quite large, with large Reynolds numbers, the friction factor is constant.*

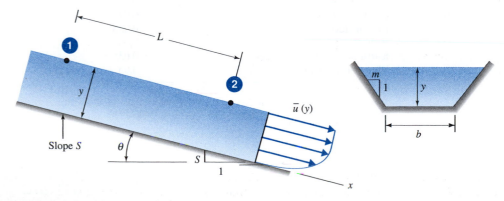

FIGURE 7.19   Uniform flow in an open channel.

where the *Chezy coefficient C* is a dimensional constant; the equation above is referred to as the *Chezy equation* named after Antoine Chezy (1718–1798). The Chezy coefficient is related to the channel roughness and the hydraulic radius (much like f is in a pipe) by

$$C = \frac{c_1}{n} R^{1/6} \tag{7.7.5}$$

where the dimensional constant c_1 has a value of 1.0 if SI units are used. The dimensionless constant n is directly related to the wall roughness; it is called the *Manning n*, named after Robert Manning (1816–1897). Values for various wall materials are listed in Table 7.3.

The flow rate, which is of primary interest in open-channel flow problems, is found to be

$$Q = \frac{c_1}{n} AR^{2/3} S^{1/2} \qquad c_1 = 1.0 \text{ for S.I. units} \tag{7.7.6}$$

This is the *Chezy–Manning equation*.

For smooth-surfaced channels, the use of the Chezy–Manning equation is discouraged since it implicitly assumes a rough wall. Calculations for smooth-surfaced channels such as glass or plastic should be based on the Darcy–Weisbach relation with variable f; see Section 7.6.3.

TABLE 7.3   Average Values[a] of the Manning n

Wall material	Manning n
Planed wood	0.012
Unplaned wood	0.013
Finished concrete	0.012
Unfinished concrete	0.014
Sewer pipe	0.013
Brick	0.016
Cast iron, wrought iron	0.015
Concrete pipe	0.015
Riveted steel	0.017
Earth, straight	0.022
Corrugated metal flumes	0.025
Rubble	0.03
Earth with stones and weeds	0.035
Mountain streams	0.05

[a]The values in this table result in flow rates too large for hydraulic radii greater than about 3 m. The Manning n should be increased by 10 to 15% for such large conduits.

Example 7.19

The depth of water at 15°C flowing in a 3.6 m-wide rectangular, finished concrete channel is measured to be 1.2 m. The slope is measured to be 0.0016. Estimate the flow rate using (a) the Chezy–Manning equation and (b) the Darcy–Weisbach equation.

Solution

The hydraulic radius is calculated to be

$$R = \frac{A}{P} = \frac{yb}{2y + b} = \frac{1.2 \times 3.6}{2 \times 1.2 + 3.6} = 0.72 \text{ m}$$

(a) Using the Chezy–Manning equation, with $n = 0.012$ from Table 7.3 and $c = 1.0$, we have

$$Q = \frac{1.0}{n} \, AR^{2/3}S^{1/2}$$

$$= \frac{1.0}{0.012} \times (1.2 \times 3.6) \times 0.72^{2/3} \times 0.0016^{1/2} = 11.56 \text{ m}^3/\text{s}$$

(b) The relative roughness is, using a low value $e = 0.45$ mm (it is finished concrete) shown on the Moody diagram:

$$\frac{e}{4R} = \frac{0.45 \times 10^{-3}}{4 \times 0.72} = 0.000156$$

Assuming a completely turbulent flow, the Moody diagram gives the friction factor as

$$f = 0.013$$

The Darcy–Weisbach equation (7.7.3) then yields the velocity as follows:

$$\therefore V = \left(\frac{8RgS}{f}\right)^{1/2}$$

$$= \left(\frac{8 \times 0.72 \times 9.81 \times 0.0016}{0.013}\right)^{1/2} = 2.64 \text{ m/s}$$

The flow rate is calculated as

$$Q = VA = 2.64 \times 1.2 \times 3.6 = 11.40 \text{ m}^3/\text{s}$$

These two values are within 1.5%, an acceptable engineering tolerance for this type of problem. That found using the Moody diagram is considered to be more accurate, however.

Example 7.20

A 1.0-m-diameter concrete pipe transports 20°C water at a depth of 0.4 m. If the slope is 0.001, find the flow rate using (a) Chezy–Manning equation and (b) the Darcy–Weisbach equation.

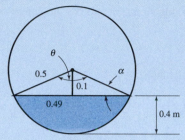

FIGURE E7.20

Solution

From the sketch of the pipe in Fig. E7.20 the following are calculated:

$$\alpha = \sin^{-1}\frac{0.1}{0.5} = 11.54°$$

$$\therefore \theta = 156.9°$$

$$\therefore A = \pi \times 0.5^2 \times \frac{156.9}{360} - 0.49 \times 0.1 = 0.2933 \text{ m}^2$$

$$P = 2\pi \times 0.5 \times \frac{156.9}{360} = 1.369 \text{ m}$$

The hydraulic radius is found, using the above calculations, to be

$$R = \frac{A}{P} = \frac{0.2933}{1.369} = 0.2142 \text{ m}$$

(a) The Chezy–Manning equation yields, with n from Table 7.3 and $c_1 = 1.0$,

$$Q = \frac{1.0}{n} AR^{2/3}S^{1/2} = \frac{1.0}{0.013} \times 0.2933 \times 0.2142^{2/3} \times 0.001^{1/2} = 0.25 \text{ m}^3/\text{s}$$

We used an intermediate value for n since sufficient information is not given.

(b) The relative roughness is, using a relatively rough value for concrete pipe (see Fig. 7.13) of $e = 2.0$ mm,

$$\frac{e}{4R} = \frac{2}{4 \times 214.2} = 0.0023$$

Assuming completely turbulent flow, the Moody diagram yields

$$f = 0.025$$

The Darcy–Weisbach equation (7.7.3) then gives the following:

$$\therefore V = \left(\frac{8RgS}{f}\right)^{1/2} = \left(\frac{8 \times 0.2142 \times 9.81 \times 0.001}{0.025}\right)^{1/2} = 0.820 \text{ m/s}$$

The flow rate is

$$Q = VA = 0.820 \times 0.2933 = 0.24 \text{ m}^3/\text{s}$$

This is within 4% of the result above, an acceptable tolerance for this type of problem. The second method, which is more difficult to apply, is considered to be more accurate, however.

7.8 SUMMARY

The laminar entrance lengths for a pipe and wide channel are, respectively,

$$\frac{L_E}{D} = 0.065 \text{ Re} \qquad \frac{L_E}{h} = 0.04 \text{ Re} \qquad (7.8.1)$$

For a high-Reynolds-number turbulent pipe flow the entrance length is

$$\frac{L_E}{D} = 120 \qquad (7.8.2)$$

For laminar flow in a pipe and a wide channel the pressure and friction factor are, respectively,

$$\Delta p = \frac{8\mu V L}{r_0^2} \qquad f = \frac{64}{\text{Re}} \text{ pipe} \qquad (7.8.3)$$

$$\Delta p = \frac{12\mu V L}{a^2} \qquad f = \frac{48}{\text{Re}} \text{ channel} \qquad (7.8.4)$$

where a is the channel height.

The torque required to rotate an inner cylinder with the outer cylinder fixed is

$$T = \frac{4\pi\mu r_1^2 r_2^2 L w_1}{r_2^2 - r_1^2} \qquad (7.8.5)$$

In a turbulent flow the head loss is calculated using

$$h_L = f \frac{L}{D} \frac{V^2}{2g} \qquad (7.8.6)$$

where f is found using the Moody Diagram of Fig. 7.13. Minor losses are included using

$$h_L = K \frac{V^2}{2g} \qquad (7.8.7)$$

where many loss coefficients K are listed in Table 7.2.

To include a pump in a piping system when the flow rate is not known, it is necessary to have pump characteristic curves, such as those of Fig. E7.18. Example 7.18 illustrates the procedure.

The flow rate in an open channel is most often estimated using the equation

$$Q = \frac{c_1}{n} A R^{2/3} S^{1/2} \qquad c_1 = 1.0 \text{ for S.I. units} \qquad (7.8.8)$$

where n is found in Table 7.3.

PROBLEMS

Laminar or Turbulent Flow

7.1 Calculate the maximum average velocity V with which 20°C water can flow in a pipe in the laminar state if the critical Reynolds number ($\text{Re} = VD/\nu$) at which transition occurs is 2000; the pipe diameter is:
   (a) 2 m   **(b)** 2 cm   **(c)** 2 mm

7.2 Water at 20°C flows in a wide river. Using a critical Reynolds number ($\text{Re} = Vh/\nu$) at which transition occurs of 1500, calculate the average velocity V that will result in a laminar flow if the depth h of the river is:
   (a) 4 m   **(b)** 1 m   **(c)** 0.3 m

7.3 Water at 10°C is flowing in a thin sheet in a parking lot at a depth of 5 mm. with an average velocity of 0.5 m/s. Is the flow laminar or turbulent?

7.4 Water is flowing, apparently quite placidly, in a river that is 20 m wide and 1.4 m deep. A leaf floating in the river is observed to move 1 m in 2 s. Is the flow laminar or turbulent? See Problem 7.2 for the definition of the Reynolds number.

7.5 Flow occurs in a 2-cm-diameter pipe. What is the maximum velocity that can occur for water at 20°C for a laminar flow if:
   (a) $\text{Re} = 2000$?   **(b)** $\text{Re} = 40\,000$?

Entrance and Developed Flow

7.6 Calculate the laminar entrance length in a 4-cm-diameter pipe if 2×10^{-4} m³/s of water is flowing at:
   (a) 10°C   **(b)** 20°C
   (c) 40°C   **(d)** 80°C

7.7 A laminar flow is to occur in an experimental facility with 20°C air flowing through a 4-cm-diameter pipe. Calculate the average velocity, the inviscid core length, and the entrance length if the Reynolds number is:
   (a) 1000   **(b)** 80 000

7.8 A 6-cm-diameter pipe originates in a tank and delivers 0.025 m³/s of water at 20°C to a receiver 50 m away. Is the assumption of developed flow acceptable?

7.9 A laboratory experiment is designed to create a laminar flow in a 2-mm-diameter tube shown in Fig P.7.9. Water flows from a reservoir through the tube. If 18 L is collected in 2 hr, can the entrance length be neglected?

7.10 Air at 23°C is used as the working fluid in a parallel-plate research project. If the plates are 1.2 cm apart, how long is the longest possible entrance length for laminar flow? What is the shortest entrance length?

7.11 Air at 25°C can exist in either the laminar state or the turbulent state (a trip wire near the entrance is used to make it turbulent) for flow in a 6-cm-diameter pipe in a research lab. If the average velocity is 5 m/s, compare the length of the entrance region for the laminar flow with that of the turbulent flow.

7.12 Water at 20°C flows with an average velocity of 0.2 m/s from a reservoir through a 4-cm-diameter pipe. Estimate the inviscid core length and the entrance length if the flow is:
   (a) Laminar
   (b) Turbulent

7.13 Draw an incremental control volume with length Δx and radius r_0 and show that for a laminar flow $(\Delta p/\Delta x)_{\text{entrance}} > (\Delta p/\Delta x)_{\text{developed}}$.

7.14 Explain the pressure variations observed for a turbulent flow in Fig. 7.3 for:
   (a) A high-Re flow ($\text{Re} > 300\,000$)
   (b) A low-Re flow ($\text{Re} \approx 10\,000$)
   (c) An intermediate-Re flow

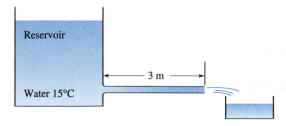

Reservoir

3 m

Water 15°C

FIGURE P7.9

Laminar Flow in a Pipe

7.15 Define $p_k = p + \gamma h$ as the kinetic pressure and write Eq. 7.3.5 or 7.3.11 in terms of p_k. Can we let $dp_x/dx = \Delta p_k/L$, where L is the length over which Δp_k is measured? If so, express $u(r)$ in terms of $\Delta p_k/L$.

7.16 Verify that Eq. 7.3.13 is, in fact, correct.

7.17 A parabolic profile is measured in a pipe flow. The flow is:

I. Laminar	**A.**	I and III
II. Developed	**B.**	I, II, and III
III. Steady	**C.**	I, II, and IV
IV. Symmetric	**D.**	I, II, IV, and IV

7.18 A pressure drop of 480 Pa occurs over a section of 2-cm-diameter pipe transporting water at 20°C. Determine the length of the horizontal section if the Reynolds number is 1600. Also, find the shear stress at the wall and the friction factor.

7.19 Find the angle θ of the 10-mm-diameter pipe of Fig. P7.19 in which water at 40°C is flowing with Re = 1500 such that no pressure drop occurs. Also, find the flow rate.

FIGURE P7.19

7.20 A liquid is pumped through a 2-cm-diameter pipe at a flow rate of 12 L/min. Calculate the pressure drop in a 10-m horizontal section if the liquid is:
(a) SAE-10W oil at 20°C
(b) Water at 20°C
(c) Glycerine at 40°C
Is the assumption of laminar flow acceptable?

7.21 A liquid flows with no pressure drop in a vertical 2-cm-diameter pipe. Find the flow rate if, assuming a laminar flow, the liquid is:
(a) Water at 5°C
(b) SAE-30W oil at 25°C
(c) Glycerine at 20°C
Is the assumption of laminar flow acceptable?

7.22 A laminar flow is to exist in a pipe transporting 3.5 L/s of SAE-10W oil at 20°C. What is the maxi-

mum allowable diameter? What is the pressure drop over 10 m of horizontal pipe for this diameter?

7.23 Estimate the flow rate through the smooth pipe shown in Fig. P7.23. How long is the entrance region? Assume a laminar flow.

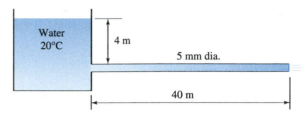

FIGURE P7.23

7.24 A manufacturer of small-diameter tubes wishes to know if the diameters are, in fact, accurate. An experimental setup, like that in Fig. P7.23, is used with a 4-m-long horizontal tube transporting 20°C water with a head of 4 m. If 3.4 L of water is collected in 60 min, what is the inside diameter of the tube, neglecting the effect of the entrance region? Is the effect of entrance region actually negligible?

7.25 Air at 20°C flows in a horizontal 2-cm-diameter pipe. Calculate the maximum pressure drop in a 10-m section for a laminar flow. Assume that $\rho = 1.23$ kg/m³.

7.26 Water at 20°C flows in the 4-mm-diameter pipe of Fig. P7.26. The pressure rise over the 10-m section is 6 kPa. Find the Reynolds number of the flow and the wall shear stress. Assume laminar flow.

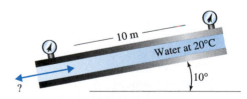

FIGURE P7.26

7.27 A research experiment requires a laminar flow of air at 20°C through a 10-cm-diameter pipe at a Reynolds number of 40 000. What maximum velocity is to be expected? What would be the pressure drop over a 10-m horizontal length of developed flow? How long would the entrance region be? Use $\rho = 1.2$ kg/m³.

7.28 Calculate the radius where a pitot probe must be placed in the laminar liquid flow of Fig. P7.28 so that the flow rate is given by $\pi R^2 \sqrt{2gH}$.

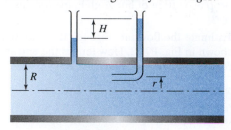

FIGURE P7.28

7.29 A laminar flow of water at 20°C occurs in a vertical 2-mm-diameter pipe. Calculate the flow rate if the pressure is constant. Is it reasonable to assume a laminar flow?

7.30 Find the radius in a developed laminar flow in a pipe where:
 (a) The velocity is equal to the average velocity.
 (b) The shear stress is equal to one-half the wall shear stress.

7.31 Find the ratio of the total flow rate through a pipe of radius r_0 to the flow rate through an annulus with inner and outer radii of $r_0/2$ and r_0. Assume developed laminar flow with the same pressure gradient.

7.32 A laminar flow of water at 15°C is obtained in a research lab at Re = 20,000 in a horizontal 5-cm-diameter pipe. Calculate the head loss in a 10-m section of developed flow, the wall shear stress, and the length of the entrance region.

7.33 Water at 20°C flows between the two concentric horizontal pipes of Fig. P7.33 with diameters of 2 cm and 3 cm. A pressure drop of 100 Pa is measured over a 10-m section of developed laminar flow. Find the flow rate and the shear stress on the inner pipe.

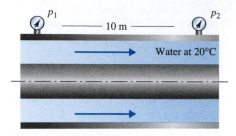

FIGURE P7.33

7.34 Air at 20°C is to flow in the annulus between two concentric, horizontal pipes, with respective diameters of 2 cm and 3 cm, such that a pressure drop of 10 Pa occurs over a 10-m length. Find the average velocity and the shear stress at the inner pipe. Assume a developed laminar flow.

7.35 Fluid flows in the annulus between two concentric horizontal pipes. The inner pipe is maintained at a higher temperature than the outer pipe so that the viscosity in the annulus cannot be assumed to be constant but $\mu = \mu(T)$. What differential equation would be solved to yield $u(r)$ assuming a developed laminar flow?

7.36 Show that the velocity distribution of Example 7.2 approaches that of pipe flow as $r_1 \to 0$ and approaches that of parallel-plate flow as $r_1 \to r_2$.

Laminar Flow Between Parallel Plates

7.37 The velocity profile between parallel plates is calculated to be Vy/s, where y is measured from the lower plate and s is the distance between the plates. We know that
 I. The flow is laminar
 II. The lower plate is moving with velocity V and the other is stationary.
 III. Lower plate is stationary and the other is moving with velocity V
 IV. The flow is steady
 A. I, III, and IV   **B.** II and III
 C. I and III   **D.** I and IV

7.38 A flow occurs in a horizontal channel 1.25 cm × 50 cm with Re = 2000. Calculate the flow rate if the fluid is:
 (a) Water at 15°C
 (b) Atmospheric air at 15°C

7.39 A board 1 m × 1 m that weighs 40 N moves down the incline shown in Fig. P7.39 with a velocity of $V = 0.2$ m/s. Estimate the viscosity of the fluid if θ is:
 (a) 20°
 (b) 30°

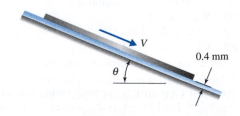

FIGURE P7.39

7.40 Water at 20°C exists between the plate and the surface of Problem 7.39. Calculate the velocity of the plate for an angle θ of:
(a) 20°
(b) 30°

7.41 Water at 20°C flows down a 20° incline at a depth of 6 mm and a width of 50 m. Calculate the flow rate and the Reynolds number assuming laminar flow. Also, find the maximum velocity and the wall shear.

7.42 Water at 20°C flows down a 100-m-wide parking lot on a slope of 0.00015 at a depth of 10 mm. Determine the flow rate and the Reynolds number assuming laminar flow. Also, calculate the friction factor and the wall shear.

7.43 A pressure drop of 50 Pa is measured over a 60-m length of a 90 cm × 2 cm rectangular horizontal channel transporting 20°C air. Calculate the maximum flow rate and the associated Reynolds number. Use $\rho = 1.2$ kg/m³.

7.44 In Fig. P7.44, a pressure difference $p_A - p_B$ is measured to be 96 kPa. Find the friction factor for the wide channel assuming laminar flow. The flow direction is unknown.

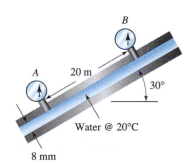

FIGURE P7.44

7.45 A slit with dimensions 0.5 mm × 100 mm exists in the 5-cm-thick side of a pressure vessel containing SAE-10W oil at 25°C and 4.0 MPa. What is the maximum flow rate that can exist from the slit? Assume developed laminar flow.

7.46 Air flows between the parallel plates as shown in Fig. P7.46. Find the pressure gradient such that:

(a) The shear stress at the upper surface is zero.
(b) The shear stress at the lower surface is zero.
(c) The flow rate is zero.
(d) The velocity at $y = 2$ mm is 4 m/s.

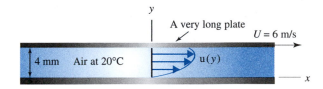

FIGURE P7.46

7.47 A pressure gradient of -20 Pa/m exists in 50°C air flowing between horizontal parallel plates spaced 6 mm apart. Find the velocity of the upper plate so that:

(a) The shear stress at the upper plate is zero.
(b) The shear stress at the lower plate is zero.
(c) The flow rate is zero.
(d) The velocity at $y = 2$ mm is 2 m/s.

7.48 Oil with $\mu = 5 \times 10^{-3}$ N·s/m² fills the concentric space between the rod and the surface shown in Fig. P7.48. Find the force F if $V = 15$ m/s. Assume that $dp/dx = 0$.

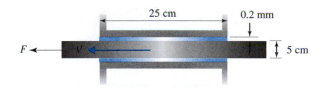

FIGURE P7.48

7.49 Calculate the torque T necessary to rotate the rod shown in Fig. P7.49 at 30 rad/s if the fluid filling the gap is SAE-10W oil at 20°C assuming a linear velocity profile.

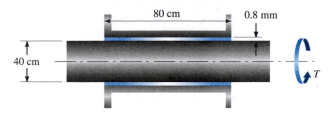

FIGURE P7.49

7.50 Oil with $\mu = 0.01$ N·s/m² fills the gap shown in Fig. P7.50. Estimate the torque necessary to rotate the disk shown assuming a linear velocity profile. Is the assumption of laminar flow valid? Use $S = 0.86$.

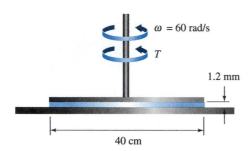

FIGURE P7.50

7.51 Approximate the torque necessary to rotate the inner 20-cm-diameter cylinder shown in Fig. P7.51. SAE-30W oil at 20°C fills the gap. Assume a linear velocity profile.

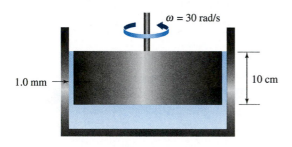

FIGURE P7.51

7.52 Find the torque needed to rotate the cone shown in Fig. P7.52 if oil with $\mu = 0.01$ N·s/m² fills the gap as shown. Assume a linear velocity profile.

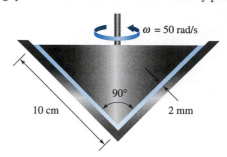

FIGURE P7.52

7.53 To create a high-Reynolds-number flow, the channel setup shown in Fig. P7.53 was proposed by Prof. John Foss of Michigan State University. It is a pressurized channel, thereby avoiding fatal leaks that always are present in a suction channel. (An upstream fan would produce blade vortices that make a high Re impossible to attain.) Estimate the power requirement of the 70%-efficient fan if the channel is 1.2 m wide and Re = 7000.

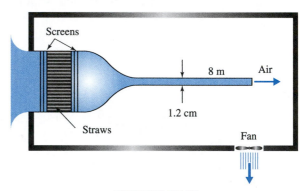

FIGURE P7.53

Laminar Flow between Rotating Cylinders

7.54 A long cylinder of radius R rotates in a large container of liquid. What is the velocity distribution in the liquid? Calculate the torque needed to rotate the 1-m-long 5-cm-diameter cylinder at 1000 rpm if the liquid is water at 15°C. Assume a laminar flow.

7.55 SAE-10W oil at 40°C fills the gap between two concentric, 40-cm-long cylinders with respective radii of 2 cm and 3 cm. What torque is necessary to rotate the inner cylinder at 3000 rpm if the outer cylinder is fixed? What power is required? Verify that Eq. 7.5.17 is, in fact, the velocity profile.

7.56 A torque of 0.015 N·m is required to rotate a 4-cm-radius cylinder inside a fixed 5-cm-radius cylinder at 40 rad/s. The concentric cylinders are 50 cm long. Calculate the viscosity of the fluid. Use $S = 0.9$. Verify that Eq. 7.5.17 is, in fact, the velocity profile.

7.57 Find an expression for the torque necessary to rotate the outer cylinder if the inner cylinder of Fig. 7.6 is fixed.

7.58 Rework Problem 7.49 using the velocity distribution of Eq. 7.5.17 and calculate the percent error assuming the linear velocity profile.

Turbulent Flow

7.59 A liquid flows in a pipe at a Reynolds number of 6000.
- **A.** The flow is laminar
- **B.** The flow is turbulent
- **C.** The flow is transitory, osscilating between laminar and turbulent
- **D.** The flow could be any of the above

7.60 Time average the differential continuity equation for an incompressible flow and show that two continuity equations result: the instantaneous continuity equation

$$\frac{\partial u'}{\partial x} + \frac{\partial v'}{\partial y} + \frac{\partial w'}{\partial z} = 0$$

and the time-average continuity equation

$$\frac{\partial \bar{u}}{\partial x} + \frac{\partial \bar{v}}{\partial y} + \frac{\partial \bar{w}}{\partial z} = 0$$

7.61 Find an expression for the difference between the time-average acceleration $\overline{Du/Dt}$ and the quantity $D\bar{u}/Dt$ using the fact that

$$\overline{u'\frac{\partial u'}{\partial x}} + \overline{v'\frac{\partial u'}{\partial y}} + \overline{w'\frac{\partial u'}{\partial z}} = \frac{\partial}{\partial x}\overline{u'^2} + \frac{\partial}{\partial y}\overline{u'v'} + \frac{\partial}{\partial z}\overline{u'w'}$$

7.62 Prove that the equation written in Problem 7.61 is indeed a fact. (*Hint:* Use the instantaneous continuity equation.)

7.63 Show that the time-averaged x-component Navier–Stokes equation results in

$$\rho\frac{\partial}{\partial y}\overline{u'v'} = -\frac{\partial \bar{p}}{\partial x} + \mu\frac{\partial^2 \bar{u}}{\partial y^2}$$

for developed flow in a wide horizontal channel. Using $\bar{\tau}_{lam} = \mu(\partial\bar{u}/\partial y)$ and $\bar{\tau}_{turb} = -\rho\overline{u'v'}$, write the time-averaged Navier–Stokes equation in terms of stresses.

7.64 The velocity components at a point in a turbulent flow are given in the following table. Find $\bar{u}$, $\bar{v}$, $\overline{u'^2}$, $\overline{v'^2}$, and $\overline{u'v'}$ at that point.

t (s)	u (m/s)	v (m/s)	t (s)	u (m/s)	v (m/s)
0.00	16.1	1.6	0.06	17.1	−1.4
0.01	25.7	−5.4	0.07	28.6	6.7
0.02	10.6	−8.6	0.08	6.7	−5.2
0.03	17.3	3.5	0.09	19.2	−8.2
0.04	5.2	4.1	0.10	21.6	1.5
0.05	10.2	−6.0			

7.65 Over a small radial distance in a developed turbulent flow, the time-average velocity is as given in the following table. The pressure drop in a 10-m horizontal section is measured to be 400 Pa. Find $\overline{u'v'}$ at $r = 20$ cm. Air with $\rho = 1.804$ kg/m^3 and $\nu = 1.5 \times 10^{-5}$ m^2/s is flowing.

r (cm)	18.0	18.9	20.7	21.6
$\bar{u}$ (m/s)	24.4	23.1	21.1	18.2

7.66 If at $r = 20$ cm for the data of Problem 7.65, we measure $\overline{u'^2} = 30$ m^2/s^2 and $\overline{v'^2} = 15$ m^2/s^2, what are the magnitudes of the eddy viscosity, the correlation coefficient, and the mixing length?

7.67 The velocity components are measured at a point in a laminar flow to be as shown in Fig. P7.67. Find $\overline{u'v'}$, η, l_m, and K_{uv} if $d\overline{u}/dy = -10\ \mathrm{s}^{-1}$ at the point.

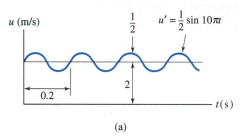

u (m/s)

$\dfrac{1}{2}$ $u' = \dfrac{1}{2}\sin 10\pi t$

2

0.2

$t(\mathrm{s})$

(a)

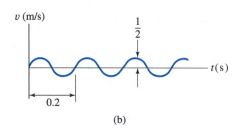

v (m/s)

$\dfrac{1}{2}$

0.2

$t(\mathrm{s})$

(b)

FIGURE P7.67

7.68 A 20-cm-diameter pipe with $e = 0.26$ mm transports water at 20°C. Determine if the pipe is smooth or rough if the average velocity is:

(a) 0.02 m/s   **(b)** 0.2 m/s   **(c)** 2 m/s

7.69 SAE-30 oil at 40°C is transported in a 10-cm-diameter pipe at an average velocity of 6 m/s. What is the largest size allowed for the roughness element if the pipe is hydraulically smooth?

7.70 Calculate the maximum velocity in the pipe of:

(a) Problem 7.68a
(b) Problem 7.68c

7.71 The velocity profile for water at 20°C in a turbulent flow in a 10-cm-diameter smooth pipe is approximated by $\overline{u} = 9.2y^{1/7}$ m/s. Find:

(a) The wall shear
(b) The velocity gradient $d\overline{u}/dy$ at the wall
(c) The pressure gradient
(d) The value of η at $r = 2.5$ cm

7.72 Water at 20°C flows in a 12.5-cm-diameter horizontal pipe at the rate of 75 L/s. Find the constant n in the exponent of Eq. 7.6.19. What is the maximum velocity?

7.73 Show that the kinetic-energy correction factor is 1.10 for $n = 5$ and 1.03 for $n = 10$ using $u = u_{max}(y/r_0)^{1/n}$ in a circular pipe.

7.74 Water at 20°C flows in a 10-cm-diameter pipe with an average velocity of 10 m/s. Using $u = u_{max}(y/r_0)^{1/n}$ with $n = 7$, plot the viscous shear and the turbulent shear as a function of r. Also, find $d\overline{p}/dx$.

7.75 SAE-10W oil at 10°C is transported in a smooth 80-cm-diameter pipe at the rate of 1.2 m³/s.
(a) Find the Reynolds number.
(b) Find the friction factor.
(c) Find the maximum velocity using Eq. 7.6.20.
(d) Find the viscous wall layer thickness.
(e) Compare part (c) with the solution using the logarithmic velocity profile.

7.76 Let the pipe in Problem 7.75 be a cast iron pipe (Fig. 7.13 provides a value for e). Estimate the maximum velocity using the logarithmic velocity profile.

7.77 A pressure drop of 10 kPa is measured with gages placed 4.5 m apart on a smooth horizontal 10-cm-diameter pipe transporting water at 80°C. Estimate:
(a) The wall shear
(b) The maximum velocity
(c) The average velocity
(d) The Reynolds number
(e) The flow rate

7.78 A 12-cm-diameter horizontal pipe transports SAE-10 oil at 10°C. Calculate the wall shear, the average velocity, and the flow rate if the pressure drop over a 10-m section of pipe is measured to be:
(a) 5 kPa   **(b)** 20 kPa   **(c)** 200 kPa

7.79 Make a linear plot (not a semilog plot) of the velocity profile of the flow in Problem 7.77 using:
(a) The log profile
(b) The power-law profile

Turbulent Flow in Pipes and Conduits

7.80 In a turbulent flow in a pipe, the head loss
- **A.** varies with the velocity squared
- **B.** is directly proportional to the flow rate
- **C.** decreases with increase with Reynolds number
- **D.** is directly proportional to the length of the pipe

7.81 The curves for the friction factor f on the Moody diagram become horizontal for sufficiently large Reynolds numbers because
- **A.** wall roughness elements project through the wall viscous layer
- **B.** the wall viscous layer completely covers the wall roughness elements
- **C.** the viscous effects become dominant in the flow
- **D.** inertial effects cease to become significant in the flow

7.82 Water at 20°C flows in an 8-cm-diameter plastic pipe with a flow rate of 20 L/S. Determine the friction factor using (a) the Moody diagram, and (b) Eq. 7.6.26.

7.83 A flow rate of 0.03 m³/s of 15°C water occurs in a 10-cm-diameter cast iron pipe. Determine the friction factor using (a) the Moody diagram, and (b) one of the equations (7.6.26)–(7.6.28).

7.84 Water at 20°C flows in a 4-cm-diameter cast iron pipe. Determine the friction factor, using the Moody diagram, if the average velocity is:
- **(a)** 0.025 m/s
- **(b)** 0.25 m/s
- **(c)** 2.5 m/s
- **(d)** 25 m/s

7.85 The pressure drop over 15 m of 2-cm-diameter galvanized iron pipe is measured to be 60 Pa. If the pipe is horizontal, estimate the flow rate of water. Use $\nu = 10^{-6}$ m²/s.
- **A.** 6.82 L/S
- **B.** 2.18 L/S
- **C.** 0.682 L/S
- **D.** 0.218 L/S

7.86 Water flows in an 8-cm-diameter pipe down a 30° incline and the pressure remains constant. Estimate the average velocity in the cast iron pipe. Use $\nu = 10^{-6}$ m²/s.
- **A.** 0.055 m/s
- **B.** 0.174 m/s
- **C.** 1.75 m/s
- **D.** 5.5 m/s

7.87 A flow rate of 0.02 m³/s occurs in a 10-cm-diameter wrought iron pipe. Using the Moody diagram, calculate the pressure drop over a 100-m horizontal section if the pipe transports:
- **(a)** Water at 20°C
- **(b)** Glycerine at 60°C
- **(c)** SAE-30W at 30°C
- **(d)** Kerosene at 10°C

Compare each answer with that using Eq. 7.6.29.

7.88 Water at 15°C flows in a 4-cm-diameter pipe with a flow rate of 2 L/s. Using the Moody diagram, determine the head loss in a 180-m section if the pipe is:
- **(a)** Cast iron
- **(b)** Galvanized iron
- **(c)** Wrought iron
- **(d)** Plastic

7.89 A mass flux of 1.2 kg/s occurs in a 10-cm-diameter plastic pipe at 20°C and 500 kPa absolute. Assume an incompressible flow and, using the Moody diagram, find the pressure drop in a 100-m section of pipe if the fluid flowing is:
- **(a)** Air
- **(b)** Carbon dioxide
- **(c)** Hydrogen

Compare each answer with that using Eq. 7.6.29.

7.90 SAE-30W oil flows at the rate of 0.08 m³/s in a 15-cm-diameter horizontal galvanized iron pipe. Find the pressure drop in 100 m if the temperature of the oil is:
- **(a)** 0°C
- **(b)** 30°C
- **(c)** 60°C
- **(d)** 90°C

Compare each answer with that using Eq. 7.6.29.

7.91 Select the material, listed in Fig. 7.13, from which each of the following pipes is probably constructed. Each 5-cm-diameter pipe is tested with water at 20°C using a flow rate of 400 L/min. The following pressure drops are measured over a 10-m length of horizontal pipe:
- **(a)** Pipe 1: 36 kPa
- **(b)** Pipe 2: 24 kPa
- **(c)** Pipe 3: 19 kPa

7.92 Water at 10°C flows up a 30° incline in a 6-cm-diameter plastic pipe with a flow rate of 10 L/s. Find the pressure change over a 100-m length of pipe.

7.93 Water at 40°C flows in a horizontal section of 5-cm-diameter wrought iron pipe with a flow rate of 0.02 m³/s. Does the pipe behave as a smooth pipe, or is the roughness significant?

7.94 An 80-cm-diameter concrete pipe is to transport storm water at 20°C at a rate of 5 m³/s. What pressure drop is to be expected over a 100-m section of horizontal pipe?

7.95 A pressure drop of 500 kPa is not to be exceeded over a 200-m length of horizontal 10-cm-diameter cast iron pipe. Calculate the maximum flow rate if the fluid is:
  (a) Water at 20°C
  (b) Glycerine at 20°C
  (c) SAE-10W oil at 20°C
  (d) Kerosene at 20°C

7.96 A liquid with a density of 900 kg/m³ flows straight down in a 6-cm-diameter cast iron pipe. Estimate the pressure rise over 20 m of pipe if the average velocity is 4 m/s. Assume $\nu = 8 \times 10^{-6}$ m²/s.
  A. 250 kPa
  B. 100 kPa
  C. 77 kPa
  D. 10.2 kPa

7.97 A pressure drop of 200 kPa is not to be exceeded over a 100-m length of horizontal 4-cm-diameter pipe. Estimate the maximum flow rate if water at 20°C is transported and the pipe is:
  (a) Cast iron
  (b) Wrought iron
  (c) Plastic

7.98 Neglecting all losses except that due to wall friction, estimate the flow rate through the pipe shown in Fig. P7.98 if the diameter is:
  (a) 4 cm
  (b) 8 cm
  (c) 12 cm
  (d) 16 cm

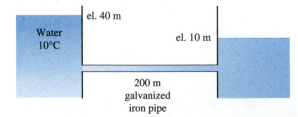

el. 40 m
Water 10°C
el. 10 m
200 m galvanized iron pipe

FIGURE P7.98

7.99 A pressure drop of 400 Pa is allowable in gas flow in a 400-m horizontal section of 12-cm-diameter wrought iron pipe. If the temperature and pressure are 40°C and 200 kPa absolute, find the maximum mass flux if the gas is:
  (a) Air   **(b)** Carbon dioxide
  (c) Hydrogen

7.100 A pressure drop of 220 kPa is not to be exceeded over a 180-m length of horizontal 1.2-m-diameter concrete pipe transporting water at 15°C. What flow rate can be accommodated? Use:
  (a) The Moody diagram
  (b) Equation 7.6.30

7.101 Estimate the size of plastic tubing that should be selected if 0.002 m³/s of fluid is to be transported such that the pressure drop does not exceed 200 kPa in a 100-m horizontal section. The fluid is:
  (a) Water at 20°C
  (b) Glycerine at 60°C
  (c) Kerosene at 20°C
  (d) SAE-10W oil at 40°C

7.102 Select a concrete pipe size that will transport 5 m³/s of 20°C water so that the head loss does not exceed 20 m in a 300-m horizontal pipe section. Use:
  (a) The Moody diagram
  (b) Equation 7.6.31

7.103 A farmer wishes to siphon 10°C water from a lake 1200 m from a field positioned a distance of 3 m below the surface of the lake. What size of drawn tubing should be selected if 400 L of water is desired each minute? Use:
  (a) The Moody diagram
  (b) Equation 7.6.31

Neglect all losses except that due to wall friction. Is the exiting kinetic energy also negligible?

7.104 The flow of water occurs in a square cast iron conduit, 4 cm on a side. If the horizontal conduit transports 0.02 m³/s, estimate the pressure drop over 40 m. Use $\nu = 10^{-6}$ m²/s.
  A. 162 Pa   **B.** 703 Pa
  C. 1390 Pa   **D.** 1590 Pa

7.105 Atmospheric air at 30°C is to be transported through a square sheet metal (smooth) conduit at the rate of 4 m³/s. What should be the conduit dimensions so that the head loss does not exceed 10 m over a horizontal length of 200 m?

7.106 Water at 20°C is transported through a 2 cm × 4 cm smooth conduit and experiences a pressure drop of 80 Pa over a 2-m horizontal length. What is the flow rate?

7.107 A plastic conduit 4 cm × 10 cm transports water at 20°C. If a pressure drop of 100 Pa is measured by gages spaced 5 m apart on a horizontal section, find the flow rate.

Minor Losses

7.109 A new design of a valve is to be tested. Which of the following parameters is the most important if liquid benzene flows through the valve?
  A. Froude number
  B. Reynolds number
  C. Mach number
  D. Euler number

7.110 If the loss coefficient for a sudden expansion is based on the exit velocity V_2, determined the loss coefficient in terms of A_1 and A_2.

7.111 Explain, with reference to Eq. 3.4.15, why the high- and low-pressure regions of Fig. 7.14 exist. Also, sketch a velocity profile from the inside corner of the bend to the outside of the bend along a line from B to C as sketched in Fig. P7.111. Explain why a secondary flow results after the bend.

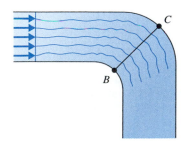

FIGURE P7.111

7.112 For each system shown in Fig. P7.112, find p_2 if $Q = 0.02$ m³/s of air at 20°C and $p_1 = 50$ kPa.

7.108 An open rectangular 1.2-m-wide concrete (use $e = 1.5$ mm) channel transports 20°C water from a reservoir to a location 10 000 m away. Using the Moody diagram, estimate the flow rate if the channel is on a slope of 0.0015 and the water depth is:
  (a) 0.3 m   **(b)** 0.6 m
  (c) 0.9 m

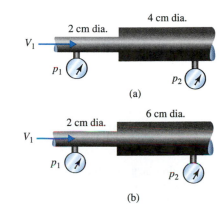

FIGURE P7.112

7.113 Replace the sudden enlargement of Problem 7.112 with a 20° expansion angle and rework the problem.

7.114 For each system shown in Fig. P7.114, estimate the loss coefficient based on V_2 using the data of Fig. 7.16.

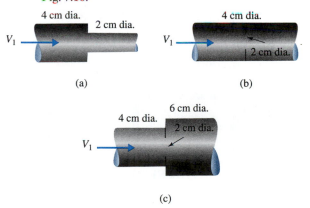

FIGURE P7.114

7.115 The flow rate is measured to be 3.5 L/s from the pipe shown in Fig. P7.115. Find the loss coefficient of the valve. Neglect wall friction.

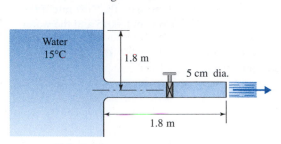

FIGURE P7.115

7.116 The flow rate is measured to be 6 L/s in the pipe shown in Fig. P7.116. Find the loss coefficient of the valve if H is:

(a)   4 cm   (b)   8 cm

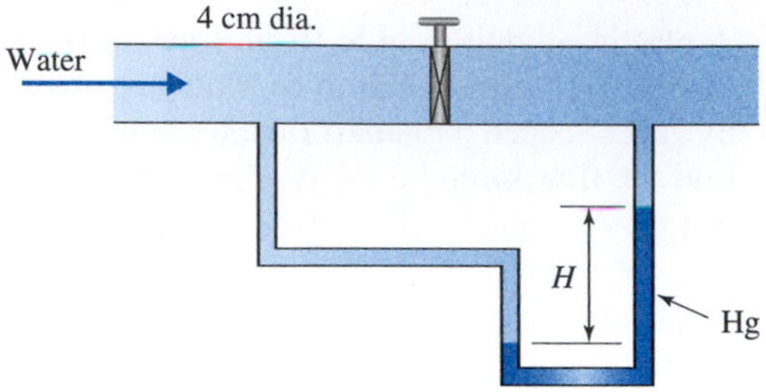

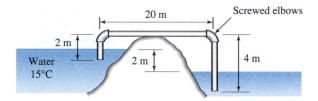

FIGURE P7.116

Simple Piping Systems

7.117 A water system is to be installed in a community that is quite hilly. The design engineer must be certain that:

A.   the hydraulic grade line must always be above the pipe line.

B.   the energy grade line must always be above the pipe line.

C.   the stagmation pressure must remain positive in the pipe line

D.   the elevation of the pipe must not be allowed to be negative.

7.118 Find the flow rate from the pipe shown in Fig. P7.118. Sketch the EGL and the HGL.

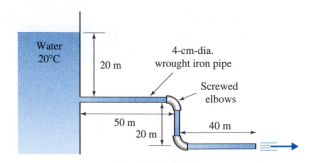

FIGURE P7.118

7.119 Water at 20°C flows from a 10-cm-diameter 360-m length of cast iron horizontal pipe that is attached to a reservoir with a square-edged entrance. A screwed globe valve that controls the flow is half

open. Find the flow rate if the reservoir elevation above the pipe exit is:

(a)   5 m   (b)   10 m   (c)   20 m

7.120 Estimate the flow rate to be expected through the plastic siphon shown in Fig. P7.120 if the diameter is:

(a)   4 cm   (b)   8 cm   (c)   12 cm

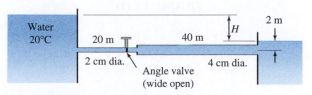

FIGURE P7.120

7.121 For the cast-iron piping shown in Fig. P7.121, calculate the flow rate and minimum pressure and sketch the HGL and the EGL if:

(a) $H = 10$ m

(b) $H = 20$ m

(c) $H = 30$ m

FIGURE P7.121

7.122 Water at 35°C flows from a bathtub through a 2.5-cm-diameter plastic pipe to a large air-filled drain pipe. There are two screwed elbows in the 10 m of pipe. If the drain pipe is 0.8 m below the water level in the tub, estimate how long it will take to collect 10 L of water.

7.123 A 3-cm-diameter plastic tube with screwed elbows is used to siphon the water as shown in Fig. P7.123. Estimate the maximum height H for which the siphon will operate.

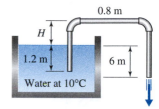

FIGURE P7.123

7.124 A 12-L lawn sprayer is filled with 8 L of 20°C water. It is 1.2 m high and has an 8-mm-diameter copper tube ($e \approx 0$) that reaches the bottom (it is a little short). A 1.2-m-long 5-mm-diameter smooth hose is attached to the copper tube. The hose ends with a 2-mm-diameter nozzle. If the sprayer is pressurized to 100 kPa, estimate the initial velocity exiting the nozzle.

7.125 What is the maximum flow rate through the pipe shown in Fig. P7.125 if the elevation difference of the reservoir surfaces is:
(a) 80 m?
(b) 150 m?
(c) 200 m?

7.126 A 9.4-mm-diameter 60-m-long plastic pipe transports 20°C water from a spring to a pond 3 m below, similar to the situation shown in Problem 7.125. It is observed that the water alternates between a relatively fast moving stream to a relatively slow moving stream. Explain this phenomenon with supporting calculations.

7.127 Water at 15°C is to be pumped through 250 m of cast iron pipe from a reservoir to a device that is 10 m above the reservoir surface. It is to enter the device at 220 kPa. Screwed components include two elbows, a square-edged entrance, and an angle valve. If the flow rate is to be 20 L/s, what pump power is needed (assume 80% efficiency) if the pipe diameter is:
(a) 4 cm? **(b)** 8 cm? **(c)** 12 cm?

7.128 What pump power (85% efficient) is needed for a flow rate of 0.01 m³/s in the pipe shown in Fig. P7.128? What is the greatest distance from the left reservoir that the pump can be located?

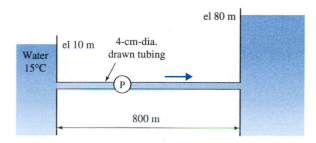

FIGURE P7.128

7.129 A flow rate of 2 m³/s exists in the pipe shown in Fig. P7.129. What is the expected power output of

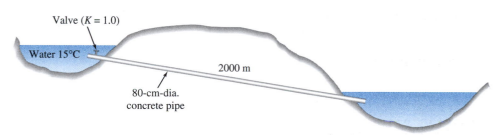

FIGURE P7.125

the turbine (85% efficient) if the elevation difference of the reservoir surfaces is:
(a) 20 m? **(b)** 60 m? **(c)** 100 m?

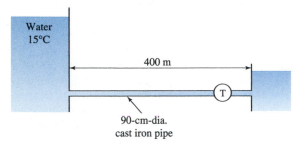

Water 15°C

400 m

T

90-cm-dia. cast iron pipe

FIGURE P7.129

7.130 What pump power (75% efficient) is needed in the piping system shown in Fig. P7.130? What is the greatest distance from the reservoir that the pump can be located?

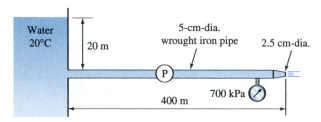

Water 20°C

20 m

5-cm-dia. wrought iron pipe

2.5 cm-dia.

P

700 kPa

400 m

FIGURE P7.130

7.131 The pump shown in Fig. P7.131 has the characteristic curves shown in Example 7.18.
(a) Estimate the flow rate and the power required by the pump.
(b) Sketch the EGL and the HGL.
(c) If cavitation is possible, determine the maximum distance from the reservoir to locate the pump.

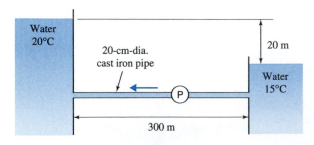

Water 20°C

20-cm-dia. cast iron pipe

20 m

Water 15°C

P

300 m

FIGURE P7.131

7.132 Reverse the flow direction in Problem 7.131 and redo the problem.

7.133 A turbine replaces the pump of Problem 7.131. Estimate the power output if $\eta_T = 0.88$. The turbine's characteristic curve is $H_T = 0.8Q$, where H_T is measured in meters and Q in L/s.

7.134 The pump shown in Fig. P7.134 has characteristic curves shown in Example 7.18. Estimate the flow rate and:
(a) Calculate the pump power requirement.
(b) Calculate the pressure at the pump inlet.
(c) Calculate the pressure at the pump outlet.
(d) Sketch the EGL and the HGL.

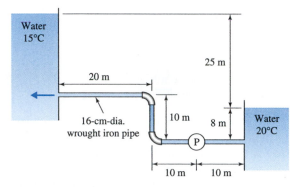

Water 15°C

25 m

20 m

16-cm-dia. wrought iron pipe

10 m

8 m

Water 20°C

P

10 m 10 m

FIGURE P7.134

7.135 Reverse the flow direction in Problem 7.134 and redo the problem.

7.136 A turbine with a characteristic curve shown in Fig. P7.136 is inserted in the pipeline. Calculate the turbine power output. Assume that $\eta_T = 0.90$.

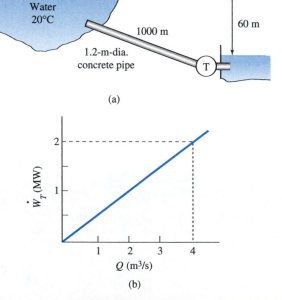

Water 20°C

1000 m

60 m

1.2-m-dia. concrete pipe

T

(a)

$\dot{W}_T$ (MW)

Q (m³/s)

(b)

FIGURE P7.136

Open-Channel Flow

7.137 Water flows in a 2.4-m-wide, rectanglular, finished concrete channel at a depth of 80 cm. If the slope is 0.002, the flow rate is nearest
A. 2.2 m³/s
B. 3.4 m³/s
C. 4.6 m³/s
D. 6.2 m³/s

7.138 Using a control volume surrounding a finite length of the liquid in a channel flowing at constant depth, find the average shear stress on the walls if water flows in a 3-m-wide rectangular channel at a depth of 1.8 m on a 0.001 slope.

7.139 Using the approach mentioned in Problem 7.138, determine the average shear stress on the portion of the wall of a 40-cm-diameter circular conduit in contact with the water, which is flowing at a constant depth of 10 cm. The slope is 0.0016.

7.140 Calculate the flow rate in a 2-m-wide rectangular, planed wood channel on a slope of 0.001 if the depth is 60 cm. Use:
(a) The Chezy–Manning equation
(b) The Darcy–Weisbach equation

7.141 Water flows in a 2-m-diameter finished concrete conduit on a 0.0012 slope. Predict the flow rate if the depth of flow is:
(a) 2 m minus a little
(b) 1.9 m
(c) 1 m
(d) 0.5 m
(e) 0.2 m

7.142 For the channel shown in Fig. P7.142, find the flow rate and average velocity if $S = 0.001$ using:
(a) The Chezy–Manning equation
(b) The Darcy–Weisbach equation

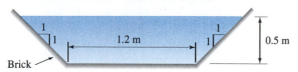

FIGURE P7.142

7.143 At what depth will 5 m³/s of water flow in a 2-m-wide rectangular brick channel with $S = 0.001$? Use:
(a) The Chezy–Manning equation
(b) The Darcy–Weisbach equation

7.144 The cross section of a straight river is approximated as shown in Fig. P7.144. At what depth will 100 m³/s of water flow? The slope is 0.001.

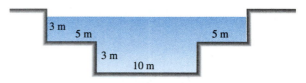

FIGURE P7.144

7.145 For the channel shown in Fig. P7.145, if $S = 0.0016$, find the flow depth if $Q = 10$ m³/s and the channel is constructed with:
(a) Planed wood
(b) Brick

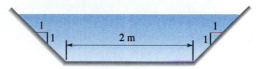

FIGURE P7.145

7.146 Water flows in a 1.2-m-diameter sewer pipe at a flow rate of 0.75 m³/s. Estimate the depth if the slope is 0.001.

7.147 Water flows in a 80-cm-diameter sewer pipe at a flow rate of 0.2 m³/s. Determine the depth if the slope is 0.001.

A blimp is a lighter-than-air-vessel that is streamlined to reduce the drag forces encountered during its forward motion. Large blimps, perhaps 1000 m long, could be used as cruise ships to tour the globe; sea sickness would be avoided! (Courtesy of The Goodyear Tire & Rubber Company)

8

External Flows

Outline

Chapter Objectives

The objectives of this chapter are to:

▲ Discuss separated and attached flows.

▲ Introduce the lift and drag coefficients.

▲ Determine the drag on a variety of bodies.

▲ Study the influence of vortex shedding and streamlining.

▲ Determine when cavitation is present.

▲ Calculate the lift and drag on airfoils.

▲ Superimpose several simple potential flows to construct a flow of interest.

▲ Analyze both laminar and turbulent boundary layers on a flat plate.

▲ Provide numerous examples and problems demonstrating how the various quantities of interest are determined for the many exernal flows discussed in this chapter.

8.1 INTRODUCTION

The study of external flows is of particular importance to the aeronautical engineer in the analysis of airflow around the various components of an aircraft. In fact, much of the present knowledge of external flows has been obtained from studies motivated by such aerodynamic problems. There is, however, substantial interest by other engineers in external flows; the flow of fluid around turbine blades, automobiles, buildings, athletic stadiums, smokestacks, spray droplets, bridge abutments, submarine pipelines, river sediment, and red blood cells suggest a variety of phenomena that can be understood only from the perspective of external flows.

It is a difficult task to determine the flow field external to a body and the pressure distribution on the surface of a body, even for the simplest geometry. To discuss this subject, consider low-Reynolds-number flows (Re < 5, or so) and high-Reynolds-number flows (Re > 1000). Low-Reynolds-number flows, called *creeping flows* or *Stokes flows*, seldom occur in engineering applications (flow around spray droplets and red blood cells, lubrication in small gaps, and flow in porous media would be exceptions) and are not presented in this book; they are left to the specialist. We will direct our attention to high-Reynolds-number flows only. We do, however, show a Stokes flow in Fig. 8.1.

High-Reynolds-number flows can be subdivided into three major categories: (1) incompressible immersed flows involving such objects as automobiles, helicopters, submarines, low-speed aircraft, take-off and landing of commercial aircraft, buildings, and turbine blades; (2) flows of liquids that involve a free surface as experienced by a ship or a bridge abutment; and (3) compressible flows involving high-speed objects ($V > 100$ m/s) such as aircraft, missiles, and bullets. We will focus our attention on the first category of flows in this chapter and

KEY CONCEPT *Low-Reynolds-number flows seldom occur in engineering applications.*

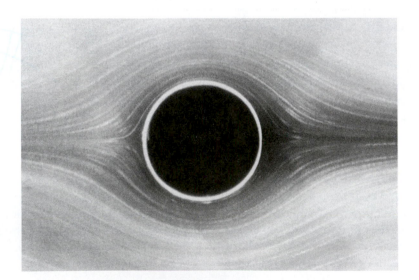

FIGURE 8.1 Flow past a circular cylinder at Re = 0.16. The flow is from left to right. It resembles superficially the pattern of potential flow. The flow of water is shown by aluminum dust. (Courtesy of The Parabolic Press, Stanford, California. Reprinted with permission.)

consider cases in which the object is far from a solid boundary or other objects. The flow becomes significantly influenced by the presence of a boundary or another object, as shown in Fig. 8.2; in part (d) the slender object must be at least five body lengths below the free surface before free surface effects can be neglected. Flows such as those shown in Fig. 8.2 are not included in an introductory presentation.

High-Reynolds-number incompressible immersed flows are divided into two categories: flows around blunt bodies and flows around streamlined bodies, as displayed in Fig. 8.3. The boundary layer (see Section 3.3.2) near the stagnation point is a laminar boundary layer, and for a sufficiently large Reynolds number, undergoes transition downstream to a turbulent boundary layer, as shown; the flow may separate from the body and form a **separated region**, a region of recirculating flow, as shown for the blunt body, or it simply leaves the streamlined body at the trailing edge (there may be a small separated region here). The **wake**, which is characterized by a *velocity defect* is a growing (diffusion) region, and trails the body, as shown. The boundaries of the wake, the separated region, and the turbulent boundary layer are quite time dependent; in the sketch the time-average location of the wake is shown by the dashed lines. Shear stresses due to viscosity are concentrated in the thin boundary layer, the separated region, and the wake; outside these regions the flow is approximated by an inviscid flow.

From the figure it could be assumed that the separated region does not exchange mass with the free stream since mass does not cross a streamline. When viewed instantaneously, however, the separation streamline is highly time dependent, and due to this unsteady character, the separated region is able to exchange mass slowly with the free stream.

Separated region: *A region of recirculating flow.*

Wake: *A region of velocity defect that grows due to diffusion.*

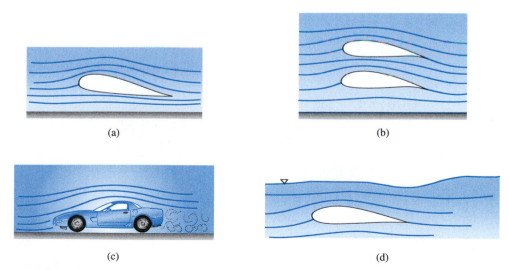

FIGURE 8.2 Examples of complicated immersed flows: (a) flow near a solid boundary; (b) flow between two turbine blades; (c) flow around an automobile; (d) flow near a free surface.

lbl = Laminar boundary layer
tbl = Turbulent boundary layer

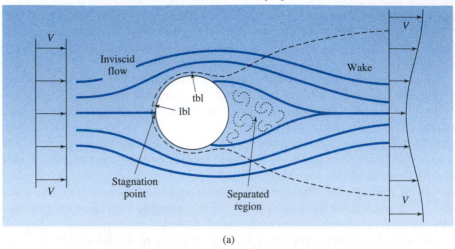

(a)

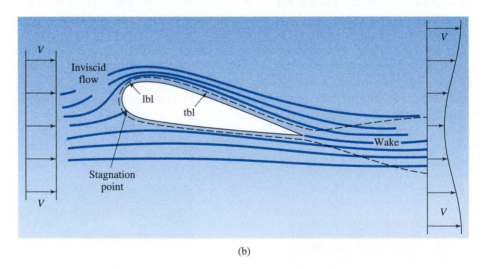

(b)

FIGURE 8.3 Flow around a blunt body and a streamlined body.

Drag: *The force the flow exerts in the direction of the flow.*

Lift: *The force the flow exerts normal to the direction of flow.*

Several comments should be made regarding the separated region and the wake. The separated region eventually closes; the wake keeps diffusing into the main flow and eventually disappears as its area becomes exceedingly large (the fluid regains the free-stream velocity). Time-average streamlines do not enter a separated region; they do enter a wake. The separated region is always submerged within the wake.

Flow around a blunt object is usually treated empirically, as was done for a turbulent flow in a conduit. We follow this procedure here. We are interested primarily in the **drag**, the force the flow exerts on the body in the direction of the flow. The **lift**, which acts normal to the direction of flow, will be of interest for airfoil shapes, as presented in Section 8.4. The actual details of the flow field are seldom of interest and are not presented in this introductory text. We present the drag F_D and lift F_L as dimensionless coefficients: the *drag coefficient* and *lift coefficient*, defined as

$$C_D = \frac{F_D}{\frac{1}{2}\rho V^2 A} \qquad C_L = \frac{F_L}{\frac{1}{2}\rho V^2 A} \qquad\qquad (8.1.1)$$

where A is most often the projected area (projected on a plane normal to the direction of the flow); for airfoil shapes, the area is based upon the chord (see Fig. 8.4). Drag coefficients for several common shapes are presented in Section 8.3.1. Since the drag on a blunt object is dominated by the flow in the separated region, there is little interest in studying the boundary layer growth on the front part of a blunt body and the associated viscous shear at the wall. Hence interest is focused on the empirical data that provide the drag coefficient.

KEY CONCEPT   *The drag on a blunt object is dominated by the flow in the separated region.*

Flow around a streamlined body — the separated region is insignificantly small or nonexistent — provides the motivation for detailed study of the laminar and turbulent boundary layers. A boundary layer that develops on a plane streamlined surface, such as an airfoil, is usually sufficiently thin that the curvature of the surface can be ignored and the problem can be treated as a boundary layer developing on a flat plate with a nonzero pressure gradient. We will provide a detailed study of flow on a flat plate with zero pressure gradient; once that problem is understood, the influence of a pressure gradient can be discussed. If the flow in the boundary layer on a streamlined body can be determined, the drag can be calculated, since the drag is a result of the shear stress and pressure force acting on the body surface.

Outside the boundary layer there exists an inviscid, *free-stream flow*, as shown in Figs. 8.3 and 8.4. Initially, we will assume that the free-stream flow is known. Before the velocity profile in the boundary layer can be determined, it is

KEY CONCEPT   *The inviscid, free-stream flow exists outside the boundary layer.*

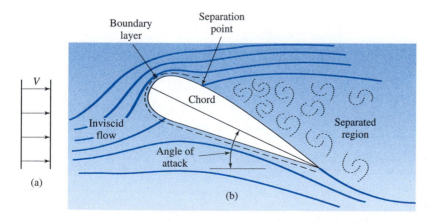

FIGURE 8.4   Streamlined body that is stalled.

necessary that the inviscid flow solution be known. It is found by completely ignoring the boundary layer, since it is so thin, and solving the appropriate inviscid equations. The inviscid flow solution is then used to provide the lift on the body, and the two quantities used in the boundary layer flow solution: the pressure gradient and the velocity at the boundary. With the inviscid flow known and the boundary layer flow determined, quantities of interest can be obtained in the flow about a streamlined body.

8.2 SEPARATION

Before we present the empirical data associated with the flow around blunt bodies, the general nature of separation will be discussed. Separation occurs when the main stream flow leaves the body, resulting in a separated region of flow, as sketched in Fig. 8.3a. When separation occurs on a streamlined body near the forward portion of an airfoil, as it will with a sufficiently large angle of attack (the angle the oncoming flow makes with the **chord**, a line connecting the trailing edge with the nose), the flow situation is referred to as **stall**, as shown in Fig. 8.4. Stall is highly undesirable on aircraft at cruise conditions and leads to inefficiencies when it occurs on turbine blades. It is used, however, to provide the high drag needed when landing an aircraft, or in certain maneuvers by stunt planes. For blunt bodies, however, separation is unavoidable at high Reynolds numbers and its effect must be understood.

The location of the separation point is dependent primarily on the geometry of the body; if the body has an abrupt change in geometry, such as that shown in Fig. 8.5, separation will occur at, or near, the abrupt change; however, it will also occur upstream on the flat surface as shown. In addition, reattachment will occur at some location, as shown. Let us establish the criterion used in predicting the location of the separation point on a surface with no abrupt geometry change.

Consider the flow on the flat surface just before the step of Fig. 8.5. The region near the forward separation point is enlarged and shown in Fig. 8.6; the y-coordinate is normal to the wall and the x-coordinate is measured along the wall. Downstream of the separation point the x-component velocity near the wall is in the negative x-direction and thus at the wall $\partial u/\partial y$ must be negative.

Chord: *A line connecting the trailing edge with the nose.*

Stall: *Flow condition where separation occurs on a streamlined body near the forward portion.*

KEY CONCEPT *For blunt bodies, separation is unavoidable at high Reynolds numbers.*

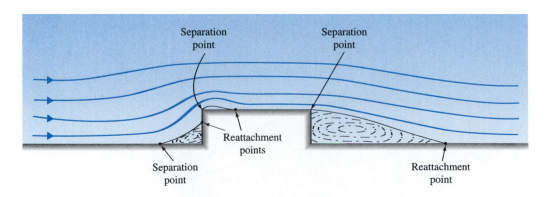

FIGURE 8.5 Separation due to abrupt geometry changes.

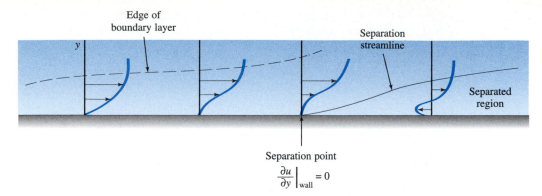

FIGURE 8.6 Flow separation on a flat surface due to an adverse pressure gradient.

Upstream of the separation point the x-component velocity near the wall is in the positive x-direction, demanding that $\partial u/\partial y$ at the wall be positive. Hence we conclude that the separation point is defined as that point where $(\partial u/\partial y)_{\text{wall}} = 0$.

Observe that separation on the flat surface occurs as the flow is approaching a stagnation region where the velocity is low and the pressure is high. As the flow approaches the stagnation region the pressure increases, that is, $\partial p/\partial x > 0$; the pressure gradient is positive. Since separation is often undesirable, a positive pressure gradient is called an *adverse pressure gradient*; a negative gradient is a *favorable pressure gradient*. In general, the effect of an adverse pressure gradient results in decreasing velocities in the streamwise direction; if an adverse pressure gradient acts on a surface over a sufficient distance, separation may result. This is true even if the surface is a flat plate, such as the wall of a diffuser. See Section 8.6.7 for further discussion.

In addition to geometry and pressure gradient, several other parameters influence separation. These include the Reynolds number as a very important parameter, with the wall roughness, the *free-stream fluctuation intensity*[1] (the intensity of the disturbances that exist away from the boundary), and the wall temperature having less but occasionally significant influence.[2] Visualize, for example, flow around a sphere; at sufficiently low Reynolds numbers no separation will occur. As the Reynolds number increases to a particular value, separation will occur over a small area at the rear; this area will become larger and larger as the Reynolds number increases until at a sufficiently large Reynolds number no additional increase in separation area will be observed. The boundary layer before separation will still be laminar. An interesting phenomenon takes place as the boundary layer before separation becomes turbulent; there is a sudden movement of the separation point to the rear of the sphere, resulting in a substantial reduction in the separation area and thus the drag. This phenomenon is explained by comparing the velocity profile of a laminar boundary layer with that of a turbulent boundary layer, as displayed in Fig. 8.7. As was true in a pipe flow, the turbulent profile has a much larger gradient near the wall (much larger wall

KEY CONCEPT *The separation point is defined as that point where $(\partial u/\partial y)_{\text{wall}} = 0$.*

KEY CONCEPT *As the boundary layer before separation becomes turbulent, the separation point moves to the rear.*

---

[1] Free-stream fluctuation intensity is defined as $\sqrt{\overline{u'^2}}/V$, where u' is the fluctuation. A value of 0.001 is quite low and 0.1 is quite high.

[2] If a fluid is flowing past a body that is rigidly supported, the vibration level of the support system will also influence the separation phenomenon. External sound waves can also be significant.

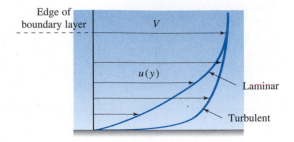

FIGURE 8.7 Comparison of laminar and turbulent velocity profiles.

stress) and thus the momentum of the fluid near the wall is substantially larger in the turbulent boundary layer. For a given geometry a greater distance is required to reduce the velocity near the wall to zero, resulting in the movement of the separation point to the rear, as is quite obvious in Fig. 8.8, where both spheres are moving with the same velocity (the sphere in (b) has sandpaper attached in the nose region). In Fig. 8.8a it is observed that separation occurs on the front half of the sphere, in a region of favorable pressure gradient. This separation is due to the centrifugal effects as the fluid moves around the sphere. This phenomenon of drag reduction is observed in the drop in the drag coefficient curves for a sphere and a cylinder, presented in the following section.

8.3 FLOW AROUND IMMERSED BODIES

8.3.1 Drag Coefficients

From our study of dimensional analysis we know that for a steady, incompressible flow in which gravity, thermal, and surface tension effects are neglected, the primary flow parameter that influences the flow is the Reynolds number; other occasionally important parameters include the relative wall roughness and the free-stream velocity fluctuation intensity.

The drag coefficient curves for two bodies that do not exhibit sudden geometric changes will be presented; the drag coefficients for the smooth sphere and the long smooth cylinder are shown in Fig. 8.9 over a large range of Reynolds numbers. At Re < 1 creeping flow with no separation results. For the sphere, this creeping flow problem has been solved, with the result that

$$C_D = \frac{24}{\text{Re}} \qquad \text{Re} < 1 \tag{8.3.1}$$

Separation is observed at Re $\simeq$ 10 over a very small area on the rear of the body. The separated area increases as the Reynolds number increases until Re $\simeq$ 1000, where the separated region ceases to enlarge; during this growth of the separated region the drag coefficient decreases. At Re = 1000, 95% of the drag is due to

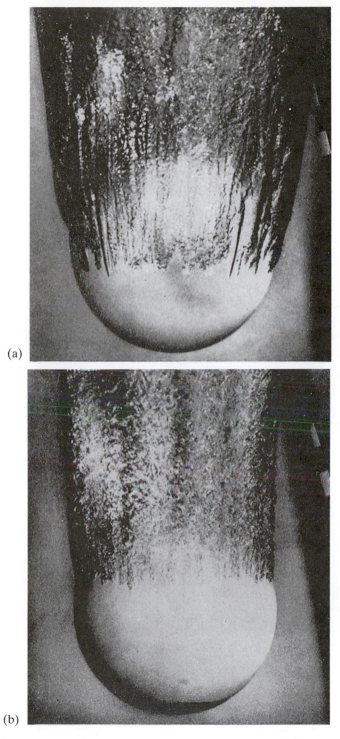

(a)

(b)

FIGURE 8.8 Effect of boundary layer transition on separation: (a) laminar boundary layer before separation; (b) turbulent boundary layer before separation. (U.S. Navy photographs.)

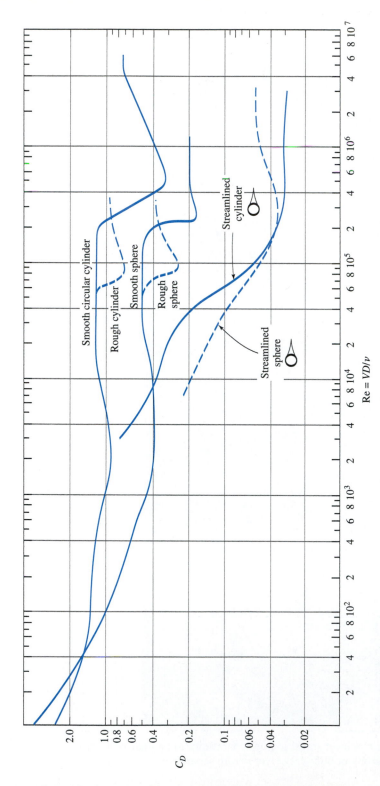

FIGURE 8.9 Drag coefficients for flow around a long cylinder and a sphere. (See E. Achenbach, *J. Fluid Mech.*, Vol. 46, 1971, and Vol. 54, 1972.)

form drag (the drag force due to the pressure acting on the body) and 5% is due to frictional drag (the drag force due to the shear stresses acting on the body).

The drag coefficient curve is relatively flat for smooth bodies over the range $10^3 < \text{Re} < 2 \times 10^5$. The boundary layer before the point of separation is laminar and the separated region is as shown in Fig. 8.8a. At $\text{Re} \simeq 2 \times 10^5$, for a smooth surface and with low free-stream fluctuation intensity, the boundary layer before separation undergoes transition to a turbulent state and the increased momentum in the boundary layer "pushes" the separation back, as shown in Fig. 8.8b, with a substantial decrease (a 60 to 80% drop) in the drag. If the surface is rough (dimples on a golf ball) or the free stream has high free-stream fluctuation intensity, the drop in the C_D curve may occur at $\text{Re} \simeq 8 \times 10^4$. Since a lower drag is usually desirable, surface roughness is often added; the dimples on the golf ball may increase the flight distance by 50 to 100%.

After the sudden drop in drag, the C_D curve is observed to again increase with increased Reynolds number. Experimental data are not readily available for $\text{Re} > 10^6$ for a sphere and $\text{Re} > 6 \times 10^7$ for a cylinder; however, a value of $C_D = 0.2$ for a sphere at large Reynolds number appears acceptable. Some engineers use $C_D = 0.4$ for cylinders at large Reynolds numbers; however, the data presented here suggests that that is too low. Additional experimental data is needed.

For cylinders of finite length and for elliptic cylinders, the drag coefficients are presented in Table 8.1. The finite-length cylinders are assumed to have two free ends. If one end is fixed to a solid surface, its length must be doubled when using Table 8.1. Blunt objects with sudden geometry changes have separated regions that are relatively insensitive to the Reynolds number; the drag coefficients for some common shapes are given in Table 8.2.

> **KEY CONCEPT** *Dimples on the golf ball may increase the flight distance by 50 to 100%.*

TABLE 8.1 Drag Coefficients of Finite-Length Circular Cylinders[a] with Free Ends[b] and of Infinite-Length Elliptic Cylinders

Circular cylinder		Elliptic cylinder[c]		
$\dfrac{Length}{Diameter}$	$\dfrac{C_D}{C_{D\infty}}$	$\dfrac{Major\ axis}{Minor\ axis}$	Re	C_D
∞	1	2	4×10^4	0.6
40	0.82	4	10^5	0.46
20	0.76	4	2.5×10^4 to 10^5	0.32
10	0.68	8	2.5×10^4	0.29
5	0.62	8	2×10^5	0.20
3	0.62			
2	0.57			
1	0.53			

[a] $C_{D\infty}$ is the drag coefficient for the infinite-length circular cyulinder obtained in Fig. 8.8.
[b] If one end is fixed to a solid surface, double the length of the cylinder.
[c] Flow is in the direction of the major axis.

TABLE 8.2   Drag Coefficients for Various Blunt Objects

Object		Re	C_D
Square cylinder → $\square\, w$ over w $L/w = \begin{cases} \alpha \\ 1 \text{ (cube)} \end{cases}$		$> 10^4$ $> 10^4$	2.0 1.1
rounded corners $L/w = \alpha$ ($r = 0.2w$)		$= 10^5$	1.2
Rectangular plates → $\vert\, w$ $L/w = \begin{cases} \alpha \\ 20 \\ 5 \\ 1 \end{cases}$		$> 10^3$ $> 10^3$ $> 10^3$ $> 10^3$	2.0 1.5 1.2 1.1
Circular cylinder → $\square\, D$ (L) $L/D = \begin{cases} 0.1 \text{ (disk)} \\ 4 \\ 7 \end{cases}$		$> 10^3$ $> 10^3$ $> 10^3$	1.1 0.9 1.0
Semicircular → D cylinder → C		$> 10^4$ $> 10^4$	2.2 1.2
Semicircular → $)$ shell → $($		$= 2 \times 10^4$ $= 2 \times 10^4$	2.3 1.1
Equilateral cylinders → $\triangleleft$ → $\triangleright$	2.0 1	$> 10^4$ $> 10^4$	2.0 1.4
Cone → $\triangleleft$ (α) $\alpha = \begin{cases} 30° \\ 60° \\ 90° \end{cases}$		$> 10^4$ $> 10^4$ $> 10^4$	0.6 0.8 1.2
Solid hemisphere → $\blacktriangleright$ → $\blacktriangleleft$		$> 10^4$ $> 10^4$	1.2 0.4
Hollow hemisphere → D → C		$> 10^4$ $> 10^4$	1.4 0.4
Parachute		$> 10^7$	1.4

Automobile
   1920
   Modern, with square corners
   Modern, with rounded corners
	—
	 —
	 —
		$> 10^5$
		 $> 10^5$
		 $> 10^5$
			0.80
			 0.30
			 0.29

Van		$> 10^5$	0.42

Bicycle, upright rider
   racing, bent over
   racing, drafting
			1.1
			 0.9
			 0.5

Semitruck, standard
   with streamlined deflector
   with deflector and gap seal
			0.96
			 0.76
			 0.70

Example 8.1

A square sign, 3 m × 3 m, is attached to the top of a 18-m-high pole which is 30 cm in diameter (Fig. E8.1). Approximate the maximum moment that must be resisted by the base for a wind speed of 30 m/s.

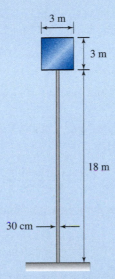

3 m

3 m

18 m

30 cm →|←

FIGURE E8.1

Solution

The maximum force F_1 acting on the sign occurs when the wind is normal to the sign; it is

$$F_1 = C_D \times \tfrac{1}{2}\rho V^2 A$$
$$= 1.1 \times \tfrac{1}{2} \times 1.23 \times 30^2 \times 3^2 = 5480 \text{ N}$$

where C_D is found in Table 8.2 and we use the standard value $\rho = 1.23 \text{ kg/m}^3$ since it was not given. The force F_2 acting on the cylindrical pole is (using the projected area as $A = 18 \times 0.3 \text{ m}^2$)

$$F_2 = C_D \times \tfrac{1}{2}\rho V^2 A$$
$$= 0.8 \times \tfrac{1}{2} \times 1.23 \times 30^2 \times (18 \times 0.3) = 2391 \text{ N}$$

where C_D is found from Fig. 8.8 with Re $= 30 \times 0.3/(1.51 \times 10^{-5}) = 5.96 \times 10^5$, assuming a high-intensity fluctuation level (i.e., a rough cylinder); since neither end is free, we do not use the multiplication factor of Table 8.1.

The resisting moment that must be supplied by the supporting base is

$$M = d_1 F_1 + d_2 F_2$$
$$= 19.5 \times 5480 + 9 \times 2391 = 128380 \text{ N·m}$$

assuming that the forces act at the centers of their respective areas.

Example 8.2

Determine the terminal velocity of a 30-cm-diameter smooth sphere ($S = 1.02$) if it is dropped in (a) air at 20°C and (b) water at 20°C.

Solution

(a) When terminal velocity is reached by a falling object, the weight of the object is balanced by the drag force acting on the object. Using $\Sigma F = 0$ and Eq. 8.1.1, we have

$$W = F_D$$

$$\therefore \gamma_{\text{sphere}} \times \text{volume} = C_D \times \tfrac{1}{2}\rho V^2 A$$

The volume of a sphere is $\frac{4}{3}\pi R^3$ and $\gamma_{\text{sphere}} = S\gamma_{\text{water}}$; the projected area is $A = \pi R^2$. Thus

$$S\gamma_{\text{water}} \times \tfrac{4}{3}\pi R^3 = C_D \times \tfrac{1}{2}\rho V^2 \pi R^2$$

The velocity can now be expressed as

$$V = \left(\frac{8RS\gamma_{\text{water}}}{3\rho C_D}\right)^{1/2} = \left(\frac{8 \times 0.15 \times 1.02 \times 9800}{3 \times 1.20 \times C_D}\right)^{1/2} = \frac{57.7}{\sqrt{C_D}}$$

Let us assume that the Reynolds number will be quite large and use C_D from Fig. 8.8 as $C_D = 0.2$. Then

$$V = \frac{57.7}{\sqrt{0.2}} = 129 \text{ m/s}$$

We must check the Reynolds number to verify the C_D value assumed. It is

$$\text{Re} = \frac{VD}{\nu} = \frac{129 \times 0.3}{1.6 \times 10^{-5}} = 2.42 \times 10^6$$

This is beyond the end of the curve where data are unavailable; we will assume that the drag coefficient is unchanged at 0.2, so the terminal velocity is 129 m/s.

(b) For the sphere falling in water, we must include the buoyancy force B acting in the same direction as the drag force F_D. Hence the summation of forces yields

$$W = F_D + B$$

$$\therefore \gamma_{\text{sphere}} \times \text{volume} = C_D \times \tfrac{1}{2}\rho V^2 A + \gamma_{\text{water}} \times \text{volume}$$

This gives

$$(S - 1)\gamma_{\text{water}} \times \tfrac{4}{3}\pi R^3 = C_D \times \tfrac{1}{2}\rho V^2 \pi R^2$$

Using $\rho = 1000$ kg/m^3, there results

$$V = \left(\frac{8R(S - 1)\gamma_{\text{water}}}{3\rho C_D}\right)^{1/2} = \left(\frac{8 \times 0.15 \times 0.02 \times 9800}{3 \times 1000 \times C_D}\right)^{1/2} = \frac{0.28}{\sqrt{C_D}}$$

We anticipate the Reynolds number being lower than in part (a), so let's assume that it is in the range $2 \times 10^4 < \text{Re} < 2 \times 10^5$. Then $C_D = 0.5$ and there results

$$V = 0.40 \text{ m/s}$$

This gives a Reynolds number of

$$\text{Re} = \frac{VD}{\nu} = \frac{0.40 \times 0.3}{10^{-6}} = 1.2 \times 10^5$$

This is in the required range, so the terminal velocity is expected to be 0.40 m/s. Of course, if the sphere were roughened (sand glued to the surface), the C_D value would be less and the velocity would be greater.

8.3.2 Vortex Shedding

Long blunt objects, such as circular cylinders, exhibit a particularly interesting phenomenon when placed normal to a fluid flow; vortices or eddies (regions of circulating fluid) are shed from the object, regularly and alternately from opposite sides, as shown in Fig. 8.10. The resulting flow downstream is often referred to as a *Kármán vortex street* named after Theodor von Kármán (1881–1963). The vortices are shed in the Reynolds number range $40 < \text{Re} < 10\,000$, and are accompanied by turbulence above $\text{Re} = 300$. Photographs of low- and high-Reynolds-number vortex shedding are presented in Fig. 8.11.

Dimensional analysis may be applied to find an expression for the shedding frequency. For high-Reynolds-number flows, that is, flows with insignificant viscous forces, the shedding frequency f, in hertz, depends only on the velocity and diameter. Thus $f = f(V, D)$. Using dimensional analysis we can show that

> **KEY CONCEPT** *Vortices are shed regularly and alternately from opposite sides from circular cylinders.*

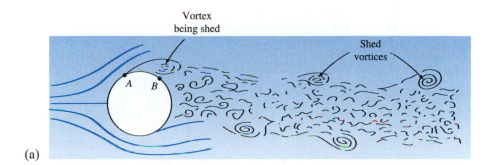

(a)

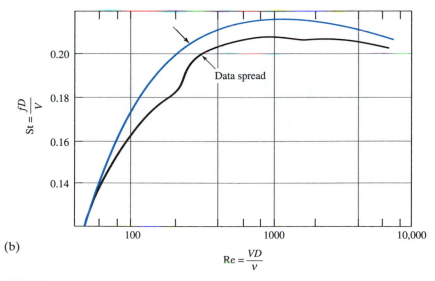

(b)

FIGURE 8.10 Vortex shedding from a cylinder: (a) vortex shedding; (b) Strouhal number versus Reynolds number. (From *NACA Rep. 1191*, by A. Roshko, 1954.)

fD/V = const. The shedding frequency, expressed as a dimensionless quantity, is expressed as the Strouhal number,

$$\text{St} = \frac{fD}{V} \tag{8.3.2}$$

From the experimental results of Fig. 8.10, we observe that the Strouhal number is essentially constant (0.21) over the range $300 < \text{Re} < 10\,000$; hence, the frequency is directly proportional to the velocity over this relatively large Reynolds number range.

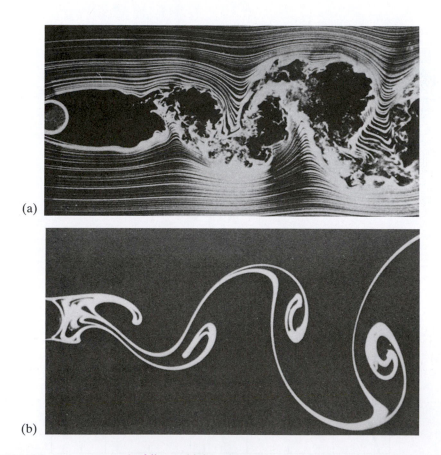

(a)

(b)

FIGURE 8.11   Vortex shedding at high and low Reynolds numbers: (a) Re = 10 000 (photograph by Thomas Corke and Hassan Nagib); (b) Re = 140 (photograph by Sadatoshi Taneda).

Example 8.3

The velocity of a slow-moving, 30°C air stream is to be measured using a cylinder and a pressure tap located between points A and B on the cylinder in Fig. 8.10. The velocity range is expected to be $0.1 < V < 1$ m/s. What size cylinder should be selected and what frequency would be observed by the pressure-measuring device for $V = 1$ m/s?

Solution

The Reynolds number should be in the vortex shedding range, say 4000. For the maximum velocity the diameter would be found as follows:

$$4000 = \frac{VD}{\nu}$$

$$= \frac{1.0 \times D}{1.6 \times 10^{-5}}$$

$$\therefore D = 0.064 \text{ m} \qquad \text{Select } D = 6 \text{ cm}$$

At $V = 0.1$ m/s the Reynolds number is $0.1 \times 0.06/1.6 \times 10^{-5} = 375$. Vortex shedding would occur, so this is acceptable. The expected vortex shedding frequency at $V = 1.0$ m/s is found using a Strouhal number from Fig. 8.10 of 0.21. Hence

$$0.21 = \frac{fD}{V}$$

$$= \frac{f \times 0.06}{1.0}$$

$$\therefore f = 3.5 \text{ hertz}$$

The engineer or architect must be very careful when designing structures, such as towers and bridges, that shed vortices. When a vortex is shed, a small force is applied to the structure; if the frequency of shedding is close to the natural frequency[3] (or one of the harmonics) of the structure, the phenomenon of resonance may occur in which the response to the applied force is multiplied by

KEY CONCEPT *If the phenomenon of resonance occurs, the response to an applied force is multiplied by a large factor.*

---

[3] The natural frequency is that frequency with which a structure vibrates when given a "knock."

a large factor. For example, when resonance occurs on a television tower, the deflection of the tower due to the applied force may become so large that the supporting cables fail, leading to collapse of the structure. This has occurred many times, leading to severe damage and numerous deaths and injuries. The collapse of the Tacoma Narrows suspension bridge is undoubtedly the most spectacular failure due to vortex shedding. "Galloping" power lines, in which a power line alternates between the usual catenary and an inverted catenary, is another example that may lead to significant damage; this may occur when a power line has iced up, providing a much larger cross-sectional area for the wind.

8.3.3   Streamlining

If the flow is to remain attached to the surface of a blunt object, such as a cylinder or a sphere, it must move into regions of higher and higher pressure as it progresses to the rear stagnation point. At sufficiently high Reynolds numbers (Re > 10) the slow-moving boundary layer flow near the surface is unable to make its way into the high-pressure region near the rear stagnation point, so it separates from the object. **Streamlining** reduces the high pressure at the rear of the object so that the slow-moving flow near the surface is able to negotiate its way into a slightly higher-pressure region. The fluid may not be able to make it all the way to the trailing edge of the streamlined object, but the separation region will be reduced to only a small percentage of the initial separated region on the blunt object. The included angle at the trailing edge must not be greater than about 20° or the separation region will be too large and the effect of streamlining will be negated. Drag coefficients for streamlined cylinders and spheres are displayed in Fig. 8.9.

Streamlining: *Reducing the high pressure at the rear of the object allowing slow-moving surface flow to move rearward.*

> **KEY CONCEPT** *The angle at the trailing edge must not be greater than about 20° for streamlining to be effective.*

When a body is streamlined, the surface area is increased substantially. This eliminates the majority of the pressure drag but increases the shear drag on the surface. To minimize drag, the idea is to minimize the sum of the pressure drag and the shear drag. Consequently, the streamlined body cannot be so long that the shear drag is larger than the pressure drag plus the shear drag for a shorter body. An optimization procedure is required. Such a procedure would lead to a thickness-to-chord length ratio of about 0.25 for a strut.

Obviously, for a low-Reynolds-number flow (Re < 10) the drag is due primarily to shear drag and thus streamlining is unnecessary; it would undoubtedly lead to increased drag since the surface area would be increased.

Finally, it should be pointed out that another advantage of streamlining is that the periodic shedding of vortices is usually eliminated. The vibrations produced by vortex shedding are often undesirable, so streamlining not only decreases drag but can eliminate the vibrations.

Example 8.4

A strut on a stunt plane traveling at 60 m/s is 4 cm in diameter and 24 cm long. Calculate the drag force acting on the strut as a circular cylinder and as a streamlined strut as shown in Fig. E8.4. Would you expect vortex shedding from the circular cylinder?

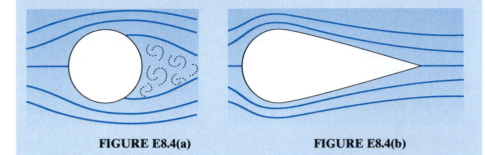

FIGURE E8.4(a)   FIGURE E8.4(b)

Solution

The Reynolds number associated with the cylinder and the streamlined strut is, assuming that $T = 20°C$,

$$Re = \frac{VD}{v}$$

$$= \frac{60 \times 0.04}{1.5 \times 10^{-5}} = 1.6 \times 10^5$$

Assuming a smooth surface as in (a), the drag coefficient is $C_D = 1.2$ from Fig. 8.9. The drag force is then

$$F_D = C_D \times \tfrac{1}{2}\rho V^2 A$$

$$= 1.2 \times \tfrac{1}{2} \times 1.20 \times 60^2 \times (0.24 \times 0.04) = 24.9 \text{ N}$$

For the streamlined strut of (b), Fig. 8.9 yields $C_D = 0.04$. The drag force is

$$F_D = C_D \times \tfrac{1}{2}\rho V^2 A$$

$$= 0.04 \times \tfrac{1}{2} \times 1.20 \times 60^2 \times (0.24 \times 0.04) = 0.82 \text{ N}$$

This is a reduction of 97% in the drag, a rather substantial percentage.

Vortex shedding is not to be expected on the circular cylinder; the Reynolds number is too high.

8.3.4 Cavitation

Cavitation: *A change of phase from liquid to vapor which occurs whenever the local pressure is less than the vapor pressure.*

Cavitation is a very rapid change of phase from liquid to vapor which occurs in a liquid whenever the local pressure is equal to or less than the vapor pressure. The first appearance of cavitation is at the position of lowest pressure in the field of flow. Four types of cavitation have been identified:

1. *Traveling cavitation*, which exists when vapor bubbles or cavities are formed, are swept downstream, and collapse.
2. *Fixed cavitation*, which exists when a fixed cavity of vapor exists as a separated region. The separated region may reattach to the body or the separated region may enclose the rear of the body and be closed by the main flow, in which case it is referred to as *supercavitation*.
3. *Vortex cavitation*, which is found in the high-velocity, and thus low-pressure, core of a vortex, often observed in the tip vortex leaving a propeller.
4. *Vibratory cavitation*, which may exist when a pressure wave moves in a liquid. A pressure wave consists of a pressure pulse, which has a high pressure followed by a low pressure. The low-pressure part of the wave (or vibration) can result in cavitation.

The first type of cavitation, in which vapor bubbles are formed and collapse, is associated with potential damage. The instantaneous pressures resulting from the collapse can be extremely high (perhaps 1400 MPa) and may cause damage to stainless steel components as happens on the propellers of ships.

KEY CONCEPT *The high instantaneous pressures may cause damage to stainless steel components.*

Cavitation occurs whenever the *cavitation number* σ, defined by

$$\sigma = \frac{p_\infty - p_v}{\frac{1}{2}\rho V^2} \tag{8.3.3}$$

is less than the critical cavitation number σ_{crit}, which depends on the geometry of the body and the Reynolds number. Here p_∞ is the absolute pressure in the undisturbed free stream and p_v is the vapor pressure. As σ decreases below σ_{crit}, the cavitation increases in intensity, moving from traveling cavitation to fixed cavitation to supercavitation.

TABLE 8.3 Drag Coefficients for Zero Cavitation Number for Blunt Objects

Two-dimensional body				Axisymmetric body		
Geometry		θ	$C_D(0)$	Geometry	θ	$C_D(0)$
Flat plate		—	0.88	Disk	—	0.8
Circular cylinder		—	0.50	Sphere	—	0.30
Wedge		120	0.74	Cone	120	0.64
		90	0.64		90	0.52
		60	0.49		60	0.38
		30	0.28		30	0.20

The drag coefficient of a body is dependent on the cavitation number and for small cavitation numbers is given by

$$C_D(\sigma) = C_D(0)(1 + \sigma) \tag{8.3.4}$$

where some values of $C_D(0)$ for common shapes are listed in Table 8.3 for $Re \simeq 10^5$.

The hydrofoil, an airfoil-type body that is used to lift a vessel out of the water, is a shape that is invariably associated with cavitation. Drag and lift coefficients and critical cavitation numbers are given in Table 8.4 for a typical hydrofoil with $10^5 < Re < 10^6$, where the Reynolds number is based on the chord length and the area used with C_D and C_L is the chord times the length.

TABLE 8.4 Drag and Lift Coefficients and Critical Cavitation Number for a Typical Hydrofoil

Angle (°)	Lift coefficient C_L	Drag coefficient C_D	Critical cavitation number σ_{crit}
−2	0.2	0.014	0.5
0	0.4	0.014	0.6
2	0.6	0.015	0.7
4	0.8	0.018	0.8
6	0.95	0.022	1.2
8	1.10	0.03	1.8
10	1.22	0.04	2.5

Example 8.5

A hydrofoil is to operate 50 cm below the surface of 15°C water at an angle of attack of 8° and travel at 15 m/s. If its chord length is 60 cm and it is 1.8 m long, calculate its lift and drag. Is cavitation present?

Solution
The absolute pressure p_∞ is

$$p_\infty = \gamma h + p_{atm}$$
$$= 9810 \times \left(\frac{50}{100}\right) + 101 \times 10^3 = 105.9 \times 10^3 \text{ Pa} = 105.9 \text{ kPa absolute}$$

The vapor pressure is $p_v = 1.76$ kPa, so

$$\sigma = \frac{p_\infty - p_v}{\frac{1}{2}\rho V^2}$$

$$= \frac{(105.9 - 1.76) \times 10^3}{\frac{1}{2} \times 1000 \times 15^2} = 0.925$$

(continued)

Answering the last question first, we see that this is less than 1.8, hence cavitation exists.

The lift force is, finding C_L in Table 8.4,

$$F_L = C_L \times \tfrac{1}{2}\rho V^2 A$$

$$= 1\,1 \times \tfrac{1}{2} \times 1000 \times 15^2 \times (0.6 \times 1.8) = 133650 \text{ N}$$

The drag force is, finding C_D in Table 8.4,

$$F_D = C_D \times \tfrac{1}{2}\rho V^2 A$$

$$= 0.03 \times \tfrac{1}{2} \times 1000 \times 15^2 \times (0.6 \times 1.8) = 3645 \text{ N}$$

8.3.5   Added Mass

The previous sections in this chapter have dealt with bodies moving at constant velocity. In this section we consider bodies accelerating from rest in a fluid. When a body accelerates, we say that an unbalanced force acts on the body; not only does the body accelerate but so does some of the fluid surrounding the body. The acceleration of the surrounding fluid requires an added force over and above the force required to accelerate only the body. A relatively simple way of accounting for the fluid mass being accelerated is to add a mass, called the *added mass* m_a, to the mass of the body. Summing forces in the direction of motion for a symmetrical body moving in the direction of its axis of symmetry we have, for horizontal motion,

$$F - F_D = (m + m_a)\frac{dV_B}{dt} \tag{8.3.5}$$

where V_B is the velocity of the body and F_D is the drag force. For an initial acceleration from rest F_D would be zero.

The added mass is related to the mass of the fluid m_f displaced by the body by the relationship

$$m_a = km_f \tag{8.3.6}$$

where k is an added mass coefficient. For a sphere $k = 0.5$; for an ellipsoid with major axis twice the minor axis and moving in the direction of the major axis, $k = 0.2$; for a long cylinder moving normal to its axis, $k = 1.0$. These values were calculated for inviscid flows and thus are applicable for motions starting from rest so that the viscous forces are negligible.

For dense bodies accelerating in the atmosphere the added mass is negligibly small and is typically ignored. Masses accelerating from rest in a liquid are more influenced by the added mass and it must usually be accounted for. Offshore structures that are subjected to oscillatory wave motions experience periodic forces the determination of which must include the added mass effect.

Example 8.6

A sphere with specific gravity 2.5 is released from rest in water. Calculate its initial acceleration. What is the percentage error if the added mass is ignored?

Solution

The summation of forces in the vertical direction, with zero drag, is

$$W - B = (m + m_a) \frac{dV_B}{dt}$$

where B is the buoyant force. Substituting in the appropriate quantities gives, letting V = sphere volume,

$$S\gamma_{water}V - \gamma_{water}V = (\rho_{water}SV + 0.5\rho_{water}V) \frac{dV_B}{dt}$$

This gives

$$g(S - 1) = (S + 0.5) \frac{dV_B}{dt}$$

Hence

$$\frac{dV_B}{dt} = \frac{g(S - 1)}{S + 0.5} = \frac{9.8(2.5 - 1)}{2.5 + 0.5} = 4.90 \text{ m/s}^2$$

If we ignored the added mass, the acceleration would be

$$\frac{dV_B}{dt} = \frac{g(S - 1)}{S} = \frac{9.8(2.5 - 1)}{2.5} = 5.88 \text{ m/s}^2$$

This is an error of 20%.

8.4 LIFT AND DRAG ON AIRFOILS

Separation occurs on a blunt body, such as a cylinder, due to the strong adverse pressure gradient in the boundary layer on the rear of the body. An airfoil is a streamlined body designed to reduce the adverse pressure gradient so that separation will not occur, usually with a small angle of attack, as shown in Fig. 8.12. Without separation the drag is due primarily to the wall shear stress, which results from viscous effects in the boundary layer.

The boundary layer on an airfoil is very thin, and thus it can be ignored when solving for the flow field (the streamline pattern and the pressure distribution) surrounding the airfoil. Since the boundary layer is so thin, the pressure on the wall is not significantly influenced by the boundary layer's existence. Hence the lift on an airfoil can be approximated by integrating the pressure distribution as given by the inviscid flow solution on the wall. In the next section we will demonstrate how this is done; in this section we simply give empirical results.

The drag on an airfoil can be predicted by solving the boundary layer equations (simplified Navier–Stokes equations) for the shear stress on the wall, and performing the appropriate integration. The inviscid flow field must be known

KEY CONCEPT *The lift on an airfoil can be approximated by integrating the inviscid flow solution.*

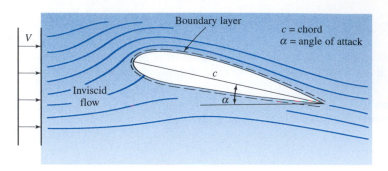

FIGURE 8.12 Flow around an airfoil at an angle of attack.

before the boundary layer equations can be solved since the pressure gradient on the wall and the inviscid flow velocity at the wall are needed as inputs in solving for the boundary layer flow. Boundary layer calculations will be presented in Section 8.6; in this section we present the empirical results for the drag.

The drag coefficient presented may seem quite low compared with the coefficients of the preceding section. For airfoils a much larger projected area is used, namely, the plan area, which is the chord c (see Fig. 8.12) times the length L of the airfoil. Thus the drag and lift coefficients are defined as

$$
C_D = \frac{F_D}{\frac{1}{2}\rho V^2 cL} \qquad\qquad C_L = \frac{F_L}{\frac{1}{2}\rho V^2 cL} \qquad\qquad (8.4.1)
$$

For a typical airfoil the lift and drag coefficients are given in Fig. 8.13. For a specially designed airfoil the drag coefficient may be as low as 0.0035, but the maximum lift coefficient is about 1.5. The design lift coefficient (cruise condition) is about 0.3, which is near the minimum drag coefficient condition. This corresponds to an angle of attack of about 2°, far from the stall condition of about 16°.

Conventional airfoils are not symmetric; hence there is a positive lift coefficient at zero angle of attack. The lift is directly proportional to the angle of attack but deviates from the straight-line function just before stall. The drag coefficient also increases linearly up to an angle of attack of about five degrees for a conventional airfoil; then it increases in a nonlinear relation with angle of attack.

To take off and land at relatively low speeds, it is necessary to attain significantly higher lift coefficients than the maximum of 1.7 of Fig. 8.13. Or if a relatively low lift coefficient is to be accepted, the area $c \times L$ must be enlarged. Both are actually accomplished. Flaps are moved out from a section of each airfoil, resulting in an increased chord, and the angle of attack of the flap is also increased. Slots are used to move high-pressure air from the underside into the relatively low momentum boundary layer flow on the top side, as shown in Fig. 8.14; this prevents separation from the flap, thereby maintaining high lift. The lift coefficient can reach 2.5 with a single-slotted flap and 3.2 with a double-slotted flap. On some modern aircraft there may be three flaps in series with three slots along with a nose flap, to ensure that the boundary layer does not separate from the upper surface of the airfoil.

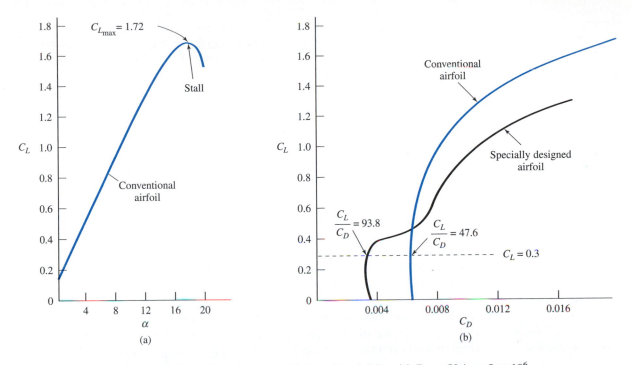

FIGURE 8.13 Lift and drag coefficients for airfoils with $Re = Vc/\nu \approx 9 \times 10^6$.

The total lift on an aircraft is supplied primarily by the airfoil. The effective length of the airfoil when calculating the lift is taken to be the tip-to-tip distance, the **wingspan**, since the fuselage acts to produce the lift of the midsection of the airfoil. The drag calculation must include the shear acting on the airfoil, the fuselage, and the tail section.

The drag coefficient is essentially constant on airfoils up to a Mach number of about 0.75. Then a sudden rise occurs until the Mach number reaches unity; see Fig. 8.15. The drag coefficient then slowly falls. Obviously, the condition of M = 1 is to be avoided. Thus aircraft either fly at M < 0.75 or M > 1.5 or so, to avoid the high drag coefficients near M = 1. Near M = 1 there are also regions of flow that oscillate from subsonic to supersonic. Such oscillations create forces that are best avoided.

Wingspan: *The effective length of an airfoil is the tip-to-tip distance.*

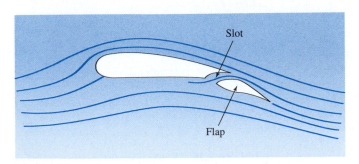

FIGURE 8.14 Flapped airfoil with slot for separation control.

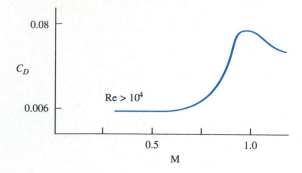

FIGURE 8.15 Drag coefficient as a function of Mach number (speed) for a typical unswept airfoil.

It is useful to use swept-back airfoils since it is the component of velocity normal to the leading edge of the airfoil that must be used in calculating the Mach number in Fig. 8.15. Cruise speeds at M = 0.8 with swept-back wings are not uncommon. It should be pointed out, though, that fuel consumption depends on power required, and power is drag force times velocity; hence fuel consumption depends on the velocity cubed since the drag force depends on the velocity squared, assuming all other parameters are constant. A lower velocity results in a fuel savings even though the engines must operate longer when traveling a fixed distance.

KEY CONCEPT *The component of velocity normal to the leading edge is used in calculating the Mach number.*

A final comment on airfoils regards the influence of a finite airfoil. To understand flow around a finite airfoil we make reference to a vortex. Fluid particles rotate about the center of a vortex as they travel along in the flow field. There is a high pressure on the bottom and a low pressure on the top side of the airfoil sketched in Fig. 8.16 with a model airfoil shown in Fig. 8.17. This results in a movement of air from the bottom side to the top side around the ends of the airfoil, as shown, resulting in a strong tip vortex. Distributed vortices are also shed all along the airfoil and they all collect into two large trailing vortices. On a clear day, the two trailing vortices may show up as visible white streaks of water vapor behind a high-flying aircraft. The trailing vortices persist for a considerable distance (perhaps 15 km) behind a large aircraft, and their 90-m/s velocities can

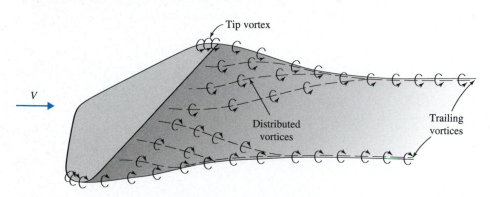

FIGURE 8.16 Trailing vortex.

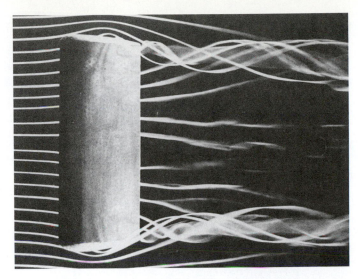

FIGURE 8.17   Trailing vortices from a rectangular wing. The flow remains attached over the entire wing surface. The centers of the vortex cores leave the trailing edge at the tips. The model is tested in a smoke tunnel at Reynolds number 100 000. (Courtesy of The Parabolic Press, Stanford, California. Reprinted with permission.)

cause a small aircraft to flip over. Also, the trailing vortices induce a downwash, a downward velocity component, that must be accounted for in the design of the aircraft. The tail section is located up high to minimize the effect of this downwash.

KEY CONCEPT   *Trailing vortices behind a large aircraft can cause a small aircraft to lose control.*

Example 8.7

A light airplane weighs 10 000 N, its wingspan measures 12 m, its chord measures 1.8 m, and a payload of 2000 N is anticipated. Predict (a) the takeoff speed if an angle of attack of 8° is desired, (b) the stall speed of the conventional airfoil, and (c) the power required by the airfoil during cruise at 50 m/s.

Solution
(a) The lift on an airplane is equal to its weight. With the payload the total weight is 12 000 N; hence the lift-coefficient equation (8.4.1) gives the following:

$$V = \left(\frac{F_L}{\frac{1}{2}\rho C_L cL} \right)^{1/2}$$

$$= \left(\frac{12\ 000}{\frac{1}{2} \times 1.20 \times 1.0 \times 1.8 \times 12} \right)^{1/2} = 30.4 \text{ m/s}$$

where we have used $C_L = 1.0$ at $\alpha = 8°$ from Fig. 8.13, and $\rho = 1.2 \text{ kg/m}^3$ since light aircraft take off at ground level.

(continued)

(b) The stall speed is found using a maximum lift coefficient of 1.72 from Fig. 8.13.

$$V_{stall} = \left(\frac{F_L}{\frac{1}{2}\rho C_L cL} \right)^{1/2}$$

$$= \left(\frac{12\,000}{\frac{1}{2} \times 1.20 \times 1.72 \times 1.8 \times 12} \right)^{1/2} = 23.2 \text{ m/s}$$

(c) The power demanded by the airfoil during cruise is equal to the drag force times the velocity. The design lift coefficient is assumed equal to 0.3 and thus from Fig. 8.13, assuming a conventional airfoil, $C_D = 0.0063$. This gives

$$F_D = \frac{1}{2}\rho V^2 cLC_D$$

$$= \frac{1}{2} \times 1.20 \times 50^2 \times 1.8 \times 12 \times 0.0063 = 204 \text{ N}$$

The power is then

$$\text{power} = F_D \times V$$

$$= 204 \times 50 = 10\,200 \text{ W} \qquad \text{or } 13.7 \text{ hp}$$

The total power would be significantly larger since the drag on the fuselage and tail section must be included.

8.5 POTENTIAL FLOW THEORY

8.5.1 Basic Flow Equations

Inviscid flow exists outside the boundary layer and the wake in high-Reynolds-number flow around bodies. For an airfoil the boundary layer is quite thin, and the inviscid flow provides a good approximation to the actual flow; it is used to predict the pressure distribution on the surface, thereby giving a good estimate of the lift. It will also provide us with the velocity to be used as a boundary condition in the boundary layer solution of Section 8.6; from such a solution we can estimate the drag and predict possible points of separation. Consequently, the inviscid flow solution is very important in our study of external flows. Obviously, if we use the empirical results of previous sections, the details of the inviscid flow solution are unnecessary. If, on the other hand, we wish to predict quantities such as the lift and drag, and locate possible points of separation, by using the required differential equations, the inviscid flow solution is essential.

KEY CONCEPT *The inviscid flow solution is very important in our study of external flows.*

Consider a velocity field that is given by the gradient of a scalar function ϕ, that is,

$$\mathbf{V} = \nabla\phi \tag{8.5.1}$$

Potential flow: *A flow with zero vorticity.*

in which ϕ is called a *velocity potential function*. Such a velocity field is called a **potential flow** (or an *irrotational flow*) and possesses the property that the vorticity $\boldsymbol{\omega}$, which is the curl of the velocity vector, is zero; this is expressed by

$$\boldsymbol{\omega} = \boldsymbol{\nabla} \times \mathbf{V} = 0 \tag{8.5.2}$$

The fact that the vorticity is zero for a potential flow can easily be shown by letting $\mathbf{V} = \boldsymbol{\nabla}\phi$ and expanding Eq. 8.5.2 in rectangular coordinates. A fluid particle that does not possess vorticity (i.e., is not rotating) cannot obtain vorticity without the action of viscosity; normal pressure forces and body forces that act through the mass center cannot impart rotation to a fluid particle. This result is also observed from the *vorticity equation*, which is obtained by taking the curl of the Navier–Stokes equation (5.3.17) (see Section 5.3.4 if the details are of interest); the vorticity equation is

KEY CONCEPT   *The vorticity is zero for a potential flow.*

$$\frac{D\boldsymbol{\omega}}{Dt} = (\boldsymbol{\omega} \cdot \boldsymbol{\nabla})\,\mathbf{V} + \nu\nabla^2\boldsymbol{\omega} \tag{8.5.3}$$

Note that if $\boldsymbol{\omega} = 0$, the only way that $D\boldsymbol{\omega}/Dt$ can be nonzero is for viscous effects to act through the last term. If viscous effects are absent, as in an inviscid flow, then $D\boldsymbol{\omega}/Dt = 0$ and vorticity must remain zero.

With the velocity given by the gradient of a scalar function, the differential continuity equation, for an incompressible flow, gives

$$\boldsymbol{\nabla} \cdot \boldsymbol{\nabla}\phi = \nabla^2\phi = 0 \tag{8.5.4}$$

and is known as *Laplace's equation*, named after Pierre S. Laplace (1749–1827). In rectangular coordinates this is

$$\frac{\partial^2\phi}{\partial x^2} + \frac{\partial^2\phi}{\partial y^2} + \frac{\partial^2\phi}{\partial z^2} = 0 \tag{8.5.5}$$

With the appropriate boundary conditions this equation could be solved. However, three-dimensional problems are quite difficult, so we focus our attention on plane flows in which the velocity components u and v depend on x and y. This is acceptable for two-dimensional airfoils and other plane flows, such as flow around cylinders.

Before we attempt a solution to Eq. 8.5.5, let us define another scalar function that will aid us in our study of plane fluid flows. The continuity equation

$$\frac{\partial u}{\partial x} + \frac{\partial v}{\partial y} = 0 \tag{8.5.6}$$

motivates the definition. If we let

$$u = \frac{\partial\psi}{\partial y} \qquad \text{and} \qquad v = -\frac{\partial\psi}{\partial x} \tag{8.5.7}$$

we observe that the continuity equation is automatically satisfied; the scalar function $\psi(x, y)$ is called a **stream function**. By using the mathematical description of a streamline, $\mathbf{V} \times d\mathbf{r} = 0$, we see that, for a plane flow, $u\,dy - v\,dx = 0$. Substituting from Eqs. 8.5.7, this becomes

$$\frac{\partial \psi}{\partial y}\,dy + \frac{\partial \psi}{\partial x}\,dx = 0 \tag{8.5.8}$$

This is, of course, $d\psi = 0$. Thus ψ is constant along a streamline. Example 8.9 will show that the difference $(\psi_2 - \psi_1)$ between any two streamlines is equal to the flow rate per unit depth between the two streamlines.

The vorticity vector for a plane flow has only a z-component since $w = 0$ and there is no variation with z. The vorticity is

$$\omega_z = (\mathbf{\nabla} \times \mathbf{V})_z = \frac{\partial v}{\partial x} - \frac{\partial u}{\partial y} \tag{8.5.9}$$

For our potential flow we demand that the vorticity be zero, so Eq. 8.5.9 gives, using Eqs. 8.5.7,

$$\frac{\partial^2 \psi}{\partial x^2} + \frac{\partial^2 \psi}{\partial y^2} = 0 \tag{8.5.10}$$

So we see that both the stream function ψ and the potential function ϕ satisfy Laplace's equation for this plane flow.

Rather than attempt a solution of Laplace's equation for a particular flow of interest, we will use a different technique; we will identify some relatively simple functions that satisfy Laplace's equation and then superimpose these simple functions to create flows of interest. It is possible to generate any plane flow desired using this technique. Hence we will not actually solve Laplace's equation.

Before we present some simple functions, some additional observations concerning ϕ and ψ will be made. Using Eqs. 8.5.1 and 8.5.7, we see that

$$u = \frac{\partial \phi}{\partial x} = \frac{\partial \psi}{\partial y} \qquad v = \frac{\partial \phi}{\partial y} = -\frac{\partial \psi}{\partial x} \tag{8.5.11}$$

These relationships between the derivatives of ϕ and ψ are the famous *Cauchy–Riemann equations*, named after Augustin L. Cauchy (1789–1857) and Georg F. Riemann (1826–1866), from the theory of complex variables. The functions ϕ and ψ are harmonic functions since they satisfy Laplace's equation and form an *analytic function* $(\phi + i\psi)$ called the *complex velocity potential*. The theory of complex variables with all its powerful theorems is thus applicable to this restricted class of problems: namely, plane, incompressible, potential flows. An example will show that the streamlines and constant-potential lines intersect each other at right angles.

Example 8.8

A scalar potential function is given by $\phi = A \tan^{-1}(y/x)$. Find the stream function $\psi(x, y)$.

Solution

The relationship between ϕ and ψ is given by Eq. 8.5.11. We have

$$\frac{\partial \psi}{\partial y} = \frac{\partial \phi}{\partial x} = \frac{\partial}{\partial x}\left(A \tan^{-1}\frac{y}{x}\right) = -\frac{Ay}{x^2 + y^2}$$

This can be integrated as follows:

$$\int \frac{\partial \psi}{\partial y}\, dy = -A \int \frac{y}{x^2 + y^2}\, dy$$

$$\therefore \psi = -\frac{A}{2}\ln(x^2 + y^2) + f(x)$$

A function of x rather than a constant must be added since partial derivatives are being used. Now, let us differentiate this expression with respect to x. There results

$$\frac{\partial \psi}{\partial x} = -A\frac{x}{x^2 + y^2} + \frac{df}{dx}$$

This must equal $(-\partial \phi/\partial y)$ as demanded by Eq. 8.5.11; that is,

$$-\frac{Ax}{x^2 + y^2} + \frac{df}{dx} = -\frac{Ax}{x^2 + y^2}$$

Thus

$$\frac{df}{dx} = 0 \qquad \text{or} \qquad f = C$$

Since ϕ and ψ are used to find the velocity components by differentiation, the constant is of no concern; it is usually set equal to zero. Hence

$$\psi = -\frac{A}{2}\ln(x^2 + y^2)$$

Example 8.9

Show that the difference in the stream function between any two streamlines is equal to the flow rate per unit depth between the two streamlines. The flow rate per unit depth is denoted by q.

Solution

Consider the flow between two streamlines infinitesimally close, as shown in Fig. E8.9a. The flow rate per unit depth through the elemental area is, referring to Fig. E8.9b,

(continued)

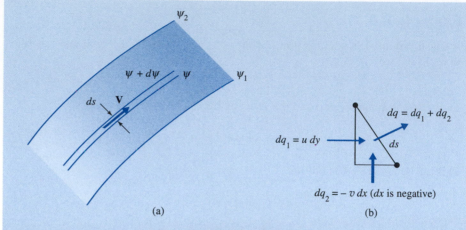

FIGURE E8.9

$$dq = dq_1 + dq_2$$

$$= u\,dy - v\,dx$$

$$= \frac{\partial \psi}{\partial y}\,dy + \frac{\partial \psi}{\partial x}\,dx = d\psi$$

If this is integrated between two streamlines with $\psi = \psi_1$ and $\psi = \psi_2$, there results

$$q = \psi_2 - \psi_1$$

thereby proving the statement of the example.

Example 8.10

Show that the streamlines and equipotential lines of a plane, incompressible, potential flow intersect one another at right angles.

Solution

If, at a point, the slope of a streamline is the negative reciprocal of the slope of an equipotential line, the two lines are perpendicular to each other. The slope of a stream-line (see Fig. E8.10a) is given by

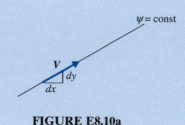

FIGURE E8.10a

$$\frac{dy}{dx}\bigg|_{\psi = \text{const.}} = \frac{v}{u}$$

The slope of an equipotential line is found from

$$d\phi = \frac{\partial \phi}{\partial x} dx + \frac{\partial \phi}{\partial y} dy = 0$$

since ϕ = constant along an equipotential line. This gives

$$\frac{dy}{dx}\bigg|_{\phi = \text{const}} = -\frac{\partial \phi / \partial x}{\partial \phi / \partial y} = -\frac{u}{v}$$

Hence we see that the slope of the streamline is the negative reciprocal of the slope of the equipotential line, that is,

$$\frac{dy}{dx}\bigg|_{\phi = \text{const}} = -\left(\frac{dy}{dx}\bigg|_{\psi = \text{const.}}\right)^{-1}$$

Thus, whenever the streamlines intersect the equipotential lines, they must do so at right angles. A sketch of the streamlines and equipotential lines (equally spaced at large distances from the body), known as a *flow net*, is shown in Fig. E8.10b for flow over a weir. Such a carefully constructed sketch can be used to approximate the velocities at points of interest in an inviscid flow. Pressures can then be found using Bernoulli's equation.

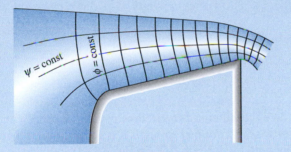

FIGURE E8.10b

8.5.2   Simple Solutions

Now, let us identify some relatively simple functions that satisfy Laplace's equation. However, before we do, it is often more convenient to use polar coordinates. Laplace's equation, the continuity equation, and the velocity components take the following forms:

$$\nabla^2 \psi = \frac{1}{r} \frac{\partial}{\partial r}\left(r \frac{\partial \psi}{\partial r}\right) + \frac{1}{r^2} \frac{\partial^2 \psi}{\partial \theta^2} = 0 \qquad (8.5.12)$$

$$\frac{1}{r} \frac{\partial}{\partial r}(r v_r) + \frac{1}{r} \frac{\partial v_\theta}{\partial \theta} = 0 \qquad (8.5.13)$$

$$v_r = \frac{1}{r}\frac{\partial \psi}{\partial \theta} = \frac{\partial \phi}{\partial r} \qquad v_\theta = -\frac{\partial \psi}{\partial r} = \frac{1}{r}\frac{\partial \phi}{\partial \theta} \qquad (8.5.14)$$

We will introduce the names of four simple flows, sketched in Fig. 8.18, and their corresponding functions. Each function obviously satisfies Laplace's equation. The names and functions are:

Uniform flow: $\psi = U_\infty y$ $\phi = U_\infty x$ (8.5.15)

Line source: $\psi = \dfrac{q}{2\pi}\theta$ $\phi = \dfrac{q}{2\pi}\ln r$ (8.5.16)

Irrotational vortex: $\psi = \dfrac{\Gamma}{2\pi}\ln r$ $\phi = \dfrac{\Gamma}{2\pi}\theta$ (8.5.17)

Doublet: $\psi = -\dfrac{\mu \sin \theta}{r}$ $\phi = -\dfrac{\mu \cos \theta}{r}$ (8.5.18)

The uniform flow velocity U_∞ is assumed to be in the x-direction; if a y-component is desired an appropriate term is simply added. The *source strength q* is the volume rate of flow per unit depth issuing from the source; a negative value would create a *sink*. The *vortex strength* Γ is the *circulation* about the origin, defined by

$$\Gamma = \oint_L \mathbf{V} \cdot d\mathbf{s} \qquad (8.5.19)$$

where L must be a closed curve (a circle is usually used) around the origin and clockwise is positive. The *doublet strength* μ is for a doublet oriented in the negative x-direction; note the large arrow (in Fig. 8.18d) showing the direction of the doublet. Doublets oriented in other directions are seldom of interest and are not considered here.

Of the four flows presented above, the doublet is rather mysterious; it can be visualized as a source and a sink of equal strength separated by a very small distance. Its usefulness is in the creation of certain other flows of interest. The irrotational vortex is encountered when water swirls down a drain or into the turbine of a hydropower dam, or more spectacularly, in a tornado.

The velocity components for the four simple flows are easily shown, using Eqs. 8.5.11 and 8.5.14 for rectangular and polar coordinates, to be as follows:

KEY CONCEPT *The doublet can be visualized as a source and sink of equal strength separated by a very small distance.*

Uniform flow: $u = U_\infty$ $v = 0$

$v_r = U_\infty \cos \theta$ $v_\theta = -U_\infty \sin \theta$ (8.5.20)

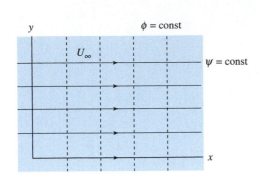

(a) Uniform flow in x-direction

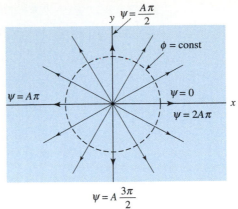

(b) Line source

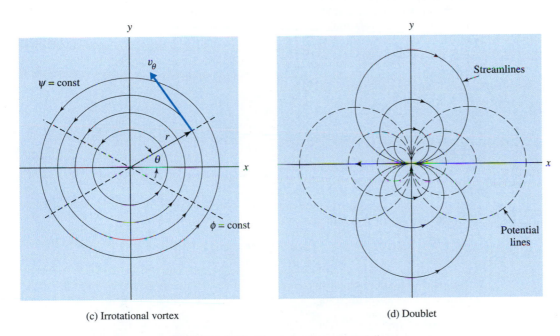

(c) Irrotational vortex

(d) Doublet

FIGURE 8.18   Four simple potential flows.

Line source: $\qquad v_r = \dfrac{q}{2\pi r} \qquad\qquad v_\theta = 0$ $\qquad$ (8.5.21)

$$u = \frac{q}{2\pi}\frac{x}{x^2 + y^2} \qquad v = \frac{q}{2\pi}\frac{y}{x^2 + y^2}$$

Irrotational vortex: $\quad v_r = 0 \qquad\qquad\qquad v_\theta = -\dfrac{\Gamma}{2\pi r}$

$$u = -\frac{\Gamma}{2\pi}\frac{y}{x^2 + y^2} \qquad v = \frac{\Gamma}{2\pi}\frac{x}{x^2 + y^2}$$

$$(8.5.22)$$

Doublet: $\qquad\qquad v_r = -\dfrac{\mu \cos\theta}{r^2} \qquad v_\theta = -\dfrac{\mu \sin\theta}{r^2}$

$$u = -\mu\frac{x^2 - y^2}{(x^2 + y^2)^2} \qquad v = -\mu\frac{2xy}{(x^2 + y^2)^2}$$

$$(8.5.23)$$

Example 8.11

The pressure far from an irrotational vortex (a simplified tornado) in the atmosphere is zero gage. If the velocity at $r = 20$ m is 20 m/s, estimate the velocity and the pressure at $r = 2$ m. (The irrotational vortex ceases to be a good model for a tornado when r is small. In the "eye" of the tornado the motion is approximated by rigid-body motion.)

Solution

For an irrotational vortex, we know that

$$v_\theta = -\frac{\Gamma}{2\pi r}$$

Hence

$$\Gamma = -2\pi r v_\theta$$
$$= -2\pi \times 20 \times 20 = -800\pi \text{ m}^2/\text{s}$$

The velocity at $r = 2$ m is then

$$v_\theta = -\frac{-800\pi}{2\pi \times 2} = 200 \text{ m/s}$$

Bernoulli's equation for this incompressible, inviscid, steady flow then gives the pressure as follows assuming a stagnant atmosphere away from the tornado:

$$\overset{0}{\cancel{p_\infty}} + \overset{0}{\cancel{\frac{U_\infty^2}{2}}}\rho = p + \frac{v_\theta^2}{2}\rho$$
$$\therefore p = -\tfrac{1}{2}\rho v_\theta^2$$
$$= -\tfrac{1}{2} \times 1.20 \times 200^2 = -24\,000 \text{ Pa}$$

The negative sign denotes a vacuum. It is this vacuum that causes roofs of buildings to blow off during a tornado.

8.5.3 Superposition

The simple flows presented in Section 8.5.2 are of particular interest because they can be superimposed with each other to form more complicated flows of engineering importance. In fact, the most complicated incompressible, plane flow can be constructed using these simple flows. For example, suppose that the flow around an airfoil with a slotted flap is desired. We could divide the surface of the airfoil into a relatively large number (say, 200) of panels, locate a source or a sink (or alternatively, a doublet) at the center of each panel, add a uniform flow and an irrotational vortex, and then by adjusting[4] the panel source strengths, the desired inviscid flow could be created. The development of the model and the computer routine necessary to perform the calculations are considered beyond the scope of this book.

> **KEY CONCEPT** *The most complicated incompressible, plane flow can be constructed using these simple flows.*

In this section we demonstrate superposition by creating flow around a circular cylinder with and without circulation. First, superimpose a uniform flow and a doublet; there results

$$\psi = U_\infty y - \frac{\mu \sin \theta}{r} \tag{8.5.24}$$

The velocity component v_r is (let $y = r \sin \theta$)

$$v_r = \frac{1}{r} \frac{\partial \psi}{\partial \theta}$$

$$= U_\infty \cos \theta - \frac{\mu}{r^2} \cos \theta \tag{8.5.25}$$

Let us ask the question: Is there a radius r_c for which $v_r = 0$? If we set $v_r = 0$ we find that

$$r_c = \sqrt{\frac{\mu}{U_\infty}} \tag{8.5.26}$$

At this radius v_r is identically zero for all θ and thus the circle $r = r_c$ must be a streamline.

The stagnation points are found by setting $v_\theta = 0$ on the circle $r = r_c$. There results

$$v_\theta = - \frac{\partial \psi}{\partial r}$$

$$= -U_\infty \sin \theta - \frac{\mu \sin \theta}{r_c^2} = -2U_\infty \sin \theta = 0 \tag{8.5.27}$$

---

[4] The source strengths are adjusted so that the normal component of velocity at the center of each panel is zero. The vortex strength is adjusted so that the rear stagnation point occurs at the trailing edge.

Thus we see that $v_\theta = 0$ at $\theta = 0°$ and $180°$. The flow is as sketched in Fig. 8.19a. We only have an interest in the flow external to the circular streamline $r = r_c$.

If the pressure distribution were desired on the cylinder, Bernoulli's equation could be used between the stagnation point where $V = 0$ and $p = p_0$ and some arbitrary point on the cylinder to yield

$$p_c = p_0 - \rho \frac{v_\theta^2}{2}$$
$$= p_0 - 2\rho U_\infty^2 \sin^2\theta \qquad (8.5.28)$$

This provides a symmetric pressure distribution that yields zero drag and zero lift. The zero-lift prediction is acceptable for a real flow, but the zero-drag result is obviously unacceptable. This may suggest that we ignore the inviscid flow solution; however, compared with the actual flow situation of Fig. 8.19b, the measured pressure distribution on the cylinder up to the point of separation is very

KEY CONCEPT *The pressure distribution up to the point of separation is very nearly the same as that predicted by potential flow.*

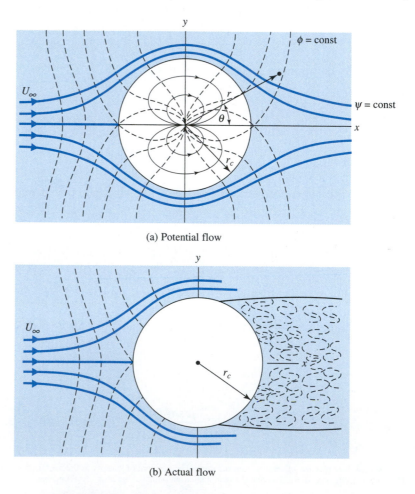

(a) Potential flow

(b) Actual flow

FIGURE 8.19 Flow around a circular cylinder.

nearly the same as that predicted by the potential flow solution. Hence the potential flow solution is quite useful to us, even for blunt bodies that experience a separated flow. At low Reynolds numbers, viscous effects are not confined to a thin boundary layer, so potential flow theory is not useful.

Let us now consider flow around a rotating cylinder. This is accomplished by adding an irrotational vortex to the doublet and uniform flow so that

$$\psi = U_\infty y - \frac{\mu \sin \theta}{r} + \frac{\Gamma}{2\pi} \ln r \qquad (8.5.29)$$

Since the vortex flow, consisting of circular streamlines, does not influence the velocity component v_r the cylinder $r = r_c$ remains unchanged. The stagnation points, however, are changed and are located by setting $v_\theta = 0$ on $r = r_c$; that is, letting $\mu = U_\infty r_c^2$,

$$v_\theta = -\frac{\partial \psi}{\partial r}$$

$$= -2U_\infty \sin \theta - \frac{\Gamma}{2\pi r_c} = 0 \qquad (8.5.30)$$

This gives the location of the stagnation points as shown in Fig. 8.20. In (a) the stagnation points are on the cylinder where $r = r_c$, but in (b) the circulation is sufficiently large that a single stagnation point is off the cylinder with $\theta = 270°$.

Bernoulli's equation gives the pressure distribution as

$$p_c = p_0 - \rho \frac{U_\infty^2}{2} \left(2 \sin \theta + \frac{\Gamma}{2\pi r_c U_\infty} \right)^2 \qquad (8.5.31)$$

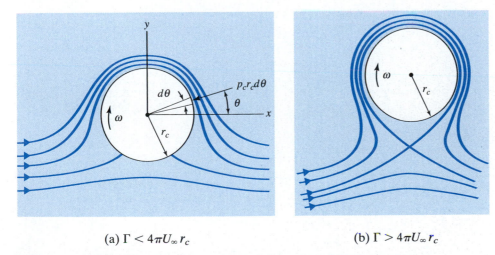

(a) $\Gamma < 4\pi U_\infty r_c$ (b) $\Gamma > 4\pi U_\infty r_c$

FIGURE 8.20 Flow around a circular cylinder with circulation.

KEY CONCEPT *The expression $\rho U_\infty \Gamma$ provides an excellent approximation to the lift for all cylinders.*

This can be integrated to give drag = 0 and the lift per unit length as

$$F_L = -\int_0^{2\pi} p_c \sin\theta \, r_c d\theta$$

$$= \rho U_\infty \Gamma \qquad (8.5.32)$$

This expression for the lift provides an excellent approximation to the lift for all cylinders, including the airfoil. Along with the zero-drag conclusion, it forms the *Kutta-Joukowsky theorem*.

Other superpositions of the simple flows are included in the problems.

Example 8.12

An 20-cm-diameter cylinder rotates clockwise at 1000 rpm in a 15°C-atmospheric airstream flowing at 4.5 m/s. Locate any stagnation points and find the minimum pressure on the cylinder.

Solution

The circulation is calculated (see Eq. 8.5.19) to be

$$\Gamma = \oint_L \mathbf{V} \cdot d\mathbf{s}$$

$$= 2\pi r_c^2 \omega = 2\pi \times \left(\frac{10}{100}\right)^2 \times \frac{1000 \times 2\pi}{60} = 6.58 \text{ m}^2/\text{s}$$

This is greater than $4\pi U_\infty r_c = 4\pi \times 4.5 \times \left(\frac{10}{100}\right) = 5.65 \text{ m}^2/\text{s}$; hence the stagnation point is off the cylinder (see Fig. 8.20b) at

$$\theta = 270° \qquad r_0 = -\frac{\Gamma}{4\pi U_\infty \sin 270°} = \frac{6.58}{4\pi \times 4.5 \times (-1)} = 0.116 \text{ m}$$

Only one stagnation point exists.

The minimum pressure is located on the top of the cylinder where $\theta = 90°$. Using Bernoulli's equation from the free stream to that point, we have, letting $p_\infty = 0$,

$$\overset{0}{\cancel{p_\infty}} + \rho \frac{U_\infty^2}{2} = p_{min} + \frac{\rho}{2}(v_\theta)_{max}^2$$

$$\therefore p_{min} = \frac{\rho}{2}[U_\infty^2 - (v_\theta)_{max}^2] = \frac{\rho}{2}\left[U_\infty^2 - \left(-2U_\infty \sin 90° - \frac{\Gamma}{2\pi r_c}\right)^2\right]$$

$$= \frac{1.23}{2}\left[4.5^2 - \left(2 \times 4.5 + \frac{6.58}{2\pi \times 10/100}\right)^2\right] = -245.6 \text{ Pa}$$

8.6 BOUNDARY LAYER THEORY

8.6.1 General Background

In our study of high-Reynolds-number external flows we have observed that the viscous effects are confined to a thin layer of fluid, a boundary layer, next to the body and to the wake downstream of the body. For a streamlined body such as an airfoil, a good approximation to the drag can be obtained by integrating the viscous shear stress on the wall. To predict the wall shear, the velocity gradient at the wall must be known. This requires a complete solution of the flow field (i.e., a solution of the Navier–Stokes equations) inside the boundary layer. Such a solution also allows us to predict locations of possible separation. In this section we derive the integral and differential equations and provide solution techniques for a boundary layer flow on a flat plate with a zero pressure gradient; this simplified flow has many applications. Nonzero-pressure-gradient flows on flat plates and curved surfaces are beyond the scope of this introductory presentation.

Let us now discuss some of the characteristics of a boundary layer. The edge of the boundary layer, with thickness designated by $\delta(x)$, cannot be observed in an actual flow; we arbitrarily define it to be that locus of points where the velocity is equal to 99% of the free-stream velocity [the free-stream velocity is the inviscid flow wall velocity $U(x)$], as shown in Fig. 8.21. Since the boundary layer is so thin, the pressure in the boundary layer is assumed to be the pressure $p(x)$ at the wall as predicted by the inviscid flow solution.

The boundary layer begins as a laminar flow with zero thickness at the leading edge of a flat plate, as shown in Fig. 8.22, or with some finite thickness at the stagnation point of a blunt object or an airfoil (see Fig. 8.3). After a distance x_T, which depends on the free-stream velocity, the viscosity, the pressure gradient, the wall roughness, the free-stream fluctuation level, and wall rigidity, the laminar flow undergoes a transition process that results, after a short distance, in a turbulent flow, as sketched. For flow over a flat plate with zero pressure gradient, this transition process occurs when $U_\infty x_T/\nu = 3 \times 10^5$ for flow on rough plates or with high free-stream fluctuation intensity ($\sqrt{\overline{u'^2}}/U_\infty \simeq 0.1$), or $U_\infty x_T/\nu = 5 \times 10^5$

> **KEY CONCEPT** *The edge of the boundary layer cannot be observed in an actual flow.*

> **KEY CONCEPT** *The pressure in the boundary layer is the pressure at the wall of the inviscid flow solution.*

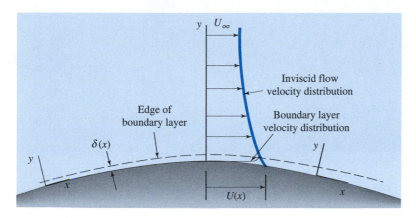

FIGURE 8.21 Boundary layer on a curved surface.

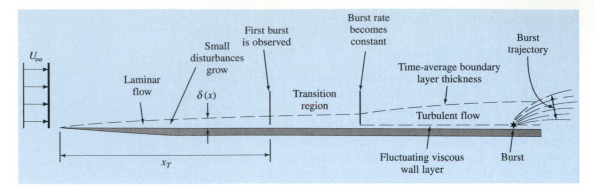

FIGURE 8.22 Boundary layer with transition.

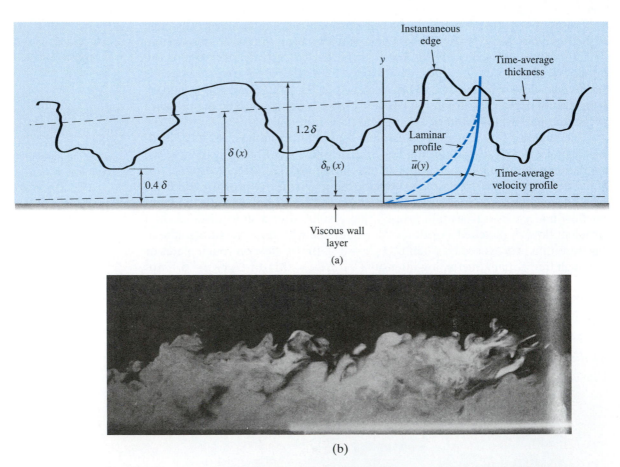

FIGURE 8.23 Turbulent boundary layer: (a) nomenclature sketch; (b) streamwise slice of the boundary layer. (Photograph by R. E. Falco.)

for flow on smooth rigid plates with low free-stream fluctuation intensity. For extremely low fluctuation levels in research labs, laminar flows have been observed on smooth rigid plates with carefully designed leading edges up to $U_\infty x_T/\nu = 10^6$.

The quantity $U_\infty x/\nu$ is the *local Reynolds number* and $U_\infty x_T/\nu$ is the *critical Reynolds number*. For a smooth rigid flat plate and a very low free-stream fluctuation level, a zero-pressure-gradient flow becomes unstable (i.e., small disturbances will grow) at a local Reynolds number of about 6×10^4. The small disturbances grow initially as a two-dimensional wave, then a three-dimensional wave, and finally burst into a turbulent spot; the initial burst forms the beginning of the transition region. The transition region is relatively short and is usually ignored in calculations. The flow up to x_T is assumed to be laminar, and the flow after x_T is considered to be turbulent.

The turbulent boundary layer thickens much more rapidly than the laminar layer. It also has a substantially greater wall shear. A sketch of a turbulent boundary layer with its submerged viscous wall layer is shown in Fig. 8.23a, and an actual photograph in Fig. 8.23b. The time-average thickness is $\delta(x)$ and the time-average viscous wall layer thickness is $\delta_\nu(x)$. Both layers are actually quite time dependent. The instantaneous boundary layer thickness varies between 0.4δ and 1.2δ, as shown. The turbulent profile has a greater slope at the wall than does a laminar profile with the same boundary layer thickness, as can be observed in Fig. 8.23a.

Finally, we should emphasize that the boundary layer is quite thin. A thick boundary layer is drawn to scale in Fig. 8.24. We have assumed a laminar flow up to x_T and a turbulent flow thereafter. For higher velocities the boundary layer thickness decreases. If we assumed that $U_\infty = 100$ m/s, the boundary layer would hardly be noticed drawn to the same scale, yet all the viscous effects are confined in that thin layer; the velocity is brought to rest with very large gradients. Viscous dissipative effects in this thin layer are large enough to cause sufficiently high temperatures that satellites burn up on reentry.

8.6.2   Von Kármán Integral Equation

From the velocity profile in the boundary layer of Fig. 8.24 it is observed that the velocity goes from $u = 0.99U_\infty$ at $y = \delta$ to $u = 0$ at $y = 0$ over a very short distance (the thickness of the boundary layer). Hence it is not surprising that we can

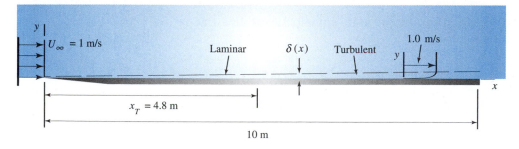

FIGURE 8.24   Boundary layer in air with $\text{Re}_{\text{crit}} = 3 \times 10^5$.

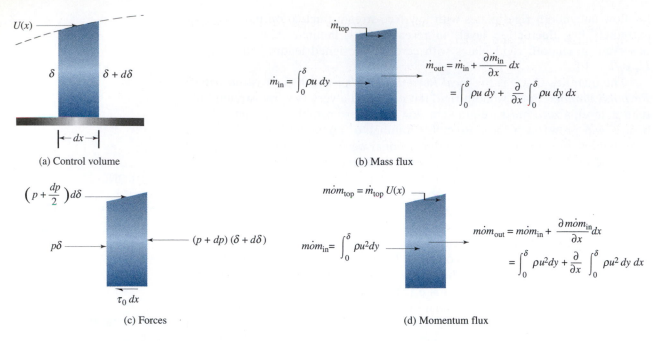

FIGURE 8.25 Control volume for a boundary layer with variable $U(x)$.

approximate the velocity profile, for both laminar and turbulent flow, with a good deal of accuracy. If the velocity profile can be assumed known, the continuity and momentum integral equations will enable us to predict the boundary layer thickness and the wall shear and thus the drag. So let us develop the integral equations for the boundary layer.

Consider an infinitesimal control volume, shown in Fig. 8.25a. The integral continuity equation allows us to find $\dot{m}_{\text{top}}$ (see Fig. 8.25b). It is, assuming unit depth,

$$\dot{m}_{\text{top}} = \dot{m}_{\text{out}} - \dot{m}_{\text{in}}$$

$$= \frac{\partial}{\partial x} \int_0^\delta \rho u \, dy \, dx \qquad (8.6.1)$$

The integral momentum equation takes the form

$$\Sigma F_x = \dot{m\dot om}_{\text{out}} - \dot{m\dot om}_{\text{in}} - \dot{m\dot om}_{\text{top}} \qquad (8.6.2)$$

where $\dot{m\dot om}$ represents the momentum flux in the x-direction. Referring to Figs. 8.25c and 8.25d this becomes, neglecting high-order terms,

$$-\delta \, dp - \tau_0 dx = \frac{\partial}{\partial x} \int_0^\delta \rho u^2 \, dy \, dx - \left(\frac{\partial}{\partial x} \int_0^\delta \rho u \, dy \, dx \right) U(x) \qquad (8.6.3)$$

where $\dot{m}_{\text{top}}$ is given in Eq. 8.6.1. We have not assumed that $U(x)$ is constant. Divide through by $-dx$ and obtain

$$\tau_0 + \delta \frac{dp}{dx} = U(x) \frac{d}{dx} \int_0^\delta \rho u \, dy - \frac{d}{dx} \int_0^\delta \rho u^2 \, dy \qquad (8.6.4)$$

where we have used ordinary derivatives since the integrals are only functions of x. This equation is often referred to as the *von Kármán integral equation*.

For flow over a flat plate with zero pressure gradient, so that $dp/dx = 0$ and $U(x) = U_\infty$, von Kármán's integral equation takes the simplified form

$$\tau_0 = \frac{d}{dx} \int_0^\delta \rho u U_\infty dy - \frac{d}{dx} \int_0^\delta \rho u^2 \, dy$$

$$= \frac{d}{dx} \int_0^\delta \rho u (U_\infty - u) \, dy \qquad (8.6.5)$$

If the velocity profile can be assumed, this equation along with $\tau_0(x) = \mu \partial u/\partial y|_{y=0}$ allows us to solve for both $\delta(x)$ and $\tau_0(x)$. This will be demonstrated in the following sections for both a laminar and a turbulent boundary layer.

Before we do this, however, there are two additional lengths often used in boundary layer theory. They are the *displacement thickness* δ_d and the *momentum thickness* θ, defined by

$$\delta_d = \frac{1}{U} \int_0^\delta (U - u) \, dy \qquad (8.6.6)$$

$$\theta = \frac{1}{U^2} \int_0^\delta u(U - u) \, dy \qquad (8.6.7)$$

The displacement thickness is the displacement of the streamlines in the free stream as a result of velocity deficits in the boundary layer, as can be shown by continuity considerations. The momentum layer thickness is the equivalent thickness of a fluid layer with velocity U with momentum equal to the momentum lost due to friction; the momentum thickness is often used as a characteristic length in turbulent boundary layer studies. The end-of-chapter problems will demonstrate the use of δ_d and θ. It should be noted, though, that von Kármán's integral equation (8.6.5) takes the form, assuming that $\rho = \text{const.}$,

$$\tau_0 = \rho U_\infty^2 \frac{d\theta}{dx} \qquad (8.6.8)$$

8.6.3   Approximate Solution to the Laminar Boundary Layer

It is possible to use the von Kármán integral equation and obtain a fairly accurate approximation to the laminar boundary layer on a flat plate with zero pressure gradient. We have four conditions that a proposed velocity profile should satisfy:

$$
\begin{aligned}
u &= 0 && \text{at} \quad y = 0 \\
u &= U_\infty && \text{at} \quad y = \delta \\
\frac{\partial u}{\partial y} &= 0 && \text{at} \quad y = \delta \\
\frac{\partial^2 u}{\partial y^2} &= 0 && \text{at} \quad y = 0
\end{aligned}
\tag{8.6.9}
$$

The first three of these conditions are obvious from a sketch of a velocity profile; the fourth condition comes from the x-component Navier–Stokes equation (5.3.14) since $u = v = 0$ at the wall, $\partial^2 u/\partial x^2 = 0$ on the wall, and $dp/dx = 0$ for the steady flow on the flat plate under consideration.

A cubic polynomial can satisfy the four conditions above; let us assume that

$$
\frac{u}{U_\infty} = A + By + Cy^2 + Dy^3
\tag{8.6.10}
$$

where A, B, C, and D can be functions of x. Using the four conditions, we see that

$$
A = 0, \quad B = \frac{3}{2\delta}, \quad C = 0, \quad D = -\frac{1}{2\delta^3}
\tag{8.6.11}
$$

Hence a good approximation for the velocity profile in a laminar flow is

$$
\frac{u}{U_\infty} = \frac{3}{2}\frac{y}{\delta} + \frac{1}{2}\left(\frac{y}{\delta}\right)^3
\tag{8.6.12}
$$

Let us now use this velocity profile to find $\delta(x)$ and $\tau_0(x)$. Von Kármán's integral equation (8.6.5) gives

$$
\begin{aligned}
\tau_0 &= \frac{d}{dx}\int_0^\delta \rho\left(\frac{3y}{2\delta} - \frac{y^3}{2\delta^3}\right)\left(1 - \frac{3y}{2\delta} - \frac{y^3}{2\delta^3}\right)U_\infty^2\,dy \\
&= 0.139\rho U_\infty^2 \frac{d\delta}{dx}
\end{aligned}
\tag{8.6.13}
$$

At the wall we know that $\tau_0 = \mu\,\partial u/\partial y\,|_{y=0}$, or using the cubic profile (8.6.12),

$$
\tau_0 = \mu\left(U_\infty \frac{3}{2\delta}\right)
\tag{8.6.14}
$$

Equating the foregoing expressions for $\tau_0(x)$, we find that

$$\delta \, d\delta = \frac{\frac{3}{2}\mu U_\infty}{0.139\rho U_\infty^2} \, dx = 10.8 \, \frac{\nu}{U_\infty} \, dx \qquad (8.6.15)$$

Using $\delta = 0$ at $x = 0$ (the leading edge), Eq. 8.6.15 is integrated to give

$$\delta = 4.65 \, \sqrt{\frac{\nu x}{U_\infty}} = 4.65 \, \frac{x}{\sqrt{\text{Re}_x}} \qquad (8.6.16)$$

where Re_x is the local Reynolds number. This is substituted back into Eq. 8.6.14, giving the wall shear as

$$\tau_0 = 0.323\rho U_\infty^2 \, \sqrt{\frac{\nu}{x U_\infty}}$$

$$= \frac{0.323\rho U_\infty^2}{\sqrt{\text{Re}_x}} \qquad (8.6.17)$$

The shearing stress is made dimensionless by dividing by $\frac{1}{2}\rho U_\infty^2$. The **local skin friction coefficient** c_f results; it is

Local skin friction coefficient: *A dimensionless wall shearing stress.*

$$c_f = \frac{\tau_0}{\frac{1}{2}\rho \, U_\infty^2}$$

$$= \frac{0.646}{\sqrt{U_\infty x/\nu}} = \frac{0.646}{\sqrt{\text{Re}_x}} \qquad (8.6.18)$$

If the wall shear is integrated over the length L, there results, per unit width,

$$F_D = \int_0^L \tau_0 \, dx = 0.646\rho U_\infty \, \sqrt{U_\infty L \nu}$$

$$= \frac{0.646\rho U_\infty^2 L}{\sqrt{\text{Re}_L}} \qquad (8.6.19)$$

or in terms of the **skin friction coefficient** C_f,

Skin friction coefficient: *A dimensionless drag force.*

$$C_f = \frac{F_D}{\frac{1}{2}\rho \, U_\infty^2 L}$$

$$= \frac{1.29}{\sqrt{U_\infty L/\nu}} = \frac{1.29}{\sqrt{\text{Re}_L}} \qquad (8.6.20)$$

where Re_L is the Reynolds number at the end of the flat plate.

Note that the shearing stress τ_0 becomes unbounded as $x \to 0$. Hence we would not expect $\tau_0(x)$ to be a very good approximation to the wall shear near the leading edge. The expression for the drag is however acceptable.

The results above are quite good when compared with results from a solution of the differential equations (refer to Section 8.6.6). The boundary layer thickness is 7% too low; the constant in Eq. 8.6.16 should be 5 if an exact solution were obtained. The wall shear is less than 3% too low; the constant in Eq. 8.6.17 should be 0.332.

Example 8.13

Assume a parabolic velocity profile and calculate the boundary layer thickness and the wall shear. Compare with those calculated above.

Solution

The parabolic velocity profile is assumed to be

$$\frac{u}{U_\infty} = A + By + Cy^2$$

The fourth condition, which would be impossible to satisfy, of (8.6.9) is omitted; this leaves

$$0 = A$$
$$1 = A + B\delta + C\delta^2$$
$$0 = B + 2C\delta$$

A simultaneous solution provides

$$A = 0, \qquad B = \frac{2}{\delta}, \qquad C = -\frac{1}{\delta^2}$$

The velocity profile is then

$$\frac{u}{U_\infty} = 2\frac{y}{\delta} - \frac{y^2}{\delta^2}$$

This is substituted into von Kármán's integral equation (8.6.5) to obtain

$$\tau_0 = \frac{d}{dx}\int_0^\delta \rho U_\infty^2 \left(2\frac{y}{\delta} - \frac{y^2}{\delta^2}\right)\left(1 - \frac{2y}{\delta} + \frac{y^2}{\delta^2}\right) dy$$
$$= \frac{2}{15}\rho U_\infty^2 \frac{d\delta}{dx}$$

We also use $\tau_0 = \mu\, \partial u/\partial y \,|_{y=0}$; that is,

$$\tau_0 = \mu U_\infty \frac{2}{\delta}$$

Equating the two expressions above, we obtain

$$\delta \, d\delta = 15 \frac{\nu}{U_\infty} dx$$

Using $\delta = 0$ at $x = 0$, this is integrated to

$$\delta = 5.48 \sqrt{\frac{\nu x}{U_\infty}}$$

This is 18% higher than the value using the cubic but only 10% higher than the more accurate result of $5 \sqrt{\nu x/U_\infty}$.

The wall shear is found to be

$$\tau_0 = \frac{2\mu U_\infty}{\delta}$$

$$= 0.365\rho U_\infty^2 \sqrt{\frac{\nu}{x U_\infty}}$$

This is 13% higher than the value using the cubic and 10% higher than the more accurate value of $0.332 \, \rho U_\infty^2 \sqrt{\nu/x U_\infty}$. Because the boundary layer is so thin, there is little difference between a cubic and a parabola or the actual profile; refer to the profile in Fig. 8.24.

8.6.4 Turbulent Boundary Layer: Power-Law Form

For turbulent boundary layer flow we have two methods for obtaining the information desired. Both methods utilize experimental data but the one that we present in this section is the simpler of the two. The second method, to be presented in the next section, provides us with more information than we usually desire for most applications; however, it is more accurate.

In the method to be presented first we fit the data for the velocity profile with a power-law equation. The power-law form is

$$\frac{\bar{u}}{U_\infty} = \left(\frac{y}{\delta}\right)^{1/n} \qquad n = \begin{cases} 7 & \mathrm{Re}_x < 10^7 \\ 8 & 10^7 < \mathrm{Re}_x < 10^8 \\ 9 & 10^8 < \mathrm{Re}_x < 10^9 \end{cases} \qquad (8.6.21)$$

where

$$\mathrm{Re}_x = \frac{U_\infty x}{\nu} \qquad (8.6.22)$$

Von Kármán's integral equation can now be applied following the steps used for laminar flow, except when the shear stress at the wall is evaluated. The power-law form (8.6.21) yields $(\partial \bar{u}/\partial y)_{y=0} = \infty$; hence the profile gives poor results near the wall, especially for the shear stress. So rather than using $\tau_0 = (\mu \, \partial \bar{u}/\partial y)_{y=0}$, we use an empirical relation; the *Blasius formula*, named after Paul R. H. Blasius

KEY CONCEPT *The power-law form gives poor results at the wall.*

(1883–1970), which relates the local skin friction coefficient to the boundary layer thickness, is

$$c_f = 0.046 \left(\frac{\nu}{U_\infty \delta} \right)^{1/4} \tag{8.6.23}$$

or, relating τ_0 to c_f using the definition of c_f in Eq. 8.6.18, the shear stress relation is

$$\tau_0 = 0.023 \, \rho U_\infty^2 \left(\frac{\nu}{U_\infty \delta} \right)^{1/4} \tag{8.6.24}$$

Von Kármán's integral equation provides us with a second expression for τ_0. Substitute the velocity profile of (8.6.21) with $\mathrm{Re}_x < 10^7$ into Eq. 8.6.5 and obtain

$$\tau_0 = \frac{d}{dx} \int_0^\delta \rho U_\infty^2 \left(\frac{y}{\delta} \right)^{1/7} \left[1 - \left(\frac{y}{\delta} \right)^{1/7} \right] dy$$

$$= \frac{7}{72} \rho U_\infty^2 \frac{d\delta}{dx} \tag{8.6.25}$$

Combining the two expressions above for τ_0, we find that

$$\delta^{1/4} \, d\delta = 0.237 \left(\frac{\nu}{U_\infty} \right)^{1/4} dx \tag{8.6.26}$$

Assuming a turbulent flow from the leading edge (the laminar portion is often quite short, i.e., $L \gg x_T$), there results

$$\delta = 0.38x \left(\frac{\nu}{U_\infty x} \right)^{1/5}$$

$$= 0.38x \mathrm{Re}_x^{-1/5} \qquad \mathrm{Re}_x < 10^7 \tag{8.6.27}$$

Substituting this expression for δ back into Eq. 8.6.23, we find that

$$c_f = 0.059 \, \mathrm{Re}_x^{1/5} \qquad \mathrm{Re}_x < 10^7 \tag{8.6.28}$$

and, performing the required integration, there results, with $n = 7$,

$$C_f = 0.073 \, \mathrm{Re}_L^{-1/5} \qquad \mathrm{Re}_L < 10^7 \tag{8.6.29}$$

where

$$\mathrm{Re}_L = \frac{U_\infty L}{\nu}$$

The relations above can be stretched to $\mathrm{Re}_x \simeq 10^8$ without substantial error.

If L is not much larger than x_T, say $L = 3x_T$, then there is a significant laminar portion on the leading part of a flat plate, and the skin friction coefficient can be modified as

KEY CONCEPT   *There may be a significant laminar portion on the leading edge of a flat plate.*

$$C_f = 0.073\mathrm{Re}_L^{-1/5} - 1700\mathrm{Re}_L^{-1} \qquad \mathrm{Re}_L < 10^7 \qquad (8.6.30)$$

This relationship is based on transition occurring at $\mathrm{Re}_{\mathrm{crit}} = 5 \times 10^5$. If $\mathrm{Re}_{\mathrm{crit}} = 3 \times 10^5$, the constant of 1700 is replaced by 1060; if $\mathrm{Re}_{\mathrm{crit}} = 6 \times 10^5$, it becomes 2080.

Finally, the displacement thickness and momentum thickness can be evaluated, using $n = 7$, to be

$$\delta_d = 0.048x\mathrm{Re}_x^{-1/5} \qquad (8.6.31)$$

$$\theta = 0.037x\mathrm{Re}_x^{-1/5}$$

Example 8.14

Estimate the boundary layer thickness at the end of a 4-m-long flat surface if the free-stream velocity is $U_\infty = 5$ m/s. Use atmospheric air at 30°C. Also, predict the drag force if the surface is 5 m wide. (a) Neglect the laminar portion of the flow and (b) account for the laminar portion using $\mathrm{Re}_{\mathrm{crit}} = 5 \times 10^5$.

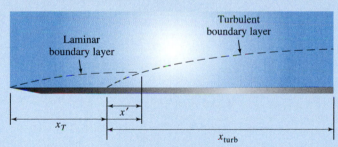

FIGURE E8.14

Solution
(a) Let us first assume turbulent flow from the leading edge. The boundary layer thickness is given by Eq. 8.6.27. It is

$$\delta = 0.38x\mathrm{Re}_x^{-1/5}$$

$$= 0.38 \times 4 \times \left(\frac{5 \times 4}{1.6 \times 10^{-5}}\right)^{-1/5} = 0.0917 \text{ m}$$

The drag force is, using Eq. 8.6.29,

$$F_D = C_f \times \tfrac{1}{2}\rho U_\infty^2 Lw$$

$$= 0.073\left(\frac{5 \times 4}{1.6 \times 10^{-5}}\right)^{-1/5} \times \frac{1}{2} \times 1.16 \times 5^2 \times 4 \times 5 = 1.28 \text{ N}$$

(continued)

The predictions above assume that $\mathrm{Re}_L < 10^7$. The Reynolds number is

$$\mathrm{Re}_L = \frac{5 \times 4}{1.6 \times 10^{-5}} = 1.25 \times 10^6$$

Hence the calculations are acceptable.

(b) Now let us account for the laminar portion of the boundary layer. Referring to Fig. E8.14, the distance x_T is found as follows:

$$\mathrm{Re}_{\mathrm{crit}} = 5 \times 10^5 = \frac{U_\infty x_T}{\nu}$$

$$\therefore x_T = 5 \times 10^5 \times 1.6 \times \frac{10^{-5}}{5} = 1.6 \text{ m}$$

The boundary layer thickness at x_T is, replacing the constant of 4.65 in Eq. 8.6.16 with the more accurate value of 5.0,

$$\delta = 5 \sqrt{\frac{x\nu}{U_\infty}}$$

$$= 5 \sqrt{\frac{1.6 \times 1.6 \times 10^{-5}}{5}} = 0.0113 \text{ m}$$

The location of the fictitious origin of the turbulent flow (see Fig. E8.14) is found using Eq. 8.6.27 to be

$$x'^{4/5} = \frac{\delta}{0.38} \left(\frac{U_\infty}{\nu} \right)^{1/5}$$

$$\therefore x' = \left(\frac{0.0113}{0.38} \right)^{5/4} \left(\frac{5}{1.6 \times 10^{-5}} \right)^{1/4} = 0.292 \text{ m}$$

The distance x_{turb} is then $x_{\mathrm{turb}} = 4 - 1.6 + 0.292 = 2.69$ m. Using Eq. 8.6.27, the thickness at the end of the surface is

$$\delta = 0.38x \left(\frac{\nu}{U_\infty x} \right)^{1/5}$$

$$= 0.38 \times 2.69 \times \left(\frac{1.6 \times 10^{-5}}{5 \times 2.69} \right)^{1/5} = 0.067 \text{ m}$$

The value of part (a) is 37% too high when compared with this more accurate value.

The more accurate drag force is found using Eq. 8.6.30 to be

$$F_D = C_f \times \tfrac{1}{2} \rho U_\infty^2 Lw$$

$$= [0.073 \, \mathrm{Re}_L^{-1/5} - 1700 \mathrm{Re}_L^{-1}] \times \frac{1}{2} \rho U_\infty^2 Lw$$

$$= \left[0.073 \left(\frac{5 \times 4}{1.6 \times 10^{-5}} \right)^{-1/5} - 1700 \left(\frac{5 \times 4}{1.6 \times 10^{-5}} \right)^{-1} \right] \times \frac{1}{2} \times 1.16 \times 5^2 \times 4 \times 5$$

$$= 0.88 \text{ N}$$

The prediction of part (a) is 45% too high. For relatively short surfaces it is obvious that significant errors result if the thinner laminar portion with its smaller shear stress is neglected.

8.6.5   Turbulent Boundary Layer: Empirical Form

The second method of predicting turbulent flow quantities on a flat plate with zero pressure gradient is based entirely on data. It is more accurate than the power-law form but also more complicated. The time-average turbulent velocity profile can be divided into two regions, the *inner region* and the *outer region*, as shown in Fig. 8.26. The inner region is characterized by the *self-similar* (the dimensionless dependent variable depends on only one dimensionless independent variable) relation,

$$\frac{\overline{u}}{u_\tau} = f\left(\frac{u_\tau y}{\nu}\right) \tag{8.6.32}$$

in which u_τ is the *shear velocity*, given by[5]

$$u_\tau = \sqrt{\frac{\tau_0}{\rho}} \tag{8.6.33}$$

KEY CONCEPT u_τ *is a ficticious velocity called the shear velocity.*

The velocity profile in the outer region is given by the self-similar relation

$$\frac{U_\infty - \overline{u}}{u_\tau} = f\left(\frac{y}{\delta}\right) \tag{8.6.34}$$

where $(U_\infty - \overline{u})$ is called the *velocity defect*.

The inner region has three distinct zones; the viscous wall layer, the buffer zone, and the turbulent zone, as shown in Fig. 8.26b. The highly fluctuating *viscous wall layer* has a linear time-average profile given by

KEY CONCEPT *The viscous wall layer is highly fluctuating.*

$$\frac{\overline{u}}{u_\tau} = \frac{u_\tau y}{\nu} \tag{8.6.35}$$

The quantity ν/u_τ is the characteristic length in the turbulent inner region; hence the dimensionless distance from the wall is denoted by

$$y^* = \frac{u_\tau y}{\nu} \tag{8.6.36}$$

The viscous wall layer is very thin, extending up to $y^* \simeq 5$. A logarithmic profile exists from $y^* \simeq 50$ to $y/\delta \simeq 0.15$. In this self-similar *turbulent zone*,

---

[5] The shear velocity is a fictitious velocity and is defined because the quantity $\sqrt{\tau_0/\rho}$ occurs often in empirical relations in turbulent boundary layer flows.

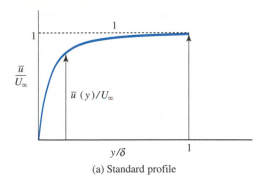

(a) Standard profile

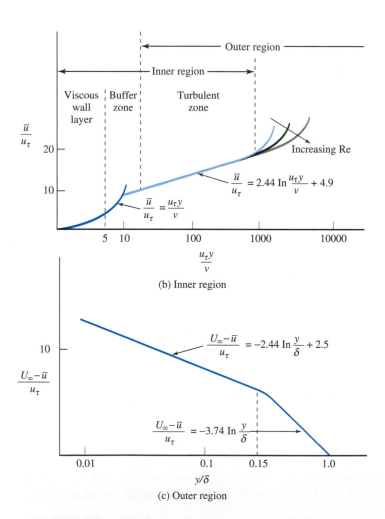

(b) Inner region

(c) Outer region

FIGURE 8.26 Velocity profile in a turbulent boundary layer.

$$\frac{\overline{u}}{u_\tau} = 2.44 \ln \frac{u_\tau y}{\nu} + 4.9 \qquad 50 < \frac{u_\tau y}{\nu} \qquad \frac{y}{\delta} < 0.15 \qquad (8.6.37)$$

The location of the outer edge of the turbulent zone is strongly dependent on the Reynolds number. The value of $u_\tau y/\nu$ that locates the outer edge increases as the Reynolds number increases, as shown. A *buffer zone*, with no specified velocity profile, connects the two self-similar zones.

The outer region relates the velocity defect to y/δ. In the turbulent zone the velocity-defect profile is sketched in Fig. 8.26c and is

$$\frac{U_\infty - \overline{u}}{u_\tau} = -2.44 \ln \frac{y}{\delta} + 2.5 \qquad 50 < \frac{u_\tau y}{\nu} \qquad \frac{y}{\delta} < 0.15 \qquad (8.6.38)$$

Between $y/\delta = 0.15$ and $y/\delta = 1$ researchers fit the data with several relations; the one selected here is

$$\frac{U_\infty - \overline{u}}{u_\tau} = -3.74 \ln \frac{y}{\delta} \qquad \frac{y}{\delta} > 0.15 \qquad (8.6.39)$$

The equations above involve the shear velocity u_τ, which depends on the wall shear τ_0. The equation that is valid at the wall is Eq. 8.6.35, which, using $\tau_0 = \mu \, \partial u/\partial y|_{y=0}$, simply provides us with an identity. It does not allow us to calculate τ_0. Hence a relationship is necessary to provide us with τ_0 (or equally c_f). There are several such relationships used; one that gives excellent results is

$$c_f = \frac{0.455}{(\ln 0.06 \mathrm{Re}_x)^2} \qquad (8.6.40)$$

This local skin friction coefficient allows us to determine τ_0 and thus u_τ at any location of interest. Knowing u_τ, the velocity profiles can be used to calculate quantities of interest.

Assuming turbulent flow from the leading edge, the shear stress can be integrated to yield the drag. Then the skin friction coefficient becomes

KEY CONCEPT *This local skin friction coefficient allows us to determine τ_0 and thus u_τ.*

$$C_f = \frac{0.523}{(\ln 0.06 \mathrm{Re}_L)^2} \qquad (8.6.41)$$

This relation is very good and can be used up to $\mathrm{Re}_L = 10^9$ with an error of 2% or less. Even at $\mathrm{Re}_L = 10^{10}$ the error is about 4%. To account for a laminar portion, the same term included in Eq. 8.6.30 can be subtracted from Eq. 8.6.41.

In concluding this section, a very useful relationship can be obtained by combining the two logarithmic profiles for the common turbulent zone. Substitute Eq. 8.6.37 into Eq. 8.6.38 and obtain

$$\frac{U_\infty}{u_\tau} = 2.44 \ln \frac{u_\tau \delta}{\nu} + 7.4 \tag{8.6.42}$$

This equation allows an easy calculation of δ knowing u_τ.

Example 8.15

Estimate the thickness δ_ν of the viscous wall layer and the boundary layer thickness at the end of a 4.5-m-long flat plate if $U_\infty = 30$ m/s in 15°C–atmospheric air. Also, calculate the drag force on one side if the plate is 3 m wide. Use the empirical data.

Solution

To find the viscous wall layer thickness we must know the shear velocity and hence the wall shear. The wall shear, using Eq. 8.6.40, and the shear velocity at $x = 15$ ft are

$$\tau_0 = \tfrac{1}{2}\rho U_\infty^2 c_f$$

$$= \frac{1}{2}\rho U_\infty^2 \frac{0.455}{(\ln 0.06 \mathrm{Re}_x)^2}$$

$$= \frac{1}{2} \times 1.23 \times 30^2 \frac{0.455}{\left(\ln 0.06 \dfrac{30 \times 4.5}{1.47 \times 10^{-5}}\right)^2} = 1.44 \text{ Pa}$$

$$u_\tau = \sqrt{\frac{\tau_0}{\rho}} = \sqrt{\frac{1.44}{1.23}} = 1.08 \text{ m / s}$$

The viscous wall layer thickness is determined using Eq. 8.6.36 with $y^* = 5$ as follows:

$$\frac{u_\tau \delta_\nu}{\nu} = 5$$

$$\therefore \delta_\nu = \frac{5\nu}{u_\tau} = \frac{5 \times 1.47 \times 10^{-5}}{1.08} = 6.8 \times 10^{-5} \text{ m}$$

The boundary layer thickness is found using Eq. 8.6.42:

$$\frac{U_\infty}{u_\tau} = 2.44 \ln \frac{u_\tau \delta}{\nu} + 7.4$$

$$\therefore \frac{30}{1.08} = 2.44 \ln \frac{1.08 \times \delta}{1.47 \times 10^{-5}} + 7.4 \qquad \therefore \delta = 0.067 \text{ m}$$

The drag force is calculated using Eq. 8.6.41 to be

$$F_D = C_f \times \tfrac{1}{2}\rho U_\infty^2 Lw$$

$$= \frac{0.523}{(\ln 0.06 \mathrm{Re}_L)^2} \times \tfrac{1}{2}\rho U_\infty^2 Lw$$

$$= \frac{0.523}{\left(\ln 0.06 \dfrac{30 \times 4.5}{1.47 \times 10^{-5}}\right)^2} \times \tfrac{1}{2} \times 1.23 \times 30^2 \times 4.5 \times 3 = 22.4\ \mathrm{N}$$

The laminar portion of the boundary layer has been neglected.

Example 8.16

Estimate the maximum boundary layer thickness and the drag on the side of a ship that measures 40 m long with a submerged depth of 8 m assuming the side of the ship is approximated as a flat plate. The ship travels at 10 m/s. (a) Use the empirical methods and (b) compare with the results using the power law model.

Solution

(a) The boundary layer thickness is found from Eq. 8.6.42. First we must find τ_0 from Eq. 8.6.40 and then u_τ as follows:

$$\tau_0 = \frac{1}{2}\rho U_\infty^2 \frac{0.455}{(\ln 0.06 \mathrm{Re}_L)^2}$$

$$= \frac{1}{2} \times 1000 \times 10^2 \frac{0.455}{\left(\ln 0.06 \dfrac{10 \times 40}{10^{-6}}\right)^2} = 78.8\ \mathrm{Pa}$$

$$\therefore u_\tau = \sqrt{\frac{\tau_0}{\rho}}$$

$$= \sqrt{\frac{78.8}{1000}} = 0.28\ \mathrm{m/s}$$

The maximum boundary layer thickness is found using Eq. 8.6.42:

$$\frac{U_\infty}{u_\tau} = 2.44 \ln \frac{u_\tau \delta}{\nu} + 7.4$$

$$\therefore \frac{10}{0.28} = 2.44 \ln \frac{0.28\delta}{10^{-6}} + 7.4 \qquad \therefore \delta = 0.39\ \mathrm{m}$$

The drag is

$$F_D = C_f \times \tfrac{1}{2}\rho U_\infty^2 Lw$$

$$= \frac{0.523}{\left(\ln 0.06 \dfrac{10 \times 40}{10^{-6}}\right)^2} \times \tfrac{1}{2} \times 1000 \times 10^2 \times 40 \times 8 = 29\,000\ \mathrm{N}$$

(continued)

(b) First, calculate the Reynolds number: Re $= 10 \times 40/10^{-6} = 4 \times 10^8$. We select $n = 9$. Equation (8.6.25) becomes

$$\tau_0 = \frac{d}{dx} \int_0^\delta \rho U_\infty^2 \left(\frac{y}{\delta}\right)^{1/9} \left[1 - \left(\frac{y}{\delta}\right)^{1/9}\right] dy$$

$$= \frac{9}{110} \rho U_\infty^2 \frac{d\delta}{dx}$$

Equating this to the τ_0 of Eq. 8.6.24, we find that

$$\delta^{1/4} \, d\delta = 0.281 \, (\nu/U_\infty)^{1/4} \, dx$$

Assume $\delta = 0$ at $x = 0$ and integrate. This provides

$$\delta = 0.433x \, \mathrm{Re}_x^{-1/5}$$

$$= 0.433(40) \left(\frac{10 \times 40}{10^{-6}}\right)^{-1/5} = 0.33 \text{ m}$$

This value is 15% too low.
    The drag force is found to be

$$F_D = 0.071 \mathrm{Re}_L^{-1/5} \times \tfrac{1}{2} \rho U_\infty^2 \, Lw$$

$$= 0.071 \left(\frac{10 \times 40}{10^{-6}}\right)^{-1/5} \times \frac{1}{2} \times 1000 \times 10^2 \times 40 \times 8 = 21\,600 \text{ N}$$

This value is 25% too low. Obviously, the power-law equations are in significant error.

8.6.6 Laminar Boundary Layer Equations

The solution presented in Section 8.6.3 for the laminar boundary layer was an approximate solution using a cubic polynomial to approximate the velocity profile. In this section we simplify the Navier–Stokes equations and present a more accurate solution for the laminar boundary layer on a flat plate with a zero pressure gradient.

The x-component Navier–Stokes equation for a steady, incompressible, plane flow is (see Eq. 5.3.14 and ignore the gravity term)

$$u \frac{\partial u}{\partial x} + v \frac{\partial u}{\partial y} = -\frac{1}{\rho} \frac{\partial p}{\partial x} + \nu \left(\frac{\partial^2 u}{\partial x^2} + \frac{\partial^2 u}{\partial y^2}\right) \tag{8.6.43}$$

KEY CONCEPT *There is no pressure variation in the y-direction in the boundary layer.*

In boundary layer theory the boundary layer is assumed to be very thin (see Fig. 8.22), so there is no pressure variation in the y-direction in the boundary layer, that is, $p = p(x)$. In addition (this is a very important point), the pressure $p(x)$ is given by the inviscid flow solution as the wall pressure; hence the pressure is not an unknown. This leaves only two unknowns, u and v. Equation 8.6.43 provides us with one equation and the continuity equation

$$\frac{\partial u}{\partial x} + \frac{\partial v}{\partial y} = 0 \tag{8.6.44}$$

provides us with the other. The y-component Navier–Stokes equation is not of use in boundary layer theory since all the terms are negligibly small (obviously, $v \ll u$ as inferred from Fig. 8.24).

In addition to the simplification that a known pressure gradient provides, it is obvious that $\partial^2 u/\partial x^2$ is much, much less than the large gradients that exist in the y-direction (refer to the sketch of Fig. 8.24); consequently, neglecting $\partial^2 u/\partial x^2$, the boundary layer equation that must be solved is

$$u \frac{\partial u}{\partial x} + v \frac{\partial u}{\partial y} = -\frac{1}{\rho} \frac{dp}{dx} + \nu \frac{\partial^2 u}{\partial y^2} \tag{8.6.45}$$

where the pressure gradient dp/dx is assumed to be known from the inviscid flow solution. This is often referred to as the *Prandtl boundary layer equation*, named after Ludwig Prandtl (1875–1953). Neither term on the left can be neglected; the y-component v may be small, but the velocity gradient $\partial u/\partial y$ is obviously quite large; hence the product must be retained.

Let us focus on flow over a flat plate with zero pressure gradient. In addition, let us introduce the stream function:

$$u = \frac{\partial \psi}{\partial y} \qquad v = -\frac{\partial \psi}{\partial x} \tag{8.6.46}$$

The boundary layer equation becomes

$$\frac{\partial \psi}{\partial y} \frac{\partial^2 \psi}{\partial x \partial y} - \frac{\partial \psi}{\partial x} \frac{\partial^2 \psi}{\partial y^2} = \nu \frac{\partial^3 \psi}{\partial y^3} \tag{8.6.47}$$

In this form the x and y dependence cannot be separated. If we transform this equation (such transformations are selected by trial and error and experience) by letting

$$\xi = x \qquad \eta = y \sqrt{\frac{U_\infty}{\nu x}} \tag{8.6.48}$$

there results

$$-\frac{1}{2\xi} \left(\frac{\partial \psi}{\partial \eta} \right)^2 + \frac{\partial \psi}{\partial \eta} \frac{\partial^2 \psi}{\partial \xi \partial \eta} - \frac{\partial \psi}{\partial \xi} \frac{\partial^2 \psi}{\partial \eta^2} = \nu \frac{\partial^3 \psi}{\partial \eta^3} \sqrt{\frac{U_\infty}{\nu \xi}} \tag{8.6.49}$$

This equation may appear to be a more difficult equation to solve than Eq. 8.6.47, but by observing the position of ξ in this equation, we separate variables by letting

$$\psi(\xi, \eta) = \sqrt{U_\infty \nu \xi} \, F(\eta) \tag{8.6.50}$$

The velocity components can then be shown to be, using Eqs. 8.6.48 and 8.6.50,

$$u = \frac{\partial \psi}{\partial y} = U_\infty F'(\eta)$$

$$v = -\frac{\partial \psi}{\partial x} = \frac{1}{2}\sqrt{\frac{\nu U_\infty}{x}}\,(\eta F' - F) \tag{8.6.51}$$

Substitute Eq. 8.6.50 into Eq. 8.6.49 and an ordinary nonlinear, differential equation results; it is

$$F\frac{d^2 F}{d\eta^2} + 2\frac{d^3 F}{d\eta^3} = 0 \tag{8.6.52}$$

This equation replaces the partial differential equation (8.6.47). Now let us state the boundary conditions.

The boundary conditions $[u(x, 0) = 0, v(x, 0) = 0$ and $u(x, y > \delta) = U_\infty]$ take the form

$$F = F' = 0 \text{ at } \eta = 0 \qquad F' = 1 \text{ at large } \eta \tag{8.6.53}$$

The boundary value problem, consisting of the ordinary differential equation (8.6.52) and boundary conditions (8.6.53), can now be solved numerically using a computer. The results are tabulated in Table 8.5. The last two columns are used to provide v and τ_0, respectively.

Defining the boundary layer thickness to be the point where $u = 0.99U_\infty$, we see from Table 8.5 that this occurs where $\eta = 5$. Hence, letting $\eta = 5$ and $y = \delta$ in Eq. 8.6.48, we have

$$\delta = 5\sqrt{\frac{\nu x}{U_\infty}} \tag{8.6.54}$$

Using

$$\frac{\partial u}{\partial y} = \frac{\partial u}{\partial \eta}\frac{\partial \eta}{\partial y} = U_\infty F''\sqrt{\frac{U_\infty}{\nu x}} \tag{8.6.55}$$

the wall shear in a laminar boundary layer with $dp/dx = 0$ is

$$\tau_0 = \mu\frac{\partial u}{\partial y}\bigg|_{y=0} = 0.332\,\rho U_\infty^2 \sqrt{\frac{\nu}{xU_\infty}} \tag{8.6.56}$$

TABLE 8.5   Solution for the Laminar Boundary Layer with $dp/dx = 0$

$\eta = y\sqrt{\dfrac{U_\infty}{\nu x}}$	F	$F' = u/U_\infty$	$\frac{1}{2}(\eta F' - F)$	F''
0	0	0	0	0.3321
1	0.1656	0.3298	0.0821	0.3230
2	0.6500	0.6298	0.3005	0.2668
3	1.397	0.8461	0.5708	0.1614
4	2.306	0.9555	0.7581	0.0642
5	3.283	0.9916	0.8379	0.0159
6	4.280	0.9990	0.8572	0.0024
7	5.279	0.9999	0.8604	0.0002
8	6.279	1.0000	0.8605	0.0000

The local skin friction coefficient is

$$c_f = \frac{0.664}{\sqrt{\mathrm{Re}_x}} \qquad (8.6.57)$$

and the skin friction coefficient is

$$C_f = \frac{1.33}{\sqrt{\mathrm{Re}_L}} \qquad (8.6.58)$$

Numerically integrating Eqs. 8.6.6 and 8.6.7, the displacement and momentum thicknesses are found to be

$$\delta_d = 1.72\sqrt{\frac{\nu x}{U_\infty}} \qquad \theta = 0.644\sqrt{\frac{\nu x}{U_\infty}} \qquad (8.6.59)$$

Example 8.17

Atmospheric air at 30°C flows over a 8-m-long 2-m-wide flat plate at 2 m/s. Assume that laminar flow exists in the boundary layer over the entire length. At $x = 8$ m, calculate (a) the maximum value of v, (b) the wall shear, and (c) the flow rate through the layer. (d) Also, calculate the drag force on the plate.

(continued)

Solution

(a) The y-component of velocity has been assumed to be small in boundary layer theory. Its maximum value at $x = 8$ m is found using 8.6.51 to be

$$v = \sqrt{\frac{\nu U_\infty}{x}} \times \frac{1}{2}(\eta F' - F)$$

$$= \sqrt{\frac{1.6 \times 10^{-5} \times 2}{8}} \times 0.86 = 0.00172 \text{ m/s}$$

Compare this with $U_\infty = 2$ m/s.

(b) The wall shear at $x = 8$ m is found using Eq. 8.6.56 to be

$$\tau_0 = 0.332 \rho U_\infty^2 \sqrt{\frac{\nu}{U_\infty x}}$$

$$= 0.332 \times 1.16 \times 2^2 \sqrt{\frac{1.6 \times 10^{-5}}{2 \times 8}} = 0.00154 \text{ Pa}$$

(c) The flow rate through the boundary layer at $x = 8$ m is given by

$$Q = \int_0^\delta u \times 2 dy = 2\sqrt{\frac{\nu x}{U_\infty}} \int_0^5 U_\infty F' \, d\eta$$

where we have substituted for u and y from Eqs. 8.6.51 and 8.6.48. Recognizing that $\int F' \, d\eta = F$, the flow rate is

$$Q = 2 U_\infty \sqrt{\frac{\nu x}{U_\infty}} [F(5) - F(\cancel{0})^0]$$

$$= 2 \times 2 \sqrt{\frac{1.6 \times 10^{-5} \times 8}{2}} \times 3.28 = 0.105 \text{ m}^3/\text{s}$$

(d) The drag force is determined to be

$$F_D = \tfrac{1}{2} \rho U_\infty^2 L w C_f$$

$$= \frac{1}{2} \times 1.16 \times 2^2 \times 8 \times 2 \times \frac{1.33}{\sqrt{2 \times 8/1.6 \times 10^{-5}}} = 0.049 \text{ N}$$

8.6.7 Pressure Gradient Effects

In the foregoing sections we have concentrated our study of boundary layers on the flat plate with a zero pressure gradient. This is the simplest boundary layer flow and allows us to model many flows of engineering interest. The inclusion of a pressure gradient, even though relatively low, markedly alters the boundary layer flow. In fact, a strong negative pressure gradient (such as flow in a contraction) may relaminarize a turbulent boundary layer; that is, the turbulence production in the viscous wall layer which sustains the turbulence ceases and a laminar boundary layer is reestablished. A positive pressure gradient quickly causes the boundary layer to thicken and eventually to separate. These two effects are shown in the photographs of Fig. 8.27.

The flow about any plane body with curvature, such as an airfoil, can be modeled as flow over a flat plate with a nonzero pressure gradient. The boundary layer thickness is so much smaller than the radius of curvature that the additional

KEY CONCEPT *A strong negative pressure gradient may relaminarize a turbulent boundary layer.*

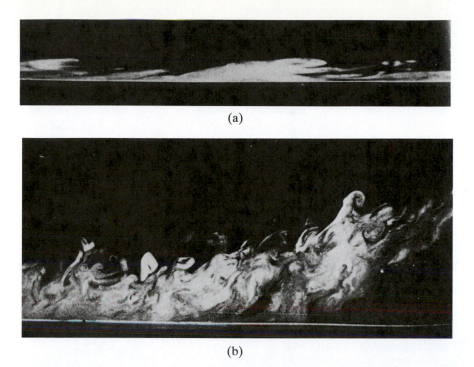

(a)

(b)

FIGURE 8.27   Influence of a strong pressure gradient on a turbulent flow: (a) a strong negative pressure gradient may relaminarize a flow; (b) a strong positive pressure gradient causes a strong boundary layer to thicken. (Photographs by R. E. Falco.)

curvature terms drop out of the differential equations. The inviscid flow solution at the wall provides the pressure gradient dp/dx and the velocity $U(x)$ at the edge of the boundary layer. For axisymmetric flows, such as flow over the nose of an aircraft, boundary layer equations in cylindrical coordinates must be utilized.

The pressure gradient determines the value of the second derivative $\partial^2 u/\partial y^2$ at the wall. From the boundary layer equation (8.6.45) at the wall, $u = v = 0$, so that

$$\frac{dp}{dx} = \mu \left. \frac{\partial^2 u}{\partial y^2} \right|_{y=0} \tag{8.6.60}$$

for either a laminar or a turbulent boundary layer flow. For a zero pressure gradient the second derivative is zero at the wall; then since the first derivative is greatest at the wall and decreases as y increases, the second derivative must be negative for positive y. The profiles are sketched in Fig. 8.28a.

For a negative (favorable) pressure gradient the slope of the velocity profile near the wall is relatively large with a negative second derivative at the wall and throughout the layer. The momentum near the wall is larger than that of the zero pressure gradient flow, as shown in Fig. 8.28b, and thus there is a reduced tendency for the flow to separate. Turbulence production is discouraged and relaminarization may occur for a sufficiently large negative pressure gradient.

If a positive (unfavorable) pressure gradient is imposed on the flow the second derivative at the wall will be positive and the flow will be as sketched in part

KEY CONCEPT   *For a negative pressure gradient, there is a reduced tendency for the flow to separate.*

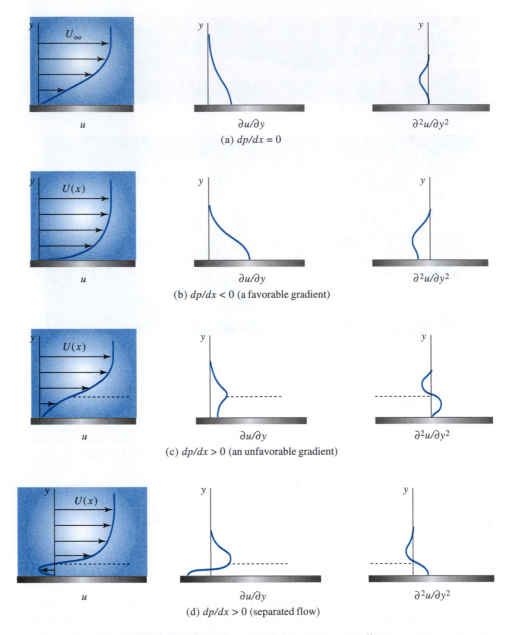

FIGURE 8.28 Influence of the pressure gradient.

(c) or (d). If the unfavorable pressure gradient acts over a sufficient distance, part (d) will probably represent the flow situation with the flow separated from the surface. Near the wall the higher pressure downstream will drive the low momentum flow near the wall in the upstream direction, resulting in flow reversal, as shown. The point at which $\partial u/\partial y = 0$ at the wall locates the point of separation.

The problem of a laminar boundary layer with a pressure gradient can be solved using conventional numerical techniques with a computer. The procedure is relatively simple using the simplified boundary layer equation (8.6.45) with a known pressure gradient. For a turbulent flow the Reynolds stress term must be included; much work is being done to develop models of turbulent quantities that will result in acceptable numerical solutions. Experimental results are often necessary for turbulent flow problems, as was the situation for internal flows.

8.7 SUMMARY

The drag and lift coefficients are defined as

$$C_D = \frac{\text{Drag}}{\frac{1}{2}\rho V^2 A} \qquad C_L = \frac{\text{Lift}}{\frac{1}{2}\rho V^2 A} \tag{8.7.1}$$

where the area is the projected area for blunt objects, and the chord times the length for an airfoil.

Vortex shedding occurs from a cylinder whenever the Reynold's number is in the range $300 < \text{Re} < 10\ 000$. The frequency of shedding is found from the Strouhal number

$$\text{St} = \frac{fD}{V} \tag{8.7.2}$$

where f is the frequency in hertz.

Plane potential flows are constructed by superimposing the following simple flows:

$$
\begin{aligned}
&\text{Uniform flow:} && \psi = U_\infty y \\
&\text{Line source:} && \psi = \frac{q}{2\pi\theta} \\
&\text{Irrotational vortex:} && \psi = \frac{\Gamma}{2\pi} \ln r \\
&\text{Doublet:} && \psi = -\frac{\mu}{r} \sin\theta
\end{aligned}
\tag{8.7.3}
$$

The stream function for the rotating cylinder is given by

$$\psi_{\text{cylinder}} = U_\infty y - \frac{\mu}{r}\sin\theta + \frac{\Gamma}{2\pi}\ln r \tag{8.7.4}$$

where the cylinder radius is

$$r_c = \sqrt{\frac{\mu}{U_\infty}} \tag{8.7.5}$$

The velocity components are

$$u = \frac{\partial \psi}{\partial y}, \qquad v = -\frac{\partial \psi}{\partial x}$$

$$v_r = \frac{1}{r}\frac{\partial \psi}{\partial \theta}, \qquad v_\theta = -\frac{\partial \psi}{\partial r} \tag{8.7.6}$$

For a laminar boundary layer on a flat plate with zero pressure gradient the exact solution provides

$$\delta = 5\sqrt{\frac{\nu x}{U_\infty}}, \qquad c_f = 0.664\sqrt{\frac{\nu}{xU_\infty}}, \qquad C_f = 1.33\sqrt{\frac{\nu}{LU_\infty}} \tag{8.7.7}$$

For a turbulent flow from the leading edge, the power-law profile with $\eta = 7$ gives

$$\delta = 0.38x\left(\frac{\nu}{xU_\infty}\right)^{1/5}, \qquad c_f = 0.059\left(\frac{xU_\infty}{\nu}\right)^{1/5}, \qquad C_f = 0.073\left(\frac{\nu}{LU_\infty}\right)^{1/5} \tag{8.7.8}$$

where the wall shear and drag force per unit width are, respectively,

$$\tau_0 = \frac{1}{2}c_f \rho U_\infty^2, \qquad F_D = \frac{1}{2}C_f \rho U_\infty^2 L \tag{8.7.9}$$

PROBLEMS

Separated Flows

8.1 Sketch the flow over an airfoil at a large angle of attack for both attached flow and separated flow. Also, sketch the expected pressure distributions on the top and bottom surfaces for both flows. Identify regions of favorable and unfavorable pressure gradients.

8.2 A spherical particle is moving in 20°C atmospheric air at a velocity of 20 m/s. What must its diameter be for Re = 5 and Re = 10^5? Sketch the expected flow field at these Reynolds numbers. Identify all regions of the flow.

8.3 Sketch the flow that is expected over a semitruck (a tractor and trailer) where the trailer is substantially higher than the tractor with and without an air deflector attached to the roof of the tractor. Sketch a side view indicating any separated regions, boundary layers, and the wake.

8.4 Air is blowing by a long, rectangular building with the wind blowing parallel to the long sides. Sketch the top view showing the separated flow regions, the inviscid flow region, the boundary layers, and the wake.

8.5 The drag force on a streamlined shape is due primarily to:
- **A.** The wake.
- **B.** The component of the pressure force acting in the flow direction.
- **C.** The shear stress.
- **D.** The separated region near the trailing edge.

8.6 A golf ball has dimples to increase its flight distance. Select the best reason that accounts for the longer flight distance of a dimpled ball compared to that of a smooth ball.
- **A.** The shearing stress is smaller on the dimpled ball.
- **B.** The dimpled ball has an effective smaller diameter.
- **C.** The wake on the dimpled ball is smaller.
- **D.** The pressure on the front of the smooth ball is larger.

8.7 Flood waters at 10°C flow over an 8-mm diameter fence wire with a velocity of 0.8 m/s. Which of the following is true?
- **A.** It is a Stokes flow with no separation.
- **B.** The separated region covers most of the rear of the wire.
- **C.** The drag is due to the relatively low pressure in the separated region.
- **D.** The separated region covers only a small area at the rear of the wire.

8.8 A 2-cm-diameter sphere is to move with Re = 5. At what velocity does it travel if it is submerged in:
- **(a)** Water at 15°C?
- **(b)** Water at 80°C?
- **(c)** Standard air at 15°C?

8.9 Air at 20°C flows around a cylindrical body at a velocity of 20 m/s. Calculate the Reynolds number if the body is:
- **(a)** A 6-m-diameter smokestack.
- **(b)** A 6-cm-diameter flagpole.
- **(c)** A 6-mm-diameter wire.

Use Re = VD/ν. Would a separated flow be expected?

8.10 The pressure distribution on the front of a 2-m-diameter disk (Fig. P8.10) is approximated by $p(r) = p_0(1 - r^2)$. If $V = 20$ m/s in this 20°C–atmospheric airflow, estimate the drag force and the drag coefficient for this disk. Assume that the pressure on the back side is zero.

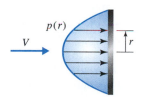

FIGURE P8.10

8.11 A flat plate 30 cm × 30 cm acts as a hydrofoil. If it is oriented at an angle of attack of 10°, estimate the lift and the drag if the pressure on the bottom side is 20 kPa and on the top side there is a vacuum of 10 kPa; neglect the effect of shear stress. Also, estimate the lift and drag coefficients if the hydrofoil velocity is 5 m/s. Use the surface area of the plate in the definition of the lift and drag coefficients.

8.12 The symmetric airfoil shown in Fig. P8.12 flies at an altitude of 12 000 m with an angle of attack of 5°. If $p_l = 26$ kPa and $p_u = 8$ kPa, estimate the lift and drag coefficients neglecting shear stresses.

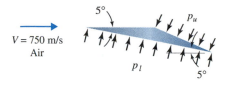

FIGURE P8.12

8.13 If the drag coefficient for a 10-cm-diameter sphere is given by $C_D = 1.0$, calculate the drag if the sphere is falling in the atmosphere:
- **(a)** At sea level.
- **(b)** At 30 000 m.
- **(c)** In 10°C water.

8.14 Calculate the drag on a smooth 50-cm-diameter sphere when subjected to a 20°C–atmospheric airflow of:
- **(a)** 6 m/s
- **(b)** 15 m/s

8.15 The drag on a 10-m-diameter spherical water storage tank in an 80 km/hr wind is approximately:
- **A.** 6300 N
- **B.** 4700 N
- **C.** 3200 N
- **D.** 2300 N

8.16 A 4-m-long smooth cylinder experiences a drag of 60 N when subjected to an atmospheric air speed of 40 m/s. Estimate the diameter of the cylinder.
- **A.** 127 mm
- **B.** 63 mm
- **C.** 26 mm
- **D.** 4.1 mm

8.17 A 4.45-cm-diameter golf ball is roughened to reduce the drag during its flight. If the Reynolds number at which the sudden drop occurs is reduced from 3×10^5 to 6×10^4 by the roughening (dimples), would you expect this to lengthen the flight of a golf ball significantly? Justify your reasoning with appropriate calculations.

8.18 A 10-cm-diameter smooth sphere experiences a drag of 2.5 N when placed in standard 15°C air.
- **(a)** What is the velocity of the airstream?
- **(b)** At what increased speed will the sphere experience the same drag?

8.19 A smooth 20-cm-diameter sphere experiences a drag of 4.2 N when placed in a 20°C water channel. Calculate the drag coefficient and the Reynolds number.

8.20 A 2-m-diameter smokestack stands 60 m high. It is designed to resist a 40-m/s wind. At this speed, what total force would be expected, and what moment would the base be required to resist? Assume atmospheric–20°C air.

8.21 A flagpole is composed of three sections: a 5-cm-diameter top section that is 10 m long, a 7.5-cm-diameter middle section 15 m long, and a 10-cm-diameter bottom section 20 m long. Calculate the total force acting on the flagpole and the resisting moment provided by the base when subjected to a 25-m/s wind speed. Make the calculations for:
   (a) A winter day at −30°C.
   (b) A summer day at 35°C.

8.22 A drag force of 50 N is desired at Re = 10^5 on a 2-m-long cylinder in a 15°C–atmospheric airflow. What velocity should be selected, and what should be the cylinder diameter?

8.23 A 20-m-high structure is 2 m in diameter at the top and 8 m in diameter at the bottom as shown in Fig. P8.23. If the diameter varies linearly with height, estimate the total drag force due to a 30-m/s wind. Use atmospheric air at 20°C.

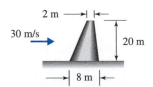

FIGURE P8.23

8.24 A steel sphere ($S = 7.82$) is dropped in water at 20°C. Calculate the terminal velocity if the diameter of the sphere is:
   (a) 10 cm   **(c)** 1 cm
   (b) 5 cm   **(d)** 2 mm

8.25 Estimate the terminal velocity of a 50-cm-diameter sphere as it falls in a 15°C atmosphere near the Earth if it has a specific gravity of:
   (a) 0.005   **(b)** 0.02   **(c)** 1.0

8.26 Estimate the terminal velocity of a skydiver by making reasonable approximations of the arms, legs, head, and body. Assume air at 20°C.

8.27 Assuming that the drag on a modern automobile at high speeds is due primarily to form drag, estimate the power (horsepower) needed by an automobile with a 3.2-m² cross-sectional area to travel at:
   (a) 80 km/h   **(b)** 90 km/h
   (c) 100 km/h

8.28 The 2 m × 3 m sign shown in Fig. P8.28 weighs 400 N. What wind speed is required to blow over the sign?

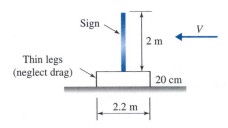

FIGURE P8.28

8.29 Calculate the drag force on a 60-cm-diameter 6-m-long cylinder if 20°C-air is blowing normal to its axis at 40 km/h and the cylinder:
   (a) Is a section of a very long cylinder.
   (b) Has both ends free.
   (c) Is fixed to the ground with the top free.

8.30 An 80-kg paratrooper leaps from an elevation of 3000 m. Estimate the paratrooper's landing speed if she:
   (a) Curls up as tightly as possible.
   (b) Uses an 8-m-diameter lightweight parachute.
   (c) Uses a 2-m-diameter safety parachute.

8.31 A semitruck travels 200 000 km each year at an average speed of 90 km/h. Estimate the savings if a streamlined deflector is added to reduce the drag coefficient. Fuel costs $0.40 per liter and the truck without the deflector averages 1.2 km per liter of fuel.

8.32 A rectangular car carrier has a cross section 2 m × 1 m. Estimate the minimum added power (horsepower) required to travel at 100 km/h because of the car carrier.

8.33 Assume that the velocity at the corners of an automobile where the rear-view mirror is located is 1.6 times the automobile's velocity. How much horsepower is required by the two 10-cm-diameter rear-view mirrors for an automobile speed of 100 km/h?

8.34   Atmospheric air at 25°C blows normal to a 4-m-long section of a cone, 30 cm in diameter at one end and 2 m in diameter at the other end, with a velocity of 20 m/s. Predict the drag on the object. Assume that $C_D = 0.4$ for each cylindrical element of the object.

8.35   An 80-cm-diameter balloon (Fig. P8.35) that weighs 0.5 N is filled with 20°C helium at 20 kPa. Neglecting the weight of the string, estimate V if α equals:
   (a)   80°
   (b)   70°
   (c)   60°
   (d)   50°

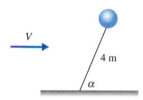

FIGURE P8.35

8.36   For the newly planted tree shown in Fig. P8.36, the soil–ball interface is capable of resisting a moment of 5000 N·m. Predict the minimum wind speed that might possibly tip the tree over. Assume that $C_D = 0.4$ for a cylinder in this air flow.

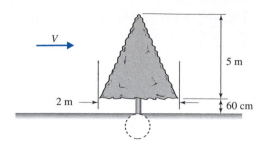

FIGURE P8.36

8.37   A sign 1.2 m × 0.6 m is attached to the top of a pizza delivery car. The car travels 10 hours a day, 6 days a week. Estimate the added cost in a year that the sign adds to the fuel used by the car. The average speed of the car is 40 km/h, fuel costs $0.30 per liter, the engine/drive train is 30% efficient, and the fuel contains 12 000 kJ/kg.

8.38   A bike rider is able to travel at an average speed of 40 km/h while riding upright. It is determined that the rider's projected area is 0.56 m². If the rider assumes a racing position so that his projected area is 0.40 m², estimate his increased speed if his drag coefficient decreases 20%, assuming the same energy expenditure.

8.39   An automobile with a cross-sectional area of 3 m² is powered by a 40-hp engine. Estimate the maximum possible speed if the drive train is 90% efficient. (The engine is rated by the power produced before the transmission.)

Vortex Shedding

8.40   Vortices are shed from a 2-cm-diameter cylinder due to a 4 m/s air stream. How far apart would you expect the vortices to be downstream of the cylinder?
   A.   44 cm
   B.   23 cm
   C.   9 cm
   D.   4 cm

8.41   Over what range of velocities would you expect vortex shedding from a telephone wire 3 mm in diameter? Could you hear any of the vorticies being shed? (Human beings can hear frequencies between 20 and 20 000 Hz.)

8.42   A wire is being towed through 15°C water normal to its axis at a speed of 2 m/s. What diameter (both large and small) could the wire have so that vortex shedding would not occur?

8.43   It is quite difficult to measure low velocities. To determine the velocity in a low-speed airflow, the vorticies being shed from a 10-cm-diameter cylinder are observed to occur at 0.2 Hz. Estimate the airspeed if the temperature is 20°C.

8.44   Motion pictures show that vortices are shed from a 2-m-diameter cylinder at 0.002 Hz while it is moving in 20°C water. What is the cylinder velocity?

8.45 The cables supporting a suspended bridge (Fig. P8.45) have a natural frequency of $T/(\pi\rho L^2 d^2)$ hertz, where T is the tension, ρ the cable density, d its diameter, and L its length. The vortices shed from the cables may lead to resonance and possible failure. A certain 1.6-cm-diam- eter steel cable is subjected to a force of 30 000 N. What length cable would result in resonance in a 10-m/s wind? (*Note:* The third and fifth harmonics may also lead to resonance. Calculate all three lengths.)

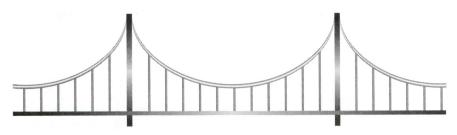

FIGURE P8.45

Streamlining

8.46 Streamlining reduces drag primarily by:
- **A.** Reducing the wall shear.
- **B.** Reducing the pressure in the stagnation region.
- **C.** Reducing the separated flow area.
- **D.** Eliminating the wake.

8.47 A 15-cm-diameter smokestack on a semitruck extends 2 m straight up into the free stream. Estimate the horsepower needed because of the stack for a speed of 100 kmph. If the stack were streamlined, estimate the reduced horsepower.

8.48 A wind speed of 3 m/s blows normal to an 8-cm-diameter smooth cylinder that is 2 m long. Calculate the drag force. The cylinder is now streamlined. What is the percentage reduction in the drag? Assume that $T = 20°C$.

8.49 Water flows by an 80-cm-diameter cylinder that protrudes 2 m up from the bottom of a river. For an average water speed of 2 m/s, estimate the drag on the cylinder. If the cylinder were streamlined, what would be the percentage reduction in drag?

8.50 Circular 2-cm-diameter tubes are used as supports on an ultralight airplane, designed to fly at 50 km/hr. If there are 20 linear meters of the tubes, estimate the horsepower needed by the tubes. If the tubes were streamlined, estimate the reduced horsepower required of the tubes.

8.51 A bike racer is able to ride 50 km/hr at top speed. Estimate the drag force just due to his head. If he wore a streamlined, tight-fitting helmet, estimate the reduced drag force.

Cavitation

8.52 The critical cavitation number for a streamlined strut is 0.7. Find the maximum velocity of the body to which the strut is attached if cavitation is to be avoided. The body is traveling 5 m beneath a water surface.

8.53 A lift force of 200 kN is desired at a speed of 12 m/s on a hydrofoil, designed to operate at a depth of 40 cm. It has a 40-cm chord and is 10 m long. Calculate the angle of attack and the drag force. Is cavitation present under these conditions?

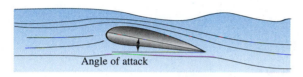

FIGURE P8.53

8.54 A hydrofoil, designed to operate at a depth of 40 cm, has a 40-cm-chord and is 10 m long. A lift force of 200 kN is desired at a speed of 10 m/s. Calculate the angle of attack and the drag force. Is cavitation present under these conditions?

8.55 A body resembling a sphere has a diameter of approximately 0.8 m. It is towed at a speed of 20 m/s, 5 m below the water surface. Estimate the drag acting on the body.

8.56 A 2200-kg mine sweeper is designed to ride above the water with hydrofoils at the four corners providing the lift. If the hydrofoils have a 40-cm chord length, what total hydrofoil length is required if they are to operate 60 cm below the surface at a 6° angle of attack? The vehicle is to travel at 50 m/s.

Added Mass

8.57 A 40-cm-diameter sphere, which weighs 400 N, is released from rest while submerged in water. Calculate its initial acceleration:
(a) Neglecting the added mass.
(b) Including the added mass.

8.58 A submersible, whose length is twice its maximum diameter, resembles an ellipsoid. If its added mass is ignored, what is the percentage error in a calculation of its initial acceleration if its specific gravity is 1.2?

Lift and Drag on Airfoils

8.59 Estimate the takeoff speed needed for a 1200-kg aircraft (including payload) if the angle of attack at takeoff is to be 10°. The effective wing area (chord times length) is 16 m².
A. 22 m/s   **B.** 33 m/s
C. 44 m/s   **D.** 55 m/s

8.60 An aircraft with a mass, including payload, of 1000 kg is designed to cruise at a speed of 80 m/s at 10 km. The effective wing area is approximately 15 m². Determine the lift coefficient and the angle of attack. What power is required by the airfoil during cruise? Assume a conventional airfoil.

8.61 A 1500-kg aircraft is designed to carry a payload of 3000 N when cruising at 80 m/s at 10 km. The effective wing area is 20 m². Assuming a conventional airfoil, calculate:
(a) The takeoff speed if an angle of attack of 10° is desired.
(b) The stall speed when landing.
(c) The power required at cruise if 45% of the power is needed to move the airfoil.

8.62 In Problem 8.61 it was necessary to assume that takeoff was at sea level. What would be the take-off speed in Wyoming, where the elevation is 2000 m?

8.63 The aircraft of Problem 8.61 is flown at 2 km rather than 10 km. Estimate the percentage increase or decrease in power required at cruise.

8.64 An additional load of 6000 N is added to the aircraft of Problem 8.61. Estimate the takeoff speed if the angle of attack remains at 10°.

8.65 Estimate the minimum landing speed for a 250 000-kg aircraft in an emergency situation if the angle of attack is selected to be near stall and:
(a) No slotted flaps are used.
(b) One slotted flap is used.
(c) Two slotted flaps are used.

Its wingspan is 60 m and the airfoil chord averages 8 m.

8.66 In Problem 8.65 it is assumed that the air-craft lands at standard sea-level conditions since neither elevation nor temperature is given. Calculate the percentage increase or decrease in the emergency landing speed if the aircraft is to land:

(a) In Denver where the elevation is 1600 m.
(b) At sea level when the temperature is very cold at $-40°C$.
(c) At sea level when the temperature is hot at $50°C$.

8.67 A proposed aircraft is to resemble a huge airfoil, a flying wing (Fig. P8.67). Its wingspan will be 200 m and its chord will average 30 m. Estimate, assuming a conventional airfoil, the total mass of the aircraft, including payload, for a design speed of 800 km/h at an elevation of 8 km. Also, calculate the power requirement.

FIGURE P8.67

Vorticity, Velocity Potential, and Stream Function

8.68 Take the curl of the Navier–Stokes equation and show that the vorticity equation (8.5.3) results. (See Eq. 5.3.20.)

8.69 Write the three component vorticity equations contained in Eq. 8.5.3 using rectangular coordinates. Use $\boldsymbol{\omega} = \omega_x\hat{\mathbf{i}} + \omega_y\hat{\mathbf{j}} + \omega_z\hat{\mathbf{k}}$.

8.70 Simplify the vorticity equation 8.5.3 for a plane flow ($w = 0$ and $\partial/\partial z = 0$). Use $\boldsymbol{\omega} = (\omega_x, \omega_y, \omega_z)$. What conclusion can be made about the magnitude of ω_z in an inviscid, plane flow (such as flow through a short contraction) that contains vorticity?

8.71 Determine which of the following flows are irrotational and incompressible and find the velocity potential function, should one exist for each incompressible flow.
(a) $\mathbf{V} = 10x\hat{\mathbf{i}} + 20y\hat{\mathbf{j}}$
(b) $\mathbf{V} = 8y\hat{\mathbf{i}} + 8x\hat{\mathbf{j}} - 6z\hat{\mathbf{k}}$
(c) $\mathbf{V} = (x\hat{\mathbf{i}} + y\hat{\mathbf{j}})/\sqrt{x^2 + y^2}$
(d) $\mathbf{V} = (x\hat{\mathbf{i}} + y\hat{\mathbf{j}})/(x^2 + y^2)$

8.72 An attempt is to be made to solve Laplace's equation for flow around a circular cylinder of radius r_c oriented in the center of a channel of height $2h$. The velocity profile far from the cylinder is uniform. State the necessary boundary conditions. Assume that $\psi = 0$ at $y = -h$. The origin of the coordinate system is located at the center of the cylinder.

8.73 State the stream function and velocity potential corresponding to a uniform velocity of $100\hat{\mathbf{i}} + 50\hat{\mathbf{j}}$ using rectangular coordinates.

8.74 A flow is represented by the stream function $\psi = 40 \tan^{-1}(y/x)$.
(a) Express the stream function in polar form.
(b) Is this an incompressible flow? Show why.
(c) Determine the velocity potential.
(d) Find the radius where the acceleration is -10 m/s^2.

8.75 The stream function for a flow is $\psi = 20 \ln (x^2 + y^2)$ m²/s. Determine the complex velocity potential for this incompressible flow. If the pressure at a large distance from the origin is 20 kPa, what is the pressure at the point (0, 20 cm) if water is flowing?

8.76 A stream function is given by

$$\psi = 10y - \frac{10y}{x^2 + y^2}$$

(a) Show that this satisfies $\nabla^2 \psi = 0$.
(b) Find the velocity potential $\phi (x, y)$.
(c) Assuming water to be flowing, find the pressure along the x-axis if $p = 50$ kPa at $x = -\infty$.
(d) Locate any stagnation points.

8.77 The velocity potential for a flow is

$$\phi = 10x + 5 \ln (x^2 + y^2)$$

(a) Show that this function satisfies Laplace's equation.
(b) Find the stream function $\psi (x, y)$.
(c) Assume that water is flowing and find the pressure along the x-axis if $p = 100$ kPa at $x = -\infty$.
(d) Locate any stagnation points.
(e) Find the acceleration at $x = -2$ m, $y = 0$.

8.78 The velocity profile in a wide 0.2-m-high channel is given by $u(y) = y - y^2/0.2$. Determine the stream function for this flow. Calculate the flow rate by integrating the velocity profile and by using $\Delta \psi$. Explain why a velocity potential does not exist by referring to Eq. 8.5.2.

Superposition of Simple Flows

8.79 The body formed by superimposing a source at the origin of strength 0.5π m²/s and a uniform flow of 9 m/s is shown in Fig. P8.79.
(a) Locate any stagnation points.
(b) Find the y-intercept y_B of the body.
(c) Find the thickness of the body at $x = \infty$.
(d) Find u at $x = -30$ cm, $y = 0$.

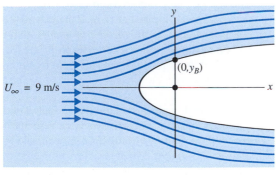

FIGURE P8.79

8.80 A source with strength π m²/s and a sink of equal strength are located at (−1 m, 0) and (1 m, 0), respectively. They are combined with a uniform flow $U_\infty = 10$ m/s to form a *Rankine oval*. Calculate the length and maximum thickness of the oval. If $p = 10$ kPa at $x = -\infty$, find the minimum pressure if water is flowing.

8.81 An oval is formed from a source and sink of strengths 2π m²/s located at (−1, 0) and (1, 0), respectively, combined with a uniform flow of 2 m/s. Locate any stagnation points, and find the velocity at (−4, 0) and (0, 4). Distances are in meters.

8.82 Two sources of strength 2π m²/s are located at (0, 1) and (0, −1), respectively. Sketch the resulting flow and locate any stagnation points. Find the velocity at (1, 1). Distances are in meters.

8.83   The two sources of Problem 8.82 are superimposed with a uniform flow. Sketch the flow, locate any stagnation points, and find the y-intercept of the body formed if:
   (a) $U_\infty = 10$ m/s.
   (b) $U_\infty = 1$ m/s.
   (c) $U_\infty = 0.2$ m/s.

8.84   A doublet of strength 60 m³/s is superimposed with a uniform flow of 8 m/s of water. Calculate:
   (a)   The radius of the resulting cylinder.
   (b)   The pressure increase from $x = -\infty$ to the stagnation point.
   (c)   The velocity $v_\theta(\theta)$ on the cylinder.
   (d)   The pressure decrease from the stagnation point to the point of minimum pressure on the cylinder.

8.85   A sink of strength 4π m²/s is superimposed with a vortex of strength 20π m²/s.
   (a)   Sketch a pathline of a particle initially occupying the point $(x = 0, y = 1$ m$)$. Use tangents at every 45° with straight-line extensions.
   (b)   Calculate the acceleration at $(0, 1)$.
   (c)   If $p(10, 10) = 20$ kPa, what is $p\,(0, 0.1)$ if atmospheric air is flowing?

8.86   The cylinder shown in Fig. P8.86 is formed by combining a doublet of strength 40 m³/s with a uniform flow of 10 m/s.
   (a)   Sketch the velocity along the y-axis from the cylinder to $y = \infty$.
   (b)   Calculate the velocity at $(x = -4$ m, $y = 3$ m$)$.
   (c)   Calculate the drag coefficient for the cylinder assuming potential flow over the front half and constant pressure over the back half.

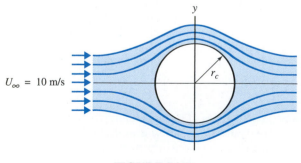

FIGURE P8.86

8.87   A 2-m-diameter cylinder is placed in a uniform water flow of 4 m/s.
   (a)   Sketch the velocity along the x-axis from the cylinder to $x = -\infty$.
   (b)   Find v_θ on the front half of the cylinder.
   (c)   Find $p(\theta)$ on the front half of the cylinder if $p = 50$ kPa at $x = -\infty$.
   (d)   Estimate the drag force on a 1-m length of the cylinder if the pressure over the rear half is constant and equal to the value at $\theta = 90°$.

8.88   Superimpose a free stream $U_\infty = 9$ m/s, a doublet $\mu = 11.5$ m³/s, and a vortex $\Gamma = 90$ m²/s. Locate any stagnation points and predict the minimum and maximum pressure on the surface of the cylinder if $p = 0$ at $x = -\infty$ and atmospheric air is flowing.

8.89   A 0.8-m-diameter cylinder is placed in a 20-m/s atmospheric airflow. At what rotational speed should the cylinder rotate so that only one stagnation point exists on its surface? Calculate the minimum pressure acting on the cylinder if $p = 0$ at $x = -\infty$.

8.90   A 1.2-m-diameter cylinder rotates at 120 rpm in a 3-m/s atmospheric airstream. Locate any stagnation points and calculate the minimum and maximum pressures on the cylinder if $p = 0$ at $x = -\infty$.

8.91   The circulation around a 20-m-long airfoil (measured tip to tip) is calculated to have a value of 1500 m²/s. Estimate the lift generated by the airfoil if the aircraft is flying at 10,000 m with a speed of 100 m/s. Assume the flow to be incompressible.

8.92   The velocity field due to a source of strength 2π m²/s per meter, located at $(2$ m, 2 m$)$ in a 90° corner, is desired. Use the *method of images*, that is, add one or more sources at the appropriate locations, and determine the velocity field by finding $u\,(x, y)$ and $v\,(x, y)$.

8.93   Flow through a porous medium is modeled with Laplace's equation and an associated velocity potential function. Natural gas can be stored in certain underground rock structures for use at a later time. A well is placed next to an impervious rock formation, as shown in Fig. P8.93. If the well is to extract 0.2 m³/s per meter, predict the velocity to be expected at the point (4 m, 3 m). See Problem 8.92 for the method of images.

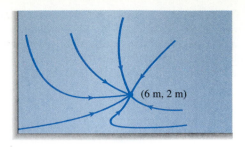

(6 m, 2 m)

FIGURE P8.93

Boundary Layers

8.94 How far from the leading edge can turbulence be expected on an airfoil traveling at 100 m/s if the elevation is:

(a) 0 m? **(b)** 4000 m? **(c)** 10,000 m?

Use $Re_{crit} = 6 \times 10^5$ and assume a flat plate with zero pressure gradient.

8.95 The boundary layer on a flat plate with zero pressure gradient is to be studied in a wind tunnel. How far from the leading edge would you expect to find turbulent flow if $U_\infty = 10$ m/s and:

(a) The plate is held rigid with high free-stream disturbance level?

(b) The plate is held rigid with low free-stream disturbance level?

(c) The plate is vibrated with low free-stream disturbance level?

(d) The plate is vibrated with high free-stream disturbance level?

(e) At what distance would you expect a small disturbance to grow for the flow of part (b)?

8.96 Repeat Problem 8.95 but place the flat plate in a water channel.

8.97 A laminar region is desired to be at least 2 m long on a smooth rigid flat plate. A wind tunnel and a water channel are available. What maximum speed can be selected for each? Assume low free-stream fluctuation intensity.

8.98 Determine the pressure $p(x)$ in the boundary layer and the velocity $U(x)$ at the edge of the boundary layer that would be present on the front of the cylinder of Problem 8.86. Let the pressure at the stagnation point be 20 kPa with water flowing. Let x be measured from the stagnation point; see Figure 8.19.

8.99 A boundary layer would develop as shown in Fig. P8.99 from the front stagnation point of Problem 8.87. Determine $U(x)$ and $p(x)$ that would be needed to solve for the boundary layer growth on the cylinder front. Measure x from the stagnation point.

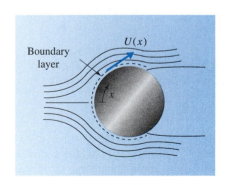

$U(x)$

Boundary layer

x

FIGURE P8.99

8.100 Assuming inviscid uniform flow of air through the contraction shown in Figure P8.100, estimate $U(x)$ and dp/dx, which are necessary to solve for the boundary layer growth on the flat plate. Assume a one-dimensional flow with $\rho = 1.0$ kg/m^3.

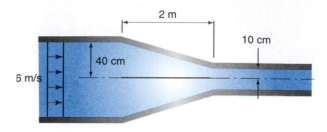

FIGURE P8.100

Von Kármán Integral Equation

8.101 Provide the detailed steps to the final form of Eq. 8.6.1 and Eq. 8.6.3. Refer to Figure 8.23.

8.102 Show that the von Kármán integral equation 8.6.4 can be put in the form

$$\tau_0 = -\delta \frac{dp}{dx} + \rho \frac{d}{dx} \int_0^\delta u(U - u)\, dy - \rho \frac{dU}{dx} \int_0^\delta u\, dy$$

Note that the quantity $\int_0^\delta u\, dy$ is only a function of x.

8.103 Show that the von Kármán integral equation of Problem 8.102 can be written as

$$\tau_0 = \rho \frac{d}{dx}(\theta U^2) + \rho \delta_d U \frac{dU}{dx}$$

To accomplish this, we must show that Bernoulli's equation $p + \rho U^2/2 = \text{const.}$ can be differentiated to yield

$$\frac{dp}{dx} = -\rho U \frac{dU}{dx} = -\frac{\rho}{\delta} \frac{dU}{dx} \int_0^\delta U\, dy$$

8.104 Assume that $u = U_\infty \sin(\pi y/2\delta)$ in a zero pressure gradient boundary layer. Calculate:
   (a) $\delta(x)$.
   (b) $\tau_0(x)$.
   (c) v at $y = \delta$ and $x = 3$ m.

8.105 Assume a linear velocity profile and find $\delta(x)$ and $\tau_0(x)$. Compute the percentage error when compared with the exact expressions for a laminar flow. Use $dp/dx = 0$.

8.106 A boundary layer profile is approximated with:

$$u = 3\, U_\infty \frac{y}{\delta} \qquad\qquad 0 < y \le \delta/6$$

$$u = U_\infty\left(\frac{y}{\delta} + \frac{1}{3}\right) \qquad\qquad \delta/6 < y \le \delta/2$$

$$u = U_\infty\left(\frac{y}{3\delta} + \frac{2}{3}\right) \qquad\qquad \delta/2 < y \le \delta$$

Determine $\delta(x)$ and $\tau_0(x)$; compute the percentage error when compared with the exact expressions for a laminar flow.

8.107 If the walls in a wind-tunnel test section are parallel, the velocity in the center portion of the tunnel will accelerate as shown in Fig. P8.107. To maintain a constant tunnel velocity so that $dp/dx = 0$, show that the walls should be displaced outward a distance $\delta_d(x)$. If a wind tunnel were square, how far should one wall be displaced outward for $dp/dx = 0$?

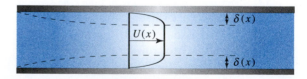

FIGURE P8.107

8.108 The velocity profile at a given x-location in the boundary layer (Fig. P8.108) is assumed to be

$$u(y) = 10\left(2\frac{y}{\delta} - \frac{y^2}{\delta^2}\right)$$

A streamline is 2 cm from the flat plate at the leading edge. How far is it from the plate when $x = 3$ m (i.e., what is h)? Also, calculate the dis- placement thickness at $x = 3$ m. Compare the displacement thickness to $(h - 2)$ cm.

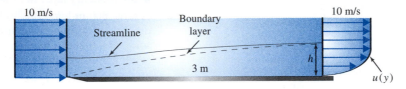

FIGURE P8.108

8.109 It is desired that the test section in a wind tunnel experience a zero pressure gradient. If the test section has a square cross section, what should be the equation of the displacement of one of the walls (three walls will be straight and parallel or perpendicular) if the 30°C–air is pressurized to 160 kPa absolute? Assume a boundary layer profile of $u/U_\infty = 2y/\delta - y^2/\delta^2$. Assume that $y = 0$ at $x = 0$ of the displacement equation $y(x)$.

8.110 Find δ_d and θ for a laminar boundary layer assuming:
(a) A cubic profile.
(b) A parabolic profile.
(c) That $u = U_\infty \sin(\pi y/2\delta)$.
Compute percentage errors when compared with the exact values of $\delta_d = 1.72 \sqrt{\nu x/U_\infty}$ and $\theta = 0.644 \sqrt{\nu x/U_\infty}$.

8.111 A laminar flow is maintained in a boundary layer on a 6-m-long 5-m-wide flat plate with 15°C–atmospheric air flowing at 4 m/s. Assuming parabolic profile, calculate:
(a) δ at $x = 6$ m.
(b) τ_0 at $x = 6$ m.
(c) The drag force on one side.
(d) v at $y = \delta$ and $x = 3$ m.

8.112 Work Problem 8.111, but assume a cubic profile.

Laminar and Turbulent Boundary Layers

8.113 Atmospheric air at 20°C flows at 10 m/s over a 2-m-long 4-m-wide flat plate. Calculate the maximum boundary layer thickness and the drag force on one side assuming:
(a) Laminar flow over the entire length.
(b) Turbulent flow over the entire length.

8.114 Fluid flows over a flat plate at 20 m/s. Determine δ and τ_0 at $x = 6$ m if the fluid is:
(a) Atmospheric air at 20°C.
(b) Water at 20°C.

Neglect the laminar portion.

8.115 Assume a turbulent velocity profile $\bar{u} = U_\infty(y/\delta)^{1/7}$. Does this profile satisfy the conditions at $y = \delta$? Can it give the shear stress at the wall? Plot both a cubic laminar profile and the one-seventh power law profile on the same graph assuming the same boundary layer thickness.

8.116 Estimate the drag on one side of a 4-m-long 5-m-wide flat plate if atmospheric air at 15°C is flowing with a velocity of 6 m/s. Assume that:
(a) $Re_{crit} = 3 \times 10^5$
(b) $Re_{crit} = 5 \times 10^5$
(c) $Re_{crit} = 6 \times 10^5$

8.117 A 1-m-long flat plate with a sharp leading edge that is 2 m wide is towed parallel to itself in 20°C water at 1.2 m/s. Estimate the total drag if:
- **(a)** $Re_{crit} = 3 \times 10^5$
- **(b)** $Re_{crit} = 6 \times 10^5$
- **(c)** $Re_{crit} = 9 \times 10^5$

8.118 Air moving at 60 km/h is considered to have a zero boundary layer thickness at a distance of 100 km from shore. At the beach, estimate the boundary layer thickness and the wall shear using:
- **(a)** A one-seventh power law.
- **(b)** Empirical data.

Use $T = 20°C$.

8.119 For the conditions of Problem 8.118, calculate:
- **(a)** The thickness of the viscous wall layer.
- **(b)** The displacement thickness at the beach.

8.120 Atmospheric air at 15°C flows over a flat plate at 100 m/s. At $x = 6$ m, estimate:
- **(a)** The local skin friction coefficient.
- **(b)** The wall shear.
- **(c)** The viscous wall layer thickness.
- **(d)** The boundary layer thickness.

8.121 Water at 20°C flows over a flat plate at 10 m/s. At $x = 3$ m, estimate:
- **(a)** The viscous wall layer thickness.
- **(b)** The velocity at the edge of the viscous wall layer.
- **(c)** The value of y at the outer edge of the turbulent zone.
- **(d)** The boundary layer thickness.

8.122 Estimate the total shear drag on a ship traveling at 10 m/s if the sides are flat plates 10 m × 100 m with zero pressure gradients. What is the maximum boundary layer thickness?

8.123 A large 600-m-long, 100-m-diameter, cigar-shaped dirigible is planned for cruises for retired professors. As a first estimate the drag is calculated assuming it to be a flat plate with zero pressure gradient, with the drag on the nose and rear areas neglected.
- **(a)** Estimate the power required of each of four engines if it is to cruise at 15 m/s.
- **(b)** Estimate the payload if one half of its volume is filled with helium, and its engines, equipment, and structure have a mass of 1.2×10^6 kg.

FIGURE P8.123

Laminar Boundary Layer Equations

8.124 Assuming that $dp/dx = 0$, show that Eq. 8.6.47 follows from Eq. 8.6.45.

8.125 Recalling from calculus that

$$\frac{\partial \psi}{\partial y} = \frac{\partial \psi}{\partial \xi} \frac{\partial \xi}{\partial y} + \frac{\partial \psi}{\partial \eta} \frac{\partial \eta}{\partial y}$$

so that

$$\frac{\partial^2 \psi}{\partial y^2} = \frac{\partial}{\partial y}\left(\frac{\partial \psi}{\partial y}\right) = \frac{\partial(\partial \psi/\partial y)}{\partial \xi} \frac{\partial \xi}{\partial y} + \frac{\partial(\partial \psi/\partial y)}{\partial \eta} \frac{\partial \eta}{\partial y}$$

show that Eq. 8.6.49 follow from Eq. 8.6.47.

8.126 Show that Eqs. 8.6.51 follow from the preceding equations.

8.127 Solve Eq. 8.6.52 with the appropriate boundary conditions, using a third-order Runge–Kutta scheme (or any other appropriate numerical algorithm) and verify the results of Table 8.5. (This was a Ph.D. thesis project for Blasius before the advent of the computer!)

8.128 A laminar boundary layer exists on a flat plate with atmospheric air at 20°C moving at 5 m/s. At $x = 2$ m, find:
- **(a)** The wall shear.
- **(b)** The boundary layer thickness.
- **(c)** The maximum value of v.
- **(d)** The flow rate through the boundary layer.

8.129 A laminar boundary layer exists on a flat plate with atmospheric air at 15°C moving at 5 m/s. At $x = 2$ m, find:
- **(a)** The wall shear.
- **(b)** The boundary layer thickness.
- **(c)** The maximum value of v.
- **(d)** The flow rate through the boundary layer.

8.130 Water at 20°C flows over a flat plate with zero pressure gradient at 5 m/s. At $x = 2$ m, find:
   (a) The wall shear.
   (b) The boundary layer thickness.
   (c) The flow rate through the boundary layer.

8.131 If, when we defined the boundary layer thickness, we defined δ to be that y-location where $u = 0.999U_\infty$, estimate the thickness of the boundary layer of:
   (a) Problem 8.128.   **(b)** Problem 8.129.

8.132 Find the y-location where $u = 0.5U_\infty$ for the boundary layer of Problem 8.128. What is the value of v at that y-location? What is the shear stress there?

8.133 Assume that v remains unchanged between $y = \delta$ and $y = 10\delta$. Sketch $u(y)$ for $0 \le y < 10\delta$ for a flat plate with zero pressure gradient. Now, assume that $v = 0$ at $y = 10\delta$. Again sketch $u(y)$. Explain by referring to appropriate equations.

8.134 Sketch the boundary layer velocity profile near the end of the flat plate of Problem 8.111 and show a Blasius profile of the same thickness on the same sketch.

Pressure Gradient Effects

8.135 Sketch the velocity profiles near and normal to the cylinder's surface at each of the points indicated in Fig. P8.135. The flow separates at C.

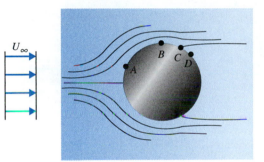

FIGURE P8.135

8.136 Sketch the expected boundary layer profiles at each of the points indicated in Fig. P8.136, showing relative thicknesses. The flow undergoes transition to turbulence just after point A. It separates at D. Show all profiles on the same plot. Indicate the sign of the pressure gradient at each point.

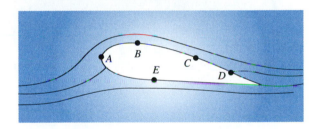

FIGURE P8.136

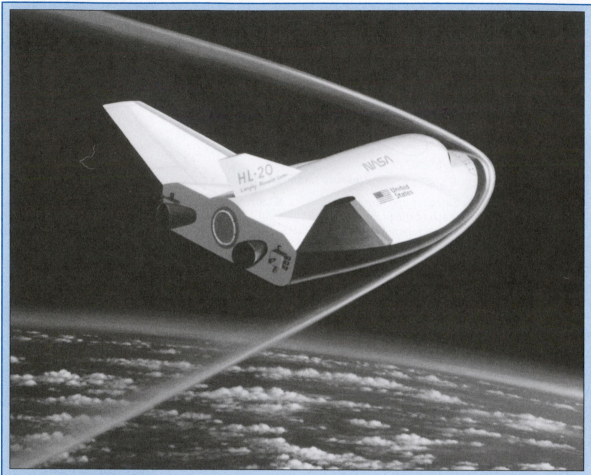

A new-generation space vehicle is rendered by an artist as it re-enters the earth's atmosphere. A shock wave is shown to be generated at the leading edge of the vessel, signifying that it is traveling faster than the speed of sound. (U.S. National Aeronautics and Space Administration)

9

Compressible Flow

Outline

Chapter Objectives

The objectives of this chapter are to:

▲ Present the various equations needed to solve the uniform-flow problems of a compressible gas flow.

▲ Apply the basic equations to isentropic flow through a nozzle.

▲ Introduce the normal shock wave.

▲ Solve for velocities and pressures when a shock wave exists in a converging-diverging nozzle.

▲ Analyze the supersonic flow of steam through a nozzle.

▲ Study the oblique shock wave needed to turn a supersonic flow through an angle.

▲ Calculate the angle through which expansion waves can turn a flow around a convex corner.

▲ Provide numerous examples and problems that illustrate isentropic flow, flow through normal and oblique shock waves, and the turning of a supersonic flow around a convex corner.

9.1 INTRODUCTION

Compressible flow: *The flow of gas in which the density changes significantly between points on a streamline.*

In this chapter we consider flows of gases in which the density changes significantly between points on a streamline; such flows are called **compressible flows**. We will consider those problems that can be solved using the integral equations: the continuity equation, the energy equation, and for some problems, the momentum equation.

KEY CONCEPT *Not all gas flows are compressible flows.*

Not all gas flows are compressible flows, neither are all compressible flows gas flows. At low speeds, less than a Mach number (M = $V/\sqrt{kRT}$) of about 0.3, gas flows may be treated as incompressible flows. This is justified because the density variations caused by the flow are negligible (less than about 3%). Incompressible gas flows occur in a large number of situations of engineering interest; many of these have been considered in earlier chapters. There are many flows, however, in which the density variations must be accounted for. Included among these are airflows around commercial and military aircraft, airflow through jet engines, and the flow of a gas in compressors and turbines.

There are examples of compressible effects important in liquid flows; water hammer and compression waves due to underwater blasts are examples of compressible liquid flows. Compressibility of rock accounts for the propagation through the earth's surface of longitudinal waves due to an earthquake. In this chapter we are concerned with compressibility effects in gas flows.

Let us introduce the effects of compressibility into the simplest of flow situations. The velocity at a given streamwise location in a conduit is assumed to be uniform and hence does not vary normal to the flow direction. For this simple uniform flow we recall that the continuity equation takes the form (see Eq. 4.4.5)

$$\dot{m} = \rho_1 A_1 V_1 = \rho_2 A_2 V_2 \tag{9.1.1}$$

The momentum equation for the uniform, compressible flow takes the form (see Eq. 4.6.6)

$$\Sigma \mathbf{F} = \dot{m}(\mathbf{V}_2 - \mathbf{V}_1) \tag{9.1.2}$$

The energy equation, neglecting potential energy changes, is written (see Eqs. 4.5.17 and 4.5.18)

$$\frac{\dot{Q} - \dot{W}_s}{\dot{m}} = \frac{V_2^2 - V_1^2}{2} + h_2 - h_1 \tag{9.1.3}$$

where we have used $h = \tilde{u} + p/\rho$. Assuming an ideal gas with constant specific heats, the energy equation takes the form

$$\frac{\dot{Q} - \dot{W}_s}{\dot{m}} = \frac{V_2^2 - V_1^2}{2} + c_p(T_2 - T_1) \tag{9.1.4}$$

or

$$\frac{\dot{Q} - \dot{W}_s}{\dot{m}} = \frac{V_2^2 - V_1^2}{2} + \frac{k}{k-1}\left(\frac{p_2}{\rho_2} - \frac{p_1}{\rho_1}\right) \qquad (9.1.5)$$

where we have used the thermodynamic relations

$$h_2 - h_1 = c_p(T_2 - T_1) \qquad c_p = R + c_v \qquad k = \frac{c_p}{c_v} \qquad (9.1.6)$$

and the ideal-gas law

$$p = \rho R T \qquad (9.1.7)$$

Should we have an interest in calculating the entropy change between two sections, we will use the definition of entropy as

$$\Delta S = \int \left. \frac{\delta Q}{T}\right|_{\text{reversible}} \qquad (9.1.8)$$

where δQ represents the differential heat transfer. Using the first law, this becomes, for an ideal gas with constant specific heat,

$$\Delta s = c_p \ln \frac{T_2}{T_1} - R \ln \frac{p_2}{p_1} \qquad (9.1.9)$$

If a process is adiabatic ($Q = 0$) and reversible (no losses) Eq. 9.1.8 shows that the entropy change is zero (i.e., the flow is *isentropic*). If the flow is isentropic, the relationship above can be used, along with the ideal-gas law and Eqs. 9.1.6b and c, to show that

$$\frac{T_2}{T_1} = \left(\frac{p_2}{p_1}\right)^{(k-1)/k} \qquad \frac{p_2}{p_1} = \left(\frac{\rho_2}{\rho_1}\right)^{k} \qquad (9.1.10)$$

Note: Temperatures and pressures must be measured on absolute scales.

Let us now introduce a parameter, the Mach number, that will be of special interest throughout our study of compressible flow.

9.2 SPEED OF SOUND AND THE MACH NUMBER

The speed of sound is the speed at which a pressure disturbance of small amplitude travels through a fluid. It is analogous to the small ripple, a gravity wave, that travels radially outward when a pebble is dropped into a pond. To determine the speed of sound consider a small pressure disturbance, called a **sound wave**, to be passing through a pipe, as shown in Fig. 9.1. It travels with a velocity c relative to a stationary observer as shown in part (a); the pressure, density, and temperature will change by the small amounts Δp, $\Delta \rho$, and ΔT, respectively. There will also be an induced velocity ΔV in the fluid immediately behind the sound wave. To simplify the problem we will create a steady flow by having the observer travel at the speed of the wave so that the sound wave appears to be stationary. The flow will then approach the wave from the right, with the speed of sound c, as shown in Fig. 9.1b. The flow properties will all change across the wave and the velocity in the downstream flow will be expressed as $c + \Delta V$, where ΔV is the small change in velocity. If the flow speed is reduced, the wave would move to the right, and if the flow speed is increased it would move to the left. For the flow speed equal to the speed of sound, $V = c$, the sound wave would be stationary, as shown.

Let us apply the continuity equation and the energy equation to a small control volume enclosing the sound wave, shown in Fig. 9.1c. The continuity equation (9.1.1) takes the form

$$\rho A c = (\rho + \Delta \rho)A(c + \Delta V) \tag{9.2.1}$$

This can be rewritten as

$$\rho \Delta V = -c \Delta \rho \tag{9.2.2}$$

where we have neglected the higher-order term $\Delta \rho\, \Delta V$; that is, $\Delta \rho$ represents a small percentage change in ρ so that $\Delta \rho \ll \rho$.

The streamwise-component momentum equation yields

$$pA - (p + \Delta p)A = \rho A c(c + \Delta V - c) \tag{9.2.3}$$

which simplifies to, neglecting higher-order terms,

$$-\Delta p = \rho c\, \Delta V \tag{9.2.4}$$

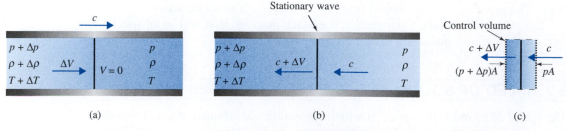

FIGURE 9.1 Sound wave: (a) stationary observer; (b) observer moving with the wave; (c) control volume enclosing the wave.

Combining this with Eq. 9.2.2 results in

$$c = \sqrt{\frac{\Delta p}{\Delta \rho}} \qquad (9.2.5)$$

Since the changes Δp and $\Delta \rho$ are quite small, we can write

$$\frac{\Delta p}{\Delta \rho} \simeq \frac{dp}{d\rho} \qquad (9.2.6)$$

Small-amplitude, moderate-frequency waves (up to about 18 000 Hz) travel with no change in entropy (isentropically), so that

$$\frac{p}{\rho^k} = \text{const.} \qquad (9.2.7)$$

This can be differentiated to give

$$\frac{dp}{d\rho} = k\frac{p}{\rho} \qquad (9.2.8)$$

Using this in Eq. 9.2.5, the speed of sound c is given by

$$c = \sqrt{\frac{kp}{\rho}} \qquad (9.2.9)$$

or, using the ideal-gas law,

$$c = \sqrt{kRT} \qquad (9.2.10)$$

At high frequency, sound waves generate friction and the process ceases to be isentropic. It is better approximated by an isothermal process. For an ideal gas an isothermal approximation would lead to

$$c = \sqrt{RT} \qquad (9.2.11)$$

For small waves traveling through liquids or solids the *bulk modulus* is used; it has dimensions of pressure and is equal to $\rho dp/d\rho$. For water it has a nominal value of 2110 MPa. It does vary slightly with temperature and pressure. Using Eq. 9.2.5, this leads to a speed of propagation of 1450 m/s for a small-amplitude pressure wave in water.

An important quantity used in the study of compressible flow is the dimensionless velocity called the *Mach number*, introduced in Eq. 3.3.3 as

KEY CONCEPT *A small-amplitude pressure wave in water travels 1450 m/s.*

$$M = \frac{V}{c} \qquad (9.2.12)$$

If M < 1, the flow is a *subsonic* flow, and if M > 1, it is a *supersonic* flow.

If a source of sound waves is at a fixed location, the waves travel radially away from the source with the speed of sound c. Figure 9.2a shows the position of the sound waves after a time increment Δt and after multiples of Δt. Part (b) shows a source moving with a velocity V which is less than the speed of sound. Note that the sound waves always propagate ahead of the source so that an airplane traveling at a speed less than the speed of sound will always "announce" its approach. This is not true, however, for an object traveling at a speed greater than the speed of sound, as shown in Fig. 9.2c. The region outside the cone is a *zone of silence*, so that an approaching object moving at a supersonic speed could not be heard until it passed overhead and the *Mach cone*, the cone shown, intercepted the observer. From the figure the angle α of the Mach cone is given by

$$\alpha = \sin^{-1}\frac{c}{V} = \sin^{-1}\frac{1}{M} \qquad (9.2.13)$$

The discussion above is limited to small-amplitude sound waves, often called *Mach waves*. They are formed on the needle nose of an aircraft, or on the leading edge of an airfoil if the leading edge is sufficiently sharp. If the nose is blunt or if the leading edge is not sufficiently sharp, a supersonic aircraft will produce a large-amplitude wave called a **shock wave**. The shock wave will also form a zone of silence, but the initial angle at the source created by the shock wave will be larger than that of a Mach wave. Shock waves will be considered in subsequent sections.

Shock wave: *A large-amplitude wave which may be created by a blunt object.*

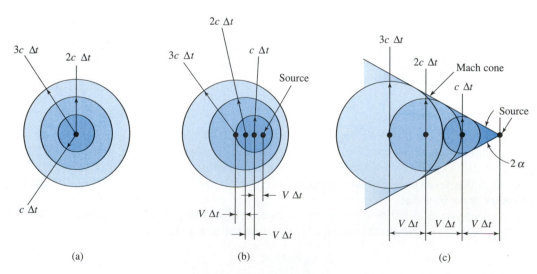

FIGURE 9.2 Sound waves propagating from a noise source: (a) stationary source; (b) moving source: $V < c$; (c) moving source, $V > c$.

Example 9.1

A needle-nose projectile traveling at a speed with M = 3 passes 200 m above the observer of Fig. E9.1. Calculate the projectile's velocity and determine how far beyond the observer the projectile will first be heard.

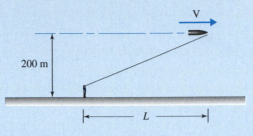

FIGURE E9.1

Solution

At a Mach number of 3 the velocity is

$$V = Mc = M\sqrt{kRT}$$

$$= 3\sqrt{1.4 \times 287 \times 288} = 1021 \text{ m/s}$$

where a standard temperature of 15°C has been assumed since no temperature is given. Using h as the height and L as the distance beyond the observer (refer to Fig. 9.2c), we have

$$\sin \alpha = \frac{h}{\sqrt{L^2 + h^2}} = \frac{1}{M}$$

With the information given,

$$\frac{200}{\sqrt{L^2 + 200^2}} = \frac{1}{3}$$

giving

$$L = 566 \text{ m}$$

Note: The units on kRT are $\dfrac{\text{N} \cdot \text{m}}{\text{kg} \cdot \text{K}} \times \text{K} = \dfrac{\text{N} \cdot \text{m}}{\text{N} \cdot \text{s}^2/\text{m}} = \dfrac{\text{m}^2}{\text{s}^2}$. The quantity k is dimensionless.

9.3 ISENTROPIC NOZZLE FLOW

There are many applications where gas flows through a section of tube or conduit that has a changing area in which a steady, uniform, isentropic flow is a good approximation to the actual flow situation. The diffuser near the front of a jet engine, exhaust gases passing through the blades of a turbine, the nozzles on a rocket engine, a broken natural gas line, and gas flow measuring devices are all examples of situations that can be modeled with a steady, uniform, isentropic

KEY CONCEPT *Steady, uniform, isentropic flow is a good approximation to many flow situations.*

flow. Consider the flow through the infinitesimal control volume shown in Fig. 9.3. With a changing area the continuity equation

$$\rho AV = \text{const.} \tag{9.3.1}$$

applied between two sections a distance dx apart takes the form

$$\rho AV = (\rho + d\rho)(A + dA)(V + dV) \tag{9.3.2}$$

Keeping only first-order terms in the differential quantities, Eq. 9.3.2 can be put in the form

$$\frac{dV}{V} + \frac{dA}{A} + \frac{d\rho}{\rho} = 0 \tag{9.3.3}$$

The energy equation can be written (see Eq. 9.1.5)

$$\frac{V^2}{2} + \frac{k}{k-1}\frac{p}{\rho} = \text{const.} \tag{9.3.4}$$

For the present application we have

$$\frac{V^2}{2} + \frac{k}{k-1}\frac{p}{\rho} = \frac{(V+dV)^2}{2} + \frac{k}{k-1}\frac{p+dp}{\rho+d\rho} \tag{9.3.5}$$

or again, retaining only first-order terms,

$$V\,dV + \frac{k}{k-1}\frac{\rho\,dp - p\,d\rho}{\rho^2} = 0 \tag{9.3.6}$$

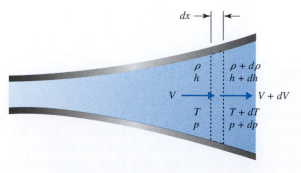

FIGURE 9.3 Uniform, isentropic flow.

For an isentropic process we use Eq. 9.2.8 and there results

$$V \, dV + k \frac{p}{\rho^2} \, d\rho = 0 \qquad (9.3.7)$$

Substituting for $d\rho/\rho$ from Eq. 9.3.3, the above becomes

$$\frac{dV}{V} \left(\frac{\rho V^2}{kp} - 1 \right) = \frac{dA}{A} \qquad (9.3.8)$$

In terms of the speed of sound this is written as

$$\frac{dV}{V} \left(\frac{V^2}{c^2} - 1 \right) = \frac{dA}{A} \qquad (9.3.9)$$

Introduce the Mach number and we have the very important relationship

$$\frac{dV}{V} (M^2 - 1) = \frac{dA}{A} \qquad (9.3.10)$$

for an isentropic uniform flow in a changing area.

From Eq. 9.3.10 we can make the following observations:

1. If the area is increasing, $dA > 0$, and $M < 1$, we see that dV must be negative, that is, $dV < 0$. The flow is decelerating for this subsonic flow.
2. If the area is increasing and $M > 1$, we see that $dV > 0$; hence the flow is accelerating in the diverging section for this supersonic flow.
3. If the area is decreasing and $M < 1$, then $dV > 0$, resulting in an accelerating flow.
4. If the area is decreasing and $M > 1$, then $dV < 0$, indicating a decelerating flow.
5. At a throat where $dA = 0$, either $dV = 0$ or $M = 1$, or possibly both.

If we define a **nozzle** as a device that accelerates the flow, we see that observations 2 and 3 describe a nozzle and observations 1 and 4 describe a **diffuser**, a device that decelerates the flow. The supersonic flow leads to rather surprising results: an accelerating flow in an enlarging area and a decelerating flow in a decreasing area. This is, in fact, the situation encountered in a traffic flow on a freeway; hence a supersonic flow might be used to model a traffic flow.

Note that the observations above prohibit a supersonic flow in a converging section attached to a reservoir. If a supersonic flow is to be generated by releasing a gas from a reservoir, there must be a converging section in which a subsonic flow accelerates to the throat where $M = 1$ followed by a diverging section in which the flow continues to accelerate with $M > 1$, as shown in Fig. 9.4. This is the type of nozzle observed on rockets used to place satellites into orbit.

Nozzle: *A device that accelerates the flow.*

Diffuser: *A device that decelerates the flow.*

The isentropic flow in the nozzle will now be considered in more detail. The energy equation between the reservoir where $V_0 = 0$ and any section can be written in the form

$$c_p T_0 = \frac{V^2}{2} + c_p T \qquad (9.3.11)$$

Stagnation quantities:
Quantities with a zero subscript at a location where $V = 0$.

Quantities with a zero subscript are often called **stagnation quantities** since they occur at a location where $V = 0$. Recognizing that $V = Mc$, $c_p/c_v = k$, $c_p = c_v + R$, and $c = \sqrt{kRT}$, we can write the energy equation as

$$\frac{T_0}{T} = 1 + \frac{k-1}{2} M^2 \qquad (9.3.12)$$

For our isentropic flow the pressure and density ratios are expressed as

$$\frac{p_0}{p} = \left(1 + \frac{k-1}{2} M^2\right)^{k/(k-1)}$$

$$\frac{\rho_0}{\rho} = \left(1 + \frac{k-1}{2} M^2\right)^{1/(k-1)} \qquad (9.3.13)$$

If a supersonic flow occurs downstream of the throat, then $M = 1$ at the throat and denoting this *critical area* with an asterisk (*) superscript, we have the following critical ratios:

$$\frac{T^*}{T_0} = \frac{2}{k+1}$$

$$\frac{p^*}{p_0} = \left(\frac{2}{k+1}\right)^{k/(k-1)}$$

$$\frac{\rho^*}{\rho_0} = \left(\frac{2}{k+1}\right)^{1/(k-1)} \qquad (9.3.14)$$

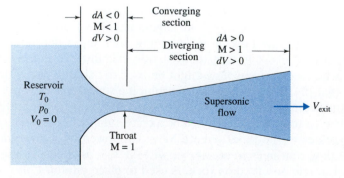

FIGURE 9.4 Supersonic nozzle.

We often make reference to the critical area even though an actual throat does not occur; we can imagine a throat occurring and call it the critical area. In fact, the isentropic flow table, Table D.1 in Appendix D, includes just such an area. For air with $k = 1.4$ the critical values are

KEY CONCEPT   *We reference the critical area even though an actual throat does not occur.*

$$p^* = 0.5283 \, p_0 \qquad T^* = 0.8333 \, T_0 \qquad \rho^* = 0.6340 \, \rho_0 \qquad (9.3.15)$$

We can determine an expression for the mass flux through the nozzle from the equation

$$\dot{m} = \rho A V$$
$$= \frac{p}{RT} A M \sqrt{kRT} = \frac{p}{\sqrt{T}} \sqrt{\frac{k}{R}} \, A M \qquad (9.3.16)$$

Using Eqs. 9.3.12 and 9.3.13, this can be expressed as

$$\dot{m} = \frac{p_0 \left(1 + \dfrac{k-1}{2} M^2\right)^{k/(1-k)}}{\sqrt{T_0} \left(1 + \dfrac{k-1}{2} M^2\right)^{-1/2}} \sqrt{\frac{k}{R}} \, A M$$

$$= p_0 \sqrt{\frac{k}{RT_0}} \, M A \left(1 + \frac{k-1}{2} M^2\right)^{(k+1)/2(1-k)} \qquad (9.3.17)$$

If we choose the critical area where $M^* = 1$, we see that

$$\dot{m} = p_0 A^* \sqrt{\frac{k}{RT_0}} \left(\frac{k+1}{2}\right)^{(k+1)/2(1-k)} \qquad (9.3.18)$$

This shows that the mass flux in the nozzle is only dependent on the reservoir conditions and the critical area A^*. By combining Eqs. 9.3.17 and 9.3.18, the area ratio A/A^* can be written in terms of the Mach number as

$$\frac{A}{A^*} = \frac{1}{M} \left[\frac{2 + (k-1)M^2}{k+1}\right]^{(k+1)/2(k-1)} \qquad (9.3.19)$$

This ratio is included in Table D.1 for airflow.

As a final consideration in our study of isentropic nozzle flow, we will present the influence of reservoir pressure and receiver pressure on the mass flux. First, the converging nozzle will be presented; then the converging–diverging nozzle will follow. The converging nozzle is assumed to be attached to a reservoir, as shown in Fig. 9.5a, with fixed conditions; the pressure in the receiver can be lowered to provide an increasing mass flux through the nozzle, as shown by the left curve in Fig. 9.5b. When the receiver pressure p_r reaches the critical pressure (for

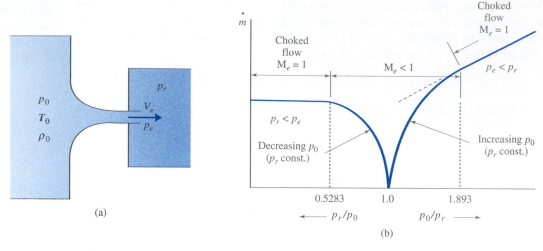

FIGURE 9.5 Converging nozzle.

air $p_r = 0.5283\, p_0$) the Mach number M_e at the throat (the exit) is unity. As p_r is reduced below this critical value the mass flux will not increase, and the condition of *choked flow* occurs as shown by the left curve in Fig. 9.5b. If the throat is a critical area, that is, $M = 1$ at the throat, then $\dot{m}$ is only dependent on the throat area A^* and the reservoir conditions, as indicated by Eq. 9.3.18. Hence, reducing p_r below the critical pressure has no effect on the upstream flow. This is reasonable since disturbances travel at the speed of sound; if p_r is reduced the disturbances that would travel upstream thereby changing conditions cannot do so since the velocity of the stream at the exit is equal to the speed of sound thereby preventing any disturbances from propagating upstream.

If, in the converging nozzle, we keep p_r constant and increase the reservoir pressure (keep T_0 constant also), a choked flow again occurs when $M_e = 1$; however, when p_0 is increased still further, we see from Eq. 9.3.18 that the mass flux will increase, as shown by the right curve of Fig. 9.5b. The pressure p_e is equal to p_r until the Mach number M_e is just equal to unity. This begins the condition of choked flow. The exit pressure p_e for the choked flow condition will be greater than the receiver pressure p_r, a condition that occurs when a gas line ruptures.

A note may be in order explaining how it is possible for the flow exit pressure p_e to exceed the receiver pressure p_r. If $p_e > p_r$, the flow exiting the nozzle is able to turn rather sharply, causing a flow pattern sketched in Fig. 9.6. This possible flow situation will be studied in Section 9.8.

Now, consider the converging–diverging nozzle, shown in Fig. 9.7, with reservoir and receiver as indicated. We will only present the condition of constant reservoir pressure and reduced receiver pressure. For this nozzle we sketch the pressure ratio p/p_0 as a function of location in the nozzle for various receiver pressure ratios p_r/p_0. If $p_r/p_0 = 1$, no flow occurs, corresponding to curve A. If p_r is reduced a small amount, curve B results and a subsonic flow exists throughout the nozzle with a minimum pressure occurring at the throat. As the pressure is reduced still further, a pressure is reached that will result in the Mach number at

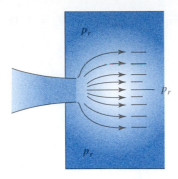

FIGURE 9.6   Nozzle exit flow for $p_e > p_r$.

the throat just being unity, as sketched by curve C; the flow remains subsonic throughout, however. Another particular receiver pressure exists, considerably below that of curve C, that will also produce an isentropic flow throughout; it results in curve D. Any receiver pressure in between these two particular pressures will result in a nonisentropic flow in the nozzle; a shock wave, to be studied later, occurs which renders our assumption of isentropic flow invalid. If the receiver pressure is below that associated with curve D, we again find that the nozzle exit pressure p_e is greater than the receiver pressure p_r. The mass flux in the nozzle increases from curve A to curve C; but as the receiver pressure is reduced below that of curve C, no increase in mass flux can occur since the throat condition will remain unchanged. The mass flux is given by Eq. 9.3.17.

Final notes regard nozzle and diffuser effectiveness. The purpose of a nozzle is to convert enthalpy (which can be thought of as stored energy) into kinetic

KEY CONCEPT   *The purpose of a nozzle is to convert stored energy into kinetic energy.*

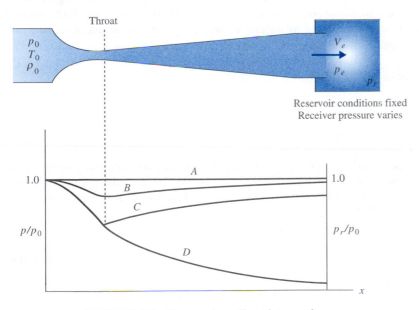

FIGURE 9.7   Converging–diverging nozzle.

energy as described by Eq. 9.1.3 with $\dot{Q} = \dot{W}_s = 0$. The efficiency η_N of a nozzle is defined as

$$\eta_N = \frac{(\Delta KE)_{actual}}{(\Delta KE)_{isentropic}}$$
$$= \frac{h_0 - h_e}{h_0 - h_{es}} \tag{9.3.20}$$

where h_e is the actual exit enthalpy and h_{es} is the isentropic exit enthalpy. Efficiencies are between 90 and 99% with larger nozzles having the higher percentages because the viscous wall effects that account for most of the losses are relatively small with the larger nozzles.

The purpose of a diffuser is to slow down the fluid and recover the pressure. For a diffuser we define the *pressure recovery factor* C_p to be

$$C_p = \frac{\Delta p_{actual}}{\Delta p_{isentropic}} \tag{9.3.21}$$

Such factors vary between 40% when the flow actually separates from the wall (the included angle should be less than about 10° for the subsonic section to avoid this separation) to 85%. Vanes can be installed in wide-angled diffusers such that each passage between vanes expands at an angle of 10° or less. Viscous effects are significantly greater in the diffuser than in the nozzle because of the thicker viscous wall layers.

The reader may think that the flow in the supersonic diverging nozzle may also tend to separate from the wall; this is not the case. Expansion fans, a phenomenon provided by nature, to be studied in a subsequent section, allow the supersonic flow to turn rather sharp angles so that supersonic nozzles are constructed with large included angles, such as those attached to the rockets that propel satellites into orbit.

Example 9.2

Air exits from a reservoir maintained at 20°C and 500 kPa absolute into a receiver maintained at (a) 300 kPa absolute and (b) 200 kPa absolute. Estimate the mass flux if the exit area is 10 cm². Use the equations first and then the isentropic flow table, Table D.1. Refer to Fig. 9.5.

Solution

To estimate the mass flux we will assume isentropic flow. For air the receiver pressure that would result in $M_e = 1$ is

$$p_r = 0.5283 \, p_0 = 0.5283 \times 500 = 264.2 \text{ kPa}$$

For part (a) $M_e < 1$ since $p_r > 264.2$ kPa, and for part (b) choked flow occurs and $M_e = 1$ since $p_r < 264.2$ kPa.

(a) To find the exit Mach number Eq. 9.3.13 gives

$$1 + \frac{k-1}{2} M^2 = \left(\frac{p_0}{p}\right)^{(k-1)/k}$$

or

$$M_e^2 = \frac{2}{k-1} \left(\frac{p_0}{p}\right)^{(k-1)/k} - \frac{2}{k-1}$$

$$= \frac{2}{0.4} \left(\frac{500}{300}\right)^{0.2857} - \frac{2}{0.4} = 0.7857 \quad \therefore M_e = 0.8864$$

The mass flux is given by Eq. 9.3.17 and is found to be

$$\dot{m} = p_0 \sqrt{\frac{k}{RT_0}} MA \left(1 + \frac{k-1}{2} M^2\right)^{(k+1)/2(1-k)}$$

$$= 500\,000 \sqrt{\frac{1.4}{287 \times 293}} \times 0.8864 \times 0.001 \left(1 + \frac{0.4}{2} \times 0.8864^2\right)^{-2.4/0.8}$$

$$= 1.167 \text{ kg/s}$$

(b) Choked flow occurs and thus $M_e = 1$ and Eq. 9.3.18 yields

$$\dot{m} = p_0 A^* \sqrt{\frac{k}{RT_0}} \left(\frac{k+1}{2}\right)^{(k+1)/2(1-k)}$$

$$= 500\,000 \times 0.001 \sqrt{\frac{1.4}{287 \times 293}} \left(\frac{2.4}{2}\right)^{-2.4/0.8} = 1.181 \text{ kg/s}$$

Now, let us use the isentropic flow table (Table D.1) and solve parts (a) and (b).

(a) For a pressure ratio of $p/p_0 = 300/500 = 0.6$, we interpolate to find

$$M_e = \frac{0.6041 - 0.6}{0.6041 - 0.5913} \times 0.02 + 0.88 = 0.886$$

$$\frac{T_e}{T_0} = \frac{0.6041 - 0.6}{0.6041 - 0.5913} (0.8606 - 0.8659) + 0.8659 = 0.864$$

$$T_e = 0.864 \times 293 = 253 \text{ K}$$

The velocity and density are, respectively,

$$V = Mc = 0.886 \sqrt{1.4 \times 287 \times 253} = 282 \text{ m/s}$$

$$\rho = \frac{p}{RT} = \frac{300}{0.287 \times 253} = 4.13 \text{ kg/m}^3$$

The mass flux is then

$$\dot{m} = \rho A V$$

$$= 4.13 \times 0.001 \times 282 = 1.165 \text{ kg/s}$$

(b) For choked flow we know that $M_e = 1$. The table gives

(continued)

$$\frac{T_e}{T_0} = 0.8333 \qquad \text{and} \qquad \frac{p_e}{p_0} = 0.5283$$

Thus the temperature, velocity, and density are, respectively,

$$T = 0.8333 \times 293 = 244.2 \text{ K}$$

$$V = \text{M}c = 1 \times \sqrt{1.4 \times 287 \times 244.2} = 313.2 \text{ m/s}$$

$$\rho = \frac{p}{RT} = \frac{0.5283 \times 500}{0.287 \times 244.2} = 3.769 \text{ kg/m}^3$$

The mass flux is calculated to be

$$\dot{m} = \rho A V$$

$$= 3.769 \times 0.001 \times 313.2 = 1.180 \text{ kg/s}$$

The results using the equations are essentially the same as those using the tables.

Example 9.3

A converging–diverging nozzle, with an exit area of 40 cm^2 and a throat area of 10 cm^2, is attached to a reservoir with $T = 20°C$ and $p = 500$ kPa absolute. Determine the two exit pressures that result in M = 1 at the throat for an isentropic flow. Also, determine the associated exit temperatures and velocities.

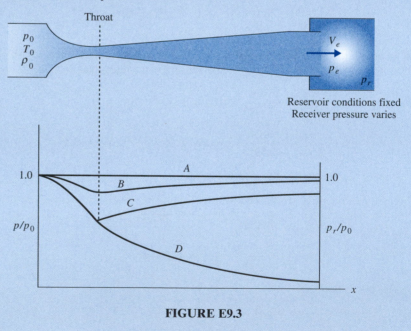

FIGURE E9.3

Solution

The exit pressures we seek are associated with curves C and D of Fig. E9.3. The area ratio is

$$\frac{A}{A^*} = \frac{40}{10} = 4$$

We could solve Eq. 9.3.19 for M using a trial-and-error technique; however, it is simpler to use the isentropic flow table, Table D.1. There are two entries for $A/A^* = 4$. Interpolation gives

$$\left(\frac{p}{p_0}\right)_C = \frac{4.182 - 4.0}{4.182 - 3.673}(0.9823 - 0.9864) + 0.9864 = 0.9849$$

$$\left(\frac{p}{p_0}\right)_D = \frac{4.0 - 3.999}{4.076 - 3.999}(0.02891 - 0.02980) + 0.02980 = 0.02979$$

Hence the two exit pressures that will result in isentropic flow are

$$p_C = 492.4 \text{ kPa} \quad \text{and} \quad p_D = 14.9 \text{ kPa}$$

Note the very small pressure difference (7.6 kPa) between receiver and reservoir necessary to create the flow condition of curve C of Fig. E9.3.

The exit temperature ratios and Mach numbers are interpolated to be

$$\left(\frac{T}{T_0}\right)_C = 0.3576(0.9949 - 0.9961) + 0.9961 = 0.9957$$

$$\left(\frac{T}{T_0}\right)_D = 0.01299(0.3633 - 0.3665) + 0.3665 = 0.3665$$

$$M_C = 0.3576 \times 0.02 + 0.14 = 0.147$$

$$M_D = 0.01299 \times 0.02 + 2.94 = 2.94$$

The exit temperatures associated with curves C and D are thus

$$T_C = 0.9957 \times 293 = 291.7 \text{ K}$$

$$T_D = 0.3665 \times 293 = 107.4 \text{ K}$$

The exit velocities are found from $V = Mc$ to be

$$V_C = 0.147 \sqrt{1.4 \times 287 \times 291.7} = 50.3 \text{ m/s}$$

$$V_D = 2.94 \sqrt{1.4 \times 287 \times 107.4} = 611 \text{ m/s}$$

Example 9.4

Gas flows are assumed to be incompressible flows at Mach numbers less than about 0.3. Determine the error involved in calculating the stagnation pressure for an airflow with $M = 0.3$.

Solution

For an incompressible airflow the energy equation (4.4.20) with no losses would give

$$p_0 = p_1 + \rho \frac{V_1^2}{2}$$

where $V_0 = 0$ at the stagnation point.

The isentropic flow equation (9.3.13), with $k = 1.4$, gives

$$p_0 = p_1(1 + 0.2M_1^2)^{3.5}$$

Use the binomial theorem $(1 + x)^n = 1 + nx + n(n - 1)x^2/2! + \cdots$ and express this as, letting $x = 0.2M_1^2$,

$$p_0 = p_1(1 + 0.7M_1^2 + 0.175M_1^4 + \cdots)$$

This can be written as (see Eqs. 9.2.9 and 9.2.12)

$$p_0 - p_1 = p_1M_1^2(0.7 + 0.175M_1^2 + \cdots)$$

$$= \frac{\rho_1 V_1^2}{1.4}(0.7 + 0.175M_1^2 + \cdots)$$

$$= \rho_1 \frac{V_1^2}{2}(1 + 0.25M_1^2 + \cdots)$$

Substituting $M_1 = 0.3$, we see that

$$p_0 - p_1 = \rho_1 \frac{V_1^2}{2}(1 + 0.0225 + \cdots)$$

Comparing this with the incompressible flow equation, we see that the error is only slightly greater than 2%. Hence it is reasonable to approximate a gas flow below $M = 0.3$ ($V \cong 100$ m/s for air at standard conditions) with an incompressible flow for many engineering applications.

9.4 NORMAL SHOCK WAVE

Small-amplitude disturbances travel at the speed of sound, as was established in a preceding section. In this section a large-amplitude disturbance will be studied. Its speed of propagation and its effect on other flow properties, such as pressure and temperature, will be considered. Large-amplitude disturbances occur in a number of situations. Examples include the flow in a gun barrel ahead of the projectile, the exit flow from a rocket or jet engine nozzle, the airflow around a supersonic aircraft, and the expanding front due to an explosion. Such large disturbances propagating through a gas are called *shock waves*. They can be oriented normal to the flow or at oblique angles. In this section we consider only the normal shock wave that occurs in a tube or directly in front of a blunt object; the photograph in Fig. 9.8 shows the shock wave in front of a sphere.

The property changes that occur across a shock wave take place over an extremely short distance. For usual conditions the distance is only several mean free paths of the molecules, on the order of 10^{-4} mm. Phenomena such as viscous dissipation and heat conduction that occur within the shock wave will not be studied here. We will treat the shock wave as a zero-thickness discontinuity in the flow and allow our integral control volume equations to relate the quantities of interest.

Consider a normal shock wave moving with velocity V_1. We can make it stationary by moving the flow in a tube at velocity V_1, as shown in Fig. 9.9. The continuity equation, recognizing that $A_1 = A_2$, is

$$\rho_1 V_1 = \rho_2 V_2 \tag{9.4.1}$$

The energy equation (9.1.5), with $\dot{Q} = \dot{W}_s = 0$, is

$$\frac{V_2^2 - V_1^2}{2} + \frac{k}{k-1}\left(\frac{p_2}{\rho_2} - \frac{p_1}{\rho_1}\right) = 0 \tag{9.4.2}$$

KEY CONCEPT *The changes that occur across a shock wave take place over an extremely short distance.*

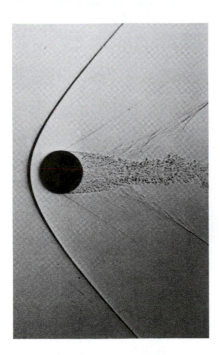

FIGURE 9.8 A shock wave is observed in front of a sphere at M = 1.53. (Photograph by A. C. Charters.)

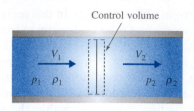

FIGURE 9.9 Stationary shock wave in a tube.

The momentum equation (9.1.2), with only pressure forces, becomes

$$p_1 - p_2 = \rho_1 V_1 (V_2 - V_1) \tag{9.4.3}$$

where the areas have divided out. These three equations allow us to determine three unknown quantities; if ρ_1, V_1, and p_1 are known, we can find ρ_2, V_2, p_2, and subsequently T_2 and M_2, using the appropriate equations.

It is convenient, however, to express the equations in terms of the Mach numbers M_1 and M_2. This results in a set of equations that are simpler to solve than solving the three simultaneous equations listed above. To do this we write Eq. 9.4.3, using $\rho_1 V_1 = \rho_2 V_2$, in the form

$$p_1 \left(1 + \frac{\rho_1 V_1^2}{p_1} \right) = p_2 \left(1 + \frac{\rho_2 V_2^2}{p_2} \right) \tag{9.4.4}$$

Introducing $M^2 = V^2 \rho / pk$, the momentum equation becomes

$$\frac{p_2}{p_1} = \frac{1 + k M_1^2}{1 + k M_2^2} \tag{9.4.5}$$

Similarly, the energy equation (9.4.2), with $p = \rho R T$, is written as

$$T_1 \left(1 + \frac{k-1}{kRT_1} \frac{V_1^2}{2} \right) = T_2 \left(1 + \frac{k-1}{kRT_2} \frac{V_2^2}{2} \right) \tag{9.4.6}$$

or, substituting $M^2 = V^2 / kRT$.

$$\frac{T_2}{T_1} = \frac{1 + \dfrac{k-1}{2} M_1^2}{1 + \dfrac{k-1}{2} M_2^2} \tag{9.4.7}$$

If we substitute $\rho = p/RT$ into the continuity equation (9.4.1), we have

$$\frac{p_1 V_1}{RT_1} = \frac{p_2 V_2}{RT_2} \tag{9.4.8}$$

which becomes, using $V = M\sqrt{kRT}$,

$$\frac{p_2}{p_1}\frac{M_2}{M_1}\sqrt{\frac{T_1}{T_2}} = 1 \tag{9.4.9}$$

Substituting for the pressure and temperature ratios from Eqs. 9.4.5 and 9.4.7, the continuity equation takes the form

$$\frac{M_1\left(1 + \dfrac{k-1}{2}M_1^2\right)^{1/2}}{1 + kM_1^2} = \frac{M_2\left(1 + \dfrac{k-1}{2}M_2^2\right)^{1/2}}{1 + kM_2^2} \tag{9.4.10}$$

Hence the downstream Mach number is related to the upstream Mach number by

$$M_2^2 = \frac{M_1^2 + \dfrac{2}{k-1}}{\dfrac{2k}{k-1}M_1^2 - 1} \tag{9.4.11}$$

This allows us to express the pressure and temperature ratios in terms of M_1 only. The momentum equation (9.4.5) takes the form

$$\frac{p_2}{p_1} = \frac{2k}{k+1}M_1^2 - \frac{k-1}{k+1} \tag{9.4.12}$$

and the energy equation becomes

$$\frac{T_2}{T_1} = \frac{\left(1 + \dfrac{k-1}{2}M_1^2\right)\left(\dfrac{2k}{k-1}M_1^2 - 1\right)}{\dfrac{(k+1)^2}{2(k-1)}M_1^2} \tag{9.4.13}$$

For air, with $k = 1.4$, the preceding three equations reduce to

$$M_2^2 = \frac{M_1^2 + 5}{7M_1^2 - 1}$$

$$\frac{p_2}{p_1} = \frac{7M_1^2 - 1}{6}$$ (9.4.14)

$$\frac{T_2}{T_1} = \frac{(M_1^2 + 5)(7M_1^2 - 1)}{36M_1^2}$$

From the first of these three equations, we observe:

- If $M_1 = 1$, then $M_2 = 1$ and no shock wave exists.
- If $M_1 > 1$, then $M_2 < 1$ and the normal shock wave converts a supersonic flow into a subsonic flow.
- If $M_1 < 1$, then $M_2 > 1$ and a subsonic flow appears to be converted into a supersonic flow by the presence of a normal shock wave. This possibility is eliminated by the second law since it would demand a decrease in entropy by a process in an isolated system, an impossibility.

The impossibility stated above is observed by considering the entropy increase, given by

$$
\begin{aligned}
s_2 - s_1 &= c_p \ln \frac{T_2}{T_1} - R \ln \frac{p_2}{p_1} \\
&= c_p \ln \frac{2 + (k-1)M_1^2}{2 + (k-1)M_2^2} - R \ln \frac{1 + kM_1^2}{1 + kM_2^2}
\end{aligned}
$$ (9.4.15)

For air, with $k = 1.4$, this is plotted in Fig. 9.10, relating M_2 to M_1, with Eq. 9.4.11. Note the impossible negative entropy change whenever $M_1 < 1$.

The relationship among the thermodynamic properties, listed in the equations above, can be demonstrated graphically with reference to the T–s diagram

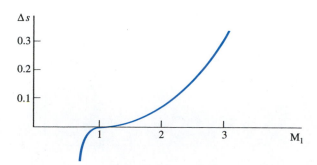

FIGURE 9.10   Entropy change for a normal shock in air.

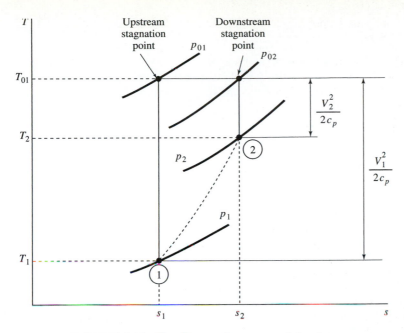

FIGURE 9.11 T–s diagram for a normal shock wave.

in Fig. 9.11. The conditions upstream of the normal shock are designated by state 1, and downstream by state 2. Note the dashed line from state 1 to state 2 for the irreversible process that occurs inside the shock wave. The energy equation, with $\dot{Q} = \dot{W}_s = 0$, can be written as

$$\frac{V_1^2}{2c_p} + T_1 = \frac{V_2^2}{2c_p} + T_2 \qquad (9.4.16)$$

The stagnation temperature is defined as the temperature that would exist if the flow is brought to rest isentropically. Thus the energy equation gives

$$T_{01} = T_{02} \qquad (9.4.17)$$

as shown in the figure. The substantial decrease in stagnation pressure, $p_{02} < p_{01}$, is also observed in Fig. 9.11. If the entropy increases from state 1 to state 2, as it must, the stagnation pressure p_{02} must decrease, as shown, if we are to maintain $T_{01} = T_{02}$.

Gas tables are available that give the pressure ratio, the temperature ratio, the downstream Mach number, and the stagnation pressure ratio as a function of the upstream Mach number. Table D.2 is such a table for $k = 1.4$ and includes the ratios given by Eq. 9.4.14. Note that M_2 is always less than unity, p_2 is always greater than p_1, T_2 is always greater than T_1, and p_{02} is always less than p_{01}.

Example 9.5

A normal shock wave passes through stagnant air at 15°C and atmospheric pressure of 80 kPa with a speed of 450 m/s. Calculate the pressure and temperature downstream of the shock wave. Use (a) the equations and (b) the gas tables.

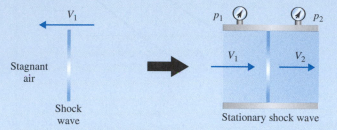

FIGURE E9.5

Solution

We consider the shock wave to be stationary with $V_1 = 450$ m/s and $p_1 = 80$ kPa. (a) To use the simplified equations (9.4.14) we must know the upstream Mach number. It is

$$M_1 = \frac{V_1}{c_1} = \frac{V_1}{\sqrt{kRT_1}}$$

$$= \frac{450}{\sqrt{1.4 \times 287 \times 288}} = 1.323$$

The pressure and temperature are then found to be

$$p_2 = \frac{p_1(7M_1^2 - 1)}{6}$$

$$= \frac{80 \times 10^3 \left(7 \times 1.323^2 - 1\right)}{6} = 150 \text{ kPa}$$

$$T_2 = \frac{T_1(M_1^2 + 5)(7M_1^2 - 1)}{36M_1^2}$$

$$= \frac{288 \left(1.323^2 + 5\right)\left(7 \times 1.323^2 - 1\right)}{36 \times 1.323^2} = 347.2 \text{ K}$$

(b) From part (a), we use $M_1 = 1.323$. Interpolation in Table D.2 yields

$$\frac{p_2}{p_1} = \frac{1.323 - 1.32}{1.34 - 1.32}\left(1.928 - 1.866\right) + 1.866 = 1.8753$$

$$\frac{T_2}{T_1} = \frac{1.323 - 1.32}{1.34 - 1.32}\left(1.216 - 1.204\right) + 1.204 = 1.2058$$

Using the information given, we have

$$p_2 = 80 \times 10^3 \times 1.8753 = 150 \times 10^3 \text{ Pa} = 150 \text{ kPa}$$

$$T_2 = 288 \times 1.2058 = 347.2 \text{ K}$$

Example 9.6

A normal shock wave propagates through otherwise stagnant air at standard conditions at a speed of 700 m/s. Determine the speed induced in the air immediately behind the shock wave as shown in Fig. E9.6.

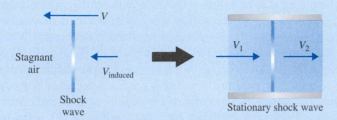

FIGURE E9.6

Solution

For standard conditions the temperature is 15°C. The upstream Mach number is thus

$$M_1 = \frac{V_1}{c_1} = \frac{V_1}{\sqrt{kRT_1}} = \frac{700}{\sqrt{1.4 \times 287 \times 288}} = 2.06$$

Using Eq. 9.4.14, we find that

$$M_2 = \left(\frac{M_1^2 + 5}{7M_1^2 - 1}\right)^{1/2} = \left(\frac{2.06^2 + 5}{7 \times 2.06^2 - 1}\right)^{1/2} = 0.567$$

$$T_2 = \frac{T_1(M_1^2 + 5)(7M_1^2 - 1)}{36M_1^2}$$

$$= \frac{288(2.06^2 + 5)(7 \times 2.06^2 - 1)}{36 \times 2.06^2} = 500.2 \text{ K}$$

This allows us to calculate

$$V_2 = M_2 c_2 = 0.567 \sqrt{1.4 \times 287 \times 500.2} = 254.4 \text{ m/s}$$

This velocity assumes a flow with the shock wave stationary and the air approaching the shock wave at 700 m/s. If we superimpose a velocity of 700 m/s moving opposite to V_1, we find the induced velocity to be

$$V_{induced} = V_2 - V_1$$
$$= 254.4 - 700 = -446 \text{ m/s}$$

where the negative sign means that the induced velocity would be moving to the left if V_1 is to the right. The induced velocity would be in the same direction as the propagation of the shock wave. Such large induced velocities are responsible for much of the damage away from the bomb center caused by explosions of high-power bombs.

9.5 SHOCK WAVES IN CONVERGING–DIVERGING NOZZLES

The converging–diverging nozzle has already been presented for an isentropic flow; for receiver-to-reservoir pressure ratios between those of curves C and D of Figs. 9.7 and 9.12, shock waves exist in the flow either inside or outside the nozzle. If $p_r/p_0 = a$ (locate a on the vertical axis on the right of Fig. 9.12), a normal shock wave would exist at an internal location in the diverging portion of the nozzle. Usually, the location of the shock wave is prescribed in student problems since to locate the shock a trial-and-error solution is necessary. When $p_r/p_0 = b$ the normal shock wave is located in the exit plane of the nozzle. For a pressure ratio less than b but greater than e, two types of oblique shock wave patterns are observed, one with a central normal shock wave, as sketched for $p_r/p_0 = c$, and one with only oblique waves, as sketched for $p_r/p_0 = d$. These pressure ratios that result in oblique shock waves will not be considered here. As we move from d to e, the oblique shock waves become weaker and weaker until the isentropic flow is again realized at $p_r/p_0 = e$ with all shock waves absent. For pressure ratios below e a very complicated flow exists. The flow turns the corner at the nozzle exit rather abruptly due to expansion waves (isentropic waves to be considered in a subsequent section), then turns back due to the same expansion waves, resulting in a billowing out of the exhaust flow, as is visible from high-altitude satellite rocket engines. Let us work some examples now for the converging–diverging nozzle; no new equations are necessary.

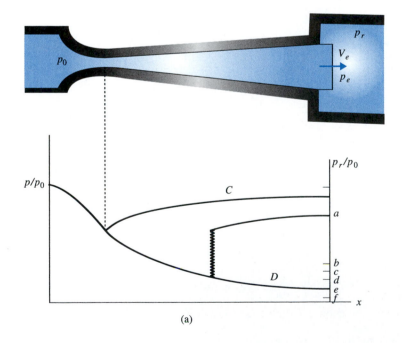

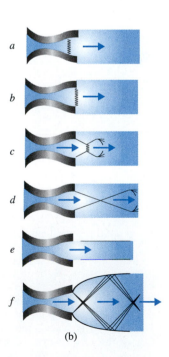

FIGURE 9.12 Converging–diverging nozzle.

Example 9.7

A converging–diverging nozzle has a throat diameter of 5 cm and an exit diameter of 10 cm. The reservoir is the laboratory, maintained at atmospheric conditions of 20°C and 90 kPa absolute. Air is constantly pumped from a receiver so that a normal shock wave stands across the exit plane of the nozzle. Determine the receiver pressure and the mass flux.

Solution

Isentropic flow occurs from the reservoir, to the throat, to the exit plane in front of the normal shock wave at state 1. Supersonic flow occurs downstream of the throat making the throat the critical area. Hence

$$\frac{A_1}{A^*} = \frac{10^2}{5^2} = 4$$

Interpolation in the isentropic flow table (Table D.1) gives

$$M_1 = 2.94 \qquad \frac{p_1}{p_0} = 0.0298$$

Hence the pressure in front of the normal shock is

$$p_1 = p_0 \times 0.0298$$
$$= 90 \times 0.0298 = 2.68 \text{ kPa}$$

From the normal shock table (Table D.2), using $M_1 = 2.94$, we find that

$$\frac{p_2}{p_1} = 9.918$$
$$\therefore p_2 = 9.918 \times 2.68 = 26.6 \text{ kPa}$$

This is the receiver pressure needed to orient the shock across the exit plane as shown for $p_r/p_0 = b$ in Fig. 9.12.

To find the mass flux through the nozzle, we need only consider the throat. Recognizing that $M_t = 1$, so that $V_t = c_t$, we can write

$$\dot{m} = \rho_t A_t V_t = \frac{p_t}{RT_t} A_t \sqrt{kRT_t} = p_t A_t \sqrt{\frac{k}{RT_t}}$$

The isentropic flow table yields

$$\frac{p_t}{p_0} = 0.5283 \qquad \frac{T_t}{T_0} = 0.8333$$

Thus the mass flux becomes

$$\dot{m} = (0.5283 \times 90\,000) \times \frac{\pi \times 0.05^2}{4} \sqrt{\frac{1.4}{287 \times (0.8333 \times 293)}}$$

$$= 0.417 \text{ kg/s}$$

Remember, the pressure must be measured in pascals in the equation above.

Example 9.8

Air flows from a reservoir at 20°C and 200 kPa absolute through a 5-cm-diameter throat and exits from a 10-cm-diameter nozzle. Calculate the pressure needed to locate a normal shock wave at a position where the diameter is 7.5 cm.

FIGURE E9.8

Solution

We will use the gas tables for this flow represented by the curve established by $p_r/p_0 = a$ in Fig. 9.12. The throat is a critical area since for a supersonic flow $M_t = 1$. The area ratio is

$$\frac{A_1}{A^*} = \frac{7.5^2}{5^2} = 2.25$$

For this area ratio we find, from Table D.1, that

$$M_1 = 2.33$$

Then from Table D.2, for this Mach number, we obtain

$$M_2 = 0.531 \qquad \frac{p_{02}}{p_{01}} = 0.570$$

The reservoir pressure $p_0 = p_{01}$. Thus

$$p_{02} = 0.570 \times 200 = 114 \text{ kPa}$$

Isentropic flow occurs from state 2 immediately after the normal shock wave to the exit. Hence, for $M_2 = 0.531$, we find from Table D.1 that

$$\frac{A_2}{A^*} = 1.285$$

so that, if A_e is the exit area,

$$\frac{A_e}{A^*} = \frac{A_2}{A^*} \times \frac{A_e}{A_2} = 1.285 \times \frac{10^2}{7.5^2} = 2.284$$

The Mach number and pressure ratio corresponding to this area ratio are

$$M_e = 0.265 \qquad \frac{p_e}{p_{0e}} = 0.952$$

For our isentropic flow between the shock and the exit, we know that $p_{02} = p_{0e}$; thus

$$p_e = p_{02} \times 0.952 = 114 \times 0.952 = 109 \text{ kPa}$$

Note the usefulness of the critical area ratio in obtaining the desired results.

Example 9.9

A pitot probe, the device used to measure the stagnation pressure in a flow, is inserted into an airstream and measures 300 kPa absolute, as shown in Fig. E9.9. The pressure in the flow is measured to be 75 kPa absolute. If the temperature at the stagnation point of the probe is measured as 150°C, determine the free-stream velocity V.

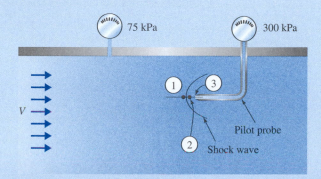

FIGURE E9.9

Solution

When a blunt object is placed in a supersonic flow a detached shock wave forms around the object, as it does around the front of the pitot probe shown. The flow that meets the front of the pitot probe at the stagnation point passes through a normal shock wave from state 1 to state 2; the subsonic flow at state 2 then decelerates isentropically to state 3, the stagnation point.

For the isentropic flow from state 2 to state 3 we can use Eq. 9.3.13,

$$\frac{p_3}{p_2} = \left(1 + \frac{k-1}{2}\,\mathrm{M}_2^2\right)^{k/(k-1)}$$

Across the normal shock we know that (see Eq. 9.4.12)

$$\frac{p_2}{p_1} = \frac{2k}{k+1}\,\mathrm{M}_1^2 - \frac{k-1}{k+1}$$

Also, the Mach numbers are related by Eq. 9.4.11,

$$\mathrm{M}_2^2 = \frac{(k-1)\mathrm{M}_1^2 + 2}{2k\mathrm{M}_1^2 - k + 1}$$

(continued)

The three equations above can be combined, with some algebraic manipulation, to yield the *Rayleigh-pitot-tube formula* for supersonic flows, namely,

$$\frac{p_3}{p_1} = \frac{\left(\dfrac{k+1}{2}M_1^2\right)^{k/(k-1)}}{\left(\dfrac{2kM_1^2}{k+1} - \dfrac{k-1}{k+1}\right)^{1/(k-1)}}$$

Substituting $k = 1.4$, $p_3 = 300$ kPa, and $p_1 = 75$ kPa. we have

$$\frac{300}{75} = \frac{(1.2M_1^2)^{3.5}}{(1.167M_1^2 - 0.1667)^{2.5}}$$

This can be solved by trial and error to give

$$M_1 = 1.65$$

The Mach number after the normal shock is interpolated from the shock table to be

$$M_2 = 0.654$$

Using the isentropic flow table with this Mach number, we interpolate the temperature at state 2 to be

$$T_2 = T_3 \times 0.921$$

$$= 423 \times 0.921 = 389.6 \text{ K}$$

The temperature in front of the normal shock is found by using the normal shock table as follows:

$$\frac{T_2}{T_1} = 1.423$$

$$\therefore T_1 = \frac{T_2}{1.423} = 274 \text{ K}$$

Finally, the velocity before the normal shock is given by

$$V_1 = M_1 c_1 = M_1 \sqrt{kRT_1}$$

$$= 1.65 \sqrt{1.4 \times 287 \times 274} = 547 \text{ m/s}$$

9.6 VAPOR FLOW THROUGH A NOZZLE

The flow of vapor through a nozzle forms a very important engineering problem. High-pressure steam flows through the nozzles of turbines in electrical generating plants; in this section we present the technique for analyzing such a problem. We recall, however, that vapor that is not substantially superheated does not behave very well as an ideal gas; the vapor tables must be consulted since c_p and c_v cannot be assumed to be constant. Consider the problem of a superheated vapor entering the nozzle of Fig. 9.12. The flow would be isentropic unless a shock wave were encountered, and the expansion would be as sketched on the T–s diagram of Fig. 9.13. Assume that the flow initiates in a reservoir with stagnation conditions, flows through a throat indicated by state t, and exits out a diverging section to the exit at state e.

Note that the exit state could very likely be in the quality region with the possibility of condensation of liquid droplets; however, for sufficiently high flow velocities there may not be sufficient time for the formation of the droplets and the associated heat transfer process. This produces a situation called *supersaturation* and a condition of *metastable equilibrium* exists; that is, the nonequilibrium state e is reached rather than the equilibrium state e'. To estimate the temperature of the metastable state e, we assume that the steam behaves as an ideal gas, so that $Tp^{k/(k-1)} = $ const. If the exit state is sufficiently far into the quality region, a *condensation shock* will be experienced, a phenomenon not considered in this book.

Because of supersaturation it is possible to model the isentropic flow of a vapor through a nozzle with acceptable accuracy by considering the ratio of specific heats to be constant. For steam $k = 1.3$ gives acceptable results over a considerable range of temperatures. The critical pressure ratio given by Eq. 9.3.14 for steam becomes

$$\frac{p^*}{p_0} = \left(\frac{2}{k+1}\right)^{k/(k-1)} = 0.546 \qquad (9.6.1)$$

If saturated steam enters the nozzle, some liquid droplets will be formed and entrained by the vapor; provided that a condensation shock does not exist, a good approximation to the critical pressure ratio is 0.577 corresponding to $k = 1.14$.

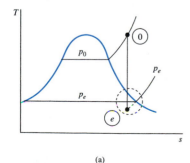

(a)

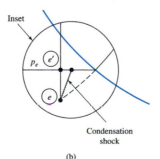

(b)

FIGURE 9.13   Isentropic expansion of a vapor.

Example 9.10

Steam is to be expanded isentropically from reservoir conditions of 300°C and 800 kPa absolute to an exit condition of 100 kPa absolute. If supersonic flow is desired, calculate the necessary throat and exit diameters if a mass flux of 2 kg//s is demanded.

Solution

From the steam tables (found in any thermodynamics textbook) we find that

$$s_0 = s_e = 7.2336 \text{ kJ/kg} \cdot \text{K}$$

$$h_0 = 3056.4 \text{ kJ/kg}$$

To estimate the temperature of the metastable exit state, we use

$$T_e = T_0 \left(\frac{p_e}{p_0}\right)^{(k-1)/k} = 593 \left(\frac{100}{800}\right)^{0.3/1.3} = 367 \text{ K} \qquad \text{or} \qquad 94°\text{C}$$

Using the steam tables at this temperature, we interpolate, using the exit quality x_e, to find that

$$7.2336 = 1.239 + 6.90 x_e$$

$$\therefore x_e = 0.968$$

Thus we have the enthalpy and specific volume at the exit:

$$h_e = 394 + 0.968 \times 2273 = 2594 \text{ kJ/kg}$$

$$v_e = 0.001 + 0.968 \times (2.06 - 0.001) = 1.99 \text{ m}^3/\text{kg}$$

Using the energy equation, the exit velocity is estimated as follows:

$$\cancel{\frac{V_0^2}{2}}^{0} + h_0 = \frac{V_e^2}{2} + h_e$$

$$\therefore V_e = \sqrt{2(h_0 - h_e)} = \sqrt{2(3056 - 2594) \times 1000} = 961 \text{ m/s}$$

where the 1000 converts kJ to J. From the definition of mass flux we have

$$m_e = \rho_e A_e V_e$$

$$2 = \frac{1}{1.99} \times \frac{\pi d_e^2}{4} \times 961$$

$$\therefore d_e = 0.0726 \text{ m} \qquad \text{or} \qquad 7.26 \text{ cm}$$

To determine the diameter of the throat, we recognize the throat to be the critical area; thus Eq. 9.6.1 gives

$$p^* = 0.546 \, p_0 = 437 \text{ kPa}$$

Using this pressure and $s^* = s_0 = 7.2336 \text{ kJ/kg} \cdot \text{K}$ we could use the steam tables to find h^* and v^*; the energy equation would then allow us to find V^* and thus d_t. However, a simpler, approximate technique, assuming constant specific heats, is to use Eq. 9.3.18 with $k = 1.3$ and obtain the following:

$$\dot{m} = p_0 A^* \sqrt{\frac{1.3}{RT_0}} \left(\frac{2.3}{2}\right)^{2.3/-0.6}$$

$$2 = 800\,000 \frac{\pi d_t^2}{4} \sqrt{\frac{1.3}{462 \times 573}} \times 0.585$$

$$\therefore d_t = 0.049 \text{ m} \quad \text{or} \quad 4.9 \text{ cm}$$

This is reasonable since we have already assumed constant specific heats in predicting T_e. Obviously, the above is approximate; using $k = 1.3$ does, though, give reasonable predictions.

9.7 OBLIQUE SHOCK WAVE

In this section we investigate the oblique shock wave, a finite-amplitude wave that is not normal to the incoming flow. The flow approaching the oblique shock wave will be assumed to be in the x-direction. After the oblique shock the velocity vector will have a component normal to the flow direction. We will continue to assume that the flow before and after the oblique shock is uniform and steady.

Oblique shock waves form on the leading edge of a supersonic airfoil or in an abrupt corner, as sketched in Fig. 9.14. Oblique shock waves may also be found on axisymmetric bodies such as a nose cone or a bullet traveling at supersonic speeds. In this book, we consider only plane flows.

The function of the oblique shock wave is to turns the flow so that the velocity vector V_2 is parallel to the plane wall. The angle between the two velocity vectors introduces another variable into our analysis. The problem remains solvable, however, with the additional tangential momentum equation.

KEY CONCEPT *The oblique shock wave turns the flow so that the velocity vector is parallel to the plane wall.*

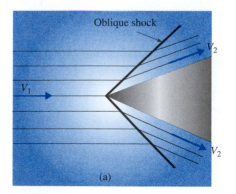
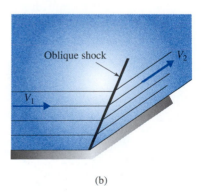

FIGURE 9.14 Oblique shock waves in a supersonic flow: (a) flow over a symmetrical wedge; (b) flow in a corner.

To analyze the oblique shock wave, consider a control volume enclosing a portion of the shock as shown in Fig. 9.15. The velocity vector upstream is assumed to be in the x-direction only; the oblique shock wave makes an angle β with the upstream velocity vector and turns the flow through the *deflection angle* or *wedge angle* θ so that V_2 is parallel to the wall. The components of the velocity vectors are shown normal and tangential to the oblique shock wave. The tangential components do not cause fluid to flow through the shock; hence the continuity equation, with $A_1 = A_2$, gives

$$\rho_1 V_{1n} = \rho_2 V_{2n} \tag{9.7.1}$$

The pressure forces act normal to the oblique shock and produce no tangential components. Thus the momentum equation expressed in the tangential direction requires that the tangential momentum into the control volume equal the tangential momentum leaving the control volume, that is,

$$\dot{m}_1 V_{1t} = \dot{m}_2 V_{2t} \tag{9.7.2}$$

or, using $\dot{m}_1 = \dot{m}_2$, we demand that

$$V_{1t} = V_{2t} \tag{9.7.3}$$

The normal momentum equation takes the form

$$p_1 - p_2 = \rho_2 V_{2n}^2 - \rho_1 V_{1n}^2 \tag{9.7.4}$$

The energy equation, using $V^2 = V_n^2 + V_t^2$, may be written as

$$\frac{V_{1n}^2}{2} + \frac{k}{k-1}\frac{p_1}{\rho_1} = \frac{V_{2n}^2}{2} + \frac{k}{k-1}\frac{p_2}{\rho_2} \tag{9.7.5}$$

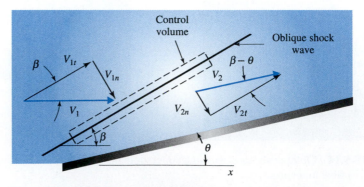

FIGURE 9.15 Control volume enclosing a small portion of an oblique shock wave.

where the tangential component terms have been canceled from both sides. Note that the tangential components of the two velocity vectors do not enter the continuity, normal momentum, or energy equations, the three equations used in the solution of the normal shock wave. Hence we may substitute V_{1n} and V_{2n} for V_1 and V_2, respectively, of the normal shock wave equations and obtain a solution. Either the normal shock wave equations or the normal shock wave table (Table D.2) may be used. Of course, we also replace M_1 and M_2 with M_{1n} and M_{2n}, respectively.

> **KEY CONCEPT** *The tangential components of the two velocity vectors do not enter the equations.*

It is useful to relate the oblique shock angle β to the deflection angle θ. Using the continuity equation (9.7.1), with reference to Fig. 9.15, there results

$$\frac{\rho_2}{\rho_1} = \frac{V_{1n}}{V_{2n}} = \frac{V_{1t}\tan\beta}{V_{2t}\tan(\beta - \theta)} = \frac{\tan\beta}{\tan(\beta - \theta)} \tag{9.7.6}$$

From the normal shock wave equations (9.4.12) and (9.4.13) we can find the density ratio to be

$$\frac{\rho_2}{\rho_1} = \frac{p_2 T_1}{p_1 T_2} = \frac{(k+1)M_{1n}^2}{(k-1)M_{1n}^2 + 2} \tag{9.7.7}$$

Substituting this into Eq. 9.7.6 gives

$$\tan(\beta - \theta) = \frac{\tan\beta}{k+1}\left[k - 1 + \frac{2}{M_1^2 \sin^2\beta}\right] \tag{9.7.8}$$

For a flow with a given M_1, this equation relates the oblique shock angle β to the wedge or corner angle θ. The three variables β, θ, and M_1 in the equation above are often plotted as in Fig. 9.16. We can observe several phenomena by studying the figure.

- For a specified upstream Mach number M_1 and a given wedge angle θ there are two possible oblique shock angles β, the large one corresponding to a "strong" shock and the smaller corresponding to a "weak" shock.
- For a given wedge angle θ there is a minimum Mach number for which there is only one oblique shock angle β.
- For a given wedge angle θ, if M_1 is less than the minimum for that particular curve, no oblique shock wave exists and the shock wave becomes detached, as shown in Fig. 9.17. Also, for a given M_1 there is a sufficiently large θ that will result in a detached shock wave.

The pressure rise across the oblique shock wave determines whether a weak shock or a strong shock occurs. For a relatively small pressure rise a weak shock will occur with $M_2 > 1$. If the pressure rise is relatively large a strong shock

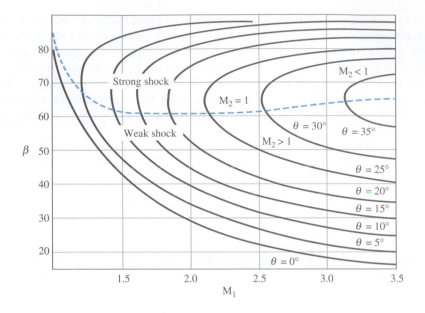

FIGURE 9.16 Oblique shock wave relationships for $k = 1.4$.

occurs with $M_2 < 1$. Note that for the detached shocks around bodies a normal shock exists for the stagnation streamline; this is followed away from the stagnation point by the strong oblique shock, then the weak oblique shock, and eventually a Mach wave. For blunt bodies moving at supersonic speeds the shock wave is always detached.

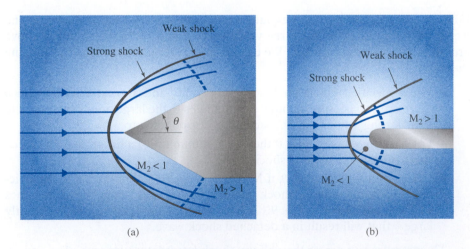

FIGURE 9.17 Detached shock waves: (a) flow around a wedge; (b) flow around a blunt object.

Example 9.11

Air flows over a wedge with $M_1 = 3$ as shown in Fig. E9.11. A weak shock reflects from the wall. Determine the values of M_3 and β_3 for the reflected wave.

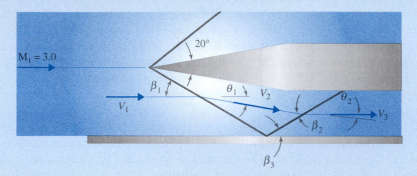

FIGURE E9.11

Solution

From Fig. 9.16 with $\theta_1 = 10°$ and $M_1 = 3.0$, we find for the weak shock that $\beta_1 = 27.5°$. This yields

$$M_{1n} = 3 \sin 27.5° = 1.39$$

From the shock table we interpolate to find

$$M_{2n} = 0.744 = M_2 \sin(27.5° - 10°)$$
$$\therefore M_2 = 2.48$$

The reflected shock must again turn the flow through an angle of 10°, that is, $\theta_2 = 10°$. For this wedge angle and $M_2 = 2.48$ from Fig. 9.16 for a weak shock, we see that $\beta_2 = 33°$. This results in

$$M_{2n} = 2.48 \sin 33° = 1.35$$

From the shock table

$$M_{3n} = 0.762 = M_3 \sin 23°$$
$$\therefore M_3 = 1.95$$

The desired angle is calculated to be

$$\beta_3 = \beta_2 - 10° = 23°$$

Note that Fig. 9.16 does not allow precise calculations. Equation 9.7.8 could be used, by trial and error, to improve the accuracy of the β's and hence the quantities that follow.

9.8 ISENTROPIC EXPANSION WAVES

In this section we consider the supersonic flow around a convex corner, as shown in Fig. 9.18. Let us first attempt to accomplish such a flow with the finite amplitude wave of Fig. 9.18a. The flow must turn the angle θ so that V_2 is parallel to the wall. The tangential component must be conserved because of momentum conservation. This would result in $V_2 > V_1$, as is obvious from the sketch. This would be the situation if a subsonic flow, $M_{1n} < 1$, could experience a finite increase to a supersonic flow, $M_{2n} > 1$. This, of course, is impossible because of the second law, as was noted in the discussion associated with Fig. 9.10. Consequently, we consider the turning of a finite wave around a convex corner is an impossibility.

Consider a second possible mechanism that would allow the flow to turn the corner, a fan composed of an infinite number of Mach waves, emanating from the corner, as shown in Fig. 9.18b. The second law would not be violated with such a mechanism since each Mach wave is an isentropic wave. We will determine the effect of a single Mach wave on the flow and then integrate to obtain the total effect. Figure 9.19 shows the infinitesimal velocity change due to a single Mach wave. For the control volume enclosing the Mach wave, we know that the tangential momentum is conserved; thus the tangential velocity component remains unchanged as shown, as in the oblique shock wave. From the triangles in the figure we can write

$$V_t = V \cos \mu = (V + dV) \cos(\mu + d\theta) \qquad (9.8.1)$$

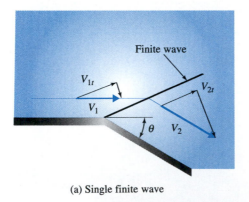

(a) Single finite wave

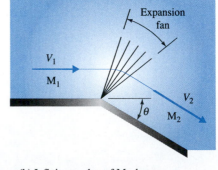

(b) Infinite number of Mach waves

FIGURE 9.18 Supersonic flow around a convex corner.

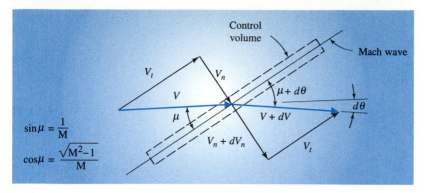

FIGURE 9.19 Mach wave.

$$V \sin \mu \, d\theta = \cos \mu \, dV \tag{9.8.2}$$

Substituting $\sin \mu = 1/M$ (see Eq. 9.2.13) and $\cos \mu = (\sqrt{M^2 - 1})/M$, we have

$$d\theta = \sqrt{M^2 - 1} \, \frac{dV}{V} \tag{9.8.3}$$

The relationship $V = M \sqrt{kRT}$ can be differentiated and rearranged to give

$$\frac{dV}{V} = \frac{dM}{M} + \frac{1}{2}\frac{dT}{T} \tag{9.8.4}$$

The energy equation, in the form $V^2/2 + kRT/(k-1) = $ const., may also be differentiated to yield

$$\frac{dV}{V} + \frac{1}{(k-1)M^2}\frac{dT}{T} = 0 \tag{9.8.5}$$

Eliminating dT/T by combining the two preceding equations, there results

$$\frac{dV}{V} = \frac{2}{2 + (k-1)M^2}\frac{dM}{M} \tag{9.8.6}$$

---

[1] Recall the trigonometric identity, $\cos(\alpha + \beta) = \cos \alpha \cos \beta - \sin \alpha \sin \beta$. Then using $\cos d\theta = 1$ and $\sin d\theta = d\theta$, we have $\cos(\mu + d\theta) = \cos \mu - d\theta \sin \mu$.

This can be substituted into Eq. 9.8.3, allowing a relationship between θ and M to be obtained. We have

$$d\theta = \frac{2\sqrt{M^2 - 1}}{2 + (k-1)M^2} \frac{dM}{M} \tag{9.8.7}$$

This can be integrated, using $\theta = 0$ at $M = 1$, to provide a relationship between the resulting Mach number (M_2 in Fig. 9.18) and the angle, provided that the incoming Mach number is unity; the relationship is

$$\theta = \left(\frac{k+1}{k-1}\right)^{1/2} \tan^{-1}\left[\frac{k-1}{k+1}(M^2 - 1)\right]^{1/2} - \tan^{-1}(M^2 - 1)^{1/2} \tag{9.8.8}$$

Prandtl–Meyer function:
The angle θ through which the supersonic flow turns.

The angle θ, which is a function of M, is the **Prandtl–Meyer function**. It is tabulated for $k = 1.4$ in Table D.3 so that trial-and-error solutions to Eq. 9.8.8 for M are not necessary. Other changes that may be desired, such as the pressure or temperature changes, can be found from the isentropic-flow equations. The collection of Mach waves that turn the flow are referred to as *expansion waves*.

KEY CONCEPT *The flow remains attached to the wall as it turns the corner, even for large angles.*

We will see, in working the examples and problems, that both the Mach number and the velocity increase as supersonic flow turns the convex corner. The flow remains attached to the wall as it turns the corner, even for large angles, a phenomenon not observed in a subsonic flow; a subsonic flow would separate from the abrupt corner, even for small angles. If we substitute $M = \infty$ in Eq. 9.8.8, we find the angle of maximum turning to be $\theta = 130.5°$. This would mean that both the temperature and pressure would be absolute zero; obviously, the gas would become a liquid before this would be possible. The angle of 130.5° is, however, an upper limit. The point is, rather large turning angles are possible in supersonic flows, angles that may exceed 90°. This introduces a design constraint on the exhaust nozzle of rocket engines that exhaust into the vacuum of space; the exhaust gases may turn through an angle so that impingement on the spacecraft body would occur if not designed properly.

KEY CONCEPT *Rather large turning angles are possible in supersonic flows.*

Example 9.12

Air at a Mach number of 2.0 and a temperature and pressure of 500°C and 200 kPa absolute, respectively, flows around a corner with a convex angle of 20° (Fig. E9.12). Find M_2, p_2, T_2, and V_2.

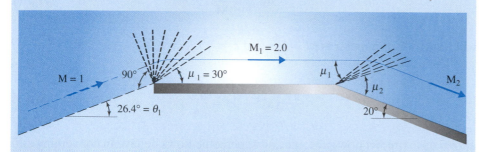

FIGURE E9.12

Solution

Table D.3 uses M = 1 as a reference condition; thus we visualize the flow as originating from a flow with M = 1 and turning through the angle θ_1 to M_1 = 2, as shown in the sketch. From the table we find, adding an additional 20° to the deflection angle, that $\theta_2 = 46.4°$. This would be equivalent to the flow at M = 1 turning a convex corner with $\theta = 46.4°$. Since the flow is isentropic, we can simply superimpose in this manner. Now, for an angle of $\theta = 46.4°$ from the table, we find that

$$M_2 = 2.83$$

From the isentropic flow table (Table D.1) we find that

$$p_2 = p_1 \frac{p_0}{p_1} \frac{p_2}{p_0}$$

$$= 200 \times \frac{1}{0.1278} \times 0.0352 = 55.1 \text{ kPa}$$

$$T_2 = T_1 \frac{T_0}{T_1} \frac{T_2}{T_0}$$

$$= 773 \times \frac{1}{0.5556} \times 0.3844 = 534.8 \text{ K} \qquad \text{or} \qquad 261.8°\text{C}$$

The velocity V_2 is found to be

$$V_2 = M_2 \sqrt{kRT_2}$$

$$= 2.83 \sqrt{1.4 \times 287 \times 534.8} = 1312 \text{ m/s}$$

9.9 SUMMARY

The zone of silence of a supersonic object that produces only Mach waves exists outside a cone with included angle α found from

$$\sin \alpha = \frac{1}{M} \tag{9.9.1}$$

where the Mach number is $M = V/c$ and $c = \sqrt{kRT}$.

The relationship

$$\frac{dV}{V}(M^2 - 1) = \frac{dA}{A} \tag{9.9.2}$$

allows us to predict how subsonic and supersonic flows behave in converging and diverging nozzles. The mass flux through a nozzle with throat area A^* where $M^* = 1$ is given by

$$\dot{m} = p_0 A^* \sqrt{\frac{k}{RT_0}} \left(\frac{k+1}{2}\right)^{(k+1)/2(1-k)} \tag{9.9.3}$$

where p_0 and T_0 are the reservoir conditions. Temperature, pressure and velocity in an isentropic flow are found from

$$\frac{T_0}{T} = 1 + \frac{k-1}{2}M^2, \quad \frac{p_0}{p} = \left(1 + \frac{k-1}{2}M^2\right)^{k/(k-1)}, \quad \frac{V^2}{2} = c_p(T_0 - T) \tag{9.9.4}$$

The flow variables across a normal shock are found from continuity, momentum, and energy:

$$\rho_1 V_1 = \rho_2 V_2, \qquad p_1 - p_2 = \rho_1 V_1(V_2 - V_1),$$
$$\frac{V_2^2 - V_1^2}{2} + \frac{k}{k-1}\left(\frac{p_2}{\rho_2} - \frac{p_1}{\rho_1}\right) = 0 \tag{9.9.5}$$

Rather than solve the above equations, we often use the normal-shock flow Table D.2.

Across an oblique shock the tangential velocity component does not change. The normal velocity component V_n simply replaces the velocity V in the normal-shock wave equations above, and Table D.2 may be used with M_n replacing M. The wedge angle θ through which the flow is turned is related to the oblique shock angle β by

$$\tan(\beta - \theta) = \frac{\tan \beta}{k+1}\left[k - 1 + \frac{2}{M_1^2 \sin^2 \beta}\right] \tag{9.9.6}$$

or Fig. 9.16 may be used to avoid a trial-and-error solution if β is desired.

In an expansion wave the angle θ through which a flow with M = 1 can be turned is the Prandtl–Meyer function.

$$\theta = \left(\frac{k+1}{k-1}\right)^{1/2} \tan^{-1}\left[\frac{k-1}{k+1}(M^2 - 1)\right]^{1/2} - \tan^{-1}(M^2 - 1)^{1/2} \quad (9.9.7)$$

If we know θ and desire M a trial-and-error solution is required. This is avoided by using Table D.3.

PROBLEMS

9.1 In thermodynamics c_p for air is often used as 0.24 kcal/kg·K. From Table B.4 c_p is 1.004 kJ/kg·K. Show these two values to be equal. Also, calculate c_v in both sets of units. (*Note:* The value of c_p used in thermodynamics and in fluid mechanics is the same when using the SI system of units.)

9.2 Show that $c_p = Rk/(k - 1)$.

9.3 Show that Eqs. 9.1.10 follow from Eqs. 9.1.9 and 9.1.7.

9.4 Verify the various forms of the energy equation in Eqs. 9.1.3, 9.1.4, and 9.1.5.

Speed of Sound

9.5 Show that Eq. 9.2.8 follows from Eq. 9.2.7.

9.6 Show that the velocity of sound in a high-frequency wave (i.e., a dog whistle) travels with speed $c = \sqrt{RT}$ if the process is assumed to be isothermal.

9.7 Show that for a small adiabatic disturbance in a steady flow, the energy equation takes the form $\Delta h = -c\,\Delta V$.

9.8 Verify that the speed of propagation of a small wave through water is about 1450 m/s.

9.9 Two rocks are slammed together on one side of a lake. An observer on the other side with his head under water "hears" the disturbance 0.6 s later. How far is it across the lake?

9.10 You and a friend are standing 10 m apart in waist-deep water. You slam two rocks together under water. How long after the rocks are slammed together does your friend hear the interaction if your friend's head is underwater?

9.11 Calculate the Mach number for an aircraft if it is flying at:
   (a) Sea level at 200 m/s
   (b) 5000 m at 200 m/s
   (c) 10 000 m at 200 m/s
   (d) 20,000 m at 200 m/s
   (e) 35 000 at 200 m/s

9.12 A wood chopper is chopping wood some distance away. You carefully observe, using your numerical wrist timer, that it takes 1.21 s for the sound of the axe to reach your ears. Calculate the distance you are from the chopper if the temperature is −10°C.

9.13 You see a bolt of lightning and 2 s later you hear the sound of thunder. How far away did the lightning bolt strike?

9.14 A needle-nosed projectile passes over you on a military testing field at a speed of 1000 m/s (Fig. P9.14). You know that it has an elevation of 1000 m where $T = -10°C$. How long after it passes overhead will you hear its sound? How far away will it be? Calculate its Mach number.

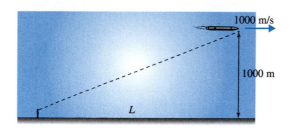

FIGURE P9.14

9.15 A special camera is capable of showing the Mach angle of a sharp-nosed bullet passing through the test section of a wind tunnel. If the Mach angle is measured to be 22°, calculate the speed of the bullet in m/s.

Assume standard conditions.

9.16 A small-amplitude wave passes through the standard atmosphere at sea level with a pressure rise of 15 Pa. Estimate the associated induced velocity and temperature rise.

Isentropic Flow

9.17 Provide all the steps needed to proceed from:
    (a)   Eq. 9.3.2 to Eq. 9.3.3
    (b)   Eq. 9.3.5 to Eq. 9.3.10
    (c)   Eq. 9.3.11 to Eq. 9.3.12
    (d)   Eq. 9.3.16 to Eq. 9.3.18
    (e)   Eq. 9.3.17 to Eq. 9.3.19

9.18 A pitot probe, an instrument that measures stagnation pressure, is used to determine the velocity of an airplane. Such a probe, attached to an aircraft, measures 10 kPa. Determine the speed of the aircraft if it is flying at an altitude of:
    (a)   3000 m   **(b)**   10 000 m

Assume an isentropic process from the free stream to the stagnation point.

9.19 A pitot probe, used to measure the stagnation pressure, indicates a pressure of 4 kPa at the nose of a surface vehicle traveling in 15°C–atmospheric air. Calculate its speed assuming:
    (a)   Isentropic flow
    (b)   Incompressible flow

Compute the percent error for part (b).

9.20 A converging nozzle with an exit diameter of 2 cm is attached to a reservoir maintained at 25°C and 200 kPa absolute. Using equations only, determine the mass flux of air if the receiver pressure is:
    (a)   100 kPa absolute
    (b)   130 kPa absolute

9.21 The converging nozzle with exit diameter of 3 cm of Fig. P9.21 is attached to a reservoir maintained at 20°C and 200 kPa. Using equations only, determine the mass flux of air if the receiver pressure is:
    (a)   75 kPa   **(b)**   150 kPa

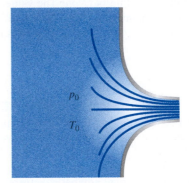

FIGURE P9.21

9.22 Rework Problem 9.20 using the isentropic flow table.

9.23 Rework Problem 9.21 using the isentropic flow table.

9.24 Air flows from a reservoir ($T_0 = 30°C, p_0 = 400$ kPa absolute) out a converging nozzle with a 10-cm-diameter exit. What exit pressure would just result in $M_e = 1$? Determine the mass flux for this condition. Use the isentropic flow table.

9.25 Air flows from a converging nozzle attached to a reservoir with $T_0 = 10°C$. What reservoir pressure is necessary to just cause $M_e = 1$ if the 6-cm-diameter nozzle exits to atmospheric pressure? Calculate the mass flux for this condition.

9.26 Air flows from a converging nozzle attached to a reservoir with $T_0 = 5°C$. If the 7-cm-diameter nozzle exits to atmospheric pressure, what reservoir pressure is necessary to just cause $M_e = 1$? Calculate the mass flux for this condition. Now double the reservoir pressure and determine the increased mass flux.

9.27 A 25-cm-diameter air line is pressurized to 500 kPa absolute and suddenly bursts (Fig. P9.27). The exit area is later measured to be 30 cm². If 6 min elapsed before the 10°C air was turned off, estimate the cubic meters of air that was lost.

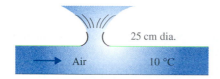

FIGURE P9.27

9.28 A converging nozzle is attached to a reservoir containing helium with $T_0 = 27°C$ and $p_0 = 200$ kPa absolute. Determine the receiver pressure that will just give $M_e = 1$. Now attach a diverging section with a 15-cm-diameter exit to the 6-cm-diameter throat. What is the highest receiver pressure that will give $M_t = 1$?

9.29 A venturi tube is used to measure the mass flux of air in a pipe by reducing the diameter from 10 cm to 5 cm and then back to 10 cm. The pressure at the inlet is 300 kPa and at the minimum diameter section it is 240 kPa. If the upstream temperature is 20°C, determine the mass flux.

9.30 The mass flux of air flowing through the pipe of Fig. P9.30 is desired. The pipe is reduced in diameter from 10 cm to 5 cm and then back to 10 cm. The pressure at the inlet is 300 kPa and at the minimum diameter section it is 250 kPa. If the upstream temperature is 15°C, determine the mass flux.

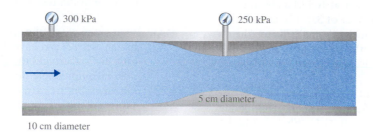

FIGURE P9.30

9.31 Air flows from a reservoir maintained at 30°C and 200 kPa absolute through a converging–diverging nozzle that has a 10-cm-diameter throat. Determine the diameter where M = 3. Use equations only.

9.32 Air flows in a converging–diverging nozzle from a reservoir maintained at 20°C and 500 kPa absolute. The throat and exit diameters are 5 cm and 15 cm, respectively. What two receiver pressures will result in M = 1 at the throat if isentropic flow occurs throughout? Use equations only.

9.33 Rework Problem 9.32 using the isentropic flow table.

9.34 Air flows from a nozzle with a mass flux of 15 kg/s. If $T_0 = 320°C$, $p_0 = 800$ kPa, and $p_e = 105$ kPa, calculate the throat and exit diameters for an isentropic flow. Also, determine the exit velocity.

9.35 Air flows from a reservoir, maintained at 20°C and 2 MPa absolute, and exits from a nozzle with $M_e = 4$. The receiver pressure is then raised until the flow is just subsonic throughout the entire nozzle. Calculate this receiver pressure.

9.36 Air at 30°C flows in a 10-cm-diameter pipe at a velocity of 150 m/s. A venturi tube is used to measure the flow rate. What should be the minimum diameter of the tube so that supersonic flow does not occur?

9.37 For a nozzle efficiency of 96%, rework Problem 9.24.

9.38 Nitrogen enters a diffuser at 100 kPa absolute and 100°C with a Mach number of 3.0. The mass flux is 10 kg/s and the exit velocity is small. Sketch the diffuser, then determine the throat area and the exit pressure and temperature, assuming isentropic flow.

9.39 Nitrogen enters a diffuser at 105 kPa absolute and 95°C with a Mach number of 3.0. The mass flux is 3 kg/s and the exit velocity is small. Sketch the diffuser, then determine the throat area and the exit pressure and temperature, assuming isentropic flow.

9.40 A rocket has a mass of 80 000 kg and is to be lifted vertically from a platform with six nozzles exiting exhaust gases with $T_e = 1000°C$. What should the exit velocity be from each 50-cm-diameter nozzle if the exhaust gases are assumed to be carbon dioxide?

9.41 A 100-kg man straps a small air-breathing jet engine on his back and lifts off the ground vertically (Fig. P9.41). The engine has an exit area of 200 cm². With what velocity must the 600°C exhaust gases exit the engine?

FIGURE P9.41

9.42 A converging–diverging nozzle is bolted into a reservoir at a diameter of 40 cm. The throat and exit diameters are 5 cm and 10 cm, respectively. If $T_0 = 27°C$ and $p_e = 100$ kPa absolute and isentropic flow of air occurs throughout the supersonic nozzle, calculate the force necessary to hold the nozzle onto the reservoir.

9.43 What is the maximum speed in **(a)** m/s and **(b)** mph that an aircraft can have during take-off or landing if the airflow around the aircraft is to be modeled as an incompressible flow? Allow a 3% error in the pressure from free stream to the stagnation point. Assume standard conditions.

Normal Shock

9.44 The pressure, temperature, and velocity before a normal shock wave are 80 kPa absolute, 10°C, and 1000 m/s, respectively. Calculate M_1, M_2, p_2, T_2, and ρ_2 for air. Use:
(a) Basic equations
(b) The normal shock table

9.45 The pressure, temperature, and velocity before a normal shock wave are 80 kPa and 1000 m/s, respectively. Calculate M_1, M_2, p_2, T_2, and ρ_2 for air. Use:
(a) Basic equations
(b) The normal shock table

9.46 Derive the *Rankine–Hugoniot relationship*,

$$\frac{p_2}{p_1} = \frac{(k+1)\,p_2/p_1 + k - 1}{(k-1)\,p_2/p_1 + k + 1}$$

which relates the density ratio to the pressure ratio across a normal shock wave. Find the limiting density ratio for air across a strong shock for which $p_2/p_1 \gg 1$.

9.47 An explosion occurs just above the surface of the Earth, producing a shock wave that travels radially outward. At a given location it has a Mach number of 2.0. Determine the pressure just behind the shock and the induced velocity.

9.48 Air at 200 kPa absolute and 20°C passes through a normal shock wave with strength so that $M_2 = 0.5$. Calculate V_1, p_2, and ρ_2.

9.49 Air at 150 kPa and 15°C passes through a normal shock wave with strength so that $M_2 = 0.5$. Calculate V_1, p_2, and ρ_2.

9.50 A blunt object travels at 1000 m/s at an elevation of 10 000 m. The flow approaching the stagnation point passes through a normal shock wave and then decelerates isentropically to the stagnation point. Calculate p_0 and T_0 at the stagnation point.

9.51 A pitot probe is inserted in an airflow in a pipe in which $p = 800$ kPa absolute, $T = 40°C$, and $M = 3.0$ (Fig. P9.51). What pressure does it measure?

9.52 Air flows from a 25°C reservoir to the atmosphere through a nozzle with a 5-cm-diameter throat and a 10-cm-diameter exit. What reservoir pressure will just result in M = 1 at the throat? Also, calculate the mass flux. Maintaining this reservoir pressure, reduce the throat diameter to 4 cm and determine the resulting mass flux. Sketch the pressure distribution as in Fig. 9.12.

9.53 Air flows from a 20°C reservoir to the atmosphere through a nozzle with a 5-cm-diameter throat and a 10-cm-diameter exit. What reservoir pressure is necessary to locate a normal shock wave at the exit? Also, calculate the velocity and pressure at the throat, before the shock, and after the shock.

9.54 Air flows from a 15°C reservoir to the atmosphere through a nozzle with a 5-cm-diameter throat and a 10-cm-diameter exit. What reservoir pressure is necessary to locate a normal shock wave at the exit? Also, calculate the velocity and pressure at the throat, before the shock, and after the shock.

9.55 Air flows from a reservoir maintained at 25°C and 500 kPa absolute. It flows through a nozzle with throat and exit diameters of 5 cm and 10 cm, respectively. What receiver pressure is necessary to locate a normal shock wave at a location where the diameter is 8 cm? Also, determine the velocity before the shock and at the exit.

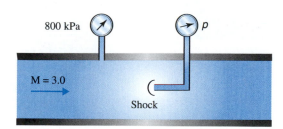

800 kPa

p

M = 3.0

Shock

FIGURE P9.51

Vapor Flow

9.56 Steam flows at a rate of 4 kg/s from reservoir conditions of 400°C and 1.2 MPa absolute through a converging–diverging nozzle to the atmosphere. Determine the throat and exit diameters if supersonic, isentropic flow exists throughout the diverging section.

9.57 Steam flows from reservoir conditions of 350°C and 1000 kPa absolute to the atmosphere at the rate of 15 kg/s. Estimate the converging nozzle exit diameter.

9.58 Steam flows from reservoir conditions of 350°C and 1.2 MPa to the atmosphere at the rate of 4.0 kg/s. Estimate the converging nozzle exit diameter.

9.59 A header supplies steam at 400°C and 1.2 MPa absolute to a set of nozzles with throat diameters of 1.5 cm. The nozzles exhaust to a pressure of 120 kPa absolute. If the flow is approximately isentropic, calculate the mass flux and the exit temperature.

Oblique Shock Wave

9.60 An airflow with velocity, temperature, and pressure of 800 m/s, 30°C, and 40 kPa absolute, respectively, is turned with an oblique shock wave emanating from the wall, which makes an abrupt 20° corner.
   (a) Find the downstream Mach number, pressure, and velocity for a weak shock.
   (b) Find the downstream Mach number, pressure, and velocity for a strong shock.
   (c) If the concave corner angle were 35°, sketch the corner flow situation.

9.61 Two oblique shocks intersect as shown in Fig. P9.61. Determine the angle of the reflected shocks if the airflow must leave parallel to its original direction. Also find M_3.

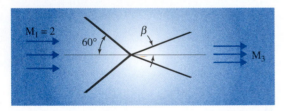

FIGURE P9.61

9.62 An oblique shock wave at a 35° angle is reflected from a plane wall. The upstream Mach number M_1 is 3.5 and $T_1 = 30°C$. Find V_3 after the reflected oblique shock wave for the airflow.

9.63 An oblique shock wave at a 35° angle is reflected from a plane wall. The upstream Mach number M_1 is 3.5 and $T_1 = 0°C$. Find V_3 after the reflected oblique shock wave for the airflow.

9.64 A supersonic inlet can be designed to have a normal shock wave oriented at the inlet, or a wedge can be used to provide a weak oblique shock wave, as shown in Fig. P9.64. Compare the pressure p_3 from the flow shown to the pressure that would exist behind a normal shock wave with no oblique shock.

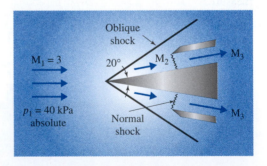

FIGURE P9.64

Expansion Waves

9.65 A supersonic airflow with $M_1 = 3$, $T_1 = -20°C$, and $p_1 = 20$ kPa absolute turns a convex corner of $25°$. Calculate M_2, p_2, T_2, and V_2 after the expansion fan. Also calculate the included angle of the fan.

9.66 A supersonic airflow with $M_1 = 2$, $T_1 = 0°C$, and $p_1 = 20$ kPa absolute turns a convex corner. If $M_2 = 4$, what angle θ should the corner have? Also calculate T_2 and V_2.

9.67 A supersonic airflow with $M_1 = 2$, $T_1 = 0°C$, and $p_1 = 35$ kPa absolute turns a convex corner. If $M_2 = 4$, what angle θ should the corner have? Also calculate T_2 and V_2.

9.68 The flat plate shown in Fig. P9.68 is used as an airfoil at an angle of attack of $5°$. Oblique shock waves and expansion fans allow the air to remain attached to the plate with the flow behind the airfoil parallel to the original direction. Calculate:

(a) The pressures on the upper and lower sides of the plate.

(b) The downstream Mach numbers M_{2u} and M_{2l}.

(c) The lift coefficient defined by $C_L = \text{lift}/(\frac{1}{2}\rho_1 V_1^2 A)$. Note that $\rho_1 V_1^2 = kM_1^2 p_1$.

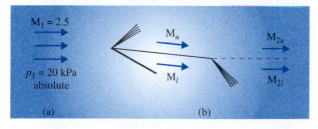

FIGURE P9.68

9.69 The supersonic airfoil shown in Fig. P9.69 is to fly at zero angle of attack. Calculate the drag coefficient $C_D = \text{drag}/(\frac{1}{2}\rho_1 V_1^2 A)$. Note that $\rho_1 V_1^2 = kM_1^2 p_1$.

FIGURE P9.69

9.70 The airfoil in Problem 9.69 flies at an angle of attack of $5°$. Determine the lift and drag coefficients, C_L and C_D. See Problem 9.68 and 9.69 for definitions of C_L and C_D.

The Wahluke Branch Canal, part of the Columbia Basin Project, is an example of an engineered canal. Water is delivered over a distance of many kilometers. (U.S. Bureau of Reclamation)

10

Flow in Open Channels

Outline

Chapter Objectives

The objectives of this chapter are to:

▲ Describe various types of open-channel, free-surface flows

▲ Apply Chezy-Manning equation with various geometric cross-sections for uniform flow applications

▲ Derive energy and momentum principles for rapidly varied flow situations

▲ Drive the differential equation for nonuniform gradually varied flow

▲ Present the method of profile synthesis — the qualitative description of open-channel flow, including establishment of controls and classification of water surface profiles

▲ Develop and present methods to numerically compute gradually varied flow

▲ Present numerous examples of uniform, rapidly varied, and gradually varied flow

▲ Detail several examples of profile synthesis and numerical analysis of complex open-channel flow with emphasis on design application

10.1 INTRODUCTION

Free-surface flow is probably the most commonly occurring flow phenomenon that we encounter on the surface of the earth. Ocean waves, river currents, and overland flow of rainfall are examples that occur in nature. Human-induced situations include flows in canals and culverts, drainage over impervious materials, such as roofs and parking lots, and wave motion in harbors.

In all of these situations, the flow is characterized by an interface between the air and the upper layer of water, which is termed the **free surface**. At the free surface, the pressure is constant, and for nearly all situations, it is atmospheric. In such a case, the hydraulic grade line and the liquid free surface coincide. In engineering practice, most open channels convey water as the fluid. However, the principles developed and implemented in this chapter are also applicable for other liquids that flow with a free surface.

Generally, the elevation of the free surface does not remain constant; it can vary along with the fluid velocities. Another complexity is that the flow is often three-dimensional. Fortunately, there are many instances in which two-dimensional and even one-dimensional simplifications can be made. The flow patterns in estuaries[1] and lakes under certain circumstances can be treated as two-dimensional in the horizontal plane, averaged vertically over the depth. River and channel flows are usually treated as one-dimensional with respect to the position coordinate along the streambed. In this chapter we restrict our consideration to one-dimensional flows.

Figure 10.1a shows a representative centerline velocity distribution in a channel. The velocity profile is three-dimensional at a given cross section (Fig. 10.1b). The boundary shear stress is nonuniform; at the free surface the shear

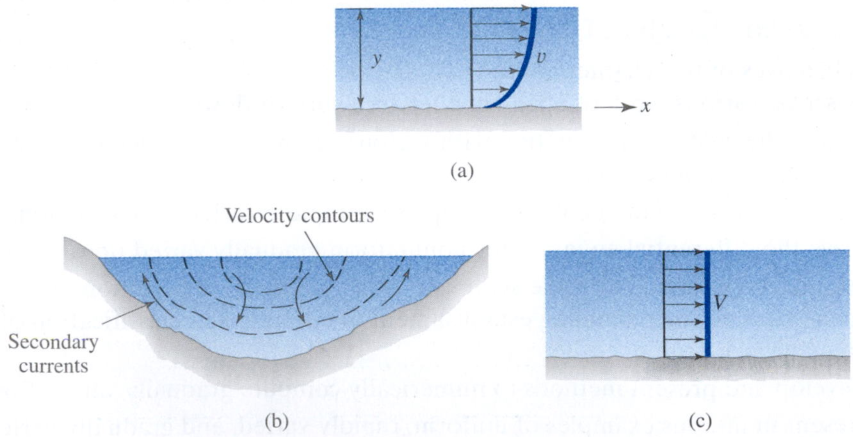

Free surface: *The interface between the air and the upper layer of water.*

FIGURE 10.1 Free-surface flow: (a) centerline velocity distribution; (b) cross section; (c) one-dimensional model.

[1]An estuary is the lower course of a river that is influenced by the ocean tides.

stress is negligible, yet the shear stress varies around the wetted perimeter. In some circumstances the presence of secondary currents will force the maximum velocity to occur slightly below the free surface. By convention, y is defined as the depth from the deepest location to the free surface; note that y is not a coordinate. The mean velocity is given by the relation

$$V = \frac{1}{A}\int_A v\, dA \qquad (10.1.1)$$

In the one-dimensional model, we assume the velocity to be equal to V everywhere at a given cross section. This model provides excellent results and is used widely. Flows in open channels are most likely to be turbulent, and the velocity profile can be assumed to be approximately constant, as in Fig. 10.1c, without significant error. Hence the one-dimensional model is employed.

10.2 OPEN-CHANNEL FLOWS

10.2.1 Classification of Free-Surface Flows

Flow in a channel is characterized by the mean velocity, even though a velocity profile exists at a given section, as shown in Fig. 10.1. The flow is classified as a combination of steady or unsteady, and uniform or nonuniform. Steady flow signifies that the mean velocity V, as well as the depth y, is independent of time, whereas unsteady flow necessitates that time be considered as an independent variable. Uniform flow implies that V and y are independent of the position coordinate in the direction of flow; nonuniform flow signifies that V and y vary in magnitude along that coordinate. The possible combinations are shown in Table 10.1; the position coordinate is designated as x.

Uniform flow is the situation where terminal velocity has been reached in a channel of constant cross section; not only is the mean velocity constant, but the depth is invariant as well. Steady, nonuniform flow is a common occurrence in rivers and in man-made channels. In those situations, it will be found to occur in two ways. In relatively short reaches, called *transitions*, there is a rapid change

KEY CONCEPT *In relatively short reaches, called transitions, there is a rapid change in depth and velocity.*

TABLE 10.1 Combinations of One-Dimensional Free-Surface Flows

Type of flow	Average velocity	Depth
Steady, uniform	$V=$ const.	$y=$ const.
Steady, nonuniform	$V=V(x)$	$y=y(x)$
Unsteady, uniform	$V=V(t)$	$y=y(t)$
Unsteady, nonuniform	$V=V(x,t)$	$y=y(x,t)$

Rapidly varied flow: *A rapid change in depth and velocity.*

in depth and velocity; such flow is termed **rapidly varied flow**. Examples are the hydraulic jump (shown in Example 4.12), flow entering a steep channel from a lake or reservoir, flow close to a free outfall from a channel, and flow in the vicinity of an obstruction such as a bridge pier or a sluice gate.

Along more extensive reaches of channel the velocity and depth may not vary rapidly, but rather, change in a slow manner. Here the water surface can be considered continuous and the regime is called **gradually varied flow**. Examples of gradually varied steady flow are the backwater created by a dam placed in a river, and the drawdown of a water surface as flow approaches a falls. Figure 10.2 illustrates how both rapidly varied flow (RVF) and gradually varied flow (GVF) can occur simultaneously in a reach of channel. Note that the vertical scale is larger than the horizontal scale; such scale distortion is common when representing open-channel flow situations.

Gradually varied flow: *In extensive reaches of a channel, the velocity and depth change in a slow manner.*

Unsteady uniform flow rarely occurs, but unsteady nonuniform flow is common. Flood waves in rivers, hydraulic bores, and regulated flow in canals are all examples of the latter category. In many situations, these flows may be considered to behave sufficiently like steady uniform flow or steady nonuniform flow to justify treating them as such. Unsteady flows are beyond the scope of a fundamental treatment and will not be presented in this book; the one exception is the hydraulic bore — a moving hydraulic jump — which can be analyzed in a quasi-steady fashion.

10.2.2 Significance of Froude Number

The primary mechanism for sustaining flow in an open channel is gravitational force. For example, the elevation difference between two reservoirs will cause water to flow through a canal that connects them. The parameter that represents this gravitational effect is the *Froude number,*

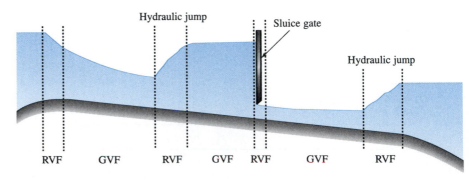

FIGURE 10.2 Steady nonuniform flow in a channel.

$$\text{Fr} = \frac{V}{\sqrt{gL}} \qquad\qquad (10.2.1)$$

which has been observed in Chapter 6 to be the ratio of inertial force to gravity force. In the context of open-channel flow, V is the mean cross-sectional velocity, and L is a representative length parameter. For a channel of rectangular cross section, L becomes the depth y of the flow.

The Froude number plays the dominant role in open-channel flow analysis. It appears in a number of relations that will be developed later in this chapter. Furthermore, by knowing its magnitude, one can ascertain significant characteristics regarding the flow regime. For example, if $\text{Fr} > 1$, the flow possesses a relatively high velocity and shallow depth; on the other hand, when $\text{Fr} < 1$, the velocity is relatively low and the depth is relatively deep. Except in the vicinity of rapids, cascades, and waterfalls, most rivers experience a Froude number less than unity. Constructed channels may be designed for Froude numbers to be greater or less than unity, or to vary from greater than unity to less than unity along the length of the channel.

10.2.3   Hydrostatic Pressure Distribution

Consider a channel in which the flow is nearly horizontal, as shown in Fig. 10.3. In this case, there is little or no vertical acceleration of fluid within the reach, and the streamlines remain nearly parallel. Such a condition is common in many open-channel flows, and indeed, if slight variations exist, the streamlines are assumed to behave as if they are parallel. Since the vertical accelerations are nearly zero, one can conclude that in the vertical direction the pressure distribution is hydrostatic. As a result, the sum $(p + \gamma z)$ remains a constant at any depth, and the hydraulic grade line coincides with the water surface. It is customary in open-channel flow to designate z as the elevation of the channel bottom, and y as the depth of flow. Since at the channel bottom $p/\gamma = y$, the hydraulic grade line

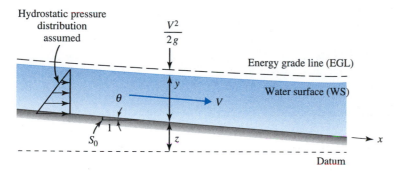

FIGURE 10.3   Reach of open-channel flow.

is given by the sum $(y + z)$. The concepts in this chapter are developed assuming a hydrostatic pressure distribution.

10.3 UNIFORM FLOW

Before we study nonuniform flow, let us focus our attention on the simpler condition of uniform flow. Such a flow is rare, but if it were to occur in a channel, the depth and velocity would not vary along its length, or in other words, terminal conditions would have been reached. In addition to uniform flow, this section covers the cross-sectional geometry of open channels; the formulations will apply to both uniform and nonuniform flow. Some of that material is covered in Section 7.7; it is reviewed here for completeness.

KEY CONCEPT *Channel cross sections can be considered regular or irregular.*

10.3.1 Channel Geometry

Regular section: *Section whose shape does not vary along the length of the channel.*

Irregular section: *Section whose shape will have changes in its geometry.*

Channel cross sections can be considered to be either regular or irregular. A **regular section** is one whose shape does not vary along the length of the channel, whereas an **irregular section** will have changes in its geometry. In this chapter we consider mostly regular channel shapes; three common geometries are shown in Fig. 10.4.

The simplest channel shape is a rectangular section. The cross-sectional area is given by

$$A = by \tag{10.3.1}$$

in which b is the width of the channel bottom (see Fig. 10.4a). Additional parameters of importance for open-channel flow are the wetted perimeter, the hydraulic radius, and the width of the free surface. The **wetted perimeter** P is the length of the line of contact between the liquid and the channel; for a rectangular channel it is

Wetted perimeter: *The length of the line of contact between the liquid and the channel.*

$$P = b + 2y \tag{10.3.2}$$

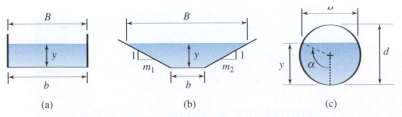

FIGURE 10.4   Representative regular cross sections: (a) rectangular; (b) trapezoidal; (c) circular.

The **hydraulic radius** R is the area divided by the wetted perimeter, that is,

Hydraulic radius: *The area divided by the wetted perimeter.*

$$R = \frac{A}{P} = \frac{by}{b + 2y} \tag{10.3.3}$$

The free-surface width B is equal to the bottom width b for a rectangular section.

A trapezoidal section (Fig. 10.4b) has the added feature of sloped side walls. If m_1 is the ratio of the horizontal to vertical change of the wall on one side, and m_2 is the corresponding quantity on the other wall, the area, wetted perimeter, and free-surface width are given as

$$A = by + \tfrac{1}{2}y^2(m_1 + m_2) \tag{10.3.4}$$

$$P = b + y\left(\sqrt{1 + m_1^2} + \sqrt{1 + m_2^2}\right) \tag{10.3.5}$$

$$B = b + y(m_1 + m_2) \tag{10.3.6}$$

Note that the rectangular section is embodied in the trapezoidal definition, since for vertical side walls m_1 and m_2 are zero, and the relations for A, P, and B become identical. In addition, if b is set equal to zero, Eqs. 10.3.4 to 10.3.6 describe the geometry of a triangular-shaped channel.

The circular cross section is an important one to consider, since many free-surface flows in drainage and sewer systems are conveyed in circular conduits. If d is the conduit diameter, the area, wetted perimeter, and free-surface width are given by

$$A = \frac{d^2}{4}(\alpha - \sin\alpha\cos\alpha) \tag{10.3.7}$$

$$P = \alpha d \tag{10.3.8}$$

$$B = d\sin\alpha \tag{10.3.9}$$

where

$$\alpha = \cos^{-1}\left(1 - 2\frac{y}{d}\right) \tag{10.3.10}$$

The angle α is defined in Fig. 10.4c.

A generalized cross-sectional geometry, such as that shown in Fig. 10.5a, can be expressed in functional form as $A(y)$, $P(y)$, $R(y)$, and $B(y)$. The functional representations encompass all of the analytical forms given above, and they also may be used to describe an irregular channel. For example, at a river section, one can describe the area and wetted perimeter in tabular form and utilize techniques

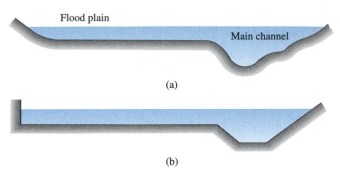

FIGURE 10.5 Generalized section representation: (a) actual cross section; (b) composite cross section.

such as curve fitting or interpolation to extract the numerical information as functions of the depth. Such procedures are useful for computer-based analyses.

A composite section is one made up of several subsections; usually these subsections are of analytic form. The example shown in Fig. 10.5b consists of a main channel and a floodplain. The main channel is approximated by a trapezoid and the floodplain by a rectangle. One could derive analytical expressions for such a composite section; however, it may be more useful to consider the functional forms for the geometric parameters. Note that the functions will be discontinuous at depths where the two sections are matched.

Most of the theoretical developments in this chapter focus on cross sections that are rectangular. Such an assumption allows one to simplify the mathematics associated with open-channel flow analysis. Even though the equations will be simplified relative to more complicated geometries, the physical understanding of the phenomena and conclusions reached will apply to most generalized prismatic cross sections. A clear distinction will be made between rectangular and other types of geometry when various developments and concepts are presented.

10.3.2 Equation for Uniform Flow

KEY CONCEPT *Uniform flow occurs in a channel when the depth and velocity do not vary along its length.*

Uniform flow occurs in a channel when the depth and velocity do not vary along its length, that is, when terminal conditions have been reached in the channel. Under such conditions, the energy grade line, water surface, and channel bottom are all parallel. It was seen in Section 7.7 that uniform flow can be predicted by an equation of the form

$$V = C \sqrt{RS_0} \qquad (10.3.11)$$

in which S_0 is the slope of the channel bottom and C is the Chezy coefficient, which is independent of the Reynolds number since the flow is considered completely turbulent. It has become common engineering practice to relate C to the channel roughness and the hydraulic radius by use of the *Manning relation*

$$C = \frac{c_1}{n} R^{1/6} \tag{10.3.12}$$

where $c_1 = 1$ for SI units. Combining Eqs. 10.3.11 and 10.3.12 with the definition of discharge results in the *Chezy–Manning equation*

$$Q = \frac{c_1}{n} A R^{2/3} \sqrt{S_0} \tag{10.3.13}$$

Values of the Manning coefficient n are given in Table 7.3.

The depth associated with uniform flow is designated y_0; it is called either *uniform depth* or *normal depth*. Uniform flow rarely occurs in rivers because of the irregularity of the geometry. In manmade channels it is not always present since the presence of controls such as sluice gates, weirs, or outfalls will cause the flow to become gradually varied. It is, however, necessary to determine y_0 when analyzing gradually varied flow conditions, since it provides in part a basis for evaluating the type of water surface that may exist in the channel. The design of gravity flow sewer networks is often based on assuming uniform flow and the use of Eq. 10.3.13, even though much of the time the flow in such systems may be nonuniform.

An examination of Eq. 10.3.13 reveals that it can be solved explicitly for Q, n, or S_0. Examples 7.18 and 7.19 provide illustrations. Use of a trial solution or equation solver is necessary when it is required to find y_0 with the remaining parameters given.

> **KEY CONCEPT** *The depth associated with uniform flow is called either uniform or normal depth.*

Example 10.1

Water is flowing at a rate of 4.5 m³/s in a trapezoidal channel (Fig. 10.4b) whose bottom width is 2.4 m and side slopes are 1 vertical to 2 horizontal. Compute y_0 if $n = 0.012$ and $S_0 = 0.0001$.

Solution
Given geometrical data are $b = 2.4$ m and $m_1 = m_2 = 2$. Rearrange Eq. 10.3.13, noting that $R = A/P$ and $c_1 = 1$:

$$\frac{A^{5/3}}{P^{2/3}} = \frac{Qn}{\sqrt{S_0}}$$

Substituting in the known data and trapezoidal geometry, one has

$$\frac{[2.4y_0 + \frac{1}{2} y_0^2(2 + 2)]^{5/3}}{[2.4 + y_0(2\sqrt{1 + 2^2})]^{2/3}} = \frac{4.5 \times 0.012}{\sqrt{0.0001}}$$

Solving for y_0, either by trial or by use of computational software, yields $y_0 = 1.28$ m.

Example 10.2

Uniform flow occasionally occurs in a 5-m-diameter circular concrete conduit (Fig 10.4c), but the depth of flow can vary. The Manning coefficient is $n = 0.013$, and the channel slope is $S_0 = 0.0005$. Plot the discharge-depth curve.

Solution

Mathcad is employed to generate the curve. Note that the solution is generalized, so that any diameter, Manning coefficient, or channel slope can be entered into the algorithm. Equations 10.3.7 and 10.3.8 are used to define the area and wetted perimeter, respectively. A MATLAB solution is shown in Appendix E, Fig E.1.

Input diameter, **Manning** coefficient, and channel slope:

$$d := 5 \qquad n := 0.013 \qquad S_0 := 0.0005 \qquad c_1 := 1.0$$

Define geometric functions:

$$\alpha(y) := acos\left(1 - 2\cdot\frac{y}{d}\right)$$

$$A(y) := \frac{d^2}{4}\cdot(\alpha(y) - sin(\alpha(y))\cdot cos(\alpha(y)))$$

$$P(y) := \alpha(y)\cdot d$$

$$R(y) := \frac{A(y)}{P(y)}$$

Define discharge function, (i.e., Manning's equation):

$$Q(y) := \frac{c_1}{n}\cdot A(y)\cdot R(y)^{\frac{2}{3}}\cdot\sqrt{S_0}$$

Plot depth versus discharge:

$$y := 0, 0.01 .. d$$

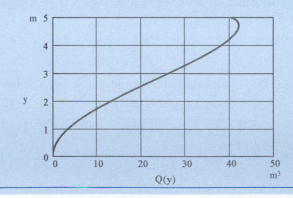

10.3.3 Most Efficient Section

Design of a channel to convey uniform flow typically consists of selecting or specifying the appropriate geometrical cross section provided that Q, n, and S_0 are known. Once chosen, an optimum sizing of the cross section can be based on the criteria of minimum resistance to flow. The flow resistance per unit length equals surface shear stress times wetted perimeter. Using a control volume for uniform flow, and assuming a small slope S_0 so that $\sin \theta \simeq S_0$ (see Fig. 10.3), one can show that the resistance per unit length is

$$\tau_0 P = \gamma A S_0 \qquad (10.3.14)$$

Hence a least resistance criterion is equivalent to requiring a minimum cross-sectional area with respect to the parameters defining the area. Furthermore, we have to satisfy the Chezy–Manning equation. Since Q, n, and S_0 are given and $R = A/P$, Eq. 10.3.13 can be written as

$$P = cA^{5/2} \qquad (10.3.15)$$

in which c is a constant. As an example, consider a rectangular channel with width b and depth y. The best hydraulic cross section is obtained by rewriting Eq. 10.3.15 such that A becomes a function of b only. Therefore, we express P in terms of A and b, using $A = by$ and $P = b + 2y$:

$$P = b + \frac{2A}{b} \qquad (10.3.16)$$

Substitution of this equation into Eq. 10.3.15 yields

$$b + \frac{2A}{b} = cA^{5/2} \qquad (10.3.17)$$

Now, differentiate Eq. 10.3.17 with respect to b, remembering that A is dependent on b:

$$1 + \frac{2}{b}\frac{dA}{db} - \frac{2A}{b^2} = \frac{5}{2} cA^{3/2}\frac{dA}{db} \qquad (10.3.18)$$

Set $dA/db = 0$, since the objective is to find the value of b that minimizes A. The result is

$$\frac{2A}{b^2} = 1 \qquad (10.3.19)$$

or, using $A = by$,

$$b = 2y \qquad (10.3.20)$$

Thus, if the width of a rectangular channel is twice the depth of the flowing water, the water will flow most efficiently.

For a trapezoidal cross section, it is simpler to begin with Eq. 10.3.15, eliminate b, and express P as a function of A, y, and m, where $m = m_1 = m_2$. Then by considering A as a function of m and y, the minimum of A is found by setting the gradient vector of A to zero. The result is $m = \sqrt{3}/3$ or a side slope angle of 60° with the horizontal. The resulting hexagonal shape is a trapezoid that best approximates a semicircle. The evaluation is left as an exercise for the reader.

The optimum criterion based on flow resistance cannot always be used, and in fact may be less significant than other design factors. Additional aspects to be considered are the type of excavation, and if the channel is unlined, the slope stability of the side walls and the possibility of erosion of the bed.

10.4 ENERGY CONCEPTS IN OPEN-CHANNEL FLOW

Total energy: *The sum of the vertical distance to the channel bottom measured from a horizontal datum, the depth of flow, and the kinetic energy.*

The energy at any position along the channel is the sum of the vertical distance measured from a horizontal datum z, the depth of flow y, and the kinetic energy $V^2/2g$. That sum defines the energy grade line and is termed the **total energy** H:

$$H = z + y + \frac{V^2}{2g} \qquad (10.4.1)$$

The kinetic-energy-correction factor associated with the $V^2/2g$ term is assumed to be unity (see Section 4.4.4); this is common practice for most prismatic channels of simple geometry since the velocity profiles are nearly uniform for the turbulent flows involved. The fundamental energy equation, developed in Section 4.4, states that losses will occur for a real fluid between any two sections of the channel, and hence the total energy will not remain constant. The energy balance is given simply by the relation

$$H_1 = H_2 + h_L \qquad (10.4.2)$$

in which h_L is the head loss. The only manner in which energy can be added to an open-channel flow system is for mechanical pumping or lifting of the liquid to take place. Equation 10.4.2 is applicable for rapidly varied as well as gradually varied flow situations; it will be used in conjunction with the momentum and continuity equations in a number of applications.

10.4.1 Specific Energy

Specific energy:
Measurement of energy relative to the bottom of the channel.

It is convenient in open-channel flow to measure the energy relative to the bottom of the channel; it provides a useful means to analyze complex flow situations. Such a measure is termed **specific energy** and is designated as E:

$$E = y + \frac{V^2}{2g} \qquad (10.4.3)$$

Specific energy then is the sum of the flow depth y and kinetic energy head $V^2/2g$.

Rectangular Sections.   For a rectangular section, the specific energy can be expressed as a function of the depth y. The **specific discharge** q is defined as the total discharge divided by the channel width, that is,

Specific discharge: The total discharge divided by the channel width (valid only for a rectangular channel).

$$q = \frac{Q}{b} = Vy \qquad (10.4.4)$$

The specific energy for a rectangular channel can thus be put in the form

$$E = y + \frac{q^2}{2gy^2} \qquad (10.4.5)$$

This E–y relation is shown in Fig. 10.6a. We may observe that a specific discharge requires at least a minimum energy; this minimum energy is referred to as *critical energy*, E_c. The corresponding depth y_c is called the **critical depth**. If the specific energy is greater than E_c, two depths are possible; those depths are referred to as **alternate depths**. For constant q, Eq. 10.4.5 is a cubic equation in y for a given value of E which is greater than E_c. The two positive solutions of y are the alternate depths.[2] Another way to express Eq. 10.4.5 is to consider E constant and vary q. Equation 10.4.5 can be solved for q as

Critical depth: Critical depth is the depth for which specific energy is a minimum.

Alternate depth: The two depths of flow that are possible for a given specific energy and discharge.

$$q = \sqrt{2gy^2\,(E - y)} \qquad (10.4.6)$$

This relation is shown in Fig. 10.6b. This form of the energy relation is useful for analyzing flows in which the specific energy remains nearly constant throughout the transition region; examples are a change in the width of a channel and the variation of depth with discharge at the entrance to a channel. Note that maximum unit discharge, q_{max}, occurs at critical depth.

The critical depth y_c can be evaluated from Eq. 10.4.5 by setting the derivative of E with respect to y equal to zero:

$$\frac{dE}{dy} = 1 - \frac{q^2}{gy^3} = 0 \qquad (10.4.7)$$

[2]The E-y curve falls between the two asymptotes $E = y$ and $y = 0$. Another curve defined by the relation exists for negative y; it is not considered since it has no physical meaning for open-channel flow.

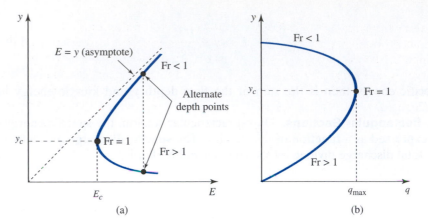

FIGURE 10.6 Variation of specific energy and specific discharge with depth: (a) E versus y for constant q; (b) q versus y for constant E.

Since $q = Vy$, the condition of minimum E is

$$1 - \frac{V^2}{gy} = 1 - Fr^2 = 0 \tag{10.4.8}$$

where the Froude number in a rectangular channel is

$$Fr = \frac{V}{\sqrt{gy}} = \frac{q}{\sqrt{gy^3}} \tag{10.4.9}$$

Thus from Eq. 10.4.8 the Froude number is equal to unity for minimum energy. Solving for the depth in Eq. 10.4.7 in terms of q, we have

$$y = y_c = \left(\frac{q^2}{g}\right)^{1/3} \tag{10.4.10}$$

This relation gives the critical depth in terms of the specific discharge. Note that at critical flow conditions, E_c can conveniently be expressed by combining Eqs. 10.4.5 and 10.4.10 to eliminate q, resulting in

$$E_c = \tfrac{3}{2}y_c \tag{10.4.11}$$

On the E–y curve, for any depth greater than y_c, the flow is relatively slow or tranquil, and $Fr < 1$; such a state is termed *subcritical flow*. Conversely, for a depth less than critical, the flow is relatively rapid or shooting, Fr > 1, and the regime is one of *supercritical flow*. The E–y diagram is a representation of the change in specific energy as the depth is varied, given a constant specific discharge. It is possible for q to vary, as when the width of a rectangular section changes in a transition region. As q increases, the E–y curve shifts to the right in Fig. 10.6a. It is left as an exercise for the reader to show that for a given specific energy, maximization of q of Eq. 10.4.6 will produce critical conditions at maximum discharge (see Fig. 10.6b).

Generalized Cross Section.   For a generalized section the specific energy can be written in terms of the total discharge Q and the cross-sectional area A:

$$E = y + \frac{Q^2}{2gA^2} \tag{10.4.12}$$

The minimum-energy condition is obtained by differentiating Eq. 10.4.12 with respect to y to obtain

$$\frac{dE}{dy} = 1 - \frac{Q^2}{gA^3} \frac{dA}{dy} \tag{10.4.13}$$

For incremental changes in depth, the corresponding change in area is $dA = B\, dy$. Thus setting Eq. 10.4.13 to zero, the minimum-energy condition becomes

$$1 - \frac{Q^2 B}{gA^3} = 0 \tag{10.4.14}$$

By analogy to Eq. 10.4.8, the second term in Eq. 10.4.14 is the square of the Froude number since

$$\text{Fr} = \sqrt{\frac{Q^2 B}{gA^3}} = \frac{Q/A}{\sqrt{gA/B}} = \frac{V}{\sqrt{gA/B}} \tag{10.4.15}$$

The ratio A/B is termed the *hydraulic depth*, and its use allows the definition of the Froude number to be generalized. The reader can easily verify that A/B is equal to y for a rectangular channel.

The specific energy diagram of Fig. 10.6 provides a useful means of visualizing the solution to a transition problem. Even though one probably would solve the problem numerically, an assessment of the graphical solution may provide useful physical insight as well as prevent the incorrect root from being selected.

Example 10.3

Water is flowing in a triangular channel with $m_1 = m_2 = 1.0$ at a discharge of $Q = 3$ m³/s. If the water depth is 2.5 m, determine the specific energy, Froude number, hydraulic depth, and alternate depth.

Solution

Recognizing that $b = 0$, the flow area and top width are computed from Eqs. 10.3.4 and 10.3.6 as follows:

$$A = \tfrac{1}{2}y^2(m_1 + m_2)$$
$$= \tfrac{1}{2} \times 2.5^2 \times (1 + 1) = 6.25 \text{ m}^2$$

$$B = (m_1 + m_2)y$$
$$= (1 + 1) \times 2.5 = 5.0 \text{ m}$$

Using Eqs. 10.4.12 and 10.4.15, E and Fr are found to be

$$E = y + \frac{Q^2}{2gA^2}$$

$$= 2.5 + \frac{3^2}{2 \times 9.81 \times 6.25^2} = 2.51 \text{ m}$$

$$\text{Fr} = \sqrt{\frac{Q^2 B}{gA^3}}$$

$$= \sqrt{\frac{3^2 \times 5}{9.81 \times 6.25^3}} = 0.137$$

The hydraulic depth is

$$\frac{A}{B} = \frac{6.25}{5.0} = 1.25 \text{ m}$$

The alternate depth is calculated using the energy equation. Recognizing that $A = y^2$, we have

$$2.51 = y + \frac{3^2}{2 \times 9.81 \times (y^2)^2}$$

$$= y + \frac{0.459}{y^4}$$

Solving, $y = 0.71$ m.

10.4.2 Use of the Energy Equation in Transitions

As mentioned in Section 10.2.1, a transition is a relatively short reach of channel where the depth and velocity change, creating a nonuniform, rapidly varied flow situation. The mechanism for such changes in the flow is usually an alteration of

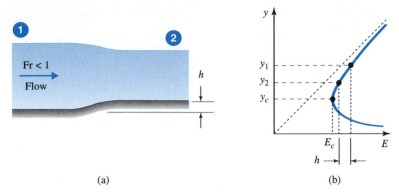

(a)   (b)

FIGURE 10.7   Channel constriction: (a) raised channel bottom;
(b) specific energy diagram.

one or more geometric parameters of the channel.[3] Within such regions, the
energy equation can be used effectively to analyze the flow in a transition or to
aid in the design of a transition. Two applications are given in this section
to demonstrate the methodology; other types of transitions can be treated
similarly.

Channel Constriction.   Consider a rectangular channel whose bottom is
raised by a distance h over a short region (Fig. 10.7a). The change in depth in the
transition can be analyzed by use of the energy equation, and as a first approximation, one can neglect losses. Assume that the specific energy upstream of the
transition is known. Recognizing that $H = E + z$, Eq. 10.4.2 is applied from location 1 to the end of the transition region, location 2:

$$E_1 = E_2 + h \tag{10.4.16}$$

The depth y_2 at the end of the transition can be visualized by inspection of the
E–y diagram (Fig. 10.7b). If the flow at location 1 is subcritical, y_1 is located on
the upper leg as shown in the specific energy diagram. The magnitude of h is
selected to be relatively small such that $y_2 > y_c$, hence the flow at location 2 is
similarly subcritical; however, as h is increased still further, a state of minimum
energy is ultimately reached in the transition. The condition of minimum energy
is sometimes referred to as a *choking condition* or as *choked flow*. Once choked
flow occurs, as h increases, the variations in depth and velocity are no longer
localized in the vicinity of the transition. Influences may be observed for significant distances both upstream and downstream of the transition.

The reader should verify that in a transition region, a narrowing of the channel width will create a situation similar to an elevation of the channel bottom.
The most general transition region is one that possesses a change in both width
and in bottom elevation.

> **KEY CONCEPT**   *The
> condition of choked flow or
> a choking condition implies
> that minimum specific
> energy exists within the
> transition.*

---

[3]One notable exception to this is the hydraulic jump, which requires use of the momentum principle and will be
considered in Section 10.5.2.

Example 10.4

A rectangular channel 3 m wide is conveying water at a depth $y_1 = 1.55$ m, and velocity $V_1 = 1.83$ m/s. The flow enters a transition region as shown in Fig. E10.4, in which the bottom elevation is raised by $h = 0.20$ m. Determine the depth and velocity in the transition, and the value of h for choking to occur.

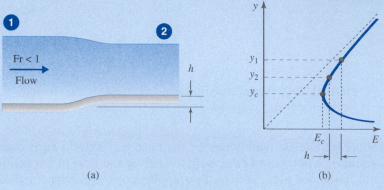

(a)   (b)

FIGURE E10.4

Solution

Use Eq. 10.4.4 to find the specific discharge to be

$$q = V_1 y_1$$
$$= 1.83 \times 1.55 = 2.84 \ \text{m}^2/\text{s}$$

The Froude number at location 1 is

$$\text{Fr} = \frac{V_1}{\sqrt{gy_1}}$$

$$= \frac{1.83}{\sqrt{9.81 \times 1.55}} = 0.47$$

which is less than unity. Hence the flow at location 1 is subcritical. The specific energy at location 1 is found, using Eq. 10.4.3, to be

$$E_1 = y_1 + \frac{V_1^2}{2g}$$

$$= 1.55 + \frac{1.83^2}{2 \times 9.81} = 1.72 \ \text{m}$$

The specific energy at location 2 is found, using Eq. 10.4.16, to be

$$E_2 = E_1 - h$$
$$= 1.72 - 0.20 = 1.52 \ \text{m}$$

If $E_2 > E_c$, it is possible to find the depth y_2. Therefore, E_c is calculated first. From Eqs. 10.4.10 and 10.4.11 the critical conditions are

$$y_c = \left(\frac{q^2}{g}\right)^{1/3} = \left(\frac{2.84^2}{9.81}\right)^{1/3} = 0.94 \text{ m}$$

$$E_c = \frac{3y_c}{2} = 3 \times \frac{0.94}{2} = 1.41 \text{ m}$$

Hence, since $E_2 > E_c$, we can proceed with calculating y_2. The depth y_2 can be evaluated by substituting known values into Eq. 10.4.5:

$$1.52 = y_2 + \frac{2.84^2}{2 \times 9.81 \times y_2^2}$$

The solution is

$$y_2 = 1.26 \text{ m}$$

$$\therefore V_2 = \frac{q}{y_2}$$

$$= \frac{2.84}{1.26} = 2.25 \text{ m/s}$$

The flow at location 2 is subcritical since there is no way in which the flow can become supercritical in the transition with the given geometry.

The value of h for critical flow to appear at location 2 is determined by setting $E_2 = E_c$ in Eq. 10.4.16:

$$h = E_1 - E_c = 1.72 - 1.40 = 0.31 \text{ m}$$

Channel Entrance with Critical Flow.   Consider flow entering a channel from a lake or reservoir over a short rounded crest (see Fig. 10.8). If the channel slope is steep, the flow will discharge freely into the channel and supercritical flow will occur downstream of the entrance region. Upstream of the crest the flow may be considered as subcritical. Since the flow regime at the crest changes from subcritical to supercritical, the flow at the crest must be critical. To determine the discharge, we assume that rapidly varied flow takes place over the crest in conjunction with the critical flow condition at the crest. An example illustrates the procedure.

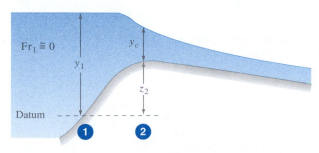

FIGURE 10.8   Outflow from a reservoir with critical flow at the channel entrance.

Example 10.5

Water flows freely from a reservoir into a trapezoidal channel with bottom width $b = 5.0$ m and side slope parameters $m_1 = m_2 = 2.0$. The elevation of the water surface in the reservoir is 2.3 m above the entrance crest. Assuming negligible losses in the transition and a negligible velocity in the reservoir upstream of the entrance, find the critical depth at the transition and the discharge into the channel.

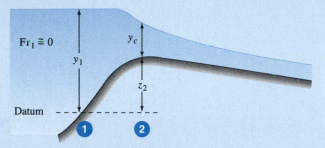

FIGURE E10.5

Solution

The total energy at location 1 in Fig. E10.5 is y_1 since the kinetic energy in the reservoir is negligible ($V_1 \approx 0$). Equating the total energies at locations 1 and 2 gives

$$y_1 = E_2 + z_2$$

Since critical conditions occur at location 2, Eqs. 10.4.12 and 10.4.14 can be combined to eliminate the discharge, with the result

$$E_2 = y_c + \frac{A}{2B}$$

Elimination of E_2 in the two equations yields the expression

$$y_1 - z_2 = y_c + \frac{A}{2B} = y_c + \frac{by_c + \frac{1}{2}(m_1 + m_2)y_c^2}{2[b + (m_1 + m_2)y_c]}$$

or, with the given data, the expression becomes

$$2.3 = y_c + \frac{5y_c + \frac{1}{2}(2 + 2)y_c^2}{2[5 + (2 + 2)y_c]}$$

The relation above is a quadratic in y_c. The positive root is chosen, which gives

$$y_c = 1.70 \text{ m}$$

Subsequently, one can find that $A = 14.28$ m^2 and $B = 11.80$ m. Use Eq. 10.4.14 to find the discharge to be

$$Q = \sqrt{\frac{gA^3}{B}}$$

$$= \sqrt{\frac{9.8 \times 14.3^3}{11.8}} = 49.3 \text{ m}^3/\text{s}$$

Energy Losses.   Energy losses in expansions and contractions are known to be relatively small when the flow is subcritical; however, it may be necessary under certain circumstances that the losses be considered. Equation 10.4.2 includes a loss term that can account for transitional losses. The following experimentally derived formulas can be employed. For a channel expansion use

$$h_L = K_e \left(\frac{V_1^2}{2g} - \frac{V_2^2}{2g} \right) \tag{10.4.17}$$

and for a channel contraction use

$$h_L = K_c \left(\frac{V_2^2}{2g} - \frac{V_1^2}{2g} \right) \tag{10.4.18}$$

In Eq. 10.4.17, K_e is an expansion coefficient; it has been suggested (King and Brater, 1963) to use $K_e = 1.0$ for sudden or abrupt expansions, and $K_e = 0.2$ for well-designed or rounded expansions. For the contraction coefficient K_c in Eq. 10.4.18, use $K_c = 0.5$ for sudden contractions, and $K_c = 0.10$ for well-designed contractions. When the flow is supercritical, significant standing-wave patterns can be generated throughout and downstream of the transition region; for such flows, proper design requires that wave mechanics be considered (Chow, 1959; Henderson, 1966).

10.4.3   Flow Measurement

The most common means of metering discharge in an open channel is to use a weir. Basically, a **weir** is a device placed in the channel that forces the flow through an opening or aperture designed to measure the discharge. Specialized weirs have been designed for specific needs; in this section two fundamental types — broad-crested and sharp-crested — will be presented.

A properly designed weir will exhibit subcritical flow upstream of the structure, and the flow will converge and accelerate to a critical condition near the top or *crest* of the weir. As a result, a correlation between discharge and a depth upstream of the weir can be made. The downstream overflow is termed the nappe, which usually discharges freely into the atmosphere. There are a number of factors that affect the performance of a weir; significant among them are the three-dimensional flow pattern, the effects of turbulence, frictional resistance, surface tension, and the amount of ventilation beneath the nappe. The simplified derivations presented here are based on the Bernoulli equation; the other effects can be accounted for by modifying the ideal discharge with a *discharge coefficient, C_d*; the actual discharge is the ideal discharge multiplied by the discharge coefficient. When possible, it is advantageous to calibrate a particular weir in place in order to obtain the desired accuracy.

Weir: *A device placed in a channel that forces the flow through an opening or aperture, often designed to measure the discharge.*

KEY CONCEPT   *Flow will converge and accelerate to a critical condition near the crest of the weir.*

Broad-Crested Weir. A broad-crested weir is shown in Fig. 10.9. It has sufficient elevation above the channel bottom to choke the flow, and it is long enough so that the overflow streamlines become parallel, resulting in a hydrostatic pressure distribution. Then, at some position, say location 2, a critical flow condition exists.

Assume a rectangular horizontal channel and let the height of the weir be h. Location 1 is a point upstream of the weir where the flow is relatively undisturbed, and Y is the vertical distance from the top of the weir to the free surface at that point. Applying Bernoulli's equation from location 1 to location 2 at the free surface and neglecting the kinetic energy head at location 1 ($V_1 \approx 0$), one has the result

$$h + Y = h + y_c + \frac{V_c^2}{2g} \tag{10.4.19}$$

Solving for V_c,

$$V_c = \sqrt{2g(Y - y_c)} \tag{10.4.20}$$

For a weir whose breadth normal to the flow is b, the ideal discharge is

$$Q = b y_c V_c = b y_c \sqrt{2g(Y - y_c)} \tag{10.4.21}$$

Recognizing that $Y = E_c$, Eq. 10.4.11 is used to relate y_c to Y, and when substituted into Eq. 10.4.21, the result is

$$Q = \frac{2}{3} \sqrt{\frac{2}{3}g}\, b Y^{3/2} \tag{10.4.22}$$

For a properly rounded upstream edge on the weir, Eq. 10.4.22 is accurate to within several percent of the actual flow; hence a discharge coefficient is not applied.

Sharp-Crested Weir. A sharp-crested weir is a vertical plate placed normal to the flow containing a sharp-edge crest so that the nappe will behave as a free jet. Figure 10.10 shows a rectangular weir with a horizontal crest extending across the entire width of the channel. Because of the presence of the side walls, lateral contractions are not present.

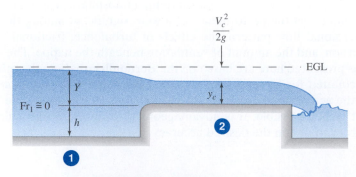

FIGURE 10.9 Broad-crested weir.

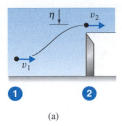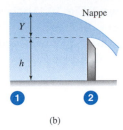

FIGURE 10.10 Rectangular sharp-crested weir: (a) ideal flow; (b) actual flow.

Let us define an idealized flow situation: The flow in the vertical plane does not contract as it passes over the crest so that the streamlines are parallel, atmospheric pressure is present in the nappe, and uniform flow exists at location 1 with negligible kinetic energy ($v_1 \approx 0$). The Bernoulli equation is applied along a representative streamline (Fig. 10.10a) and solved for v_2, the local velocity in the nappe:

$$v_2 = \sqrt{2g\eta} \qquad (10.4.23)$$

If b is the width of the crest normal to the flow, the ideal discharge is given as

$$Q = b\int_0^Y v_2\, d\eta = b\int_0^Y \sqrt{2g\eta}\, d\eta = b\frac{2}{3}\sqrt{2g}\; Y^{3/2} \qquad (10.4.24)$$

Experiments have shown that the magnitude of the exponent is nearly correct but that a discharge coefficient C_d must be applied to accurately predict the real flow, shown in Fig. 10.10b:

$$Q = C_d\frac{2}{3}\sqrt{2g}\; b\; Y^{3/2} \qquad (10.4.25)$$

The discharge coefficient accounts for the effect of contraction, velocity of approach, viscosity, and surface tension. An experimentally (Chow, 1959) obtained formula for C_d has been given in the form

$$C_d = 0.61 + 0.08\frac{Y}{h} \qquad (10.4.26)$$

Normally, for small Y/h ratio, $C_d = 0.61$. If the weir crest does not extend to the side walls but allows for end contractions to appear, as in Fig. 10.11a, the effective width of the weir can be approximated by $(b - 0.2Y)$.

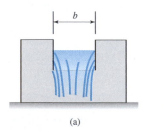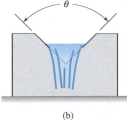

FIGURE 10.11 Contracted rectangular and V-notch weirs: (a) rectangular; (b) V-notch.

The V-notch weir (Fig. 10.11b) is more accurate than the rectangular weir for measurement of low discharge. In a manner similar to the development of the rectangular weir relation, the idealized discharge is found by integrating the local velocity throughout the nappe above the crest. A discharge coefficient is applied to yield

$$Q = C_d \frac{8}{15} \sqrt{2g} \left(\tan \frac{\theta}{2} \right) Y^{5/2} \qquad (10.4.27)$$

For use with water, and for θ varying from 22.5° to 120°, experiments (King and Brater, 1963) have shown that a value of $C_d = 0.58$ is acceptable for engineering calculations. Derivation of Eq. 10.4.27 is left as an exercise.

Example 10.6

Determine the discharge of water over a rectangular sharp-crested weir, $b = 1.25$ m, $Y = 0.35$ m, $h = 1.47$ m, with side walls and with end contractions. If a 90° V-notch weir were to replace the rectangular weir, what would be the required Y for a similar discharge?

Solution
For the rectangular weir, using Eq. 10.4.26, the discharge coefficient is

$$C_d = 0.61 + 0.08 \frac{Y}{h} = 0.61 + 0.08 \times \frac{0.35}{1.47} = 0.63$$

Substitute into Eq. 10.4.25 and calculate

$$Q = C_d \frac{2}{3} \sqrt{2g} \, b Y^{3/2}$$
$$= 0.63 \times \frac{2}{3} \times \sqrt{2 \times 9.81} \times 1.25 \times 0.35^{3/2}$$
$$= 0.48 \text{ m}^3/\text{s}$$

With end contractions the effective width of the weir is reduced by 0.2Y, resulting in

$$Q = C_d \frac{2}{3} \sqrt{2g} \, (b - 0.2Y) \, Y^{3/2}$$
$$= 0.63 \times \frac{2}{3} \times \sqrt{2 \times 9.81} \times (1.25 - 0.2 \times 0.35) \times 0.35^{3/2}$$
$$= 0.45 \text{ m}^3/\text{s}$$

With a discharge of $Q = 0.48$ m³/s, use Eq. 10.4.27 to find Y for the 90° V-notch weir:

$$Y = \left[\frac{Q}{C_d \times \frac{8}{15} \times \sqrt{2g} \, \tan(\theta/2)} \right]^{2/5}$$

$$= \left[\frac{0.482}{0.58 \times \frac{8}{15} \times \sqrt{2 \times 9.81} \times \tan 45°} \right]^{2/5} = 0.66 \text{ m}$$

Additional Methods of Flow Measurement. Other types of weirs include those whose faces are inclined in the upstream and downstream direction (triangular, trapezoidal, irregular). In addition, the spillway section of a dam can be considered as a weir with a rounded crest. King and Brater (1963) give details on the selection and use of these types.

A special type of open flume is one in which the geometry of the throat is constricted in such a manner to choke the flow, thereby creating critical flow followed by a hydraulic jump. When manufactured using a particular standardized section, the flume is called a **Parshall flume** (Fig. 10.12). Extensive calibrations have established reliable empirical formulas to predict the discharge. For throat widths from 0.3 to 2.4 m the discharge is given by the formula

Parshall flume: *An open flume where the throat is constricted to choke the flow to create critical flow followed by a hydraulic jump.*

$$Q = 0.1133BH^{1.522B^{0.026}} \qquad (10.4.28)$$

where $H = H'/0.3048$ and $B = B'/0.3048$
in which H' is the depth measured at the upstream location shown in Fig. 10.12. Note that H' and B' are measured in meters, and Q is in cubic meter per second.

In a natural section of a river it may be impractical to place a weir; in that case stream gaging can be used to measure the discharge. A control location is established upstream of the gaging site, and for a given depth, or *stage*, of the river, the two-dimensional velocity profile is measured using current meters. Subsequently, the profile is numerically integrated to yield the discharge. A series of such measurements will produce a stage-discharge curve, which can then be employed to estimate the discharge by measuring the river stage.

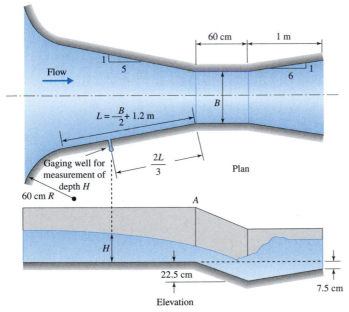

FIGURE 10.12 Parshall flume. (HENDERSON, OPEN CHANNEL FLOW, 1st,©1966. Electronically reproduced by permission of Pearson Education, Inc., Upper Saddle River, New Jersey.)

10.5 MOMENTUM CONCEPTS IN OPEN-CHANNEL FLOW

In the preceding section we have seen how the energy equation is applied to rapidly varied flow situations, and in particular, how it is utilized to analyze flows in transition regions. The momentum equation is also employed to study certain phenomena in those situations. When used in conjunction with the energy and continuity relations, the momentum equation gives the user a concise means of analyzing nearly all significant transition problems, including problems that involve a hydraulic jump.

10.5.1 Momentum Equation

Consider the open-channel reach with supercritical flow upstream of a submerged obstacle, as depicted in Fig. 10.13a. It represents a rapidly varied flow with an abrupt change in depth but no change in width. In general, such a change may result from an obstacle in the flow or from a hydraulic jump. The upstream flow is supercritical and the downstream flow is subcritical. Other flow regimes may also be considered; for example, the conditions could be subcritical throughout the entire control volume. Each situation should be approached as a unique formulation.

The generalized flow situation may be used to develop the equation of motion for transition regions. The control volume corresponding to Fig. 10.13a is shown in Fig. 10.13b. The pressure distribution is assumed hydrostatic, and the resultant hydrostatic forces are given by $\gamma A \bar{y}$, where the distance $\bar{y}$ to the centroid of the cross-sectional area is measured from the free surface. The submerged obstacle imparts a force F on the control volume with a direction opposite to the direction of flow. The linear momentum equation, presented in Section 4.5, is applied to the control volume in the x-direction to give

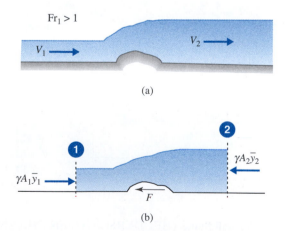

(a)

(b)

FIGURE 10.13 Channel flow over an obstacle: (a) idealized flow; (b) control volume.

$$\gamma A_1 \bar{y}_1 - \gamma A_2 \bar{y}_2 - F = \rho Q \, (V_2 - V_1) \qquad (10.5.1)$$

Note that frictional forces have not been included in Eq. 10.5.1; they are usually quite small relative to the other terms, so they can be ignored. Similarly, gravitational forces in the flow direction are negligible for the small channel slopes considered. Equation 10.5.1 can be rearranged in the form

$$M_1 - M_2 = \frac{F}{\gamma} \qquad (10.5.2)$$

in which M_1 and M_2 are terms that contain the hydrostatic force and the momentum flux at locations 1 and 2, respectively. The quantity M is called the *momentum function*, and for a general prismatic section it is given by

$$M = A\bar{y} + \frac{Q^2}{gA} \qquad (10.5.3)$$

For a rectangular section, $A\bar{y} = by^2/2$, and the momentum function is

$$M = \frac{by^2}{2} + \frac{bq^2}{gy} = b\left(\frac{y^2}{2} + \frac{q^2}{gy}\right) \qquad (10.5.4)$$

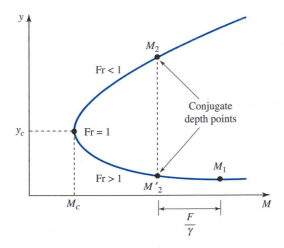

FIGURE 10.14   Variation of the momentum function with depth.

Equation 10.5.4 is sketched in Fig. 10.14. Two positive roots y exist for a given M and q; they are termed either **conjugate** or **sequent** depths. The upper leg of the curve $(y > y_c)$ applies to subcritical flow and the lower leg $(y < y_c)$ to supercritical flow. The values of M_1 and M_2 for the example of Fig. 10.13a are shown, indicating supercritical flow upstream and subcritical flow downstream of the submerged obstacle. The horizontal distance between M_1 and M_2 is equal to F/γ. The downstream flow would be supercritical if no hydraulic jump occurred; the corresponding value of the momentum function is indicated by M_2'.

The depth associated with a minimum M is found by differentiating M with respect to y in Eq. 10.5.3:

$$\frac{dM}{dy} = A - \frac{BQ^2}{gA^2} = 0 \qquad (10.5.5)$$

Note[4] that $d(A\bar{y})/dy = A$. The condition for minimum M is thus

$$Q^2B = gA^3 \qquad (10.5.6)$$

This result is identical to Eq. 10.4.14. Hence the condition of minimum M is equivalent to that of minimum energy: critical flow with a Froude number equal to unity.

Momentum Equation Applied to a Transition Region. Quite often the momentum equation is applied in situations where it is desired to determine the resultant force acting at a specific location, or to find the change in depth or velocity when there is a significant undefined loss across the transition region. It is important to remember that the energy and continuity equations are also at our disposal, and we must determine which relations are necessary. In some instances, along with the continuity equation, both energy and momentum equations must be applied. An example is given below to illustrate the technique.

---

[4]This can be observed from the definition $\bar{y} = (1/A) \int (y - \eta) \, dA$. Differentiation of $(\bar{y}A)$ using Leibnitz's rule from calculus yields

$$\frac{d}{dy}(\bar{y}A) = \frac{d}{dy}\int_0^y (y - \eta)w(\eta)d\eta$$

$$= (y - \eta)w(\eta)\Big|_{\eta=y} + \int_0^y w(\eta)d\eta$$

$$= 0 + A = A$$

Example 10.7

In a rectangular 5-m-wide channel, water is discharging at 14.0 m³/s (Fig. E10.7). Find the force exerted on the sluice gate when $y_1 = 2$ m and $y_2 = 0.5$ m.

FIGURE E10.7

Solution

Using Eq. 10.5.3, the momentum functions at 1 and 2 are

$$M_1 = A_1\bar{y}_1 + \frac{Q^2}{gA_1}$$

$$= 5 \times 2 \times 1 + \frac{(14)^2}{9.81 \times 5 \times 2} = 12.0 \text{ m}^3$$

$$M_2 = A_2\bar{y}_2 + \frac{Q^2}{gA_2}$$

$$= 5 \times 0.5 \times 0.25 + \frac{(14)^2}{9.81 \times 5 \times 0.5} = 8.62 \text{ m}^3$$

The resultant force acting on the fluid control volume is determined, using Eq. 10.5.2, to be

$$F = \gamma(M_1 - M_2)$$

$$= 9800 \times (12.0 - 8.62) = 33\ 100 \text{ N}$$

Hence the force on the gate acts in the downstream sense with a magnitude of 33.1 kN.

10.5.2 Hydraulic Jump

A **hydraulic jump** is a phenomenon in which fluid flowing at a supercritical state will abruptly undergo a transition to a subcritical state. Boundary conditions upstream and downstream of the jump will dictate its strength as well as its location. An idealized hydraulic jump is shown in Fig. 10.15. The strength of the jump varies widely, as shown in Table 10.2, with relatively mild disturbances occurring at one extreme, and significant separation and eddy formation taking place at the other. As a result, the energy loss associated with the jump is considered to be unknown, so the energy equation is not used in the initial analysis. Assuming no

Hydraulic jump: *A phenomenon where fluid flowing at a supercritical state will abruptly undergo a transition to a subcritical state.*

friction along the bottom and no submerged obstacle, that is, setting F equal to zero, Eq. 10.5.2 shows that $M_1 = M_2$. For a rectangular section, Eq. 10.5.4 can be substituted into the relation, allowing one to obtain

$$\frac{q^2}{g}\left(\frac{1}{y_1} - \frac{1}{y_2}\right) = \frac{1}{2}(y_2^2 - y_1^2) \qquad (10.5.7)$$

Rearranging, factoring, and noting that $Fr_1^2 = q^2/(gy_1^3)$ there results

$$Fr_1^2 = \frac{1}{2}\frac{y_2}{y_1}\left(\frac{y_2}{y_1} + 1\right) \qquad (10.5.8)$$

This equation is dimensionless and it relates the Froude number upstream of the jump with the ratio of the downstream to upstream depths. It can be seen that Eq. 10.5.8 is quadratic with respect to y_2/y_1 provided that Fr_1 is known. Solving for y_2/y_1, one obtains

$$\frac{y_2}{y_1} = \frac{1}{2}\left(\sqrt{1 + 8Fr_1^2} - 1\right) \qquad (10.5.9)$$

The positive sign in front of the radical has been chosen to yield a physically meaningful solution. It is worth noting that Eq. 10.5.9 is also valid if the subscripts on the depths and Froude number are reversed:

$$\frac{y_1}{y_2} = \frac{1}{2}\left(\sqrt{1 + 8Fr_2^2} - 1\right) \qquad (10.5.10)$$

The theoretical energy loss associated with a hydraulic jump in a rectangular channel can be determined once the depths and flows at locations 1 and 2 are known. The energy equation is applied from 1 to 2, including the head loss h_j

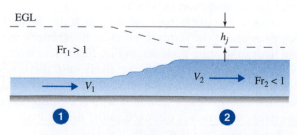

FIGURE 10.15   Idealized hydraulic jump.

across the jump. Combining that relation with Eq. 10.5.7 and the continuity equation, one can show after some algebra that

$$h_j = \frac{(y_2 - y_1)^3}{4y_1 y_2} \qquad (10.5.11)$$

Equations 10.5.8 through 10.5.11 are useful forms for solving most hydraulic jump problems in a rectangular channel.

Table 10.2 shows the various forms that a jump may assume relative to the upstream Froude number. A steady, well-established jump, with $4.5 < Fr < 9.0$, is often used as an energy dissipator downstream of a dam or spillway. It is characterized by the existence of breaking waves and rollers accompanied by a submerged jet with significant turbulence and dissipation of energy in the main body of the jump; downstream, the water surface is relatively smooth. For Froude numbers outside the range 4.5 to 9.0, less desirable jumps exist and may create undesirable downstream surface waves. The length of a jump is the distance from the front face to just downstream where smooth water exists; a steady jump has an approximate length of six times the upstream depth.

TABLE 10.2 Hydraulic Jumps in Horizontal Rectangular Channels

Upstream Fr	Type	Description	
1.0–1.7	Undular	Ruffled or undular water surface; surface rollers form near Fr = 1.7	
1.7–2.5	Weak	Prevailing smooth flow; low energy loss	
2.5–4.5	Oscillating	Intermittent jets from bottom to surface, causing persistent downstream waves	
4.5–9.0	Steady	Stable and well-balanced; energy dissipation contained in main body of jump	
>9.0	Strong	Effective, but with rough, wavy surface downstream	

Source: Adapted with permission from Chow, 1959.

Example 10.8

A hydraulic jump is situated in a 4-m-wide rectangular channel. The discharge in the channel is 7.5 m^3/s, and the depth upstream of the jump is 0.20 m. Determine the depth downstream of the jump, the upstream and downstream Froude numbers, and the rate of energy dissipated by the jump.

Solution
Find the unit discharge and upstream Froude number:

$$q = \frac{Q}{b}$$

$$= \frac{7.5}{4} = 1.88 \text{ m}^2/\text{s}$$

$$\text{Fr}_1 = \frac{q}{\sqrt{gy_1^3}}$$

$$= \frac{1.88}{\sqrt{9.81 \times 0.20^3}} = 6.71$$

The downstream depth is computed, using Eq. 10.5.9, to be

$$y_2 = \frac{y_1}{2}\left(\sqrt{1 + 8\text{Fr}_1^2} - 1\right)$$

$$= \frac{0.20}{2}\left(\sqrt{1 + 8 \times 6.71^2} - 1\right) = 1.80 \text{ m}$$

The downstream Froude number is

$$\text{Fr}_2 = \frac{q}{\sqrt{gy_2^3}}$$

$$= \frac{1.88}{\sqrt{9.81 \times 1.80^3}} = 0.25$$

The head loss in the jump is given by Eq. 10.5.11:

$$h_j = \frac{(y_2 - y_1)^3}{4y_1 y_2}$$

$$= \frac{(1.80 - 0.20)^3}{4 \times 0.20 \times 1.80} = 2.84 \text{ m}$$

Hence the rate of energy dissipation in the jump is

$$\gamma Q h_j = 9800 \times 7.5 \times 2.84 = 2.09 \times 10^5 \text{ W} \quad \text{or} \quad 209 \text{ kW}$$

KEY CONCEPT *A translating hydraulic jump is a positive surge wave, maintaining a stable front as it propagates into an undisturbed region.*

Translating Hydraulic Jump.   A translating hydraulic jump, alternately termed a *surge* or a *hydraulic bore,* is shown in Fig. 10.16a; it is termed a positive surge wave in the sense that it maintains a stable front as it propagates at speed *w* into an undisturbed region. Such a wave can be generated by abruptly lower-

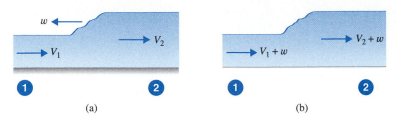

FIGURE 10.16   Translating hydraulic jump: (a) front moving upstream; (b) front appears stationary by superposition.

ing a downstream gate or by rapidly releasing water at an upstream location into a channel. This unsteady rapidly varied flow situation can be conveniently analyzed as a steady-state problem by superposing the bore speed w in the opposite sense on the control volume (Fig. 10.16b). The front appears stationary and the relative velocities at locations 1 and 2 are equal to $V_1 + w$ and $V_2 + w$, respectively.

Assume a bore translating in a rectangular, frictionless, horizontal channel. Equation 10.5.9 can be applied by substituting $V_1 + w$ for V_1 in the definition of Fr_1 to obtain

$$\frac{y_2}{y_1} = \frac{1}{2}\left[\sqrt{1 + 8\frac{(V_1 + w)^2}{gy_1}} - 1\right] \tag{10.5.12}$$

The continuity relation applied to the control volume of Fig. 10.16b is

$$y_1(V_1 + w) = y_2(V_2 + w) \tag{10.5.13}$$

Equations 10.5.12 and 10.5.13 contain five parameters: y_1, y_2, V_1, V_2, and w. Three of them must be known to solve for the remaining two. Depending on which variables are unknown, the solution of Eqs. 10.5.12 and 10.5.13 will be either explicit or based on a trial procedure.

Example 10.9

Water flows in a rectangular channel with a velocity of 2.5 m/s and a depth of 1.5 m. A gate is suddenly completely closed, forming a surge that travels upstream. Find the speed of the surge and the depth behind the surge.

Solution
Since the gate is closed, the downstream velocity is $V_2 = 0$. Equations 10.5.12 and 10.5.13 contain the two unknowns w and y_2. Combining them to eliminate y_2 and substituting $V_1 = 2.5$ and $y_1 = 1.5$ results in the relation

$$2.5 + w = \frac{w}{2}\left[\sqrt{1 + 8\frac{(2.5 + w)^2}{9.81 \times 1.5}} - 1\right]$$

(continued)

Solving yields $w = 3.41$ m/s. Use Eq. 10.5.13 to compute y_2:

$$y_2 = y_1 \frac{V_1 + w}{V_2 + w}$$

$$= 1.5 \times \frac{2.5 + 3.41}{3.41} = 2.60 \text{ m}$$

Drag on Submerged Objects. If an object is submerged in the flow it is possible to describe the drag force in the manner

$$F = C_D A_\rho \frac{V^2}{2} \qquad (10.5.14)$$

in which C_D is the drag coefficient and A is the projected area normal to the flow. A discussion of drag coefficients for various immersed objects is given in Section 8.3.1. In open channel flow, since a free surface is present, the drag coefficient must be modified to account for wave drag as well as drag due to friction and separation. Examples of submerged objects in open channel flow include pipelines and bridge piers; in these situations, a hydraulic jump may not be present.

A design application is illustrated in Fig. 10.17. Baffle blocks are devices placed in a reach of channel known as a *stilling basin* to stabilize the location of a hydraulic jump and to aid in the dissipation of flow energy. Typically they are used for Fr_1 greater than 4.5. Example 10.9 demonstrates how baffle blocks reduce the magnitude of the momentum function downstream of a hydraulic jump, resulting in reduced downstream depth and increased energy dissipation. In practice, Eq. 10.5.14 is seldom employed to analyze or design a stilling basin, since other factors such as additional appurtenances, approach velocity, scour, and cavitation must also be considered. Instead, design standards have been established based on observations of existing basins and laboratory model studies (U.S. Department of Interior, 1974; Roberson et al., 1988).

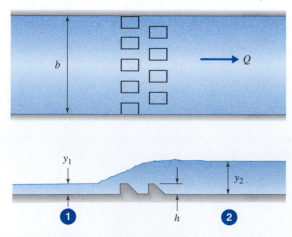

FIGURE 10.17 Stilling basin with baffle blocks.

Example 10.10

In the flow situation presented in Example 10.8, a series of baffle blocks is placed in the channel as shown in Fig. E10.10. Laboratory experimentation has shown that the arrangement has an effective drag coefficient of 0.25, provided that the blocks are submerged in the flow. If the blocks are 0.15 m high, and if the discharge and upstream depth remain the same, determine the depth downstream of the jump and the rate of energy dissipated by the jump.

FIGURE E10.10

Solution

It is necessary to use Eq. 10.5.2 since obstacles (i.e., the baffle blocks) are placed within the control volume. The upstream velocity is

$$V_1 = \frac{Q}{A_1}$$

$$= \frac{7.5}{4 \times 0.2} = 9.38 \text{ m/s}$$

The force F due to the presence of the baffle blocks is computed using Eq. 10.5.14:

$$F = C_D A \rho \frac{V_1^2}{2}$$

$$= 0.25 \times (4 \times 0.15) \times 1000 \times \frac{9.38^2}{2} = 6600 \text{ N}$$

Note that the frontal area is the width of the channel multiplied by the height of the blocks. Substituting known conditions into Eq. 10.5.2, making use of Eq. 10.5.4 which defines M for a rectangular channel, and noting that $q = 7.5/4 = 1.88$ m²/s, we find

$$b\left(\frac{y_1^2}{2} + \frac{q^2}{gy_1}\right) - b\left(\frac{y_2^2}{2} + \frac{q^2}{gy_2}\right) = \frac{F}{\gamma}$$

$$4\left(\frac{0.2^2}{2} + \frac{1.88^2}{9.81 \times 0.2}\right) - 4\left(\frac{y_2^2}{2} + \frac{1.88^2}{9.81 y_2}\right) = \frac{6600}{9800}$$

(continued)

The relation reduces to

$$y_2^2 + \frac{0.721}{y_2} = 3.31$$

The solution for y_2 is approximately 1.70 m. The change in specific energy between locations 1 and 2 is

$$E_1 - E_2 = y_1 + \frac{q^2}{2gy_1^2} - \left(y_2 + \frac{q^2}{2gy_2^2}\right)$$

$$= 0.2 + \frac{1.88^2}{2 \times 9.81 \times 0.2^2} - \left(1.70 + \frac{1.88^2}{2 \times 9.81 \times 1.70^2}\right)$$

$$= 2.94 \text{ m}$$

The rate of energy dissipation, therefore, is

$$\gamma Q(E_1 - E_2) = 9800 \times 7.5 \times 2.94$$

$$= 2.16 \times 10^5 \text{ W} \quad \text{or} \quad 216 \text{ kW}$$

10.5.3 Numerical Solution of the Momentum Equation

For nonrectangular channels the momentum relation can be used directly to analyze the hydraulic jump or other problems requiring the momentum equation; the technique is demonstrated as follows. Consider a trapezoidal channel with conditions known at location 1 upstream of the jump. Consequently, M_1 and F are evaluated as constants and Eq. 10.5.2 can be written in the form

$$M_2 - M_1 + \frac{F}{\gamma} = 0 \tag{10.5.15}$$

in which M_2 is a function of y_2. Introducing the trapezoidal geometry at location 2, with $m_1 = m_2 = m$, the relation above can be written as

$$\frac{y_2^2}{6}(2my_2 + 3b) + \frac{Q^2}{g(by_2 + my_2^2)} - M_1 + \frac{F}{\gamma} = 0 \tag{10.5.16}$$

This can be solved for y_2 by a suitable numerical technique such as interval halving, false position, or Newton's method, which are discussed in any book on numerical methods (see, e.g., Chapra and Canale, 1998). Note that by setting $F = 0$, it becomes the relation for finding the conditions downstream of a hydraulic jump, and that by additionally letting $m = 0$, a rectangular hydraulic jump problem can be solved.

Example 10.11

A hydraulic jump occurs in a triangular channel with $m_1 = m_2 = 2.5$. The discharge is 20 m³/s and $y_c = 1.67$ m. Upstream of the jump the following parameters are given: $y_1 = 0.75$ m, $Fr_1 = 7.42$, and $M_1 = 29.35$ m³. Determine the conjugate depth y_2 downstream of the jump.

Solution

Use Eq. 10.5.16 with $F = 0$:

$$\frac{y_2^2}{6} \times 2 \times 2.5 y_2 + \frac{20^2}{9.81 \times 2.5 y_2^2} - 29.35 = 0$$

The relation reduces to

$$f(y_2) = y_2^3 + \frac{19.58}{y_2^2} - 35.23 = 0$$

The false position method is chosen to find y_2. The first step is to set the upper and lower limits, termed y_u and y_l. Since $Fr_1 > 1$, and $y_2 > y_c$, an appropriate lower limit is $y_l = y_c = 1.67$ m. The upper limit is assumed to be 5 m.

Iteration	y_u	y_l	$f(y_u)$	$f(y_l)$	y_r	$f(y_r)$	Sign of $f(y_l) \times f(y_r)$	ε
1	5	1.67	90.55	−23.55	2.357	−18.61	−	
2	5	2.357	90.55	−18.61	2.808	−10.61	−	
3	5	2.808	90.55	−10.61	3.038	−5.076	−	
4	5	3.038	90.55	−5.067	3.142	−2.231	−	
5	5	3.142	90.55	−2.231	3.187	−0.944	−	
6	5	3.187	90.55	−0.944	3.205	−0.393	−	
7	5	3.205	90.55	−0.393	3.213	−0.162	−	2.5×10^{-3}
8	5	3.213	90.55	−0.162	3.216	−0.0672	−	9.4×10^{-4}
9	5	3.216	90.55	−0.0672	3.218	−0.0277	−	4.9×10^{-4}

The solution is tabulated above. In each iteration a new estimate y_r of the root is made:

$$y_r = \frac{y_u f(y_l) - y_l f(y_u)}{f(y_l) - f(y_u)}$$

The product $f(y_l) \times f(y_r)$ is formed to determine on which subinterval the root will be found. If $f(y_l) \times f(y_r) < 0$, then $y_u = y_r$; otherwise, $y_l = y_r$. It is required that initially $f(y_u)$ and $f(y_l)$ be of opposite sign. Iterations continue until a relative error ε, defined by

$$\varepsilon = \left| \frac{y_r^{new} - y_r^{old}}{y_r^{old}} \right|$$

is less than a specified value, which in the example is 0.0005. The result after nine iterations is $y_2 = 3.22$ m, rounded off to three significant figures.

Using Mathcad or MATLAB, one finds the solution to be less time consuming, see Appendix E, Figs. E.2 and E.3.

10.6 NONUNIFORM GRADUALLY VARIED FLOW

The evaluation of many open-channel flow situations must include accurate analyses of relatively long reaches where the depth and velocity may vary but do not exhibit rapid or sudden changes. In the preceding two sections we emphasized nonuniform, rapidly varied flow phenomena that occur over relatively short reaches, or transitions, in open channels. Attention is now focused on nonuniform, gradually varied flow, where the water surface is continuously smooth. One significant difference between the two is that for rapidly varied flow, losses may often be neglected without severe consequences, whereas for gradually varied flow, it is necessary to include losses due to shear stress distributed along the channel length. The shear stress is the primary mechanism that resists the flow.

Gradually varied flow is a type of steady, nonuniform flow in which y and V do not exhibit sudden or rapid changes, but instead vary so gradually that the water surface can be considered continuous. As a result, it is possible to develop a differential equation that can describe the incremental variation of y with respect to x, the distance along the channel. An analysis of this relation will enable one to predict the various trends that the water surface profile may assume based on the channel geometry, magnitude of the discharge, and the known boundary conditions. Numerical evaluation of the same equation will provide engineering design criteria.

10.6.1 Differential Equation for Gradually Varied Flow

A representative nonuniform gradually varied flow is shown in Fig. 10.18. Over the incremental distance Δx, the depth and velocity are known to change slowly. The slope of the energy grade line is designated as S. In contrast to uniform flow, the slopes of the energy grade line, water surface, and channel bottom are no longer parallel. Since the changes in y and V are gradual, the energy loss over the incremental length Δx can be represented by the Chezy–Manning equation. This means that Eq. 10.3.13, which is valid for uniform flow, can also be used to evaluate S for a gradually varied flow situation, and that the roughness coefficients presented in Table 7.3 are applicable. Additional assumptions include a regular

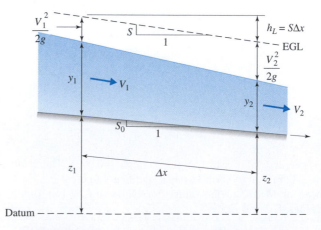

FIGURE 10.18 Nonuniform gradually varied flow.

cross section, small channel slope, hydrostatic pressure distribution, and one-dimensional flow.

The energy equation is applied from location 1 to location 2, with the loss term h_L given by $S\Delta x$. If the total energy at location 2 is expressed as the energy at location 1 plus the incremental change in energy over the distance Δx, Eq. 10.4.2 becomes

$$H_1 = H_2 + S\,\Delta x = H_1 + \frac{dH}{dx}\Delta x + S\,\Delta x \qquad (10.6.1)$$

Substitute $H = y + z + V^2/2g$ and $dz/dx = -S_0$ into this relation and rearrange to find

$$S - S_0 = -\frac{d}{dx}\left(y + \frac{V^2}{2g}\right) \qquad (10.6.2)$$

The right-hand term is $-dE/dx$, and it is transformed to

$$\frac{dE}{dx} = \frac{dE}{dy}\frac{dy}{dx} = (1 - \text{Fr}^2)\frac{dy}{dx} \qquad (10.6.3)$$

(Recall from Eqs. 10.4.13 and 10.4.15 that $dE/dy = 1 - Q^2B/gA^3 = 1 - \text{Fr}^2$.) Finally, upon substitution into the energy relation and solving for the slope of the water surface, dy/dx, one finds that

$$\frac{dy}{dx} = \frac{S_0 - S}{1 - \text{Fr}^2} \qquad (10.6.4)$$

This is the differential equation for gradually varied flow and is valid for any regular channel shape.

Example 10.12

Using an appropriate control volume for gradually varied flow, show that the slope S of the energy grade line is equivalent to $\tau_0/\gamma R$.

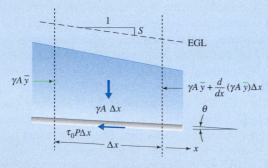

FIGURE E10.12 *(continued)*

Solution

The control volume is shown in Fig. E10.12. The resultant force acting on the control volume is due to the incremental change in hydrostatic pressure $[\gamma d(A\bar{y})/dx]\,\Delta x$, the component of weight in the x-direction $\gamma A \sin\theta \Delta x$, and the resistance term $\tau_0 P \Delta x$. Using the momentum equation

$$\Sigma F_x = \dot{m}(V_{2x} - V_{1x})$$

with $V_{2x} - V_{1x} = (dV/dx)\,\Delta x$ results in

$$-\gamma \frac{d}{dx}(A\bar{y})\,\Delta x + \gamma A \sin\theta\,\Delta x - \tau_0 P\,\Delta x = \rho V A \frac{dV}{dx}\Delta x$$

This relation can be simplified by noting that

$$\frac{d(A\bar{y})}{dx} = \frac{d(A\bar{y})}{dy}\frac{dy}{dx} = A\frac{dy}{dx}$$

and $P = A/R$. Substitute and divide the equation by $\gamma A\,\Delta x$, the weight of the control volume, and find that

$$-\frac{dy}{dx} + \sin\theta - \frac{\tau_0}{\gamma R} = \frac{V}{g}\frac{dV}{dx}$$

Since $\sin\theta \simeq S_0$ for small θ, the equation above can be rearranged in the form

$$\frac{\tau_0}{\gamma R} - S_0 = -\frac{dy}{dx} - \frac{V}{g}\frac{dV}{dx}$$

$$= -\frac{d}{dx}\left(y + \frac{V^2}{2g}\right)$$

Upon comparison with Eq. 10.6.2, it is seen that the right-hand side is equivalent to $S - S_0$, and consequently,

$$\frac{\tau_0}{\gamma R} - S_0 = S - S_0$$

or

$$S = \frac{\tau_0}{\gamma R}$$

10.6.2   Water Surface Profiles

It is possible to identify a series of water surface profiles based on an evaluation (Bakhmeteff, 1932) of Eq. 10.6.4. Essential to the development is the determination of normal and critical depths. Note that y_0 and y_c are uniquely determined once the channel properties and discharge are established. Table 10.3 shows the classification of the water surface profiles. Associated with y_c is a critical slope S_c, which is found by substituting y_c into the Chezy–Manning equation and solving for the slope. The channel slope can be designated as mild, steep, or critical, depending on whether S_0 less than, greater than, or equal to S_c, respectively. A horizontal slope exists when $S_0 = 0$ and an adverse slope occurs when $S_0 < 0$. Inspection of Table 10.3 shows that there are 12 possible profiles. Each profile is classified by a letter/number combination. The letter refers to the channel slope: M for mild, S for steep, C for critical, H for horizontal, and A for adverse. The numerical subscript designates the range of y relative to y_0 and y_c. Flow can occur at depths above or below y_c and at depths above or below y_0.

The variation of y with respect to x for each profile in Table 10.3 can now be found. For a given Q, n, S_0, and channel geometry, the analysis reduces to the determination of how S and Fr vary with y. An inspection of the Chezy–Manning equation will reveal that S decreases with increasing y; similarly, the Froude number decreases as y increases. The numerator in Eq. 10.6.4 will assume the following inequalities: $(S_0 - S) > 0$ for $y > y_0$, and $(S_0 - S) < 0$ for $y < y_0$. In addition, the denominator varies in the manner $(1 - \text{Fr}^2) > 0$ for $y > y_c$ and $(1 - \text{Fr}^2) < 0$ for $y < y_c$. With these criteria, the sign of dy/dx can be evaluated. In addition, with the use of Eq. 10.6.2, the sign of dE/dx is revealed.

TABLE 10.3   Classification of Surface Profiles

Channel slope	Profile type	Depth range	Fr	$\dfrac{dy}{dx}$	$\dfrac{dE}{dx}$	
Mild $S_0 < S_c$ $y_0 > y_c$	M_1	$y > y_0 > y_c$	<1	>0	>0	
	M_2	$y_0 > y > y_c$	<1	<0	<0	
	M_3	$y_0 > y_c > y$	>1	>0	<0	
Steep $S_0 > S_c$ $y_0 < y_c$	S_1	$y > y_c > y_0$	<1	>0	>0	
	S_2	$y_c > y > y_0$	>1	<0	>0	
	S_3	$y_c > y_0 > y$	>1	>0	<0	
Critical $S_0 = S_c$ $y_0 = y_c$	C_1	$y > y_c$ or y_0	<1	>0	>0	
	C_3	y_c or $y_0 > y$	>1	>0	<0	
Horizontal $S_0 = 0$ $y_0 \to \infty$	H_2	$y > y_c$	<1	<0	<0	
	H_3	$y_c > y$	>1	>0	<0	
Adverse $S_0 < 0$ y_0 undefined	A_2	$y > y_c$	<1	<0	<0	
	A_3	$y_c > y$	>1	>0	<0	

The boundaries of the profiles can be established as follows[5]:

1. As y tends to y_0, S tends to S_0. Hence dy/dx approaches zero, or in other words, the water surface approaches y_0 asymptotically. This applies to curves M_1, M_2, S_2, and S_3.

2. As y becomes large, the velocity becomes small and S and Fr tend to zero, so that dy/dx approaches S_0. Hence the surface approaches a horizontal asymptote; curves M_1, S_1, and C_1 are of this type.

3. As y approaches y_c, dy/dx becomes infinite, a limit that is never reached. For supercritical flow, as the M_3, H_3, and A_3 curves approach y_c, a hydraulic jump will form and create a discontinuity in the water surface; at the beginning of the curve, rapid acceleration occurs with nonparallel streamlines. When the flow is subcritical (M_2, H_2, A_2), rapid drawdown takes place close to y_c, and the streamlines are no longer parallel. For all of these situations, Eq. 10.6.4 is invalid, since the flow is no longer one-dimensional.

For the profiles shown in Table 10.3, there is no physical significance to the theoretical limit of y approaching zero, since a finite depth is necessary for the existence of flow. It is left as an exercise to show that for the critical slope condition, $dy/dx > S_c$ as y approaches y_c from either a C_1 or C_2 profile, and that dy/dx approaches a horizontal asymptote as y becomes very large.

Example 10.13

By assuming a wide rectangular channel, develop the right-hand side of Eq. 10.6.4 to show how dy/dx varies with y.

Solution

For a wide rectangular channel, assume that $b \gg y$, so that the wetted perimeter is approximated by $P \simeq b$. The hydraulic radius then becomes $R = A/P \simeq (by)/b = y$. Noting that $Q = qb$, the Chezy–Manning equation, used to evaluate S, simplifies to

$$S = \frac{(qbn)^2}{(by^{5/3})^2} = \frac{(qn)^2}{y^{10/3}}$$

It is assumed that in the Chezy–Manning equation $c_1 = 1$. For a rectangular section the square of the Froude number is

$$\text{Fr}^2 = \frac{q^2}{gy^3}$$

Substituting into Eq. 10.6.4 gives

$$\frac{dy}{dx} = \frac{S_0 - (qn)^2/y^{10/3}}{1 - q^2/(gy)^3}$$

---

[5]Note that x is measured parallel to the channel bottom, and that y is measured vertically from the channel bottom; hence dy/dx and dE/dx are evaluated relative to the channel bottom and not to a horizontal datum.

Since $(qn)^2 y_0{}^{-10/3} = S_0$ and $Fr_c^2 = q^2/(gy_c^3) = 1$, the relation can be written as

$$\frac{dy}{dx} = S_0 \frac{1 - (y_0/y)^{10/3}}{1 - (y_c/y)^3}$$

This equation can be used as an alternative to Eq. 10.6.4 to evaluate the water surface profiles shown in Table 10.3.

10.6.3   Controls and Critical Flow

The existence of the various profiles shown in Table 10.3 depend on the boundary conditions that are specified at given locations in the channel. Quite often, a control will define the boundary condition. A *control* exists when a depth-discharge relationship can be established at a section. The manner in which the control section affects the water surface away from its specific location can be studied by examining flow near the critical state in a rectangular channel.

Equation 10.5.12 is the expression for a surge of finite magnitude translating in the upstream direction in a rectangular channel. As y_2 approaches y_1, the magnitude of the surge becomes infinitesimal; these waves may be generated by the presence of control structures and other transition sections that tend to "disturb" the flow. In Eq. 10.5.12, one can replace y_1 and y_2 by y, and V_1 by V. In addition, let the wave travel in the downstream direction, so that the equation becomes

$$1 = \frac{1}{2}\left[\sqrt{1 + 8\frac{(V-w)^2}{gy}} - 1\right] \tag{10.6.5}$$

Solving for w there results

$$w = V \pm \sqrt{gy} = V \pm c \tag{10.6.6}$$

in which $c = \sqrt{gy}$, termed the *celerity*. The celerity is the speed at which an infinitesimal wave will travel into an undisturbed region, that is, a region with zero velocity. For a nonrectangular channel, $c = \sqrt{gA/B}$. Thus if a disturbance is created at a midstream location in a channel, two infinitesimal waves are generated. One wave front will tend to propagate upstream at the speed $V - c$ and the second will move downstream at the speed $V + c$. These observations are made relative to an observer in a fixed position, that is, to one who is standing on the shore observing the wave motion.

In a rectangular channel with critical flow conditions, $Fr = V/\sqrt{gy} = 1$, or $V = \sqrt{gy} = c$. Hence at critical flow, the first wave front generated by the disturbance would not move upstream but would appear stationary and become a so-called "standing wave." The opposite front would be swept downstream at the speed $2c$. If $Fr < 1$, the first wave would travel upstream, and the opposite wave would travel downstream at a speed less than $2c$. For $Fr > 1$ in the channel, since $V > c$, both waves are swept downstream. Thus, since it contains a mechanism that disturbs the flow, a control will influence upstream conditions only when the

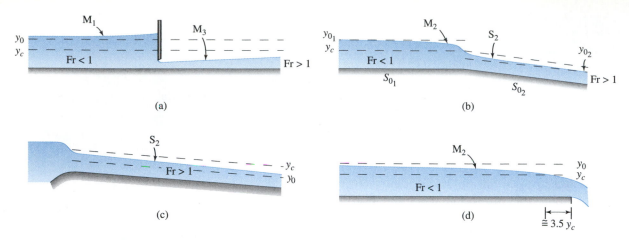

FIGURE 10.19 Representative controls: (a) sluice gate; (b) change in slope from mild (S_{01}) to steep (S_{02}); (c) entrance to a steep channel; (d) free outfall.

flow is subcritical (Fr < 1). Similarly, when the flow is supercritical (Fr > 1), the control can influence only the downstream flow conditions. This is illustrated in Figure 10.19a, where a sluice gate is positioned in a channel with subcritical flow upstream and supercritical flow downstream. An M_1 profile is generated above the gate and an M_3 profile exists below it. Any movement of the gate would influence the nature of the two profiles; lowering the gate would lengthen their range, and raising it would produce the opposite effect.

The profiles shown in Table 10.3 are all influenced by the presence of controls. In Fig. 10.19, several controls are shown which create a variety of profiles. Critical depth often is associated with an effective control. Examples include sluice gates, weirs, dams, and flumes, all of which force critical flow to occur somewhere in the transition region. In addition, a control can be located at a break in channel slope from mild upstream to steep downstream (Fig. 10.19b). The entrance to a steep channel (Fig. 10.19c) is an example of an upstream control, with critical flow occurring at the crest. Critical flow will occur a short distance upstream from a free outfall on a mild slope (Fig. 10.19d). Other controls, not shown in Fig. 10.19, are a channel constriction acting as a choke, and for a mild slope, the existence of uniform flow at some location.

10.6.4 Profile Synthesis

Identification of controls and their interaction with the possible profiles is a requirement for successful understanding and correct design and analysis of open-channel flow. Since controls are essentially transition sections, the rapidly varied flow principles presented in Sections 10.4 and 10.5 can be used to determine the necessary depth–discharge relations. Once the controls have been identified, the profiles can be selected and the range of influence of the controls established.

As an example, consider the situation shown in Fig. 10.20. Flow enters the steep channel from a reservoir, so that critical flow exists at the channel entrance.

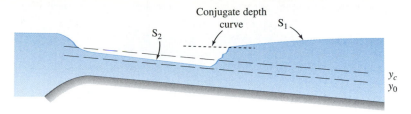

FIGURE 10.20   Example of profile synthesis.

Both the magnitude of the discharge and the S_2 profile are influenced by the depth at the entrance; therefore, the depth acts as a control. At the downstream end of the channel, the lower reservoir acts as a control to establish an S_1 profile that projects upstream. At some interior location, a hydraulic jump occurs to allow the flow to pass from a supercritical to a subcritical state. The location of the jump can be found by plotting a curve of the depth conjugate to the S_2 profile and finding its point of intersection with the S_1 curve. A lowering of the lower reservoir elevation would cause the jump to move downstream and ultimately be swept out of the channel. Increasing the lower reservoir elevation would move the jump upstream; if it were to move into the entrance region, the upstream control would no longer exist. Additional examples of profile synthesis are given in Section 10.7.

Example 10.14

In a rectangular channel, $b = 3$ m, $n = 0.015$, $S_0 = 0.0005$, and $Q = 5$ m³/s. At the entrance to the channel, flow issues from a sluice gate at a depth of 0.15 m. The channel is sufficiently long that uniform flow conditions are established away from the entrance region, Fig. E10.14a. Find the nature of the water surface profile in the vicinity of the entrance.

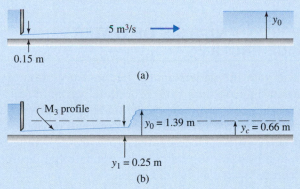

FIGURE E10.14

Solution
First find y_0 and y_c to determine the type of channel. To find y_0, follow the method shown in Example 10.1. Substitute known data into the Chezy–Manning equation:

(continued)

$$\frac{(3y_0)^{5/3}}{(3 + 2y_0)^{2/3}} = \frac{0.015 \times 5}{\sqrt{0.0005}} = 3.354$$

Solving gives $y_0 = 1.39$ m. Next, the critical depth is computed to be

$$y_c = \left(\frac{q^2}{g}\right)^{1/3}$$

$$= \left[\frac{(5/3)^2}{9.81}\right]^{1/3} = 0.66 \text{ m}$$

Since $y_0 > y_c$, a mild slope condition exists. The gate is a control and there will be an M_3 profile beginning at the entrance, terminated by a hydraulic jump. Downstream of the jump, the condition of uniform flow acts as a control, so at that location the depth is y_0, and the Froude number is

$$\text{Fr}_0 = \frac{q}{\sqrt{gy_0^3}}$$

$$= \frac{5/3}{\sqrt{9.81 \times 1.39^3}} = 0.325$$

Using Eq. 10.5.10, the depth before the jump is

$$y_1 = \frac{y_0}{2}\left(\sqrt{1 + 8\text{Fr}_0^2} - 1\right)$$

$$= \frac{1.39}{2}\left(\sqrt{1 + 8 \times 0.325^2} - 1\right) = 0.25 \text{ m}$$

The depths y_c and y_1 have been calculated to two significant figures, since the Manning coefficient is known to only two significant figures. See Fig. E10.14b.

10.7 NUMERICAL ANALYSIS OF WATER SURFACE PROFILES

Section 10.6 dealt with the understanding and interpretation of various aspects of gradually varied flow. An important procedure was outlined: A water surface profile can be synthesized, or predicted, by making use of relevant information about the channel geometry, roughness, and flow, and by determining or assuming the appropriate controls. Once a satisfactory profile synthesis has been conducted, we are in a position to numerically calculate the desired water surface profiles and accompanying energy gradelines.

Regardless of the type of method chosen to numerically evaluate gradually varied flow, analysis of a water surface profile on a reach of channel with constant slope usually follows these steps:

1. The channel geometry, channel slope S_0, roughness coefficient n, and discharge Q are given or assumed.

2. Determine normal depth y_0 and critical depth y_c.
3. Establish the controls (i.e., the depth of flow) at the upstream and downstream ends of the channel reach.
4. Integrate Eq. 10.6.4 to find y and subsequently E as functions of x, allowing for the possibility of a hydraulic jump to occur within the reach.

For a prismatic channel, y_0 can be evaluated by applying the Chezy–Manning equation and finding the root of the function

$$\frac{Qn}{c_1 A R^{2/3} \sqrt{S_0}} - 1 = 0 \qquad (10.7.1)$$

Similarly, y_c is found from application of the Froude number in the form

$$\frac{Q^2 B}{g A^3} - 1 = 0 \qquad (10.7.2)$$

Numerical methods, such as the false position or interval halving (Chapra and Canale, 1988), or computational software can be used to solve Eqs. 10.7.1 and 10.7.2.

The Chezy–Manning equation, given here in the form of Eq. 10.7.1, is used to represent the depth–discharge variation for nonuniform as well as for uniform flow. Thus one can replace S_0 by the slope S of the energy grade line in Eq. 10.7.1 and solve for S in the manner

$$S(y) = \frac{Q^2 n^2}{c_1^2 \, [A(y)]^2 \, [R(y)]^{4/3}} \qquad (10.7.3)$$

Since Q and n are either given or assumed, the right-hand side of Eq. 10.7.3 is a function of y alone.

Two numerical methods are presented in this section to compute water surface profiles along with normal and critical depths. The first, termed the *standard step method*, is the most widely used, and the second employs an accurate numerical integration scheme. Solutions will be illustrated using Excel, Mathcad, and MATLAB software. The third method is analytical, and uses Eq. 10.6.4 in integrated form, assuming simplified geometric properties of the channel cross section.

10.7.1 Standard Step Method

To develop a numerical procedure for solving gradually varied flow problems, we use Eq. 10.6.2; it is applied over the reach of channel shown in Fig. 10.21, resulting in

$$\frac{dE}{dx} = \frac{d}{dx}\left(y + \frac{V^2}{2g}\right)$$
$$= S_0 - S(y)$$

(10.7.4)

For small changes in y and V between x_i and x_{i+1}, Eq. 10.7.4 can be approximated by

$$E_{i+1} - E_i = \int_{x_i}^{x_{i+1}} [S_0 - S(y)]\, dx$$
$$\simeq (x_{i+1} - x_i)[S_0 - S(y_m)]$$

(10.7.5)

in which $y_m = (y_{i+1} + y_i)/2$. Equation 10.7.5 is solved for x_{i+1} to yield

$$x_{i+1} = x_i + \frac{E_{i+1} - E_i}{S_0 - S(y_m)}$$

(10.7.6)

The calculations take place stepwise, beginning at a control point or other location where the depth is known. Assume that it is desired to generate a water surface profile along the channel, and compute values of x_i, y_i, and E_i, for $i = 1, \ldots, k$, where k is the location at the opposite end of the reach. Other than x_k, it is not necessary that location x_i be fixed, so it can take on any value. Beginning at location i, the evaluation of conditions at location $i + 1$ proceeds as follows:

1. Choose y_{i+1}.
2. Compute E_{i+1} from Eq. 10.4.12, y_m knowing y_i and y_{i+1}, and $S(y_m)$ from Eq. 10.7.3.
3. Compute x_{i+1} from Eq. 10.7.6.
4. At location k, trial values of y_k are assumed until Eq. 10.7.6 is satisfied for the known value of x_k.

The standard step method can be carried out using spreadsheet analysis.

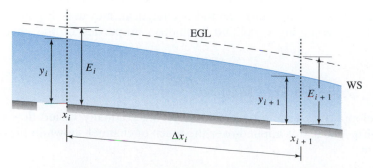

FIGURE 10.21   Notation for computing gradually varied flow.

Example 10.15

Water is flowing at $Q = 22$ m^3/s in a long trapezoidal channel, $b = 7.5$ m, $m_1 = m_2 = 2.5$. A free overfall is located at the downstream end of the channel, where $x = 2000$ m. For $n = 0.015$, $S_0 = 0.0006$, find the water surface profile and energy grade line for a distance of approximately 800 m upstream from the free outfall.

Solution

Equations 10.7.1 and 10.7.2 are used to evaluate y_0 and y_c by substituting in known data:

$$\frac{22 \times 0.015 \times [7.5 + 2y_0 \sqrt{1 + (2.5)^2}]^{2/3}}{[7.5y_0 + \frac{1}{2} y_0^2(2.5 + 2.5)]^{5/3} \sqrt{0.0006}} - 1 = 0$$

$$\frac{(22)^2 \times [7.5 + y_c(2.5 + 2.5)]}{9.8 \times [7.5y_c + \frac{1}{2} y_c^2(2.5 + 2.5)]^3} - 1 = 0$$

The roots of these equations can be found using a routine such as Excel Solver®; the solutions are $y_0 = 1.29$ m and $y_c = 0.86$ m. Hence the channel is a mild type, and control will be close to the free overfall at the downstream end of the channel. Without any serious loss of accuracy, one can assume that critical conditions will exist at the free overfall. Referring to Table 10.3, the profile upstream of the overfall will be of type M$_2$. An Excel spreadsheet solution is shown in Table E10.15. The upper part shows the values of critical and normal depths found by using Solver. In the residual column are very small numbers that should be close to zero, see the two above equations. The lower part of the table shows the step method solution. Calculations proceed from station 1 to station 5 in a straightforward manner, with arbitrary values of depth selected and placed in the y column. The beginning value of x (2000 m) is placed in the first cell of the x column, and the remaining distances are computed as explained on page 522. At station 6, different values of y are chosen until the distance is close to the desired value of 1200 m; a depth of 1.27 m results in a distance of 1230 m, which is sufficient. The spreadsheet equations for computing normal and critical depths, and for the step method, are provided in Appendix E. In addition, a MATLAB solution to this problem is shown in Appendix E, Fig. E.5.

TABLE E10.15

	Depth [m]	Residual
Critical	0.865	1.087E-06
Normal	1.292	1.812E-06

Station	y [m]	A [m^2]	V [m/s]	E [m]	y_m [m]	$S(y_m)$	Δx [m]	x [m]
1	0.865	8.358	2.632	1.218				2000
2	0.950	9.381	2.345	1.230	0.908	2.165E-03	-8	1992
3	1.050	10.631	2.069	1.268	1.000	1.527E-03	-41	1951
4	1.150	11.931	1.844	1.323	1.100	1.081E-03	-114	1837
5	1.250	13.281	1.656	1.390	1.200	7.866E-04	-357	1480
6	1.270	13.557	1.623	1.404	1.260	6.574E-04	-250	1230

Example 10.16

A trapezoidal channel 300 m long conveys water flowing at $Q = 25$ m^3/s. The inverse side slopes of the cross section are $m_1 = m_2 = 2.5$, the bottom width is 5 m, the bottom slope is $S_0 = 0.005$, and the Manning coefficient is $n = 0.013$. Compute the water surface profile and energy grade line if the upstream depth $y_u = 0.55$ m and the downstream depth $y_d = 2.15$ m.

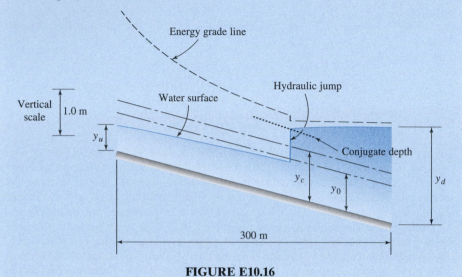

FIGURE E10.16

Solution

An Excel spreadsheet solution using the step method is shown in Table E10.16. First, input data is entered, including the gravitational constant and the corresponding constant c_1 in the Manning equation. Since the two inverse side slopes are equal, a single value m is entered. Then critical and normal depths are evaluated using Solver. Since $y_c > y_0$, the slope is steep. One control point is at the upstream end, and an S_3 profile exists in the upper reach of the channel. The profile is computed from station 1 beginning with y_u to station 7 where the depth is 0.85 m and the distance is 288 m. Then an S_1 profile is calculated from station 9 at the end of the channel beginning with y_d, and terminating when y reaches critical depth. A hydraulic jump is located approximately 190 m downstream from the channel entrance. It is located by plotting the curve of depth conjugate to the S_3 profile and finding its intersection with the S_1 profile as shown in Figure E10.16. The conjugate depths y_{cj} are computed using Excel Solver.

TABLE E10.16

$Q =$ 25	$n =$ 0.013		$S_0 =$ 0.005	$L =$ 300	$g =$ 9.81
$b =$ 5	$m =$ 2.5		$y_u =$ 0.55	$y_d =$ 2.15	$c_1 =$ 1.00

	Depth	Residual
$y_c =$	1.124	-2.26E-07
$y_0 =$	0.864	4.363E-07

Station	y	A	V	E	y_m	$S(y_m)$	Δx	x	y_{cl}	Residual
1	0.550	3.506	7.130	3.141						
2	0.600	3.900	6.410	2.694	0.575	2.189E-02	26	26	1.901	8.81E-07
3	0.650	4.306	5.806	2.368	0.625	1.624E-02	29	56	1.802	5.19E-07
4	0.700	4.725	5.291	2.127	0.675	1.231E-02	33	88	1.711	1.68E-07
5	0.750	5.156	4.848	1.948	0.725	9.500E-03	40	128	1.628	9.30E-07
6	0.800	5.600	4.464	1.816	0.775	7.449E-03	54	182	1.550	3.85E-07
7	0.850	6.056	4.128	1.719	0.825	5.923E-03	105	288	1.477	5.07E-07
9	2.150	22.306	1.121	2.214				300		
10	2.050	20.756	1.204	2.124	2.100	1.575E-04	-19	281		
11	1.950	19.256	1.298	2.036	2.000	1.924E-04	-18	263		
12	1.850	17.806	1.404	1.950	1.900	2.371E-04	-18	245		
13	1.750	16.406	1.524	1.868	1.800	2.951E-04	-17	228		
14	1.650	15.056	1.660	1.791	1.700	3.713E-04	-17	211		
15	1.550	13.756	1.817	1.718	1.600	4.728E-04	-16	195		
16	1.450	12.506	1.999	1.654	1.500	6.102E-04	-15	180		
17	1.350	11.306	2.211	1.599	1.400	7.998E-04	-13	167		
18	1.250	10.156	2.462	1.559	1.300	1.067E-03	-10	157		
19	1.124	8.778	2.848	1.537	1.187	1.514E-03	-6	151		

10.7.2 Numerical Integration Method

There are a number of numerical methods available to solve Eq. 10.6.4 for channels with regular cross sections, any of which would provide a solution with sufficient accuracy for design and analysis purposes. Worthy of mention are integration techniques using trapezoidal or Simpson's rules, and solution of the differential form using a Runge–Kutta procedure. The underlying theory of these methods and algorithms for their implementation can be found in a number of texts on numerical methods. McBean and Perkins (1975) discuss convergence problems associated with use of the trapezoidal rule applied to gradually varied flow.

A useful integration scheme is the two-point Gauss–Legendre quadrature (Chapra and Canale, 1998). It is particularly well suited for computer use and is usually more accurate than either the trapezoidal or Simpson's rule methods. Equation 10.6.4 is written in the integral form

$$x_{i+1} = x_i + \int_{y_i}^{y_{i+1}} \frac{1 - \mathrm{Fr}^2}{S_0 - S} \, dy$$

$$= x_i + \int_{y_i}^{y_{i+1}} G(y) \, dy \tag{10.7.7}$$

The last integral is approximated by the Gauss–Legendre formula, giving the result

$$\int_{y_i}^{y_{i+1}} G(y) \, dy = \frac{y_{i+1} - y_i}{2} \left[G\left(\frac{y_{i+1} + y_i - \sqrt{3}/3(y_{i+1} - y_i)}{2} \right) \right.$$

$$\left. + G\left(\frac{y_{i+1} + y_i + \sqrt{3}/3(y_{i+1} - y_i)}{2} \right) \right] \tag{10.7.8}$$

The following example illustrates use of the algorithm.

Example 10.17

Rework Example 10.15 using Gauss–Legendre quadrature.

Solution

A Mathcad solution is shown below. After the known data is entered and functions are defined, normal and critical depths are computed. Subsequently, the depth is iterated to find corresponding values of distance using Eqs. 10.7.7 and 10.7.8.

Input data:

$$Q := 22 \qquad M := 2.5 \qquad b := 7.5 \qquad L := 2000$$

$$S_0 := 0.0006 \qquad n := 0.015 \qquad g := 9.81$$

Define functions:

$$A(y) := b \cdot y + M \cdot y^2 \qquad B(y) := b + 2 \cdot M \cdot y \qquad P(y) := b + 2 \cdot y \cdot \sqrt{1 + M^2}$$

$$S(y) := \frac{Q^2 \cdot n^2}{A(y)^{3.3333} \cdot P(y)^{-1.3333}} \qquad Fr(y) := \sqrt{\frac{Q^2 \cdot B(y)}{g \cdot A(y)^3}} \qquad G(y) := \frac{1 - Fr(y)^2}{S_0 - S(y)}$$

$$E(y) := y + \frac{Q^2}{2 \cdot g \cdot A(y)^2}$$

Compute normal and critical depths:

$$y_n := root\left(S(y) - S_0, y, 0.01, 5\right) \qquad\qquad y_n = 1.292$$

$$y_c := root(Fr(y) - 1, y, 0.01, 5) \qquad\qquad y_c = 0.865$$

Compute S2 profile from upstream location:

$$x_0 := 2000 \qquad y_0 := y_c \qquad \Delta y := 0.10 \qquad N := 4$$

$$i := 1, 2 .. N$$

$$y_i := y_{i-1} + \Delta y$$

$$x_i := x_{i-1} + \frac{\Delta y}{2} \cdot \left(G\left(\frac{y_i + y_{i-1} - \frac{\sqrt{3}}{3} \cdot \Delta y}{2} \right) + G\left(\frac{y_i + y_{i-1} + \frac{\sqrt{3}}{3} \cdot \Delta y}{2} \right) \right)$$

Computed depth, distance and specific energy:

$$y = \begin{pmatrix} 0.865 \\ 0.965 \\ 1.065 \\ 1.165 \\ 1.265 \end{pmatrix} \qquad x = \begin{pmatrix} 2 \times 10^3 \\ 1.988 \times 10^3 \\ 1.938 \times 10^3 \\ 1.798 \times 10^3 \\ 1.257 \times 10^3 \end{pmatrix} \qquad E(y) = \begin{pmatrix} 1.218 \\ 1.235 \\ 1.275 \\ 1.333 \\ 1.4 \end{pmatrix}$$

10.7.3   Irregular Channels

The method demonstrated in Example 10.15 works well with regular channels. When calculating the profile for a reach of river channel in which the cross-sectional data are irregular and usually given at fixed locations, a trial solution is necessary at each location. Henderson (1966) outlines a method of computation that includes corrections for composite sections and eddy losses that occur at bends and expansions. For a composite section, the slope of the energy grade line is given by

$$S = \frac{Q^2}{(\Sigma \, K_i)^2} \tag{10.7.9}$$

in which K_i is termed the *conveyance* of subsection i:

$$K_i = \left(\frac{c_1 A R^{2/3}}{n}\right)_i \tag{10.7.10}$$

A kinetic-energy coefficient must be applied to the kinetic-energy term; for a composite section use

$$\alpha = \frac{(\Sigma \, A_i)^2}{(\Sigma \, K_i)^3} \, \Sigma \left(\frac{K_i^3}{A_i^2}\right) \tag{10.7.11}$$

With such complexity, it is most convenient to use the step method procedure in computer-coded form. The United States Army Corps of Engineers has developed the algorithm "HEC-RAS River Analysis System" for use in natural channels. The solution is based on the standard step method and is available for use on a personal computer. The program and documentation can be downloaded from the Website *www.hec.usace.army.mil*.

10.7.4   Direct Integration Methods

Equation 10.6.4 can be integrated in a straightforward manner if one assumes a wide rectangular, horizontal channel. A more general integration procedure is available for nonhorizontal prismatic channels of various shapes (Chow, 1959). Assume that in the Chezy–Manning equation the product $A^2 R^{4/3}$ is proportional to y^N, and in the Froude number the ratio A^3/B is proportional to y^M, where M and N are constants. Then Eq. 10.6.4 can be written

$$dx = \frac{1}{S_0}\left[\frac{1 - (y_c/y)^M}{1 - (y_0/y)^N}\right] dy \tag{10.7.12}$$

The solution of Eq. 10.7.12 is

$$x = \frac{y_0}{S_0}\left[u - F(u, N) + \left(\frac{y_c}{y_0}\right)^M \frac{J}{N}F(v, J)\right] \qquad (10.7.13)$$

in which $u = y/y_0$, $v = u^{N/J}$, $J = N/(N - M + 1)$, and F is the varied flow function

$$F(u, N) = \int_0^u \frac{d\eta}{1 - \eta^N} \qquad (10.7.14)$$

Equation 10.7.14 can be evaluated numerically and the results are presented as the varied flow function in Appendix E, Table E.1. Values of $N = 2\frac{1}{2}$, 3, and $3\frac{1}{3}$ are commonly used since they are consistent with a wide rectangular channel assumption (see Example 10.13). Note that the function F may have any desired constant of integration assigned to it. In the varied flow function table the constant is adjusted so that $F(0, N) = F(\infty, N) = 0$. Since integration from one location to another is direct, no intermediate calculations are necessary. This method of evaluation is in direct contrast to the numerical methods, where intermediate steps cannot be avoided without the loss of precision.

Example 10.18

A wide rectangular channel conveys a discharge of $q = 3.72$ m³/s per meter width on a slope of $S_0 = 0.001$. At a given location the depth is 3 m. Determine the distance upstream where the depth is 2.5 m. The Manning coefficient is 0.025.

Solution

From Example 10.13, for a wide rectangular channel we found that $N = 3.33$ and $M = 3$. Therefore, one can calculate J to be

$$J = \frac{N}{N - M + 1} = \frac{3.33}{3.33 - 3 + 1} = 2.5$$

Also from Example 10.13, we can determine y_0 in the manner

$$y_0 = \left(\frac{q^2 n^2}{S_0}\right)^{3/10} = \left(\frac{3.72^2 \times 0.025^2}{0.001}\right)^{3/10} = 1.91 \text{ m}$$

Furthermore,

$$y_c = \left(\frac{q^2}{g}\right)^{1/3} = \left(\frac{3.72^2}{9.81}\right)^{1/3} = 1.12 \text{ m}$$

Substitute these values into Eq. 10.7.13 and simplify:

$$x = \frac{1.91}{0.001}\left[u - F(u, 3.33) + \left(\frac{1.12}{1.91}\right)^3 \times \frac{2.5}{3.33}F(v, 2.5)\right]$$

$$= 1910[u - F(u, 3.33) + 0.151F(v, 2.5)]$$

Since $y_0 > y_c$, and at the downstream location $y > y_0$, the profile is an M_1 curve. At the downstream location where the depth $y = 3$ m,

(continued)

$$u = \frac{y}{y_0} = \frac{3}{1.91} = 1.57 \quad \text{and} \quad v = u^{N/J} = 1.57^{3.33/2.5} = 1.825$$

From Appendix E, Table E.1, making use of linear interpolation between recorded values we find that

$$F(1.57, 3.33) = 0.166 \quad \text{and} \quad F(1.825, 2.5) = 0.300$$

Considering x as a distance measured from an arbitrary datum, we find that

$$x = 1910(1.57 - 0.166 + 0.151 \times 0.300) = 2763 \text{ m}$$

To determine the distance upstream of the weir where the depth is 2.5 m, we perform the following calculations in a manner analogous to those at the downstream location:

$$u = \frac{2.5}{1.91} = 1.31 \quad \text{and} \quad v = 1.31^{3.33/2.5} = 1.43$$

$$F(1.31, 3.33) = 0.578 \quad \text{and} \quad F(1.43, 2.5) = 0.474$$

$$x = 1910(1.31 - 0.578 + 0.151 \times 0.474) = 1536 \text{ m}$$

Hence the distance between the two locations, from a depth of 3 m to where the depth is 2.5 m, is $2763 - 1536 = 1227$ m, or approximately 1230 m.

10.8 SUMMARY

As opposed to pipe flows, open channel flow is complicated by the presence of a free surface. In addition to velocity, the depth of flow is variable, so that even though it may be steady, the flow in a channel is generally nonuniform. In Section 10.3, as well in Chapter 7, we have introduced uniform flow as a limiting case — one that does not often occur but nonetheless is significant. Friction losses in open channel flow are described mathematically by the Chezy–Manning equation, which relates velocity or discharge to the geometric and hydraulic properties of the channel cross-section. Energy and momentum principles were introduced in Sections 10.4 and 10.5 and applied to rapidly varied nonuniform flow that occurs at channel transitions and at a hydraulic jump. Critical and conjugate depths have been derived along with the definition of the Froude number. We have seen that specific energy is a useful concept for dealing with transition analysis. Much of the development in this chapter has used a rectangular cross-section; this was done primarily for mathematical expediency. The guidelines that were developed and conclusions made using the rectangular section apply to nonrectangular sections as well. Table 10.4 summarizes several of the formulas developed in the chapter.

Gradually varied nonuniform flow occurs where the depth and velocity vary continuously with distance along a channel. In Section 10.6, application of the energy equation to gradually varied flow resulted in a mathematical description of the water surface profile. We have presented a classification of the various types of flow regimes that occur along a channel reach. In addition we have shown how the understanding of rapidly varied and gradually varied flow enable one to predict the manner in which water surfaces will behave, a task that is required prior to numerical computation of water surface profiles. Several numerical methods were introduced in Section 10.7 to integrate the gradually varied flow equation, and examples of solutions were provided using spreadsheets, computational software, and approximate integrals. Perhaps the most useful means of solution is the spreadsheet, in which a generalized algorithm can be readily adapted for use in different problems.

TABLE 10.4   Formulas for Rectangular and General Sections

Section	Fr	y_c	E	M
Rectangular	$\dfrac{q}{\sqrt{gy^3}}$	$\left(\dfrac{q^2}{g}\right)^{1/3}$	$y + \dfrac{q^2}{2gy^2}$	$\dfrac{by^2}{2} + \dfrac{q^2}{gy}$
General	$\dfrac{Q/A}{\sqrt{gA/B}}$	$\dfrac{Q^2 B}{gA^3} = 1$	$y + \dfrac{Q^2}{2gA^2}$	$A\bar{y} + \dfrac{Q^2}{gA}$

PROBLEMS

10.1 Give an example of each of the following types of flow for both closed-conduit and free-surface conditions. Include a sketch with each representation:
(a) Steady, uniform
(b) Unsteady, nonuniform
(c) Steady, nonuniform
(d) Unsteady, uniform

10.2 Water is flowing with a free surface at 4 m³/s in a circular conduit. Find the diameter d such that the critical depth $y_c = 0.3d$. Under critical flow conditions, $Q^2 B/(gA^3) = 1$.

10.3 Classify the following flows as steady or unsteady, and uniform or nonuniform:
(a) Water flowing through a logjam in a river with the observer standing on the shore.
(b) Water flowing through a river rapids with the observer on a raft traveling with the current.
(c) A flood wave traveling in a river which is generated by a gentle rainfall/runoff event, with the observer standing on the bank.
(d) A moving hydraulic jump in a prismatic channel, with the observer running alongside the bank at the same speed as the jump.

Uniform Flow

10.4 For each regular channel section, construct the curves R versus y and $AR^{2/3}$ versus y:

(a) Circular, $d = 2$ m.

(b) Trapezoidal, $b = 3$ m, $m_1 = m_2 = 2.5$.

(c) Composite (see Fig. P10.4).

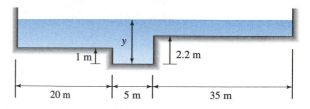

FIGURE P10.4

10.5 Determine the most efficient section based on flow resistance for a trapezoidal channel. Assume equal side slopes, that is, $m_1 = m_2$.

10.6 Determine the uniform depth y_0, area A, and wetted perimeter P for the following conditions:

    (a) Trapezoidal channel, $m_1 = 2.5$, $m_2 = 3.5$, $b = 4.5$ m, $Q = 35$ m³/s, $n = 0.015$, $S_0 = 0.00035$.

    (b) Circular channel, $d = 2.1$ m, $Q = 4.25$ m³/s, $n = 0.012$, $S_0 = 0.001$.

    (c) Trapezoidal channel, $Q = 15$ m³/s, $n = 0.011$, $S_0 = 0.0013$, "most efficient" cross section.

    (d) Rectangular channel, $b = 7.5$ m, $Q = 35$ m³/s, $n = 0.020$, $S_0 = 0.0004$.

10.7 Uniform flow occurs in a trapezoidal channel with $m_1 = m_2 = 1.75$. The channel is to convey a discharge of 4.0 m³/s, with an average velocity of 1.4 m/s. If $n = 0.015$ and $S_0 = 0.001$, find the bottom width and the uniform depth.

10.8 A channel cross section, commonly called a *gutter*, forms at the side of a street next to a curb during rainfall conditions (Fig. P10.8). The slope along the roadway is $S_0 = 0.0005$, and the Manning roughness coefficient is $n = 0.015$. Assuming that uniform flow conditions occur:

    (a) Determine the discharge if the depth of flow is $y_0 = 12$ cm.

    (b) If $Q = 80$ L/s, what is the flow depth y_0?

FIGURE P10.8

10.9 A sewer pipe of circular cross section is made of concrete ($n = 0.014$). It is to convey water at uniform flow conditions such that the pipe flows half full, that is, $y = d/2$, where $d = $ diameter of the pipe. Evaluate the following conditions:

    (a) Given $S_0 = 0.0003$ and $Q = 1.75$ m³/s, find d.

    (b) Given $S_0 = 0.00005$ and $d = 1.3$ m, find Q.

    (c) Given $d = 0.75$ m and $Q = 0.45$ m³/s, find S_0.

Energy Concepts

10.10 For a given specific energy in a rectangular channel, show that critical flow exists when the specific discharge is at its maximum value q_{max}.

10.11 The specific energy relation for a rectangular section, Eq. 10.4.5, can be made dimensionless by normalizing it with respect to critical depth y_c. Prepare such a plot with y/y_c as the ordinate and E/y_c as the abscissa. Locate the minimum point mathematically.

10.12 Water is flowing in a rectangular channel at a velocity of 3 m/s and a depth of 2.5 m. Determine the changes in water surface elevation for the following alterations in the channel bottom:

    (a) An increase (upward step) of 20 cm, neglecting losses.

    (b) A "well-designed" increase of 15 cm.

    (c) The maximum increase allowable for the specified upstream flow conditions to remain unchanged, neglecting losses.

    (d) A "well-designed" decrease (downward step) of 20 cm.

10.13 Water is flowing in a rectangular channel whose width is 5 m. The depth of flow is 2 m and the discharge is 25 m³/s. Determine the changes in depth for the following alterations in the channel width:

    (a) An increase of 50 cm, neglecting losses.

    (b) A decrease of 25 cm, assuming a "well-designed" transition.

10.14 Flow in a rectangular channel of width b takes place with a known depth y and velocity V. Downstream of this location, there is an upward step of height h. Neglecting losses, determine the change in width that must take place simultaneously for critical flow to occur within the transition. Given:

$b = 3$ m, $y = 3$ m,
$V = 3$ m/s, $h = 70$ cm

10.15 Water is flowing in a rectangular channel 2.0 m wide. At a transition section the channel bottom is lowered by $h = 0.1$ m for a short distance, and then is raised back to the original elevation (Fig. P10.15). If $y = 1.22$ m and $Q = 4.8$ m³/s, then, with losses neglected:

(a) Find the change in channel width necessary to maintain a horizontal water surface through the transition.

(b) What change in width would cause critical flow to occur in the transition?

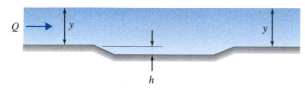

FIGURE P10.15

10.16 Water flows at a depth of 2.15 m and a unit discharge of 5.5 m²/s in a rectangular channel. Energy losses can be neglected.

(a) What is the maximum height h of a raised bottom that will permit the flow to pass over it without increasing the upstream depth?

(b) Show the solution on an E–y diagram.

(c) Sketch the water surface and energy grade line.

(d) If the channel bottom is raised greater than h, discuss a type of change that may take place upstream of the transition.

10.17 A lake discharges into a steep channel. At the channel entrance the lake level is 2.5 m above the channel bottom. Neglecting losses, find the discharge for the following geometries:

(a) Rectangular section, $b = 4$ m.

(b) Trapezoidal section, $b = 3$ m, $m_1 = m_2 = 2.5$.

(c) Circular section, $d = 3.5$ m.

10.18 Steep flow conditions exist at the outlet from a reservoir into an open canal. The canal width is

not yet fixed, but the depth y at the entrance and the discharge Q are known. Determine the necessary widths for the following geometries:

(a) Rectangular section, $y = 1$ m, $Q = 18$ m³/s

(b) Trapezoidal section, $m_1 = m_2 = 3$, $y = 1$ m, $Q = 18$ m³/s

10.19 Rapidly varied flow occurs over a rectangular sill (Fig. P10.19), with free outfall conditions at location C. Throughout the reach losses can be neglected.

(a) Compute the discharge.

(b) Evaluate the depths at locations A, B, and C.

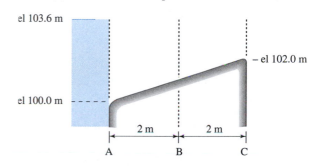

FIGURE P10.19

10.20 Prepare a computer algorithm that evaluates the critical depth of a regular channel using the false position technique or a method of your choice. Verify the program with the following data:

(a) Circular section, $Q = 5$ m³/s, $d = 2.5$ m.

(b) Rectangular section, $Q = 3.54$ m³/s, $b = 3.66$ m.

(c) Trapezoidal section, $Q = 120$ m³/s, $b = 10$ m, $m_1 = m_2 = 5$.

(d) Triangular section, $Q = 7.08$ m³/s, $m_1 = 2.5$, $m_2 = 3.0$.

10.21 For the channel cross section shown in Fig. P10.21, determine the critical depth if the discharge is 16.5 m³/s.

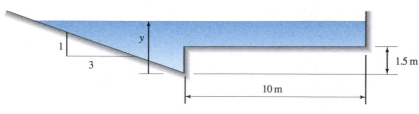

FIGURE P10.21

10.22 For the composite channel section shown in Fig. P10.22, find the critical depth y_c if:
(a) $Q = 55$ m³/s. **(b)** $Q = 3.5$ m³/s.

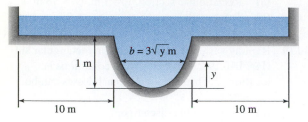

FIGURE P10.22

10.23 A parabolic channel (Fig. P10.23) is described by the function $\eta = 0.1x^2$. If $Q = 25$ m³/s and $y = 2.00$ m, find the alternate depth.

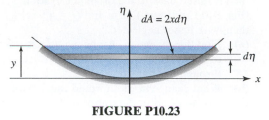

FIGURE P10.23

10.24 Prepare a computer algorithm that determines the alternate depth, given the geometry of a prismatic channel, the discharge, and the depth of flow. Use the interval-halving technique or a solution of your choice.

10.25 Derive the relation for a V-notch weir that expresses Q as a function of Y, Eq. 10.4.27.

10.26 Debris blocks the entire width of a river as shown in Fig. P10.26. The cross section of the river is approximately rectangular, where $b = 15$ m, $Q = 22.5$ m³/s, $y_0 = 1.2$ m, and $h = 0.2$ m.
(a) Neglecting losses, analyze the rapidly varied water surface in the vicinity of the debris.
(b) Sketch the energy grade line and water surface.

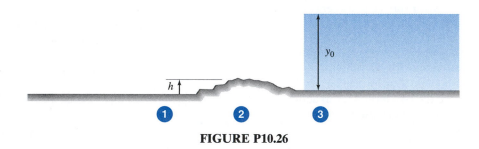

FIGURE P10.26

10.27 Develop the expression for the theoretical discharge for the weir shown in Fig. P10.27.

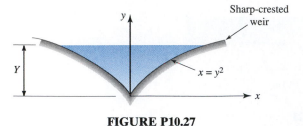

FIGURE P10.27

10.28 A Parshall flume is placed in a small stream to measure the discharge. It has a width of 90 cm at its throat.

(a) Compute the discharge if the measured upstream depth is 36.5 cm.

(b) Prepare a plot of Q versus H where H ranges from 0.1 m to 0.5 m.

10.29 Water is flowing at a given depth and discharge in a 3.5-m-wide rectangular channel. A 60-cm-high submerged raised section is placed in the channel as shown in Fig. P10.29. The raised section spans the entire width of the channel, and is designed such that losses can be neglected. Calculate the water surface and energy grade line:

(a) When $y = 1.5$ m and $Q = 4$ m³/s.
(b) When $y = 1.0$ m and $Q = 3$ m³/s.
(c) Under what given conditions, (a) or (b) above, would the raised section act similar to a broad crested weir (explain your answer in words only)?

FIGURE P10.29

10.30 Design a broad-crested weir to convey a river discharge that varies between Q_1 and Q_2. The maximum water depth upstream of the weir is not to exceed y_2 and the minimum depth is not to be less than y_1. Given:
$Q_1 = 0.15$ m³/s, $Q_2 = 30$ m³/s, $y_1 = 1.05$ m, $y_2 = 1.75$ m

10.31 Design a canal to deliver water between the two reservoirs shown in Fig. P10.31. The horizontal distance between the reservoirs is L. It is required that uniform flow exist throughout the reach. The water surface in reservoir A is to be no more than the distance h above the bottom of the canal at the entrance. The canal is rectangular in cross section, and made of concrete. Neglect entrance and exit losses, and assume uniform, subcritical flow conditions at the channel entrance. Given:
$El_A = 501.8$ m, $El_B = 500.2$ m,
$L = 1500$ m, $h = 2$ m, $b = 2.5$ m.

FIGURE P10.31

Momentum Concepts

10.32 The momentum function for a rectangular section, Eq. 10.5.4, can be made dimensionless by normalizing it with respect to $b(y_c)^2$. Prepare such a plot using y/y_c as ordinate and $M/(by_c)^2$ as abscissa.

10.33 Rework Problem 10.24 to evaluate the conjugate depth in place of the alternate depth.

10.34 Flow occurs in a wide rectangular channel at a depth y. A construction project over this channel requires that cofferdams spaced at distance w on centerlines be placed in the channel (Fig. P10.34). Assuming that the cofferdams create no energy losses between locations 1 and 2, determine the maximum permissible diameter of the cofferdams without creating upstream backwater effects (i.e., without increasing the upstream depth), and the resultant drag force on each cofferdam (assume that $C_D = 0.15$). Given:
$q = 1.5$ m²/s, $y = 1.8$ m, $w = 6$ m

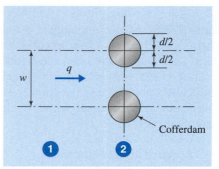

FIGURE P10.34

10.35 A temporary pipeline (diameter = 200 mm $C_D = 0.30$) is to be placed on a river bed normal to the direction of flow. The river is approximately rectangular in section, and is 100 m wide. The depth of flow downstream of the pipe is 2.5 m and the mean velocity is 3 m/s. Neglecting bed slope and resistance, determine the following:

   (a) Flow conditions (i.e., depth and velocity) upstream of the pipe once it is in place.

   (b) Resultant drag force on the pipe.

10.36 A hydraulic jump occurs over a sill located in a triangular channel with water flowing as shown in Fig. P10.36. The inverse side slopes are $m_1 = m_2 = 3$, the drag coefficient $C_D = 0.40$, and the height of the sill $h = 0.3$ m. Determine the discharge Q if $y_1 = 0.50$ m and $y_2 = 1.8$ m.

FIGURE P10.36

10.37 Water is flowing in a rectangular channel at a depth of 1.6 m and a velocity of 0.85 m/s. At a downstream location the discharge is suddenly reduced to zero, causing a surge to propagate upstream. Find the depth and velocity behind the surge, and the speed of the surge wave.

10.38 In a rectangular channel, water is flowing at a depth y and a velocity V. At a downstream location, the discharge is suddenly reduced by 60% (i.e., to 40% of the original value), causing a surge to propagate upstream. Determine the depth and velocity behind the surge, and the speed of the surge wave. Given: $y = 1.5$ m, $V = 1$ m/s

10.39 Water enters a reach of rectangular channel where $y_1 = 0.5$ m, $b = 7.5$ m, and $Q = 20$ m³/s (Fig. P10.39). It is desired that a hydraulic jump occur upstream (location 2) of the sill and on the sill critical conditions exist (location 3). Other than across the jump, losses can be neglected. Determine the following:

   (a) Depths at locations 2 and 3.

   (b) Required height of the sill, h.

   (c) Resultant force acting on the sill.

   (d) Sketch the water surface and energy grade line.

   (e) Describe the nature and character of the jump.

FIGURE P10.39

10.40 Water is flowing as shown in Fig. P10.40 under the sluice gate in a horizontal rectangular channel that is 5 m wide. The depths y_1 and y_2 are 2.5 m and 10 cm, respectively. The horizontal distances between locations 1, 2, and 3 are sufficiently short that rapidly varied flow conditions can be assumed to occur. Determine the following:

   (a) The discharge.

   (b) The depth downstream of the jump at location 3.

   (c) The power lost in the hydraulic jump.

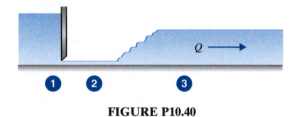

FIGURE P10.40

10.41 A transition section is located at the entrance to a rectangular channel as shown in Fig. P10.41. At location 1, the depth is sufficiently large so that the velocity is negligible. If $b = 3$ m, $\text{Fr}_3 = 0.75$, and $Q = 5.55$ m³/s, determine the following:

   (a) The water surface elevation relative to the datum at locations 1, 2, and 3. Sketch the water surface and energy grade lines between locations 1 and 3.

   (b) The resultant horizontal force acting on the channel bottom between locations 2 and 3.

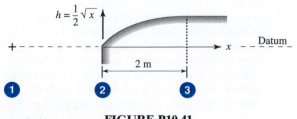

FIGURE P10.41

10.42 Water is flowing in a trapezoidal channel, $b = 5$ m, $m_1 = m_2 = 3$. A stationary hydraulic jump occurs, with the upstream depth $y_1 = 1.1$ m and the discharge $Q = 60$ m^3/s. Find the downstream depth y_2 and the power dissipated by the jump.

10.43 A 4-m-wide rectangular channel contains a rectangular step whose width is the same as the channel. The given flow conditions are $y_1 = 0.5$ m, $V_1 = 8$ m/s, and $y_2 = 2$ m.
   (a) What is the height h of the step if $C_D = 1.2$?
   (b) What would be y_2 if the step were not present?

10.44 A rectangular channel (Fig. P10.44) has a sudden upward step of 0.17 m. A hydraulic jump occurs above the step. At location 2, downstream of the

jump, the depth is $y_2 = 1.5$ m, and the Froude number there is Fr$_2 = 0.40$. If the width of the channel is 5 m and the drag coefficient on the step is $C_D = 0.35$, find the discharge Q and the depth y_1 upstream of the jump.

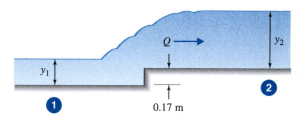

FIGURE P10.44

Nonuniform Gradually Varied Flow

10.45 The channel whose bottom profile is shown in Fig. P10.45 has a rectangular cross section, with $b = 8$ m, $n = 0.014$, and $S_0 = 0.004$. Determine the following:
   (a) The discharge in the channel.
   (b) Sketch the water surface and energy grade line.
   (c) The possible existence of a hydraulic jump.
   (*Hint:* Assume critical flow conditions at the channel entrance.)

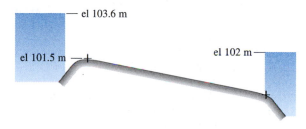

FIGURE P10.45

10.46 Uniform flow occurs in a long rectangular channel of width b, $S_0 = 0.0163$ and $n = 0.012$. The channel is altered by raising the channel bottom by Δz for a short distance. The alteration is intended to act as a "smooth transition region" so that losses can be considered negligible. Determine what changes take place due to the presence of the transition. In your analysis include all relevant rapidly varied flow calculations, identify any gradually varied flow profiles, and sketch the water surface and energy grade line. Given: $Q = 0.35$ m^3/s, $b = 1.80$ m, $\Delta z = 100$ mm.

10.47 Rework Problem 10.46 if $S_0 = 0.0013$.

10.48 Flow of water takes place in a rectangular channel with $b = 4$ m, $n = 0.012$, $L = 500$ m, $S_0 = 0.00087$. The discharge is 33 m^3/s. At the entrance to the channel, the depth is 0.68 m, and at the downstream end a free-outfall condition exists. Perform a profile synthesis to determine the nature of the water surface and energy grade line.

10.49 A flow of 8.5 m^3/s occurs in a long rectangular channel 3 m wide with $y_0 = 1.54$ m. There is a smooth constriction in the channel to 1.8 m width.
   (a) Determine the depths to be expected in and just upstream of the constriction, neglecting losses. Show the solution on an $E-y$ diagram.
   (b) Classify the gradually varied flow profile upstream of the constriction.

10.50 Water is discharging at 20 m^3/s in a triangular channel with $m_1 = 3.5$, $m_2 = 2.5$, $S_0 = 0.001$, $n = 0.014$. At the entrance to the channel the depth is 0.50 m, and at a distance 300 m downstream the depth is 2.5 m. Determine the nature of the water surface over the 300-m reach.

10.51 A long channel has a change in bottom slope 500 m from its downstream end; upstream of the transition the slope is $S_{01} = 0.0003$, and downstream of it the slope is $S_{02} = 0.005$. The downstream reach of channel terminates at a reservoir whose elevation is 3 m above the channel bottom at that location. The upper reach is quite long, and very far upstream from the transition normal flow conditions exist. The channel section is trapezoidal, with a bottom width of 3 m, $m_1 = m_2 = 1.8$, $n = 0.012$, and a discharge of 17.5 m³/s. Determine the nature of the water surface and energy grade line from a location far upstream of the transition to the end of the channel.

10.52 A rectangular channel with $b = 4$ m has a change in slope such that the normal depths are $y_{01} = 0.93$ m and $y_{02} = 1.42$ m. At locations far upstream and far downstream of the transition, flow occurs at the respective uniform depths that serve as controls. The discharge is 15 m³/s. Find the variation of the water surface and energy grade line throughout the region.

10.53 The partial water profile shown in Fig. P10.53 is for a rectangular channel of width $b = 3$ m, in which water is flowing at a discharge of $Q = 5$ m³/s.
  (a) Does a hydraulic jump occur in the channel? If so, is it located upstream or downstream of location A?
  (b) Sketch the water surface and energy grade line, and identify any known water surface profiles.

$y_0 = 0.4$ m
(uniform depth)

1.6 m

A   (horizontal slope)

FIGURE P10.53

10.54 A very long rectangular channel (Fig. P10.54) has normal flow conditions at locations A and C, and a change in the bottom slope occurs at location B.
  (a) Somewhere between locations A and C a hydraulic jump will take place. Explain why this statement is true.
  (b) Classify and sketch the two possible water surface profiles that exist between locations A and C.
  (c) By appropriate reasoning and calculation, determine whether the jump will occur upstream of location B, or downstream of location B.

10.55 A long rectangular channel (Fig. P10.55) has normal flow conditions at locations A and D, respectively, far upstream and downstream of location C. At C a change in slope occurs as shown.
  (a) Classify the water surface, and sketch the water surface and the energy grade line between A and D.
  (b) Determine the height h of a transition such that normal flow conditions occur between A and B, which is a short distance upstream of A.
  (c) Explain what type of profile now exists downstream of B.

$y_{0_1} = 1.0$ m $y_{0_2} = 1.8$ m $q = 5.75$ m²/s

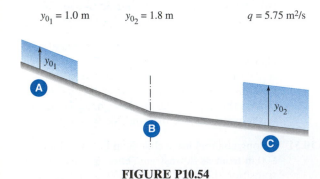

FIGURE P10.54

$y_{0_1} = 1.6$ m $y_{0_2} = 0.8$ m $q = 4$ m²/s

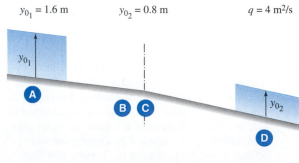

FIGURE P10.55

10.56 Water is flowing at a discharge Q in a circular conduit of diameter d and length L, with $S_0 = 0.001$, and $n = 0.015$. At the inlet the water depth is y_0 and at the outlet a free overfall exists. Determine the nature of the water surface in the conduit. Given:

$Q = 2.5$ m^3/s, $L = 500$ m, $d = 2.5$ m, $y_0 = 0.4$ m

10.57 With the results from Problem 10.51, compute the water surface profile and energy grade line using a numerical method of your choice.

10.58 A rectangular channel ($b = 5$ m, $S_0 = -0.001$) conveys flow at a rate of $Q = 3$ m^3/s. At a given location the depth is $y_1 = 2$ m and downstream of that location the depth is $y_2 = 1.8$ m. The Manning coefficient is $n = 0.02$.

(a) Determine the distance between locations 1 and 2.

(b) What type of water surface profile exists in the reach?

10.59 A rectangular channel (Fig. P10.59) has the following properties: $b = 1$ m, $S_0 = 0.0002$, $n = 0.017$. A 120° V-notch weir is placed in the channel to measure the discharge, with the bottom notch of the weir located 0.5 m above the channel floor. Upstream of the weir the depth Y is measured to be 0.37 m.

(a) What is the discharge in the channel?

(b) If uniform flow conditions exist very far upstream of the weir, classify the nature of the water surface.

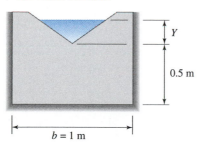

FIGURE P10.59

10.60 Gradually varied flow occurs over a reach of rectangular channel with $n = 0.013$, $S_0 = 0.005$, and $b = 2.5$ m. At location 1 the depth is 1.05 m and at location 2 the depth is 1.2 m. If the two locations are 50 m apart, find the discharge and identify the profile type.

10.61 Over a portion of a very long rectangular channel of width b (Fig. P10.61), uniform flow occurs with discharge Q and depth y_0. At the end of the channel the flow is terminated by a free outfall. In addition, at the outlet the channel width is reduced to b_1 m for a very short distance. Determine and describe as completely as possible the changes that take place in the water surface between locations A and B. Use an E–y diagram for illustration. Given:

$Q = 5.5$ m^3/s, $y_0 = 0.5$ m, $b = 3$ m, $b_1 = 1.5$ m.

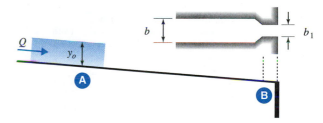

FIGURE P10.61

10.62 A sluice gate is placed in a long rectangular channel, $b = 4$ m, $n = 0.014$, and $S_0 = 0.0008$. The depth upstream of the gate is $y_1 = 1.85$ m and downstream of the gate the depth is $y_2 = 0.35$ m. Neglecting losses, identify and compute the water surface profiles on either side of the gate.

10.63 With the following data,[7] identify the profiles and plot the water surface and energy grade line to scale.

$Q =$	35	$n =$	0.011	$S_0 =$	0.001	$L =$	200	$g =$	9.81
$b =$	5	$m =$	2.5	$y_u =$	0.80	$y_d =$	2.00	$c_1 =$	1.00

	Depth	Residual
$y_c =$	1.356	-3.44E-07
$y_0 =$	1.442	-4.23E-07

Station	y	A	V	E	y_m	$S(y_m)$	Δx	x	y_{cj}	Residual
1	0.800	5.600	6.250	2.791						
2	0.850	6.056	5.779	2.552	0.825	8.311E-03	33	33	2.044	-5.98E-07
3	0.900	6.525	5.364	2.366	0.875	6.689E-03	33	65	1.961	-3.09E-07
4	0.950	7.006	4.996	2.222	0.925	5.443E-03	33	98	1.883	8.24E-07
5	1.000	7.500	4.667	2.110	0.975	4.472E-03	32	130	1.809	-7.46E-07
6	1.050	8.006	4.372	2.024	1.025	3.706E-03	32	162	1.740	-9.44E-08
7	1.100	8.525	4.106	1.959	1.075	3.097E-03	31	193	1.674	2.14E-07
8	1.150	9.056	3.865	1.911	1.125	2.606E-03	30	223	1.611	4.94E-07
7	2.000	20.000	1.750	2.156				200		
8	1.975	19.627	1.783	2.137	1.988	2.770E-04	-26	174		
9	1.950	19.256	1.818	2.118	1.963	2.916E-04	-26	147		
10	1.925	18.889	1.853	2.100	1.938	3.073E-04	-27	121		
11	1.900	18.525	1.889	2.082	1.913	3.240E-04	-27	94		
12	1.875	18.164	1.927	2.064	1.888	3.417E-04	-27	67		
13	1.850	17.806	1.966	2.047	1.863	3.607E-04	-27	40		
14	1.825	17.452	2.006	2.030	1.838	3.810E-04	-27	13		

[7]The data was generated by Excel spreadsheet.

10.64 With the following data,[8] identify the profiles and plot the water surface and energy grade line to scale.

(a)

$Q =$	3.5	$n =$	0.013	$S_0 =$	0.005	$L =$	100	$g =$	9.81
$b =$	2.5	$m =$	0	$y_u =$	0.30	$y_d =$	0.95	$c_1 =$	1.00

	Depth	Residual
$y_c =$	0.585	-8.58E-07
$y_0 =$	0.508	-1.42E-07

Station	y	A	V	E	y_m	$S(y_m)$	Δx	x	y_{cj}	Residual
1	0.300	0.750	4.667	1.410						
2	0.325	0.813	4.308	1.271	0.313	2.154E-02	8	8	0.985	6.45E-07
3	0.350	0.875	4.000	1.165	0.338	1.702E-02	9	17	0.932	-4.60E-07
4	0.375	0.938	3.733	1.085	0.363	1.369E-02	9	26	0.884	6.34E-07
5	0.400	1.000	3.500	1.024	0.388	1.119E-02	10	36	0.840	3.55E-07
6	0.425	1.063	3.294	0.978	0.413	9.272E-03	11	47	0.799	2.00E-07
7	0.450	1.125	3.111	0.943	0.438	7.774E-03	13	60	0.762	4.40E-07
8	0.475	1.188	2.947	0.918	0.463	6.587E-03	16	76	0.727	9.05E-07
9	0.500	1.250	2.800	0.900	0.488	5.635E-03	29	104	0.694	2.22E-07
10	0.950	7.006	0.500	0.963				100		
11	0.850	6.056	0.578	0.867	0.900	9.698E-04	-24	76		
12	0.800	5.600	0.625	0.820	0.825	1.236E-03	-13	64		
13	0.750	5.156	0.679	0.773	0.775	1.474E-03	-13	51		
14	0.700	4.725	0.741	0.728	0.725	1.781E-03	-14	36		
15	0.650	4.306	0.813	0.684	0.675	2.183E-03	-16	21		
16	0.600	3.900	0.897	0.641	0.625	2.725E-03	-19	2		

(b)

$Q =$	125	$n =$	0.013	$S_0 =$	0.005	$L =$	300	$g =$	32.2
$b =$	8	$m =$	0	$y_u =$	1.00	$y_d =$	3.00	$c_1 =$	1.49

	Depth	Residual
$y_c =$	1.965	-1.04E-07
$y_0 =$	1.710	-4.30E-04

Station	y	A	V	E	y_m	$S(y_m)$	Δx	x	y_{cj}	Residual
1	1.000	8.000	15.625	4.791						
2	1.100	8.800	14.205	4.233	1.050	2.155E-02	34	34	3.311	-3.50E-07
3	1.200	9.600	13.021	3.833	1.150	1.634E-02	35	69	3.102	8.20E-07
4	1.300	10.400	12.019	3.543	1.250	1.269E-02	38	107	2.914	5.79E-07
5	1.400	11.200	11.161	3.334	1.350	1.007E-02	41	148	2.744	3.35E-07
6	1.500	12.000	10.417	3.185	1.450	8.135E-03	48	195	2.589	2.31E-07
7	1.600	12.800	9.766	3.081	1.550	6.674E-03	62	258	2.447	3.00E-07
8	1.700	13.600	9.191	3.012	1.650	5.549E-03	126	384	2.317	2.58E-07
9	3.000	37.500	3.333	3.173				300		
10	2.800	33.600	3.720	3.015	2.900	1.105E-03	-40	260		
11	2.600	29.900	4.181	2.871	2.700	1.349E-03	-39	220		
12	2.400	26.400	4.735	2.748	2.500	1.674E-03	-37	183		
13	2.200	23.100	5.411	2.655	2.300	2.121E-03	-32	151		
14	2.000	20.000	6.250	2.607	2.100	2.751E-03	-21	129		
15	1.965	19.478	6.417	2.605	1.983	3.247E-03	-1	128		

[8]The data was generated by Excel spreadsheet.

10.65 Consider a backwater curve situated upstream of a dam in a rectangular reservoir (Fig. P10.65). Given data are: normal depth $y_0 = 3.92$ m, $Q = 125$ m³/s, $S_0 = 0.0004$, $n = 0.025$, $y_A = 11$ m, $y_B = 12$ m.

(a) Identify the type of profile that exists in the reservoir.

(b) Compute the distance between A and B.

(c) Sketch the water surface and energy grade line.

FIGURE P10.65

10.66 A rectangular channel has a change in slope as shown in Fig. P10.66. The channel is 3.66 m wide, with $n = 0.017$, $S_{02} = 0.00228$, and $Q = 15.38$ m³/s.

(a) Determine the depth that must exist in the downstream channel for a hydraulic jump to terminate at uniform flow conditions.

(b) If $y_{01} = 0.6$ m, calculate the length to the jump, L_j, using several increments of depth in a step calculation.

(c) Sketch the water surface and energy grade line, and identify all gradually varied flow profiles.

10.67 In the three channels shown in Fig. P10.67, normal depths and critical depths are indicated by dashed (- - -) and dotted (· · ·) lines, respectively. Sketch one possible composite profile for each system, and identify all gradually varied surfaces.

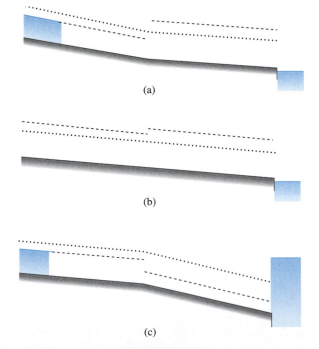

(a)

(b)

(c)

FIGURE P10.67

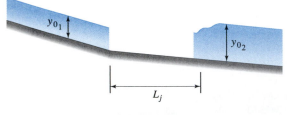

FIGURE P10.66

10.68 A design flow of $Q = 19$ m³/s occurs in a river that can be approximated as a wide rectangular channel with the following properties: $b = 20$ m, $S_0 = 0.0005$, $n = 0.016$. A small dam is to be placed in the channel as part of a flood control plan (Fig. P10.68). The height of the dam is $h = 6$ m, and the depth of flow at the toe of the dam is $y_{toe} = 0.10$ m. A concrete spillway apron ($n = 0.014$) is to be situated downstream of the dam. The purpose of the apron is to keep any hydraulic jump that may occur downstream of the dam off the river bed for the given design conditions to prevent erosion of the river bed. The depth behind the jump at the end of the apron is $y_2 = 0.75$ m. Determine the following:

(a) Power in kilowatts dissipated by the jump.

(b) Design length of the apron.

(c) The gradually-varied-flow profiles upstream of the dam, and between the toe and the hydraulic jump.

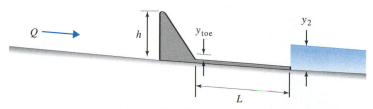

FIGURE P10.68

10.69 A very wide rectangular channel has the following properties: $S_0 = 0.0005$, $n = 0.017$, $q = 1.2$ m²/s. At some location, say location A, the depth is 0.65 m, and at location B, the depth is 0.90 m.

(a) Is location B upstream or downstream of location A?

(b) Using the varied flow function determine the distance between locations A and B.

10.70 An estuary to a river empties to the ocean. The estuary can be approximated as a very wide channel with a depth of 7 m at its outlet, and with $S_0 = 0.0001$, $n = 0.015$. If the discharge into the ocean from the estuary is $q = 1.5$ m³/s per meter width, determine the distances upstream from the ocean where the depth is equal to 6, 5, 4, and 2 m.

10.71 A river cross section can be approximated by two rectangles as shown in Fig. P10.71. The channel properties and dimensions are: $b_1 = 150$ m, $z_1 = 4$ m, $n_1 = 0.03$, $b_2 = 5$ m, $z_2 = 0$ m, and $n_2 = 0.02$. If the slope of the energy grade line is $S = 0.0005$ and the depth $y = 5$ m, find:

(a) The energy coefficient for the composite section.

(b) The discharge.

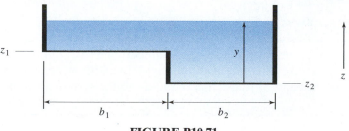

FIGURE P10.71

10.72 Consider the same river cross section of Problem 10.71. In the table below, data are provided for two locations x along the river:

At the downstream location $x = 400$ m, the depth $y = 3.0$ m when the discharge $Q = 280$ m³/s. Using a trial-and-error procedure, determine the depth y at the upstream location $x = 0$.

x (m)	b_1 (m)	b_2 (m)	z_1 (m)	z_2 (m)	n_1	n_2
0	115	107	16.7	15.0	0.05	0.03
400	149	91	17.7	15.1	0.05	0.03

In a complex of industrial piping such as this one, engineering design requires that steady flows be analyzed for correct pipe sizing and placement of pumps. In addition, analysis is occasionally undertaken to mitigate unsteady, or transient, excitations to the system. (Marafona/Shutterstock)

11

Flows in Piping Systems

Outline

Chapter Objectives

The objectives of this chapter are to:

▲ Compare empirical frictional loss equations in piping

▲ Introduce the concept of hydraulic grade line as a variable for piping analysis

▲ Describe ad hoc methods of computing discharges and pressures in simple pipe systems

▲ Present linearized solutions (i.e., Hardy Cross method) for pipe networks using spreadsheet analysis

▲ Show use of computational software for pipe network analysis

▲ Introduce simplified analysis for unsteady flows in simple pipe systems

11.1 INTRODUCTION

Internal flows in pipelines and ducts are commonly encountered in all parts of our industrialized society. From delivering potable water to transporting chemicals and other industrial liquids, engineers have designed and constructed untold kilometers of relatively large-scale piping systems. Smaller piping units are also in abundance: in hydraulic controls, in heating and air conditioning systems, and in cardiovascular and pulmonary flow systems, to name only a few. These flows can be either steady or unsteady, uniform or nonuniform. The fluid can be either incompressible or compressible, and the piping material can be elastic, inelastic, or perhaps, viscoelastic. In this chapter we concentrate primarily on incompressible, steady flows in rigid piping. The piping system may be relatively simple, such that the variables can be solved rather easily using a calculator, or it may be sufficiently complicated so it is more convenient to use computational software.

Piping systems are considered to be composed of elements and components. Basically, pipe **elements** are reaches of constant-diameter piping and the **components** consist of valves, tees, bends, reducers, or any other device that may create a loss to the system. In addition to components and elements, pumps add energy to the system and turbines extract energy. The elements and components are linked at junctions. Figure 11.1 illustrates several types of piping systems.

Following a discussion of piping losses, several pipe systems, including series, branch, and parallel configurations, are analyzed. Attention is then directed to comprehensive network systems, where several methods of solution are presented. Most of the piping problems analyzed are those where the discharge is the unknown variable; this type of problem is classified as category 2 in Section 7.6.3.

Finally, a brief introduction to unsteady flow in pipelines is presented. This topic is one that has been important for many years. It is taking on increasing importance as the demand for the construction of more cost-effective piping continues, and the manner in which piping, valves, pumps, and other components interact and perform become more sophisticated. We address only the fundamental aspects of unsteady flows, focusing on two simple flow assumptions in a single pipe of constant diameter: inelastic pipe material and incompressible liquid versus elastic pipe material and compressible liquid.

Elements: *Reaches of constant-diameter piping.*

Components: *Valves, tees, bends, reducers, or any other device that creates a loss to the system.*

11.2 LOSSES IN PIPING SYSTEMS

Losses can be divided into two categories: (a) those due to wall shear in pipe elements, and (b) those due to piping components. The former are distributed along the length of pipe elements. The latter are treated as discrete discontinuities in the hydraulic grade line and the energy grade line and are commonly referred to as minor losses; they are due primarily to separated or secondary flows.

The fundamental mechanics of wall shear and development of the empirical relations relating to pipe losses are treated in Chapter 7. Minor losses are covered in detail in Section 7.6.4 and are not discussed further here. The following material focuses on the treatment of losses in the analysis of piping systems.

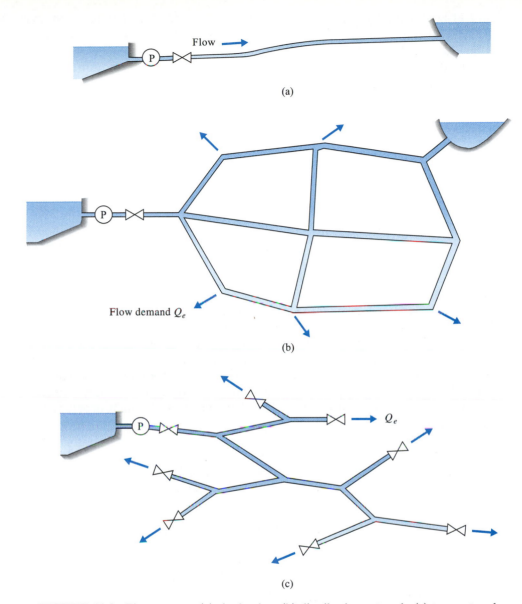

FIGURE 11.1   Pipe systems: (a) single pipe; (b) distribution network; (c) tree network.

11.2.1   Frictional Losses in Pipe Elements

It is convenient to express the pipe element frictional loss in the exponential form

$$h_L = RQ^\beta \qquad (11.2.1)$$

in which h_L is the head loss over length L of pipe, R is the *resistance coefficient*, Q is the discharge in the pipe, and β is an exponent. Depending on the formulation chosen, the resistance coefficient may be a function of pipe roughness, Reynolds number, or length and diameter of the pipe element. In particular, the Darcy–Weisbach relation, Eq. 7.6.23, can be substituted into Eq. 11.2.1. Then $\beta = 2$, and the resulting expression for R is

$$R = \frac{fL}{2gDA^2}$$

$$= \frac{8fL}{g\pi^2 D^5} \tag{11.2.2}$$

KEY CONCEPT *Frictional losses in piping are commonly evaluated using the Darcy–Weisbach or Hazen–Williams equation. The Darcy–Weisbach formulation provides a more accurate estimation.*

where f is the friction factor. The characteristics of f for commercial pipe flows are developed in Section 7.6.3. Specifically, the Moody diagram, Fig. 7.13, presents a comprehensive picture of just how the friction factor varies over a wide range of Reynolds numbers and for a range of relative roughnesses. For pipe network analysis, it is convenient to express the behavior of f using approximate, equivalent empirical formulas in which the friction factor can be obtained directly in terms of the Reynolds number and relative roughness. A number of relations have been developed and shown to be reasonably accurate for engineering calculations (Benedict, 1980). In particular, the formulas of Swamee and Jain (1976) are presented in Section 7.6.3 and are shown to accurately represent the Colebrook relation, Eq. 7.6.28. The friction factor formula developed by Swamee and Jain is

$$f = 1.325 \left\{ \ln\left[0.27 \left(\frac{e}{D}\right) + 5.74 \left(\frac{1}{\mathrm{Re}}\right)^{0.9} \right] \right\}^{-2} \tag{11.2.3}$$

Combining Eqs. 11.2.2 and 11.2.3, one finds that

$$R = 1.07 \left(\frac{L}{gD^5}\right) \left\{ \ln\left[0.27 \left(\frac{e}{D}\right) + 5.74 \left(\frac{1}{\mathrm{Re}}\right)^{0.9} \right] \right\}^{-2} \tag{11.2.4}$$

Equations 11.2.3 and 11.2.4 are valid over the ranges $0.01 > e/D > 10^{-8}$, and $10^8 > \mathrm{Re} > 5000$. The fully rough regime, where Re has a negligible effect on f, begins at a Reynolds number given by

$$\mathrm{Re} = \frac{200D}{e\sqrt{f}} \tag{11.2.5}$$

For values of Re greater than this, the friction factor is a function only of e/D, and is given by

$$f = 1.325 \left\{ \ln \left[0.27 \left(\frac{e}{D} \right) \right] \right\}^{-2} \qquad (11.2.6)$$

Two additional expressions for pipe frictional losses, which have found wide use, are the Hazen–Williams and Chezy–Manning formulas. For water flow, the value of R in Eq. 11.2.1 for the Hazen–Williams relation is

$$R = \frac{K_1 L}{C^\beta D^m} \qquad (11.2.7)$$

in which the exponents are $\beta = 1.85$, $m = 4.87$, and C is the Hazen–Williams coefficient dependent only on the roughness. In SI units K_1 has a magnitude of 10.59. Values of the Hazen–Williams roughness coefficient C are given in Table 11.1.

The Chezy–Manning equation is more commonly associated with open-channel flow. However, in sewage and drainage systems in particular, it has been applied to conduits flowing under *surcharge*, that is, under pressurized conditions. The Chezy–Manning equation was introduced in Section 7.7. For a circular pipe flowing full, Eq. 7.7.6 can be substituted into Eq. 11.2.1 and solved for R:

$$R = \frac{10.29 n^2 L}{K_2 D^{5.33}} \qquad (11.2.8)$$

in which n is the Manning roughness coefficient and $K_2 = 1$ for SI units. In Eq. 11.2.1, the exponent $\beta = 2$.

An advantage to using Eq. 11.2.7 or Eq. 11.2.8 as opposed to Eq. 11.2.4 is that in the first two, C and n are dependent on roughness only, while in the last, f depends on the Reynolds number as well as the relative roughness. However, Eq. 11.2.4 is recommended since it provides a more precise representation of pipe frictional losses. Note that the Hazen–Williams and Chezy–Manning relations are dimensionally inhomogeneous,[1] whereas the Swamee–Jain equation is dimensionally homogeneous and contains the two parameters e/D and Re, which appropriately influence the losses.

The limitations of the Hazen–Williams and Chezy–Manning formulas are demonstrated as follows. Beginning with Eq. 11.2.1, the head loss h_L based on the Darcy–Weisbach resistance coefficient, Eq. 11.2.2, with the exponent $\beta = 2$ can be equated with h_L based on the Hazen–Williams coefficient, Eq. 11.2.7, with $\beta = 1.85$. Introducing the Reynolds number to eliminate Q and solving for the friction factor f results in

$$f = \frac{1.28 g K_1}{C^{1.85} D^{0.02} (\text{Re } \nu)^{0.15}} \qquad (11.2.9)$$

---

[1] In a dimensionally inhomogeneous equation the constants in the equation have units assigned to them. In a dimensionally homogeneous equation the constants are dimensionless.

TABLE 11.1 Nominal Values of the Hazen–Williams Coefficient C

Type of pipe	C
Extremely smooth; asbestos-cement	140
New or smooth cast iron; concrete	130
Wood stave; newly welded steel	120
Average cast iron; newly riveted steel; vitrified clay	110
Cast iron or riveted steel after some years of use	95–100
Deteriorated old pipes	60–80

With SI units and for water at 20°C, Eq. 11.2.9 reduces to

$$f = \frac{1056}{C^{1.85} D^{0.02} \text{Re}^{0.15}} \tag{11.2.10}$$

Note that f is weakly dependent upon D in this equation. In a similar fashion, Eq. 11.2.8 can be substituted into Eq. 11.2.1 with $\beta = 2$ and equated to h_L based on the Darcy–Weisbach formulation to yield

$$f = \frac{124.5 n^2}{D^{0.33}} \tag{11.2.11}$$

Comparisons between the original Colebrook formula (Eq. 7.6.28), the formula of Swamee and Jain (Eq. 11.2.3), and the equivalent expressions for the Hazen–Williams and Manning coefficients (Eqs. 11.2.10 and 11.2.11) are shown in Fig. 11.2 for a 1-m-diameter concrete pipe flowing full with water ($C = 130$,

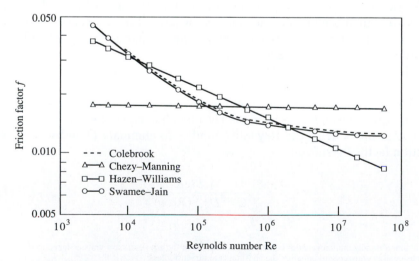

FIGURE 11.2 Comparison of several approximate formulas with the original Colebrook formula.

$n = 0.012$, $e/D = 0.00015$). It is evident that the Hazen–Williams and Chezy–Manning relations are valid over a limited range of Re, and that Eq. 11.2.3 provides a more versatile and accurate estimate of pipe losses. In the remainder of this chapter most of the frictional losses will be treated using the Darcy–Weisbach formulation, Eqs. 11.2.2 to 11.2.6. Any exceptions will be noted.

Example 11.1

A pipeline is conveying 0.05 m³/s of water, at 30°C. The length of the line is 300 m, and the diameter is 0.25 m. Estimate the head loss due to friction using the three formulas (a) Darcy–Weisbach ($e = 0.5$ mm), (b) Hazen–Williams ($C = 110$), and (c) Chezy–Manning ($n = 0.012$).

Solution

The kinematic viscosity of water at 30°C is $\nu = 0.804 \times 10^{-6}$ m²/s. The relative roughness, velocity, and Reynolds number are computed to be

$$\frac{e}{D} = 0.002 \qquad V = 1.02 \text{ m/s} \qquad \text{Re} = 3.17 \times 10^5$$

(a) Substitute values into Eq. 11.2.4:

$$R = 1.07 \frac{300}{9.81 \times (0.25)^5}$$

$$\times \{\ln[0.27 \times 0.002 + 5.74 \, (3.17 \times 10^5)^{-0.9}]\}^{-2} = 610$$

Then, with Eq. 11.2.1, the friction loss based on the Darcy–Weisbach formulation is

$$h_L = RQ^2$$

$$= 610 \, (0.05)^2 = 1.52 \text{ m}$$

(b) For the Hazen–Williams formulation, use Eq. 11.2.7:

$$R = \frac{10.59 \times 300}{(110)^{1.85}(0.25)^{4.87}} = 454$$

$$h_L = 454 \, (0.05)^{1.85} = 1.78 \text{ m}$$

(c) Use Eq. 11.2.8 for the Chezy–Manning formulation:

$$R = \frac{10.29(0.012)^2(300)}{1 \times (0.25)^{5.33}} = 719$$

$$h_L = 719 \, (0.05)^2 = 1.80 \text{ m}$$

Part (a) with $h_L = 1.52$ m is the most accurate. The Hazen–Williams result is 17% too high, and the Chezy–Manning result is 18% too high.

11.3 SIMPLE PIPE SYSTEMS

The analysis of a single pipeline is presented in Section 7.6.4; it would be appropriate here for the reader to review the three categories of pipe problems in that section. For more complex piping systems, the methodology is similar. With relatively simple networks, such as series, parallel and branching systems, ad hoc solutions can be developed that are suitable for use with calculators, spreadsheet algorithms, or computational software. Such approaches are relevant since they make use of the solver's ingenuity and require an understanding of the nature of the flow and piezometric head distributions for the particular piping arrangement. For systems of greater complexity, an alternative means of solving such problems is the Hardy Cross method, presented in Section 11.4.

The fundamental principle in the ad hoc approach is to identify all of the unknowns and write an equivalent number of independent equations to be solved. Subsequently, the system is simplified by eliminating as many unknowns as possible and reducing the problem to a series of single pipe problems — either category 1 or category 2, Section 7.6.3.

11.3.1 Series Piping

Consider the series system shown in Fig. 11.3. It consists of N pipe elements and a specified number of minor-loss components ΣK associated with each ith pipe element. A single minor loss is described in Section 7.6.5 to be equal to $h_L = KV^2/2g$. It is convenient here to express the minor loss in terms of the discharge rather than the velocity, so that $h_L = KQ^2/2gA^2$. For many flow situations, it is common practice to neglect the kinetic-energy terms at the inlet and outlet; they would be significant only if the velocities were relatively high. Assuming that the exponent in Eq. 11.2.1 is $\beta = 2$, the energy equation applied from location A to location B of Fig. 11.3 is

$$\left(\frac{p}{\gamma} + z\right)_A - \left(\frac{p}{\gamma} + z\right)_B = \left(R_1 + \frac{\Sigma K}{2gA_1^2}\right)Q_1^2 + \left(R_2 + \frac{\Sigma K}{2gA_2^2}\right)Q_2^2$$

$$+ \cdots + \left(R_N + \frac{\Sigma K}{2gA_N^2}\right)Q_N^2 \qquad (11.3.1)$$

$$= \sum_{i=1}^{N}\left(R_i + \frac{\Sigma K}{2gA_i^2}\right)Q_i^2$$

in which R_i is the resistance coefficient for pipe i.

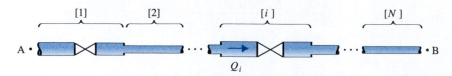

FIGURE 11.3   Series piping system.

The statement of continuity for the series system is that the discharge in every element is identical, or

$$Q_1 = Q_2 = \cdots = Q_i = \cdots = Q_N = Q \qquad (11.3.2)$$

Replacing Q_i with Q, Eq. 11.3.1 becomes

$$\left(\frac{p}{\gamma}+z\right)_A - \left(\frac{p}{\gamma}+z\right)_B = \left[\sum_{i=1}^{N}\left(R_i + \frac{\Sigma K}{2gA_i^2}\right)\right]Q^2 \qquad (11.3.3)$$

Equation 11.2.4 can be substituted for R_i, or alternatively, one may use Eq. 11.2.3 or the Moody diagram, Fig. 7.13, to obtain values of the friction factor. The reader should recognize that in a series system, the discharge remains constant from one pipe element to another, and the losses are accumulative, that is, they are the sum of the minor loss components and the pipe frictional losses.

For a category 1 problem, the right-hand side of Eq. 11.3.3 is known and the solution is straightforward. For a category 2 problem, in which Q is unknown, a trial-and-error solution is required, since the Reynolds number, in terms of the unknown discharge (Re $= 4Q/\pi\nu D$), is present in the friction factor relation. Note that if flow in the fully rough zone is assumed, f is independent of Q and Eq. 11.3.3 reduces to a quadratic in Q. A category 3 problem is not encountered in this type of analysis. A series piping system with minor losses and constant-diameter piping is applied to a category 1 problem in Example 7.13. The solution for a somewhat more complex series system is illustrated in Example 11.2.

Example 11.2

For the system shown in Fig. E11.2, find the required power to pump 100 L/s of liquid ($S = 0.85$, $\nu = 10^{-5}$ m²/s). The pump is operating at an efficiency $\eta = 0.75$. Pertinent data are given in the figure.

Line 1: $L = 10$ m, $D = 0.20$ m, $e = 0.05$ mm, $K_1 = 0.5$, $K_v = 2$
Line 2: $L = 500$ m, $D = 0.25$ m, $e = 0.05$ mm, $K_e = 0.25$, $K_2 = 1$

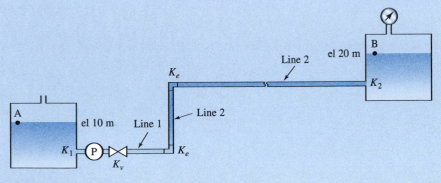

FIGURE E11.2

(continued)

Solution

This is a category 1 problem. The energy relation, Eq. 11.3.3, for the system is

$$\left(\frac{p}{\gamma} + z\right)_A + H_P = \left(\frac{p}{\gamma} + z\right)_B + \left[R_1 + \frac{K_1 + K_v}{2gA_1^2} + R_2 + \frac{2K_e + K_2}{2gA_2^2}\right]Q^2$$

The resistance coefficients, R_1 and R_2, are calculated with Eq. 11.2.4 after first evaluating Re and e/D:

$$\mathrm{Re}_1 = \frac{4Q}{\pi D_1 \nu} = \frac{4 \times 0.10}{\pi \times 0.20 \times 10^{-5}} = 6.37 \times 10^4 \qquad \left(\frac{e}{D}\right)_1 = \frac{0.05}{200} = 0.00025$$

$$\mathrm{Re}_2 = \frac{4Q}{\pi D_2 \nu} = \frac{4 \times 0.10}{\pi \times 0.25 \times 10^{-5}} = 5.09 \times 10^4 \qquad \left(\frac{e}{D}\right)_2 = \frac{0.05}{250} = 0.0002$$

$$R_1 = 1.07\left(\frac{10}{9.81 \times (0.20)^5}\right)$$

$$\times \{\ln[0.27 \times 0.00025 + 5.74 \times (6.37 \times 10^4)^{-0.9}]\}^{-2} = 53.4$$

$$R_2 = 1.07\left(\frac{500}{9.81 \times (0.25)^5}\right)$$

$$\times \{\ln[0.27 \times 0.0002 + 5.74 \times (5.09 \times 10^4)^{-0.9}]\}^{-2} = 904$$

The minor loss coefficient terms are calculated to be

$$\frac{K_1 + K_v}{2gA_1^2} = \frac{0.5 + 2}{2 \times 9.81 \times [\pi/4 \times (0.20)^2]^2} = 129.1$$

$$\frac{2K_e + K_2}{2gA_2^2} = \frac{2 \times 0.25 + 1}{2 \times 9.81 \times [\pi/4 \times (0.25)^2]^2} = 31.7$$

Substitute these values into the energy equation and obtain

$$0 + 10 + H_P = \frac{200 \times 10^3}{0.85 \times 9800} + 20 + (53.4 + 129.1 + 904 + 31.7)(0.1)^2$$

This relation reduces to

$$10 + H_P = 24 + 20 + 11.2$$

Solving for the head across the pump gives $H_P = 45.2$ m. The required input power is

$$\dot{W}_P = \frac{\gamma Q H_P}{\eta}$$

$$= \frac{(9800 \times 0.85) \times 0.10 \times 45.2}{0.75}$$

$$= 5.0 \times 10^4 \mathrm{W} \qquad \text{or} \qquad 50 \text{ kW}$$

11.3.2 Parallel Piping

A parallel piping arrangement is shown in Fig. 11.4; it is essentially an arrangement of N pipe elements joined at A and B with ΣK minor loss components associated with each pipe element i. The continuity equation applied at either location A or B is given by

$$Q = \sum_{i=1}^{N} Q_i \qquad (11.3.4)$$

The algebraic sum of the energy grade line around any defined loop must be
zero. As in the case of series piping, it is customary to assume that
$V^2/2g \ll (p/\gamma + z)$. Hence, for any pipe element i, the energy equation from
location A to B is

$$\left(\frac{p}{\gamma} + z\right)_A - \left(\frac{p}{\gamma} + z\right)_B = \left(R_i + \frac{\Sigma K}{2gA_i^2}\right) Q_i^2 \qquad i = 1, \ldots, N \qquad (11.3.5)$$

The unknowns in Eqs. 11.3.4 and 11.3.5 are the discharges Q_i and the difference
in piezometric head between A and B; the discharge Q into the system is known.
It is possible to convert the minor loss terms using an equivalent length as
defined in Section 7.6.4. For each pipe element i the equivalent length L_e for ΣK
minor loss components is

$$(L_e)_i = \frac{D_i}{f_i} \Sigma K \qquad (11.3.6)$$

Thus Eq. 11.3.5 simplifies to the form

$$\left(\frac{P}{\gamma} + z\right)_A - \left(\frac{P}{\gamma} + z\right)_B = \overline{R}_i Q_i^2 \qquad (11.3.7)$$

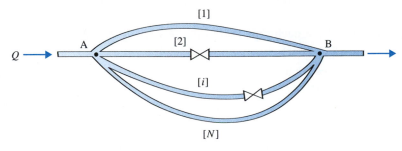

FIGURE 11.4 Parallel piping system.

in which the modified pipe resistance coefficient $\overline{R}_i$ is given by

$$\overline{R}_i = \frac{8f_i[L_i + (L_e)_i]}{g\pi^2 D_i^5} \tag{11.3.8}$$

Note that the right-hand side of Eq. 11.3.8 is equivalent to the term $R_i + \Sigma K/(2gA_i^2)$.

A solution employing the method of successive substitution is developed in the following manner. Define the variable W to be the change in hydraulic grade line between A and B; that is, $W = (p/\gamma + z)_A - (p/\gamma + z)_B$. Then Eq. 11.3.7 can be solved for Q_i in terms of W as

$$Q_i = \sqrt{\frac{W}{\overline{R}_i}} \tag{11.3.9}$$

Equations 11.3.4 and 11.3.9 are combined to eliminate the unknown discharges Q_i, resulting in

$$Q = \sum_{i=1}^{N} \sqrt{\frac{W}{\overline{R}_i}} = \sqrt{W} \sum_{i=1}^{N} \frac{1}{\sqrt{\overline{R}_i}} \tag{11.3.10}$$

The remaining unknown W is taken out of the summation sign since it is the same in all pipes. Solving for W in Eq. 11.3.10 results in

$$W = \left(\frac{Q}{\sum_{i=1}^{N} \frac{1}{\sqrt{\overline{R}_i}}} \right)^2 \tag{11.3.11}$$

An iterative procedure can be formulated to solve for W and the discharges Q_i as follows:

1. Assume flows in each line to be in the completely rough zone, and compute an initial estimate of the friction factors in each line using Eq. 11.2.6.
2. Compute $\overline{R}_i$ for each pipe and evaluate W with Eq. 11.3.11.
3. Compute Q_i in each pipe with Eq. 11.3.9.
4. Update the estimates of the friction factors in each line using the current values of Q_i and Eq. 11.2.3.
5. Repeat steps 2 to 4 until the unknowns W and Q_i do not vary according to a desired tolerance.

Note that if friction factors are in the completely rough zone so that they are independent of the discharge and therefore constant, steps 4 and 5 are unnecessary and a solution results on the first iteration. The technique is illustrated in Example 11.3.

Example 11.3

Find the distribution of flow and the drop in hydraulic grade line for the three-parallel-pipe arrangement shown in Fig. E11.3. Use variable friction factors with $\nu = 10^{-6}$ m²/s. The total water discharge is $Q = 0.020$ m³/s.

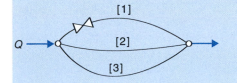

Pipe	L (m)	D (m)	e (mm)	ΣK
1	100	0.05	0.1	10
2	150	0.075	0.2	3
3	200	0.085	0.1	2

FIGURE E11.3

Solution

The initial estimates of f are based on Eq. 11.2.6. Preliminary calculations provide the following:

Pipe	e/D	f (Eq. 11.2.6)	L_e (Eq. 11.3.6)	$\overline{R}$ (Eq. 11.3.8)
1	0.002	0.023	21.7	7.40×10^5
2	0.0027	0.025	9.0	1.38×10^5
3	0.0012	0.021	8.1	8.14×10^4

Apply Eq. 11.3.11, and the first estimate of W is

$$W = \left[\frac{0.020}{(7.40 \times 10^5)^{-1/2} + (1.38 \times 10^5)^{-1/2} + (8.14 \times 10^4)^{-1/2}} \right]^2$$

$$= 7.39 \text{ m}$$

Then with Eq. 11.3.9, estimates of Q_i are made:

$$Q_1 = \left(\frac{7.39}{7.40 \times 10^5} \right)^{1/2} = 0.00316 \text{ m}^3/\text{s}$$

$$Q_2 = \left(\frac{7.39}{1.38 \times 10^5} \right)^{1/2} = 0.00732 \text{ m}^3/\text{s}$$

$$Q_3 = \left(\frac{7.39}{8.14 \times 10^4} \right)^{1/2} = 0.00953 \text{ m}^3/\text{s}$$

A continuity check is made using Eq. 11.3.4:

$$\sum_{i=1}^{3} Q_i = 0.00316 + 0.00732 + 0.00953 = 0.0200 \text{ m}^3/\text{s}$$

Even though the sum satisfies the solution, another iteration will be carried out to study the convergence of the solution technique. First, the $\overline{R}$-values are updated using Eq. 11.2.3 to evaluate the friction factors.

(continued)

Pipe	$Re = \dfrac{4Q}{\pi D \nu}$	f (Eq. 11.2.3)	$\overline{R}$ (Eq. 11.3.8)
1	8.05×10^4	0.026	8.20×10^5
2	1.24×10^5	0.027	1.49×10^5
3	1.43×10^5	0.022	8.51×10^4

Evaluating W using Eq. 11.3.11 yields $W = 7.88$ m, and application of Eq. 11.3.9 gives the new estimates of Q_i:

$$Q_1 = 0.00310 \text{ m}^3/\text{s} \qquad Q_2 = 0.00727 \text{ m}^3/\text{s} \qquad Q_3 = 0.00962 \text{ m}^3/\text{s}$$

A continuity check shows that

$$\sum_{i=1}^{3} Q_i = 0.0200 \text{ m}^3/\text{s}$$

which is the same as in the first iteration.

A solution of Example 11.3 using Mathcad is shown in Appendix F, Fig. F1.

11.3.3   Branch Piping

The branching network, illustrated in Fig. 11.5a, is made up of three elements connected to a single junction. In contrast to the parallel system shown in Fig. 11.4, no closed loops exist. In the analysis, one assumes the direction of flow in each element; then the energy equation for each element is written using an equivalent length to account for minor losses:

$$\left(\frac{p}{\gamma} + z\right)_A - \left(\frac{p}{\gamma} + z\right)_B = \overline{R}_1 Q_1^2 \tag{11.3.12}$$

$$\left(\frac{p}{\gamma} + z\right)_B - \left(\frac{p}{\gamma} + z\right)_C = \overline{R}_2 Q_2^2 \tag{11.3.13}$$

$$\left(\frac{p}{\gamma} + z\right)_B - \left(\frac{p}{\gamma} + z\right)_D = \overline{R}_3 Q_3^2 \tag{11.3.14}$$

The piezometric heads at locations A, C, and D are considered known. The unknowns are the piezometric head at B and the discharges Q_1, Q_2, and Q_3. The additional relation is the continuity balance at location B, which is

$$Q_1 - Q_2 - Q_3 = 0 \tag{11.3.15}$$

Thus there are four equations with four unknowns. One convenient ad hoc method of solution is outlined below and illustrated in Example 11.4:

1. Assume a discharge Q_1 in element 1 (with or without a pump). Establish the piezometric head H at the junction by solving Eq. 11.3.12 (a category 1 problem).

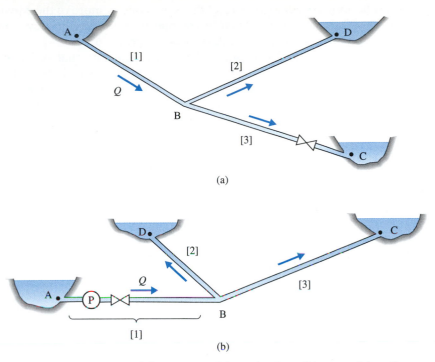

FIGURE 11.5 Branch piping systems: (a) gravity flow; (b) pump-driven flow.

2. Compute the discharge Q_i in the remaining branches using Eqs. 11.3.13 and 11.3.14 (a category 2 problem).
3. Substitute the Q_i into Eq. 11.3.15 to check for continuity balance. Generally, the flow imbalance at the junction ΔQ will be nonzero. In Eq. 11.3.15, $\Delta Q = Q_1 - Q_2 - Q_3$.
4. Adjust the flow Q_1 in element 1 and repeat steps 2 and 3 until ΔQ is within desired limits.

If a pump exists in pipe 1 (see Fig. 11.5b), Eq. 11.3.12 is altered in the manner

$$\left(\frac{p}{\gamma} + z\right)_A - \left(\frac{p}{\gamma} + z\right)_B + H_P = \overline{R}_1 Q_1^2 \qquad (11.3.16)$$

An additional unknown, namely, the pump head H_P, is introduced. The additional necessary relationship is the head-discharge curve for the pump; see, for example, Fig. 7.18. The solution can proceed in a manner similar to that described in Example 11.5. It may be convenient to follow the solution graphically by plotting the assumed discharge in line 1 versus either the piezometric head at B or the imbalance of flow at B. Such a step is helpful to determine in what manner the discharge in line 1 should be altered for the next iteration.

An alternative method of solution for a single branching system is to eliminate all of the variables except the piezometric head $H (= p/\gamma + z)$ at the junction, location B in Fig. 11.5. Then a numerical solving technique can be employed. An additional requirement is to assume the direction of flow in each pipe. It may become necessary to correct the sign in one or more of the equations if during the solution, H moves from below one of the reservoir elevations to above it, or vice versa. Example 11.4 illustrates the procedure.

Examples 11.4 and 11.5 represent a level of complexity that one may not want to exceed employing a calculator-based solution. To include additional pumps, reservoirs or pipes would make either ad hoc or simplified numerical approaches too cumbersome. For such systems, the more generalized network analysis described in Section 11.4 is recommended.

Example 11.4

For the three-branch piping system shown in Fig. E11.4 we have the following data:

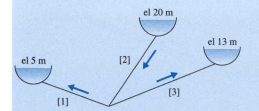

Pipe	L (m)	D (m)	f	ΣK
1	500	0.10	0.025	3
2	750	0.15	0.020	2
3	1000	0.13	0.018	7

FIGURE E11.4

Determine the flow rates Q_i and the piezometric head H at the junction. Assume constant friction factors.

Solution
The equivalent lengths and resistance coefficients are

$$(L_e)_1 = \frac{0.10}{0.025} \times 3 = 12 \text{ m} \qquad \overline{R}_1 = \frac{8 \times 0.025 \times 512}{9.81 \times \pi^2 \times (0.10)^5} = 1.06 \times 10^5 \text{ s}^2/\text{m}^5$$

$$(L_e)_2 = \frac{0.15}{0.020} \times 2 = 15 \text{ m} \qquad \overline{R}_2 = \frac{8 \times 0.020 \times 765}{9.81 \times \pi^2 \times (0.15)^5} = 1.66 \times 10^4 \text{ s}^2/\text{m}^5$$

$$(L_e)_3 = \frac{0.13}{0.018} \times 7 = 51 \text{ m} \qquad \overline{R}_3 = \frac{8 \times 0.018 \times 1051}{9.81 \times \pi^2 \times (0.13)^5} = 4.21 \times 10^4 \text{ s}^2/\text{m}^5$$

With the flow directions assumed as shown, the energy equation is written for each pipe and solved for the unknown discharge:

$$Q_1 = \left(\frac{H-5}{\overline{R}_1}\right)^{1/2} \qquad Q_2 = \left(\frac{20-H}{\overline{R}_2}\right)^{1/2} \qquad Q_3 = \left(\frac{H-13}{\overline{R}_3}\right)^{1/2}$$

The continuity equation is $-Q_1 + Q_2 - Q_3 = 0$. Eliminating Q_1, Q_2, and Q_3 with the energy relations results in an algebraic equation in terms of H:

$$w(H) = -\left(\frac{H-5}{1.06 \times 10^5}\right)^{1/2} + \left(\frac{20-H}{1.66 \times 10^4}\right)^{1/2} - \left(\frac{H-13}{4.21 \times 10^4}\right)^{1/2} = 0$$

Even though this can be solved as a quadratic, the method of false position is chosen to compute H, which would be required if the friction factors varied. The procedure is presented in Example 10.10. In the present example the recurrence formula is

$$H_r = \frac{H_l w(H_u) - H_u w(H_l)}{w(H_u) - w(H_l)}$$

The solution is shown in the table below. Note that with the initial guesses of H_l and H_u, the sign conventions in w require that $20 > H > 13$. Iteration continues until the convergence criterion shown in the last column becomes less than the arbitrary value 0.005.

Iteration	H_u	H_l	$w(H_u)$	$w(H_l)$	H_r	$w(H_r)$	Sign of $w(H_l) \times w(H_r)$	$\varepsilon = \left\lvert \dfrac{H_r^{new} - H_r^{old}}{H_r^{old}} \right\rvert$
1	18	13	−0.01100	0.01185	15.59	−0.00154	+	
2	15.59	13	−0.00154	0.01185	15.29	−0.000384	+	0.019
3	15.29	13	−0.000384	0.01185	15.22	−0.000112	+	0.0046

Hence $H \simeq 15.2$ m. The discharges are now computed:

$$Q_1 = \left(\frac{15.2 - 5}{1.06 \times 10^5}\right)^{1/2} = 0.0098 \text{ m}^3/\text{s}$$

$$Q_2 = \left(\frac{20 - 15.2}{1.66 \times 10^4}\right)^{1/2} = 0.0170 \text{ m}^3/\text{s}$$

$$Q_3 = \left(\frac{15.2 - 13}{4.21 \times 10^4}\right)^{1/2} = 0.0072 \text{ m}^3/\text{s}$$

Note that continuity is satisfied.

Example 11.5

For the system shown in Fig. E11.5, determine the flow distribution Q_i of water and the piezometric head H at the junction. The fluid power input by the pump is constant, equal to $\gamma Q H_P = 20$ kW. Assume constant friction factors.

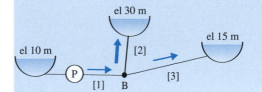

Pipe	L (m)	D (m)	f	ΣK
1	50	0.15	0.02	2
2	100	0.10	0.015	1
3	300	0.10	0.025	1

FIGURE E11.5

Solution
The equivalent lengths and resistance coefficients are computed from Eqs. 11.3.6 and 11.3.8 to be

(continued)

$$(L_e)_1 = \frac{0.15}{0.02} \times 2 = 15 \text{ m} \qquad \overline{R}_1 = \frac{8 \times 0.02 \times 65}{9.81 \times \pi^2 \times (0.15)^5} = 1.42 \times 10^3 \text{ s}^2/\text{m}^5$$

$$(L_e)_2 = \frac{0.10}{0.015} \times 1 = 6.7 \text{ m} \qquad \overline{R}_2 = \frac{8 \times 0.015 \times 106.7}{9.81 \times \pi^2 \times (0.10)^5} = 1.32 \times 10^4 \text{ s}^2/\text{m}^5$$

$$(L_e)_3 = \frac{0.10}{0.025} \times 1 = 4 \text{ m} \qquad \overline{R}_3 = \frac{8 \times 0.025 \times 304}{9.81 \times \pi^2 \times (0.10)^5} = 6.28 \times 10^4 \text{ s}^2/\text{m}^5$$

Assume flow directions as shown. The energy equation for pipe 1 from the reservoir to the junction B is

$$z_1 + H_P = H + \overline{R}_1 Q_1^2$$

in which H is the piezometric head at B. Substituting in known parameters and solving for H results in

$$H = 10 + \frac{20 \times 10^3}{9800 Q_1} - 1.42 \times 10^3 Q_1^2$$

$$= 10 + \frac{2.04}{Q_1} - 1420 Q_1^2$$

An iterative solution is shown in the accompanying table. For each iteration, a value of Q_1 is estimated. Then H is calculated and Q_2 and Q_3 are evaluated from the relations

$$Q_2 = \left(\frac{H - z_2}{\overline{R}_2}\right)^{1/2} = \left(\frac{H - 30}{1.32 \times 10^4}\right)^{1/2}$$

$$Q_3 = \left(\frac{H - z_3}{\overline{R}_3}\right)^{1/2} = \left(\frac{H - 15}{6.28 \times 10^4}\right)^{1/2}$$

In the last column of the table, a continuity balance is employed to check the accuracy of the estimate of Q_1. The third estimate of Q_1 is based on a linear interpolation by setting $\Sigma Q = 0$ and using values of Q_1 and ΔQ from the first two iterations.

Iteration	Q_1	H	Q_2	Q_3	$\Delta Q = Q_1 - Q_2 - Q_3$
1	0.050	47.25	0.0362	0.0227	−0.0089
2	0.055	42.80	0.0311	0.0210	+0.0029
3	0.054	43.64	0.0322	0.0214	+0.0004

The approximate solution is $H = 43.6$ m, $Q_1 = 54$ L/s, $Q_2 = 32$ L/s, and $Q_3 = 21$ L/s. If greater precision is desired, a solution should be employed similar to that shown in Example 11.4.

Solutions for Examples 11.4 and 11.5 using Mathcad and MATLAB are provided in Appendix F, Figs. F2 to F5. In those solutions note that the term $\overline{R}Q^2$ is replaced by $\overline{R}Q|Q|$, which automatically accounts for changes in flow direction. In addition to the computational software solutions in Appendix F, the more general pipe network solutions described in Section 11.4 can be applied to the problems described in this section.

11.4 ANALYSIS OF PIPE NETWORKS

Piping systems more complicated than those considered in Section 11.3 are best analyzed by formulating the solution for a network. Before considering a generalized set of network equations, it is worthwhile to examine a specific example of a piping network to observe the degree of complexity involved.

Figure 11.6a shows a relatively simple network consisting of seven pipes, two reservoirs, and one pump. The hydraulic grade lines at A and F are assumed known; these locations are termed *fixed-grade nodes*. Outflow demands are present at nodes C and D. Nodes C and D, along with nodes B and E are called

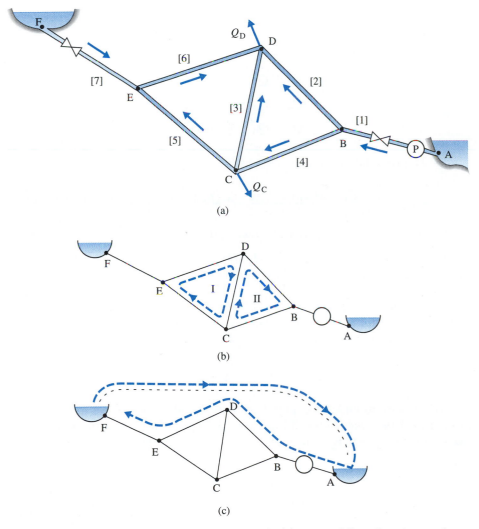

(a)

(b)

(c)

FIGURE 11.6 Representative piping network: (a) assumed flow directions and numbering scheme; (b) designated interior loops; (c) path between two fixed-grade nodes.

interior nodes or *junctions*. Flow directions, even though not initially known, are assumed to be in the directions shown. The system equations are given as follows:

1. Energy balance for each pipe (seven equations):

$$H_A - H_B + H_P(Q_1) = \overline{R}_1 Q_1^2 \qquad H_C - H_E = \overline{R}_5 Q_5^2$$
$$H_B - H_D = \overline{R}_2 Q_2^2 \qquad H_E - H_D = \overline{R}_6 Q_6^2$$
$$H_C - H_D = \overline{R}_3 Q_3^2 \qquad H_F - H_E = \overline{R}_7 Q_7^2 \tag{11.4.1}$$
$$H_B - H_C = \overline{R}_4 Q_4^2$$

2. Continuity balance for each interior node (four equations):

$$Q_1 - Q_2 - Q_4 = 0$$
$$Q_2 + Q_3 + Q_6 = Q_D$$
$$Q_4 - Q_3 - Q_5 = Q_C \tag{11.4.2}$$
$$Q_5 - Q_6 + Q_7 = 0$$

3. Approximation of pump curve (one equation):

$$H_P(Q_1) = a_0 + a_1 Q_1 + a_2 Q_1^2 \tag{11.4.3}$$

in which a_0, a_1, and a_2 are known constants.

The unknowns are $Q_1, \ldots, Q_7, H_B, H_C, H_D, H_E$, and H_P. Thus there are 12 unknowns and 12 equations to solve simultaneously. Since the energy equations and the pump equation are nonlinear, it is necessary to resort to some type of approximate numerical solution. The 12 equations can be reduced in number by combining the energy equations along special paths. Let the drop in hydraulic grade line for any pipe element i be designated as W_i. Then

$$W_i = \overline{R}_i Q_1^2 \tag{11.4.4}$$

For the system under consideration, two closed paths, or *interior loops*, can be identified (see Fig. 11.6b). Flow is considered positive in a clockwise sense around each loop. Energy balances, written around loops I and II, are

$$W_6 - W_3 + W_5 = 0$$
$$W_3 - W_2 + W_4 = 0 \tag{11.4.5}$$

To account for the flow in pipes 1 and 7, a path can be defined along nodes A, B, D, E, and F in Fig. 11.6c. Then with the addition of the pump head, the energy balance from A to F is

$$H_A + H_P - W_1 - W_2 + W_6 + W_7 = H_F \qquad (11.4.6)$$

Note that the path energy equation connects two fixed-grade nodes. Such a path is sometimes termed a *pseudoloop*, since an imaginary pipe with infinite resistance, or no flow, can be considered to connect the two reservoirs. The imaginary pipe is shown by a dotted line in Fig. 11.6c. Substituting the pump equation and the friction equation into the energy relations above results in the following reduced set of equations:

$$-\bar{R}_3 Q_3^2 + \bar{R}_5 Q_5^2 + \bar{R}_6 Q_6^2 = 0$$

$$-\bar{R}_2 Q_2^2 + \bar{R}_3 Q_3^2 + \bar{R}_4 Q_4^2 = 0$$

$$-\bar{R}_1 Q_1^2 + (a_0 + a_1 Q_1 + a_2 Q_1^2) - \bar{R}_2 Q_2^2 + \bar{R}_6 Q_6^2 + \bar{R}_7 Q_7^2 + H_A - H_F = 0$$

$$Q_1 - Q_2 - Q_4 = 0$$

$$Q_2 + Q_3 + Q_6 = Q_D$$

$$Q_4 - Q_3 - Q_5 = Q_C$$

$$Q_5 - Q_6 + Q_7 = 0$$

$$(11.4.7)$$

There are now seven unknowns $(Q_1, \dots, Q_7)$ and seven equations to solve. The energy relations are nonlinear since the loss terms and the pump head are represented as polynomials with respect to the discharges.

11.4.1 Generalized Network Equations

Networks of piping such as those shown in Fig. 11.6 can be represented by the following equations.

1. Continuity at the *j*th interior node:

$$\Sigma(\pm)_j Q_j - Q_e = 0 \qquad (11.4.8)$$

   in which the subscript *j* refers to the pipes connected to a node, and Q_e is the external demand. The algebraic plus or minus sign convention pertains to the assumed flow direction: Use the positive sign for flow into the junction, and the negative sign for flow out of the junction.

2. Energy balance around an interior loop:

$$\Sigma(\pm)_i W_i = 0 \qquad (11.4.9)$$

   in which the subscript *i* pertains to the pipes that make up the loop. There will be a relation for each of the loops. Here it is assumed that there are no pumps located in the interior of the network. The plus sign is used if the flow in the element is positive in the clockwise sense; otherwise, the minus sign is employed.

3. Energy balance along a unique path or pseudoloop connecting two fixed-grade nodes:

$$\Sigma(\pm)_i[W_i - (H_P)_i] + \Delta H = 0 \qquad (11.4.10)$$

where ΔH is the difference in magnitude of the two fixed-grade nodes in the path ordered in a clockwise fashion across the imaginary pipe in the pseudoloop. The term $(H_P)_i$ is the head across a pump that could exist in the ith pipe element. If F is the number of fixed-grade nodes, there will be $(F - 1)$ unique path equations. The plus and minus signs in Eq. 11.4.10 follow the same argument given for Eq. 11.4.9.

Let P be the number of pipe elements in the network, J the number of interior nodes, and L the number of interior loops. Then the following relation will hold if the network is properly defined:

$$P = J + L + F - 1 \qquad (11.4.11)$$

In Fig. 11.6 of the introductory example, $J = 4$, $L = 2$, $F = 2$, so that $P = 4 + 2 + 2 - 1 = 7$.

An additional necessary formulation is the relation between discharge and loss in each pipe; it is

$$W = RQ^\beta + \frac{\Sigma K}{2gA^2}Q^2 \qquad (11.4.12)$$

If the minor losses can be defined in terms of an equivalent length, Eq. 11.4.12 can be replaced by[2]

$$W = \overline{R}Q^\beta \qquad (11.4.13)$$

An approximate pump head-discharge representation is given by the polynomial

$$H_P(Q) = a_0 + a_1Q + a_2Q^2 \qquad (11.4.14)$$

---

[2] In Eq. 11.4.13, the exponent β is used, since the Hazen–Williams formula is commonly employed in pipe network analysis. The equivalent length for use in the Hazen–Williams formulation is given by

$$L_e = 0.8106 \frac{\Sigma K}{K_1} D^{0.87}C^{1.85}Q^{0.15}$$

Note that it is necessary to estimate the discharge when using this equation. The Darcy–Weisbach equation with $\beta = 2$ is recommended. An alternative to using the equivalent-length concept is to account for pipe friction and minor losses separately in each line, making use of Eq. 11.4.12.

The coefficients a_0, a_1, and a_2 are assumed known; typically, they can be found by substituting three known data points from a specified pump curve and solving the three resulting equations simultaneously. In place of defining the pump head-discharge curve, an alternative means of including a pump in a line is to specify the useful power the pump puts into the system. The useful, or actual, power $\dot{W}_f$ is assumed to be constant and allows H_P to be represented in the manner

$$H_P(Q) = \frac{\dot{W}_f}{\gamma Q} \tag{11.4.15}$$

This equation is particularly useful when the specific operating characteristics of a pump are unknown.

11.4.2   Linearization of System Energy Equations

Equation 11.4.10 is a general relation that can be applied to any path or closed loop in a network. If it is applied to a closed loop, ΔH is set equal to zero, and if no pump exists in the path or loop, $(H_P)_i$ is equal to zero. Note that Eq. 11.4.9 can be considered to be a subset of Eq. 11.4.10. In the following development, Eq. 11.4.10 will be used to represent any loop or path in the network.

Define the function $\phi(Q)$ to contain the nonlinear terms $W(Q)$ and $H_P(Q)$ in the form

$$\begin{aligned}\phi(Q) &= W(Q) - H_P(Q) \\ &= \overline{R}Q^\beta - H_P(Q)\end{aligned} \tag{11.4.16}$$

Equation 11.4.16 can be expanded in a Taylor series as

$$\phi(Q) = \phi(Q_0) + \frac{d\phi}{dQ}\bigg|_{Q_0}(Q - Q_0) + \frac{d^2\phi}{dQ^2}\bigg|_{Q_0}\frac{(Q - Q_0)^2}{2} + \cdots \tag{11.4.17}$$

in which Q_0 is an estimate of Q. To approximate $\phi(Q)$ accurately, Q_0 should be chosen so that the difference $(Q - Q_0)$ is numerically small. Retaining the first two terms on the right-hand side of Eq. 11.4.17, and using Eq. 11.4.16, yields

$$\phi(Q) \simeq \overline{R}Q_0^\beta - H_P(Q_0) + \left[\beta\overline{R}Q_0^{\beta-1} - \frac{dH_P}{dQ}\bigg|_{Q_0}\right](Q - Q_0) \tag{11.4.18}$$

Note that the approximation to $\phi(Q)$ is now linear with respect to Q.

The parameter G is introduced as

$$G = \beta\overline{R}Q_0^{\beta-1} - \frac{dH_P}{dQ}\bigg|_{Q_0} \tag{11.4.19}$$

Using Eq. 11.4.14 to represent the pump head, Eq. 11.4.19 becomes

$$G = \beta \bar{R} Q_0^{\beta-1} - (a_1 + 2a_2 Q_0) \qquad (11.4.20)$$

Alternatively, with Eq. 11.4.15 substituted into Eq. 11.4.19, one has

$$G = \beta \bar{R} Q_0^{\beta-1} + \frac{\dot{W}_f}{\gamma Q_0^2} \qquad (11.4.21)$$

Substituting Eq. 11.4.19 into Eq. 11.4.18 gives

$$\begin{aligned} \phi(Q) &= \bar{R} Q_0^{\beta} - H_P(Q_0) + (Q - Q_0)G \\ &= W_0 - H_{P0} + (Q - Q_0)G \end{aligned} \qquad (11.4.22)$$

in which $W_0 = W(Q_0)$ and $H_{P0} = H_P(Q_0)$. Finally, Eq. 11.4.22 is substituted into Eq. 11.4.10 to produce the linearized loop or path energy equation

$$\Sigma(\pm)_i[(W_0)_i - (H_{P0})_i] + \Sigma[Q_i - (Q_0)_i]G_i + \Delta H = 0 \qquad (11.4.23)$$

The second term does not contain the plus or minus sign since G_i is a monotonically increasing function of the flow correction. Since Q in the relations above can assume positive or negative values, Q_0^{β} and $Q_0^{\beta-1}$ are often replaced by $Q_0|Q_0|^{\beta-1}$ and $|Q_0|^{\beta-1}$, respectively, in solution algorithms. This is done, for example, in Eq. 11.4.23, which forms the basis for the Hardy Cross solution outlined below.

11.4.3 Hardy Cross Method

In Eq. 11.4.23, let $(Q_0)_i$ be the estimates of discharge from the previous iteration, and let Q_i be the new estimates of the discharge. Define a flow adjustment ΔQ for each loop to be

$$\Delta Q = Q_i - (Q_0)_i \qquad (11.4.24)$$

The adjustment is applied independently to all pipes in a given loop. Hence Eq. 11.4.23 can be written as

$$\Sigma(\pm)_i[(W_0)_i - (H_{P0})_i] + \Delta Q \Sigma G_i + \Delta H = 0 \qquad (11.4.25)$$

Solving for ΔQ, one has

$$\Delta Q = \frac{-\Sigma(\pm)_i[(W_0)_i - (H_{P0})_i] - \Delta H}{\Sigma G_i} \qquad (11.4.26)$$

It is necessary for the algebraic sign of Q to be positive in the direction of normal pump operation; otherwise, the pump curve will not be represented properly and

Eq. 11.4.26 will be invalid. Furthermore, it is important that the discharge Q through the pump remain within the limits of the data used to generate the curve.

For a closed loop in which no pumps or fixed-grade nodes are present, Eq. 11.4.26 reduces simply to the form

$$\Delta Q = \frac{-\Sigma(\pm)_i (W_0)_i}{\Sigma\, G_i} \qquad (11.4.27)$$

The Hardy Cross iterative solution is outlined in the following steps:

1. Assume an initial estimate of the flow distribution in the network that satisfies continuity, Eq. 11.4.8. The closer the initial estimates are to the correct values, the fewer will be the iterations required for convergence. One guideline to use is that in a pipe element, as $\overline{R}$ increases, Q decreases.

2. For each loop or path, evaluate ΔQ with Eq. 11.4.26 or 11.4.27. The numerators should approach zero as the loops or paths become balanced.

3. Update the flows in each pipe in all loops and paths, that is, from Eq. 11.4.24.

$$Q_i = (Q_0)_i + \Sigma \Delta Q \qquad (11.4.28)$$

The term $\Sigma\Delta Q$ is used, since a given pipe may belong to more than one loop; hence the correction will be the sum of corrections from all loops to which the pipe is common.

4. Repeat steps 2 and 3 until a desired accuracy is attained. One possible criterion to use is

$$\frac{\Sigma|Q_i - (Q_0)_i|}{\Sigma|Q_i|} \le \varepsilon \qquad (11.4.29)$$

in which ε is an arbitrarily small number. Typically, $0.001 < \varepsilon < 0.005$.

The Hardy Cross method of analysis is a simplified version of the method of successive approximations applied to a set of linearized equations. It does not require the inversion of a matrix; hence it can be used to solve relatively small networks using either a calculator or spreadsheet algorithm on a personal computer. The continuity relations, Eq. 11.4.8, are in a sense "decoupled" from the solution of the energy relations, Eq. 11.4.26. Continuity is satisfied initially with assumed flows and remains satisfied throughout the solution process. Essentially, one computes a correction ΔQ to the flows Q_i in each loop separately, and then applies ΔQ to the entire network to bring flow through the loops into closer balance. In effect this is a superposition type of solution. Since the flow corrections ΔQ are applied to each loop independently, convergence may not be rapid.

Example 11.6

For the piping system shown in Fig. E11.6a, determine the flow distribution and piezometric heads at the junctions using the Hardy Cross method of solution. Assume that losses are proportional to Q^2.

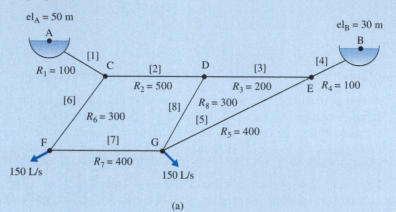

(a)

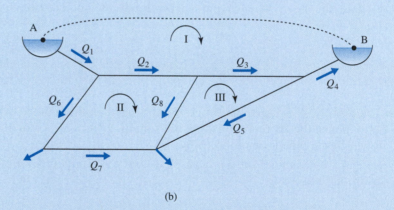

(b)

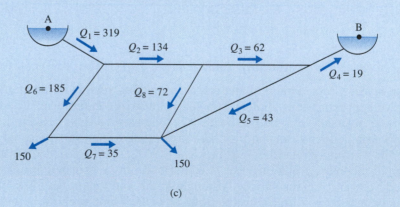

(c)

FIGURE E11.6

Solution

There are five junctions ($J = 5$), eight pipes ($P = 8$), and two fixed-grade nodes ($F = 2$). Hence the number of closed loops is $L = 8 - 5 - 2 + 1 = 2$, plus one pseudoloop. The three loops and assumed flow directions are shown in Fig. E11.6b. Equation 11.4.26 is applied to loop I:

$$\Delta Q_{\mathrm{I}} = \frac{-\left(\pm W_4 \pm W_3 \pm W_2 \pm W_1\right) - (z_{\mathrm{A}} - z_{\mathrm{B}})}{G_4 + G_3 + G_2 + G_1}$$

Apply Eq. 11.4.27 to loops II and III:

$$\Delta Q_{\mathrm{II}} = \frac{-\left(\pm W_2 \pm W_8 \pm W_7 \pm W_6\right)}{G_2 + G_8 + G_7 + G_6}$$

$$\Delta Q_{\mathrm{III}} = \frac{-\left(\pm W_3 \pm W_5 \pm W_8\right)}{G_3 + G_5 + G_8}$$

The problem is solved using an Excel spreadsheet, and the solution is shown in Table E11.6. The spreadsheet layout showing the equations used in the solution is provided in Appendix F, Fig. F.6. In the loop equations note that W and G have the correct sign attributed automatically using the relations $\pm W = \overline{R} Q |Q|$ and $G = 2\overline{R}|Q|$. In addition, the initial values of each Q take on a positive or negative sign depending on the assumed flow direction relative to the positive clockwise direction for each loop. The initial flow estimates, shown under the heading "Iteration 1," are chosen to satisfy continuity. Note that flow adjustments to each pipe element are made after all of the ΔQ's have been computed; for example, relative to loop I, $Q_2 = (Q_0)_2 + \Delta Q_{\mathrm{I}} - \Delta Q_{\mathrm{II}}$. After four iterations the absolute magnitude of the ΔQ's are all less than 0.001, and the left-hand side of Eq. 11.4.29 is 0.0051.

The values of Q_i after four iterations are shown in Fig. E11.6c, along with the final flow directions. Flow rates are in liters per second. The piezometric heads are evaluated by computing the energy drop along designated paths, beginning at a known fixed-grade node, in this case, node A:

$$H_{\mathrm{C}} = H_{\mathrm{A}} - R_1 Q_1^2 = 50 - 100(0.319)^2 = 39.8 \text{ m}$$

$$H_{\mathrm{D}} = H_{\mathrm{C}} - R_2 Q_2^2 = 39.8 - 500(0.134)^2 = 30.8 \text{ m}$$

$$H_{\mathrm{E}} = H_{\mathrm{D}} - R_3 Q_3^2 = 30.8 - 200(0.062)^2 = 30.0 \text{ m}$$

$$H_{\mathrm{F}} = H_{\mathrm{C}} - R_6 Q_6^2 = 39.8 - 300(0.185)^2 = 29.5 \text{ m}$$

$$H_{\mathrm{G}} = H_{\mathrm{D}} - R_8 Q_8^2 = 30.8 - 300(0.072)^2 = 29.2 \text{ m}$$

Note that there is negligible loss in element four.

(continued)

		Iteration 1			Iteration 2			Iteration 3			Iteration 4			
	R	Q	RQ|Q|	2R|Q|	Q	RQ|Q|	2R|Q|	Q	RQ|Q|	2R|Q|	Q	RQ|Q|	2R|Q|	Q
Loop 1 ΔH			20.000			20.000			20.000			20.000		
Pipe 4	100	-0.020	-0.040	4.000	-0.022	-0.051	4.495	-0.019	-0.038	3.899	-0.020	-0.039	3.968	-0.019
Pipe 3	200	-0.060	-0.720	24.000	-0.064	-0.809	25.440	-0.062	-0.770	24.817	-0.063	-0.787	25.098	-0.062
Pipe 2	500	-0.130	-8.450	130.000	-0.137	-9.433	137.352	-0.133	-8.907	133.466	-0.135	-9.150	135.276	-0.134
Pipe 1	100	-0.320	-10.240	64.000	-0.322	-10.399	64.495	-0.319	-10.208	63.899	-0.320	-10.230	63.968	-0.319
			0.550	222.000		-0.691	231.783		0.078	226.081		-0.206	228.309	
			ΔQ =	-2.48E-03		ΔQ =	2.98E-03		ΔQ =	-3.44E-04		ΔQ =	9.03E-04	
Loop 2 Pipe 2	500	0.130	8.450	130.000	0.137	9.433	137.352	0.133	8.907	133.466	0.135	9.150	135.276	0.134
Pipe 8	300	0.070	1.470	42.000	0.074	1.632	44.251	0.071	1.530	42.853	0.073	1.578	43.519	0.072
Pipe 7	400	-0.040	-0.640	32.000	-0.035	-0.494	28.101	-0.036	-0.519	28.823	-0.035	-0.478	27.650	-0.035
Pipe 6	300	-0.190	-10.830	114.000	-0.185	-10.281	111.075	-0.186	-10.382	111.617	-0.185	-10.219	110.738	-0.185
			-1.550	318.000		0.290	320.779		-0.464	316.759		0.031	317.183	
			ΔQ =	4.87E-03		ΔQ =	-9.03E-04		ΔQ =	1.47E-03		ΔQ =	-9.83E-05	
Loop 3 Pipe 3	200	0.060	0.720	24.000	0.064	0.809	25.440	0.062	0.770	24.817	0.063	0.787	25.098	0.062
Pipe 5	400	0.040	0.640	32.000	0.041	0.676	32.898	0.043	0.724	34.039	0.043	0.736	34.325	0.043
Pipe 8	300	-0.070	-1.470	42.000	-0.074	-1.632	44.251	-0.071	-1.530	42.853	-0.073	-1.578	43.519	-0.072
			-0.110	98.000		-0.146	102.589		-0.036	101.710		-0.054	102.941	
			ΔQ =	1.12E-03		ΔQ =	1.43E-03		ΔQ =	3.57E-04		ΔQ =	5.29E-04	

A Mathcad solution for Example 11.6 is given in Appendix F, Fig. F.7.

11.4.4   Network Analysis Using Generalized Computer Software

Spreadsheets and computational software such as Mathcad or MATLAB can be used to solve for the flow distribution in small networks, as illustrated in Example 11.6. However, for larger networks that contain more than several loops, the task of programming becomes cumbersome, especially since a unique description is required for each system accompanied by a large number of equations to solve. An advantage of using these methods of solution is that the user is applying the fundamental relations directly, and can easily discern convergence of the solution. On the other hand, generalized computer codes have robust solution algorithms that allow a wide variety of network systems to be analyzed, and provide user-friendly input/output schemes. They have become an indispensable tool for many engineers and practitioners in the water-supply industry as well as finding use in other industrial applications. However, the user must become familiar with the code so that the correctness of the solution is assured.

With the advancement of software development and the increase of personal computer memory and computational speed, it is now possible to analyze large, highly complex networks with relative ease. In addition to the Hardy Cross method outlined in Section 11.4.3, other linearized methods have been used in computer codes. Herein is described briefly the use of the program EPANET 2 developed by the United States Environmental Protection Agency.

EPANET 2 is a comprehensive program that simulates both hydraulic flow and water quality in pressurized piping networks. For the hydraulic analysis, it uses a hybrid node-loop algorithm termed the *gradient method* to solve for the unknown piezometric heads and discharges. The network can contain pipes, pipe junctions, pumps, and various types of valves, reservoirs, and storage water tanks. Friction head loss is computed using the Darcy–Weisbach, Hazen–Williams, or Chezy–Manning formulas. Either pump curve data or useful power can be accommodated. Further capabilities of the program, a comprehensive description, and details of running the code are provided in the EPANET 2 Users Manual (Rossman, 2000). The source code and users manual is available from the EPA Web site *www.epa.gov*.

Example 11.7

Rework Example 11.6 using EPANET 2.

Solution
The user-defined network map is shown in Fig. E11.7, and a partial listing of the computed results are provided in Table E11.7. For coding details refer to the EPANET 2 Users Manual.

(continued)

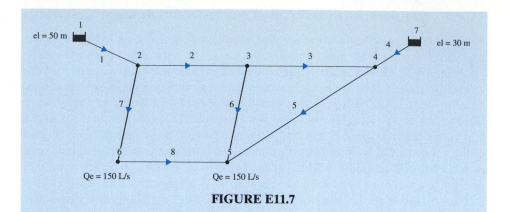

FIGURE E11.7

TABLE E11.7

Link - Node Table:

Link ID	Start Node	End Node	Length m	Diameter mm
1	1	2	66	250
2	2	3	330	250
3	3	4	130	250
4	4	7	66	250
5	4	5	260	250
6	3	5	200	250
7	2	6	200	250
8	6	5	260	250

Node Results:

Node ID	Demand LPS	Head m	Pressure m	Quality	
2	0.00	40.45	40.45	0.00	
3	0.00	30.96	30.96	0.00	
4	0.00	30.05	30.05	0.00	
5	150.00	29.16	29.16	0.00	
6	150.00	29.82	29.82	0.00	
1	-319.72	50.00	0.00	0.00	Reservoir
7	19.72	30.00	0.00	0.00	Reservoir

Link Results:

Link ID	Flow LPS	Velocity m/s	Headloss m/km	Status
1	319.72	6.52	144.71	Open
2	133.59	2.72	28.75	Open
3	62.19	1.27	6.98	Open
4	19.72	0.40	0.83	Open
5	42.47	0.87	3.44	Open
6	71.40	1.46	9.01	Open
7	186.12	3.79	53.13	Open
8	36.12	0.74	2.55	Open

In addition to Example 11.7, solutions for the two networks shown in Fig. 11.7 are provided in Appendix F, Figs. F.8 and F.9. For the branching or tree-like system shown in Fig. 11.7a, the Hazen–Williams friction formula is used with

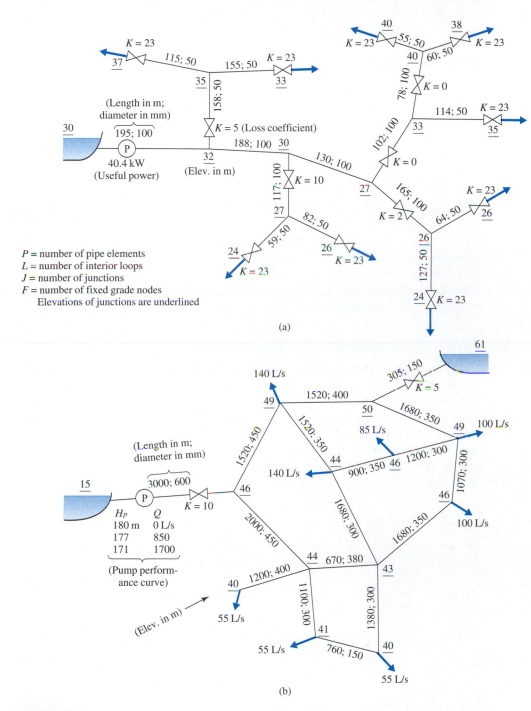

(a)

P = number of pipe elements
L = number of interior loops
J = number of junctions
F = number of fixed grade nodes
    Elevations of junctions are underlined

(b)

FIGURE 11.7   Two piping networks: (a) tree watering system: $P = 17$, $L = 0$, $J = 8$, $F = 10$; (b) water distribution system: $P = 17$, $L = 4$, $J = 12$, $F = 2$. (From Wood, 1981.)

the coefficient $C = 130$. The useful input power of the pump is 40.4 kW. For the looping network shown in Fig. 11.7b, friction losses are based on the Darcy–Weisbach relation with an absolute roughness of 0.15 mm for all pipes and a kinematic viscosity of 10^{-6} m^2/s. The pump performance curve is defined by three settings of pump head and discharge.

11.5   UNSTEADY FLOW IN PIPELINES

Unsteady, or transient flows in pipelines traditionally have been associated with hydropower piping and with long water and oil pipeline delivery systems. However, the application has broadened in recent years to include hydraulic control system operation, events that take place in power plant piping networks, fluid–structure interaction in liquid-filled piping, and pulsatile blood flow. For the unsteadiness to occur, some type of excitation to the system is necessary. Representative excitations are valve opening or closure, pump or turbine operations, pipe rupture or break, and cavitation events. In this section we examine only the most fundamental aspects of unsteady flow in piping. First we study unsteady flow in a single constant-diameter pipe assuming inelastic and incompressible conditions; that is followed by analysis of a system in which elasticity and compressibility play a significant role in the response of the pressure and velocity to an excitation. The first assumption results in the phenomenon called *surging*, while the second one leads to the phenomenon called *water hammer*.

11.5.1   Incompressible Flow in an Inelastic Pipe

Consider a single horizontal pipe of length L and diameter D (Fig. 11.8). The upstream end of the pipe is connected to a reservoir and a valve is situated at the downstream end. The upstream piezometric head is H_1, and downstream of the valve the piezometric head is H_3. Note that both H_1 and H_3 are constant time-independent reservoir elevations. The friction factor f is assumed constant, and the loss coefficient for the valve is K. There are a number of possible excitations that could be considered, but we will only look at the situation where initially there is a steady velocity V_0, the valve is then instantaneously opened to a new

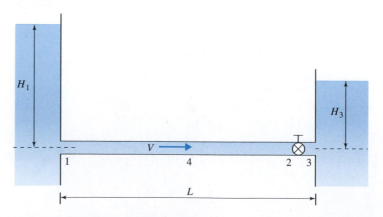

FIGURE 11.8   Horizontal pipe with a valve at the downstream end.

position, and subsequently the flow accelerates, increasing to a new steady-state velocity V_{ss}. For situations in which the valve is closed, either partially or completely, one should consider the possibility of water hammer taking place; this is treated in the next section.

In Fig. 11.8 define a control volume for the liquid in the pipe between locations 1 and 2, whose mass is ρAL. Note that location 2 is upstream of the valve. At the two ends of the control volume the pressures are p_1 and p_2, respectively, and on the surface the wall shear stress is τ_0. The conservation of momentum for that liquid volume is given by

$$A(p_1 - p_2) - \tau_0 \pi DL = \rho AL \frac{dV}{dt} \qquad (11.5.1)$$

Without serious consequences we can assume steady-flow conditions across the valve from location 2 to 3, and utilize the energy equation to provide

$$p_2 = p_3 + K \frac{\rho V^2}{2} \qquad (11.5.2)$$

It is reasonable to assume that the Darcy–Weisbach friction factor based on steady-state flow can be employed without undue error. Then, from Section 7.3.3, Eq. 7.3.19, we have the relation

$$\tau_0 = \frac{\rho f V^2}{8} \qquad (11.5.3)$$

Substituting Eqs. 11.5.2 and 11.5.3 into Eq. 11.5.1, dividing by the mass of the liquid column, and recognizing that $p_1 - p_3 = \rho g(H_1 - H_3)$, since the velocity heads are assumed to be negligible, there results

$$\frac{dV}{dt} + \left(\frac{f}{D} + \frac{K}{L} \right) \frac{V^2}{2} - g \frac{H_1 - H_3}{L} = 0 \qquad (11.5.4)$$

Equation 11.5.4 is the relation that represents incompressible unsteady flow in the pipe. The initial condition at $t = 0$ is a given velocity $V = V_0$. When the final steady-state condition is attained, $dV/dt = 0$, and that steady-state velocity, designated as V_{ss}, can be obtained by setting the derivative to zero in Eq. 11.5.4:

$$V_{ss} = \sqrt{\frac{2g(H_1 - H_3)}{fL/D + K}} \qquad (11.5.5)$$

Substituting Eq. 11.5.5 into Eq. 11.5.4, separating variables, and expressing the result in integral form, one has

$$\int_0^t dt = \frac{V_{ss}^2 L}{g(H_1 - H_3)} \int_{V_0}^V \frac{dV}{V_{ss}^2 - V^2} \qquad (11.5.6)$$

Upon integration, the resulting relation defines the velocity V relative to time t after the valve excitation:

$$t = \frac{V_{ss}L}{2g(H_1 - H_3)} \ln \frac{(V_{ss} + V)(V_{ss} - V_0)}{(V_{ss} - V)(V_{ss} + V_0)} \qquad (11.5.7)$$

There are some significant features to the solution. First, by studying Eq. 11.5.7 one concludes that infinite time is required to reach the steady-state velocity V_{ss}. In reality, V_{ss} will be reached some time sooner, because the losses have not been completely accounted for. However, it is possible to determine the time when a percentage, for example 99%, of V_{ss} has been reached that would be adequate for engineering purposes. In addition, it is possible for the fluid to be initially at rest, that is, $V_0 = 0$. Finally, remember that the solution is based on the assumptions that both liquid compressibility and pipe elasticity are ignored; the next section addresses the situation when those assumptions are invalid.

Example 11.10

A horizontal pipe 1000 m in length, with a diameter of 500 mm, and a steady velocity of 0.5 m/s, is suddenly subjected to a new piezometric head differential of 20 m when the downstream valve suddenly opens and its coefficient changes to $K = 0.2$. Assuming a friction factor of $f = 0.02$, determine the final steady-state velocity, and the time when the actual velocity is 75% of the final value.

Solution
Given are $L = 1000$ m, $D = 0.5$ m, $V_0 = 0.5$ m/s, $f = 0.02$, and $K = 0.02$. Setting $H_1 - H_3 = 20$ m, the final steady-state velocity V_{ss} is found by direct substitution into Eq. 11.5.5:

$$V_{ss} = \sqrt{\frac{2 \times 9.81 \times 20}{0.02 \times 1000/0.5 + 0.2}} = 3.12 \text{ m/s}$$

The velocity that is 75% of V_{ss} is $V = 0.75 \times 3.12 = 2.34$ m/s. Then, using Eq. 11.5.7, the time corresponding to that velocity is

$$t = \frac{3.12 \times 1000}{2 \times 9.81 \times 20} \ln \frac{(3.12 + 2.34) \times (3.12 - 0.5)}{(3.12 - 2.34) \times (3.12 + 0.5)} = 12.9 \text{ s}$$

Hence the final steady-state velocity is 3.12 m/s, and the time when 75% of that velocity is attained is approximately 13 s.

11.5.2 Compressible Flow in an Elastic Pipe

In contrast to the developments of Section 11.5.1, there are situations where the liquid is not incompressible and the piping is not rigid. Rather, the interaction between changes in momentum and applied forces cause the liquid to slightly compress and the pipe material to experience very small deformations. When this occurs, significant pressure changes can take place, and the phenomenon is termed water hammer. Water hammer is accompanied by pressure and velocity perturbations traveling at very high velocities, close to the speed of sound in the liquid. The resulting wave action occurs at relatively high frequencies. We will consider the situation in which a valve at the downstream end of a pipe closes or opens suddenly, either partially or completely, to initiate a water hammer response.

First let us develop the fundamental equations. Consider again the horizontal pipe shown in Fig. 11.8, where now the valve will close so rapidly that elastic effects cause water hammer to occur. The movement of the valve will cause an acoustic, or pressure, wave with speed a to propagate upstream. A control volume of an incremental section of liquid contained in the pipe is shown in Fig. 11.9a, where the pressure wave at a given instant is located. The presence of the wave implies that an unsteady flow is taking place within the control volume; at the entrance the velocity is V, and at the exit it is $V + \Delta V$. Steady-state conservation laws can be applied if the wavefront is made to appear stationary by an observer moving with the wavespeed (Fig. 11.9b). (Refer to Section 4.5.3 for the development of inertial reference frames that move with a constant velocity.) In contrast to Fig. 11.9a, the entrance velocity is now $V + a$, and at the exit it is $V + \Delta V + a$. In addition, the pressure, pipe area, and density at the entrance are p, A, and ρ, respectively. Due to the passage of the pressure wave, at the exit the

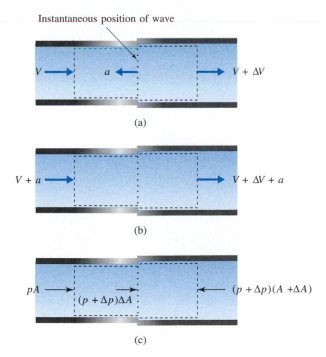

(a)

(b)

(c)

FIGURE 11.9   Pressure wave moving through a segment of horizontal pipe: (a) pressure wave moving to the left at speed a; (b) pressure wave appears stationary using the principle of superposition; (c) pressure forces acting on the control volume.

pressure, pipe area, and liquid density are altered to $p + \Delta p$, $A + \Delta A$, and $\rho + \Delta\rho$, where Δp, ΔA, and $\Delta\rho$ are the respective changes in pressure, area, and density.

Applying the conservation of mass across the control volume in Fig. 11.9b, we find that

$$0 = (\rho + \Delta\rho)(V + \Delta V + a)(A + \Delta A) - \rho(V + a)A \tag{11.5.8}$$

Neglecting frictional and gravitational forces, only pressure forces act on the control volume in the direction of flow, as shown in Fig. 11.9c. The conservation of momentum across the control volume is

$$pA + (p + \Delta p)\,\Delta A - (p + \Delta p)(A + \Delta A)$$
$$= \rho A(V + a)[V + \Delta V + a - (V + a)] \tag{11.5.9}$$

Equations 11.5.8 and 11.5.9 are expanded, and the terms containing factors of Δ^2 and Δ^3 are dropped, since they are much smaller in magnitude than the remaining ones. Then Eqs. 11.5.8 and 11.5.9 become

$$\rho A\,\Delta V + (V + a)(A\Delta\rho + \rho\Delta A) = 0 \tag{11.5.10}$$

and

$$-A\,\Delta p = \rho A(V + a)\,\Delta V \tag{11.5.11}$$

In nearly all situations $V \ll a$, so that Eq. 11.5.11 becomes

$$\Delta p = -\rho a\Delta V \tag{11.5.12}$$

Joukowsky equation: *Relates the pressure change to the pressure wave speed and the change in velocity.*

Equation 11.5.12, called the **Joukowsky equation**, relates the pressure change to the pressure wave speed and the change in velocity. Note that a velocity reduction (a negative ΔV) produces a pressure rise (a positive Δp), and a positive ΔV yields a negative Δp.

Once the wave has passed through the control volume, the altered conditions $p + \Delta p$, $V + \Delta V$, $A + \Delta A$, and $\rho + \Delta\rho$ will persist until the wave reflects from the upstream boundary; this wave action will be discussed later. First it is useful to examine the nature of the pressure pulse wave speed a. Equations 11.5.10 and 11.5.11 are combined, again recognizing that $V \ll a$, to eliminate ΔV, with the result that

$$\frac{\Delta p}{\rho a^2} = \frac{\Delta\rho}{\rho} + \frac{\Delta A}{A} \tag{11.5.13}$$

From the definition of the bulk modulus of elasticity B for a fluid, Eq. 1.5.11, we can relate the change in density to the change in pressure as $\Delta\rho/\rho = \Delta p/B$. The change in pipe area can be related to the change in pressure by considering an instantaneous, elastic response of the pipe wall to pressure changes. Assuming a circular pipe cross section of radius r, we have $\Delta A/A = 2\Delta r/r$, and the change in circumferential strain ε in the pipe wall is $\Delta\varepsilon = \Delta r/r$. For a thin-walled pipe whose thickness e is much smaller than the radius, that is, $e \ll r$, the circumferential stress is given by $\sigma = pr/e$. For small changes in r and e, $\Delta\sigma \approx (r/e)\Delta p$. The elastic modulus for the pipe wall material is the change in stress divided by the change in strain, or

$$E = \frac{\Delta\sigma}{\Delta\varepsilon} \approx \frac{(r/e)\,\Delta p}{\Delta r/r} = \frac{(2r/e)\,\Delta p}{\Delta A/A} \tag{11.5.14}$$

Solving for $\Delta A/A$, and substituting the result into Eq. 11.5.13, along with the relative change in density related to the change in pressure and bulk modulus, there results

$$\frac{\Delta\rho}{\rho a^2} = \frac{\Delta p}{B} + \frac{2r\,\Delta p}{eE} \tag{11.5.15}$$

The parameter Δp can be eliminated in Eq. 11.5.15, the diameter substituted for the radius, and the relation solved for a:

$$a = \sqrt{\frac{B/\rho}{1 + (D/e)(B/E)}} \tag{11.5.16}$$

Hence the pressure pulse wave speed is shown to be related to the properties of the liquid (ρ and B) and those of the pipe wall (D, e, and E). If the pipe is very rigid, or stiff, then the term $DB/eE \ll 1$, and Eq. 11.5.16 becomes $a = \sqrt{B/\rho}$, which is the speed of sound in an unbounded liquid (see Eq. 1.5.12). Note that the effect of pipe elasticity is to reduce the speed of the pressure wave.

The use of Eqs. 11.5.12 and 11.5.16 will provide only the magnitude of the water hammer excitation. In addition, it is necessary to understand the nature of the pressure pulse wave as it travels throughout the pipe and is reflected from the boundaries. Consistent with the development of Eqs. 11.5.12 and 11.5.16, we neglect friction to simplify the analysis. Consider the situation in which the valve is suddenly closed. Let us study in detail the sequence of events throughout one cycle of motion, as illustrated in Fig. 11.10.

In Fig. 11.10a an initial steady condition exists, the velocity is V_0, and the valve is suddenly closed at $t = 0$. Note that when friction is neglected, the hydraulic grade line appears horizontal. Subsequent to closure of the valve, the wave travels upstream (Fig. 11.10b). Behind the wave, the velocity is reduced to zero, the pressure rises by the amount Δp, the liquid has been compressed, and the pipe slightly expanded. At time L/a the wave reaches the reservoir, and an

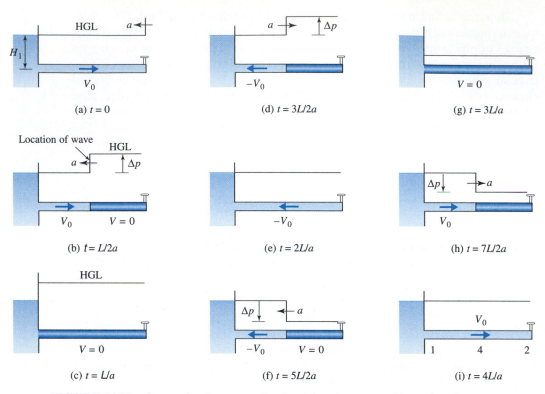

FIGURE 11.10   One cycle of wave motion in a pipe due to a sudden valve closure.

unbalanced force occurs at the pipe inlet (Fig. 11.10c). In the pipe the pressure reduces to the reservoir pressure, and the velocity suddenly reverses direction; this process begins at the upstream end and propagates downstream at the wave speed a (Fig. 11.10d).

When the wave reaches the valve at time $2L/a$, the velocity has magnitude $-V_0$ throughout the pipe (Fig. 11.10e). Adjacent to the valve, which is now closed, the velocity reduces to zero, and the pressure reduces by the amount Δp (Fig. 11.10f). The low-pressure wave travels upstream at speed a, and behind the wave the liquid is expanded and the pipe wall is contracted. Note that if the pressure behind the wave reduces to vapor pressure, cavitation will occur, and some of the liquid will vaporize. When the pressure wave reaches the reservoir at time $3L/a$, an unbalanced condition occurs again, opposite in magnitude to that at time L/a (Fig. 11.10g). Equilibrium of forces will now cause a wave to travel downstream, with the pressure increased by the amount Δp and the velocity equal to $+V_0$ behind the front (Fig. 11.10h). When the wave reaches the valve at time $4L/a$, initial steady-state conditions prevail once again throughout the pipe (Fig. 11.10i).

The process repeats itself every $4L/a$ seconds. For the ideal, frictionless situation shown here, the motion will perpetuate. The pressure waveform at the valve and at the midpoint of the pipe, and the velocity at the pipe entrance, are

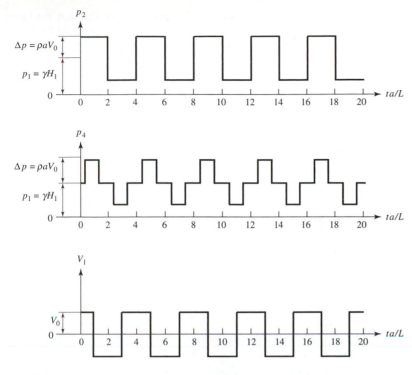

FIGURE 11.11 Pressure waveforms at the valve (p_2), pipe midpoint (p_4), and velocity waveform at the entrance to the pipe (V_1).

shown in Fig. 11.11. For a real pipe system, friction, pipe motion, and inelastic behavior of the pipe material will eventually cause the oscillation to die down and cease (see Fig. 11.12).

The pressure rise Δp predicted by Eq. 11.5.13 is based on the assumption that the valve closes instantaneously. Actually, it can be used to predict the maximum pressure rise when the valve closes with any time less than $2L/a$, the time it takes for the pressure wave to travel from the valve to the reservoir and back again. For valve closure times greater than $2L/a$, a more comprehensive analysis is required

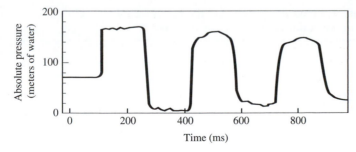

FIGURE 11.12 Pressure waveform at the valve for an actual pipe system following rapid valve closure. (Martin, C. D., Experimental Investigation of Column Separation with Rapid Valve Closure, Proceedings, 4th International Conference on Pressure Surges, BHRA Fluid Engineering, Cranfield, England, 1983, pp. 77-88. Reproduced with permission.)

(Wylie and Streeter, 1993). Keep in mind that the previous discussion pertains only to a single horizontal pipe with a reservoir at the upstream end and a valve closing instantaneously at the downstream end and one that contains a friction-less liquid. Most of the water hammer analysis performed by engineers today makes use of computer-based numerical methods for complex piping systems, incorporating a variety of excitation mechanisms such as pumps, surge sup-pressers, and various types of valves (Wylie and Streeter, 1993).

Example 11.11

A steel pipe ($E = 207 \times 10^6$ kPa, $L = 1500$ m, $D = 300$ mm, $e = 10$ mm) conveys water at 20°C. The initial velocity is $V_0 = 1$ m/s. A valve at the downstream end is closed so rapidly that the motion is considered to be instantaneous, reducing the velocity to zero. Determine the pressure pulse wave speed in the pipe, the speed of sound in an unbounded water medium, the pressure rise at the valve, the time it takes for the wave to travel from the valve to the reservoir at the upstream end, and the period of oscillation.

Solution

The density and bulk modulus of water at 20°C are found in Table B.1: $\rho = 998$ kg/m^3 and $B = 220 \times 10^7$ Pa. The pressure wave speed, a, is given by Eq. 11.15.16:

$$a = \sqrt{\dfrac{220 \times 10^7/998}{1 + \dfrac{0.3}{0.01} \times \dfrac{220 \times 10^7}{207 \times 10^9}}} = 1290 \text{ m/s}$$

The speed of sound in an unbounded water medium is found by use of Eq. 1.5.12:

$$a = \sqrt{\dfrac{220 \times 10^7}{998}} = 1485 \text{ m/s}$$

Note that the sound speed is about 15% larger than the pressure wave speed. To com-pute the pressure rise at the valve, we recognize that the reduction in velocity is $-V_0 = -1$ m/s. Using Eq. 11.15.12, we find the increase in pressure upstream of the valve to be

$$\Delta p = -998 \times 1290 \times (-1) = 1.29 \times 10^6 \text{ Pa} \quad \text{or} \quad 1290 \text{ kPa}$$

The wave travel time from the valve to the reservoir is $L/a = 1500/1290 = 1.16$ s, and the period of oscillation is $4L/a = 4 \times 1.16 = 4.65$ s.

11.6 SUMMARY

A methodology for dealing with losses in piping has been provided in Section 11.2 using the resistance coefficient as a generalized term that can include minor losses along with a chosen empirical friction loss. For accurate representation of losses, the Darcy–Weisbach friction loss is recommended, with losses propor-tional to the square of the velocity (or discharge). Equation 11.2.3 is useful for estimating the Darcy–Weisbach friction factor in an iterative, or trial, solution. In Section 11.3 simple pipe systems have been defined as those that contain one to several pipes, arranged singly, in series, parallel, or branched. Typically, a problem consists of finding the flow distribution in the piping; along with discharges the piezometric head often is an unknown. Calculator-based solutions or computational software are useful tools to analyze these systems.

For more complex piping networks, a systematic solution approach has been presented in Section 11.4. First the network equations were linearized, and then the Hardy Cross method was applied to solve for the network flows and piezometric heads. A spreadsheet solution was applied to a network using the Hardy Cross solution; it allows one to correctly apply the energy and continuity equations as well as track the convergence of the solution. Note that even the simple pipe systems presented in Section 11.3 can be solved using the Hardy Cross method. The computer program EPANET 2 was also introduced in Section 11.4. This software is useful for large systems in which spreadsheet programming becomes cumbersome.

Finally, analysis of unsteady flows in pipes has been introduced in Section 11.5. We have focused on flow that behaves either in an incompressible or compressible manner; the latter behavior is commonly called "water hammer." Water hammer is a result of acoustic waves propagating in a pipe. The waves travel at the speed of sound in the contained liquid and produce a periodic response of both pressures and velocities throughout the pipe. The Joukowsky relation, Eq. 11.5.12, gives the magnitude of the pressure wave.

PROBLEMS

Steady Flows

11.1 Using the results from Example 11.2, tabulate and plot to correct vertical scale the hydraulic and energy gradelines. Assume a 10-m length between the two elbows.

11.2 A pump is situated between two sections in a horizontal pipeline. The diameter D_1 and pressure p_1 are given at the upstream section, and D_2 and p_2 are given at the downstream section. Determine the required fluid power of the pump for the following conditions: Given:
$D_1 = 50$ mm, $p_1 = 350$ kPa, $D_2 = 80$ mm, $p_2 = 760$ kPa, $Q = 95$ L/min, $h_L = 6.6$ m, water is flowing at 20°C.

11.3 An oil ($S = 0.82$) is pumped between two storage tanks in a pipe with the following characteristics: $L = 2440$ m, $D = 200$ mm, $f = 0.02$, $\Sigma K = 12.5$. The upper tank is 32 m higher than the lower one. Using the pump data provided, determine:
(a) The oil discharge in the pipe.
(b) The power requirement for the pump.

Q (L/s)	0	15	30	45	60	75	100
H_P (m)	55	54	53	52	49	44	35
η	0	0.4	0.6	0.7	0.75	0.7	0.5

11.4 Consider the simple pumping system shown in Fig. P11.4. Assume the following parameters to be known:

Reservoir elevations, z_1, z_2

Pipe length, L

Pipe roughness, e

Sum of minor loss coefficients, ΣK

Kinematic viscosity, ν

Discharge, Q

Develop a numerical solution to find the required diameter. Use an equivalent length to represent the minor losses. Determine the diameter for the following data:

$z_2 - z_1 = 40$ m, $L = 500$ m, $e = 1$ mm, $\Sigma K = 2.5$, water at 20°C, $Q = 1.1$ m³/s, pump characteristic curve:

H_P (m)	160	158	154	148	138
Q (L/s)	0	400	800	1200	1600

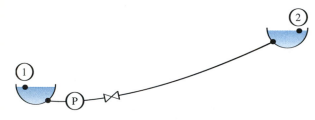

FIGURE P11.4

11.5 Gasoline is being pumped at 400 L/s in a pipeline from location A to B, as shown in Fig. P11.5. The pipe follows the topography as shown, with the highest elevation shown at location C. The only contributions to minor losses are the two valves located at the ends of the pipe. If $S = 0.81$, $\nu = 4.26 \times 10^{-7}$ m²/s, $p_v = 55.2$ kPa absolute, and $p_{atm} = 100$ kPa,
  (a) Determine the necessary power to be delivered to the system to meet the flow requirement.
  (b) What is the maximum elevation possible at location C without causing vapor pressure conditions to exist?

(c) Sketch the hydraulic grade line.

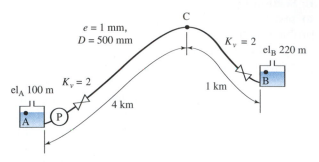

FIGURE P11.5

11.6 For the three pipes in series shown in Fig. P11.6, minor losses are proportional to the discharge squared, and the Hazen–Williams formula is used to account for friction losses. With the data given, use Newton's method to determine the discharge. Note that minor losses can be neglected in order to initially estimate Q. Given: $(p/\gamma + z)_A = 250$ m and $(p/\gamma + z)_B = 107$ m.

Pipe	L (m)	D (mm)	ΣK	C (Hazen–Williams)
1	200	200	2	100
2	150	250	3	120
3	300	300	0	90

FIGURE P11.6

11.7 A long oil pipeline consists of the three segments shown in Fig. P11.7. Each segment has a booster pump that is used primarily to overcome pipe friction. The two reservoirs are at the same elevation.
  (a) Derive a single equation to determine the discharge in the system if the resistance factors $\overline{R}_i$ for each pipe and the useful power $\dot{W}_{fi}$ for each pump are known.

(b) Determine the discharge for the data given in the figure. The unit weight of oil is $\gamma = 8830$ N/m³.

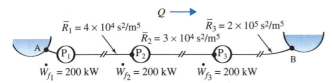

FIGURE P11.7

11.8 A liquid with a specific gravity of 0.68 is pumped from a storage tank to a free jet discharge through a pipe of length L and diameter D (Fig. P11.8). The pump provides a known amount of fluid power $\dot{W}_f$ to the liquid. Assuming a constant friction factor of 0.015, determine the discharge for the following conditions:
$z_1 = 24$ m, $p_1 = 110$ kPa, $z_2 = 18$ m, $L = 450$ m, $D = 300$ mm, $\dot{W}_f = 10$ kW.

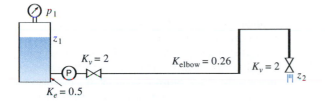

FIGURE P11.8

11.9 Water at 20°C is being pumped through the three pipes in series as shown in Fig. P11.9. The power delivered to the pump is 1920 kW, and the pump efficiency is 0.82. Compute the discharge.

Pipe	L (m)	D (mm)	e (mm)	Σ K
1	200	1500	1	2
2	300	1000	1	0
3	120	1200	1	10

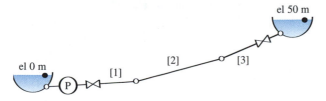

FIGURE P11.9

11.10 Find the water flow distribution in the parallel system shown in Fig. P11.10, and the required pumping power if the discharge through the pump is $Q_1 = 3$ m³/s. The pump efficiency is 0.75. Assume constant friction factors.

Pipe	L (m)	D (mm)	f	Σ K
1	100	1200	0.015	2
2	1000	1000	0.020	3
3	1500	500	0.018	2
4	800	750	0.021	4

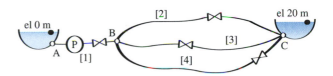

FIGURE P11.10

11.11 For the system shown in Fig. P11.11, determine the water flow distribution and the piezometric head at the junction using an ad hoc approach. Assume constant friction factors. The pump characteristic curve is $H_P = a - bQ^2$. Given :
$a = 20$ m, $b = 30$ s²/m⁵, $z_1 = 10$ m, $z_2 = 20$ m, $z_3 = 18$ m.

Pipe	L (m)	D (cm)	f	K
1	30	24	0.020	2
2	60	20	0.015	0
3	90	16	0.025	0

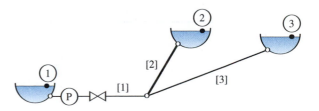

FIGURE P11.11

11.12 Work Problem 11.11 using a numerical approach:
   (a) Newton's method.
   (b) False position method.

11.13 Find the flow distribution in the parallel network shown in Fig. P11.13. Assume constant friction factors. The change in hydraulic grade line between A and B is $(p/\gamma + z)_A - (p/\gamma + z)_B = 50$ m.

Pipe	L (m)	D (mm)	e (mm)	ΣK
1	600	1000	0.1	2
2	1000	1200	0.15	0
3	550	850	0.2	4
4	800	1000	0.1	1

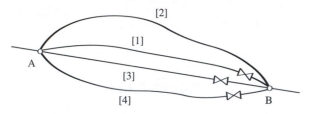

FIGURE P11.13

11.14 The water sprinkling system shown in Fig. P11.14 is applied from a large diameter pipe with constant internal pressure $p_0 = 300$ kPa. The system is positioned in a horizontal plane. Determine the flow distribution Q_1, Q_2, Q_3, Q_4 for the given data. Valve losses are included in the $\bar{R}$, values. (*Hint:* If you are clever, no trial-and-error solution is necessary!)

Pipe	$\bar{R}$ (s²/m⁵)
1	1.6×10^4
2	5.3×10^5
3	1.0×10^6
4	1.8×10^6

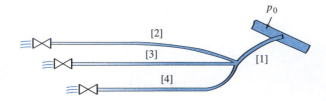

FIGURE P11.14

11.15 For the system shown in Fig. P11.15, a fluid flows from A to B. Determine in which pipe there is the greatest velocity.

Pipe	L (m)	D (mm)	f
1	2000	450	0.012
2	650	150	0.020
3	1650	300	0.015

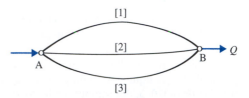

FIGURE P11.15

11.16 A proposed water irrigation system consists of one main pipe with a pump and three pipe branches (Fig. P11.16). Each branch is terminated by an orifice, and each orifice has the same elevation. It is apparent that the flow distribution can be solved by treating the piping arrangement as a branched system. However, the piping can also be treated as a parallel system in order to determine the flows.
   (a) Identify the equations and unknowns to satisfy the solution for a parallel system. Why is it possible to treat the irrigation system as a parallel piping problem?
   (b) Why would the solution to a parallel system be preferred to that of a branching system?
   (c) Determine the flow distribution and sketch the hydraulic grade line.
   (d) What part of the piping would one change to approximately double the discharge, assuming that the individual lengths and pump curve could not be altered?

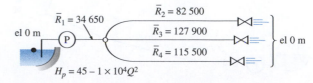

$(H_p$ in m; Q in m³/s; $\bar{R}$ in s²/m⁵, orifice loss included in $\bar{R})$

FIGURE P11.16

11.17 Determine the required head (H_P) and discharge (Q_1) of water to be handled by the pump for the system shown in Fig. P11.17. The discharge in pipe 2 is $Q_2 = 35$ L/s in the direction shown.

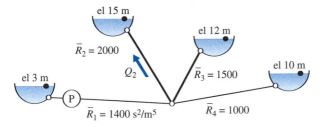

FIGURE P11.17

11.18 Determine the flow distribution of water in the parallel piping system shown in Fig. P11.18. $Q_{in} = 600$ L/min

Pipe	L (m)	D (mm)	f	Σ K
1	30	50	0.020	3
2	40	75	0.025	5
3	60	60	0.022	1

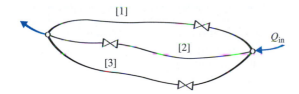

FIGURE P11.18

11.19 Water is being pumped in the piping system shown in Fig. P11.19. The pump curve is approximated by the relation $H_P = 150 - 5Q_1^2$, with H_P in meters and Q_1 in m³/s. The pump efficiency is $\eta = 0.75$. Compute the flow distribution and find the required pump power.

Pipe	$\bar{R}$ (s²/m⁵)
1	400
2	1000
3	1500

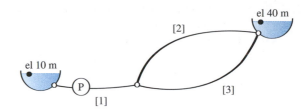

FIGURE P11.19

11.20 A pipeline consists of two pipe segments in series (Fig. P11.20). The specific gravity of the fluid is 0.81. If pump A has a constant power input of 1 MW, find the discharge, the pressure head in pumps A and B, and the required power for pump B. The minimum allowable pressure on the suction side of pump B is 150 kPa, and both pumps have an efficiency of 0.76.

Pipe	L (m)	D (mm)	Σ K	f
1	5000	750	2	0.023
2	7500	750	10	0.023

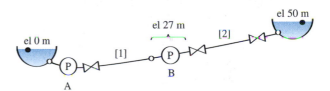

FIGURE P11.20

11.21 Determine the flow distribution of water in the system shown in Fig. P11.21. The equivalent roughness for all elements is 0.1 mm.

Pipe	L (m)	D (mm)	Σ K
1	1000	200	3
2	200	25	0
3	250	25	0
4	340	30	2
5	420	40	0
6	500	175	5

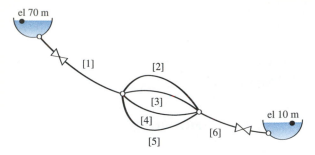

FIGURE P11.21

Pipe	L (m)	D (mm)	Σ K
1	100	350	2
2	750	200	0
3	850	200	0
4	500	200	2
5	350	250	2

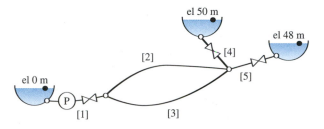

FIGURE P11.23

11.22 Determine the flow distribution of water in the branching pipe system shown in Fig. P11.22:
   (a) Without a pump in line 1.
   (b) Include a pump located in line 1 adjacent to the lower reservoir. The pump characteristic curve is given by $H_P = 250 - 0.4Q - 0.1Q^2$, with H_P in meters and Q in m³/s.

The Hazen–Williams coefficient $C = 130$ for all pipes.

Pipe	L (m)	D (mm)
1	200	500
2	600	300
3	1500	300
4	1500	400

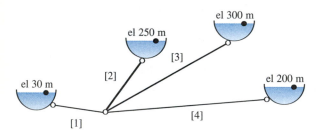

FIGURE P11.22

11.23 Determine the flow distribution of water in the system shown in Fig. P11.23. Assume constant friction factors, with $f = 0.02$. The head-discharge relation for the pump is $H_P = 60 - 10Q^2$, where H_P is in meters and the discharge is in cubic meters per second.

11.24 A pump, whose performance and efficiency curves are given in Fig. P11.24a has been selected to deliver water in a piping system. The piping consists of four pipes arranged as shown in Fig. P11.24b. Water at 15°C is being pumped from reservoir A and exits at either reservoir B or at location D, depending on whether the valves at those locations are open or closed. The pipe characteristics are shown in the accompanying table. All of the pipe diameters are 10 cm, and the friction factor in each pipe is assumed to be $f = 0.02$.
   (a) If the discharge through the pump is 20 m³/h, what is the head loss across pipe 2?
   (b) Compute the discharge in the system, assuming that the valve at location D is closed.
   (c) If the valve at location D is open and the discharge through the pump is 42 m³/h, determine the discharge in pipe 4.

Pipe	L (m)	Σ K
1	3	1
2	150	2
3	600	2
4	225	4

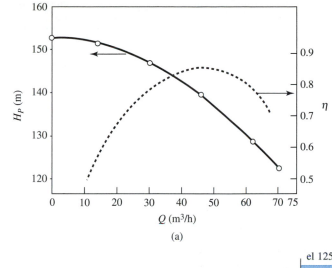

(a)

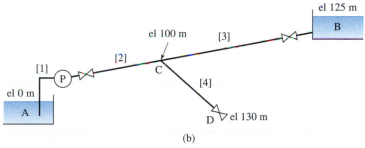

(b)

FIGURE P11.24

11.25 Refer to the system and pump curves associated with Problem 11.24. Assume that the discharge through the pump is 45 m³/h.

  (a) What is the required power for the pump?
  (b) Determine the pressure at location C.
  (c) If the valve at location D is closed, what is the gage pressure at that location?

11.26 Refer to the system and pump curves associated with Problem 11.24.

  (a) Assuming that the valve at location D is open and the head rise across the pump is 140 m, determine the discharge in pipe 3.
  (b) If the valve at location B was closed and the valve at location D was open, what is the head rise across the pump?

11.27 For the network shown in Fig. P11.27, perform the
following:
  (a) Identify all of the unknown parameters and
       write the required system equations.
       Subsequently, reduce them to one equation
       in one unknown. Assume $\beta = 2$.
  (b) Prepare the network for a Hardy Cross
       analysis by making a sketch, identifying nec-
       essary loops and paths, and writing the
       appropriate equations.

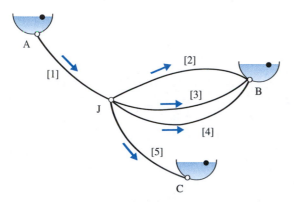

FIGURE P11.27

11.28 The branching system shown in Fig. P11.28a is to
be solved by using the Hardy Cross method.
  (a) Write the equations for ΔQ_I, ΔQ_{II}, W, and G.
  (b) Perform two iterations by filling in the
       blanks in the table shown in Fig. P11.28b.
After two iterations, sketch the hydraulic grade
line.

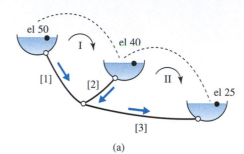

(a)

Loop	Pipe	$\bar{R}$	$2\bar{R}$	Q	W	G	Q	W	G	Q
I	1	2	4	-3.5						
	2	2	4	0						

$$\Delta Q_I = \qquad\qquad \Delta Q_I =$$

Loop	Pipe									
II	2	2	4	0						
	3	2	4	-3.5						

$$\Delta Q_{II} = \qquad\qquad \Delta Q_{II} =$$

(b)

FIGURE P11.28

11.29 Determine the flow and the hydraulic grade line
in the system shown in Fig. P11.29 using the
Hardy Cross analysis method. Use $\gamma = 9800 \text{ N/m}^3$.

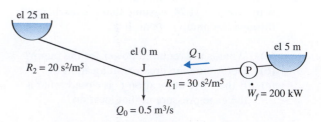

FIGURE P11.29

11.30 The solution of flow of water in a pipe network is shown in Fig. P11.30. Compute:
(a) The hydraulic grade line throughout the system.
(b) The pressure at each node.

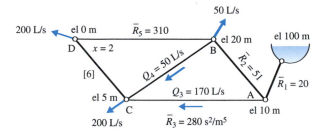

FIGURE P11.30

11.31 For the configuration shown in Problem 11.27, determine the water flow distribution and the hydraulic grade line at J. The system data are tabulated below.

Reservoir	Elevation (m)	Line	L (m)	D (cm)	f	Σ K
A	200	1	250	20	0.015	0
B	175	2	180	7.5	0.020	2
C	60	3	200	7.5	0.020	2
		4	125	7.5	0.025	3
		5	300	10	0.015	4

11.32 Water is flowing in the piping system shown in Fig. P11.32. Determine the flow distribution using the Hardy Cross method:

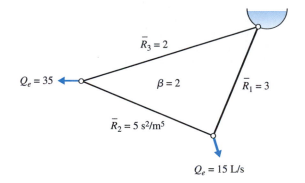

FIGURE P11.32

11.33 Determine the water flow distribution in the piping system shown in Fig. P11.33 and the pressures at the internal nodes employing the Hardy Cross method of solution.

Pipe	L (m)	D (mm)	e (mm)	Σ K
1	500	300	0.15	0
2	600	250	0.15	0
3	50	150	0.15	10
4	200	250	0.15	2
5	200	300	0.15	2

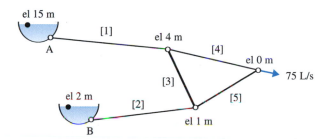

FIGURE P11.33

11.34 Solve Problem 11.9 by employing the Hardy Cross method. Use Eq. 11.2.6 to compute the friction factors.

11.35 Determine the discharge in the piping system shown in Fig. P11.35:
(a) With an exact solution.
(b) Using the Hardy Cross method.

Water is the flowing liquid and the losses are proportional to the velocity squared ($\beta = 2$). The pump characteristic curve is $H_P = 100 - 826Q^2$ (H_P in meters and Q in m³/s).

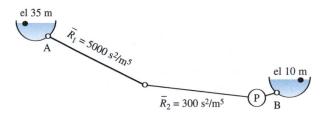

FIGURE P11.35

11.36 The piping system shown in Fig. P11.36 delivers water to two side-branch sprinklers (C and E) and to a lower reservoir (F). The water is supplied by an upper reservoir (A). The sprinklers are represented hydraulically as orifices with relatively high K values. Determine the flow distribution using the Hardy Cross method.

Line	L (m)	D (mm)	e (mm)	ΣK
1	200	100	0.1	2
2	150	50	0.1	30
3	500	100	0.1	0
4	35	50	0.1	35
5	120	100	0.1	2

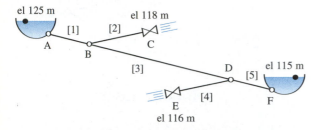

FIGURE P11.36

11.37 In Problem 11.36, if a pump supplying 10 kW of useful power was inserted near A, what changes in flow and pressure would occur?

11.38 A cooling water condenser flow system for a thermal power plant is shown in Fig. P11.38. Water is pumped from a reservoir at location A through a large-diameter pipe [1], passing through the condenser [2], and discharging through another large-diameter pipe [3] into the receiving pond, location B. The condenser consists of a large number of elevated, small-diameter tubes placed in parallel; at either end of the condenser is a large water-filled box called a header, C and C'. Water surface elevations at A and B are known, as are the lengths and diameters of pipes 1 and 3. The condenser has N identical tubes, each with the same known diameter D_2 and length L_2. The friction factor for all pipes is the same constant value, and the pump curve is approximated by the relation $H_P = a_0 + a_1 Q + a_2 Q^2 + a_3 Q^3$. Minor losses can be assumed negligible.

(a) Derive an equivalent single-pipe resistance coefficient R_2 for the condenser tubes.

(b) Write the energy equation for the entire system, using the required given variables and expressing flow resistance in terms of the resistance coefficients for pipes 1 and 3, and the equivalent resistance coefficient for pipe 2.

(c) Set up a Hardy Cross algorithm to determine the discharge through the system.

(d) If $z_A = 2$ m, $z_B = 0$ m, $L_1 = 100$ m, $D_1 = 2$ m, $L_2 = 15$ m, $D_2 = 0.025$ m, $N = 1000$, $L_3 = 200$ m, $D_3 = 2$ m, and $f = 0.02$, compute the discharge and hydraulic grade line. The coefficients for the pump curve are $a_0 = 30.4$, $a_1 = -31.8$, $a_2 = 18.6$, and $a_3 = -4.0$, where Q is in cubic meters per second.

(e) If the top of the condenser is situated at an elevation of 6 m, what are the pressures at the top of the upstream and downstream headers?

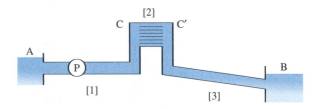

FIGURE P11.38

11.39 Determine the flow distribution in the branching or "tree-like" network shown in Fig. P11.39. Water is the flowing liquid: $\gamma = 9810$ N/m³. The source of flow is a large pipeline (location A) maintained at a constant pressure of 400 kPa. Each branch is at a location where the hydraulic grade line is known. (Courtesy of D. Wood)

11.40 The 12-pipe system shown in Fig. P11.40 represents a low-pressure region that is connected to a high-pressure system by pressure regulators set at 350 kPa. Determine the pressure and water flow distribution for the demands shown. Assume Hazen–Williams $C = 120$ for all pipes. (Courtesy of D. Wood.)

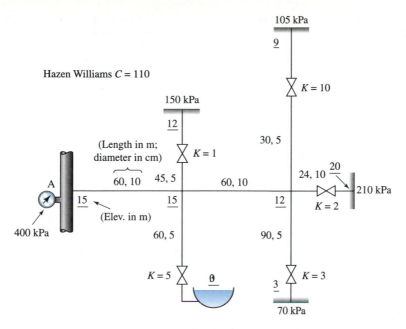

FIGURE P11.39

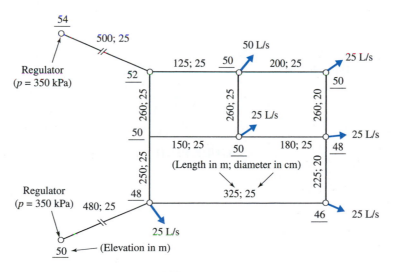

FIGURE P11.40

11.41 Determine the flow distribution for the 14-pipe water supply system shown in Fig. P11.41. The characteristic curve for the pump is represented by the following data (courtesy of D. Wood):

H_P(m)	166	132	18
Q (L/s)	0	600	1000

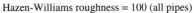

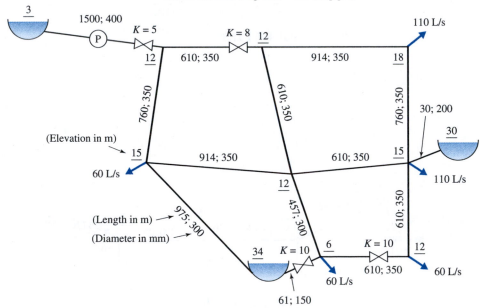

FIGURE P11.41

11.42 The layout of the required water demands for a proposed industrial complex are shown in Fig. P11.42. Design an appropriate network and determine the appropriate useful power for a pump to meet the demand needs. The design criteria are as follows:

1. The lower reservoir supplies water to the system. The upper reservoir supplies water for emergency use only, such as fire, line breaks, and so on; under normal operation, there is no flow into or out of it.

2. Under normal operation, pressures in the network can range between 550 kPa and 850 kPa.
3. Two gate valves are to be placed in each line (one at each end) to isolate it in case of a break or for needed maintenance.
4. All demands must be met in case of a single line break with minimum allowable pressures of 150 kPa.
5. Pipes are cast iron, with the following standard diameters available: 10, 15, 20, 40, 60, 75, 90, 100, 120 cm. Pipes can be placed anywhere within the region.

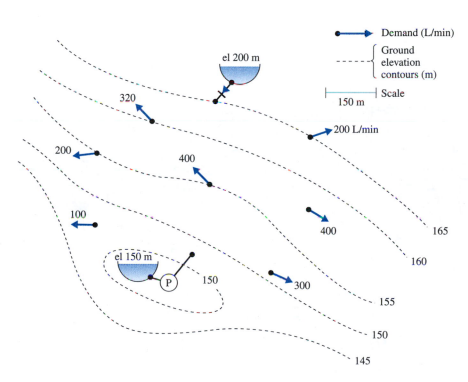

FIGURE P11.42

11.43   Design an irrigation system for part or all of the 18-hole golf course shown in Fig. P11.43. The requirements are as follows:

1.   Place the sprinkler couplings at approximately 30-m intervals along the fairway centerline. The flow rate from each sprinkler is to be 150 L/min. A maximum of two sprinklers on each fairway are to be in operation at any time.

2.   Place four sprinklers around the periphery of each green. The flow rate from each sprinkler is to be 150 L/min. A maximum of three sprinklers on each green are to be in operation at any time. As an alternative to specifying the sprinkler flow rate, specify the loss coefficient across each sprinkler (e.g., try $K_v = 35$), and with successive trials, adjust it until the desired flow rate is achieved.

3.   Water is to be pumped from the well shown. The maximum possible flow from the well is 7500 L/min. Work with useful power in order to satisfy the system demand.

4.   System pressure is to be 550 ± 75 kPa at all locations, and the design velocity in each pipe is not to exceed 1.5 m/s. Use smooth plastic piping and fittings throughout.

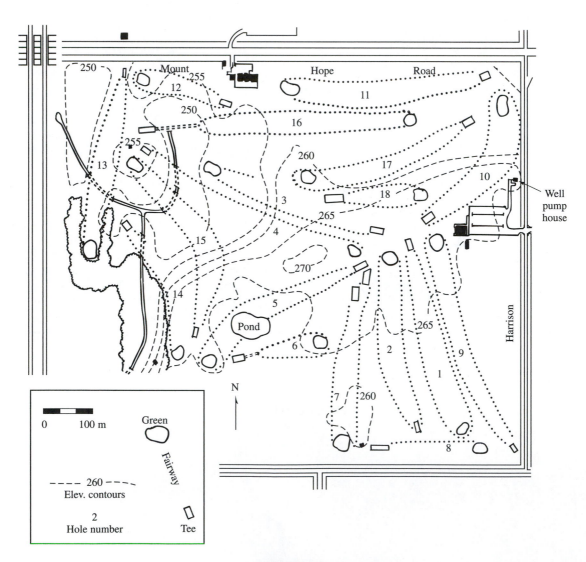

FIGURE P11.43

Unsteady Flow

11.44 An irrigation system has a nearly horizontal supply pipe that is connected to a tank at one end and a quick-opening valve at the other. The length of the pipe is 2500 m, and its diameter is 100 mm. If the elevation of water in the tank is 3 m, how long will it take for the flow to reach the 99% steady-state condition if the valve is suddenly opened from a closed position? Assume that the water is incompressible, the piping is inelastic, $f = 0.025$, and $K = 0.15$ once the valve is opened.

11.45 In Problem 11.43 plot the velocity in the pipe as a function of time.

11.46 Gasoline is supplied by gravity without pumping from a storage tank through a 800-m-long 50-mm-diameter nearly horizontal pipe into a tanker truck. There is a quick-acting valve at the end of the pipe. The difference in elevations of gasoline between the reservoir and the truck tank is 8 m. Initially, the valve is partially closed so that $K = 275$. Then the operator decides to increase the discharge by opening the valve quickly to the position where $K = 5$. Assuming an incompressible fluid and an inelastic piping system, determine the new steady-state discharge and the time it takes to reach 95% of that value. Assume that $f = 0.015$.

11.47 Water is flowing through a pipe whose diameter is 200 mm and whose length is 800 m. The pipe is made of cast iron ($E = 150$ GPa) with a thickness of 12 mm. There is a reservoir at the upstream end of the pipe and a valve at the downstream end. Under steady-state conditions the discharge is $Q = 0.05$ m³/s, when a valve at the end of the pipe is actuated very rapidly so that water hammer occurs.
   (a) How long does it take for an acoustic wave to travel from the valve to the reservoir and back to the valve?

   (b) Determine the change in pressure at the valve if the valve is opened such that the discharge is doubled.
   (c) Determine the change in pressure at the valve if the valve is closed such that the discharge is halved.

11.48 When operating the system described in Problem 11.46, the operator decides to suddenly close the valve once steady-state flow has been reached. Assuming that the gasoline is slightly compressible and the piping is elastic, determine:
   (a) The acoustic wave speed.
   (b) The pressure rise at the valve once the valve is rapidly closed in a manner to cause water hammer.
   (c) If the maximum allowable pressure in the pipe is 250 kPa, what can you conclude about the outcome of the water (gasoline) hammer activity? The pipe is made of aluminum, $E = 70$ GPa, with 2.5-mm-thick walls, and the bulk modulus of elasticity of gasoline is $B = 1.05$ GPa.

11.49 Oil with a specific gravity of $S = 0.90$ is flowing at 0.6 m³/s through a 50-cm-diameter 4000-m-long pipe. The elastic modulus of the steel pipe is $E = 200 \times 10^6$ GPa, its thickness is 10 mm and the bulk modulus of elasticity of the oil is $B = 1.5$ GPa. A valve at the downstream end of the pipe is partially closed very rapidly so that a water hammer event is initiated and a pressure wave propagates upstream. If the magnitude of the wave is not to exceed 600 kPa, determine:
   (a) The percent decrease of flow rate tolerable during the valve closure.
   (b) The time it takes the pressure wave to reach the upstream end of the pipe.

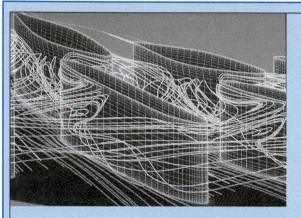
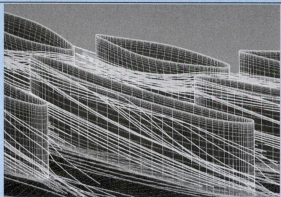

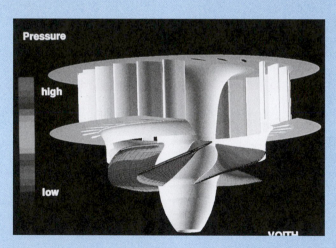

Advances in computational fluid dynamics enable engineers to predict flow patterns through complex turbomachinery. *Bottom:* Predicted pressure distribution on the blades of a Kaplan runner. *Upper:* Flow patterns through the wicked gates on a pump/turbine before (*left*) and after (*right*) improved design analysis. (Voith Hydro, Inc., York, Pennsylvania)

12

Turbomachinery

Outline

Chapter Objectives

The objectives of this chapter are to:

▲ Develop fundamental pump theory using moment of momentum principle

▲ Describe radial-flow, mixed-flow, and axial-flow pumps including presentation of prototype data and deviation from ideal conditions

▲ Introduce dimensionless pump coefficients and show how they are used in conjunction with similarity rules to develop data for turbomachinery design and analysis

▲ Demonstrate how pumps are incorporated into piping design and analysis

▲ Present introductory material on turbines: basic theory, types of turbines, and their selection and operation in conjunction with piping systems

12.1 INTRODUCTION

Turbomachines are the commonly employed devices that either supply or extract energy from a flowing fluid by means of rotating propellers or vanes. A turbopump, more commonly called a pump, adds energy to a system, with the result that the pressure is increased; it also causes flow to occur or it increases the rate of flow. A turbine extracts energy from a system and converts it to some other useful form, typically, to electric power. Pumps are essential components of piping systems which are designed to convey liquids. Similarly, turbomachines are called blowers, fans, or compressors when performing work on air or other gases in ducts. A hydroturbine, or simply turbine, is a machine that generates power from high-pressure water; relatively large conduits or tunnels deliver fluid to closed turbines in order to generate power. Steam and air turbines are also of substantial engineering importance, but such devices are considered in other courses, such as thermodynamics.

The examples above are all those of turbomachines designed to facilitate or utilize internal flows. A wind turbine, on the other hand, makes use of the surrounding external flow to convert the energy contained in the natural movement of atmospheric air into useful electrical power. As shown in Section 4.5.4, a propeller performs work on the surrounding fluid to provide thrust and propel an object along a desired path, while a stationary fan performs work to circulate air. All turbomachines are characterized as being capable of either adding or subtracting energy from the fluid by means of rotating propellers or vanes. There is no contained volume of fluid being transported, as in a positive-displacement piston pump.

The bulk of this chapter is devoted to pumps, in particular, those that convey liquids such as water or gasoline. The centrifugal, or radial-flow pump is presented in detail, and to a lesser extent, mixed-flow and axial-flow pumps are studied. Dimensional analysis, an important tool for the selection and design of turbomachines, is presented next, followed by a discussion of the proper selection and implementation of pumps in piping systems. Finally, turbines are introduced by considering their fundamental characteristics.

12.2 TURBOPUMPS

Turbopump: *Pump consisting of two parts — an impeller and the pump housing.*

A **turbopump** consists of two principal parts: an *impeller*, which imparts a rotary motion to the liquid, and the pump *housing*, or *casing*, which directs the liquid into the impeller region and transports it away under a higher pressure. Figure 12.1 shows a typical single-suction *radial-flow pump*. The impeller is mounted on a shaft and is often driven by an electric motor. The casing includes the suction and discharge nozzles and houses the impeller assembly. The portion of the casing surrounding the impeller is termed the *volute*. Liquid enters through the suction nozzle to the impeller eye and travels along the *shroud*, developing a rotary motion due to the impeller vanes. It leaves the volute casing peripherally at a higher pressure through the discharge nozzle. Some single-suction impellers are open, with the front shroud removed. Double-suction impellers have liquid entering from both sides.

KEY CONCEPT *Common types of pumps are radial flow, mixed flow, and axial flow.*

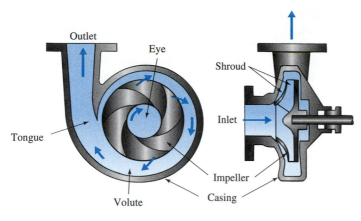

FIGURE 12.1 Single-suction pump.

In a radial-flow pump, the impeller vanes are commonly curved backward and the impeller is relatively narrow. As the impeller becomes wider, the vanes have a double curvature, becoming twisted at the suction end. Such pumps convey liquids with a lower-pressure rise than radial-flow pumps and are called *mixed-flow pumps*. At the opposite extreme from the radial-flow pump is the *axial-flow pump*; it is characterized by the flow entering and leaving the impeller region axially, parallel to the shaft axis. Typically, an axial-flow pump delivers liquid with a relatively low pressure rise. For axial-flow pumps and some mixed-flow pumps, the impellers are open; that is, there is no shroud surrounding them. Various types of impellers are shown in Fig. 12.2.

12.2.1 Radial-Flow Pumps

A representative radial-flow pump is shown in Fig. 12.1; this type of pump is commonly called a *centrifugal pump*, and it is the most commonly used pump in existence today. An elementary analysis will be performed on a radial-flow pump, which will provide a theoretical relationship between the discharge and the developed head rise. In addition, a better understanding of the manner in which momentum exchange takes place in such a turbomachine will be provided. Actual pumps operate at efficiencies less than unity; that is, they do not perform at the theoretical, idealized conditions. It is necessary, therefore, to conduct experiments so that the true operating characteristics of a turbopump can be determined.

> **KEY CONCEPT** *A radial-flow, or centrifugal, pump is designed to deliver relatively low discharge at a high head.*

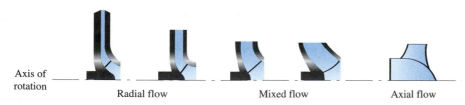

FIGURE 12.2 Various types of pump impellers.

Elementary Theory. The actual flow patterns in a turbopump are highly three-dimensional with significant viscous effects and separation patterns taking place. To construct a simplified theory for the radial-flow pump, it is necessary to neglect viscosity and to assume idealized two-dimensional flow throughout the impeller region. Figure 12.3a defines a control volume that encompasses the impeller region. Flow enters through the inlet control surface and exits through the outlet surface. Note that a series of vanes exists within the control volume, and that they are rotating about the axis with an angular speed ω.

A portion of the control volume is shown at an instant in time in Fig. 12.3b. The idealized velocity vectors are diagrammed at the inlet, location 1, and the outlet, location 2. In the velocity diagrams, V is the absolute fluid velocity, V_t is the tangential component of V, and V_n is the radial, or normal, component of V. The peripheral or circumferential speed of the blade is $u = \omega r$, where r is the radius of the control surface. The angle between V and u is α. The fluid velocity measured relative to the vane is v.

The relative velocity is assumed to be always tangent to the vane; that is, perfect guidance of the fluid throughout the control volume takes place. The angle between v and u is designated as β; since perfect guidance along the vane is assumed, β designates the blade angle as well.

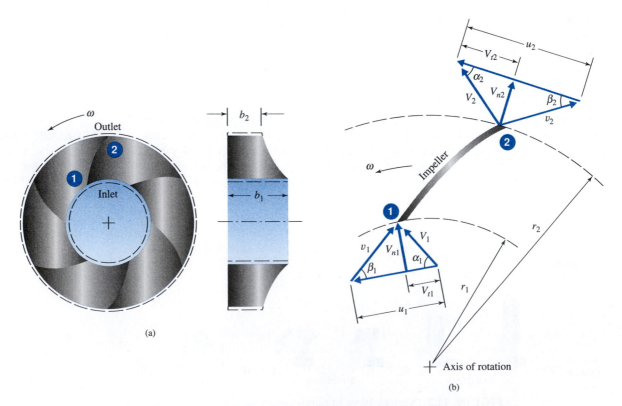

FIGURE 12.3 Idealized radial-flow impeller: (a) impeller control volume; (b) velocity diagrams at control surfaces.

The moment-of-momentum relation, Eq. 4.6.3, can be written for steady flow in the form

$$\Sigma \mathbf{M} = \int_{c.s.} \rho \mathbf{r} \times \mathbf{V}(\mathbf{V} \cdot \hat{n})\, dA \qquad (12.2.1)$$

Applied to the control volume of Fig. 12.3, this becomes

$$T = \rho Q(r_2 V_{t2} - r_1 V_{t1}) \qquad (12.2.2)$$

in which T is the torque acting on the fluid in the control volume, and the right-hand side represents the flux of angular momentum through the control volume. The power delivered to the fluid is the product of ω and T:

$$\omega T = \rho Q(u_2 V_{t2} - u_1 V_{t1}) \qquad (12.2.3)$$

From the velocity vector diagrams in Fig. 12.3b, $V_t = V \cos \alpha$, so that Eq. 12.2.3 can be written as

$$\omega T = \rho Q(u_2 V_2 \cos \alpha_2 - u_1 V_1 \cos \alpha_1) \qquad (12.2.4)$$

For the idealized situation in which there are no losses, the delivered power must be equal to $\gamma Q H_t$, in which H_t is the theoretical pressure head rise across the pump. Then there results Euler's turbomachine relation,

$$H_t = \frac{\omega T}{\gamma Q}$$
$$= \frac{u_2 V_2 \cos \alpha_2 - u_1 V_1 \cos \alpha_1}{g} \qquad (12.2.5)$$

Insight on the nature of flow through an impeller region can be obtained using Eq. 12.2.5. From the law of cosines we can write

$$v_1^2 = u_1^2 + V_1^2 - 2u_1 V_1 \cos \alpha_1$$
$$v_2^2 = u_2^2 + V_2^2 - 2u_2 V_2 \cos \alpha_2$$

These can be substituted into Eq. 12.2.5 to provide the relation

$$H_t = \frac{V_2^2 - V_1^2}{2g} + \frac{(u_2^2 - u_1^2) - (v_2^2 - v_1^2)}{2g} \qquad (12.2.6)$$

The first term on the right-hand side represents the gain in kinetic energy as the fluid passes through the impeller; the second term accounts for the increase in pressure across the impeller. This can be seen by applying the energy equation across the impeller and solving for H_t:

$$H_t = \frac{p_2 - p_1}{\gamma} + z_2 - z_1 + \frac{V_2^2 - V_1^2}{2g} \qquad (12.2.7)$$

Eliminating H_t between Eqs. 12.2.6 and 12.2.7 yields the expression

$$\frac{p_1}{\gamma} + z_1 + \frac{v_1^2 - u_1^2}{2g} = \frac{p_2}{\gamma} + z_2 + \frac{v_2^2 - u_2^2}{2g} = \text{const.} \qquad (12.2.8)$$

This relation has historically been called the Bernoulli equation in rotating coordinates. Since $z_2 - z_1$ is often much smaller than $(p_2 - p_1)/\gamma$, it can be eliminated, and thus the pressure difference is

$$p_2 - p_1 = \frac{\rho}{2}\left[(v_1^2 - u_1^2) - (v_2^2 - u_2^2)\right] \qquad (12.2.9)$$

Returning to Eq. 12.2.5, we see that a "best design" for a pump would be one in which the angular momentum entering the impeller is zero, so that maximum pressure rise can take place. Then in Fig. 12.3b, $\alpha_1 = 90°$, $V_{n1} = V_1$, and Eq. 12.2.5 becomes

$$H_t = \frac{u_2 V_2 \cos \alpha_2}{g} \qquad (12.2.10)$$

From the triangle geometry of Fig. 12.3, $V_2 \cos \alpha_2 = u_2 - V_{n2} \cot \beta_2$, so that Eq. 12.2.10 takes the form

$$H_t = \frac{u_2^2}{g} - \frac{u_2 V_{n2} \cot \beta_2}{g} \qquad (12.2.11)$$

Applying the continuity principle at the outlet region to the control volume provides the relation

$$V_{n2} = \frac{Q}{2\pi r_2 b_2}$$ (12.2.12)

in which b_2 is the width of the impeller at location 2. Introducing Eq. 12.2.12 into Eq. 12.2.11, and recalling that $u_2 = \omega r_2$, one has the relation

$$H_t = \frac{\omega^2 r_2^2}{g} - \frac{\omega \cot \beta_2}{2\pi b_2 g} Q$$ (12.2.13)

For a pump running at constant speed, Eq. 12.2.13 takes the form

$$H_t = a_0 - a_1 Q$$ (12.2.14)

in which a_0 and a_1 are constants. Equation 12.2.13 is the *theoretical head curve* and is seen to be a straight line with a slope of $-a_1$, as shown in Fig. 12.4a. The effect of the blade angle β_2 is shown in Fig. 12.4b. A forward curving blade ($\beta_2 > 90°$) can be unstable and cause pump surge, where the pump oscillates in an attempt to establish an operating point. Backward-curving vanes ($\beta_2 < 90°$) are generally preferred.

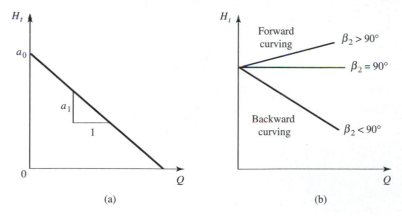

FIGURE 12.4   Ideal pump performance curves.

Example 12.1

A radial-flow pump has the following dimensions:

$$\beta_1 = 44° \qquad r_1 = 21 \text{ mm} \qquad b_1 = 11 \text{ mm}$$
$$\beta_2 = 30° \qquad r_2 = 66 \text{ mm} \qquad b_2 = 5 \text{ mm}$$

For a rotational speed of 2500 rev/min, assuming ideal conditions (frictionless flow, negligible vane thickness, perfect guidance), with $\alpha_1 = 90°$ (no prerotation), determine (a) the discharge, theoretical head, required power, and pressure rise across the impeller, and (b) the theoretical head-discharge curve. Use water as the fluid.

Solution

(a) Construct the velocity diagram at location 1 as shown in Fig. E12.1a. The rotational speed is converted to the appropriate units as

$$\omega = 2500 \frac{2\pi}{60} = 261.8 \text{ rad/s}$$

The impeller speed at r_1 is then

$$u_1 = \omega r_1 = 261.8 \times 0.021 = 5.50 \text{ m/s}$$

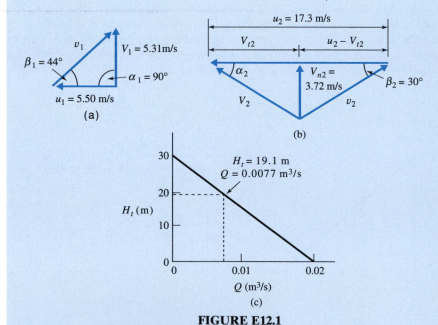

(a)

(b)

(c)

FIGURE E12.1

From the velocity diagram we see that

$$V_1 = u_1 \tan \beta_1 = 5.50 \tan 44° = 5.31 \text{ m/s}$$

and since $\alpha_1 = 90°$, $V_1 = V_{n1}$, or $V_{n1} = 5.31$ m/s. The discharge is computed to be

$$Q = 2\pi r_1 b_1 V_{n1}$$
$$= 2\pi \times 0.021 \times 0.011 \times 5.31 = 7.71 \times 10^{-3} \text{ m}^3/\text{s}$$

The normal component of velocity at location 2 is

$$V_{n2} = \frac{Q}{2\pi r_2 b_2} = \frac{7.71 \times 10^{-3}}{2\pi \times 0.066 \times 0.005} = 3.72 \text{ m/s}$$

and the impeller speed at the outlet is

$$u_2 = \omega r_2 = 261.8 \times 0.066 = 17.28 \text{ m/s}$$

The velocity diagram at location 2 is now sketched as shown in Fig. E12.1b. From the velocity diagram we see that

$$u_2 - V_{t2} = \frac{V_{n2}}{\tan \beta_2} = \frac{3.72}{\tan 30°} = 6.44 \text{ m/s}$$

Therefore,

$$V_{t2} = u_2 - 6.44 = 17.28 - 6.44 = 10.84 \text{ m/s}$$

$$\alpha_2 = \tan^{-1} \frac{V_{n2}}{V_{t2}} = \tan^{-1} \frac{3.72}{10.84} = 18.9°$$

$$V_2 = \frac{V_{t2}}{\cos \alpha_2} = \frac{10.84}{\cos 18.9°} = 11.46 \text{ m/s}$$

The theoretical head is computed with Eq. 12.2.10:

$$H_t = \frac{u_2 V_2 \cos \alpha_2}{g} = \frac{17.28 \times 11.46 \times \cos 18.9°}{9.81} = 19.1 \text{ m}$$

Hence the theoretical required power is

$$\dot{W}_p = \gamma Q H_t = 9810 \times (7.71 \times 10^{-3}) \times 19.1 = 1440 \text{ W}$$

The pressure rise is determined from the energy equation as follows:

$$p_2 - p_1 = \left(H_t + \frac{V_1^2 - V_2^2}{2g} \right) \gamma$$

$$= \left[19.1 + \frac{(5.31)^2 - (11.46)^2}{2 \times 9.81} \right] \times 9810 = 1.36 \times 10^5 \text{ Pa}$$

(b) The theoretical head-discharge curve is Eq. 12.2.13. For the present example we have

$$H_t = \frac{(\omega r_2)^2}{g} - \frac{\omega \cot \beta_2}{2\pi b_2 g} Q$$

$$= \frac{(261.8 \times 0.066)^2}{9.81} - \frac{261.8 \cot 30°}{2\pi \times 0.005 \times 9.81} Q$$

$$= 30.4 - 1471 Q$$

The curve is shown in Fig. E12.1c.

Head-Discharge Relations: Performance Curves. For real fluid flow, the theoretical head curve cannot be achieved in practice, and it is necessary to resort to experimentation to determine the actual head-discharge curve. The energy equation written across a pump from the suction side (location 1, Fig. 12.3) to the discharge side (location 2) is

$$H_P = \left(\frac{p}{\gamma} + \frac{V^2}{2g} + z\right)_2 - \left(\frac{p}{\gamma} + \frac{V^2}{2g} + z\right)_1 \qquad (12.2.15)$$
$$= H_t - h_L$$

in which H_P is the actual head across the pump, and h_L represents the losses through the pump. The actual power delivered to the fluid, designated as $\dot{W}_f$, is

$$\dot{W}_f = \gamma Q H_P \qquad (12.2.16)$$

Brake power: *Power delivered to the impeller.*

while the power delivered to the impeller is $\dot{W}_P$, often termed the **brake power**, and is given by

$$\dot{W}_P = \omega T \qquad (12.2.17)$$

If there were no losses, $\dot{W}_f$ would be equal to $\dot{W}_P$. Since in actuality $\dot{W}_f < \dot{W}_P$, the *pump efficiency*[1] η_P is defined as

$$\eta_P = \frac{\dot{W}_f}{\dot{W}_P} = \frac{\gamma Q H_P}{\omega T} \qquad (12.2.18)$$

One objective of pump design is to make the efficiency as high as possible. The theoretical head curve is compared with the actual head curve in Fig. 12.5. The difference between the two can be attributed to the following effects: (1) prerotation of the fluid before entering the impeller region, (2) separation due to imperfect guidance of fluid as it enters the impeller region, and (3) separation due to expansion in the flow passages. Leakage and high shear rates created by the differences in velocity between fluid particles in the volute and impeller contribute further to the losses. Finally, impeller blades are designed to be most efficient at the so-called design discharge; at any other discharge — that is, "off-design" — the performance will deteriorate.

---

[1] More precisely, η_P defined by Eq. 12.2.18 is termed the *overall efficiency*. It is possible to define three additional efficiencies for pumps as follows:

Hydraulic efficiency: $\eta_H = \dfrac{H_P}{H_t}$

Volumetric efficiency: $\eta_V = \dfrac{Q}{Q + Q_L}$, Q_L = leakage flow rate

Mechanical efficiency: $\eta_M = \dfrac{\gamma H_t(Q + Q_L)}{T\omega}$

These are related to the overall efficiency by $\eta_P = \eta_H\, \eta_V\, \eta_M$.

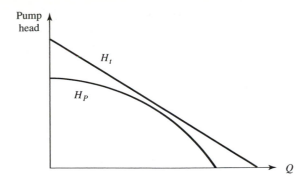

FIGURE 12.5   Comparison between theoretical and actual radial-flow pump performance curves.

A representative pump and its performance curve are shown in Fig. 12.6; this is the curve supplied by a pump manufacturer. Typically, more than one impeller diameter will be available; in the diagram, four different impellers are represented. In addition to the head-discharge performance curves, isoefficiency, power, and *net positive suction head* (*NPSH*) curves are usually provided. Net positive suction head is associated with concern for cavitation; it is discussed in Section 12.2.3.

12.2.2   Axial- and Mixed-Flow Pumps

The velocity diagram for an axial-flow pump is shown in Fig. 12.7. In the axial-flow pump, there is no radial flow and the liquid particles leave the impeller at the same radius at which they enter, so that $u_1 = u_2 = u$. Furthermore, assuming a uniform flow, continuity considerations give $V_{n1} = V_{n2} = V_n$. Equation 12.2.5, which is valid for an axial-flow pump as well as a radial-flow pump, can be combined with the identities $V_2 \cos \alpha_2 = u - V_n \cot \beta_2$ and $V_1 \cos \alpha_1 = V_n \cot \alpha_1$ to produce

KEY CONCEPT   *An axial-flow pump produces relatively large discharges at low heads.*

$$H_t = \frac{u^2}{g} - \frac{uV_n}{g} (\cot \alpha_1 + \cot \beta_2) \qquad (12.2.19)$$

This form of the turbomachine relation is useful when the ideal absolute velocity entrance angle α_1 is established by a fixed vane, or stator. If there is no prerotation, $\alpha_1 = 90°$ and the theoretical head relation, Eq. 12.2.19, becomes

$$H_t = \frac{u^2}{g} - \frac{uV_n \cot \beta_2}{g} \qquad (12.2.20)$$

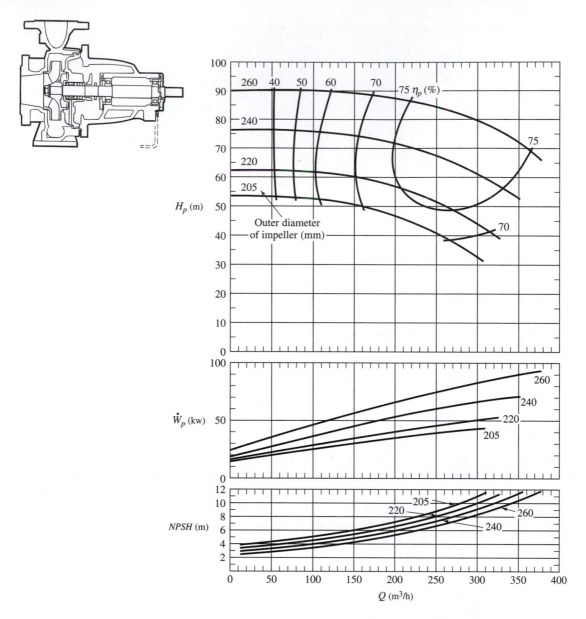

FIGURE 12.6 Radial-flow pump and performance curves for four different impellers with $N = 2900$ rpm ($\omega = 304$ rad/s). Water at 20°C is the pumped liquid. (Courtesy of Sulzer Pumps Ltd.)

This relation is identical to Eq. 12.2.11, which relates the theoretical head to the impeller outlet parameters for a radial flow pump. Note, however, that Eq. 12.2.11 is valid at the periphery of the impeller for the radial-flow pump, and that all fluid particles reach the maximum head at that location. By contrast,

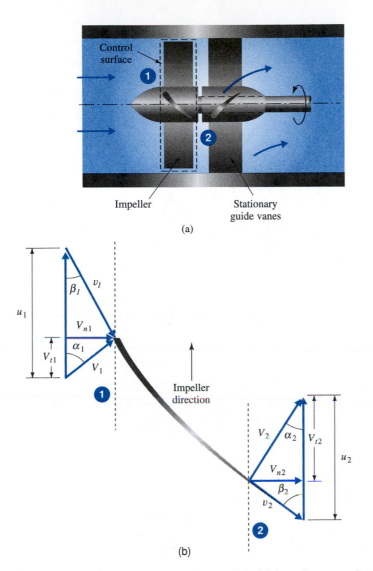

FIGURE 12.7   Idealized axial-flow impeller: (a) impeller control volume; (b) velocity diagrams at control surface.

Eq. 12.2.20 pertains only at a specified radius for the axial-flow pump, since fluid particles enter and leave the control volume at their respective radii. In this case the head varies from a minimum at the axis to a maximum at the periphery, and the total pump head is an integrated average. Axial-flow impellers are designed to maintain a constant axial velocity throughout the impeller region; this requires

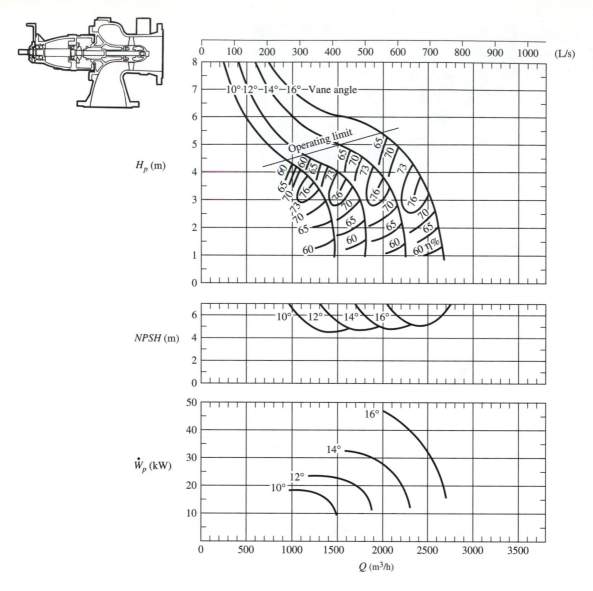

FIGURE 12.8 Axial-flow pump and performance curves for four different vane angles with $N = 880$ rpm ($\omega = 92.2$ rad/s). Propeller diameter is 500 mm. Water at 20°C is the pumped liquid. (Courtesy of Sulzer Pumps Ltd.)

that the vane angles increase gradually from the periphery to the axis, as well as from the inlet to the outlet region. Figure 12.8 shows a representative axial-flow pump, and its performance curves.

Example 12.2

An axial-flow pump is designed with a fixed guide vane, or stator blade, located upstream of the impeller. The stator imparts an angle $\alpha_1 = 75°$ to the fluid as it enters the impeller region. The impeller has a rotational speed of 500 rpm with a blade exit angle of $\beta_2 = 70°$. The control volume has an outer diameter of $D_o = 300$ mm and an inner diameter of $D_i = 150$ mm. Determine the theoretical head rise and power required if 150 L/s of liquid ($S = 0.85$) is to be pumped.

Solution

First, the normal velocity component V_n is

$$V_n = \frac{Q}{A} = \frac{Q}{(\pi/4)(D_o^2 - D_i^2)} = \frac{0.15}{(\pi/4)(0.3^2 - 0.15^2)} = 2.83 \text{ m/s}$$

The peripheral speed u of the impeller is based on an average radius:

$$u \simeq \omega \frac{D_o + D_i}{4} = 500 \left(\frac{2\pi}{60}\right) \frac{0.3 + 0.15}{4} = 5.89 \text{ m/s}$$

The theoretical head H_t is computed with Eq. 12.2.19 to be

$$H_t = \frac{u}{g}\left[u - V_n(\cot \alpha_1 + \cot \beta_2)\right] = \frac{5.89}{9.81}\left[5.89 - 2.83(\cot 75° + \cot 70°)\right] = 2.46 \text{ m}$$

Finally, for the assumed ideal conditions, the required power is

$$\dot{W}_P = \gamma Q H_t = (9810 \times 0.85) \times 0.15 \times 2.46 = 3080 \text{ W} \qquad \text{or} \qquad 3.1 \text{ kW}$$

Generally, flow in an impeller region consists of two components: the circular, vortex-like motion due to the action of the vanes, and the net flow through the impeller. The vortex motion is superposed on the radial outward flow in a radial-flow pump, and on the axial flow in an axial-flow pump. The action of the mixed-flow pump lies between these two types. The previous discussions relating to radial- and axial-flow pumps pertain to mixed-flow pumps as well. A mixed-flow pump and its characteristic curves are shown in Fig. 12.9.

As in the case for a radial-flow pump, the performance characteristics of axial- and mixed-flow pumps deviate from the idealized considerations. Generally, radial-flow pumps are designed to deliver relatively low discharges at high heads, and axial-flow pumps produce relatively large discharges at low heads. Mixed-flow pumps deliver heads and discharges between these two extremes. A method of selecting a pump appropriate for required design considerations is presented in Section 12.4.

> **KEY CONCEPT** *A mixed-flow pump delivers heads and discharges between those of radial-flow and axial-flow pumps.*

12.2.3   Cavitation in Turbomachines

Cavitation refers to conditions at certain locations within the turbomachine where the local pressure drops to the vapor pressure of the liquid, and as a result, vapor-filled cavities are formed. As the cavities are transported through the turbomachine into regions of greater pressure, they will collapse rapidly, generating

> **Cavitation:** *Condition where local pressure drops to the vapor pressure of the liquid, forming vapor-filled cavities.*

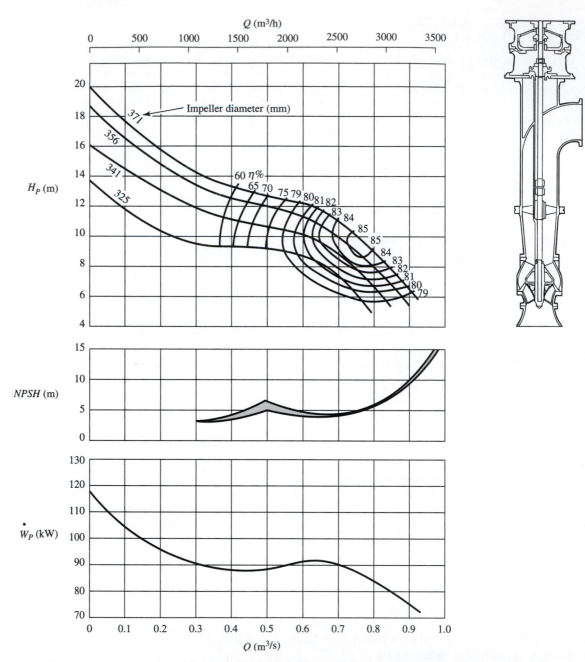

FIGURE 12.9 Mixed-flow pump and performance curves for four different impellers and vane angles with $N = 970$ rpm ($\omega = 102$ rad/s). Water at 20°C is the pumped liquid. (Courtesy of Sulzer Pumps Ltd.)

extremely high localized pressures. Those bubbles that collapse close to solid boundaries can weaken the solid surface, and after repeated collapsing, pitting, erosion, and fatigue of the surface can occur. Signs of cavitation in turbopumps include noise, vibration, and lowering of the head-discharge and efficiency

FIGURE 12.10   Cavitation bubble distribution in an impeller region. (Courtesy of Sulzer Pumps Ltd.)

curves. Regions most susceptible to damage in a turbomachine are those slightly beyond the low-pressure zones on the back side of impellers (see Fig. 12.10). In general, sudden changes in direction, sudden increases in area, and lack of streamlining are the culprits causing cavitation damage in turbopumps (Karassik et al., 1986).

The proper design of turbopumps will have minimized the possibility for cavitation to occur. Under adverse operating conditions, however, the pressures may decrease and cavitation may occur. Two parameters are used to designate the potential for cavitation; these are the cavitation number and the net positive suction head. Their interrelationship and use are now discussed.

Consider a pump operating in the manner shown in Fig. 12.11. Location 1 is on the liquid surface on the suction side, and location 2 is the point of minimum pressure within the pump. Writing the energy equation from location 1 to location 2 and using an absolute pressure datum results in

$$\frac{V_2^2}{2g} = \frac{p_{atm} - p_2}{\gamma} - \Delta z - h_L \qquad (12.2.21)$$

in which h_L is the loss between location 1 and location 2, $\Delta z = z_2 - z_1$, and the kinetic energy at location 1 is assumed negligible. The minimum allowable pressure at location 2 is the vapor pressure, p_v. Substituting this into the expression, one can say that the left-hand side of Eq. 12.2.21 represents the maximum kinetic energy head possible at location 2 when cavitation is imminent. Thus the *net positive suction head* (*NPSH*) is defined as

$$NPSH = \frac{p_{atm} - p_v}{\gamma} - \Delta z - h_L \qquad (12.2.22)$$

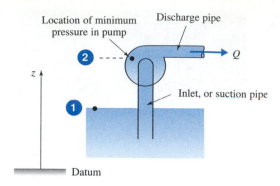

FIGURE 12.11 Cavitation setting for a pump.

The *NPSH* is also used for turbines; however, the sign of the h_L term in Eq. 12.2.22 changes, and location 1 refers to the liquid surface on the discharge side of the machine. The design requirement for a turbopump is thus established as follows:

$$NPSH \le \frac{p_{\text{atm}} - p_v}{\gamma} - \Delta z - h_L \qquad (12.2.23)$$

The performance data supplied by turbomachinery manufacturers usually includes *NPSH* curves; these are developed by testing a given family in a laboratory environment. Figures 12.6, 12.8, and 12.9 show *NPSH* curves. The *NPSH* curve enables one to specify the required maximum value of Δz to be used for a given turbomachine; note that it is necessary to estimate h_L to obtain this.

The right-hand-side of Eq. 12.2.22 can be divided by H_P, the total head across the pump to yield

$$\sigma = \frac{(p_{\text{atm}} - p_v)/\gamma - \Delta z - h_L}{H_P} \qquad (12.2.24)$$

in which σ is the *Thoma cavitation number*. This parameter is used as an alternative to the *NPSH* to establish design criteria for cavitation. A *critical cavitation number*, which signals that cavitation is imminent, is determined experimentally. Thus, for no cavitation to occur, σ must be greater than the critical cavitation number. The cavitation number is in dimensionless form, which is preferred to the dimensional form of the *NPSH*.

Example 12.3

Determine the elevation that the 240-mm-diameter pump of Fig. 12.6 can be situated above the water surface of the suction reservoir without experiencing cavitation. Water at 15°C is being pumped at 250 m³/h. Neglect losses in the system. Use $p_{atm} = 101$ kPa.

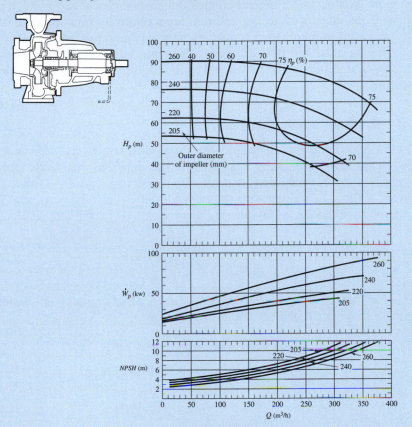

FIGURE E12.3

Solution

From Fig. E12.3, at a discharge of 250 m³/h, the *NPSH* for the 240-mm-diameter impeller is approximately 7.4 m. For a water temperature of 15°C, $p_v = 1666$ Pa absolute, and $\gamma = 9800$ N/m³. Equation 12.2.22 with $h_L = 0$ is employed to compute Δz to be

$$\Delta z = \frac{p_{atm} - p_v}{\gamma} - NPSH - h_L = \frac{101\,000 - 1666}{9800} - 7.4 - 0 = 2.74 \text{ m}$$

Thus the pump can be placed approximately 2.7 m above the suction reservoir water surface.

12.3 DIMENSIONAL ANALYSIS AND SIMILITUDE FOR TURBOMACHINERY

The development and utilization of turbomachinery in engineering practice has benefited greatly from the application of dimensional analysis, probably to an extent more than any other area of applied fluid mechanics. It has enabled pump and turbine manufacturers to test and develop a relatively small number of turbomachines, and subsequently produce a series of commercial units that cover a broad range of head and flow demands. The material developed in Chapter 6, which covers similitude and dimensional analysis, is applied to turbomachines in this section.

12.3.1 Dimensionless Coefficients

The following parameters can be considered significant ones for a turbomachine: power, rotational speed, outer diameter of the impeller, discharge, pressure change across the impeller, density of the fluid, and viscosity of the fluid. These are given in Table 12.1 along with the relevant dimensions. In functional form, the relation between these variables is given by

$$f(\dot{W}, \omega, D, Q, \Delta p, \rho, \mu) = 0 \qquad (12.3.1)$$

Using ω, ρ, and D as repeating variables, a set of convenient dimensionless groupings can be derived. They are given as follows:

$$C_{\dot{W}} = \frac{\dot{W}}{\rho \omega^3 D^5} \qquad \text{power coefficient} \qquad (12.3.2)$$

$$C_P = \frac{\Delta p}{\rho \omega^2 D^2} \qquad \text{pressure coefficient} \qquad (12.3.3)$$

$$C_Q = \frac{Q}{\omega D^3} \qquad \text{flow rate coefficient} \qquad (12.3.4)$$

$$\text{Re} = \frac{\omega D^2 \rho}{\mu} \qquad \text{Reynolds number} \qquad (12.3.5)$$

It is customary to equate Δp to γH, where H represents either the pressure head rise in the case of a pump, or the pressure head drop for a turbine. Then Eq. 12.3.3 is replaced by

$$C_H = \frac{gH}{\omega^2 D^2} \qquad \text{head coefficient} \qquad (12.3.6)$$

TABLE 12.1 Turbomachinery Parameters

Parameter	Symbol	Dimensions
Power	$\dot{W}$	ML^2/T^3
Rotational speed	ω	T^{-1}
Outer diameter of impeller	D	L
Discharge	Q	L^3/T
Pressure change	Δp	M/LT^2
Fluid density	ρ	M/L^3
Fluid viscosity	μ	M/LT

Another dimensionless parameter for a pump can be found by grouping C_Q, $C_{\dot{W}}$, and C_H to form the ratio

$$\frac{C_Q C_H}{C_{\dot{W}}} = \frac{\gamma Q H}{\dot{W}} = \eta_P \qquad \text{(pump)} \qquad (12.3.7)$$

Hence the efficiency is also a similitude parameter. For a turbine the efficiency grouping is given by

$$\frac{C_{\dot{W}}}{C_Q C_H} = \frac{\dot{W}}{\gamma Q H} = \eta_T \qquad \text{(turbine)} \qquad (12.3.8)$$

> **KEY CONCEPT** *The discharge, head, and power coefficients are used to form the dimensionless performance curves.*

in which $\dot{W}$ is the output power.[2]

The interrelationships between the dimensionless parameters are expressed in the form of dimensionless performance curves, as shown in Fig. 12.12; these are for the same turbomachine as shown in Fig. 12.6.[3] Dimensionless performance curves for the axial-flow pump of Fig. 12.8 and the mixed-flow pump of Fig. 12.9 are shown in Figs. 12.13 and 12.14, respectively. The vertical dashed lines designate the operating conditions at maximum or so-called "best" efficiency.

Note that the Reynolds number is not present as a parameter in Figs. 12.12 to 12.14, even though viscous effects are significant in turbomachinery flow. In

---

[2] It is common industrial practice to use so-called *homologous units* in place of the dimensionless coefficients. The relationships between the two groupings are as follows:

Dimensionless coefficient	C_Q	C_H	$C_{\dot{W}}$
Homologous unit	$\dfrac{Q}{ND^3}$	$\dfrac{H}{N^2D^2}$	$\dfrac{\dot{W}}{N^3D^5}$

In the homologous units, density and gravitational acceleration have been eliminated, and the rotational speed is given in revolutions per minute, designated by N. Note that the homologous units are dimensional.

[3] In the caption, Ω_P is the specific speed, which will be defined in Section 12.3.3.

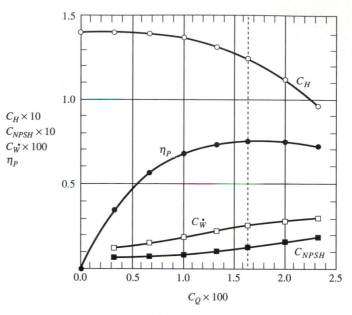

$C_H \times 10$
$C_{NPSH} \times 10$
$C_{\dot{W}} \times 100$
η_P

FIGURE 12.12 Dimensionless radial-flow pump performance curves for the pump presented in Fig. 12.6; $D = 240$ mm; $\Omega_P = 0.61$.

addition to the Reynolds number, one could include a relative roughness ratio. It is known that viscous and roughness effects will alter pump behavior; however, because of the difficulties involved, it is usual practice not to make a correlation between the Reynolds number, roughness, and pump losses. The flow regime in pump passages is usually very turbulent, and the roughness varies widely for different pumps. Furthermore, the friction losses may not be as significant as losses due to eddy formation and separation caused by diffused flow occurring in the impeller and casing. As a result, the Re effect is usually not directly represented in similitude studies. Instead, the two following performance characteristics are recognized: (1) larger pumps exhibit relatively smaller losses and higher efficiencies than smaller pumps of the same family since they have smaller roughness and clearance ratios; and (2) higher viscosity liquids will cause reductions in pump head and efficiency and an increase in power requirement for a given discharge.

A dimensionless net positive suction head coefficient, C_{NPSH}, is also presented in Figs. 12.12 to 12.14. It is formed by replacing H in Eq. 12.3.6 with the *NPSH* defined by Eq. 12.2.22.

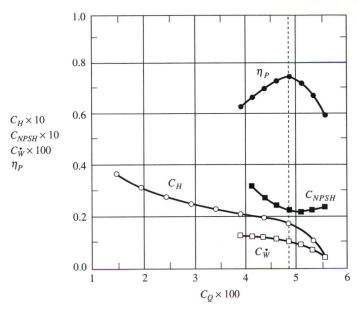

FIGURE 12.13   Dimensionless axial-flow pump performance curve for the pump presented in Fig. 12.8; $D = 500$ mm; vane angle $= 14°$; $\Omega_P = 4.5$.

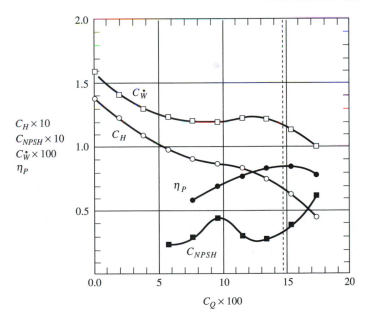

FIGURE 12.14   Dimensionless mixed-flow pump performance curve for the pump presented in Fig. 12.9; $D = 371$ mm; $\Omega_P = 3.1$.

Example 12.4

Determine the speed, size, and required power for the axial-flow pump of Fig. E12.4 to deliver 0.65 m³/s of water at a head of 2.5 m.

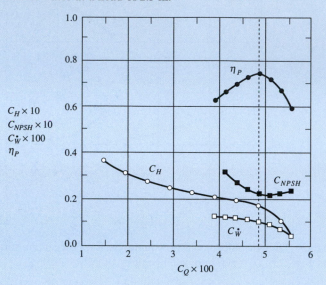

FIGURE E12.4

Solution

The pump data are obtained from Fig. E12.4. Reading from the figure we find that at the design, or maximum, efficiency of $\eta_P = 0.75$

$$C_Q = 0.048 \qquad C_H = 0.018 \qquad C_{\dot{W}} = 0.0012$$

Equations 12.3.4 and 12.3.6 are employed to determine the two unknowns ω and D. Rearrange Eq. 12.3.6 to solve for ωD:

$$\omega D = \sqrt{\frac{gH_P}{C_H}} = \sqrt{\frac{9.81 \times 2.5}{0.018}} = 36.9 \text{ m/s}$$

Equation 12.3.4 is rearranged to include ωD as a known quantity, and solved for D:

$$D = \sqrt{\frac{Q}{C_Q \omega D}} = \sqrt{\frac{0.65}{0.048 \times 36.9}} = 0.61 \text{ m}$$

Substituting $D = 0.61$ m into the relation $\omega D = 36.9$, one finds that

$$\omega = \frac{36.9}{D} = \frac{36.9}{0.61} = 60.5 \text{ rad/s} \qquad \text{or} \qquad 578 \text{ rpm}$$

The density for water is $\rho = 1000$ kg/m³. Then the pump power is determined by using Eq. 12.3.2:

$$\dot{W}_P = C_{\dot{W}} \rho \omega^3 D^5 = 0.0012 \times 1000 \times 60.5^3 \times 0.61^5 = 2.2 \times 10^4 \text{W}$$

Thus the speed, size, and required power are approximately 580 rpm, 610 mm, and 22 kW, respectively.

12.3.2 Similarity Rules

Following the principles outlined in Chapter 6 for similitude, similarity relationships between any two pumps from the same geometric family can be developed. They are:

$$(C_{\dot{W}})_1 = (C_{\dot{W}})_2$$

or

$$\frac{\dot{W}_2}{\dot{W}_1} = \frac{\rho_2}{\rho_1}\left(\frac{\omega_2}{\omega_1}\right)^3\left(\frac{D_2}{D_1}\right)^5 \tag{12.3.9}$$

$$(C_H)_1 = (C_H)_2$$

or

$$\frac{H_2}{H_1} = \frac{g_1}{g_2}\left(\frac{\omega_2}{\omega_1}\right)^2\left(\frac{D_2}{D_1}\right)^2 \tag{12.3.10}$$

$$(C_Q)_1 = (C_Q)_2$$

or

$$\frac{Q_2}{Q_1} = \frac{\omega_2}{\omega_1}\left(\frac{D_2}{D_1}\right)^3 \tag{12.3.11}$$

KEY CONCEPT *Similarity rules relate variables from a family of geometrically similar turbomachines, or are used to examine changes in variables for a given machine.*

These equations, called the *turbomachinery similarity rules*, are used to design or select a turbomachine from a family of geometrically similar units. Another use of them is to examine the effects of changing speed, fluid, or size on a given unit. One could even design a pump to deliver flow on the moon or on a space station!

In light of Eq. 12.3.7, one should also expect that $\eta_1 = \eta_2$, but as stated previously, larger pumps are more efficient than smaller ones of the same geometric family. An acceptable empirical correlation (Stepanoff, 1957) relating efficiencies to size is

$$\frac{1 - (\eta_P)_2}{1 - (\eta_P)_1} = \left(\frac{D_1}{D_2}\right)^{1/4} \tag{12.3.12}$$

This relation can also be used for turbines by replacing η_P by η_T.

Example 12.5

Determine the performance curve for the mixed-flow pump whose characteristic curve is shown in Fig. E12.5a. The required discharge is 2.5 m³/s with the pump operating at a speed of 600 rpm. At design efficiency, what are the power and the *NPSH* requirements? Water is being pumped.

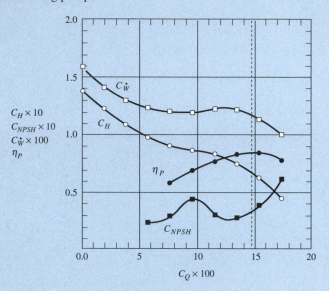

$$C_H \times 10$$
$$C_{NPSH} \times 10$$
$$C_{\dot{W}} \times 100$$
$$\eta_P$$

FIGURE E12.5a

Solution

From Fig. E12.5 the design efficiency is $\eta_P = 0.85$ and the corresponding design values for the coefficients are

$$C_Q = 0.15 \qquad C_H = 0.067 \qquad C_{\dot{W}} = 0.0115 \qquad C_{NPSH} = 0.035$$

The rotational speed in radians per second is

$$\omega = 600 \times \frac{\pi}{30} = 62.8 \text{ rad/s}$$

The pump diameter is computed with Eq. 12.3.4:

$$D = \left(\frac{Q}{\omega C_Q} \right)^{1/3}$$

$$= \left(\frac{2.5}{62.8 \times 0.15} \right)^{1/3} = 0.64 \text{ m}$$

With Eqs. 12.3.2 and 12.3.6, the required power and *NPSH* are computed:

$$\dot{W}_P = \rho \omega^3 D^5 C_{\dot{W}}$$

$$= 1000 \times 62.8^3 \times 0.64^5 \times 0.0115 = 3.1 \times 10^5 \text{W}$$

$$NPSH = \frac{\omega^2 D^2}{g} C_{NPSH}$$

$$= \frac{62.8^2 \times 0.64^2}{9.81} \times 0.035 = 5.8 \text{ m}$$

Hence the required power is 310 kW and the required *NPSH* is 5.8 m. The performance curve is constructed using Eqs. 12.3.10 and 12.3.11:

$$H_2 = \frac{\omega_2^2 D_2^2}{g}(C_H)_1$$

$$= \frac{62.8^2 \times 0.64^2}{9.81}(C_H)_1 = 165(C_H)_1$$

$$Q_2 = \omega_2 D_2^3 (C_Q)_1$$

$$= 62.8 \times 0.64^3 (C_Q)_1 = 16.5(C_Q)_1$$

Values of $(C_H)_1$ and $(C_Q)_1$ are read from Fig. 12.14, along with η_P and placed in the table shown below. Columns 4 and 5 are Q_2 and H_2 computed from the similarity equations. At best efficiency $H_2 = 10.4$ m and $Q_2 = 2.54$ m³/s. The characteristic curve is plotted in Fig. E12.5b.

$(C_Q)_1$	$(C_H)_1$	η_P	$Q_2(m^3/s)$	$H_2(m)$
0	0.138		0	22.8
0.019	0.123		0.31	20.3
0.039	0.110		0.64	18.2
0.058	0.099		0.96	16.3
0.077	0.091	0.59	1.27	15.0
0.096	0.088	0.70	1.58	14.5
0.116	0.085	0.78	1.91	14.0
0.135	0.076	0.84	2.23	12.5
0.154	0.063	0.85	2.54	10.4
0.174	0.046	0.79	2.87	7.6

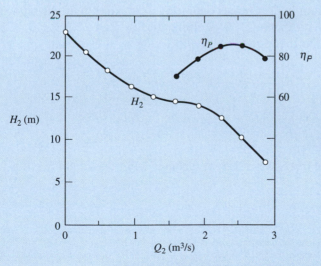

FIGURE E12.5b

12.3.3 Specific Speed

Specific speed: Dimensionless number that characterizes a turbomachine at maximum efficiency.

It is possible to correlate a turbomachine of a given family to a dimensionless number that characterizes its operation at optimum conditions. Such a number is termed the **specific speed**, and it is determined in the following manner. The specific speed Ω_P of a pump is a dimensionless parameter associated with its operation at maximum efficiency, with known ω, Q, and H_P. It is obtained by eliminating D between Eqs. 12.3.4 and 12.3.6 and expressing the rotational speed ω to the first power:

$$\Omega_P = \frac{C_Q^{1/2}}{C_H^{3/4}} = \frac{\omega Q^{1/2}}{(gH_P)^{3/4}} \tag{12.3.13}$$

In Eq. 12.3.13, ω is usually based on motor requirements, and the values of Q and H_P are those at maximum efficiency.

A preliminary pump selection can be based on the specific speed. Figure 12.15 shows how the maximum efficiency and the discharge vary with Ω_P for radial-flow pumps. The specific speed can be correlated with the type of impeller shown in Fig. 12.2; it is given as follows:

$$
\begin{aligned}
&\Omega_P < 1 && \text{radial-flow pump} \\
&1 < \Omega_P < 4 && \text{mixed-flow pump} \\
&\Omega_P > 4 && \text{axial-flow pump}
\end{aligned}
\tag{12.3.14}
$$

For a turbine, the specific speed Ω_T is a dimensionless parameter associated with a given family of turbines operating at maximum efficiency, with known ω, H_T, and $\dot{W}_T$. The diameter D is eliminated in Eqs. 12.3.2 and 12.3.6, and ω is expressed to the first power to obtain:

$$\Omega_T = \frac{C_W^{1/2}}{C_H^{5/4}} = \frac{\omega(\dot{W}_T/\rho)^{1/2}}{(gH_T)^{5/4}} \tag{12.3.15}$$

A comparison of Ω_T for different turbines will be made in Section 12.5.[4]

---

[4] In place of Eqs. 12.3.13 and 12.3.15, it has been common industrial practice to use specific speeds in dimensional form. Hence, for a pump, the dimensional, or *homologous specific speed* is $(N_s)_P = NQ^{1/2}/H^{3/4}$, and for a turbine the corresponding homologous specific speed is $(N_s)_T = N\dot{W}^{1/2}/H^{5/4}$. For these expressions, Q is in gal/min, ft³/sec, or m³/s, H is in ft or m, $\dot{W}$ is in horsepower or kW, and N is in rpm.

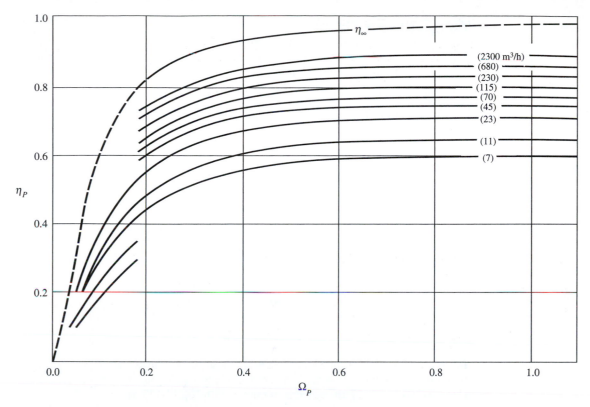

FIGURE 12.15 Maximum efficiency as a function of specific speed and discharge for radial-flow pumps. (PUMP HANDBOOK, 2ND EDITION (DURO) by Karassik. Copyright 1986 by McGraw-Hill Companies, Inc. - Books. Reproduced with permission of McGraw-Hill Companies, Inc. - Books in the format Textbook via Copyright Clearance Center.)

Another measure of cavitation is *suction specific speed*, S. Analogous to the formulation of specific speed of a pump, Eq. 12.3.13, suction specific speed for either a pump or turbine is given by

$$S = \frac{\omega Q^{1/2}}{(gNPSH)^{3/4}} \qquad (12.3.16)$$

Here, two geometrically similar units will have the same S when operating at the same flow coefficient C_Q. Conversely, equal values of S indicate similar cavitation characteristics when the units are operating differently. Design values of S are determined by experiment; when cavitation is not present, Eq. 12.3.16 is no longer valid.

Example 12.6

Select a pump to deliver 30 L/s of water with a pressure rise of 450 kPa. Assume a rotational speed not to exceed 3600 rpm.

Solution

To estimate the specific speed we need the following:

$$\omega = 3600 \times \frac{\pi}{30} = 377 \text{ rad/sec}$$

$$H_P = \frac{\Delta p}{\rho g} = \frac{450 \times 10^3}{1000 \times 9.81} = 45.9 \text{ m}$$

$$Q = 30 \text{ L/s} = 0.03 \text{ m}^3/\text{s}$$

Use Eq. 12.3.13 to find the specific speed to be

$$\Omega_P = \frac{\omega \sqrt{Q}}{(gH_P)^{3/4}}$$

$$= \frac{377 \sqrt{0.03}}{\left(9.81 \times 45.9\right)^{3/4}} = 0.67$$

From Eq. 12.3.14, this would indicate a radial-flow pump. The pump of Fig. 12.12 could be used, even though the specific speed (0.61) would result in a lower efficiency. With $\Omega_P \approx 0.61$ (Fig. 12.12), the speed is estimated with Eq. 12.3.13 as

$$\omega = \frac{\Omega_P (gH_P)^{3/4}}{\sqrt{Q}}$$

$$= \frac{0.61 \times \left(9.81 \times 45.9\right)^{3/4}}{\sqrt{0.03}} = 344 \text{ rad/sec}$$

Hence the required speed is $344 \times 30/\pi = 3285$ rpm, which does not exceed 3600 rpm. The required diameter is determined with use of Fig. 12.12, where at maximum efficiency, $C_Q = 0.0165$. This is substituted into Eq. 12.3.4 to determine the diameter:

$$D = \left(\frac{Q}{C_Q \omega}\right)^{1/3}$$

$$= \left(\frac{0.03}{0.0165 \times 344}\right)^{1/3} = 0.174 \text{ m} = 17.4 \text{ cm}$$

12.4 USE OF TURBOPUMPS IN PIPING SYSTEMS

The appropriate selection of one or more pumps to meet the flow demands of a piping system requires, in addition to a fundamental understanding of turbopumps, a hydraulic analysis of the pumps integrated into the piping system. A single pipe and pump arrangement is analyzed in Section 7.6.7. In Example 7.16 a system demand curve is obtained, and a trial solution to find the discharge and required pump head is shown. The technique is extended to simple piping systems in Section 11.3, wherein it is shown that one can employ either a pump performance curve or assume a constant power input to the pump. Section 11.4 deals with piping networks. In those systems, the pump performance curve can be represented by a polynomial relation (Eq. 11.4.14) and incorporated into the linearized energy equations, or alternatively, a constant pump power requirement can be employed (Eq. 11.4.15).

In this section we look more closely at the manner in which pumps are selected. The procedure involves not only satisfying the required flow and pressure requirements for a piping system, but also making sure that cavitation is avoided and that the most efficient pump is selected.

12.4.1 Matching Pumps to System Demand

Consider a single pipeline that contains a pump to deliver fluid between two reservoirs. The *system demand curve* is defined as

$$H_P = (z_2 - z_1) + \left(f\frac{L}{D} + \Sigma K\right)\frac{Q^2}{2gA^2} \qquad (12.4.1)$$

where atmospheric pressure is assumed at each reservoir, and the upstream and downstream reservoirs are at elevations z_1 and z_2, respectively. Note that z_2 need not be greater than z_1, and that the friction factor f can vary as the discharge (i.e., Reynolds number) varies. The term ΣK represents the minor losses in the pipe, see Section 7.6.4. Equation 12.4.1 is represented as curve (a) in Fig. 12.16. The first term on the right-hand side is the static head and the second term is the head loss due to pipe friction and minor losses. The steepness of the demand curve is

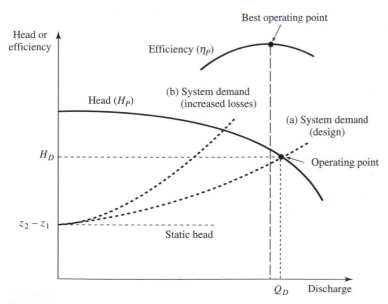

FIGURE 12.16   Pump characteristic curve and system demand curve.

dependent on the sum of the loss coefficients in the system; as the loss coefficients increase, signified by curve (b) in Fig. 12.16, the pumping head required for a given discharge is increased. Piping systems may experience short-term changes in the demand curve such as throttling of valves, and over the long term, aging of pipes may cause the loss coefficients to increase permanently. In either case the system demand curve could change from (a) to (b) as shown.

For a given design discharge, Eq. 12.3.14 allows the selection of a pump based on the specific speed. Once the type of pump is determined, an appropriate size is selected from a manufacturer's characteristic curve; a representative curve is shown in Fig. 12.16. The intersection of the characteristic curve with the desired system demand curve will provide the *design head H_D* and *design discharge Q_D*. It is desirable to have the intersection occur at or close to the point of maximum efficiency of the pump, designated as the *best operating point*.

12.4.2 Pumps in Parallel and in Series

In some instances, pumping installations may have a wide range of head or discharge requirements, so that a single pump may not meet the required range of demands. In these situations, pumps may be staged either in series or in parallel to provide operation in a more efficient manner. In this discussion, it is assumed that the pumps are placed at a single location with short lines connecting the separate units.

Where a large variation in flow demand is required, two or more pumps are placed in a parallel configuration (Fig. 12.17). Pumps are turned on individually to meet the required flow demand; in this way operation at higher efficiency can be attained. It is not necessary to have identical pumps, but individual pumps, when running in parallel, should not be operating in undesirable zones. For parallel pumping the combined characteristic curve is generated by recognizing that the head across each pump is identical, and the total discharge through the pumping system is ΣQ, the sum of the individual discharges through each pump. Note

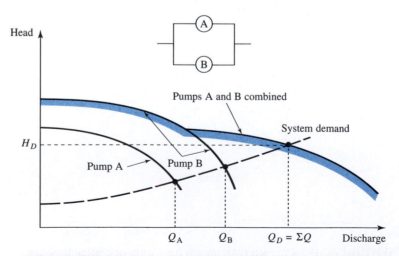

FIGURE 12.17 Characteristic curves for pumps operating in parallel.

Note the existence of three operating points in Fig. 12.17, in which pump A or pump B is used separately, or in which pumps A and B are combined. Other design operating points could be obtained by throttling the flow or by changing the pump speeds. The overall efficiency of pumps in parallel is

$$\eta_P = \frac{\gamma H_D \Sigma Q}{\Sigma \dot{W}_P}$$
(12.4.2)

in which $\Sigma \dot{W}_P$ is the sum of the individual power required by each pump.

For high head demands, pumps placed in series will produce a head rise greater than those of the individual pumps (Fig. 12.18). Since the discharge through each pump is identical, the characteristic curve is found by summing the head across each pump. Note that it is not necessary that the two pumps be identical. In Fig. 12.18 the system demand curve is such that pump A operating alone cannot deliver any liquid because its shutoff head is lower than the static system head. There are two operating points, either with pump B alone or with pumps A and B combined. The overall efficiency is

$$\eta_P = \frac{\gamma (\Sigma H_P) Q_D}{\Sigma \dot{W}_P}$$
(12.4.3)

in which ΣH_P is the sum of the individual heads across each pump.

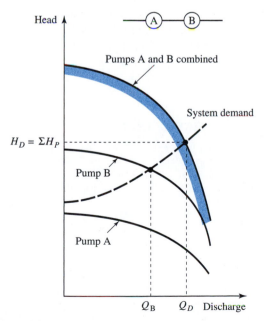

FIGURE 12.18   Characteristic curves for pumps operating in series.

Example 12.7

Water is pumped between two reservoirs in a pipeline with the following characteristics: $D = 300$ mm, $L = 70$ m, $f = 0.025$, $\Sigma K = 2.5$. The radial-flow pump characteristic curve is approximated by the formula

$$H_P = 22.9 + 10.7Q - 111Q^2$$

where H_P is in meters and Q is in m³/s.

Determine the discharge Q_D and pump head H_D for the following situations: (a) $z_2 - z_1 = 15$ m, one pump placed in operation; (b) $z_2 - z_1 = 15$ m, with two identical pumps operating in parallel; and (c) the pump layout, discharge and head for $z_2 - z_1 = 25$ m.

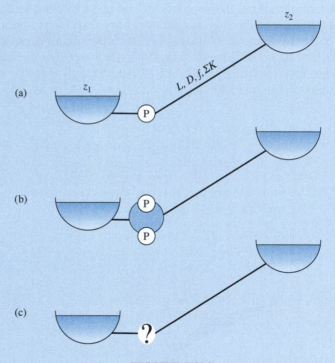

FIGURE E12.7

Solution

(a)   The system demand curve (Eq. 12.4.1) is developed first:

$$H_P = (z_2 - z_1) + \left(f\frac{L}{D} + \Sigma K\right)\frac{Q^2}{2gA^2}$$

$$= 15 + \left(\frac{0.025 \times 70}{0.3} + 2.5\right)\frac{Q^2}{2 \times 9.81(\pi/4 \times 0.3^2)^2}$$

$$= 15 + 85Q^2$$

To find the operating point, equate the pump characteristic curve to the system demand curve,

$$15 + 85Q_D^2 = 22.9 + 10.7Q_D - 111Q_D^2$$

Reduce and solve for Q_D in the manner

$$195Q_D^2 - 10.7Q_D - 7.9 = 0$$

$$\therefore Q_D = \frac{1}{2 \times 195}[10.7 + \sqrt{10.7^2 + 4 \times 195 \times 7.9}] = 0.23 \text{ m}^3/\text{s}$$

Using the system demand curve, H_D is computed as

$$H_D = 15 + 85 \times 0.23^2 = 19.5 \text{ m}$$

(b) For two pumps in parallel, the characteristic curve is

$$H_P = 22.9 + 10.7\left(\frac{Q}{2}\right) - 111\left(\frac{Q}{2}\right)^2$$

$$= 22.9 + 5.35Q - 27.75Q^2$$

Equate this to the system demand curve and solve for Q_D:

$$15 + 85Q_D^2 = 22.9 + 5.35Q_D - 27.75Q_D^2$$

$$112.8Q_D^2 - 5.35Q_D - 7.9 = 0$$

$$\therefore Q_D = \frac{1}{2 \times 112.8}[5.35 + \sqrt{5.35^2 + 4 \times 112.8 \times 7.9}] = 0.29 \text{ m}^3/\text{s}$$

The design head is calculated to be

$$H_D = 15 + 85 \times 0.29^2 = 22.2 \text{ m}$$

(c) Since $z_2 - z_1$ is greater than the single pump shutoff head (i.e., $25 > 22.9$ m), it is necessary to operate with two pumps in series. The combined pump curve is

$$H_D = 2(22.9 + 10.7Q - 111Q^2)$$

$$= 45.8 + 21.4Q - 222Q^2$$

The system demand curve is changed since $z_2 - z_1 = 25$ m. It becomes

$$H_D = 25 + 85Q^2$$

Equating the two relations above and solving for Q_D and H_P results in

$$25 + 85Q_D^2 = 45.8 + 21.4Q_D - 222Q_D^2$$

or

$$307Q_D^2 - 21.4Q_D = 20.8 = 0$$

$$\therefore Q_D = \frac{1}{2 \times 307}[21.4 + \sqrt{21.4^2 + 4 \times 307 \times 20.8}] = 0.30 \text{ m}^3/\text{s}$$

and

$$H_D = 25 + 85 \times 0.30^2 = 32.7 \text{ m}$$

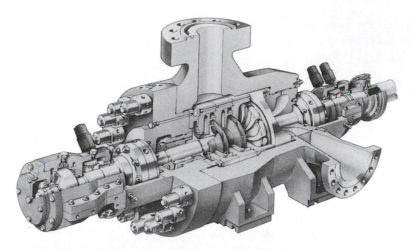

FIGURE 12.19   Four-stage centrifugal pump. (Courtesy of Sulzer Pumps Ltd.)

12.4.3   Multistage Pumps

Instead of placing several pumps in series, multistage pumps are available (Fig. 12.19). Basically, the impellers are all housed in a single casing and the outlet from one impeller stage ejects into the eye of the next. Such pumps can provide extremely high heads. For the pump shown in Fig. 12.19, the pressure head range is 1500 to 3400 m, and the discharge can vary from 4500 m^3/h down to 260 m^3/h. Up to eight stages can be selected and the maximum speed is about 8000 rpm.

12.5 TURBINES

In many parts of the world, where sufficient head and large flow rates are possible, hydroturbines are used to produce electrical power. In contrast to pumps, turbines extract useful energy from the water flowing in a piping system. The moving component of a turbine is called a **runner**, which consists of vanes or buckets that are attached to a rotating shaft. The energy available in the liquid is transferred to the shaft by means of the rotating runner, and the resulting torque transferred by the rotating shaft can drive an electric generator. Hydroturbines vary widely in size and capacity, ranging from microunits generating 5 kW to those in large hydroelectric installations that produce over 400 MW.

There are two types of turbines. The **reaction turbine** utilizes both flow energy and kinetic energy of the liquid; energy conversion takes place in an enclosed space at pressures above atmospheric conditions. Reaction turbines can be subdivided further, according to the available head, as either Francis or propeller type. The **impulse turbine** requires that the flow energy in the liquid be converted into kinetic energy by means of a nozzle before the liquid impacts on the runner; the energy is in the form of a high-velocity jet at or near atmospheric pressure. Turbines can be classified according to turbine specific speed, as shown in Fig. 12.20; the Pelton wheel is a particular type of impulse turbine (see Section 12.5.2).

KEY CONCEPT   *In contrast to pumps, turbines extract useful energy from the water flowing in a piping system.*

Runner: *Moving component of a turbine.*

Reaction turbine: *Turbine using both flow energy and kinetic energy.*

Impulse turbine: *Turbine requiring that the flow energy be converted into kinetic energy through a nozzle before the liquid impacts the runner.*

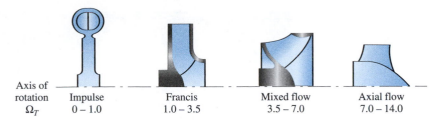

Axis of rotation Ω_T	Impulse 0 – 1.0	Francis 1.0 – 3.5	Mixed flow 3.5 – 7.0	Axial flow 7.0 – 14.0

FIGURE 12.20   Various types of turbine runners.

12.5.1 Reaction Turbines

In reaction turbines, the flow is contained in a *volute* that channels the liquid into the runner (Fig. 12.21). Adjustable *guide vanes* (also called *wicket gates*) are situated upstream of the runner; their function is to control the tangential component of velocity at the runner inlet. As a result, the fluid leaves the guide vane exit and enters the runner with an acquired angular momentum. As the fluid travels through the runner region, its angular momentum is reduced and it imparts a torque on the runner, which in turn drives the shaft to produce power. The flow exits from the runner into a diffuser, called the *draft tube*, which acts to convert the kinetic energy remaining in the liquid into flow energy.

Francis Turbine.   In the Francis turbine, the incoming flow through the guide vanes is radial, with a significant tangential velocity component at the entrance to the runner vanes (Fig. 12.22). As the fluid traverses the runner the velocity takes on an axial component while the tangential component is reduced. As it leaves the runner, the fluid velocity is primarily axial with little or no tangential component. The pressure at the runner exit is below atmospheric.

The theoretical torque delivered to the runner is developed by applying Eq. 12.2.1 to the control volume shown in Fig. 12.22; the same assumptions leading to the pump torque relation, Eq. 12.2.2, apply. The resulting relation is

$$T = \rho Q(r_1 V_{t1} - r_2 V_{t2}) \tag{12.5.1}$$

Multiplying the torque by the angular speed ω gives the power delivered to the shaft:

$$\dot{W}_T = \omega T \tag{12.5.2}$$
$$= \rho Q(u_1 V_1 \cos \alpha_1 - u_2 V_2 \cos \alpha_2)$$

The fluid power input to the turbine is given by

$$\dot{W}_f = \gamma Q H_T \tag{12.5.3}$$

in which H_T is the actual head drop across the turbine. Thus the overall efficiency is given by

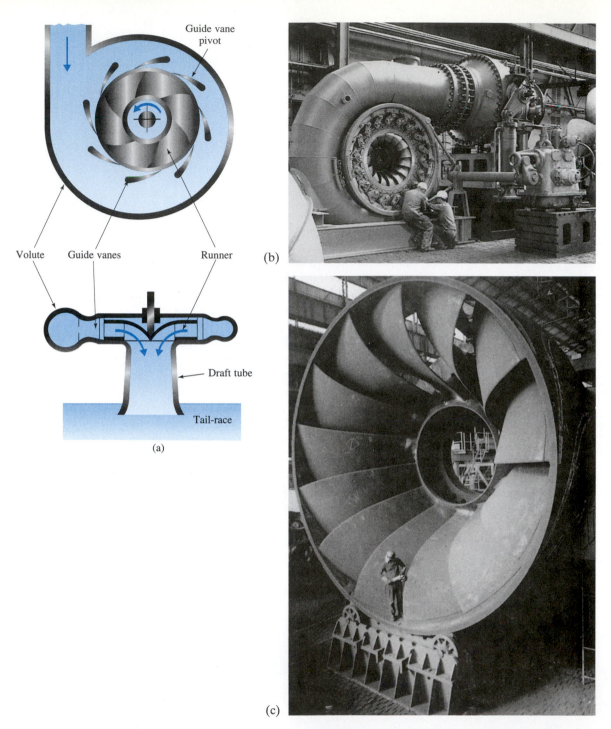

Guide vane pivot

Volute   Guide vanes   Runner   (b)

Draft tube

Tail-race

(a)

(c)

FIGURE 12.21 Reaction turbine (Francis type): (a) schematic; (b) Francis spiral turbine for the Victoria Falls power station, Australia; (c) runner for one of the turbines for the Itaipu power station, Brazil. (Courtesy of Voith Siemens Hydro Power Generation, Inc.)

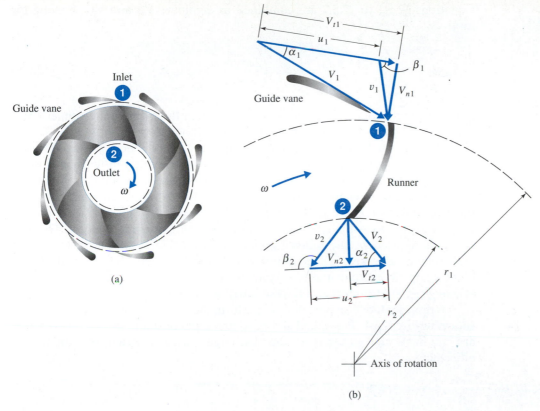

FIGURE 12.22 Idealized Francis turbine runner: (a) runner control volume: (b) velocity diagrams at control surfaces.

$$\eta_T = \frac{\dot{W}_T}{\dot{W}_f} = \frac{\omega T}{\gamma Q H_T} \qquad (12.5.4)$$

The action of the guide vanes can be described by considering the velocity vector diagrams in Fig. 12.22b. Assume perfect guidance of the fluid along the guide vane; then the tangential velocity at the entrance to the runner is

$$V_{t1} = V_{n1} \cot \alpha_1 \qquad (12.5.5)$$

From the velocity vector diagram the tangential velocity is also given by

$$V_{t1} = u_1 + V_{n1} \cot \beta_1 \qquad (12.5.6)$$

The radial velocity component can be expressed in terms of the discharge Q and the width of the runner b_1:

$$V_{n1} = \frac{Q}{2\pi r_1 b_1} \qquad (12.5.7)$$

Equations 12.5.5 to 12.5.7 are combined to eliminate V_{t1} and V_{n1}. Solving for α_1 gives

$$\alpha_1 = \cot^{-1}\left(\frac{2\pi r_1^2 b_1 \omega}{Q} + \cot \beta_1\right) \tag{12.5.8}$$

For a constant angular speed, in order to maintain the appropriate runner entry angle α_1, the guide vane angle is adjusted as Q changes.

Under normal operation, a turbine operates under a nearly constant head H_T so that the performance characteristics are viewed differently from those of a pump. For turbine operation under constant head, the important quantities are the variations of discharge, speed, and efficiency. The interrelationships among the three parameters are shown in the isoefficiency curve of Fig. 12.23. Note the drop in discharge as the speed increases at a given guide vane setting. The efficiency is reduced by the following mechanisms: (1) frictional head losses and draft tube head losses; (2) separation due to mismatch of flow entry angle with blade angle; (3) need to attain a certain turbine speed before useful power output is achieved; and (4) mechanical losses attributed to bearings, seals, and the like.

A representative dimensionless performance curve of a Francis turbine is shown in Fig. 12.24. The speed and head are kept constant, and the guide vanes are automatically adjusted as the discharge varies in order to attain peak efficiency.

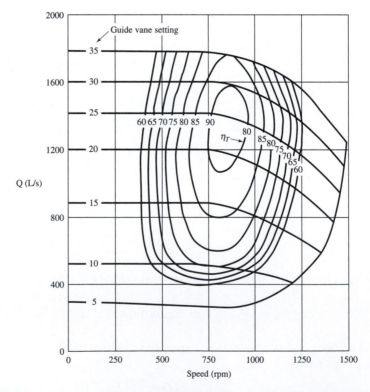

FIGURE 12.23 Isoefficiency curve for a Francis turbine: $D = 500$ mm, $H_T = 50$ m. (Courtesy of Gilbert Gilkes and Gordon, Ltd.)

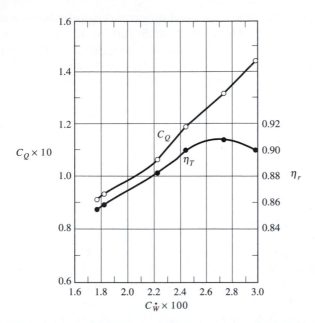

FIGURE 12.24   Performance curve for a prototype Francis turbine: $D = 1000$ mm, $\omega = 37.7$ rad/s, $\Omega_T = 1.063$, $C_H = 0.23$. (Data courtesy of Gilbert Gilkes and Gordon, Ltd.)

Example 12.8

A reaction turbine, whose runner radii are $r_1 = 300$ mm and $r_2 = 150$ mm, operates under the following conditions: $Q = 0.057$ m^3/s, $\omega = 25$ rad/s, $\alpha_1 = 30°$, $V_1 = 6$ m/s, $\alpha_2 = 80°$, and $V_2 = 3$ m/s. Assuming ideal conditions, find the torque applied to the runner, the head on the turbine, and the fluid power. Use $\rho = 1000$ kg/m^3.

Solution
The applied torque is computed using Eq. 12.5.1 to be

$$T = \rho Q(r_1 V_{t1} - r_2 V_{t2})$$
$$= \rho Q(r_1 V_1 \cos \alpha_1 - r_2 V_2 \cos \alpha_2)$$
$$= 1000 \times 0.057(0.3 \times 6 \times \cos 30° - 0.15 \times 3 \times \cos 80°)$$
$$= 84.4 \text{ N} \cdot \text{m}$$

Under ideal conditions, the power delivered to the shaft is the same as the fluid power input to the turbine (i.e., $\eta_T = 1$). Thus

$$\dot{W}_f = \dot{W}_T$$
$$= \omega T$$
$$= 25 \times 84.4 = 2110 \text{ W} \qquad \text{or} \qquad 2.11 \text{ kW}$$

The head on the turbine is found with the use of Eq. 12.5.3 to be

$$H_T = \frac{\dot{W}_f}{\gamma Q}$$
$$= \frac{2110}{9810 \times 0.057} = 3.77 \text{ m}$$

Axial-Flow Turbine.   In an axial-flow turbine the flow is parallel to the axis of rotation (Fig. 12.25). Unlike the Francis turbine, the angular momentum of the liquid remains nearly constant and the tangential component of velocity is reduced across the blade. Both fixed-blade and pivoting-blade turbines are in use; the latter type, termed a Kaplan turbine, permits the blade angle to be adjusted to accommodate changes in head. Axial-flow turbines can be installed either vertically or horizontally. They are well suited for low-head installations.

Cavitation Considerations.   The net positive suction head (*NPSH*) and the cavitation number, defined in Section 12.2.3, are applicable for turbines. In

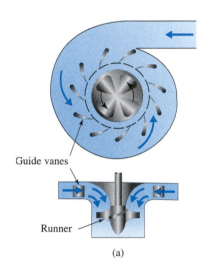

Guide vanes

Runner

(a)

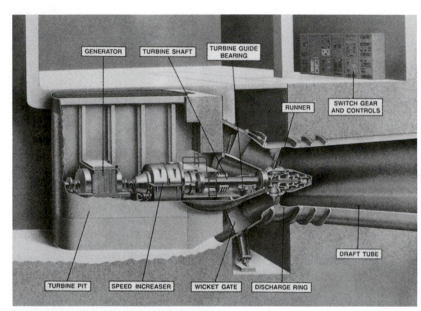

GENERATOR   TURBINE SHAFT   TURBINE GUIDE BEARING

RUNNER

SWITCH GEAR AND CONTROLS

DRAFT TUBE

TURBINE PIT   SPEED INCREASER   WICKET GATE   DISCHARGE RING

(b)

(c)

FIGURE 12.25   Axial-flow turbine (Kaplan type): (a) schematic; (b) side view; (c) turbine blades in the Altenwörth power station, Austria. (Courtesy of Voith Siemens Hydro Power Generation, Inc.)

Eqs. 12.2.21 to 12.2.23 the sign of the loss term becomes positive. The Thoma cavitation number, defined by Eq. 12.2.24 for a pump, is given in the following form for a turbine:

$$\sigma = \frac{(p_{atm} - p_v)/\gamma - \Delta z + h_L}{H_T} \qquad (12.5.9)$$

Typically, location 2 is defined at the outlet of the runner, and location 1 refers to the liquid surface at the draft tube outlet (Fig. 12.26a).

The cavitation number is commonly used to characterize the cavitation behavior of a turbine. Figure 12.26b shows a representative plot of the cavitation number versus turbine efficiency, which is obtained experimentally by the turbine manufacturer. From such a plot operational procedures related to settings of the headwater and tailwater elevations can be established, and in addition, admissible levels of the cavitation number can be determined.

12.5.2 Impulse Turbines

The Pelton wheel (Fig. 12.27) is an impulse turbine that consists of three basic components: one or more stationary inlet nozzles, a runner, and a casing. The

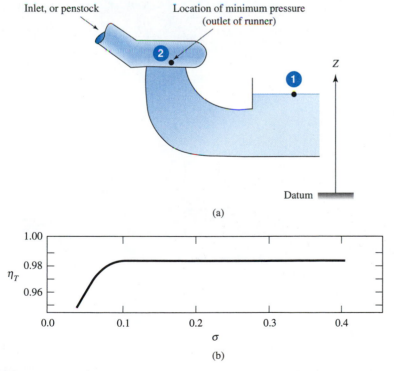

FIGURE 12.26   Cavitation considerations: (a) schematic; (b) representative cavitation number curve. (Courtesy of Voith Siemens Hydro Power Generation, Inc.)

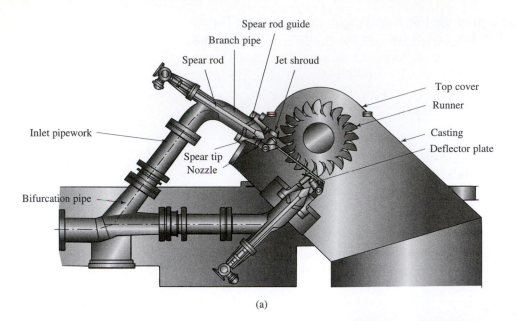

(a)

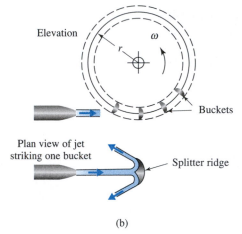

(b)

(c)

FIGURE 12.27 Impulse turbine (Pelton type): (a) typical arrangement of twin-jet machine. (Courtesy of Gilbert Gilkes and Gordon, Ltd.); (b) jet striking a bucket; (c) Pelton runner in the New Colgate power station, USA. (Courtesy Voith Siemens Hydro Power Generation, Inc.)

runner consists of multiple buckets mounted on a rotating wheel. The pressure head upstream of the nozzle is converted into kinetic energy contained in the water jet leaving the nozzle. As the jet impacts the rotating buckets, the kinetic energy is converted into a rotating torque. The buckets are shaped in a manner to divide the flow in half and turn its relative velocity vector in the horizontal plane nearly 180°; the ideal angle of 180° is not attainable since the exiting liquid must stay free of the trailing buckets.

The moment of momentum equation, illustrated in the preceding sections, can be applied to the control volume shown in Fig. 12.28. Neglecting friction, the torque delivered to the wheel by the liquid jet is

$$T = \rho Q r (V_1 - u)(1 - \cos \beta_2) \tag{12.5.10}$$

in which Q is the discharge from all jets, and $u = r\omega$, r being the wheel radius as shown in Fig. 12.27b. The power delivered by the fluid to the turbine runner is

$$\dot{W}_T = \rho Q u (V_1 - u)(1 - \cos \beta_2) \tag{12.5.11}$$

Usually, β_2 varies between 160 and 168°. Differentiation of Eq. 12.5.11 with respect to u and setting it equal to zero shows that maximum power occurs when $u = V_1/2$. The jet velocity can be given in terms of the available head H_T:

$$V_1 = C_v \sqrt{2gH_T} \tag{12.5.12}$$

The *velocity coefficient* C_v accounts for the nozzle losses; typically, $0.92 \leq C_v \leq 0.98$. The efficiency is

$$\eta_T = \frac{\dot{W}_T}{\gamma Q H_T} \tag{12.5.13}$$

Substituting Eqs. 12.5.11 and 12.5.12 into this relation and rearranging produces

$$\eta_T = 2\phi(C_v - \phi)(1 - \cos \beta_2) \tag{12.5.14}$$

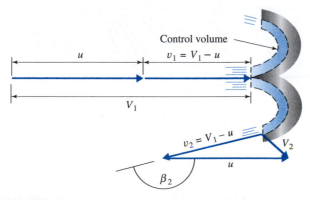

FIGURE 12.28   Velocity vector diagram for Pelton bucket.

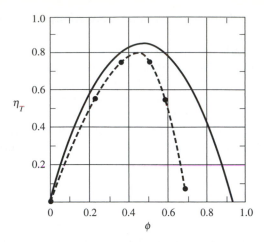

FIGURE 12.29 Speed factor versus efficiency for a laboratory-scale Pelton turbine: solid line, Eq. 12.5.14 ($C_v = 0.94$, $\beta_2 = 168°$); dashed line, experimental data.

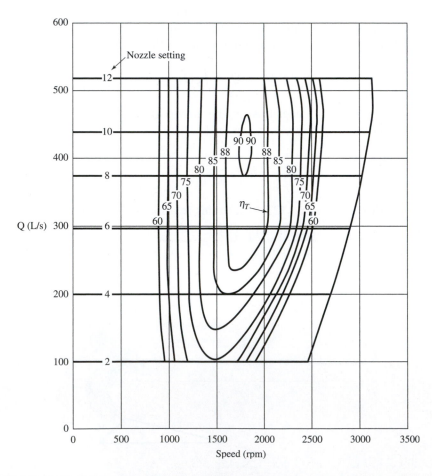

FIGURE 12.30 Isoefficiency curve for a Pelton turbine: $D = 500$ mm, $H_T = 500$ m. (Courtesy of Gilbert Gilkes and Gordon, Ltd.)

in which the *speed factor* ϕ is defined as

$$\phi = \frac{r\omega}{\sqrt{2gH_T}} \qquad\qquad (12.5.15)$$

Maximum efficiency occurs at $\phi = C_v/2$. Figure 12.29 shows Eq. 12.5.14 plotted for $C_v = 0.94$ and $\beta_2 = 168°$. In addition, data are plotted for a laboratory-scale Pelton wheel. The theoretical maximum efficiency is 0.874, while the experimental maximum efficiency is approximately 0.80. The efficiency is reduced because (1) the liquid jet does not strike the bucket with a uniform velocity, (2) losses occur when the jet strikes the bucket and splitter, (3) frictional loss is present due to the rotation of the bucket and splitter, and (4) there is frictional loss at the nozzle.

The interrelationships among Q, ω, and η_T are shown in the isoefficiency curves of Fig. 12.30. For a constant head and nozzle setting, the discharge remains constant, since the nozzle outlet is at atmospheric pressure. A representative dimensionless performance curve for a prototype Pelton turbine is shown in Fig. 12.31.

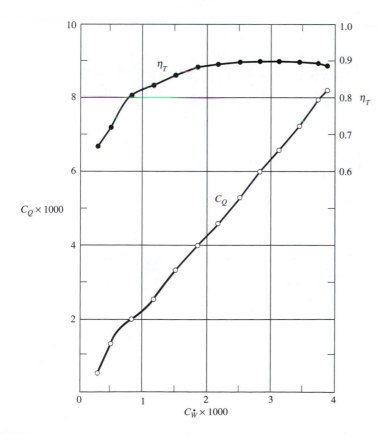

FIGURE 12.31   Performance curve for a prototype Pelton turbine: $D = 1000$ mm, $\omega = 75.4$ rad/s, $\Omega_T = 0.135$, $C_H = 0.52$. (Data courtesy of Gilbert Gilkes and Gordon, Ltd.)

Example 12.9

A Pelton turbine rotates at an angular speed of 400 rpm, developing 67.5 kW under a head of 60 m of water. The inlet pipe diameter at the base of the single nozzle is 200 mm. The operating conditions are $C_v = 0.97$, $\phi = 0.46$, and $\eta_T = 0.83$. Determine (a) the volumetric flow rate, (b) the diameter of the jet, (c) the wheel diameter, and (d) the pressure in the inlet pipe at the nozzle base.

Solution

(a) The discharge is computed from Eq. 12.5.13 to be

$$Q = \frac{\dot{W}_T}{\gamma H_T \eta_T} = \frac{67\,500}{9810 \times 60 \times 0.83} = 0.138 \text{ m}^3/\text{s}$$

(b) From Eq. 12.5.12, the velocity of the jet is

$$V_1 = C_v \sqrt{2gH_T} = 0.97 \sqrt{2 \times 9.81 \times 60} = 33.3 \text{ m/s}$$

The area of the jet is the discharge divided by V_1, or

$$A_1 = \frac{Q}{V_1} = \frac{0.138}{33.3} = 4.14 \times 10^{-3} \text{m}^2$$

Hence the jet diameter D_1 is

$$D_1 = \sqrt{\frac{4}{\pi} A_1} = \sqrt{\frac{4}{\pi} \times 4.14 \times 10^{-3}} = 0.0726 \text{ m} \qquad \text{or} \qquad 73 \text{ mm}$$

(c) Use Eq. 12.5.15 to compute the wheel diameter D to be

$$D = 2r = \frac{2\phi}{\omega} \sqrt{2gH_T}$$

$$= \frac{2 \times 0.46}{400 \times \pi/30} \sqrt{2 \times 9.81 \times 60} = 0.754 \text{ m} \qquad \text{or} \qquad 754 \text{ mm}$$

(d) The area of the inlet pipe is

$$A = \frac{\pi}{4} \times 0.20^2 = 0.0314 \text{ m}^2$$

The piezometric head just upstream of the nozzle is equal to H_T, so that the pressure at that location is

$$p = \gamma \left(H_T - \frac{Q^2}{2gA^2} \right)$$

$$= 9810 \left(60 - \frac{0.138^2}{2 \times 9.81 \times 0.0314^2} \right)$$

$$= 5.79 \times 10^5 \text{ Pa} \qquad \text{or} \qquad \text{approximately 580 kPa}$$

12.5.3   Selection and Operation of Turbines

A preliminary selection of the appropriate type of turbine for a given installation is based on the specific speed. Figure 12.20 shows how the turbine impeller varies with Ω_T. Impulse turbines normally operate most economically at heads above 300 m, but small units can be used for heads as low as 60 m. Heads up to 300 m are possible for Francis units, and propeller turbines are normally used for heads lower than 30 m. Figure 12.32 illustrates the ranges of application for a variety of hydraulic turbines.

So-called minihydro and microhydro installations have recently been subject to renewed interest, not only in developing countries, but in industrially developed countries such as the United States, where the cost of energy has increased and incentive programs have been made available to promote their use.

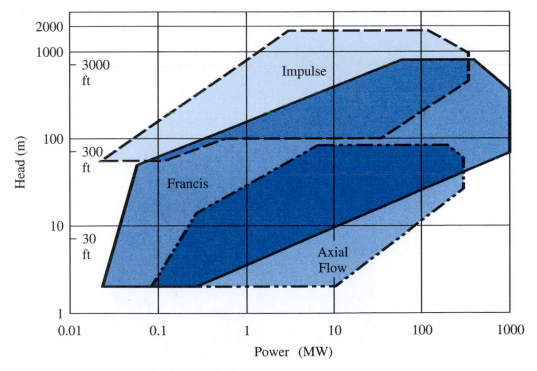

FIGURE 12.32   Application ranges for hydraulic turbines. (Courtesy of Voith Siemens Hydro Power Generation, Inc.)

Microhydro has been classified as units that possess a capacity of less than 100 kW, and *minihydro* refers to system capacity from 100 to 1000 kW (Warnick, 1984). It is becoming common practice for manufacturers to develop a standardized turbine unit. In addition to specialized small Pelton, Francis, and propeller units, the *cross-flow turbine* is employed; basically, it is an impulse turbine that possesses a higher rotational speed than that of other impulse turbines. Another alternative is to use a commercially available pump and operate it in the reverse direction; in the turbine mode, the best efficiency operating point requires a larger head and flow than when the pump operates at best efficiency in the pumping mode.

Reversible pump/turbine units are used in pumped/storage hydropower installations. Water is pumped from a lower reservoir to a higher storage reservoir during periods of low demand for conventional power plants. Subsequently, water is released from the upper reservoir, and the pumps are rotated in reverse to generate power during periods of high electric power demand. Figure 12.33 shows a representative installation consisting of a Francis reversible pump/turbine unit combined with a generator/motor. Such units require special design considerations to operate efficiently in either mode. The problem of converting the flow from the pump mode to the turbine mode, or vice versa, is a very complicated transient problem that is beyond the scope of this book. Water with enormous mass moving through the entire system must be reversed in direction; a special technique must be developed to accomplish this.

FIGURE 12.33   Runner, distributor, and top cover of the pump turbine in the Kühtai power station, Austria. (Courtesy of Voith Siemens Hydro Power Generation, Inc.)

Example 12.10

A discharge of 2100 m³/s and a head of 113 m are available for a proposed pumped-storage hydroelectric scheme. Reversible Francis pump/turbines are to be installed; in the turbine mode of operation, $\Omega_T = 2.19$, the rotational speed is 240 rpm, and the efficiency is 80%. Determine the power produced by each unit and the number of units required.

Solution

The power produced by each unit is found using the definition of specific speed, Eq. 12.3.15. Solving for the power, we have

$$\dot{W}_T = \rho \left[\frac{\Omega_T}{\omega} (gH_T)^{5/4} \right]^2$$

$$= 1000 \left[\frac{2.19}{240 \times \pi/30} (9.81 \times 113)^{5/4} \right]^2 = 3.11 \times 10^8 \text{ W}$$

From Eq. 12.5.13, the discharge in each unit is

$$Q = \frac{\dot{W}_T}{\gamma H_T \eta_T}$$

$$= \frac{3.11 \times 10^8}{9800 \times 113 \times 0.8} = 351 \text{ m}^3/\text{s}$$

The required number of units is equal to the available discharge divided by the discharge in each unit, or 2100/351 = 5.98. Hence, six units are required.

12.6 SUMMARY

This chapter focuses on pumps or turbines that supply or extract energy by means of rotating impellers or vanes. First we emphasized the radial-flow, or centrifugal, pump and made use of the moment-of-momentum principle to derive an idealized head-discharge relation. We then contrasted that with the actual head-discharge performance curve, one that is obtained experimentally and is required for use in pump analysis and design. Axial-flow and mixed-flow pumps were introduced and the differences between these and the radial-flow pump were discussed. The significance of pump efficiency and net positive suction head (relating to cavitation) diagrams were shown for the proper design and selection of pumps.

Dimensional analysis and similitude were applied to both pumps and turbines; the three significant dimensionless numbers that were developed are the power coefficient, flow coefficient, and head coefficient. These coefficients were combined to yield pump or turbine efficiency, or to produce the specific speed for a pump or turbine. It was shown that a family of pumps or turbines could be represented in dimensionless form by a single curve. The grouping of pumps — either singly, in parallel, or in series — to meet appropriate head and discharge requirements for piping was illustrated.

Finally, a brief introduction to turbines was provided, by relating fundamental theory along with the application of reaction, axial flow, and impulse turbines to various hydraulic design situations.

PROBLEMS

Elementary Theory

12.1 Determine the torque, power, and head input or output for each turbomachine shown in Fig. P12.1. Is a pump or turbine implied? The following data are in common: outer radius 300 mm, inner radius 150 mm, $Q = 0.057$ m³/s, $\rho = 1000$ kg/m³, and $\omega = 25$ rad/s.

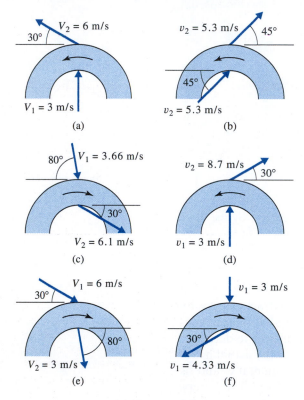

FIGURE P12.1

12.2 A centrifugal water pump rotates at 800 rpm. The impeller has uniform blade widths $b_1 = 50$ mm, $b_2 = 25$ mm, and radii $r_1 = 40$ mm, $r_2 = 125$ mm. The blade angles are $\beta_1 = 45°$, $\beta_2 = 30°$. Assuming no angular momentum of fluid at the blade entrance, determine the ideal flow rate, pressure head rise across the impeller, and the theoretical torque and power requirements.

12.3 The suction and discharge pipes of a pump are connected to a differential mercury manometer. The diameter of the suction pipe is 200 mm and the diameter of the discharge pipe is 150 mm. The

specific gravity of the pumped liquid is $S = 0.81$, and the manometer reading is $H = 600$ mm. Determine the pressure head rise across the pump if the discharge is $Q = 115$ L/s. The centerlines of the suction and discharge pipes are at the same elevation.

12.4 A centrifugal pump with the dimensions $r_2 = 6$ cm, $b_2 = 10$ mm, $\beta_2 = 60°$, is pumping kerosene ($S = 0.80$) at a rate of 8 L/s and rotating at 2000 rpm. For "best design" entry conditions, determine the theoretical pressure head rise and fluid power.

12.5 An axial-flow pump has a stator blade positioned upstream of the impeller, and it provides an angle $\alpha_1 = 60°$ to the flow. The radius of the impeller tip is 285 mm and the hub radius is 135 mm. Determine the required average blade angle at the exit of the impeller if the pump is to deliver 0.57 m³/s of water with a theoretical head rise of 2.85 m. The rotational speed is 1500 rpm.

12.6 For the system shown in Fig. P12.6, water flows through the pump at a rate of 50 L/s. The allowable *NPSH* provided by the manufacturer at that flow is 3 m. Determine the maximum height Δz above the water surface that the pump can be located to operate without cavitating. Include all losses in the suction pipe.

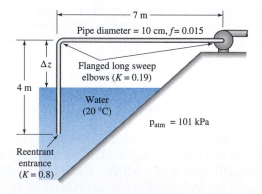

FIGURE P12.6

12.7 During a test on a water pump, no cavitation is detected when the gage pressure in the pipe at the pump inlet is −68 kPa, and the water temperature is 25°C. The inlet pipe diameter is 10 cm, the head across the pump is 35 m, and the discharge is 50 L/s.

(a) Compute the *NPSH* and the cavitation number if the atmospheric pressure is 101 kPa.

(b) If the pump is to produce the same head and discharge at a location where the atmospheric pressure is 80 kPa, what is the necessary change in elevation of the pump relative to the inlet reservoir to avoid cavitation?

Dimensional Analysis and Similitude

12.8 Consider the 371-mm-diameter pump curve of Fig. 12.9. If the pump is run at 1200 rpm to deliver 0.8 m³/s of water, find the resulting head rise and fluid power.

12.9 In Problem 12.8, how high above the suction reservoir can the pump running at 1200 rpm be located if the water temperature is 50°C and the atmospheric pressure is 101 kPa?

12.10 Prepare dimensionless performance curves for the 205-mm-diameter radial flow pump of Fig. 12.6. Compare these with the curves shown in Fig. 12.12. Can you explain why there may be differences between the two sets of curves?

12.11 If the 220-mm-diameter pump whose characteristics are shown Fig. 12.6 is operating at 3300 rpm, what will be the head and the *NPSH* when the discharge is 200 m³/h?

12.12 A pump is needed to transport 150 L/s of oil (S = 0.86) with an increase in head across the pump of 22 m when running at 1800 rpm. What type of pump is best suited for this operation?

12.13 A pump is required to deliver 0.17 m³/s of water with a head increase of 104 m when operating at maximum efficiency at 2000 rpm. Determine the type of pump best suited for this duty.

12.14 Refer to Fig. P12.14, the dimensionless performance curves for an axial-flow pump. It is desired to deliver water at a rate of 1.25 m³/s and a speed of 750 rpm. Determine:

(a) The available head, diameter, power, and *NPSH* requirements.

(b) The actual head-discharge and power-discharge curves.

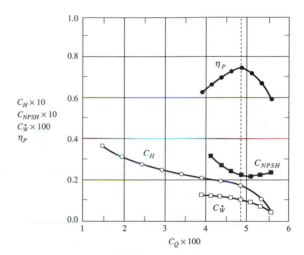

FIGURE P12.14

12.15 A pump running at 400 rpm discharges water at a rate of 85 L/s. The total head rise across the pump is 7.6 m and the efficiency is 0.7. A second pump, whose linear dimensions are two-thirds of the former, runs at 400 rpm under dynamically similar conditions; it is to be placed on a space satellite where simulated gravity produces a gravitational field which is 50% that of Earth's. What are the discharge, head rise, and power requirements for the second pump?

12.16 A pump is designed to operate under optimum conditions at 600 rpm when delivering water at 22.7 m³/min against a head of 19.5 m. What type of pump is recommended?

12.17 A pump operating under the conditions stated in Problem 12.16 has a maximum efficiency of 0.70. If the same pump is now required to deliver water at a head of 30.5 m at maximum efficiency, determine the rotational speed, the discharge, and the required power.

12.18 A manufacturer desires to double both the discharge and head on a geometrically similar pump. Determine the change in rotational speed and diameter under the new conditions.

12.19 A pump is required to move water at 200 L/s against a head of 60 m. The approximate rotational speed is 1400 rpm. Using the appropriate dimensionless performance curve in the text,

determine the actual speed and size of the pump to operate at peak efficiency.

12.20 Select the type, size, and speed of pump to deliver a liquid ($\gamma = 8830$ N/m^3) at 0.66 m^3/s with fluid power requirements of 200 kW. (*Hint:* Assume that the pump speed can vary between 1000 and 2000 rpm.)

Use of Turbopumps

12.21 For the system shown in Fig. P12.21, recommend the appropriate machine to pump 1 m^3/s of water if the impeller speed is 600 rpm.

expect the efficiency of the pump to be the value shown on the curve?

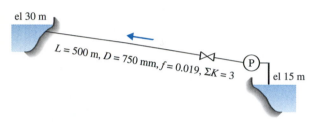

FIGURE P12.21

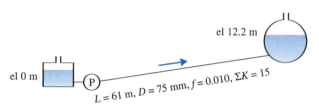

FIGURE P12.23a

12.22 The following performance data were obtained from a test on a 216-mm double-entry centrifugal pump moving water at a constant speed of 1350 rev/min:

Q (m^3/min)	H (m)	η_P
0	12.2	0
0.454	12.8	0.26
0.905	13.1	0.46
1.36	13.4	0.59
1.81	13.4	0.70
2.27	13.1	0.78
2.72	12.2	0.78
3.80	11.0	0.74

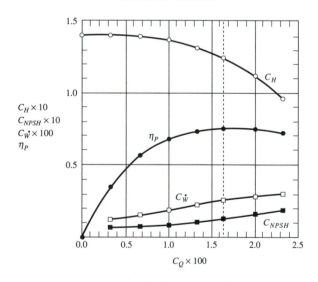

FIGURE P12.23b

Plot H versus Q, η_P versus Q, and $\dot{W}_P$ versus Q. If the pump operates in a system whose demand curve is given by $5 + Q^2$, find the discharge, head, and power required. In the demand curve, Q is given in cubic meters per minute.

12.23 An oil ($S = 0.85$) is to be pumped through a pipe as shown in Fig. P12.23a. Using the characteristic curves of Fig. P12.23b, determine the required impeller diameter, speed, and input power to deliver the oil at a rate of 14.2 L/s. Would you

12.24 With reference to the pump data in Problem 12.22, if the pump is run at 1200 rpm, find the discharge, head, and required power. Is the operating efficiency under these conditions the same as the pump running at 1350 rpm?

12.25 For the pump data presented in Problem 12.22, plot the dimensionless pump curves C_H versus C_Q, $C_{\dot{W}}$ versus C_Q, and η_P versus C_Q. What is the specific speed of the pump? (*Note:* For double-entry pumps, use one-half the discharge when computing the specific speed.)

12.26 A closed-loop flow facility (Fig. P12.26) is to be used to study water flow in a hydraulics laboratory. It is constructed of 300-mm-diameter hydraulically smooth pipe. The loss coefficient at each of the four bends is 0.1, and the total length of the loop is 14 m. The design velocity is $V = 3$ m/s. It is proposed that an axial flow pump running at about 300 rpm be used. Verify that an axial-flow pump is the correct type to employ.

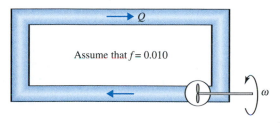

Assume that $f = 0.010$

FIGURE P12.26

12.27 In Problem 12.26, assume that the only axial pump available is one from the family shown in Fig. P12.27. If the diameter of the impeller were restricted to the pipe diameter, what would be the rotational speed and resulting head rise across the pump?

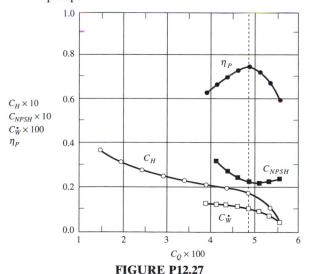

FIGURE P12.27

12.28 The 240-mm-diameter pump represented in Fig. 12.6 is used to move water in a piping system whose demand curve is $62 + 270Q^2$, where Q is in cubic meters per second. Find the discharge,

required input power, and the *NPSH* requirement for:
(a) One pump. **(b)** Two pumps in parallel.

12.29 It is desired to pump 600 m³/h of liquid ammonia ($\rho = 607$ kg/m³) at a speed of 2500 rpm and a pressure rise of 1000 kPa.
(a) Determine the type of pump best suited for this duty.
(b) Using the appropriate dimensionless pump characteristic curve, find the diameter and number of stages required to meet the pumping requirements.

12.30 Crude oil ($S = 0.86$) is to be pumped through 5 kilometer of 45-cm-diameter cast iron pipe. The elevation rise from the upstream to the downstream end is 200 m. If an available pump is the 240-mm-diameter radial-flow machine of Fig. 12.6, how many pumps in series are required to provide the most efficient operation? Find the required power.

12.31 A water supply piping network designed for a community receives water from a 25-cm-diameter well (Fig. P12.31a). The pump is situated 50 m below the 25-cm-diameter primary water main that delivers the flow to the network, and due to the available ground water, the hydraulic grade line on the suction side of the pump is 22 m above the pump. The design discharge is 30 L/s, and a pressure of 550 kPa is required in the primary water main. Making use of Fig. P12.31b, determine the size, speed, and number of stages of a pump to satisfy the design requirements, and the required horsepower. Assume that the pump impeller diameter is 20 cm and that the friction factor in the well delivery pipe is constant, $f = 0.02$.

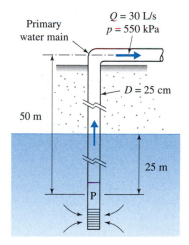

FIGURE P12.31a

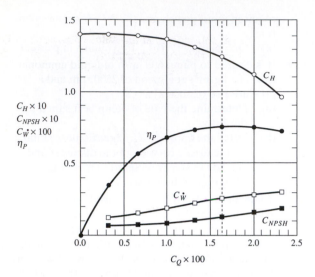

FIGURE P12.31b

12.32 A model pump (Fig. P12.32) delivers 80°C water at a speed of 2400 rpm. It begins to cavitate when the inlet pressure $p_i = 83$ kPa absolute and inlet velocity $V_i = 6$ m/s.
   (a) Compute the *NPSH* of the model pump. Neglect losses and elevation changes between the inlet section where p_i is recorded and the cavitation region in the pump.
   (b) What is the required *NPSH* of a prototype pump which is four times larger and runs at 1000 rpm?

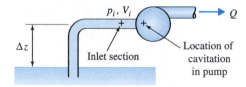

FIGURE P12.32

12.33 It is required to pump water at 1.25 m³/s from a lower to an upper reservoir. The water surface in the lower reservoir is at an elevation of 20 m, and the upper reservoir elevation is 23 m. The length of the 750-mm-diameter pipe is 60 m. The friction factor can be assumed constant, $f = 0.02$, and the entrance and exit loss coefficients are 0.5 and 1.0, respectively.

(a) What type of pump is best suited for this system if the impeller speed is to be 600 rpm?

(b) Assuming a reasonable efficiency from an appropriate pump curve, estimate the power in kilowatts required to operate the pump.

12.34 Carbon tetrachloride at 20°C is pumped from a tank opened to the atmosphere through 200 m of 50-mm-diameter smooth horizontal pipe and discharges into the atmosphere at a velocity of 3 m/s. The elevation of liquid in the tank is 3 m above the pipe. A pump is situated at the pipe inlet immediately downstream of the reservoir.

(a) Determine the speed and size of a pump to meet the system demand.

(b) What is the available net positive suction head at the pump inlet? Neglect losses in the short suction pipe connecting the pump to the tank. Assume that the atmospheric pressure is 101 kPa.

(c) Sketch the hydraulic grade line for the system.

12.35 A wind tunnel with a test section 1 m × 2 m is powered by an axial fan 1.5 m in diameter. The velocity in the test section is 20 m/s and the rotational speed of the fan is 2500 rpm. Assuming negligible losses in the tunnel and a performance curve for the fan defined by Fig. P12.35:
   (a) Calculate the power required to operate the fan.
   (b) What is the pressure rise across the fan?

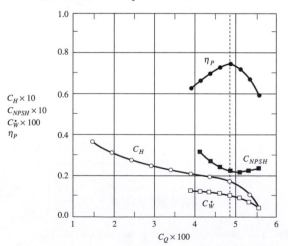

FIGURE P12.35

Turbines

12.36 A Francis turbine has the following dimensions: $r_1 = 4.5$ m, $r_2 = 2.5$ m, $b_1 = b_2 = 0.85$ m, $\beta_1 = 75°$, $\beta_2 = 100°$. The angular speed is 120 rpm, and the discharge is 150 m³/s. For no separation at the entrance to the runner, determine the guide vane angle, theoretical torque, head, and power.

12.37 A model Francis turbine at 1/5 full scale develops 3 kW at 360 rpm under a head of 1.8 m. Find the speed and power of the full-size turbine when operating under a head of 5.8 m. Assume that both units are operating at maximum efficiency.

12.38 A model is to be built for studying the performance of a turbine having a runner diameter of 1 m, maximum output of 2200 kW under a head of 50 m and a speed of 240 rpm. Determine the diameter of the model runner and its speed if the corresponding model power is 9 kW and the head is 8 m.

12.39 A hydroelectric site has available $H_T = 80$ m and $Q = 3$ m³/s. It is proposed to use a Francis turbine unit with the operating characteristics at optimum efficiency shown in Fig. P12.39. Determine the required speed and diameter of the machine.

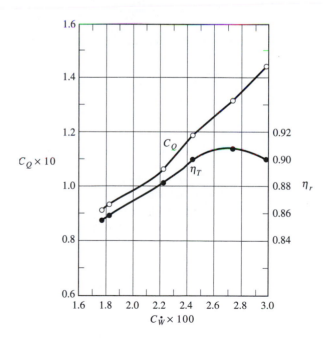

FIGURE P12.39

12.40 Determine the power output, the type of turbine, and the approximate speed for the installation shown in Fig. P12.40. Neglect all minor losses except those existing at the valve.

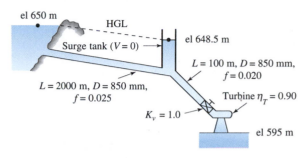

FIGURE P12.40

12.41 A water turbine is required to operate at 420 rpm under a net head of 3 m, a discharge of 0.312 m³/s, and with an efficiency of 0.9. Tests on a 1/6-scale model running at 2000 rpm are to be carried out using water. For the model turbine, determine the head drop, flow rate, output power, and expected efficiency.

12.42 A Pelton wheel develops 4.5 MW under a head of 120 m at a speed of 200 rpm. The wheel diameter is eight times the jet diameter. Use the experimental data of Fig. P12.42 at maximum efficiency to determine the required flow, wheel diameter, diameter of each jet, the number of jets required, and the specific speed.

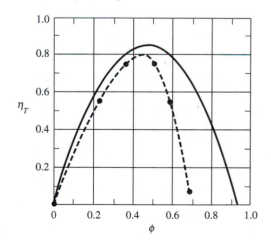

FIGURE P12.42

12.43 A reaction turbine installation typically includes a draft tube (Fig. P12.43), whose function is to convert kinetic energy of the fluid exiting the runner into flow energy. Consider a turbine operating at $Q = 85$ m³/s and $H_T = 31.8$ m. The radius of the draft tube at the outlet of the runner is 2.5 m and the diameter at the end of the tube is 5.0 m. Losses in the draft tube can be neglected.

(a) For water at 20°C, what is the pressure at the runner outlet if in Fig. P12.43, $z_2 - z_1 = 2.5$ m?

(b) What is the permissible $z_2 - z_1$ if the allowable cavitation number is $\sigma = 0.14$?

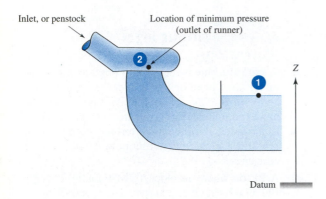

FIGURE P12.43

12.44 The Ludington pumped-storage system (Fig. P12.44) on the eastern shore of Lake Michigan consists of six penstocks delivering a combined discharge of 2100 m³/s in the power-producing mode of operation. Each turbine generates 427,300 hp at an efficiency of 0.85. What diameter penstock is required for the system to operate under design conditions? Classify the type of turbine employed.

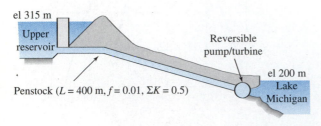

FIGURE P12.44

12.45 A Francis turbine of the same family as that represented in Fig. P12.45 is running at a speed of 480 rpm under a head of 9.5 m. Determine the runner diameter, discharge, and developed power if the turbine is operating at maximum efficiency.

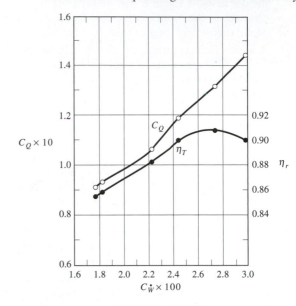

FIGURE P12.45

12.46 The available head from reservoir level to the four nozzles of a twin-runner Pelton wheel is 305 m. The length of the supply pipe, or penstock, is 3 km with a friction factor $f = 0.02$ and $\Sigma K = 2.0$. The turbine develops 10.4 MW with an efficiency of 0.85.

(a) If the head across the turbine is 95% of the available head, what is the diameter of the penstock?

(b) If $C_v = 0.98$ for each nozzle, calculate the diameter of each jet.

12.47 In a projected low-head hydroelectric scheme, 282 m³/s of water is available under a head of 3.7 m. It is proposed to use Francis turbines with specific speed 2.42, for which the rotational speed is 50 rpm. Determine the number of units required and the power to be developed by each machine. Assume an efficiency of 0.9.

12.48 Repeat Problem 12.47 if propeller turbines are substituted for the Francis units. Assume a specific speed of 4.15 and an efficiency of 0.9.

12.49 A landowner has constructed an elevated reservoir as shown in Fig. P12.49a. She estimates that 1200 L/min is available as continuous flow and desires to install a small Francis turbine to generate electricity for her own use.
  (a) What fluid power is expected to be delivered to the turbine?
  (b) If a turbine similar to the one whose characteristics are shown in Fig. P12.49b is to be installed, what are the speed and size required for this installation?
  (c) Using an appropriate empirical equation, estimate the efficiency for this installation. What power is expected to be delivered to the generator?

12.50 A proposed design for a hydroelectric project is based on a discharge of 0.25 m³/s through the penstock and turbine (Fig. P12.50). The friction factor can be assumed constant, $f = 0.015$, and the minor losses are negligible.
  (a) Determine the power in kilowatts that can be expected from the facility, assuming that the turbine efficiency is 0.85.
  (b) Show that the type of machine to be installed is a Francis turbine if the desired rotational speed is 1200 rpm.
  (c) Determine the size, actual speed, and actual power output of the selected turbine.

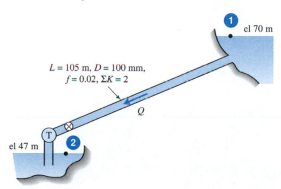

FIGURE P12.49a

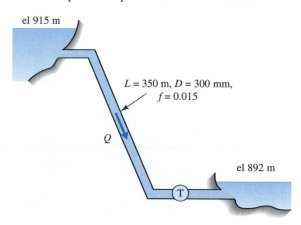

FIGURE P12.50

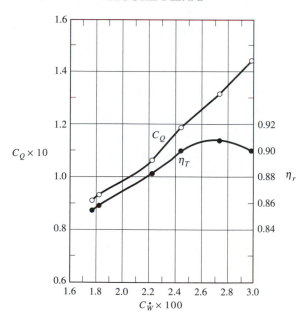

FIGURE P12.49b

Scale model of a Formula One race car in a wind tunnel. Measurements are made of flow patterns velocity distributions, and pressures on the surface of the vehicle. Such tests are employed to reduce the drag, resulting in increased speed of the prototype vehicle. (All American Racers archives)

13

Measurements in Fluid Mechanics

Outline

Chapter Objectives

The objectives of this chapter are to:

▲ Describe of how various flow parameters are measured; these include measurements taken at a local flow region—for example, pressure or velocity—as well as integrated measurements such as discharge or pressure averaged over a cross-section

▲ Present techniques and equipment employed to visualize flow fields

▲ Introduce the concepts of uncertainty analysis and error propagation resulting from collecting laboratory and field data, and demonstrate the use of regression analysis in nonlinear flow measurement

13.1 INTRODUCTION

In the engineering laboratory and in many industrial situations, the need to measure fluid properties and various flow parameters, such as pressure, velocity, and discharge, are exceedingly important. In many instances these needs are obvious. Examples include the flow rate in a pipe or irrigation channel; the contaminant or sediment load in a river; the peak pressures on the surface of a high-rise building or the flow patterns around that building; the drag on an automobile or truck traveling at high speeds; the velocity field about a commercial aircraft; the wind velocity profile above a hilly terrain; and the size and distribution of ocean or lake waves.

Fluid mechanics measurements are not exclusive to the engineer. Medical diagnosticians are interested in monitoring the functions of the cardiovascular and pulmonary systems in the human body. The movement of fluids in the agricultural, petroleum, gas, chemical, beverage, and water supply and wastewater industries require large investments annually; the uncertainties present in the measurement of these flows can have a significant impact on material and monetary considerations.

Many devices have been developed to measure flow parameters; obviously, each was designed to serve a specific purpose. Before undertaking the task of measurement, it is important to define clearly the need for measuring a particular parameter. Knowledge of the underlying fluid mechanics and the physical principles involved are essential to selecting an appropriate measuring instrument and conducting a successful measurement. In addition, we must recognize that there are differences in various types of measurements. Consider, for example, discharge readings resulting from a measurement of a velocity profile integrated over a cross section; the fluctuating velocity can be time-averaged at a specific location to produce a mean value. On the other hand, instantaneous measurements of velocity in a very small sampling volume may be required to observe the transient nature of an unsteady flow. The degree of sophistication in recording a parameter can range from a simple visual reading of a manometer or dial to high-speed digital sampling of voltages representing a number of variables.

The purpose of this chapter is to provide the reader with a basic introduction to the concepts and techniques applied by engineers who measure flow parameters either in the laboratory or in an industrial environment. Included is a general overview of experimental methods commonly used in fluid mechanics teaching and research laboratories; the treatment is not exhaustive, but references are provided for additional information. In the following two sections, methods and instrumentation employed to measure pressure, velocity, and discharge are presented. Next, a discussion of flow visualization is given, and the final section is devoted to data acquisition and methods of analyzing data.

13.2 MEASUREMENT OF LOCAL FLOW PARAMETERS

A local flow measurement means that a quantity is measured over a relatively small sample volume of the fluid. Usually, the volume is sufficiently small so that we can say that the measurement represents the magnitude of the quantity at a point in the flow field. Two significant local flow quantities are pressure and velocity; others are temperature, density, and viscosity. Only the first two are discussed in this section; the last three are not considered in this book.

13.2.1   Dynamic Response and Averaging

Flow measurements may be classified according to whether the flow is steady or unsteady. If the magnitude of a physical quantity remains constant with time, this value is referred to as the steady-state value. Conversely, if the quantity changes with time, the measurement is transient, or unsteady. Transient measurement demands a more highly specialized measuring apparatus. Most measuring instruments require a certain amount of time to respond to the physical quantity sensed. In transient measurements this response time should be much less than the time for a significant change in the physical quantity to occur. In steady-state measurements the only concern is to take a reading after the measuring system has time to respond to the flow conditions. In many situations, for example, turbulent flow, it is desired to obtain a time-average measurement of a quantity, even though the local flow conditions are unsteady; examples are Reynolds stress and turbulence intensity. In this situation, a time-averaging mechanism must be included in the measurement. In addition, an instrument must be sufficiently small to measure the spatial variation of the measured parameter, such as the wavelength of a pressure disturbance. Generally, the issue of spatial and temporal resolution must be resolved by selecting instrumentation compatible with the flow scale and with the scale of any flow disturbance; see Goldstein (1996) for further explanation.

13.2.2   Pressure

Fluid pressures are measured in many different ways. The type of instrument utilized depends on the levels of precision and detail required for the particular application. Virtually all pressure measurements are based either on the manometer principle or on the concept of pressure deforming a solid material such as a crystal, membrane, tube, or plate, and then converting that deformation to an electrical signal or mechanical readout. Pressure measurements are in either the static or dynamic mode. Time-dependent pressures result from unsteadiness in the flow and from pressure disturbances; the pressure disturbances are a result of either hydrodynamic or acoustic perturbations. Often, static pressure measurements are based on time-averaged mean values, since unsteadiness is always present when the flow is turbulent. Several representative pressure gages and transducers are described below.

Manometer.   The manometer was analyzed in Section 2.4.3. It is a simple, inexpensive instrument for measuring static pressure, and it has no moving mechanical parts. High accuracy can be attained by using an inclined tube manometer (Fig. 13.1a) or a micromanometer (Fig. 2.6). If the pressure tap is properly machined normal to the interior wall in a duct or pipe with no burrs, and the diameter is sufficiently small (usually about 1 mm diameter), the static pressure is measured accurately (Fig. 13.1b). Often the taps are placed circumferentially around the duct in the form of a piezometer ring to average out any type of flow irregularities.

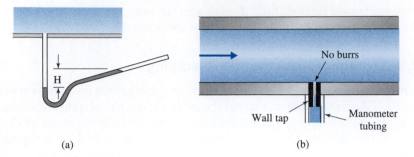

(a)   (b)

FIGURE 13.1 Manometer used to measure pressure: (a) inclined tube manometer; (b) piezometer opening.

Bourdon Gage.   An inexpensive pressure gage for reading relatively large static pressures is the Bourdon tube (Fig. 13.2). It is a curved, flattened tube, which will expand outward when subjected to internal pressure. A mechanical linkage attached to a dial gage will provide a direct reading of the pressure. When properly designed, these gages will provide good accuracy; however, they should not be used in situations where large pressure pulsations are present, since damage to the tube may result.

Pressure Transducer.   Pressure transducers are electromechanical devices that are designed to measure either steady or unsteady hydrodynamic or acoustic pressures. They consist of flexible membranes or piezoelectric elements that respond to pressure changes (Fig. 13.3). All of them require accompanying electronic circuitry, which adds to their costs. Since their output is an electric signal, they can readily be interfaced with data acquisition systems for automatic storage and retrieval of data.

The condenser microphone (Fig. 13.3a) is commonly utilized in aeroacoustics. Basically, it is a capacitor whose capacitance varies as the membrane is deflected due to applied pressure. These transducers are quite delicate, since they measure very small pressures (100 μbar) and hence are not suitable for use in liquids. The piezoelectric transducer, commonly used with liquids, has a crystalline material (e.g., quartz) that produces an electric field when deformed (Fig. 13.3b). These rugged units can be made very small (diameters down to 2.5 mm

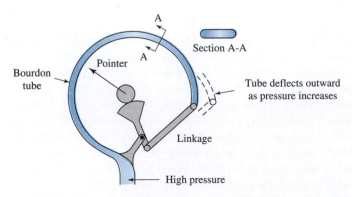

FIGURE 13.2   Bourdon tube pressure gage.

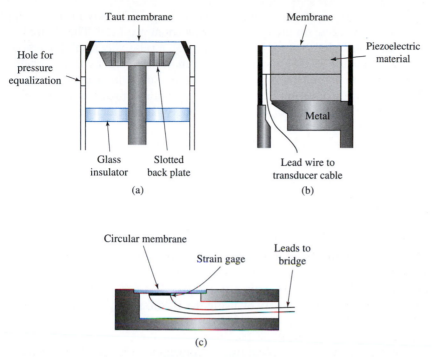

FIGURE 13.3   Pressure transducers: (a) condenser microphone; (b) piezoelectric pressure transducer; (c) strain-gage transducer.

are possible) and can be mounted flush with the wall of a pipe. They respond very rapidly and can measure a wide range of dynamic pressures (full vacuum to 20 MPa). A strain gage transducer is another type of transducer employed with liquids; it makes use of a strain gage attached to the membrane that translates its deflection into an electrical signal (Fig. 13.3c). It is subject to DC drift and sensitivity changes; however, it is rugged, reliable, and insensitive to vibration.

13.2.3   Velocity

The fluid velocity is of primary interest in a fluid flow. By measuring the velocity we can calculate the flow rate, and perhaps even piece together an image of the streamline pattern, identifying regions of separated flow, regions of stagnated flow, and other flow characteristics. There are three categories of velocity measurements: local (at a point), spatially averaged, and velocity field measurements. The discussion in this section pertains to the first two; the third is covered in Section 13.4.

Local velocities are actually measured over a small region of flow. The required precision varies, depending on the desired application. For example, time-averaged velocities measured at various points in a flow cross section may be recorded so that the velocity distribution can be integrated to provide the discharge, or flow rate. On the other hand, it may be necessary to determine the history of the time-dependent turbulent velocity components at a particular location in a region of the flow.

Instruments for velocity measurements vary considerably in their complexity and cost, depending on the type of measurement required. The desire to measure unsteady turbulent velocity components on a relatively small local scale has popularized the use of the thermal anemometer as well as the laser-doppler velocimeter. Less complicated instruments that usually measure velocity over a large spatial region are, for example, the pitot-static probe and the propeller anemometer; they are more suitable for measuring velocities that either are steady or are slowly varying with time.

Particle Velocity Measurement. Neutrally buoyant particles can be seeded in the flow to provide a visual image of the flow field; two of these, hydrogen bubbles in water and particles in air, are discussed in Section 13.4. Other tracer particles used in water are neutrally buoyant plastic beads, immiscible dyed droplets, and floats placed on a liquid surface or at a desired depth. It is important to ascertain if the particle is actually tracking the fluid motion. Very small particles that are natural impurities in the flow can be observed with appropriate optical equipment to measure velocities in many flows, especially microscopic flow fields.

Pitot-Static Probe. The pitot-static probe is analyzed in Section 3.4. It measures the local mean velocity by using the Bernoulli equation. Assuming frictionless, steady flow the velocity u is given by

$$u = \sqrt{\frac{2}{\rho}(p_T - p)} \tag{13.2.1}$$

in which p_T is the total or stagnation pressure and p is the static pressure. Equation 13.2.1 is valid only if the probe does not greatly disturb the flow, and if it is aligned with the flow direction so that the velocity u is parallel to the pitot probe. A particular design, the so-called *Prandtl tube*, is shown in Fig. 13.4. It has the static pressure holes positioned along the horizontal tube so that the reduced pressure resulting from flow past the nose is balanced by the increased pressure due to the vertical stem. Typically, the Reynolds number based on the pitot-tube diameter should be in the neighborhood of 1000, so that viscous effects are not

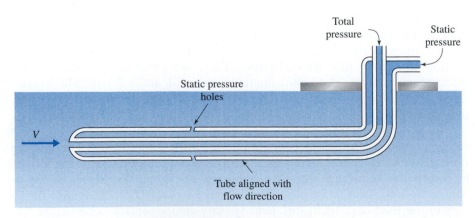

FIGURE 13.4 Pitot-static probe.

significant. If the flow is not parallel to the probe head, the measurement error is about 1% at a yaw angle of 5°, and if the yaw angle is greater than 5° the measurement error may be substantial. When the probe is placed close to a wall, the streamlines are deflected by the interaction of the probe with the wall; errors occur when the distance between the probe axis and the wall is less than two tube diameters. In a turbulent flow, the actual pressure is less than the sensed value; generally, the results are reliable if the fluctuation intensity is less than 10%.

Other probes of a similar design are the *pitot* or *impact tube*, *Preston tube*, *static tube*, and *yaw meter*. The pitot tube measures the total pressure, the static tube the static pressure, and the yaw meter, the direction of flow. The Preston tube is a special form of pitot tube that has been used for the measurement of wall shear stress in boundary layers over smooth walls; basically, it is a hypodermic needle that senses the mean velocity adjacent to the wall. See Bean (1971) for details related to calibration and design.

Cup or Propeller Anemometer. Cup or propeller anemometers are employed to measure the velocity of either gases or liquids. They are relatively large, and as a result, the measurement is averaged spatially over a relatively large area. These anemometers are used for many applications where extreme precision is not required. One type of cup anemometer, the *current meter* (Fig. 13.5a), measures the velocity in water; similar devices can measure air velocity. Cup anemometers always rotate in one sense, and thus cannot provide the flow direction. On the other hand, the *propeller anemometer* will reverse rotation

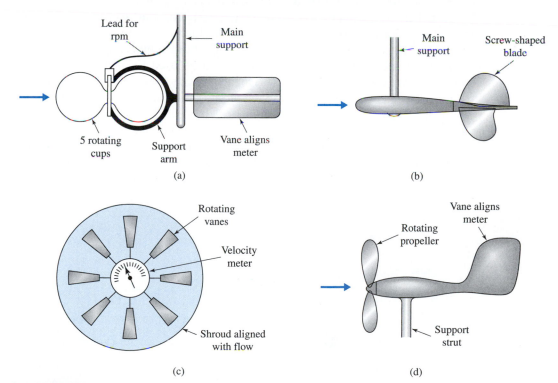

FIGURE 13.5 Anemometers: (a) current meter; (b) propeller anemometer; (c) ducted vane anemometer; (d) wind vane anemometer.

when the flow reverses (Fig. 13.5b); however, in contrast to cup anemometers, they must be aligned with the flow direction. A ducted propeller anemometer for measuring air velocities is shown in Fig. 13.5c. Unducted propellers are also employed; an example is the wind vane (Fig. 13.5d). Propellers are also placed inside pipes and used as flow meters. For all of these devices, counters or electro-mechanical readouts attached to sensors provide the rotational frequency, which is correlated to the velocity by means of a calibration test.

Thermal Anemometer. The *thermal* (hot wire) *anemometer* is employed to measure velocity in air. It consists of a small wire, typically made of tungsten, platinum, or platinum-iridium, which is insulated and mounted on supports (Fig. 13.6). The wire sensor is typically 2 mm in length and 5 μm in diameter, although smaller wires (0.2 mm long, 1 μm in diameter) have been used. When the wire is heated, the anemometer senses the changes in heat transfer as the flow speed varies; a representative wire temperature is 250°C.

The most common circuitry is one that operates in the constant-temperature mode (Fig. 13.6); earlier versions employed a constant-current arrangement. As the velocity past the probe is increased, the hot wire tends to cool, thereby attempting to lower the temperature and the resistance across the wire. The circuitry then increases the current to adjust the resistance and the probe temperature to their initial values, a process that appears to be instantaneous. Either the voltage of the amplifier or the sensor current is correlated with the velocity past the probe. The output is nonlinear with respect to velocity, and the sensitivity decreases as the velocity increases; often the signal is linearized electronically. The sensitivity as a percent of reading stays nearly constant so that a very wide range of velocities can be measured. When the fluctuation intensity is less than 20%, if the flow is incompressible, and if the fluid temperature and density remain constant, the hot-wire anemometer can be used to measure a velocity component. We can then calculate the mean velocity, fluctuation intensity, and turbulence spectrum. The single-wire anemometer senses only the normal component of velocity; if the wire is properly oriented, both the mean component and the fluctuations in the mean flow direction can be measured. Two-component measurements are made using a probe with wires crossed in an "×" fashion, to provide the correlation $\overline{u'v'}$ needed in turbulence measurements (u' and v' are the small perturbations). Three components may be measured either by rotating the ×-probe or adding a third wire.

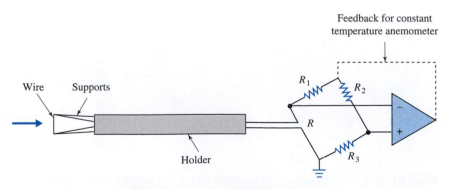

FIGURE 13.6 Thermal anemometer.

In liquids and gases that contain solid particles, the wire probe is replaced by a more rugged cylindrical film sensor. It operates on the same principle as the hot-wire probe, but it generates turbulence in its wake because of its larger size. This self-generated turbulence can limit the probe's capabilities for measuring actual fluctuation intensities in the flow field. In addition to measuring velocities, thermal anemometers have been adapted to monitor pressure and discharge.

Laser-Doppler Velocimeter. A device that can be used advantageously when it is desired not to have the probe immersed in the flow is the *laser-doppler velocimeter* (LDV). The most common LDV is the dual-beam type, shown in Fig. 13.7. The frequency of the scattered light is measured by the photodetector, which converts the receiving light to an electrical signal. A simplified way to understand the principle is as follows. At the intersection of the two beams a fringe pattern is formed with fringe spacing Δx. Light scattered from particles passing through the fringe pattern is modulated with a frequency f, which is directly proportional to the velocity component u normal to the fringes:

$$u = \frac{\Delta x}{\Delta t} = \frac{\lambda f}{2 \sin(\theta/2)} \qquad (13.2.2)$$

in which λ is the wavelength of the laser beam and θ is the beam angle. Detection of flow reversal is possible by using a moving fringe pattern. The ellipsoidal measurement volume is quite small, being on the order of 0.1 mm in diameter and 1.0 mm in length. A complete analysis of the phenomenon requires a detailed description of wave optics, which is beyond the scope of this book.

Velocities in both liquids and gases can be measured with an LDV. Liquids normally contain sufficient impurities which act as natural seeding agents, but gases usually are artificially seeded with extremely small liquid droplets or solid particles. Typically, an LDV can measure velocities ranging from 2 mm/s up to the supersonic range. The frequency response extends to 30 kHz, which enables

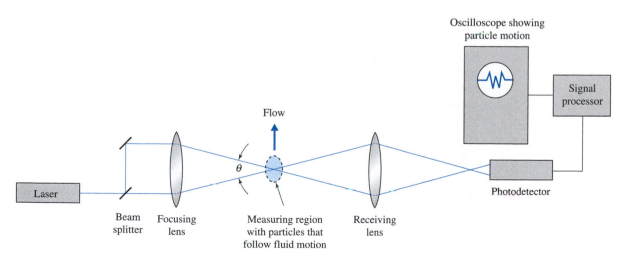

FIGURE 13.7 Dual-beam laser-doppler velocimeter.

turbulent velocity components to be measured. In contrast to the thermal anemometer, the LDV requires no calibration and measures flow reversal with relative ease. The initial cost of the LDV is high, and extensive electronic instrumentation is required to process the output signal: a counter-type signal processor is most commonly used. The LDV has been employed to measure supersonic flow, flow in pumps, natural free convection, steam and gas turbine flows, combusting flows, blood flows, and open-channel flows.

13.3 FLOW RATE MEASUREMENT

The measurement of flow rate — alternately termed discharge or volumetric flow — is one of the most common measurements made in flowing fluids. Numerous devices have been invented or adapted for the purpose of flow metering, ranging widely in sophistication, size, and accuracy. Basically, instruments for flow rate measurements can be divided into those that employ direct or "quantity" means of measurement, and those that are indirect, that is, the so-called "rate" meters. Quantity meters either weigh or measure a volume of fluid over a known time increment; examples are liquid weigh tanks, reciprocating pistons, and for gases, bellows and the liquid-sealed drum. These direct devices usually are relatively large and possess poor frequency response characteristics. However, they do provide high precision and accuracy, and as a result, are used most often as "primary" standards for the calibration of the indirect metering devices.

Indirect or rate meters consist of two components: the primary part, which is in contact with the fluid, and the secondary part, which converts the reaction of the primary part to a measurable quantity. They can be classified according to a characteristic operating principle: velocity–area measurement, pressure drop correlations, hydrodynamic drag, and the like. Rate meters are relatively low in cost, take up little space, and hence are commonly found in industrial and research laboratories. Several of the more common types are discussed below.

13.3.1 Velocity–Area Method

A number of the velocity measuring devices discussed in Section 13.2.3 may be inserted in a cross section of flow and used to measure the flow rate. Thus a thermal anemometer, a laser-doppler velocimeter, a pitot-static tube, or an anemometer can be inserted in a pipe or duct of known area and calibrated to measure the flow rate. In a manner that does not require calibration and for steady flow, a velocity distribution can be measured by systematically moving a velocity probe throughout the flow cross section, or by using a so-called "rake" where local velocities are measured simultaneously. Subsequently, a numerical integration of the velocity profile over the cross-sectional area will yield the discharge. This technique has been used in both closed conduit and open-channel flows.

13.3.2 Differential Pressure Meters

Differential pressure meters are widely used in industrial applications and laboratories because of their simplicity, reliability, ruggedness, and low cost. Three commonly used types are discussed in this section: the orifice meter, venturi meter, and flow nozzle. Their operation is based on the principle of an obstruction to flow being present in a duct or pipe, and consequently, a pressure differential will exist across the obstruction. This pressure drop can be correlated to the discharge by means of a calibration, and subsequently, the pressure-discharge curve can be used to determine the discharge by reading the differential pressure. In this section we deal only with the discharge of incompressible fluids in circular pipes.

The fundamental discharge relation for the differential pressure meter can be described in the following manner. Figure 13.8 represents a thin-plate orifice meter, which can be considered as a representative differential pressure meter. Consider steady flow to occur in a circular duct, encounter the restrictive orifice with area A_0, and issue as a downstream jet. Downstream of the restriction, the streamlines converge to form a minimal flow area A_c, termed the vena contracta. Pressure taps are located at two positions: upstream of the restriction in the undisturbed flow region (location 1) and downstream at some location in the vicinity of the vena contracta (location 2). Assuming an ideal, frictionless, incompressible fluid, the Bernoulli equation applied along the center streamline from the upstream location to the vena contracta is

$$\frac{V_1^2}{2g} + \frac{p_1}{\gamma} + z_1 = \frac{V_c^2}{2g} + \frac{p_c}{\gamma} + z_c \tag{13.3.1}$$

Similarly, the continuity equation is

$$V_1 A_1 = V_c A_c \tag{13.3.2}$$

Combining Eqs. 13.3.1 and 13.3.2, and solving for V_c yields

$$V_c = \sqrt{\frac{2g(h_1 - h_c)}{1 - (A_c/A_1)^2}} \tag{13.3.3}$$

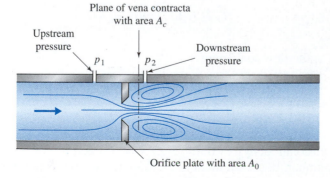

FIGURE 13.8   Flow through an orifice meter.

in which

$$h_1 = \frac{p_1}{\gamma} + z_1 \qquad h_c = \frac{p_c}{\gamma} + z_c \qquad\qquad (13.3.4)$$

The ideal discharge Q_i is equal to the area multiplied by the average velocity at the vena contracta:

$$Q_i = A_c V_c$$

$$= \frac{A_c}{\sqrt{1 - (A_c/A_1)^2}} \sqrt{2g(h_1 - h_c)} \qquad\qquad (13.3.5)$$

The actual discharge differs from the ideal for two primary reasons. Because of real fluid flow, friction causes the velocity at the centerline to be greater than the average velocity at each cross section. Second, the piezometric head h_c, evaluated at the vena contracta in the relation, is substituted with h_2, the known reading at the downstream pressure tap. Also, since the area of the vena contracta is unknown, it is convenient in Eq. 13.3.5 to replace A_c by $C_c A_0$, where C_c is the contraction coefficient. These anomalies are accounted for by introducing a discharge coefficient C_d, which is the product of the contraction coefficient and a velocity coefficient, so that the actual discharge Q is given by the relation

$$Q = \frac{C_d A_0}{\sqrt{1 - (C_c A_0/A_1)^2}} \sqrt{2g(h_1 - h_2)} \qquad\qquad (13.3.6)$$

For a circular cross section, which is typical of most differential pressure meters, it is convenient to introduce the diameter ratio

$$\beta = \sqrt{\frac{A_0}{A_1}} = \frac{D_0}{D} \qquad\qquad (13.3.7)$$

where D is the pipe diameter. A convenient way to express Eq. 13.3.6 is

$$Q = K A_0 \sqrt{2g(h_1 - h_2)} \qquad\qquad (13.3.8)$$

in which K is the flow coefficient

$$K = \frac{C_d}{\sqrt{1 - C_c^2 \beta^4}} \qquad\qquad (13.3.9)$$

A dimensional analysis would reveal that C_d and K are dependent on the Reynolds number. It is convenient to evaluate the Reynolds number either at the approach region or at the obstruction.

Orifice Meter.   A thin plate orifice meter (Fig. 13.9) is typically manufactured in the range $0.2 \leq \beta \leq 0.8$. In the figure, two means of locating the pressure taps are shown: (1) flange taps, positioned 25 mm upstream and downstream of the orifice plate, and (2) taps placed one diameter upstream and one-half diameter downstream of the plate. The second arrangement is preferred, since it is capable of sensing a larger differential pressure, and it conforms to geometric similarity laws. A third arrangement, not shown in the figure, has the pressure taps located in the pipe wall immediately upstream and downstream of the orifice; taps placed at this location have been termed corner taps.

Figure 13.10 shows experimentally determined values of the flow coefficient K for orifices as a function of β and Re_0. These data were obtained using corner taps; however, data taken with flange taps or with $D : D/2$ taps would be indistinguishable when plotted on Fig. 13.10. If greater precision is desired, numerical data for the different taps are provided in Bean (1971). Notice that for a given β, K becomes nearly constant at high Reynolds numbers, but as the Reynolds number becomes lower, K first increases to a maximum and then decreases. Maximum values of K occur at Reynolds numbers between 100 and 1000, depending on the value of β; here K is dominated by the reduced area of the vena contracta.

If the discharge is known, the Reynolds number at the orifice is known. Hence, with Fig. 13.10 one can read K directly, and subsequently determine $(h_1 - h_2)$ from Eq. 13.3.8. However, one would more likely use the figure in conjunction with Eq. 13.3.8 to determine the flow rate, given that $(h_1 - h_2)$ has been read from an attached manometer or pressure transducer. In that situation, K is not known a priori, since it depends on Re_0. One can initially estimate K based on an assumed Re_0 (usually, assume Re_0 to be large), and subsequently by trial and error improve on that estimate by successive substitution into Eq. 13.3.8.

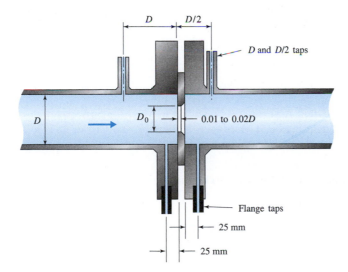

FIGURE 13.9   Details of a thin-plate orifice meter. (FLUID MECHANICS MEASUREMENTS by G. E. Mattingly. Copyright 1996 by Taylor & Francis Group LLC - Books. Reproduced with permission of Taylor & Francis Group LLC - Books in the format Textbook via Copyright Clearance Center.)

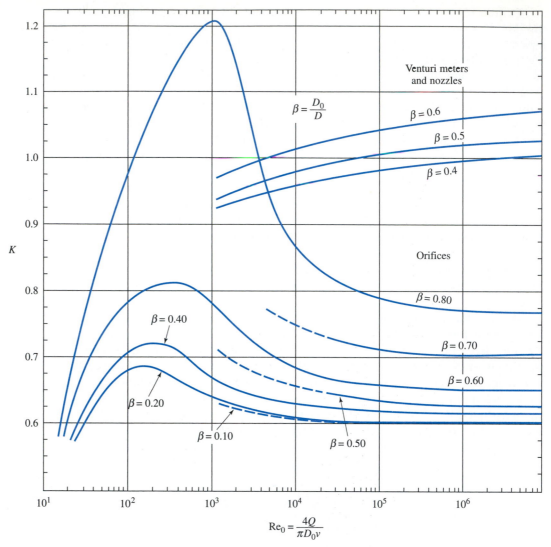

FIGURE 13.10 Flow coefficient K versus the Reynolds number for orifices, nozzles, and venturi meters. (Adapted from Engineering Fluid Mechanics, Robertson and Crowe, © 1990 John Wiley & Sons, Inc., New York. Reproduced with permission of John Wiley & Sons, Inc.)

However, if repeated readings are to be taken, it is more convenient to determine a calibration formula of the form

$$Q = C(h_1 - h_2)^m \tag{13.3.10}$$

in which C and m are constants determined by a "best fit" criterion. See Section 13.5.3 for details of one well-known criterion, the method of least squares.

The orifice has become the most widely used differential meter for measuring liquids. This is understandable, since it is relatively inexpensive and can easily be placed in an existing pipe. One disadvantage is that it produces a large head loss that cannot be recovered downstream of the orifice.

Venturi Meter.   The venturi meter has a shape that attempts to mimic the flow patterns through a streamlined obstruction in a pipe. The classical, or Herschel, type of venturi meter is rarely used today, since its dimensions are rather large, making it cumbersome to install and expensive to fabricate. It is made up of a 21° conical inlet contraction, followed by a short cylindrical throat, leading to a 7° or 8° conical exit expansion. The discharge coefficient is nearly unity. By contrast, the contemporary venturi tube, shown in Fig. 13.11, consists of a standard flow nozzle inlet section [ISA 1932 standard (Bean, 1971)] and a conical exit expansion no greater than 30°. Its recommended range of Reynolds numbers is limited from 1.5×10^5 to 2×10^6.

Equation 13.3.8 is valid for the venturi as well as for the orifice; representative values of the flow coefficient K are shown in Fig. 13.10. Because of streamlining of the flow passage, the head loss in the venturi meter is much less than in the orifice. The vena contracta is not present, and as a result the discharge coefficient C_d remains close to unity.

Flow Nozzle.   The flow nozzle is illustrated in Fig. 13.12. It consists of a standardized shape with pressure taps typically located one diameter upstream of the inlet and one-half diameter downstream. There are two standard shapes, either the long radius or the short radius. Due to the lack of an expansion section downstream of the nozzle, the total head loss is similar to that of an orifice, except that the vena contracta is nearly eliminated and the discharge coefficient is nearly unity. Similarly, when the pressure taps are located at $D:D/2$, the flow coefficient K varies with the Reynolds number in a manner nearly identical to that for the venturi meter, as shown in Fig. 13.10. The flow nozzle has an advantage over the orifice plate in that it is less susceptible to erosion and wear, and relative to the venturi meter, it is less expensive and simpler to install.

Elbow Meter.   A relatively simple meter can be fabricated by locating pressure taps at the outside and inside of a pipe elbow (Fig. 13.13). By applying the linear momentum equation to a control volume encompassing the fluid in the elbow (or Euler's equation normal to the streamlines), one can derive the flow equation

$$Q = KA \sqrt{\frac{R}{D} \frac{\Delta p}{\rho}} \qquad (13.3.11)$$

Pipe with diameter D Throat with diameter D_0 15 to 30°   Pipe with diameter D

FIGURE 13.11   Venturi meter.

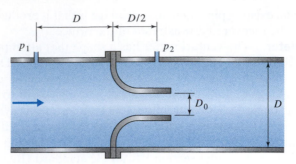

FIGURE 13.12 Flow nozzle.

in which R is the radius of curvature of the elbow and Δp is the pressure difference generated by centrifugal force across the bend. The flow coefficient K can be most accurately determined by in-place calibration of the meter. However, a relation for K has been reported (Bean, 1971) for 90° elbows with pressure taps located in a radial plane 45° from the inlet:

$$K = 1 - \frac{6.5}{\sqrt{Re}} \tag{13.3.12}$$

which is valid when $10^4 \leq Re = VD/\nu \leq 10^6$, and $R/D \geq 1.5$. The elbow meter costs much less than any of the differential pressure meters, and it usually does not add to the overall head loss of the piping system since elbows are often present in any case.

13.3.3   Other Types of Flow Meters

Turbine Meter.   The turbine meter consists of a propeller mounted inside a duct that is spun by the flowing fluid. The propeller's angular speed is correlated to the discharge; the angular rotation is measured in a manner identical to the anemometer discussed in Section 13.2.3. Accuracies up to ± 0.25% of the flow rate are attainable over a relatively large discharge range (Goldstein, 1996). Viscous effects become a limiting factor at the low end, whereas at the high end, accuracy is limited by the interaction between the blade tips and the electronic sensors. Each type of turbine meter must be calibrated individually. They are commonly employed to monitor flow rates in fuel supply lines.

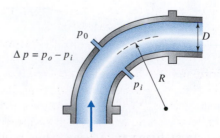

FIGURE 13.13   Elbow meter.

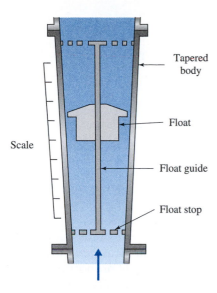

Scale

Tapered
body

Float

Float guide

Float stop

FIGURE 13.14   Rotameter.

Rotameter.   The rotameter consists of a tapered tube in which the flow is directed vertically upward (Fig. 13.14). A float moves upward or downward in response to the flow rate until a position is reached where the drag force on the float is in equilibrium with its submerged weight. Calibration consists of correlating the vertical elevation of the float with the discharge. With appropriate design, the float position can be made linearly proportional to the discharge, or if desired, another relation, such as position logarithmically proportional to discharge, can be formulated. The head loss depends on the friction loss of the tube plus the loss across the floating element. The rotameter does not provide accuracy as good as the differential pressure meters; typically it is in the range of 5% full scale.

Target Meter.   Figure 13.15 shows a target meter, which consists of a disk suspended on a support strut immersed in the flow. The strut is connected to a lever arm, or alternatively, has a strain gage bonded to its surface. Fluid drag on the disk will cause the strut to flex slightly, and the recorded drag force can be related to the discharge. The assembly must be sufficiently stiff so that the drag can be measured without movement or rotation of the disk; otherwise, the drag characteristics would be altered. One advantage of this device is that it is possible to record flow reversal by measuring the sense of the drag force; hence it can be used as a bidirectional flow meter. Drag meters are fairly rugged, and can be used to measure discharges in sediment-laden fluids.

Electromagnetic Flowmeter.   This is a nonintrusive device that consists of an arrangement of magnetic coils and electrodes encircling the pipe (Fig. 13.16). The coils are insulated from the fluid and the electrodes make contact with the fluid. Sufficient electrolytes are dissolved in the fluid so that it becomes capable of conducting an electric current. When it passes through the magnetic field generated by the coils, the liquid will create an induced voltage proportional to the flow. Electromagnetic flowmeters have been used for many applications, including blood flow and seawater measurements. Commercial models are

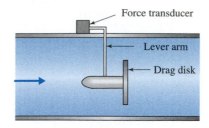

FIGURE 13.15   Target meter.

available in a wide range of diameters; they are costly and can be used only with liquids.

Acoustic Flowmeter.   The ultrasonic, or acoustic, flowmeter is based on one of two principles. The first one uses ultrasonic transmitters/receivers beamed across a flow path (Fig. 13.17). The measured differences in the travel times (or inversely, the frequencies) are directly proportional to the average velocity, and consequently, the discharge. The second type is based on the Doppler effect, in that acoustic waves are transmitted into the flow field and subsequently scattered by seeded particles or contaminants. The Doppler shift recorded between the transmitter and the receiver is then related to the discharge. Like the electromagnetic flowmeter, acoustic flowmeters are nonintrusive, and have been used in both piping and free-surface flow systems.

Vortex-Shedding Meter.   This flow-metering device consists of a single strut or series of struts placed chord-like inside a pipe normal to the flow direction. If the Reynolds number is above a threshold value, vortices appear in the near-wake regions behind the struts (see Section 8.3.2). The periodic vortex shedding that occurs behind the struts can be correlated to the Reynolds number as shown in Fig. 8.9. Different devices are employed to detect the vortex-shedding frequency. Examples include a pressure transducer mounted on the downstream surface of the strut, a thermal sensor that detects changes in heat flux on the strut surface, or a strain gage that senses oscillation of the strut. Vortex-shedding meters are can be specified to operate in flows that range up to two orders of magnitude.

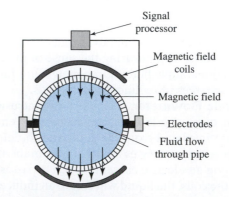

FIGURE 13.16   Electromagnetic flowmeter.

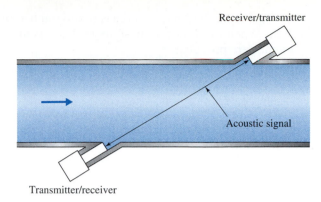

FIGURE 13.17   Acoustic flowmeter.

Coriolis-Acceleration Flowmeter.   These are meters that are based on the Coriolis acceleration principle (see Eq. 3.2.15). Typically they are used to measure liquid flows in piping. A U-shaped tube filled with the liquid to be measured is vibrated normal to the plane of the U and with the tips of the U stationary. The liquid flowing in the tube produces a torsional motion about the axis of symmetry of the tube. The amplitude of the motion is proportional to the flow rate, and its frequency can be correlated to the density of the liquid. The meter provides both flow rate and density measurements in a noninvasive manner.

Open-Channel Flowmeters.   In Section 10.4.3, flow measurement in open-channel systems is presented. Discharge relations for the broad-crested weir and sharp-crested weir are developed, and additional methods of flow measurement are discussed. The reader is referred to that section for detailed information.

13.4 FLOW VISUALIZATION

Transparent flow fields have been visualized in many different ways. This is well illustrated by the hundreds of photographs presented in the publication *An Album of Fluid Motion* (Van Dyke, 1982). The book shows flows visualized with smoke, dye, bubbles, particles, shadowgraphs, schlieren images, interferometry, and other techniques. Usually, a method is selected that best shows the flow features of interest. Examples include particles used to visualize pathlines in liquid flows around submerged objects, dye released to study the mixing process in a stream, smoke released at the upwind end of a wind tunnel to study the development of a boundary layer, and airflow patterns above a solid surface visualized by coating that surface with a viscous liquid (streaky lines are formed which coincide with the streamlines near the surface). Several of these techniques are discussed in this section.

Pathlines, streaklines, and streamlines have been defined in Chapter 3. The purpose of defining them was to provide assistance in describing the motion of a flow field. A pathline is physically generated by following the motion of an individual particle such as a bubble or small neutrally buoyant sphere over a period of time; this could be achieved by time-lapse photography or video recording. On the other hand, if a trace from smoke, a train of extremely small bubbles, or dye continuously emanating from a stationary source is photographed or recorded, a streakline is observed. When the generation of a streakline is interrupted periodically, time streaklines are produced.

13.4.1 Tracers

Tracers are fluid additives that permit the observation of flow patterns. An effective tracer does not alter the flow pattern, but is transported with the flow and is readily observable. It is important that tracers are not affected by gravitational or centrifugal forces resulting from density differences. Furthermore, their size should be at least one order of magnitude smaller than the length scale of the flow field. Several of them are discussed below.

Hydrogen Bubbles. A very thin metallic wire can be placed in water to serve as the cathode of a direct-current circuit, with any suitable conductive material acting as an anode. When a voltage is supplied to the circuit, hydrogen bubbles will be released at the cathode and oxygen bubbles at the anode. The primary reaction is the electrolysis of water in the manner $2H_2O \rightarrow 2H_2 + O_2$. Usually, the hydrogen bubbles are used as the tracer since they are smaller than the oxygen bubbles, and more of them are formed. The bubbles are transported by the flow field away from the cathode wire as a continuous sheet. If the voltage is pulsed, discrete streaklines are formed; Fig. 13.18 shows such a pattern. Standard water supplies usually contain sufficient electrolytes to sustain a current, although addition of an electrolyte such as sodium sulfate greatly increases bubble generation. Use of extremely fine wire (0.025 to 0.05 mm in diameter) with adequate velocities (Reynolds numbers based on the wire diameter should be less than 20) may create bubbles sufficiently small to negate effects of buoyancy.

Chemical Indicators. Certain organic chemicals change color when the pH of water changes. For example, thymol blue solution is yellow at pH 8.0 and

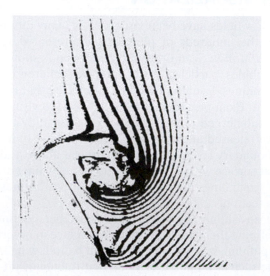

FIGURE 13.18 Hydrogen bubble streaklines showing separated flow around a rotating airfoil. (Courtesy of M. Koochesfahani.)

FIGURE 13.19   Turbulent wake behind a circular cylinder at Re = 1760. (Courtesy of R. Falco.)

blue at pH 9.2; phenol red solution is yellow at pH 6.8 and red at pH 8.2. Thus a change in pH caused by the injection of a base solution changes the color of the liquid. The lifetime of the colored water depends on the molecular diffusion of the hydrogen ions and the turbulence. An advantage of this technique is that color residues can be eliminated by changing the pH. One application is in studies of solid-liquid phase transport phenomena; if the chemical indicator is in contact with a metal surface and the metal surface is given a negative charge, the solution in immediate contact with the plate will change color.

Particles in Air.   Particles in air may be introduced as helium bubbles. By properly selecting the mixture of soap and water, one can produce helium-filled soap bubbles that are nonbuoyant; it is possible to produce bubble diameters on the order of 4 mm. Solid particles or liquid droplets could also be utilized, but they must be extremely small to avoid gravitational effects. With these small sizes one must employ an extremely strong light source to visualize the flow.

Smoke has been used successfully to study the detailed structure of complex flow phenomena. It is the most popular agent used for flow visualization in wind tunnels. One injection technique is the so-called smoke-wire method, where the smoke is generated by vaporizing oil from a fine electrically heated wire. The method can be applied to flows where the Reynolds number based on the wire diameter is less than 20. An example of such a flow structure is given in Fig. 13.19. Smoke can also be released from a small-diameter tube or "rake" to create one or more streaklines.

Velocimetry.   Recent technological advances in simultaneous measurement of two- and three-dimensional flow domains have led to the definition of a generic pulsed light velocimeter (Fig. 13.20). It consists of a pulsed light source that illuminates small fluid-entrained particles over short exposure times, and a camera that is synchronized with the light to record the location of the particles. The velocity of each marker is then given by $\Delta s / \Delta t$, where Δs is the displacement of the marker and Δt is the time between exposures. The technique is called *Particle Image Velocimetry* (PIV). The markers are usually particles that vary in size from 1 to 20 μm. An example of PIV is shown in Fig. 13.21. In addition to PIV, progress has been made in the use of molecular markers such as photochromic dyes in liquids, photochromic aerosols in gases, and molecular phosphorescence in both liquids and gases. A sample of *Molecular Tagging Velocimetry* (MTV) is shown in Fig. 13.22a and b, in which a noninvasive grid is

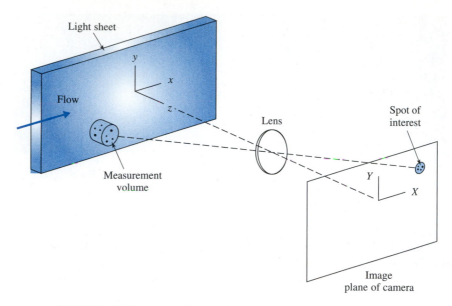

FIGURE 13.20 Pulsed light velocimeter. (Courtesy of R. Adrian.)

tagged and subsequently displaced during a brief interval. The two grids are captured with a CCD camera and analyzed to produce the velocity field shown in Fig. 13.22c. In that figure, only the left half of the velocity field is shown. The axis of symmetry is indicated by the dashed line, and the box delimits the region containing vectors derived from Fig. 13.22a and b. Velocimetry methods are particularly useful in the studies of explosions, transients in channels and ducts, flow in internal combustion engines, bubble growth and collapse, fluid-solid interaction, and the structure of turbulent flow.

Tufts and Oil Films. Flow visualization along a surface can be accomplished by attaching tufts to the surface, or if flow away from the surface is to be observed, they may be supported on wires. Surface tufts may be used to observe the transition from laminar to turbulent motion. They may also be used to study flow separation qualitatively, where the violent motion of the tufts or their tendency to point in the upstream direction identifies that separation is taking place. In air, surface flow patterns can also be visualized by distributing oil over the solid surface in such a manner that clear streaky patterns develop. Interpretation of the patterns is not always simple, particularly when flow reversals occur; the reason is that streak patterns do not indicate the flow direction.

Photography and Lighting. Many kinds of cameras are used for flow visualization, including still, stereo, and cinematic (Adrian, 1986). The video camera has become increasingly popular due to advances in resolution, and it is possible to interface certain models with a computer to obtain quantitative data. Stroboscopes, as well as laser, tungsten, and xenon lighting, are commonly employed to visualize tracers. Depending on the application, images can be produced using steady light to generate streak lines; alternatively, images can be produced using light pulses which are triggered by the camera shutter or a stroboscope to freeze the motion. If two-dimensional visualization is desired, the flow field can be illuminated by creating a narrow sheet of light directed through a

(a)

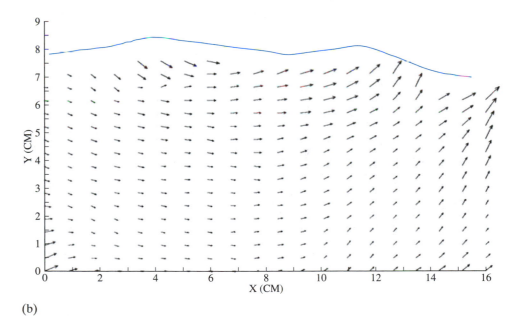

(b)

FIGURE 13.21   Particle Image Velocimetry (PIV): (a) photograph of particle pathlines;
(b) scaled velocity vectors. (Courtesy of R. Bouwmeester.)

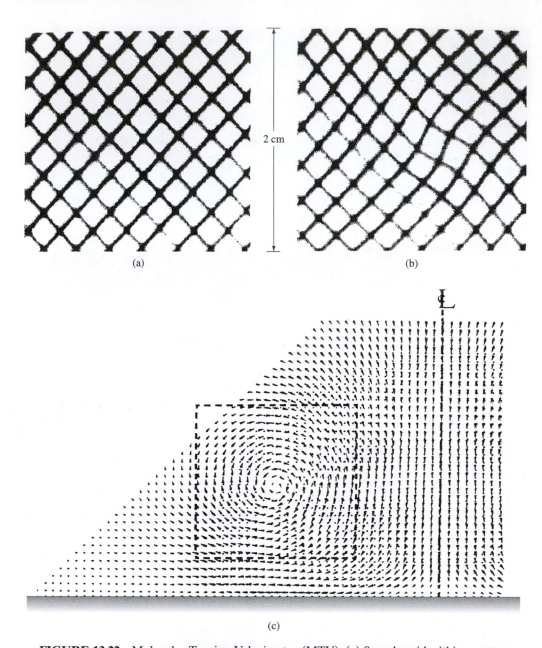

FIGURE 13.22 Molecular Tagging Velocimetry (MTV): (a) Sample grid within a vortex ring just after laser tagging; (b) displaced grid 6 ms later; (c) velocity field of the vortex ring approaching a wall. (Courtesy of C. Gendrich and M. Koochesfahani.)

window in the test section. Figure 13.23 demonstrates the technique of Laser Induced Fluorescence (LIF) to visualize a spanwise flow structure and mixing field in an unsteady wake flow field. Three-dimensional imaging can be accomplished by stereoscopic pictures or by use of holography.

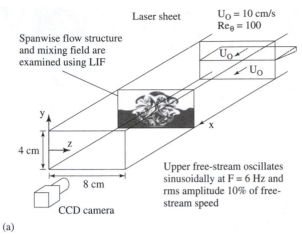

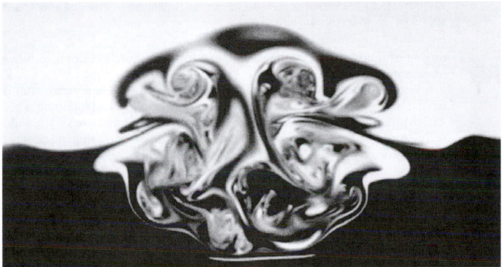

FIGURE 13.23   Laser Induced Fluorescence (LIF): (a) Experimental layout; (b) detail of spanwise flow structure measured by LIF. (Courtesy of C. MacKinnon and M. Koochesfahani.)

13.4.2 Index of Refraction Methods

The three index of refraction techniques are schlieren, shadowgraph, and interferometry. They are used in attempting to visualize flows when the density differences are either naturally or artificially altered; both gas and liquid mediums can be employed. The methods depend on the variation of the index of refraction in a transparent fluid medium and the resulting deflection of a light beam directed through the medium. Most often the flows to be studied are two-dimensional, and the light beam is directed normal to the flow field; the result is a measurement integrated over the test section. The index of refraction is a function of the thermodynamic state of the fluid and typically depends only on the density. Examples of flows that are investigated with these techniques are: high-speed flows with shock waves, convection flows, flame and combustion flow fields,

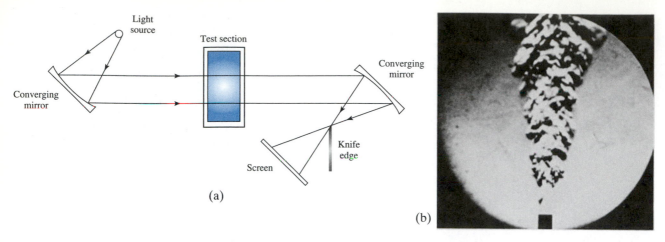

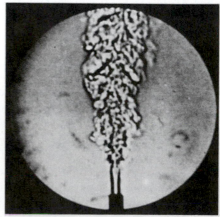

FIGURE 13.24 Schlieren system: (a) schematic; (b) helium jet entering atmospheric air. (Courtesy of R. Goldstein.)

mixing of fluids of different densities, and forced-convection flows with heat or mass transfer.

With a *schlieren* system, one measures the angle of the light beam after it has passed through the test section and emerges into the surrounding air (Fig. 13.24a); the angle is proportional to the first spatial derivative of the index of refraction. The light beam is turned in the direction of increasing index of refraction, or for most media, toward the region of higher density. A representative schlieren image is shown in Fig. 13.24b. A *shadowgraph* system measures the linear displacement of perturbed light (Fig. 13.25). In contrast to the schlieren imaging, the light pattern in a shadowgraph is determined by the second derivative of the index of refraction.

Shadowgraph and schlieren systems are usually employed for qualitative flow visualization, whereas interferometers are often used in quantitative studies.

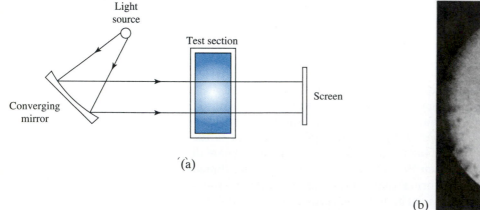

FIGURE 13.25 Shadowgraph system: (a) schematic; (b) helium jet entering atmospheric air. (Courtesy of R. Goldstein.)

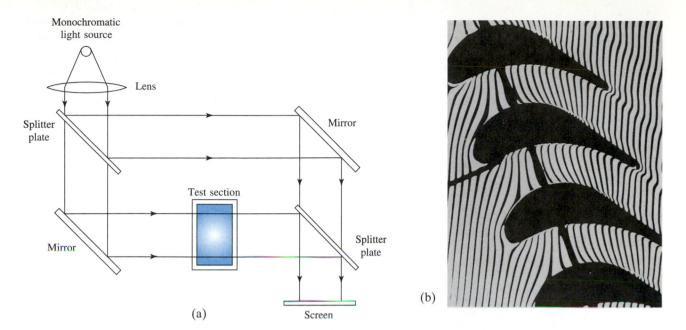

FIGURE 13.26   Mach–Zehnder interferometer: (a) schematic;(b) flow over heated gas-turbine blades. (Courtesy of R. Goldstein.)

An interfero-meter responds to differences in optical path length; since a flow field with a variable density represents an optical disturbance, the phase of transmitted light waves is changed relative to undisturbed waves. A two-beam interferometer is designed to measure that phase difference (Fig. 13.26a). When the disturbed and undisturbed light beams are recombined, interference fringes appear on the screen that are a measure of the density variation in the flow field. An example of an interferogram is shown in Fig. 13.26b.

13.5 DATA ACQUISITION AND ANALYSIS

This section begins with a brief introduction to automated acquisition of flow data using personal computers. Next, methodology to deal with the uncertainty ty of data is presented, including the manner in which those uncertainties are propagated when one flow parameter is represented indirectly as the result of measuring others. Finally, since a number of flow relations involve exponential formulations, a means of curve fitting exponential data is presented.

13.5.1   Digital Recording of Data

Some of the measurements that have been described in previous sections are made simply by the direct reading of an instrument; those readings are probably steady-state, time-averaged, or perhaps root-mean-square values. Examples include reading a manometer attached to an orifice flowmeter, monitoring a digital rpm readout of a vane anemometer, and recording the pressure indicated on a Bourdon gage. The data acquired in this manner are probably sufficiently accurate for the desired goal; indeed, any further sophistication used to collect the data could be costly and unnecessary. On the other hand, there are situations that require more complex means of data acquisition. For instance, monitoring transient pressures in a pipeline using a strain-gage transducer, or observing turbulent velocity fluctuations in a free-stream flow, require that data be recorded continuously and accurately. Another illustration is the capture of the movement of a large number of moving particles in a two-dimensional plane in order to determine an unsteady flow field. Such measurements necessitate that data be recorded automatically, rapidly, and accurately.

The advent of the personal computer has revolutionized automated acquisition of flow data. Many of the transducers in use provide a voltage output signal (an analog signal) that must be converted into a digital signal before the computer can store and process the data. Readily available today are a number of plug-in analog input/output (I/O) boards specifically made for personal computers, which can be inserted into the central processing unit bus. These I/O boards consist of input multiplexers, an input amplifier, a sample-and-hold circuit, one or more output digital-to-analog converters, and interfacing and timing circuitry. In addition, some include built-in random-access memory, read-only memory, and real-time clocks. Compatible software is available so that programming either is unnecessary or requires minimal effort.

13.5.2 Uncertainty Analysis

When flow measurements are made, it is important to realize that no recorded value of a given parameter is perfectly accurate. Instruments do not measure the so-called "true value" of the parameter, but they will provide an estimate of that value. Along with each measurement, one must attempt to answer the important question: How accurate is that measurement? The inaccuracies associated with measurements can be considered as uncertainties rather than errors. An *error* is the fixed, unchangeable difference between the true value and the recorded value, whereas an *uncertainty* is the statistical value that an error may assume for any given measurement.

The purpose of this section is to describe a technique for propagating uncertainties of flow measurements into results. It has a variety of uses, among them helping us to decide whether calculations agree with collected data or lie beyond acceptable limits. Perhaps more importantly, it can help us to improve a given experiment or measuring procedure. It is not meant to be a substitute for accurate calibrations and sound experimental techniques; rather, it should be viewed as an additional aid for obtaining good, reliable results. Only a brief introduction is presented here; the reading of Coleman and Steele (1989) and Kline (1985) is recommended for a more thorough treatment of the subject.

Uncertainty can be divided into three categories: (1) uncertainty due to calibration of instruments, (2) uncertainty due to acquisition of data, and (3) uncertainty as a result of data reduction. The total uncertainty is the sum of bias and precision. *Bias* is the systematic uncertainty that is present during a test; it is considered to remain constant during repeated measurements of a certain set of parameters. There is no statistical formulation that can be applied to estimate bias, hence its value must be based on estimates. Calibrations and independent measurements will aid in its estimation. *Precision* is alternately termed *repeatability*; it is found by taking repeated measurements from the parameter population and using the standard deviation as a precision index. For the situation in which repeated measurements are not taken (which occurs quite often for flow measurements), a single value must be used; however, less precision will be obtained. Figure 13.27 illustrates how uncertainties are defined. A sampling of data (solid line) is shown that is assumed to be distributed in a Gaussian manner. The parameter x_i is one particular measured variable, or measurand, $<x_i>$ is the average of all the measured variables, and δx_i is the uncertainty interval, or precision, associated with x_i. The uncertainty for the measurand can be given as

$$x_i = <x_i> \pm \delta x_i \qquad (13.5.1)$$

Furthermore, it is necessary to state the bias and the probability P of recording the measurand with the stated precision.

In addition to defining uncertainties in the data, it is necessary to propagate those uncertainties into the results. A two-step process is described as follows.

1. *Describe the uncertainty for the measurands.* Each measurand is assumed to be an independent observation, and it must come from a Gaussian population. Also, the precision for each measurand must be quoted at the same probability P. For example, consider an experiment to calibrate a weir, in which the measurands are the water level upstream of the weir, and the discharge. The water level is measured using a water-level gage, and the discharge is recorded by measuring the time it takes to fill a known volume. Thus the two measurands are independent. In addition, based on experience and perhaps some repeated measurements, one can reasonably state that (1) there is no bias, and (2) the recorded precision for each measurand has a probability of $P = 0.95$; that is, 20 out of every 21 readings of that measurand would yield the recorded value plus or minus the stated precision.

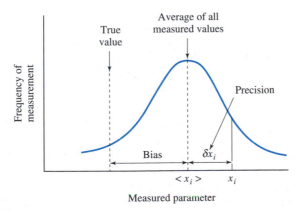

FIGURE 13.27   Uncertainties associated with a sampling of data.

2. *Compute the overall uncertainty interval.* Let the parameter R represent the result computed from n variables $x_1, x_2, \ldots, x_n$, that is, $R = R(x_1, x_2, \ldots, x_n)$. Here we assume that R is continuous and has continuous derivatives in the desired domain. In addition we assume that both the variables are uncorrelated and the uncertainty intervals are uncorrelated, that is, they are independent of one another. Then the overall uncertainty in the result δR can be obtained by expanding R in a Taylor series about the measurands $<x_1>, <x_2>, \ldots, <x_n>$ and obtaining the root-sum-square result

$$\delta R = \left[\left(\frac{\partial R}{\partial x_1} \delta x_1 \right)^2 + \left(\frac{\partial R}{\partial x_2} \delta x_2 \right)^2 + \cdots + \left(\frac{\partial R}{\partial x_n} \delta x_n \right)^2 \right]^{1/2} \qquad (13.5.2)$$

Situations arise when the expression for R becomes so complicated that it is difficult to obtain the partial derivatives in Eq. 13.5.2. In such a case, one can employ finite difference approximations of the partial derivatives to simplify the analysis. For example, consider a forward finite difference approximation for the first partial derivative:

$$\frac{\partial R}{\partial x_1} \approx \frac{R(x_1 + \delta x_1, x_2, \ldots, x_n) - R(x_1, x_2, \ldots, x_n)}{(x_1 + \delta x_1) - x_1} \qquad (13.5.3)$$

Here we have assumed that in the denominator the finite difference interval is equivalent to the uncertainty interval δx_1. The numerator of Eq. 13.5.3 is the difference between R evaluated at $x_1 + \delta x_1$ and x_1 with the remaining variables fixed. Expressing the other partial derivatives in Eq. 13.5.2 in a similar manner and substituting the results back into the relation we have the desired result:

$$\delta R \approx \left\{ \begin{array}{l} [R(x_1 + \delta x_1, x_2, \ldots, x_n) - R(x_1, x_2, \ldots, x_n)]^2 \\ + [R(x_1, x_2 + \delta x_2, \ldots, x_n) - R(x_1, x_2, \ldots, x_n)]^2 \\ + \cdots + [R(x_1, x_2, \ldots, x_n + \delta x_n) - R(x_1, x_2, \ldots, x_n)]^2 \end{array} \right\}^{1/2} \qquad (13.5.4)$$

Equation 13.5.4 can easily be evaluated using a spreadsheet algorithm.

Use of either Eq. 13.5.2 or 13.5.4 provides an estimate of δR. The overall uncertainty interval is an estimate of the standard deviation of the population of all possible experiments like the one being conducted.

Example 13.1

Discharge and pressure data are collected for water flow through an orifice meter in a pipe Fig. E13.1a. The orifice diameter is 35.4 mm and the diameter of the pipe is 50.8 mm. The original data are reduced to the form shown in the accompanying table. The second column is the discharge Q; the precision δQ (zero bias, $P = 0.95$) is given in the third column; and the pressure head data $\Delta h = h_1 - h_2$ are presented in the fourth column. In addition, the precision for each of the pressure measurements is determined to be the same, namely, $\delta(\Delta h) = 0.025$ m (zero bias, $P = 0.95$). The fifth and sixth columns show the Reynolds numbers and the flow coefficient K computed using Eq. 13.3.8. Determine the uncertainties in the result K as a result of the estimated uncertainties in the measurands Q and Δh.

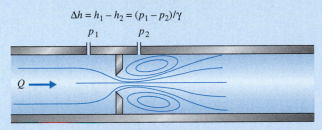

$$\Delta h = h_1 - h_2 = (p_1 - p_2)/\gamma$$

FIGURE E13.1a

Solution

Equation 13.3.8 is used in conjunction with Eq. 13.5.4 to propagate the uncertainties from Q and Δh into K. For example, consider the data for $i = 6$. Calculation of δk proceeds as follows. The orifice area is

$$A_0 = \frac{\pi}{4} 0.0354^2 = 9.84 \times 10^{-4} \text{m}^2$$

Data no. i	Q (m³/s)	δQ (m³/s)	Δh (m)	Re	K	δK
1	0.00119	0.00010	0.139	42 800	0.732	0.085
2	0.00162	0.00012	0.265	58 300	0.722	0.062
3	0.00200	0.00015	0.403	71 900	0.723	0.058
4	0.00223	0.00015	0.517	80 200	0.711	0.051
5	0.00251	0.00015	0.655	90 300	0.711	0.045
6	0.00276	0.00017	0.781	99 300	0.716	0.046
7	0.00294	0.00018	0.907	105 700	0.708	0.045
8	0.00314	0.00019	1.008	112 900	0.717	0.045
9	0.00330	0.00016	1.134	118 700	0.711	0.035
10	0.00346	0.00017	1.247	124 400	0.711	0.036
11	0.00373	0.00017	1.386	134 200	0.727	0.034
12	0.00382	0.00017	1.512	137 400	0.713	0.032
13	0.00408	0.00018	1.663	146 700	0.726	0.033
14	0.00424	0.00019	1.852	152 500	0.715	0.032
15	0.00480	0.00021	2.293	172 600	0.727	0.032

and the diameter ratio is

$$\beta = \frac{35.4}{50.8} = 0.697 \simeq 0.7$$

With Eq. 13.3.8, values of K are determined for $(Q, \Delta h)$, $(Q + \delta Q, \Delta h)$, and $(Q, \Delta h + \delta(\Delta h))$:

(continued)

$$K(Q, \Delta h) = \frac{0.00276}{9.84 \times 10^{-4} \sqrt{2 \times 9.81 \times 0.781}}$$

$$= 0.716$$

$$K(Q + \delta Q, \Delta h) = \frac{0.00276 + 0.00017}{9.84 \times 10^{-4} \sqrt{2 \times 9.81 \times 0.781}}$$

$$= 0.761$$

$$K(Q, \Delta h + \delta(\Delta h)) = \frac{0.00276}{9.84 \times 10^{-4} \sqrt{2 \times 9.81 \times (0.781 + 0.025)}}$$

$$= 0.705$$

These are then substituted into Eq. 13.5.4, with the variable K replacing the result R in that relation:

$$\delta K = \{[K(Q + \delta Q, \Delta h) - K(Q, \Delta h)]^2 + [K(Q, \Delta h + \delta(\Delta h)) - K(Q, \Delta h)]^2\}^{1/2}$$
$$= [(0.761 - 0.716)^2 + (0.705 - 0.716)^2]^{1/2}$$
$$= 0.046$$

The results for all measurands are similarly evaluated and shown in the last column in the table. In addition, they are plotted in Fig. E13.1b, where the vertical lines, or uncertainty bands, associated with each value of K have the magnitude of $2\delta K$. The bands can be interpreted as the estimated precision for K, based on zero bias and $P = 0.95$ for the two measurands. Note that the precision δK stabilizes at approximately 0.032 at the higher Reynolds numbers, and that the magnitude of δK is greater at lower Reynolds numbers. An alternative way to express this is that the uncertainty of K decreases with increasing Reynolds number. The data in the figure should be compared with the orifice curve labeled $\beta = 0.7$ in Fig. 13.10.

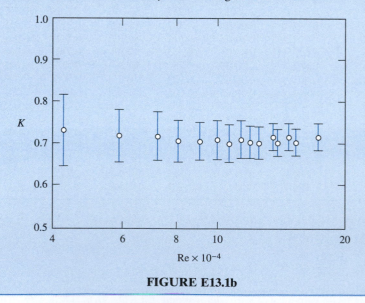

FIGURE E13.1b

A Mathcad solution of Example 13.1 is shown in Fig. 13.28.

Input data:

$$D := 0.0354 \qquad ORIGIN := 1 \qquad N := 15 \qquad \nu := 1.006 \times 10^{-6} \qquad g := 9.81$$

$$Q := \begin{pmatrix} 0.00119 \\ 0.00162 \\ 0.002 \\ 0.00223 \\ 0.00251 \\ 0.00276 \\ 0.00294 \\ 0.00314 \\ 0.0033 \\ 0.00346 \\ 0.00373 \\ 0.00382 \\ 0.00408 \\ 0.00424 \\ 0.0048 \end{pmatrix} \quad \delta Q := \begin{pmatrix} 0.00010 \\ 0.00012 \\ 0.00015 \\ 0.00015 \\ 0.00015 \\ 0.00017 \\ 0.00018 \\ 0.00019 \\ 0.00016 \\ 0.00017 \\ 0.00017 \\ 0.00017 \\ 0.00018 \\ 0.00019 \\ 0.00021 \end{pmatrix} \quad \Delta h := \begin{pmatrix} 0.139 \\ 0.265 \\ 0.403 \\ 0.517 \\ 0.655 \\ 0.781 \\ 0.907 \\ 1.008 \\ 1.134 \\ 1.247 \\ 1.386 \\ 1.512 \\ 1.663 \\ 1.852 \\ 2.293 \end{pmatrix} \quad \delta \Delta h := \begin{pmatrix} 0.025 \\ 0.025 \\ 0.025 \\ 0.025 \\ 0.025 \\ 0.025 \\ 0.025 \\ 0.025 \\ 0.025 \\ 0.025 \\ 0.025 \\ 0.025 \\ 0.025 \\ 0.025 \\ 0.025 \end{pmatrix}$$

Define function to compute K:
$$K(q, \Delta H) := \frac{q}{\left(\dfrac{\pi \cdot D^2}{4} \right)} \cdot (2 \cdot g \cdot \Delta H)^{-0.5}$$

Compute uncertainty in K:

$$i := 1 .. N$$

$$KK_i := K(Q_i, \Delta h_i)$$

$$\delta K_i := \sqrt{\left(K(Q_i + \delta Q_i, \Delta h_i) - K(Q_i, \Delta h_i) \right)^2 + \left(K(Q_i, \Delta h_i + \delta \Delta h_i) - K(Q_i, \Delta h_i) \right)^2}$$

$$Re_i := \frac{4 \cdot Q_i}{\pi \cdot \nu \cdot D}$$

FIGURE 13.28 Mathcad Solution to Example 13.1.

Plot of K versus Re:

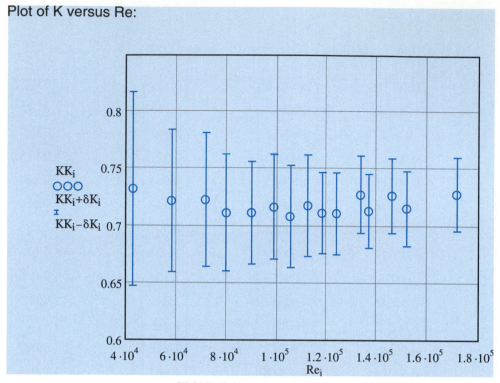

FIGURE 13.28 (*continued*)

13.5.3 Regression Analysis

A significant number of flow-measuring devices correlate the discharge with pressure drop or head according to the general formula

$$Y = CX^m \qquad (13.5.5)$$

in which Y is the derived result, X is the measurand, and C and m are constants. By taking a number of independent measurements of X and Y, estimates of C and m can be found. A systematic approach to determine C and m is to find the best estimates based on the method of least squares. If the logarithm of each side of Eq. 13.5.5 is taken, it becomes

$$\ln Y = \ln C + m \ln X \qquad (13.5.6)$$

which is of the form

$$y = b + mx \qquad (13.5.7)$$

By association we observe that Eq. 13.5.6 is linear provided that $y = \ln Y$, $b = \ln C$, and $x = \ln X$.

Consider the data set X_i, Y_i, $i = 1, \ldots, n$. The objective is to generate a straight line through the logarithms of the data (x_i, y_i) such that the errors of estimation are small. That error is defined as the difference between the observed value y_i and the corresponding value on the line; in Fig. 13.29, s_i is the error associated with the data point (x_i, y_i). The method of least squares requires that to minimize the error, the sum of the squares of the errors of all the data be as small as possible. From Eq. 13.5.5,

$$
\begin{aligned}
S &= \sum_{i=1}^{n} s_i^2 \\
&= \sum_{i=1}^{n} [y_i - (b + mx_i)]^2
\end{aligned}
\qquad (13.5.8)
$$

The parameter S is minimized by forming the partial derivatives of S with respect to b and m, setting the resulting algebraic equations to zero, and solving for b and m. It is left as an exercise for the reader to derive the results.

$$m = \frac{\sum x_i y_i - (\sum x_i \sum y_i)/n}{\sum x_i^2 - (\sum x_i)^2/n} \qquad (13.5.9)$$

$$b = \frac{\sum y_i - m \sum x_i}{n} \qquad (13.5.10)$$

$$C = e^b \qquad (13.5.11)$$

A simple computer algorithm can be prepared to evaluate m and C based on a set of input data (X_i, Y_i), or alternatively, a spreadsheet solution can be used.

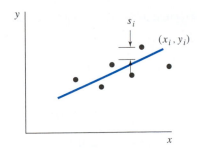

FIGURE 13.29   Method of least squares.

Example 13.2

With the data given in Example 13.1, perform a least squares regression to determine the coefficient C and exponent m in Eq. 13.5.5. Compare the result with Eq. 13.3.8.

Solution:

Equations 13.5.9 to 13.5.11 are employed to evaluate C, b, and m. In Eqs. 13.5.9 and 13.5.10, x is replaced by $\ln \Delta h = \ln(h_1 - h_2)$, and y is replaced by $\ln Q$. The calculations are tabulated in the accompanying table.

i	$x_i = \ln \Delta h$	$y_i = \ln Q$	x_i^2	$x_i y_i$
1	−1.9733	−6.7338	3.8939	13.2878
2	−1.3280	−6.4253	1.7636	8.5328
3	−0.9088	−6.2146	0.8259	5.6478
4	−0.6597	−6.1058	0.4352	4.0280
5	−0.4231	−5.9875	0.1790	2.5333
6	−0.2472	−5.8925	0.06111	1.4566
7	−0.09761	−5.8293	0.009528	0.5690
8	0.007968	−5.7635	0.00006349	−0.0459
9	0.1258	−5.7138	0.01583	−0.7188
10	0.2207	−5.6665	0.04871	−1.2506
11	0.3264	−5.5913	0.1065	−1.8250
12	0.4134	−5.5675	0.1709	−2.3016
13	0.5086	−5.5017	0.2587	−2.7982
14	0.6163	−5.4632	0.3798	−3.3670
15	0.8299	−5.3391	0.6887	−4.4309
Σ	−2.5886	−87.7954	8.8374	19.3173

Substitute the summed values into Eqs. 13.5.9 to 13.5.11:

$$m = \frac{19.3173 - (-2.5886)(-87.7954)/15}{8.8374 - (-2.5886)^2/15} = 0.497$$

$$b = \frac{-87.7954 - 0.497(-2.5886)}{15} = -5.7673$$

$$C = \exp(-5.7673) = 0.00313$$

Hence the discharge is correlated to Δh by the regression curve

$$Q = 0.00313 \, \Delta h^{0.497}$$

Based on the analysis performed in Example 13.1, the average value of K is 0.718. The area of the orifice is $A_0 = 9.84 \times 10^{-4} \text{m}^2$. Substituting these into Eq. 13.3.8, we find that

$$Q = 0.00313 \, \Delta h^{0.500}$$

Thus the two results agree quite closely. The data and regression line are plotted in Fig. E13.2.

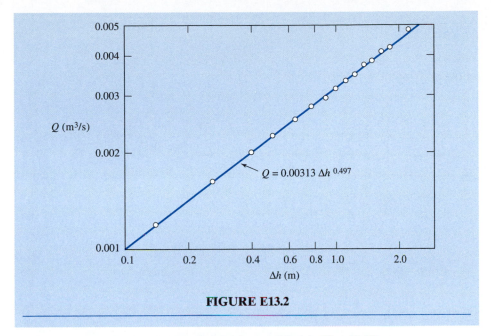

FIGURE E13.2

A Mathcad solution for Example 13.2 is provided in Fig. 13.30.

Input data:

$$n := 15$$

$$M := \begin{pmatrix} 0.00119 & 0.139 \\ 0.00162 & 0.265 \\ 0.002 & 0.403 \\ 0.00223 & 0.517 \\ 0.00251 & 0.655 \\ 0.00276 & 0.781 \\ 0.00294 & 0.907 \\ 0.00314 & 1.008 \\ 0.0033 & 1.134 \\ 0.00346 & 1.247 \\ 0.00373 & 1.386 \\ 0.00382 & 1.512 \\ 0.00408 & 1.663 \\ 0.00424 & 1.852 \\ 0.0048 & 2.293 \end{pmatrix}$$

FIGURE 13.30 Mathcad solution to Example 13.2.

Convert to logarithmic form: $\quad X := \ln\left(M^{\langle 1 \rangle}\right) \quad Y := \ln\left(M^{\langle 0 \rangle}\right)$

Compute m, b, and C:

$$m := \frac{\left(\sum\limits_{i=0}^{n-1} X_i \cdot Y_i\right) - \dfrac{1}{n} \cdot \sum\limits_{i=0}^{n-1} X_i \sum\limits_{i=0}^{n-1} Y_i}{\sum\limits_{i=0}^{n-1} (X_i)^2 - \left(\sum\limits_{i=0}^{n-1} X_i\right)^2 \cdot \dfrac{1}{n}} \qquad b := \frac{\sum\limits_{i=0}^{n-1} Y_i - m \cdot \sum\limits_{i=0}^{n-1} X_i}{n} \qquad C := \exp(b)$$

$$m = 0.496 \qquad\qquad b = -5.767 \qquad\qquad C = 3.128 \times 10^{-3}$$

Plot Q versus Δh:

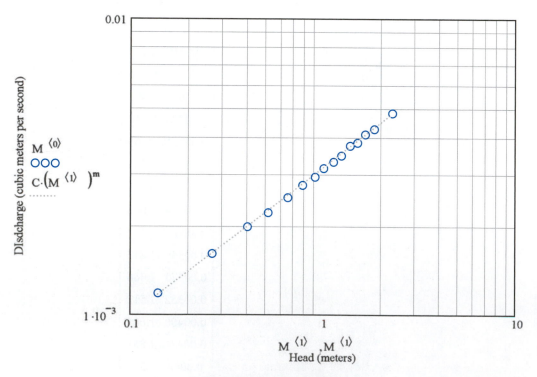

FIGURE 13.30 (*continued*)

13.6 SUMMARY

There are many devices available to measure flow parameters; we have presented some of the more common ones. The monitoring of pressure and velocity at discrete locations in flow fields were first discussed followed by descriptions of integrated flow rate or discharge measurements. Various devices are used for measurement of these parameters, ranging from purely mechanical to electromechanical.

Flow visualization is an important technique that is employed in both industrial and research laboratories to study complex flow fields, especially turbulent ones. Uses of tracers and index of refraction methods are presented. Of these methods, velocimetry techniques are rapidly gaining favor due to recent advances in computer imaging and development of laser-activated dyes.

Today, a significant number of measurements are automated and may require analog or digital interfacing with data loggers or personal computers. Regardless of the manner in which data is acquired and stored, the data taker must be aware of uncertainties that are present and unavoidable in the process of acquiring data. We have shown how the uncertainties present in measured parameters can be propagated into results using spreadsheet or computational software analyses. These techniques are useful, for example, to determine how accurate and reliable a flow meter may be, or to design an experiment to calibrate a velocity meter that will insure a desired degree of accuracy. Finally, regression analysis is applied to determine the parameters that describe an exponential relationship that correlates discharge with pressure drop or head.

PROBLEMS

13.1 A 20° inclined manometer attached to a piezometer opening is used to measure the pressure at the wall of airflow. If the reading is measured to be 4 cm of mercury, calculate the pressure at the wall.

13.2 A pitot-static probe is to be used for repeated measurements of the velocity of an atmospheric airstream. It is desired to relate the velocity V to the manometer reading h in centimeters of mercury (i.e., $V = C \sqrt{h}$). Calculate the value of the constant C for a laboratory at 20°C situated at:
  (a) Sea level.
  (b) 2000 m elevation.

13.3 A pitot-static probe attached to a manometer that has water as the manometer fluid is proposed as the measuring device for an airflow with a velocity of 8 m/s. What manometer reading is expected? Comment as to the use of the proposed device.

13.4 Water at 25°C flows through a 2.5-mm-diameter tube into a calibrated tank. If 2 liters are collected in 10 minutes, calculate the flow rate (m³/s),

the mass flux (kg/s), and the average velocity (m/s). Is the flow laminar, turbulent, or can't you tell?

13.5 A velocity traverse in a rectangular channel that measures 10 cm by 100 cm is represented by the following data. Estimate the flow rate and the average velocity assuming a symmetric plane flow.

y (cm)	0	0.5	1	2	3	4	5
u (m/s)	0	5.2	8.1	9.2	9.8	10	10

13.6 A velocity traverse in a 10-cm-diameter pipe is represented by the following data. Estimate the flow rate and the average velocity assuming a symmetric flow.

r (cm)	0	1	2	3	4	4.5
u (m/s)	10	10	9.8	9.2	8.1	5.2

13.7 The flow rate of water in a 12-cm-diameter pipe is measured with a 6-cm-diameter venturi meter to be 0.09 m³/s. What is the expected deflection on a water-mercury manometer? Assume a water temperature of 20°C.

13.8 The flow rate of 20°C water flow in a 24-cm-diameter pipe is desired. If a water-mercury manometer reads 12 cm, calculate the discharge if the manometer is connected to:
(a) A 15-cm-diameter orifice
(b) A 15-cm-diameter nozzle

13.9 Calculate the flow rate of 40°C water in the pipes shown in Fig. P13.9.

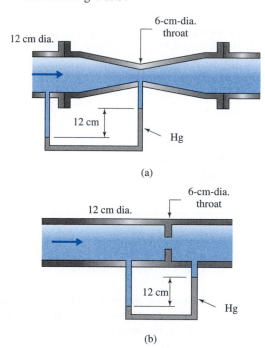

(a)

(b)

FIGURE P13.9

13.10 A pressure difference across an elbow in a water pipe is measured. Estimate the flow rate in the pipe. The pressure difference, pipe diameter, and radius of curvature are, respectively:
(a) 80 kPa, 10 cm, 20 cm
(b) 80 kPa, 5 cm, 20 cm

Use a water temperature of 20°C.

13.11 A pitot tube is used to measure the velocity of gasoline flowing in a pipe. The estimate of the velocity is given by the equation

$$v_1 = \sqrt{\frac{2}{\rho}(p_T - p_1)}$$

in which v_1 is the local velocity at location 1, ρ the density of the liquid, p_T the total pressure measured by the tube, and p_1 the static pressure at location 1. The density is estimated to be 680 ± 50 kg/m³ ($P = 0.95$, 0 bias).
(a) Assume that the two pressures are measured separately using individual pressure gages: $p_T = 102 \pm 1$ kPa ($P = 0.95$, 0 bias), $p_1 = 95 \pm 1$ kPa ($P = 0.95$, 0 bias). Estimate the uncertainty for v_1.
(b) Assume that the pressure difference $p_T - p_1$ is measured using a single differential pressure gage: $p_T - p_1 = 7 \pm 1$ kPa ($P = 0.95$, 0 bias). Estimate the uncertainty for v_1.
(c) Which arrangement, (a) or (b), is preferable?

13.12 Experimental data used to calibrate discharge of water in a venturi meter are tabulated below. The diameter of the meter throat is 33.3 mm and the diameter at the upstream tap location of the differential manometer is 54.0 mm. The precision δQ of the discharge measurements has zero bias and $P = 0.95$, and the precision for all the manometer measurements is $\delta(\Delta h) = 2$ mm Hg, with zero bias and $P = 0.95$. Note that the manometer readings are given in mm Hg. With use of Eq. 13.3.8, determine the uncertainties in K as a result of the uncertainties in the two measurands. Plot the results in a fashion similar to those in Example 13.1, and compare the curve with Fig. 13.10.

Data no. i	Q (m³/s)	δQ (m³/s)	Δh (mm Hg)	Re
1	0.00168	0.00011	13	64 200
2	0.00208	0.00014	22	79 500
3	0.00249	0.00016	32	95 100
4	0.00281	0.00019	40	107 400
5	0.00328	0.00022	54	125 300
6	0.00355	0.00024	62	135 600
7	0.00372	0.00025	70	142 100
8	0.00402	0.00023	82	153 600
9	0.00415	0.00024	91	158 600
10	0.00444	0.00025	100	169 700
11	0.00479	0.00027	113	183 000
12	0.00493	0.00028	123	188 400
13	0.00523	0.00030	134	199 800
14	0.00533	0.00035	140	203 700
15	0.00509	0.00029	140	194 500

13.13 With the numerical results obtained from Example 13.1, plot the relative uncertainty $\delta K/K$ versus Re on a linear scale. What conclusions can you draw?

13.14 Using the data from Problem 13.12, perform a least squares regression and determine the coefficient C and exponent m in Eq. 13.5.5. Compare the result with Eq. 13.3.8. Note that the manometer readings are given in mm Hg.

13.15 Establish a head–discharge relation of the form $Q = CY^m$ for an orificemeter using the data in the following table. In the equation, Y is the head causing the flow and Q is the discharge.

Data no. i	$Q \times 10^{-3}$ (m³/s)	$Y \times 10^{2}$(m)
1	3.94	11.80
2	3.68	11.52
3	3.62	11.40
4	3.23	10.88
5	2.73	10.24
6	2.65	9.75
7	2.29	9.51
8	2.21	9.33
9	2.13	9.20
10	1.97	9.05
11	1.47	8.02
12	1.23	7.44
13	1.18	7.25
14	0.90	6.28

Crude oil swirls on the surface of Prince William Sound, Alaska, near the oil tanker Exxon Valdez. In April, 1989, over 45 400 m^3 of crude oil was released into these southern Alaska waters after the tanker hit a reef. Containment booms can be seen attached to the aft of the ship. (© Jean Louis Atlan/Sygma/Corbis)

14

Environmental Fluid Mechanics

Outline

Chapter Objectives

The objectives of this chapter are to:

▲ Introduce concepts of gradient flux laws

▲ Present fundamental equations of mass and heat transport

▲ Describe concepts of advective and diffusive transport

▲ Describe concepts of turbulent diffusion and dispersion

▲ Present simplified solutions of the one-dimensional transport equation

▲ Demonstrate procedures for evaluating dispersion coefficients in the environment

14.1 INTRODUCTION

In recent decades the scope of fluid mechanics has broadened to become increasingly involved with water quality and a variety of other environmental issues. Fluid mechanics is implicitly involved because it provides the physical basis for transport processes of heat, mass, and momentum in the environment. An engineer or environmental specialist should be able to describe and quantify these processes, with due consideration given to chemical or biological effects inherent in the environment. The objective of this chapter is to provide the physical basis for the transport of mass, heat, and momentum in the environment. Emphasis is placed upon advection and diffusion in rivers, lakes, reservoirs, and coastal waters. The material presented is intended to help readers develop an intuitive sense of transport processes and serve as a guide by constructing and evaluating integrated environmental and ecosystem models.

14.2 TRANSPORT PROCESSES IN FLUIDS

The equations expressing the principles of conservation of momentum and energy to define the macroscopic properties of a fluid motion are derived in Chapter 5. The constitutive relations, examples of which are Newton's law of viscosity, Eq. 1.5.3, and Fourier's law of heat transfer, Eq. 5.4.1, were essential building blocks in describing the macroscopic properties of fluid motion. They are based on empirical evidence, and are expressed in terms of a product of the proportionality constant (dynamic viscosity, thermal conductivity) and a spatial gradient of a state variable, for example, the fluid velocity gradient or the fluid temperature gradient. The proportionality constant reflects the nature of the transporting medium while the gradient designates the spatial intensity of the state variable. Thus the transport fluxes[1] are determined by the nature of the transporting medium and the local gradients of the state variables. Momentum flux is related to the velocity gradient by fluid viscosity, heat flux is related to the temperature gradient by fluid thermal conductivity, and mass flux is related to the concentration gradient by the fluid diffusion. Since these fluxes combine a fluid property with the state variable gradient they are sometimes called **gradient flux laws**.

Gradient flux laws: *Fluxes combining fluid property with state variable gradients.*

14.2.1 Momentum Transport

Newton's law of viscosity, Eq. 1.5.3, is reexamined in terms of momentum transport in this section. Such a description is considered because it illustrates the analogy with energy and mass transport. Consider the velocity profile shown in Fig. 14.1. Suppose that the flow is laminar so that fluid particles travel along streamlines that vary smoothly over the flow domain. Obviously, the macroscopic fluid flow is in the x-direction, but it is not obvious that fluid molecules are moving in all directions. Across the plane A-B, molecules are pictured moving randomly in the vertical direction. The molecules crossing A-B from above carry

---

[1] A flux is a vector describing the net flow of a quantity along a certain axis per unit area perpendicular to the axis and per unit time.

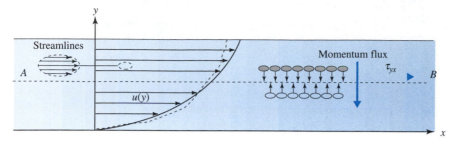

FIGURE 14.1   Momentum flux due to molecular transport.

their x-momentum with them below the plane A-B. Similarly, the molecules crossing A-B from below carry their x-momentum with them above the plane A-B. The net effect of the molecular transport is to equalize the velocities above and below plane A-B. Consequently, the velocity distribution will have a tendency to converge to that shown by the dashed line. The viscosity μ is a reflection of the net transport of x-momentum downward; that is, viscous terms transfer momentum from high values to low ones. The **momentum flux** M_{xy} moves in the direction of the negative velocity gradient, and it is equivalent to the fluid shear stress:

Momentum flux: Flux moving in the direction of the negative velocity gradient; equal to shear stress.

$$M_{xy} = -\tau_{yx} = -\mu \frac{du}{dy} = -\rho\, \nu \frac{du}{dy} \qquad (14.2.1)$$

In Eq. 14.2.1 the kinematic viscosity $\nu = \mu/\rho$ has the same dimensions (L^2/T) as the coefficient of diffusion in mass or heat transfer. As a result, ν is often interpreted as a *momentum diffusion coefficient*.

14.2.2   Heat Transport

Whenever there exists a temperature difference in a medium or between media, heat transfer must occur. Consequently, heat transport is defined as energy transfer due to a temperature difference. The three basic modes of heat transport are (a) *conduction*, (b) *convection*, and (c) *radiation*. Conduction occurs due to collisions between molecules, convection is the transport of heat in the presence of a moving fluid, and radiation is heat transport that occurs by electromagnetic waves (all surfaces of finite temperature emit energy in the form of electromagnetic waves). All three modes of heat transport occur in the environment.

Consider the two plates bounding a medium of thickness h and unit width (Fig. 14.2). The upper plate is maintained at temperature T_2 and the lower plate at temperature T_1. If T_2 is greater than T_1, heat will be transported through the medium in the negative y-direction and the steady-state temperature variation will be linear. In order to maintain the temperature distribution, heat flux q through the medium must be constant and proportional to the temperature difference divided by the thickness, that is, $q \sim (T_2 - T_1)/h$. For an incremental change in distance Δy the temperature change is ΔT, and in the limit as $\Delta y \to 0$,

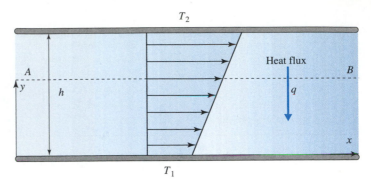

FIGURE 14.2 Temperature distribution between two fixed parallel plates.

$$q = -K \frac{dT}{dy} \tag{14.2.2}$$

in which the proportionality constant K is defined as the thermal conductivity. Equation 14.2.2 is the one-dimensional form of *Fourier's law*, and it is valid for both steady and unsteady heat transfer. In the SI system, the units of q are W/m^2 and those of K are $W/m - °C$. The thermal diffusivity α is defined in Eq. 5.4.13 to be the thermal conductivity divided by the density multiplied by the specific heat. Like kinematic viscosity, thermal diffusivity has dimensions of L^2/T. The tendency of a fluid to diffuse its temperature gradients is directly proportional to the thermal diffusivity. The thermal diffusivity can also be interpreted as a measure of the rate at which thermal energy is transported through a fluid.

Example 14.1

An insulated pipe is injected vertically into a river bed as shown. The top of the pipe is at the sediment/water interface and the bottom of the pipe is 0.2 m below the interface. Sediment temperatures are measured at both interfaces at the top and bottom of the pipe.

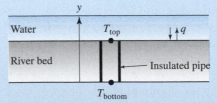

FIGURE E14.1

(a) If the upper temperature is 15°C and the lower temperature is 18°C, compute the heat flux q and the rate of heat transfer Q through the pipe.

(b) Is the river bed sediment heating or cooling the river?

(c)   What is the heat transfer rate from the sediment along the river bed if the river has an average width of 20 m?

The pipe has a diameter of 5 cm, and the thermal conductivity of the sediment is $K = 1.54$ W/m − °C.

Solution

(a)   The one-dimensional form of Fourier's law is acceptable if the sediment temperature varies only in one direction. Assume that dT/dy is a constant equal to $(T_{top} - T_{bottom})/(y_{top} - y_{bottom})$, so that Eq. 14.2.2 becomes

$$q = -K\frac{dT}{dy} = -1.54\,\frac{15 - 18}{0.2} = 23.1 \text{ W/m}^2$$

The rate of heat transfer through the pipe is the flux q multiplied by the area A normal to the direction of flux:

$$Q = qA = 23.1 \times 0.025^2 \times \pi = 0.04 \text{ W}$$

(b)   Along the river bed the heat flux is positive upward, so that the sediment heats the water in the river.

(c)   The rate of heat transfer per unit length of river is equal to q multiplied by the width of the river, or

$$q \times b = 23.1 \times 20 = 462 \text{ W/m}$$

14.2.3   Mass Transport

Whenever there exists a concentration difference of two or more substances in a fluid mixture, a transport of mass, or diffusion of mass, will take place. A complication in analyzing diffusion is that we have to deal with fluid mixtures. There are several ways of defining the concentration of a mixture. The most common way is related to the number of moles of substance, i, per unit volume. The mass of one mol of a substance is the molecular weight of that substance. For example, 180 g of substance with molecular weight of 18 contains 10 mol. The molar concentration is defined as

$$C_i = \frac{n_i}{V} \tag{14.2.3}$$

where C_i is the molar concentration of substance i (mol/m^3), n_i is the number of moles of substance i, and V is the volume of the solution (m^3). Another way of expressing concentration is defined in terms of the mass of substance i per volume of the solution. This is known as a mass concentration (kg/m^3).

Consider an infinitely long channel filled with stagnant water B into which a colored dye A is injected at time $t = 0$ (Fig. 14.3). The dye will spread outward from the point of injection to the other regions of the water where dye is not present. The dye will be transported or "diffused" from high to low concentrations. After a sufficient time t_1, when the dye concentration becomes constant

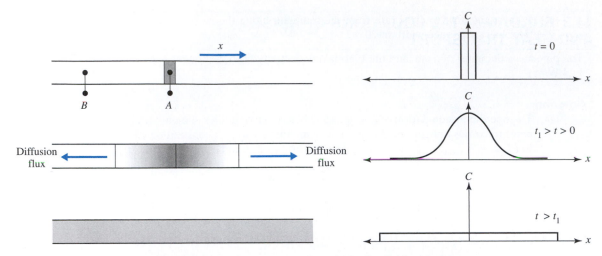

FIGURE 14.3 Spreading of dye within the fluid with time.

everywhere in the channel ($t > t_1$), there will be no net flux of dye across any section of the channel. The diffusion of dye in a binary system of dye and water occurs because of a concentration gradient of dye. Random molecular motions of the dye and water molecules lead, after a sufficient time, to a uniform mixture of dye and water. This is termed **ordinary binary diffusion**.

The mass of dye crossing a unit area per unit time in the x-direction is proportional to the gradient of dye concentration in that direction. This is **Fick's first law of diffusion**, which is stated mathematically as

$$J = -D \frac{dC}{dx} \tag{14.2.4}$$

In Eq. 14.2.4, D is the mass diffusion coefficient or mass diffusivity in a binary system, with dimension L^2/T, C is the molar concentration of diffusing dye, and the minus sign indicates that transport is from high to low concentration. This form of Fick's law is acceptable if the concentration varies only in the x-direction and if the fluid mixture is dilute, that is, the concentration of dye is low in the host fluid.

Ordinary binary diffusion:
Condition where random molecular motions of dye and water molecules, for instance, lead to a uniform mixture after a sufficient amount of time.

Fick's first law of diffusion:
The mass of dye crossing a unit time in the x-direction is proportional to the gradient of dye concentration in that direction.

14.2.4 Closure

The transport equations for momentum, energy, and mass have been introduced in this section. The equations essentially state that molecular transport of momentum, energy, and mass occurs due to the gradient of momentum concentration, energy concentration, and mass concentration, respectively. Note that these laws are linear transport laws with only the first derivatives of the momentum, energy, and mass concentration appearing in the equations. For the transport processes on a molecular level, this is acceptable since mean free paths of molecules are too small to be affected by the velocity, temperature, and mass concentration profiles.

14.3 FUNDAMENTAL EQUATIONS OF MASS AND HEAT TRANSPORT

14.3.1 Control Volume Formulation

In Section 4.2 the Reynolds transport theorem, Eq. 4.2.10, was derived. It expresses the time rate of change of an extensive property in terms of quantities that pertain to a control volume. The law of conservation of mass states that accumulation = input − output + generation − degradation. In other words: what accumulates in the control volume equals what goes in minus what goes out plus what is made there and minus what is depleted there. Let us now apply Eq. 4.2.10 to the conservation of mass of a substance that moves with a fluid. Consider a fluid mixture that contains a substance such as a chemical contaminant or a dissolved gas. Let the extensive property of the substance be its mass, and define C to be the concentration of the substance, expressed as mass per unit volume of the mixture. In addition, let $\mathbf{J}$ be the mass flux vector of the substance through the control surface, r_g be the rate of generation of the substance within the control volume, and r_d be the rate of degradation of the substance within the control volume due to a chemical reaction, for example. Then Eq. 4.2.10 becomes

$$\int_{c.v.} \frac{\partial}{\partial t} C \, dV = -\int_{c.s.} \mathbf{J} \cdot n \, dA + \int_{c.v.} r_g \, dV - \int_{c.v.} r_d \, dV \qquad (14.3.1)$$

In words, Eq. 14.3.1 states that the rate of *accumulation* of the substance within the control volume is equal to the *net mass flow* of the substance through the control surface plus the rate of *generation* of the substance within the control volume minus the rate of *degradation* of the substance within the control volume.

Equation 14.3.1 can be simplified for situations in which the concentration of substance within the control volume is uniform, that is, the fluid and substance are considered to be well mixed. For such a case, the flux of the substance through the control surface is expressed simply by the concentration multiplied by the normal component of the discharge at the surface. In many situations, however, the concentration is influenced by spatial changes in the mass flux vector so that an incremental control volume must be utilized. Consider the one-dimensional model shown in Fig. 14.4, such as would be encountered in pipe or a stream.

Here it is assumed that changes in the concentration of the substance take place only in the x-direction. The control volume has the length Δx and cross-sectional area A normal to the x-direction with flux $\mathbf{J}$ at the left area. Since the control volume is of fixed magnitude $A\Delta x$, application of Eq. 14.3.1 yields

$$\frac{\partial C}{\partial t} A \, \Delta x = -\left[-JA + \left(J + \frac{\partial J}{\partial x}\Delta x\right)A\right] + r_g A\Delta x - r_d A\Delta x \qquad (14.3.2)$$

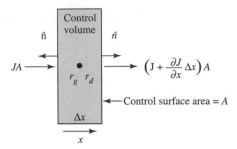

FIGURE 14.4 Rate of mass transfer of an infinitesimal fluid element.

Dividing by $A\Delta x$, Eq. 14.3.2 becomes

$$\frac{\partial C}{\partial t} + \frac{\partial J}{\partial x} = r_g - r_d \qquad (14.3.3)$$

Equation 14.3.3 is the fundamental differential mass transport equation in one dimension. The three-dimensional form of Eq. 14.3.3 can be derived in a similar manner considering the mass flux components in the three coordinate directions. The result is

$$\frac{\partial C}{\partial t} + \nabla \cdot \mathbf{J} = r_g - r_d \qquad (14.3.4)$$

in which the gradient operator is defined by Eq. 5.2.5. Before solutions of either Eq. 14.3.3 or 14.3.4 can be developed, it is necessary to explore the nature of **J** and relate how it affects the substance concentration C. First, an example.

Example 14.2

A tank with a fixed volume of 10 L has water flowing through an inlet and an outlet at a constant rate of 0.5 L/min. The water in the tank is continuously mixed by a mechanical device. The concentration of dissolved oxygen in the tank is initially zero, and the inlet flow is known to be at the saturation concentration C_s. It is desired to determine the concentration of dissolved oxygen in the tank ten minutes after initiation of flow under two conditions: (a) The tank is open to the atmosphere so that oxygen can transfer across the air–water interface; (b) The tank is completely sealed with oxygen entering the system only through the inlet. Assume that oxygen flux occurs at the air–water interface according to the relation $r = k\,(C_s - C)$. Given data are $C_s = 9$ mg/L and $k = 0.004$ sec^{-1}.

Solution

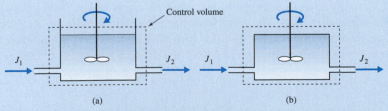

FIGURE E14.2 (a) Open tank with mass transfer at air–water interface, (b) closed tank with no mass transfer at air–water interface.

The control volumes for situations (a) and (b) are shown. The control volumes are constant, there is constant flux of oxygen at the inlet and the outlet, and oxygen transfer at the air–water interface is generation of the oxygen within the control volume, hence Eq. 14.3.1 reduces to

$$\frac{dC}{dt}\mathcal{V} = J_1A_1 - J_2A_2 + r_g\mathcal{V}$$

Substituting in the given expression for $J_1A_1 = QC_s, J_2A_2 = QC$ and $r_g = k(C_s - C)$, we can rearrange the equation in the manner

$$\frac{dC}{dt} = -\frac{Q}{\mathcal{V}}(C - C_s) + k(C_s - C) = \left(k + \frac{Q}{\mathcal{V}}\right)(C_s - C)$$

In this general situation there is surface transfer with the input concentration at location 1 being C_s. The equation can be rearranged to allow integration by separation of variables:

$$\int_0^c \frac{dC}{(C_s - C)} = \left(k + \frac{Q}{\mathcal{V}}\right)\int_0^t dt$$

The solution is

$$C = C_s\left[1 - e^{-\left(k+\frac{Q}{\mathcal{V}}\right)t}\right]$$

(a) Substitute known values of $k, Q, \mathcal{V}$, and t into the solution to find the concentration at the desired time, noting that $k = 0.004 \times 60 = 0.24$ min^{-1}, and $\frac{Q}{\mathcal{V}} = \frac{0.5}{10} = 0.05$ min^{-1}:

$$C = 9 \times [1 - e^{-(0.24+0.05)\times10}] = 85 \text{ mg/L}$$

Let's solve this problem again but with another choice for the control volume, one with its top surface below the water level. Oxygen transfer at the air–water interface is flux of oxygen to the control volume; hence Eq. 14.3.1 reduces to

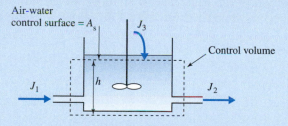

FIGURE E14.2(c)

$$\frac{dC}{dt}\mathcal{V} = J_1A_1 - J_2A_2 + J_3A_s$$

Substituting in the above expression for J_1A_1, J_2A_2, and $J_3 = k\,h(C_s - C)$; and dividing by $\mathcal{V} = A_sh$ yields

$$\frac{dC}{dt} = \left(k + \frac{Q}{\mathcal{V}}\right)(C_s - C)$$

This is the same result as given above. This example shows that there may be more than one good choice for a control volume.

(b) Beginning with the control volume formulation, we note that $r_g = 0$, which is equivalent to $k = 0$ in the solution derived for part (a). Hence

$$C = C_s[1 - e^{\frac{Q}{\mathcal{V}}t}] = 9 \times [1 - e^{-0.05\times10}] = 3.5 \text{ mg/L}$$

Thus in the open system the water reaches $8.5/9 \times 100 \approx 94\%$ saturation limit after 10 min, while at the same time in the closed system the saturation is only $3.5/9 \times 100 \approx 39\%$.

14.3.2 Generation and Degradation Processes

The flux of the substance, **J**, through the control volume leaves the molecular structure of the substance unchanged. In this section we focus our attention to processes by which a substance is transformed into other substances by chemical and biologically mediated reactions. Suppose we perform an experiment in which we determine how a substance concentration changes as a function of time in lake water. By measuring the concentration of substance in a series of well-mixed bottles filled with the lake water, we would develop data for the concentration of substance over time. When plotting the concentration of substance as a function of time, we observe an exponential decrease in concentration as schematically presented in Fig. 14.5. The mass balance of substance in the bottle is described by Eq. 14.3.3. The flux of substance in and out of the bottle control volume is zero. Thus the Eq. 14.3.3 reduces to

$$\frac{dC}{dt} = -r_d \qquad (14.3.5)$$

In words, Eq. 14.3.5 states that the rate of depletion of the substance within the control volume is equal to the rate of transformation of the substance within the control volume. *Chemical kinetics* is concerned with the velocity of reactions. Therefore, Eq. 14.3.5 is a useful measure of *reaction rate* for substance. Experimental evidence shows that the rate of reaction is influenced by energy and composition of the substance. By energy we consider the temperature, the light, and the magnetic field intensity. Ordinarily, we only take account of the

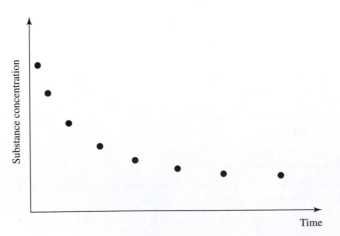

FIGURE 14.5 Decrease in pollutant concentration as a function of time in water.

temperature. The composition of substance is expressed by a function of the concentration of the reactants. Thus Eq. 14.3.5 can be represented as

$$\frac{dC_a}{dt} = -r_a = -k\,f\,(C_a, C_b, \ldots) \tag{14.3.6}$$

The minus sign indicates a decrease of substance with time, k is the temperature-dependent *rate constant*, $f(C_a, C_b, \ldots)$ is the function of the concentration of the reactants, and C_a is the concentration of the reactant "a." In this chapter we focus on a single reactant C. We can write Eq. (14.3.6) for a single reactant as

$$\frac{dC}{dt} = -k\,C^n \tag{14.3.7}$$

where "n" is the *order of the reaction*. Reactions are named to be of the first, second, and third order, with the order of the reaction being defined by the exponent n. Consideration must be paid to the units of the reaction rate constant, k. They are different for the different orders and depend on concentration units used (Table 14.1).

Equation 14.3.7 is an ordinary differential equation in which C is the dependent variable and t is the independent variable. We can integrate the equation by separation of variables and obtain an integrated form for C as a function of time, t (Table 14.1). From Table 14.1 we can see that, if a reaction obeys a zero-order model, a plot of C versus time should yield a straight line. Similarly, a plot of $\ln C$ versus time and $1/C$ versus time should yield straight lines for first-, and second-order models, respectively. The rate constant k can be determined from the slope of this line. The reaction order is determined by guessing an order and checking a goodness of fit with the data.

TABLE 14.1   Reaction Order for Irreversible Reactions

Order	Equation Form	Integrated Form	Plot Dependent	Plot Independent	Slope	Dimensions k
Zero $(n = 0)$	$\dfrac{dC}{dt} = \pm k$	$C = C_0 \pm kt$ at $t = 0\ C = C_0$	C	t	$\pm k$	$M/(L^3 T)$
First $(n = 1)$	$\dfrac{dC}{dt} = \pm kC$	$C = C_0 \exp(\pm k\,t)$ at $t = 0\ C = C_0$	$\ln C$	t	$\pm k$	$1/T$
Second $(n = 2)$	$\dfrac{dC}{dt} = \pm kC^2$	$\dfrac{1}{C} = \dfrac{1}{C_0} \pm kt$ at $t = 0\ C = C_0$	$\dfrac{1}{C}$	t	$\mp k$	$L^3/(M\,T)$

Example 14.3

The following are chlorophyll-a data from a laboratory experiment to assess a phytoplankton growth in a series of well mixed bottles.

Time (days)	0	0.25	0.5	0.75	1	1.25	1.5
Clorophylla-a concentration (mg/L)	0.052	0.068	0.091	0.120	0.145	0.182	0.235

Determine (a) the reaction order, and (b) growth rate constant.

Solution

(a) The flux of chlorophyll-a in and out of the bottle control volume is zero. We have an increase in chlorophyll-a concentration over time. Thus the Eq. 14.3.4 is

$$\frac{dC}{dt} = r_g = k\, C^n.$$

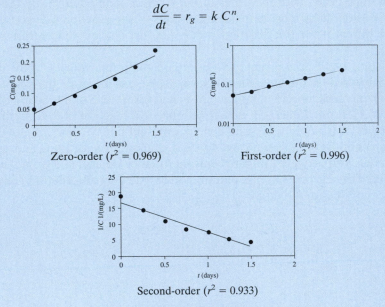

Zero-order ($r^2 = 0.969$) First-order ($r^2 = 0.996$)

Second-order ($r^2 = 0.933$)

FIGURE E14.3

Assuming zero-, first-, and second-order models, the data are plotted in Fig. E14.3. Each plot includes a straight line developed from a least-squares analysis of the data. By visual inspection we can see that the plot of ln (C) versus t, most closely approximates a straight line. In addition, the coefficient of variation, r^2, is the largest for the first-order model. Therefore, the phytoplankton growth is most closely represented by first-order reaction.

(b) The growth rate constant can be determined directly from the slope of the straight line. The best-fit line for the first-order model is

$$\ln C = -2.923 + 0.994\, t$$

where $k = 0.994$ (1/day) and $\ln C_0 = -2.923$. The resulting model is

$$C = C_0 \exp(kt) = \exp(-2.93)\exp(0.994\, t) = 0.053 \exp(0.994\, t)$$

The rate of most reactions in aquatic environments increases with temperature. This dependency is provided by an Arrhenius equation

$$k(T_2) = k(T_1)\, \theta^{T_2 - T_1} \qquad (14.3.8)$$

where k is the temperature-dependent reaction rate constant at temperature T, $\theta = e^{\frac{E}{RT_{a1}T_{a2}}}$ is a constant usually within the range 1.0–1.10, E is the activation energy (J/mole), R is the universal gas constant (8.314 J/mole K), and T_a is the absolute temperature in K. Some typical values of θ that are used in water quality modeling are summarized in Table 14.2

TABLE 14.2   Typical Values of θ Used in Water Quality Modeling

Process	θ
Biochemical oxygen demand	1.05
Bacterial respiration	1.03
Phytoplankton growth	1.06
Zooplankton grazing	1.07
Oxygen reaeration	1.03
Sediment oxygen demand	1.03

Example 14.4

A laboratory experiment was conducted to characterize the rate of bacterial growth. The first-order rate constant for bacterial growth is 0.1 day^{-1} at 20°C and 0.14 day^{-1} at 30°C. (a) What is the value of θ? (b) What is the first-order rate constant at 25°C?

Solution

(a) Rearranging the Eq. 14.3.8, we have

$$\theta = \exp\left[\frac{\ln k(T_2) - \ln k(T_1)}{T_2 - T_1}\right]$$

Substitute known values of $k(T_2), k(T_1), T_2,$ and T_1

$$\theta = \exp\left[\frac{\ln k(T_2) - \ln k(T_1)}{T_2 - T_1}\right] = \exp\left[\frac{\ln(0.14) - \ln(0.1)}{30 - 20}\right] = 1.03$$

(b) The rate constant at 25°C is

$$k(T_2) = k(T_1)\theta^{T_2 - T_1} = 0.1 \times 1.03^{25 - 20} = 0.116 \ \text{1/day}$$

14.3.3   Diffusive Transport

Diffusive transport is expressed in terms of the flux vector component (Eq. 14.2.4) that describes the net transport of mass along the x-axis per unit cross-sectional area and per unit time. However, in some cases we would like to know how the concentration of mass changes in time at a particular point in a flow domain. As an illustrative example consider dye spreading in a long channel

(Fig. 14.3). It is assumed that the channel is shallow with uniform dye concentration at a given x-location. The diffusion occurs due to the concentration gradient along the unbounded x-axis. A mass balance on the infinitesimal fluid element (Fig. 14.4) for the conservative dye, $r_g = 0$ and $r_d = 0$, reduces Eq. 14.3.3 to

$$\frac{\partial C}{\partial t} = \frac{\partial J}{\partial x} \qquad (14.3.9)$$

Recognizing that the concentration of diffusing dye depends upon time and space, the partial derivative is used in place of the ordinary derivative in Fick's first law, Eq. 14.2.4. Equation 14.3.9 when combined with Eq. 14.2.4 becomes

$$\frac{\partial C}{\partial t} = D \frac{\partial^2 C}{\partial x^2} \qquad (14.3.10)$$

This is the one-dimensional version of *Fick's second law*. The three-dimensional form of Eq. 14.3.10 can be derived in a similar manner by considering a mass balance on an infinitesimal cube including all three coordinate directions, and substance generation within the control volume. The result is

$$\frac{\partial C}{\partial t} = D\nabla^2 C + r_g - r_d \qquad (14.3.11)$$

where $\nabla^2 = \dfrac{\partial^2}{\partial x^2} + \dfrac{\partial^2}{\partial y^2} + \dfrac{\partial^2}{\partial z^2}$. Equations 14.3.10 and 14.3.11 are known as diffusion equations; Equation 14.3.10 is Eq. 14.3.11 simplified for one dimension with conservative substance ($r_g = 0$ and $r_d = 0$).

The identical approach can be applied to develop the heat diffusion equation in a liquid:

$$\rho c_v \frac{\partial T}{\partial t} = K\nabla^2 T + S \qquad (14.3.12)$$

where S is the source term associated with heat production (or dissipation) within the control volume.

14.3.4 Boundary Conditions

The diffusion equations related to mass and heat transfer outlined in this section have to be solved with due considerations given to boundary and initial conditions. For the one-dimensional models, three boundary conditions are commonly encountered in mass and heat transfer; they are applied at $x = 0$ and commonly at $x = \pm\infty$ and are summarized in Table 14.3. The simplest boundary condition is one of constant prescribed concentration. The second, more complicated one is prescribed mass flux. The third boundary condition is of practical significance in environmental problems; it consists of advective mass flux between the boundary

TABLE 14.3   Boundary Conditions for the Diffusion Equation Acting at the Control Surface

Condition at the Surface	Formulation	Graphical Description
1. Constant surface concentration	$C(0, t) = C_0$	
2. Constant surface mass flux		
(a) Finite flux	$J_0 = -D \left. \dfrac{\partial C}{\partial x} \right|_{x=0}$	
(b) No flux	$\left. \dfrac{\partial C}{\partial x} \right|_{x=0} = 0$	
3. Advection surface condition	$J_0 = -D \left. \dfrac{\partial C}{\partial x} \right|_{x=0} = k_m \left(C(0, t) - C_\infty \right)$	

and the adjacent environment. In the formulation for this condition k_m is a mass transfer coefficient.

Solutions of the diffusion equation are usually integrals of exponentials, such as error functions or exponential functions. As an illustrative example consider the dye spreading example, Fig. 14.3. In order to solve Eq. 14.3.10 we have to specify two boundary conditions on x and one initial condition. The initial condition at $t = 0$ is the pulsed input, $\left. \dfrac{\partial C}{\partial x} \right|_{x=0} = 0$, where M is the total mass of dye in the channel, A is the cross-sectional area across which the dye is diffusing, and $\delta(x)$ is the Dirac-delta function.[2] The boundary conditions are $C = 0$ at $x = \infty$ and at $x = 0$. The first boundary condition implies that far from the pulse input the dye concentration is zero; the second boundary condition implies that the dye flux has the same magnitude in the positive and negative directions. The solution of Eq. 14.3.10 with the initial and boundary conditions stated above is

$$C(x, t) = \frac{M}{A \sqrt{4\pi Dt}} \, e^{-\frac{x^2}{4Dt}} \qquad (14.3.13)$$

Equation 14.3.13 can be transformed in the following manner. Dividing by the total amount of mass M in the channel, substituting the relation

$$\sigma = \sqrt{2Dt} \qquad (14.3.14)$$

[2]This function has the features that it is infinite at $x = 0$, and zero otherwise, with the integral along the x-axis $\int_{-\infty}^{\infty} \delta(x) \, dx = 1$. The above initial condition can be shown to be appropriate by considering the simple mass balance $\int_{-\infty}^{\infty} CA \, dx = \int_{-\infty}^{\infty} \frac{M}{A} \delta(x) \, A \, dx = M$.

and replacing $AC(x, t)/M$ by $p(x)$, Eq. 14.3.13 becomes

$$p(x) = \frac{1}{\sqrt{2\pi}\,\sigma}\, e^{-\frac{x^2}{2\sigma^2}} \tag{14.3.15}$$

Equation 14.3.15 is known as a Gaussian distribution with the mean, or central tendency at $x = 0$, and a standard deviation given by σ.

It is common to designate the central tendency as the first statistical moment and the square of the standard deviation (the variance) as the second statistical moment. Equation 14.3.15 implies that the centroid of the dye cloud keeps its location at $x = 0$, and the dye spreading due to diffusion is measured by the variance. Equation 14.3.14 states that the variance σ^2 of the diffusing cloud increases linearly with time. This condition has been used to estimate the diffusion coefficient from the measured time rate of increase of the variance of the diffusing substance:

$$D = \frac{1}{2}\frac{d\sigma^2}{dt} \tag{14.3.16}$$

Example 14.5

The steady-state population of flies and mosquitoes varies in a control volume of air with elevation as shown:

Elevation (m)	50	300	600	900	1500
C (Number/m³)	630	210	110	70	20

An entomologist observes that the number of flies and mosquitoes is controlled by diffusion and settling (gradual sinking). The settling is assumed to be proportional to the concentration of insects at any elevation. Use the given data to verify this hypothesis.

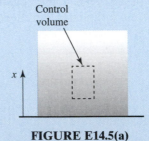

FIGURE E14.5(a)

Solution
Let $C(x)$ denote the concentration of insects at position x, the vertical direction. It is assumed that changes in concentration take place only in the x-direction. The mass bal-

ance of an infinitesimal control volume is given by Eq. 14.3.3. The rate of settling of the insects decreases the concentration of insects within the control volume and is given by $r_d = k_s C$, where k_s is the settling rate coefficient. Hence for the steady-state case Eq. 14.3.3 reduces to

$$\frac{dJ}{dx} = -r_d = -k_s C$$

The flux of insects in and out of the control volume is due to random or "diffusionlike" movement. By analogy to the random movement of the molecules, the flux is given by Fick's first law, Eq. 14.2.4, where D is the diffusivity of the insects. Thus, the above equation reduces to

$$-D\frac{d^2 C}{dx^2} + k_s C = 0$$

This equation is subject to two constant concentration boundary conditions: $C = C_0$ at $x = 0$ and $C = 0$ at $x = \infty$. The solution of the equation with the boundary conditions stated above is

$$\frac{C}{C_0} = e^{-\sqrt{\frac{k_s}{D}}x}$$

Physically this steady-state distribution presents a balance between the diffusive flux of insects in the direction of decreasing concentration (x) and the settling flux in the negative x-direction. A least-squares analysis of the data will provide the best-fit straight line. The result is

$$\ln C = -\sqrt{\frac{k_s}{D}}\, x + \ln C_0 \qquad \text{where} \quad \sqrt{\frac{k_s}{D}} = 0.002 \qquad \text{and } \ln C_0 = 6.2$$

Graphical comparison between the data denoted by symbols and the steady-state distribution is given below.

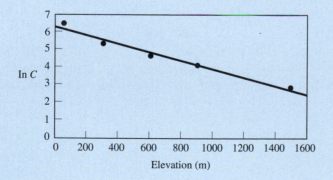

FIGURE E14.5(b)

14.3.5 Advective and Diffusive Transport

Transport of substances by molecular diffusion is an important source of motion on the microscopic scale. Molecular diffusion is extremely slow, thus its importance is significant only over short distances or over long time scales. Examples of the short distances dominated by molecular diffusion are: the diffusive sub-boundary layer thickness[3] at the sediment-water interface or air–water interface, the reaction distance between different substances, the distance over which kinetic energy is converted into heat, and the boundary layer thickness around a living cell. These short distances are integral parts of the macroscopic scales found in lakes, oceans, rivers, and subsurface aquifers. An estimate for the diffusion time scale t_{diff} over a certain distance L follows from Eq. 14.3.14:

$$t_{\text{diff}} = \frac{L^2}{2\,D} \qquad (14.3.17)$$

Consider the time it takes oxygen to diffuse across a 100 cm depth of water at a lake surface. The molecular diffusion coefficient for oxygen in water is about $D = 1.8 \times 10^{-5}$ cm²/s at 20°C. Thus the required time is $t_{\text{diff}} = 100^2/2 \times 1.8 \times 10^{-5} \approx 9$ years. If the molecular diffusion is the only transport process by which the oxygen is diffused in the water, fish and aquatic organisms will not exist in an aquatic environment. Molecular diffusion, although present, is dominated by a more rapid transport mechanism performed by the motion of the fluid itself. This kind of transport is called *advection*. Typical advection velocities are between 1 to 100 cm/s in lakes, rivers, and oceans. The transport time due to advection, for the case of constant velocity u, is $t_{\text{ad}} = \dfrac{L}{u}$. Thus, the transport time due to the advection across the 100 cm distance for the average advection velocity of the order 50 cm/s is $t_{\text{ad}} = 100/50 \approx 2$ sec.

Advective transport is expressed in terms of the flux vector along a given axis per unit area perpendicular to that axis. Thus in the x-direction the advective flux is

$$J = uC \qquad (14.3.18)$$

where u is the fluid velocity in the x-direction, and C is the substance concentration. The total flux of mass transport that results from the advection and diffusion is

$$J = uC - D\,\frac{\partial C}{\partial x} \qquad (14.3.19)$$

The diffusion flux is simply superimposed on the moving fluid. Equation 14.3.19 can be substituted into the conservation of mass, Eq. 14.3.3, with the result

---

[3]The diffusive sub-boundary layer is the zone of pure molecular transport.

$$\frac{\partial C}{\partial t} + \frac{\partial}{\partial x}(uC) = D\frac{\partial^2 C}{\partial x^2} + r_g - r_d \tag{14.3.20}$$

Equation 14.3.20 is the one-dimensional equation for mass transport; it included both advection and diffusion.

The mass transport equation in three dimensions can be obtained by considering mass flux components in the three coordinate directions, along with the conservation of mass for an incompressible fluid, Eq. 5.2.8. The result is

$$\frac{\partial C}{\partial t} + u\frac{\partial C}{\partial x} + v\frac{\partial C}{\partial y} + w\frac{\partial C}{\partial z} = D\nabla^2 C + r_g - r_d \tag{14.3.21}$$

The energy equation, sometimes called the *temperature equation* (since temperature is thermal energy per unit volume), has a form similar to the mass transport equation. The major difference is in the source term. As presented in Section 5.4, one form of the energy equation for incompressible flow is

$$\frac{\partial T}{\partial t} + u\frac{\partial T}{\partial x} + v\frac{\partial T}{\partial y} + w\frac{\partial T}{\partial z} = \frac{K}{\rho c_v}\nabla^2 T + \frac{S}{\rho c_v} \tag{14.3.22}$$

The terms in Eq. 14.3.22 are analogous to the corresponding terms in Eq. 14.3.21. The term S is associated with thermal energy production or dissipation.

A Lagrangian viewpoint can be employed to transform Eq. 14.3.20 to a form that is easier to solve analytically. Consider a coordinate system whose origin x_m moves at velocity u. The distance $\tilde{x}$ relative to the moving origin is $\tilde{x} = x - x_m = x - u\,\tilde{t}$ where $t = \tilde{t}$. Substituting $x = \tilde{x} + u\,\tilde{t}$ into Eq. 14.3.20 with $r_g = 0$ and $r_d = 0$ will produce the result[4]

$$\frac{\partial C}{\partial \tilde{t}} = D\frac{\partial^2 C}{\partial \tilde{x}^2} \tag{14.3.23}$$

which is identical to Eq. 14.3.10. Hence the governing equation for the mass transport with advection and diffusion in a fixed reference frame is the same as that with only diffusion in a stagnant fluid when viewed in a coordinate system moving at speed u. The dye spreading example shown in Fig. 14.3 in a moving fluid would result in the same solution as Eq. 14.3.13, but adjusted for the moving coordinates:

$$C(x,t) = \frac{M}{A\sqrt{4\pi Dt}}\,e^{\frac{(x-ut)^2}{4Dt}} \tag{14.3.24}$$

[4]
$$\frac{\partial}{\partial x} = \frac{\partial}{\partial \tilde{x}}\frac{\partial \tilde{x}}{\partial x} + \frac{\partial}{\partial \tilde{t}}\frac{\partial \tilde{t}}{\partial x} = \frac{\partial}{\partial \tilde{x}}1 + 0 = \frac{\partial}{\partial \tilde{x}}$$

$$\frac{\partial}{\partial t} = \frac{\partial}{\partial \tilde{x}}\frac{\partial \tilde{x}}{\partial t} + \frac{\partial}{\partial \tilde{t}}\frac{\partial \tilde{t}}{\partial t} = \frac{\partial}{\partial \tilde{x}}(-u) + \frac{\partial}{\partial \tilde{t}}1 = -u\frac{\partial}{\partial \tilde{x}} + \frac{\partial}{\partial \tilde{t}}$$

Example 14.6

The chemical benzyl chloride is discharged through a multiple diffuser into a river at a rate of $I = 1.2$ mol/h. The benzyl chloride is not conserved in the water, that is, it reacts with the river water and produces benzyl alcohol. A decrease rate of benzyl chloride concentration can be explained by a first-order reaction. From laboratory measurements the reaction rate constant is estimated to be $k = 0.046 \text{ hr}^{-1}$ at 25°C. In addition, benzyl chloride experiences air–water exchange at the water surface. Assume that transfer at the air–water interface takes the form $r = \dfrac{k_m}{H} C$, where $k_m = 6 \times 10^{-5}$ m/s is the mass transfer coefficient, and H is the mean river depth. Considering only the advective flux in a steady, one-dimensional flow, estimate the concentration of benzyl chloride 5 km downstream from the discharge point. The river has a discharge $Q_r = 25$ m³/s, an average cross-sectional area of 20 m² with a mean depth of 1 m.

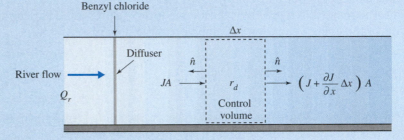

FIGURE E14.6

Solution

We will assume that the benzyl chloride is well-mixed over the river cross-sectional area, A, at the point of discharge, and it changes only along the flow direction. Thus, the concentration, C, is a cross-sectional average; lateral and vertical deviations from the average are considered to be small and negligible.

The benzyl chloride is depleted in the control volume due to the chemical reaction and transfer at the air–water interface. The rate of degradation of benzyl chloride due to the chemical reaction is $r_{d1} = kC$. The transfer at the air–water interface is $r_{d2} = k_m \dfrac{C}{H}$. The mass balance of benzyl chloride is described by Eq. 14.3.20. The mass transport equation that results only from advection in the steady-state flow reduces Eq. 14.3.20 to

$$U \frac{dC}{dx} = -r_d$$

where U is the cross-sectional average velocity; We let $D = 0$ since we are ignoring diffusion. Thus the above equation reduces to

$$\frac{dC}{dx} = -\frac{1}{U}\left(k + \frac{k_m}{H}\right)C$$

This differential equation is subject to the constant concentration boundary condition $C = C_0$ at $x = 0$. Integrating the above differential equation yields

$$C = C_0 \, e^{-\frac{1}{U}\left(k + \frac{k_m}{H}\right)x}$$

The initial concentration C_0 can be estimated from the steady-state mass balance at the discharge point:

$$C_0 = \frac{I}{Q_r} = \frac{1.2 \times \dfrac{1}{3600}}{25 \times 10^3} = 13.3 \times 10^{-9} \text{ mol/L} = 13.3 \text{ nmol/L}$$

The average cross-sectional velocity $U = Q/A = 25/20 = 1.2$ m/s. The benzyl chloride concentration 5 km downstream from the discharge point is

$$C = 13.3 \, e^{-\frac{1}{12}\left(0.046\frac{1}{3600} + \frac{6\times10^{-5}}{1}\right)5000} = 9.8 \text{ nmol/L}$$

The benzyl chloride travel time over 5 km is $t = 5000/1.2 = 4200$ sec ≈ 1.2 h. The initial concentration of the benzyl chloride will be reduced by about 26% 5 km downstream from the diffuser.

14.4 TURBULENT TRANSPORT

14.4.1 Turbulent Diffusion

Most flows encountered in the environment are turbulent, and consequently, it is important to have some understanding of the nature of turbulent flow. A brief description of turbulent flow in a pipe is presented in Sections 7.6.1 and 7.6.2. The reader is encouraged to review that material at this time. In relation to transport processes in fluids, turbulent flow has the following significant consequences:

(a)   The concentration of a substance, water temperature, and velocity components are composed of the mean and the fluctuating components (Fig. 7.7). While the mean components are usually statistically stable over several repeatable measurements, the fluctuating components are irregular and unpredictable.

(b)   A rapid rate of spreading of a concentrated substance or a concentration of high temperature are due to the three components of the velocity fluctuations in the x-, y-, and z-directions and the mean velocity.

(c)   The identifiable structures in a turbulent flow are called **eddies**. A characteristic feature of turbulence is the existence of a large range of eddy sizes. The diameter of these eddies can be as large as the flow domain scale, that is, a river, lake or reservoir depth, the thickness width of a boundary layer along the earth's surface, or they can be a few millimeters in size.

Eddies: *Identifiable structures in a turbulent flow.*

The largest eddies in turbulent flow lose much of their kinetic energy in one revolution, and they are usually unstable. However, they transfer some of the kinetic energy to a smaller scale and thus generate smaller eddies. These smaller eddies become unstable and transfer some of their kinetic energy to yet smaller eddies, and so on. Such a net transfer of the energy from larger to smaller eddies is independent of fluid viscosity and is known as an **energy cascade**. The end of the cascade is at the smallest length scale with the highest frequency. Here the kinetic

Energy cascade: *A net transfer of energy from larger to smaller eddies independent of fluid viscosity.*

energy is converted into heat by the action of molecular viscosity. The order of length scale for the smallest eddies is given by the *Kolmogorov microscale*

$$\eta_K = \left(\frac{\nu^3}{\varepsilon}\right)^{1/4} \tag{14.4.1}$$

where ε is the rate of energy dissipation per unit mass and ν is the kinematic viscosity. The scale at which the rate of formation of concentration gradients are smeared by molecular diffusion is defined by the *Batchelor scale*

$$\eta_B = \left(\frac{D^2 \nu}{\varepsilon}\right)^{1/4} \tag{14.4.2}$$

where D is the molecular diffusion coefficient. The order of length scale for the smallest eddies is given in the following example.

Example 14.7

As a preliminary part of a tracer study, Freon-12 is to be mixed in a laboratory in a container of 100 kg of water at 25°C. We will use a 200 watt hand mixer for mixing. (a) What is the scale at which velocity fluctuations will disappear? (b) What is the scale at which the concentration gradients of Freon-12 will be smeared by molecular diffusion? The molecular diffusion coefficient for Freon-12 is $D = 10 \times 10^{-10}$ m²/s.

Solution
(a) The scale at which velocity gradients are smeared by viscosity is defined by Eq. 14.4.1. The rate of energy dissipation in 100 kg of water is $\varepsilon = 200$ watt/100 kg = 2 m²/s³. The kinematic viscosity of water at 25°C is 0.905×10^{-6} m²/s (Table B.1). Thus

$$\eta_K = \left(\frac{\nu^3}{\varepsilon}\right)^{1/4} = \left(\frac{(0.905 \times 10^{-6})^3}{2}\right)^{1/4} = 2.5 \times 10^{-5} \text{ m}$$

The scale at which fluctuations of a conservative scalar disappear is defined by Eq. 14.4.2. Freon-12 has a molecular diffusion coefficient of $D = 10 \times 10^{-10}$ m²/s.

$$\eta_B = \left(\frac{D^2 \nu}{\varepsilon}\right)^{1/4} = \left(\frac{(10 \times 10^{-10})^2\, 0.905 \times 10^{-6}}{2}\right)^{1/4} = 8 \times 10^{-7} \text{ m}$$

We must now incorporate turbulence in our transport equations for mass and energy derived in the previous section. Following the approach outlined in Section 7.6.1, we separate the flow velocity and substance concentrations into mean and fluctuating components, a procedure called *Reynolds decomposition*. Although the turbulence is a three-dimensional phenomenon, for simplicity we will restrict ourselves only to the x-component:

$$u = \bar{u} + u' \quad \text{and} \quad C = \bar{C} + C' \tag{14.4.3}$$

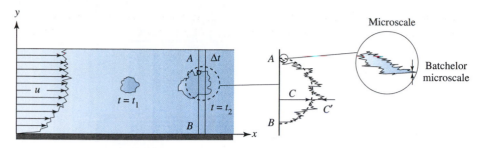

FIGURE 14.6   Spreading of dye in a turbulent flow.

where the overbar denotes a time average and the prime denotes the instantaneous fluctuation. Consider turbulent flow in a long channel into which a colored conservative dye is injected (Fig. 14.6). Turbulent velocity fluctuations in all three spatial directions will cause rapid spreading of the dye as it is carried by the mean flow in the x-direction. The fluctuations will generate large irregular dye surfaces on which molecular diffusion finally acts. Molecular diffusion is balanced by the increase in concentration gradients caused by turbulent fluctuations at the Batchelor microscale, Eq. 14.4.2. The dashed line represents the mean dye concentration $\overline{C}$ over a portion of the time record Δt measured at the fixed location A-B, and C' represents the instantaneous fluctuation.

Let us apply Reynolds' decomposition to the three-dimensional mass transport equation. After substituting Eq. 14.4.3 into Eq. 14.3.21 and aligning the x-axis with the flow so that $v = v'$ and $w = w'$, we obtain

$$\frac{\partial(\overline{C} + C')}{\partial t} + (\overline{u} + u')\frac{\partial(\overline{C} + C')}{\partial x} + v'\frac{\partial(\overline{C} + C')}{\partial y} + w\frac{\partial(\overline{C} + C')}{\partial z}$$
$$= D\frac{\partial^2(\overline{C} + C')}{\partial x^2} + (\overline{r}_g + r_g') - (\overline{r}_d + r_d') \qquad (14.4.4)$$

where $\overline{r}$ is the time average source/sink term, and r' is the fluctuation in the source term. After averaging over time we obtain

$$\overline{\frac{\partial(\overline{C} + C')}{\partial t}} + \overline{(u + u')\frac{\partial(\overline{C} + C')}{\partial x}} + \overline{v'\frac{\partial(\overline{C} + C')}{\partial y}} + \overline{w'\frac{\partial(\overline{C} + C')}{\partial z}}$$
$$= D\overline{\frac{\partial^2(\overline{C} + C')}{\partial x^2}} + \overline{r}_g + r_g' - \overline{(r_d + r_d')} \qquad (14.4.5)$$

In order to proceed further, we will use averaging rules that are provided in Chapter 7 (see Example 7.6). The averages of the four terms in Eq. 14.4.5 are

$$\overline{\frac{\partial(\overline{C} + C')}{\partial t}} = \frac{\partial \overline{C}}{\partial t} + \frac{\partial \overline{C'}}{\partial t} = \frac{\partial \overline{C}}{\partial t},$$

$$\overline{(\overline{u} + u')\frac{\partial(\overline{C} + C')}{\partial x}} = \overline{u}\frac{\partial \overline{C}}{\partial x} + \overline{u}\frac{\partial \overline{C'}}{\partial x} + \overline{u'\frac{\partial \overline{C}}{\partial x}} + \overline{u'\frac{\partial C'}{\partial x}} = \overline{u}\frac{\partial \overline{C}}{\partial x} + \overline{u'\frac{\partial C'}{\partial x}},$$

$$\overline{v'\frac{\partial(\overline{C}+C')}{\partial y}} = \overline{v'\frac{\partial\overline{C}}{\partial y}} + \overline{v'\frac{\partial C'}{\partial y}} = \overline{v'\frac{\partial C'}{\partial y}},$$

$$\overline{w'\frac{\partial(\overline{C}+C')}{\partial z}} = \overline{w'\frac{\partial\overline{C}}{\partial z}} + \overline{w'\frac{\partial C'}{\partial z}} = \overline{w'\frac{\partial C'}{\partial z}},$$

$$\overline{\frac{\partial^2(\overline{C}+C')}{\partial x^2}} = \frac{\partial^2\overline{C}}{\partial x^2} + \overline{\frac{\partial^2 C'}{\partial x^2}} = \frac{\partial^2\overline{C}}{\partial x^2}, \text{ and}$$

$$\overline{\overline{r_g}+r_g'} = \overline{\overline{r_g}}+\overline{r_g'} = \overline{r_g}$$

$$\overline{\overline{r_d}+r_d'} = \overline{\overline{r_d}}+\overline{r_d'} = \overline{r_d}$$

Collecting the averaged terms, the mean mass transport equation becomes

$$\frac{\partial\overline{C}}{\partial t} + \overline{u}\frac{\partial\overline{C}}{\partial x} + \overline{u'\frac{\partial C'}{\partial x}} + \overline{v'\frac{\partial C'}{\partial y}} + \overline{w'\frac{\partial C'}{\partial z}} = D\frac{\partial^2\overline{C}}{\partial x^2} + \overline{r_g} - \overline{r_d} \quad (14.4.6)$$

Finally, employing the continuity equation for turbulent fluctuations, namely $\frac{\partial u'}{\partial x} + \frac{\partial v'}{\partial y} + \frac{\partial w'}{\partial z} = 0$ (see Problem 7.42) and restricting ourselves only to the x component, results

$$\frac{\partial\overline{C}}{\partial t} + \overline{u}\frac{\partial\overline{C}}{\partial x} = D\frac{\partial^2\overline{C}}{\partial x^2} - \frac{\partial\overline{u'C'}}{\partial x} + \overline{r_g} - \overline{r_d} \quad (14.4.7)$$

In Eq. 14.4.7 the terms from left to right describe: the mean storage of tracer concentration, the advection of the tracer by the mean flow, the mean molecular diffusion of the tracer, the additional transport due to the turbulence, and the net mean body source for additional tracer processes. The additional transport due to the turbulence is called *turbulent diffusion* or *eddy diffusion*. It is noteworthy that this flux is advective in nature. For example, $\overline{u'C'}$ is the mean turbulent flux of tracer whose concentration is C' across the unit area normal to the velocity component u'. The turbulent flux components are clearly diffusive in nature. They tear apart a tracer cloud, and provide room for molecular diffusion to diffuse the tracer from a high to a low concentration. Note that Eq. 14.4.7, which describes the mean tracer concentration, is very similar to the basic advection-diffusion relation, Eq. 14.3.20, which we started with. The major difference between the two is the additional transport term due to the turbulence.

It is now necessary to relate the turbulent flux term to a property of the mean flow; this exercise is known as a *closure approximation*. The turbulent flux can often be represented as a gradient diffusion process, that is,

$$\overline{u'C'} = -D_{tx}\frac{\partial\overline{C}}{\partial x} \quad (14.4.8)$$

in which D_{tx} is the turbulent diffusion or eddy diffusion coefficient along the x axis. The dimension of D_{tx} is L^2/T. Note the similarity of Eq. 14.4.8 with Eq. 7.6.9, which describes the turbulent transport of momentum. Equations 7.6.9 and 14.4.8 are termed *gradient transport formulations*. In addition to the transport of momentum and a substance concentration, the gradient transport formulation can be applied to describe the flux of heat. Note that, whereas the molecular diffusion coefficient D, used in Eq. 14.2.4, is a known property of the fluid mixture, the eddy diffusion coefficient D_{tx} depends on the condition of the flow. Substituting Eq. 14.4.8 into Eq. 14.4.7 yields

$$\frac{\partial \overline{C}}{\partial t} + \overline{u}\,\frac{\partial \overline{C}}{\partial x} = D\,\frac{\partial^2 \overline{C}}{\partial x^2} + \frac{\partial}{\partial x}\left(D_{tx}\,\frac{\partial \overline{C}}{\partial x}\right) + \overline{r}_g - \overline{r}_d \qquad (14.4.9)$$

Assume that the turbulent diffusion coefficient is constant in the x-direction (i.e., the flow is developed), so that D_{tx} becomes D_t = constant. Equation 14.4.9 reduces to

$$\frac{\partial \overline{C}}{\partial t} + \overline{u}\,\frac{\partial \overline{C}}{\partial x} = (D + D_t)\,\frac{\partial^2 \overline{C}}{\partial x^2} + \overline{r}_g - \overline{r}_d \qquad (14.4.10)$$

Now compare the orders of magnitude of the turbulent diffusion and molecular diffusion coefficients in surface waters. Typical values for the turbulent diffusion coefficients in oceans, lakes, and rivers are given in Table 14.4.

Turbulent diffusion coefficients in oceans, lakes, and rivers vary by several orders of magnitude. A typical molecular diffusion coefficient is of the order of 10^{-5} cm²/s. Comparing the values in Table 14.4 with molecular diffusion, we see that turbulent diffusion is at least 100 times greater than molecular diffusion, and

TABLE 14.4   Representative Turbulent Diffusion Coefficient in Oceans, Lakes, and Rivers

Surface Water	D_t (cm²/s)
Ocean	
Vertical (mixed layer)	10 to 5000
Vertical (deep sea)	0.1 to 100
Horizontal	100 to 10^7
Lakes, reservoirs	
Vertical (mixed layer)	10 to 5000
Vertical (below thermocline)	0.001 to 0.1
Horizontal	10 to 10^6
Rivers	
Vertical	1 to 3000
Longitudinal[a]	10 to 8000
Transverse[b]	10 to 10000

[a] along the mean flow
[b] perpendicular to the mean flow

in most cases it is over 10,000 greater. Since $D_t \gg D$, Eq. 14.4.10 can be further simplified to

$$\frac{\partial \overline{C}}{\partial t} + \overline{u} \frac{\partial \overline{C}}{\partial x} = D_t \frac{\partial^2 \overline{C}}{\partial x^2} + \overline{r}_g - \overline{r}_d \qquad (14.4.11)$$

Notice that we have used the order of magnitude argument to neglect the molecular diffusion. However, it is not permissible to neglect the molecular diffusion where the turbulence is damped (e.g., near the solid boundaries) or over short distances (e.g., the diffusive sub-boundary layer thickness) the reaction distance between chemical particles, and the boundary layer thickness around a living cell.

The turbulent mass transport equation in three dimensions, subject to the mean flow in three directions, can be obtained by substituting $u = \overline{u} + u'$, $v = \overline{v} + v'$, $w = \overline{w} + w'$, $C = \overline{C} + C'$, and $r = \overline{r} + r'$ into Eq. 14.3.21 and following similar steps that lead to Eq. 14.4.11. The result is

$$\frac{\partial \overline{C}}{\partial t} + \overline{u} \frac{\partial \overline{C}}{\partial x} + \overline{v} \frac{\partial \overline{C}}{\partial y} + \overline{w} \frac{\partial \overline{C}}{\partial z} = D_t \nabla^2 \overline{C} + \overline{r}_g - \overline{r}_d \qquad (14.4.12)$$

where the turbulent diffusion coefficient D_t is assumed to be isotropic (identical in all three directions).

We can apply similar procedures to derive a heat transport equation in a turbulent flow, resulting in the relation

$$\frac{\partial \overline{T}}{\partial t} + \overline{u} \frac{\partial \overline{T}}{\partial x} + \overline{v} \frac{\partial \overline{T}}{\partial y} + \overline{w} \frac{\partial \overline{T}}{\partial z} = \alpha_t \nabla^2 \overline{T} + \frac{\overline{S}}{\rho c_v} \qquad (14.4.13)$$

where α_t is the turbulent diffusion coefficient for heat. In Eqs. 14.4.12 and 14.4.13 the overbars are often omitted with the understanding that the parameters are time-averaged values.

Example 14.8

In order to improve the water quality in a lake it is desirable to determine the movement of a substance in the lake. For that purpose, tetrachloroethene was injected and measured 5 days later in the lake. Tetrachloroethene is assumed to be conservative in water. Determine the magnitude and direction of the vertical flux of tetrachloroethene per unit time per unit area 7 m below the lake surface. The vertical profile of tetrachloroethene, measured from the lake surface at $z = 0$ (z is measured positive downward) is given below. The molecular diffusion coefficient for tetrachloroethene is $D = 10 \times 10^{-10}$ m²/s, and the vertical turbulent diffusion coefficient is $D_{tz} = 2.1 \times 10^{-5}$ m²/s.

Depth (m)	0	1	2	3	4	5	6	7	8	9	10
C (nmol/L)	2.0	3.5	5.5	5.3	4.3	3.4	3.3	3.1	2.8	2.7	2.6

Solution

Between 3 and 5 m depth there is a high decrease in concentration. This region corresponds to the thermocline[5] in the lake. Consider fluxes that originate from the molecular and turbulent diffusion at 7 m below the lake surface.

The flux of tetrachloroethene that results from molecular diffusion is given by Fick's first law (Eq. 14.2.4):

$$J = -D \left. \frac{\partial \overline{C}}{\partial z} \right|_{z=7\,m}$$

The vertical gradient at 7 m depth can be estimated using central differences[6]:

$$J = -D \, \frac{C|_{8\,m} - C|_{6\,m}}{8\,m - 6\,m} = -10 \times 10^{-10} \left(\frac{2.8 - 3.3}{2} \right) = 2.5 \times 10^{-10} \, \frac{m}{s} \, \frac{nmol}{L} \, \frac{10^3 L}{m^3}$$

$$= 2.5 \times 10^{-7} = 2.5 \times 10^{-7} \, \frac{nmol}{s \cdot m^2}$$

The flux has a positive sign, therefore the tetrachloroethene is transported along the z axis towards the lake bottom.

The flux of tetrachloroethene that results from turbulent diffusion is given by Eq. 14.4.8. For the vertical z-direction the equation becomes

$$J = \overline{w'C'} = -D_{tz} \, \frac{\partial \overline{C}}{\partial z} = -2.1 \times 10^{-5} \left(\frac{2.8 - 3.3}{2} \right) \times 10^3 = 5.2 \times 10^{-3} \, \frac{nmol}{s \cdot m^2}$$

The turbulent flux is positive, thus the tetrachloroethene is transported downward. The turbulent and molecular diffusion flux have the same direction since they are influenced by the same concentration gradient so both of the diffusion coefficients are positive. Notice that the turbulent flux is about 10^4 times higher than the molecular diffusion. We could have neglected molecular diffusion.

[5]The thermocline is the depth region with the highest temperature gradient.

[6] $\left. \frac{\partial f}{\partial x} \right|_{x_i} \approx \frac{f|_{x_{i+1}} - f|_{x_{i-1}}}{x_{i+1} - x_{i-1}}$.

14.4.2   Dispersion

Equations 14.4.12 to 14.4.13 provide the basis for studying transport processes of mass and heat in the environment. The scale of motions in a turbulent flow can range from the Kolmogorov scale (typical range 1 to 10 mm) to, for example, a lake basin scale (10 km). These equations, although simple looking, require a significant amount of information and effort to apply them to rivers, lakes, and coastal waters. Sometimes the primary interest is to obtain cross-sectional average solutions for the contaminant concentration or water temperature profiles. In order to proceed from the previous section it is necessary to simplify the equations. This simplification is obtained by integrating the transport equations over a cross-section of interest, a step that generates additional restrictions and transport terms.

As an example, consider a parallel turbulent flow bounded by two horizontal plates, as shown in Fig. 14.7. At time $t = 0$ a finite quantity of a conservative

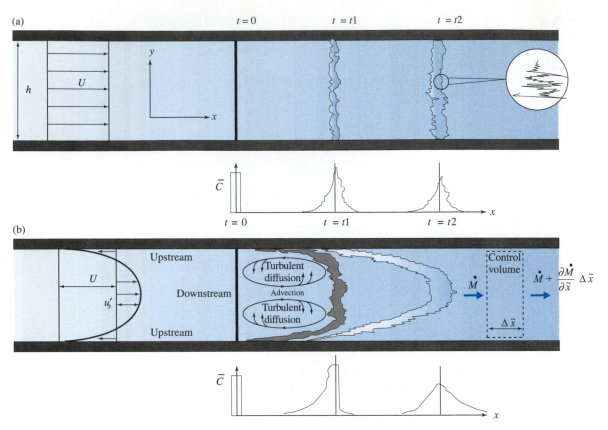

FIGURE 14.7 A mechanism of longitudinal dispersion. (a) Turbulent diffusion in a uniform flow with longitudinal distribution of the cross-sectional average concentration; (b) turbulent dispersion due to a nonuniform velocity distribution with the longitudinal distribution of the cross-sectional average concentration.

dye is introduced between the plates. Assume that the mean velocity between the plates is uniform (Fig. 14.7a). Dye carried by the mean flow spreads longitudinally due to turbulent mixing and the spread increases with time along the flow direction. Turbulent fluctuations tear apart the dye cloud somewhat like spaghetti, and thus increase the contact surface of the dye with the surrounding fluid. The concentration gradients are balanced by the vertical molecular diffusion at the Batchelor scale.

In the actual shear flow (Fig. 14.7b) adjacent fluid layers are moving with different longitudinal velocities. The cross-sectional average velocity can be obtained by integration $U = \dfrac{1}{h} \displaystyle\int_0^h u\, dy$ and the deviation of the velocity from the cross-sectional average is defined as $u_s' = u - U$. Similarly, we can define the cross-sectional average concentration[7] $\overline{C} = \dfrac{1}{h} \displaystyle\int_0^h C\, dy$ and the deviation of the concentration from the cross-sectional average as $c_s' = C - \overline{C}$. The distance

---

[7]In this section an overbar denotes a cross-sectional average concentration, with the understanding that the concentration is already time averaged.

between plates, h, is the area per unit width of flow. If one travels with the mean flow velocity, it will appear that dye at the midpoint between the plates moves downstream, while dye near the plates moves upstream. Therefore, this advective transport increases the dye concentration near the midpoint and decreases the dye concentration near the boundaries. The vertical difference in the concentration gradient should be balanced by vertical turbulent diffusion. These two mechanisms balance each other and cause a rapid rate of mixing or dispersion of the dye. Turbulent diffusion thickens the line of the dye (Fig. 14.7b, shaded area), while velocity shear determines the shape of the dye cloud. It appears that dispersion is a mechanism somewhat similar to the enhanced mixing due to turbulence described in Fig. 14.7a. The notable difference is that dispersion occurs on a larger scale in comparison to turbulent diffusion. In addition, the rate of mixing caused by dispersion is much larger than that by turbulent diffusion. Note also that dispersion caused by nonuniform velocity distributions can occur in non-turbulent flows; in this case the differential advection is balanced by the molecular diffusion.

Dye spread between two boundaries can be described by the three-dimensional relation for mass transport in a turbulent flow, Eq. 14.4.12. Taylor obtained the solution for this problem in 1953 by considering a balance between only longitudinal advection and vertical diffusion as shown in Fig. 14.7b. The major argument behind this approximation is that after the tracer has spent sufficient time in the flow, an equilibrium between longitudinal advection and turbulent diffusion becomes established. The mass balance of the dye over the control volume that moves with the mean flow velocity (Fig. 14.7b) is

$$\frac{\partial \overline{C}}{\partial t} h \Delta \tilde{x} = \dot{M} - \left(\dot{M} + \frac{\partial \dot{M}}{\partial \tilde{x}} \Delta \tilde{x} \right) \tag{14.4.14}$$

where $\overline{C}$ is the cross-sectional average concentration, and $\dot{M}$ is the total mass flux of the dye at any cross-section at $\tilde{x}$ that results from the velocity and concentration deviation from the cross-sectional average. Dividing by $h \Delta \tilde{x}$, and taking the limit as $\Delta \tilde{x} \to 0$ yields

$$\frac{\partial \overline{C}}{\partial \tilde{t}} = -\frac{1}{h} \frac{\partial \dot{M}}{\partial \tilde{x}} \tag{14.4.15}$$

The total mass flux across the cross-sectional area is

$$\dot{M} = \int_0^h u_s' c_s' \, dy = -h \, K \frac{\partial \overline{C}}{\partial \tilde{x}} \tag{14.4.16}$$

where u_s' is the deviation from the cross-sectional average velocity U, c_s' is the deviation from the cross-sectional average concentration $\overline{C}$, and K is the *longitudinal dispersion coefficient*, that is the coefficient that accounts for the effects on the cross-sectional averaged tracer concentration due to variations of velocity across the channel cross section. Note that the mean flow velocity does not transport any dye across the control volume boundaries because the volume is traveling at the mean velocity. Substituting Eq. 14.4.16 into Eq. 14.4.15 yields the diffusion equation for the cross-sectional average concentration

$$\frac{\partial \overline{C}}{\partial \tilde{t}} = K \frac{\partial^2 \overline{C}}{\partial \tilde{x}^2} \tag{14.4.17}$$

We can return to the fixed coordinate system (see footnote 4) using

$$\frac{\partial \overline{C}}{\partial t} + U \frac{\partial \overline{C}}{\partial x} = K \frac{\partial^2 \overline{C}}{\partial x^2} \tag{14.4.18}$$

This is the *one-dimensional advection-diffusion* or *dispersion equation* for mass transport. We have to be aware of Taylor's approximation, that is the equilibrium between the longitudinal advection and vertical turbulent diffusion, that we used in order to outline the dispersion equation.

Some common analytical solutions of the one-dimensional transport relation, Eq. 14.4.18, are available. The solutions, with their appropriate initial and boundary conditions are given in Table 14.6. In the first two equations the concentration varies both with respect to time and distance. In equation (1) there is a degradation term which is proportional to the concentration; this is the first-order reaction. Equations (2) and (3) do not contain a reaction term. In solutions 2 and 3 the function $\text{erfc}(\varphi)$ is the complementary error function. Values of the error function $\text{erf}(\varphi)$ are given in Table 14.5. Note that $\text{erfc}(\varphi) = 1 - \text{erf}(\varphi)$. Equation (3) is a steady-state relation in which the concentration varies with x only.

TABLE 14.5 Values of the Error Function[8]

ϕ	$\text{Erf}(\phi)$
0.0	0.0
0.1	0.1129
0.2	0.2227
0.3	0.3286
0.4	0.4284
0.5	0.5205
0.6	0.6309
0.7	0.6778
0.8	0.7421
0.9	0.7969
1.0	0.8427
1.2	0.9103
1.4	0.9523
1.6	0.9763
1.8	0.9891
2.0	0.9953
2.5	0.9996
3.0	0.99998
∞	1.00000

[8]$\text{erf}(\varphi) = \dfrac{2}{\sqrt{\pi}} \displaystyle\int_0^\varphi e^{-\zeta^2} \, d\zeta \cong 1 - \dfrac{1}{(1 + a_1\varphi + a_2\varphi^2 + a_3\varphi^3 + a_4\varphi^4)^4}$

$a_1 = 0.278393, a_2 = 0.230389, a_3 = 0.000972, a_4 = 0.078108.$

TABLE 14.6 Solutions of the One-Dimensional Transport Equation

Equation	Initial and Boundary Conditions	Solution
(1) $\dfrac{\partial C}{\partial t} + U\dfrac{\partial C}{\partial x} = K\dfrac{\partial^2 C}{\partial x^2} - kC$	Instantaneous input at $x = 0$ $C = \dfrac{M}{A}\delta(x), x = 0, t = 0$	$C(x,t) = \dfrac{M}{A\sqrt{4\pi Kt}}\exp\left[-\dfrac{(x-Ut)^2}{4Kt} - kt\right]$ Note: solution is valid when $k = 0$ or when $U = 0$
	Rectangular input at $x = 0$ $C = 0, x = \infty, t > 0$ $\dfrac{\partial C}{\partial x} = 0, x = 0, t > 0$ $C = 0, x \geq 0, t = 0$ $C = C_0, x = 0, \xi > t > 0$ $C = 0, x = 0, t > \xi$	$C(x,t) = \dfrac{C_0}{2}\exp\left(-\dfrac{kx}{U}\right)\left[G(t)\,\text{erfc}\left(\dfrac{x - Ut(1+\phi)}{\sqrt{4Kt}}\right) - G(t-\xi)\,\text{erfc}\left(\dfrac{x - U(t-\xi)(1+\phi)}{\sqrt{4K(t-\xi)}}\right)\right]$ $\phi = \dfrac{2kK}{U^2}$; $G(t-\xi) = 1$ when $t - \xi > 0$; $G(t-\xi) = 0$ when $t - \xi < 0$
(2) $\dfrac{\partial C}{\partial t} + U\dfrac{\partial C}{\partial x} = K\dfrac{\partial^2 C}{\partial x^2}$	A constant concentration C_0 is maintained at $x = 0$; $C = 0, x > 0, t = 0$ $C = 0, x = \infty, t \geq 0$ $C = C_0, x = 0, t \geq 0$ Case when $U = 0$	$C(x,t) = \dfrac{C_0}{2}\left[\exp\left(\dfrac{Ux}{K}\right)\text{erfc}\left(\dfrac{x + Ut}{\sqrt{4Kt}}\right) + \text{erfc}\left(\dfrac{x - Ut}{\sqrt{4Kt}}\right)\right]$ $C(x,t) = C_0\left[1 - \text{erf}\left(\dfrac{x}{\sqrt{4Kt}}\right)\right]$
(3) $K\dfrac{d^2C}{dx^2} - U\dfrac{dC}{dx} - kC = 0$	Continuous input of concentration C_0 at $x = 0$ $C = C_0$ at $x = 0$ $C = 0, x = \pm\infty$ Case when $U = 0$ Case when $k = 0$	$C(x) = C_0\exp\left(-x\dfrac{U}{2K}\left(1 + \sqrt{1 + \dfrac{4kK}{U^2}}\right)\right)\ x \leq 0$ $C(x) = C_0\exp\left(x\dfrac{U}{2K}\left(1 - \sqrt{1 + \dfrac{4kK}{U^2}}\right)\right)\ x \geq 0$ $C(x) = C_0\exp\left(-\sqrt{\dfrac{kx^2}{K}}\right)$ $C(x) = C_0\exp\left(-\dfrac{Ux}{K}\right)\ x \leq 0$ $C(x) = C_0\quad x \geq 0$ Note that D (molecular diffusion) or D_t (turbulent diffusion) can be substituted for K

Example 14.9

Diffusion of a dilute solution of NaCl into an equal density of water is measured in a study of one-dimensional diffusion without advection in a long laboratory channel. In these experiments turbulence is generated with the mean flow velocity $u = 0$ along the x-direction. At the midpoint of the channel $x = 0$, a barrier separated NaCl in the negative x-direction from the water solution in the positive direction. The barrier is removed at the beginning of the experiment, $t = 0$, maintaining a constant concentration C_0 of NaCl at $x = 0$. Discrete concentrations $C(x, t)$ of NaCl along the positive x-direction relative to the initial concentration C_0 is given by

	$t = 100$ sec		$t = 200$ sec			$t = 600$ sec			$t = 1200$ sec		
x (cm)	30	60	30	60	100	30	60	100	30	60	100
$\dfrac{\overline{C}}{C_0}$	0.52	0.22	0.65	0.36	0.15	0.78	0.52	0.36	0.81	0.68	0.43

Assume NaCl to be conservative in the solution. (a) What causes the enhanced rate of mixing of NaCl into the water? Should we name the transport parameter as a dispersion or a turbulent diffusion coefficient? (b) Estimate the dispersion or turbulent diffusion coefficient for this test.

Solution

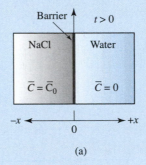

(a)

FIGURE E14.9(a)

(a) Since there is no mean velocity distribution ($u = 0$), the enhanced rate of mixing is caused by turbulent diffusion. The transport coefficient is the turbulent diffusion coefficient.

(b) The turbulent diffusion of NaCl into the water along the x-axis is given by Eq. 14.4.9. The solution of this equation subject to above test conditions, that is $u = 0$, $r_g = 0$ and $r_d = 0$, is given in Table 14.6:

$$\frac{\overline{C}}{C_0} = 1 - \mathrm{erf}\left(\frac{x}{2\sqrt{D_t\,t}}\right)$$

If we plot the above equation as an ordinate $\dfrac{\overline{C}}{C_0}$ on an arithmetic-probability scale, and $x/\sqrt{t}$ as an abscissa on a linear scale, the above equation becomes a straight line. NaCl concentration measurements are plotted in the figure below. We assume that there is a point on the abscissa at which $\frac{x}{\sqrt{t}} = \frac{\sqrt{\pi D_t}}{2}$. Substituting this in the above equation results in

$$\frac{\bar{C}}{C_0} = 1 - \mathrm{erf}\left(\frac{\sqrt{\pi D_t}}{2 \times 2 \sqrt{D_t}}\right) = 1 - \mathrm{erf}\left(\frac{\sqrt{\pi}}{4}\right) = 1 - 0.47 = 0.53$$

Thus, we can read the value of $\frac{x}{\sqrt{t}}$ that corresponds to $\frac{\bar{C}}{C_0} = 0.53$ from the plot below.

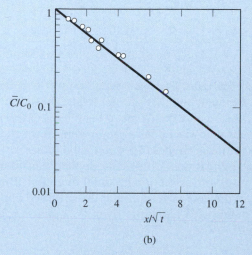

(b)

FIGURE E14.9(b)

We have

$$\frac{x}{\sqrt{t}} = 2.6 = \frac{\sqrt{\pi D_t}}{2}; D_t = 8.6 \ \mathrm{cm^2/s}$$

14.4.3 Closure

Three different forms of the advection-diffusion equations are introduced in this section. A notable difference is that these equations describe spreading of the substances at different scales. A turbulent flow (Eq. 14.4.10) spreads out the tracers into irregular surfaces resulting in enhanced vertical molecular diffusion. These two mechanisms combine and cause the enhanced spread of tracers. An analogous mechanism of stretching and vertical mixing occurs on a larger macro scale described by Eq. 14.4.18. However, in this case longitudinal advection combines with vertical turbulent diffusion. The spread of tracers on this macro scale is much greater than in the previous turbulent flow case. Measurements suggest that $K \gg D_t \gg D$, that is, the dispersion coefficient is significantly larger than the turbulent diffusion coefficient, the molecular diffusion coefficient being by far the smallest.

Equation 14.4.18 provides a basis for studying cross-sectional averaged mass and heat transport. Considering additional body source terms, such as chemical and biologically mediated reactions, and variable dispersion coefficient along the x axis, Eq. 14.4.18 becomes

$$\frac{\partial \overline{C}}{\partial t} + \frac{\partial \overline{UC}}{\partial x} = \frac{1}{A}\frac{\partial}{\partial x}\left(AK\frac{\partial \overline{C}}{\partial x}\right) + \overline{r}_g - \overline{r}_d \qquad (14.4.19)$$

in which A is the cross-sectional area. A similar procedure can be followed to derive the dispersion equation for heat transport:

$$\frac{\partial \overline{T}}{\partial t} + \frac{\partial \overline{UT}}{\partial x} = \frac{1}{A}\frac{\partial}{\partial x}\left(AE\frac{\partial \overline{T}}{\partial x}\right) + \frac{\overline{S}}{\rho\, c_v} \qquad (14.4.20)$$

where $\overline{T}$ is the cross-sectional averaged temperature, E is the longitudinal dispersion coefficient for heat, and $\overline{S}$ is the net heat generation within the control volume. Equations 14.4.19 and 14.4.20 have been used extensively for studies of mass and heat transport in rivers, lakes, reservoirs, and estuaries. In lakes and reservoirs where significant changes in water quality are anticipated along the vertical direction, the equations are usually applied considering the maximum depth of a lake or reservoir as the longitudinal direction. This implies that A is the horizontal area of the lake or reservoir at the specified depth. In rivers, where major changes are anticipated along the river, the equations are usually applied along the river reach.

14.5 EVALUATING THE TRANSPORT COEFFICIENTS IN THE ENVIRONMENT

A variety of flow patterns can develop in the ambient receiving water body depending upon its type (stream, river, pond, lake, reservoir, or coastal water), characteristics (depth, surface area, stratified or uniform, stagnant or flowing), the result of a cold or warm effluent which is characterized by its hydraulic conditions (discharge, velocity, submerged single port or multiport discharge), and quality characteristics (temperature, substance concentration, total solid content). In the previous section we focused on the development of various transport equations that are the basis for transport processes of mass and heat in the environment. It was seen that these equations may have different levels of complexity. We will now outline some methods for estimating the mixing coefficients that are used in the transport equations.

14.5.1 Open Channels

Equations 14.4.19 and 14.4.20 for longitudinal dispersion of mass and heat have been extensively used in a variety of engineering applications. These equations are very useful and practical prediction tools providing they are applied in the equilibrium zone, which is defined as the distance beyond the effluent injection

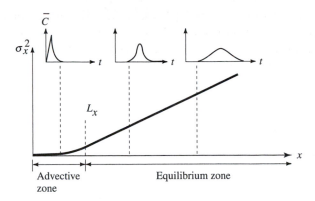

FIGURE 14.8 Schematic of the equilibrium and advective zone with cross-sectional averaged tracer concentration at different locations.

point where the longitudinal variance of the cross-sectional averaged tracer concentration increases linearly with time (Fig. 14.8).

Consider a steady, point source, side channel effluent such as that from a wastewater treatment plant or from some industrial process into a river (Fig. 14.9). The effluent interacts with the receiving river in several ways before the two become fully mixed. However, as soon as equilibrium between shear velocity and turbulent diffusion is established, the longitudinal dispersion transport equations are applicable. This means that the effluent cloud is sufficiently long so that the effluent recycles across the channel, but is not necessarily uniform across the channel. The longitudinal distance L_x between the effluent injection point and the equilibrium zone is called the *advective zone* (Fig. 14.9):

$$L_x = \alpha_x \frac{U L_t^2}{K_y} \tag{14.5.1}$$

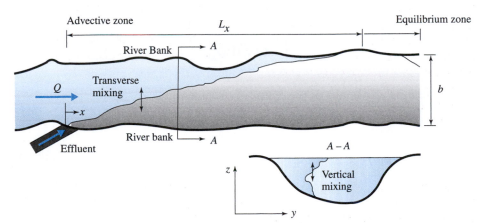

FIGURE 14.9 Mixing zones downstream from an effluent source.

In Eq. 14.5.1, α_x is an empirical constant ($\alpha_x = 0.3$ for a uniform straight channel with a transverse line source, $\alpha_x = 0.6$ for a bankside source; $\alpha_x = 5$ for a small mountain stream; and $\alpha_x = 1.6$ and 10 for a channel with dead zones that occupy 5 and 30% of the channel volume, respectively), L_t is the transverse length scale ($L_t \approx 0.5b$ for a symmetrical channel and $L_t \approx 0.7b$ for a natural channel), and K_y is the transverse dispersion coefficient that accounts for the effects on the depth-averaged tracer concentration of depth variations in the transverse velocity. A summary of approximate formulas in estimating the dispersion coefficients is given in Table 14.7. A considerable amount of uncertainty is present in the formulas given in that table. Therefore, formulations of the upper and lower bounds are given for the dispersion coefficients.

It is often possible to estimate the dispersion coefficient from measurements for the channel under consideration. Among a variety of different techniques, the method of moments appears to be popular, especially with the advent of relatively inexpensive automatic sampling stations that have a high precision of tracer detection. The basis for the method of moments is given by Eq. (14.3.16):

$$K = \frac{1}{2} \frac{d\sigma_x^2}{dt} \tag{14.5.2}$$

where σ_x^2 is the spatial variance of the diffusing particles. However, for practical applications it is much more convenient to measure temporal variance of the diffusing tracer in a river at a specified location rather than measuring spatial variance along the river. The relation between the two may be approximated in the equilibrium zone as

$$\sigma_x^2 = U^2 \sigma_t^2 \tag{14.5.3}$$

The temporal variance, which is the measure of the spread around the centroid of the cross-sectional averaged concentration, is defined as

$$\sigma_t^2(x) = \frac{\displaystyle\int_{-\infty}^{\infty} (t - \bar{t})^2 \, C(x, t) \, dt}{\displaystyle\int_{-\infty}^{\infty} C(x, t) \, dt} \tag{14.5.4}$$

where the time of passage of the centroid $\bar{t}$ is given by

$$\bar{t} = \frac{\displaystyle\int_{-\infty}^{\infty} t C(x, t) dt}{\displaystyle\int_{-\infty}^{\infty} C(x, t) dt} \tag{14.5.5}$$

TABLE 14.7 Longitudinal and Transverse Dispersion Coefficients

Coefficient	Notation	Formulation
Transverse dispersion	K_y	$K_y = \alpha_y U_* H$
		$0.3 \leq \dfrac{K_y}{HU_*} \leq 0.9$
Longitudinal dispersion	K	$K = \alpha_k \dfrac{U^2 b^2}{K_y}$
		$30 \leq \dfrac{K}{HU_*} \leq 3000$

$U_* =$ average shear velocity $(U_* = \sqrt{gHS}), g =$ acceleration due to gravity, $H =$ mean depth, $b =$ channel width, $S =$ channel slope, α_y is a constant (0.6, in meandering rivers can increase by a factor of 2), and α_k is a constant (0.0066, may vary between 0.001 to 0.016).

The denominator of Eqs. 14.5.4 and 14.5.5 is termed the zeroth statistical moment μ_0. Equation 14.5.5 is designated the first statistical moment, and Eq. 14.5.4 is the second statistical moment. Integrating Eq. (14.5.2) over a time interval t_1 to t_2 yields

$$K = \frac{1}{2} U^2 \frac{\sigma_t^2(x_2) - \sigma_t^2(x_1)}{t_2 - t_1} \tag{14.5.6}$$

where $\sigma_t^2(x)$ is the temporal variance of the cross-sectional tracer concentration at two specified locations x_1 and x_2 along the river. Hence, if tracer concentration measurements are available at two locations over time along the river in the equilibrium zone, we can estimate the dispersion coefficient.

Water quality models are often used to improve the management of waste discharge into streams. A stream water quality model must simulate the hydraulic transport; and the one-dimensional advection diffusion equation is used for this purpose. The equation requires a knowledge of parameters which are related to the physical characteristics of the reach to be modeled, particularly its geometry and flow conditions such as cross-sectional averaged velocity and the dispersion coefficient. One of the techniques used to derive the flow parameters is to use a dye-tracing experiment that follows the progress of a dye cloud along a river reach. If the cross-sectional averaged dye concentrations are measured as a function of time at the two cross-sections, the data can be used to derive the necessary parameters. This is demonstrated in the following example.

Example 14.10

Waste Load Allocation Studies in the Metropolitan area of Minneapolis/St. Paul were based on a stream water quality model. Dye (Rhodamine WT) studies were conducted on the Mississippi River. Sample data measured at station 1 and at station 2 are provided. The river flow rate is $Q = 131$ m^3/s, and the average river cross-section is $A = 1254$ m^2. (a) Determine the first and second temporal moments of the measured cross-sectional averaged concentration distribution. (b) Use the moments to predict the dispersion coefficient (K) in the river reach between sampling sites. The distance between two sampling sites is 5.3 km. The location and extent of the reaches studied follows.

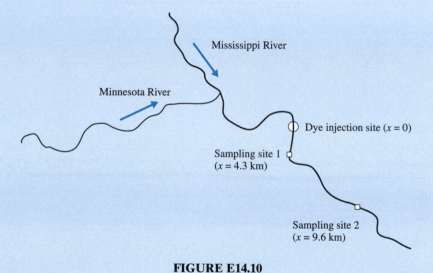

FIGURE E14.10

Mississippi River @ Station 1

	May 11													May 12		
t (hours)	0	1	3	5	7	9	11	13	15	17	19	21	23	25	27	29
C (mg/L)	0.00	0.18	0.94	12.9	19.0	8.5	2.5	1.4	0.93	0.71	0.51	0.40	0.29	0.21	0.16	0.13

Mississippi River @ Station 2

	May 11													May 12		
t (hours)	20	20.5	22	24	25.5	27	29	31	33	35	37	39	41	43	45	47
C (mg/L)	0.00	0.2	2.8	5.0	8.8	8.1	7.4	6.0	4.4	3.0	2.0	1.45	1.0	0.75	0.56	0.42

t (hours)	49	51	53	55	57
C (mg/L)	0.32	0.25	0.19	0.15	0.11

Solution

The first and second statistical moments of the measured dye concentration are determined from Eqs. 14.5.4 and 14.5.5. Numerical integration of the integrals can be conducted on a spreadsheet for the two sampling stations.

Step 1:   Estimate the zeroth statistical moment $\mu_0 \approx \sum\limits_{i=1}^{n} \left[\dfrac{C(t_i) + C(t_{i-1})}{2} \right] \Delta t,$

where n is the number of measurements and Δt is the time between successive observations.

Step 2:   Approximate the numerator of Eq. 14.5.5 using the summation

$\sum\limits_{i=1}^{n} t_i \left[\dfrac{C(t_i) + C(t_{i-1})}{2} \right] \Delta t.$ Divide this result by μ_0 to yield $\bar{t}$.

Step 3:   Approximate the numerator of Eq. 14.5.4 using the approximation

$\sum\limits_{i=1}^{n} (t_i - \bar{t})^2 \left[\dfrac{C(t_i) + C(t_{i-1})}{2} \right] \Delta t.$ Divide this by μ_0 to yield σ_t^2.

The following results are obtained:

Location	Station 1	Station 2
x (m)	4300	9600
t(hr)	8.91	31.23
σ_t^2(hr^2)	14.82	38.72

Cross-sectional average velocity:

$$U = \frac{x_2 - x_1}{t_2 - t_1} = \frac{9600 - 4300}{31.23 - 8.91} = 237.4 \text{m/hr, or } 6.6 \times 10^{-2} \text{ m/s}$$

Dispersion coefficient Eq. (14.5.6):

$$K = \frac{1}{2} U^2 \frac{\sigma_t^2(x_2) - \sigma_t^2(x_1)}{t_2 - t_1} = \frac{1}{2} 237.4^2 \frac{38.72 - 14.82}{31.23 - 8.91} = 30173 \text{ m}^2/\text{hr} = 8.4 \text{ m}^2/\text{s}$$

Example 14.11

The herbicide atrazine is frequently used on agricultural fields. Due to a road accident, 250 kg of atrazine is spilled onto the surface of a river. The river has the following characteristics: average width $b = 20$ m, mean depth $H = 1$ m, discharge $Q = 10$ m^3/s, slope $S = 0.0002$, and it is gently meandering in the study reach. (a) Estimate the length of the advective zone in which the one-dimensional longitudinal dispersion transport equation does not apply. (b) Estimate the longitudinal dispersion coefficient. (c) Estimate and plot the atrazine concentration as a function of time 18 km downstream of the injection, and (d) Estimate the peak concentration 25 km downstream of the injection. Assume that atrazine does not undergo any degradation or removal during the first 16 hours in the river. Atrazine has a molecular weight of 215.7 g/mol.

Solution

(a) First we will investigate whether we can use the one-dimensional advection-diffusion relation, Eq. 14.4.18, in order to predict atrazine concentration 18 km downstream from the injection point. To apply Eq. 14.4.18 we have to be in the equilibrium zone.

The length of the advective zone is given by Eq. 14.5.1. The transverse dispersion coefficient from Table 14.6 is $K_y = 0.6\ HU_*$, where the mean shear stress velocity is

$$U_* = (gHS)^{0.5} = (9.81 \times 1 \times 0.0002)^{0.5} = 0.044 \text{ m/s.}$$

Then

$$K_y = 0.6 \times 1 \times 0.044 = 0.026 \text{ m}^2/\text{s}$$

We will assume that $\alpha_x = 0.3$ for the transverse injection, and $L_t = 0.5\ b$ in a gently meandering channel. Hence the length of the advective zone is

$$L_x = \alpha_x \frac{UL_t^2}{K_y} = 0.3\,\frac{0.5 \times (0.5 \times 20)^2}{0.026} \approx 577 \text{ m or about 0.6 km.}$$

The equilibrium zone starts where $x \approx 0.6$ km. The corresponding advective time is

$$t = L_x/U = 576/0.5 = 0.32 \text{ hr.}$$

(b) The longitudinal dispersion coefficient is needed in order to estimate the atrazine concentration 18 km downstream. Formulations for the longitudinal dispersion coefficients are given in Table 14.7. From this table, $30 \le \dfrac{K}{HU_*} \le 3000$, which gives $1 \le K \le 132$ m^2/s. From the formula,

$$K = \alpha_k \frac{U^2 b^2}{K_y} = 0.0066\,\frac{0.5^2\ 20^2}{0.026} = 25.4 \text{ m}^2/\text{s.}$$

The dispersion coefficient falls into the expected range.

(c) The advection-diffusion with constant cross-sectional area and constant dispersion coefficient is given by Eq. 14.4.18. The solution of this equation for the impulse input M is given in Table 14.6:

$$C(x, t) = \frac{M}{A\ \sqrt{4\pi\ Kt}}\ \exp\left(\frac{-(x - Ut)^2}{4\ Kt}\right)$$

The total amount of atrazine is $M = 250$ kg, and the cross-sectional area is $A = bH = 20 \times 1 = 20$ m^2. The estimated atrazine concentration versus time at $x = 18$ km is given in Fig. E14.11.

(d) The peak concentration will occur when $x = Ut$. Therefore, the peak concentration is given by

$$C_{max} = \frac{M}{A\sqrt{4\pi K \dfrac{x}{U}}} = \frac{250 \times 10^3}{20\sqrt{4 \times \pi \times 25.4 \times \dfrac{25 \times 10^3}{0.5}}} = 3.1 \text{ g/m}^3$$

$$= \frac{3.1}{215.7 \times 10^3} \approx 1.4 \times 10^{-5} \text{ mol/L}$$

The required flow time is

$$t = \frac{x}{U} = \frac{25 \times 10^3}{0.5} \approx 14 \text{ hr.}$$

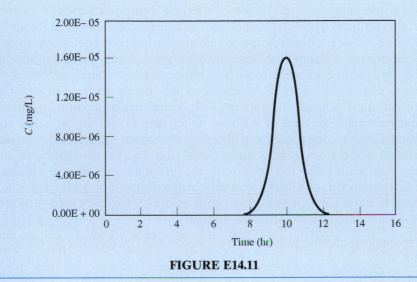

FIGURE E14.11

14.5.2   Lakes and Reservoirs

Lakes, ponds, and reservoirs, sometimes called *standing waters*, are often characterized by water movement that is very slow. The driving forces are from wind shear, solar radiation, heat losses, and inflows and outflows. Standing waters are usually vertically stratified. This implies that water density changes with a depth in a lake or reservoir. Very small density differences caused usually by temperature gradients from surface heating/cooling often control the dynamics in lakes and reservoirs. The stability of the water column, for a vertically stratified fluid, is quantified by a *stability frequency*.

$$N = \left(\frac{g}{\rho_0}\frac{d\rho_e}{dz}\right)^{0.5} \tag{14.5.7}$$

where g is the acceleration of gravity, ρ_0 is the mean water density, $\rho_e(z)$ is the density above ρ_0 (note that $\rho = \rho_0 + \rho_e$), and $\dfrac{d\,\rho_e}{dz}$ is the vertical gradient of the water density. Note in a homogenous fluid, $\rho = \rho_0$. The vertical coordinate z in Eq. 14.5.7 is defined positive from the lake surface downward. The physical meaning of stability frequency can be understood by observing the motion of a parcel of fluid which moves vertically within a stratified water column maintaining its density. If the fluid parcel is displaced upward from its stable[9] position it experiences restoring force downward; because it has a higher density than the surrounding fluid, the fluid parcel returns to the initial stable position. The time scale required for a fluid parcel to return to the initial stable position is equivalent to the inverse of the stability frequency, $t \sim N^{-1}$. The stronger the density gradient the larger the stability frequency, and the smaller the time needed for the fluid parcel to return to the stable position. Therefore, stratification slows down vertical mixing and enhances mixing along nearly horizontal surfaces (the plains of constant density) in a lake or reservoir. Typical vertical and horizontal turbulent diffusion coefficients in lakes and oceans are given in Table 14.4. Comparing the values between horizontal and vertical turbulent diffusion coefficients, horizontal diffusion is at least 100 times greater than vertical diffusion.

Density stratification due to vertical temperature gradients inhibits vertical mixing in lakes and reservoirs, and mixing in turn affects the distribution of phytoplankton, nutrients, and other water quality constituents. Quantifying turbulent transport phenomena is one of the major challenges in lake and reservoir water quality analysis. Among a variety of different techniques, the method of moments outlined in the previous section can be used. Observing the broadening of a conservative tracer distribution over time the turbulent diffusion coefficient is obtainable from Eq. 14.5.2. The vertical turbulent diffusion coefficient below a fixed depth is usually determined from the budgets of scalar quantities such as water temperature or a tracer concentration:

$$D_{tz} = -\frac{\dfrac{\partial}{\partial t} \displaystyle\int_z^{z_{max}} C(z)\, A(z)\, dz}{\dfrac{\partial C(z)}{\partial z} A(z)} \tag{14.5.8}$$

in which $C(z)$ is the measured tracer concentration, $A(z)$ is the horizontal area of a lake, t is time, and z is the downward coordinate starting at the lake surface to the lake sediment where $z = z_{max}$. The numerator of Eq. 14.5.8 represents the change in mass content with time below the specified depth, while the denominator multiplied by D_{tz} is the mean total turbulent mass flux. Water temperature can replace $C(z)$ in Eq. 14.5.8, but heat addition due to the solar radiation or heat loss to the sediments must be accounted for.

---

[9]If the density of the water is increasing with the depth from the water surface the water column is stable: a parcel of water that is displaced, say, from higher to lower depth in the water column will be denser than its surroundings and will sink back to its original position.

Example 14.12

Two vertical profiles of tetrachloroethene were measured in a small lake at a time interval of $\Delta t = 15$ days. The water temperature profile was measured during the first day of measurement. The lake has a surface area of 1 km^2 and maximum depth of 10 m. (a) Estimate the vertical turbulent diffusion coefficient at depths of 5, 6, 7, and 8 m. (b) Why should the turbulent diffusion coefficient increase with depth?

Depth z (m)	Area A (10^6 m^2)	Temperature (°C) Day 1	Concentration (μ mol/m^3) Day 1	Concentration (μ mol/m^3) Day 16
1	0.82	12.4	5.0	2.0
2	0.76	12.3	9.5	3.5
3	0.70	11.0	5.5	5.5
4	0.62	10.0	3.9	4.2
5	0.48	6.0	3.1	3.7
6	0.30	5.5	2.6	3.4
7	0.28	5.3	2.3	3.3
8	0.20	5.2	2.1	3.2
9	0.15	5.1	2.0	3.2
10	0.0	5.0	1.9	3.1

Solution

(a) The vertical turbulent diffusion coefficient is estimated from Eq. 14.5.8. Numerical integration of the integral can be carried out on a spreadsheet.

Step 1:   Approximate the numerator of Eq. 14.5.8 by $\frac{1}{\Delta t} \sum_{i=1}^{n} \Delta C_{ti} \Delta z_i A_i$, where n is the number of horizontal layers in the lake, ΔC_t is the tetrachloroethene concentration difference over the time interval Δt, Δz_i is the layer thickness, that is, the difference between two successive depths, and A_i is the average area between two successive depths. From the data the following table can be constructed:

Layer (m)	A_i (10^6 m^2)	C_{1i} (μmol/m^3)	C_{2i} (μmol/m^3)	ΔC_{12i} (μmol/m^3)	$\Delta C_{12i} A_i \Delta z$ (mol)	$\frac{1}{\Delta t} \sum \Delta C_{12i} A_i \Delta z$ (mol/d)
5–6	0.39	2.85	3.55	0.70	0.273	0.0710
6–7	0.29	2.45	3.35	0.90	0.261	0.0528
7–8	0.24	2.20	3.25	1.05	0.252	0.0354
8–9	0.17	2.05	3.20	1.15	0.195	0.0186
9–10	0.07	1.95	3.15	1.20	0.084	0.0006

Step 2:   Approximate the denominator of Eq. 14.5.8 by $\left. \frac{\Delta C_z}{\Delta z} \right|_{1\text{-}2} A(z)$, where ΔC_z is the mean concentration difference over a vertical distance Δz and over profiles 1 and 2, and $A(z)$ is the lake horizontal area at depth z. The tetrachloroethene concentration gradients are estimated using central differences (see Example 14.8). The following results are obtained:

(continued)

Depth (m)	A (10^6 m^2)	$\frac{\Delta C}{\Delta z}\big\|_1$ (μmol/m^4)	$\frac{\Delta C}{\Delta z}\big\|_2$ (μmol/m^4)	$\frac{\Delta C}{\Delta z}\big\|_{1-2}$ (μmol/m^4)	$\frac{\Delta C}{\Delta z}\big\|_{1-2} A$ (mol/m^2)
5	0.48	−0.65	−0.40	−0.52	−0.25
6	0.30	−0.40	−0.20	−0.30	−0.09
7	0.28	−0.25	−0.10	−0.17	−0.05
8	0.20	−0.15	−0.05	−0.10	−0.02

Step 3: Turbulent diffusion coefficient Eq. 14.5.8:

Depth (m)	5	6	7	8
D_{tz} (cm^2/s)	0.03	0.07	0.08	0.11

(b) We assume that the stratification in the lake remains stable over the 15 day time interval. From the water temperature profile it is evident that the water temperature changes from high, near the lake surface, to low near the bottom. The corresponding water density changes from low to high. The density difference decreases with depth below 5 m. The weaker the density gradient, the smaller the stability frequency, the longer the time needed for fluid particles to return to a stable position. Therefore, a weak stratification at large depth suppresses less vertical turbulent motion than a strong stratification at smaller depth.

14.6 SUMMARY

The transport processes of heat, mass, and momentum are related by gradient flux laws. These laws combine a fluid property with the state variable gradient. The transport of heat, mass, and momentum are from high values to low ones i.e., down the gradient. The laws expressing the principles of conservation of heat, mass, and momentum, are combined with the flux laws and provided the basis for advection-diffusion transport in the environment. Three different forms of the advection-diffusion equation are introduced. A notable difference is that these equations describe the spreading of substances at different scales. A pollutant spreading by dispersion is significantly larger than the pollutant spreading by turbulent diffusion. Although the molecular diffusion spreading is by far the smallest, it is not permissible to neglect the molecular diffusion where the turbulence is damped or over short distances such as the boundary layer thickness around a living cell, the diffusive sublayer thickness at the solid-water interface, and the reaction distance between chemical particles. Methods and procedures for estimating the mixing coefficients that are used in the transport equations of heat, mass, and momentum in the environment are presented.

REFERENCES

Taylor, G.I. (1953). Proc. Royal Soc. London, London, 219A, 186–203.

PROBLEMS

14.1 Estimate the steady state heat flux across a glass window 0.5 cm thick, one side which is maintained at 25°C and the other side at 15°C. The thermal conductivity of glass is $K = 0.86$ W/m-°C.

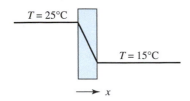

$T = 25°C$

$T = 15°C$

x

FIGURE P14.1

14.2 The thermal properties of a soil sample are tested in a laboratory. The cylindrically shaped sample has a cross-sectional area of 400 cm² and length of 40 cm. The cylinder wall is insulated.

(a) What is the basic approximation with respect to the temperature change in the soil sample for the one-dimensional form of Fourier's law?

(b) If the left end of the cylinder is kept at a constant temperature T_1, and the right end is kept at constant temperature T_2 ($T_1 > T_2$), write the expression for the heat flux per unit area in the x-direction of the soil sample.

(c) If the left end of the cylinder is kept at $T_1 = 100°C$, and the right end is kept at $T_2 = 50°C$, estimate the heat flux per unit area if the thermal conductivity is $K = 35$ W/m-°C.

(d) Write the expression for the heat flux which governs the heat flow through the soil sample if the temperature in the cylinder is changing in all three directions. Assume the thermal conductivity to be isotropic.

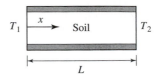

T_1 x Soil T_2

L

FIGURE P14.2

14.3 Two pure gas streams, CO_2 and air, are flowing in the same direction in a channel. The channel is divided into equal volumes by a piece of glass wall 1 cm thick. At the midpoint of the glass wall, a hole 3.14 cm² in area allows the diffusion of CO_2 into air and the diffusion of air into CO_2. The concentration of CO_2 upstream of the hole is zero in the air stream and 35 mol/m³ in the CO_2 stream. Estimate:

(a) The steady state molar flux of CO_2 into air.

(b) The mass of CO_2 in kg that passes through the opening in 2 hours. Assume the molecular diffusion coefficient in air is. CO_2 has molecular weight of MW = 44.01 g/mol.

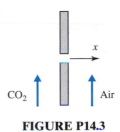

CO_2 x Air

FIGURE P14.3

14.4 An experiment was conducted to characterize the rate of a chemical reaction. Concentrations were measured at different times and the data are given below.

(a) Is this a zero-, first-, or second-order reaction?

(b) What is the value of the rate constant?

Time (h)	5	10	15	20	25	30	35	40	45	50	60
C (mg/L)	7.92	6.25	5.02	3.93	3.32	2.52	2.08	1.65	1.25	0.94	0.62

14.5 Bacterial growth has been described by a first order reaction. The growth constant is 0.12 1/day at 20°C, and θ is 1.03.
   (a) What is the value of the rate constant at 35°C?
   (b) What is the value of activation energy?

14.6 An ice-covered pond experiences a high deoxygenation rate due to a high rate of oxygen demand within the pond sediments. Prior to the freeze-over date, the dissolved oxygen (DO) concentration in the pond is 12.4 mg/L. Assume that the DO transfer occurs at the sediment-water interface according to the relation $r = kC$, where the rate constant $k = 0.1$ day^{-1} and C is the DO concentration in the well-mixed pond. The DO lower survival limit is 3 mg/L for cold-water fish.
   (a) Estimate the dissolved oxygen concentration in the pond 30 days after the freeze-over date.
   (b) How many days can fish survive in the pond? The pond has a volume of 10^4 m^3.

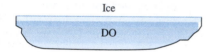

FIGURE P14.6

14.7 In order to improve the dissolved oxygen (DO) concentration in the pond of Problem 14.6, water is withdrawn from the pond and exposed to the atmosphere. The water, replenished with DO from the atmosphere, is reintroduced into the pond below the ice. The inlet and the outlet discharge is 2 L/s, and the inlet DO is $C_s = 13.8$ mg/L.
   (a) Estimate the DO concentration in the pond 30 days after the freeze-over date.
   (b) How many days can fish survive in the pond?

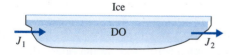

FIGURE P14.7

14.8 Over a long period of time, total phosphorus (TP) has been discharged into a lake through sewage effluents, tributaries and direct runoff at a total rate of $J_1 = \sum J_i = 10$ tons/day. The load is then decreased, at time $t = 0$, to a rate $J_1 = 1$ ton/day by diverting the component from the sewage

effluents. It is desired to determine the concentration of TP in a lake several years after the diversion assuming:
   (a) The total phosphorus to be conservative.
   (b) The total phosphorus loss to be proportional to the relation $r = kC$, where the total phosphorus loss rate $k = 0.001$ day^{-1} and C is the phosphorus concentration in the lake. The volume of the well-mixed lake is 380×10^8 m^3 with a steady outflow of 500 m^3/s.

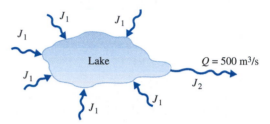

FIGURE P14.8

14.9 It is desired to simulate three-dimensional contaminant spreading in a turbulent flow. The computational domain is of size $L = 50$ m. The average dissipation in the water column is $\varepsilon = 5 \times 10^{-6}$ m^2/s^3. Assume an average water temperature of 15°C. How many computational grid points over the entire flow domain are needed in order to resolve contaminant concentration fluctuations to the scale where they are smeared by molecular diffusion? The contaminant has a molecular diffusion coefficient of 8.2×10^{-10} m^2/s.

14.10 Mixing is an important operation in many phases of wastewater treatment. Substance A is to be mixed in a container of 20 kg water at 25°C. After 30 minutes of stirring with a mixer the temperature of the mixture rises by 2°C.
   (a) What is the power output of the mixer?
   (b) What is the scale at which velocity fluctuations will disappear?
   (c) What is the scale at which the formation of concentration gradients are smeared by molecular diffusion? The mixing substance has a molecular diffusion coefficient of 7.2×10^{-6} cm^2/s and a kinematic viscosity of $\nu = 1.1 \times 10^{-6}$ m^2/s.

14.11 Nitrilotriacetic acid (NTA) is used frequently in detergents. Assume that NTA enters a small stratified lake through the sewage system. In the summer the vertical profile of the NTA concentration is measured in a lake which has a surface area of

0.2 km^2 and a maximum depth of 9 m. Estimate the magnitude and direction of the vertical flux of NTA 7 m below the lake surface. Assume a vertical turbulent diffusion coefficient of 1.2 cm^2/s and a molecular diffusion coefficient of 6.5 × 10^{-6} cm^2/s.

Depth z(m)	0	1	2	3	4	5	6	7	8	9
C (10^{-6} mol/m^3)		4.9	5.2	5.1	4.9	3.2	2.9	2.6	2.4	2.1	1.8

14.12 The temperature gradient in the atmosphere is observed to be $\dfrac{\partial \overline{T}}{\partial z} = 0.02°C/m$. The turbulent diffusion coefficient for heat is $\alpha_{tz} = 10$ m^2/s.
(a) Estimate the turbulent heat flux.
(b) Explain the sign of the heat flux. The specific heat of air at 10°C is 1003 J/kg-°C.

14.13 Turbulent diffusion of helium (He) into nitrogen (N$_2$) is measured in a long laboratory chamber. At $x = 0$ the chamber containing N$_2$ is connected to a large tank containing He. At the connection a barrier separates the He from the N$_2$. Turbulence is generated by a stack of oscillating grids moving vertically. The mean flow velocity in the long chamber is zero. At time $t = 0$ the barrier is removed and concentrations of He are measured at different times and locations along the positive x-direction. The concentration $C(x, t)$ of He along the chamber at three given times relative to the initial concentration C_0 in the tank are

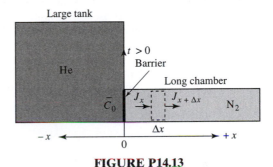

FIGURE P14.13

(a) What basic equation describes the turbulent diffusion of He in the positive x-direction?
(b) What boundary conditions are appropriate?
(c) Estimate the turbulent diffusion coefficient.

	$t = 2 \times 10^{-2}$ sec				$t = 6 \times 10^{-2}$ sec				$t = 9 \times 10^{-2}$ sec			
x (cm)	0.2	0.3	0.4	0.5	0.2	0.3	0.4	0.5	0.2	0.3	0.4	0.5
$\overline{C}/\overline{C}_0$	0.79	0.68	0.59	0.51	0.86	0.81	0.74	0.69	0.89	0.83	0.79	0.72

14.14 Aquatic herbicides are often injected into channels for weed control. A total amount of 360 kg of the acrolein herbicide (AH) is injected uniformly into the cross-section of the channel at a constant rate over a time interval of 1 hour. The decay of AH in the channel is characterized according to the relation $r = kC$, where k is the degradation rate constant 0.19 hr^{-1}, and C is the AH concentration.

(a) What is the initial AH concentration, C_0, at the injection point?
(b) Estimate and plot the AH concentration as a function of time 1 km downstream from the injection point. The channel has a mean depth of 1 m, a width of 20 m, a slope of 0.00018, and a discharge of 20 m^3/s.

14.15   A pharmaceutical company claims to continuously discharge a substance into a river at 1 m³/s with an effluent concentration of 12 mg/L. Concentration measurements are taken at locations A and B at 1 km and 3 km downstream from the discharge point. The concentrations at these locations are 2 mg/L and 1.5 mg/L, respectively. Is the pharmaceutical company's claim true? If not, what is the estimate of the discharge concentration? Assume that the substance is not conservative in the river. Assume well-mixed conditions at the discharge point. The receiving river has a discharge of 20 m³/s, a mean velocity $U = 2$ m/s, and a background concentration of the substance is $C_r = 1$ mg/L. Consider advective transport only.

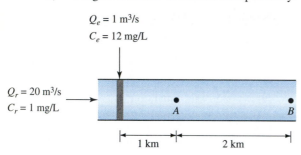

$Q_e = 1$ m³/s
$C_e = 12$ mg/L

$Q_r = 20$ m³/s
$C_r = 1$ mg/L

|← 1 km →|← 2 km →|
A   B

FIGURE P14.15

14.16   Cooling water from an industrial plant is discharged continuously at $Q_{ef} = 5$ m³/s and $T_{ef} = 45°C$ into a river with a flow of $Q_r = 20$ m³/s at $T_r = 20°C$. The cooling water is discharged through a multipoint diffuser across the entire width of the river, so that mixing is complete within a very short distance. The river is 50 m wide and 1 m deep. Downstream of the diffuser the artificially heated river loses its excess heat to the atmosphere according to the relationship $S = k(T - T_e)$, where T is the water temperature, $T_e = 19°C$ is the equilibrium water temperature (the temperature that the river reaches after a long distance below the diffuser), and k is the overall coefficient of surface heat transfer.
   (a)   Formulate a one-dimensional dispersion model for heat transport along the river, below the discharge point.
   (b)   Simplify the heat dispersion equation for steady, advective flow.
   (c)   Find an expression for the water temperature a long distance downstream from the diffuser.
   (d)   White bass, a cool water fish, have a survival temperature limit close to 23°C. How far downstream from the discharge point will white bass be unable to survive in the river? Consider advective transport only. Let $k = 100$ J/m²-s-°C and the specific heat for water $c_v = 4180$ J/kg-°C.

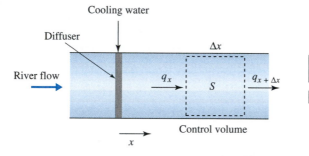

Cooling water

Diffuser

River flow

Δx

q_x S $q_{x + \Delta x}$

Control volume

x

A h

b

FIGURE P14.16

14.17   A truck transporting salt falls from a bridge into a river. The river at the point is 1.0 m deep, 50 m wide, has a discharge of 30 m³/s, and a slope $S = 0.00015$. Assume that after the accident the salt forms in a pile in the middle of the river. A saltwater plume develops downstream from the pile and lasts for several days.

   (a)   How far downstream from the pile will the salty water become "uniformly" mixed throughout the entire cross section of the channel?
   (b)   What time is required for this to occur?

14.18   Two vertical profiles of water temperature were measured in a lake at a time interval of 24 days. The lake has a surface area of 1.8 km² and a maximum depth of 20 m.

(a)   Estimate the vertical turbulent diffusion coefficient from 12 to 20 m.

(b)   Estimate the stability frequency for the same depth range.

(c)   Plot the turbulent diffusion coefficient versus stability frequency.

(d)   Why should an inverse dependency be expected? A convenient empirical expression for water density in kg/m³ as a function of water temperature in °C is

Depth z(m)	Area A (10^6 m²)	Temperature T (°C) Day 1	Temperature (°C) Day 25
0	1.82	21.85	25.65
2	1.71	20.70	25.35
4	1.46	19.60	25.00
6	1.14	17.4	20.95
8	1.00	14.8	14.15
10	0.82	10.9	11.50
11	0.75	10.25	10.85
12	0.72	9.75	10.30
13	0.66	9.25	9.75
14	0.55	9.10	9.35
15	0.52	9.00	9.15
16	0.35	8.95	9.05
17	0.28	8.85	8.95
18	0.15	8.8	8.95
19	0.12	8.7	8.92
20	0.00	8.7	8.92

$$\rho = \left(\frac{999.84 + 18.22 \times T - 0.0079 \times T^2 - 55.45 \times 10^{-6}\, T^3 + 149.76 \times 10^{-9} \times T^4}{1 + 18.16 \times 10^{-3} \times T} \right)$$

14.19   For the dye spreading example (Fig. 14.3), show that:

(a)   The centroid of the diffusing dye does not move, that is $\dfrac{dx_c}{dt} = 0$, where

$x_c = \displaystyle\int_{-\infty}^{+\infty} x\, P(x)\, dx$ is the coordinate of the centroid of the diffusing dye, and

$\displaystyle\int_{-\infty}^{+\infty} P(x)\, dx = 1.0$ is the normalized total amount of dye in the channel.

(b)   The variance of the concentration distribution increases linearly with time according to $\sigma^2 = 2\,Dt$. (*Hint:* Normalize the diffusion equation, Eq. 14.3.6 by the total amount of dye in the channel $M = \displaystyle\int_{-\infty}^{+\infty} C\, A\, dx$, and integrate by parts.)

14.20   In a stratified lake the vertical turbulent diffusion coefficient is measured from the vertical spreading of an injected tracer, sulfurhexafluoride (SF_6). The lake has a surface area of 5 km² and a maximum depth of 34 m. SF_6 is injected at a depth of 17 m along a 3 km distance. Vertical profiles of SF_6 were measured over 30 days. Estimated vertical variance of the tracer SF_6 as a function of time after tracer release is given below.

(a)   Estimate the vertical turbulent diffusion coefficient.

(b)   What is the possible explanation for the finite variance quantity at time $t = 0$?

σ_t^2 (10^4 cm²)	0.45	7.1	8.9	9.5	12.8	15.6	18.3	19.9
Time (day)	2	6	8	10	15	20	25	30

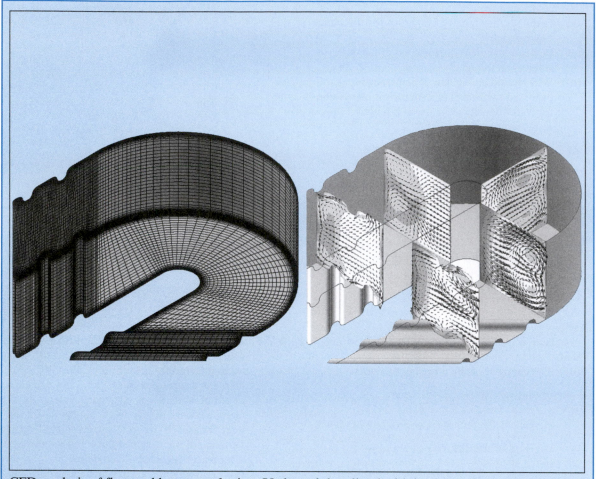

CFD analysis of flow and heat transfer in a U-shaped duct lined with inclined ribs. Left is grid. Right is solution at selected cross sections. Gray scales represent normalized temperature distribution. Darker is hotter; lighter is cooler. (Courtesy of T.I-P. Shih.)

15

Computational Fluid Dynamics

Outline

Chapter Objectives

The objectives of this chapter are to:

▲ Present an introduction to finite-difference and finite-volume methods.

▲ Present ideas on consistency, numerical stability, convergence, and errors.

▲ Present an introduction to grid generation for flow problems with complicated geometries.

▲ Present methods for compressible flows.

▲ Present the artificial compressibility method for incompressible flows.

15.1 INTRODUCTION

The equations governing fluid-flow problems are the *continuity*, the *Navier–Stokes*, and the *energy equations*. These equations, derived in Chapter 5, form a system of coupled quasi-linear *partial differential equations* (PDEs). Because of the nonlinear terms in these PDEs, analytical methods can yield very few solutions. In general, analytical solutions are possible only if these PDEs can be made linear, either because nonlinear terms naturally drop out (e.g., fully developed flows in ducts and flows that are irrotational everywhere) or because nonlinear terms are small when compared to other terms so that they can be neglected (e.g., flows where the Reynolds number is less than unity). Yih (1969) and Schlichting & Gersten (2000) describe most of the better-known analytical solutions. If the nonlinearities in the governing PDEs cannot be neglected, which is the situation for most engineering flows, then numerical methods are needed to obtain solutions.

Computational fluid dynamics, or simply CFD, is concerned with obtaining numerical solutions to fluid-flow problems by using the computer. The advent of high-speed and large-memory computers has enabled CFD to obtain solutions to many flow problems, including those that are compressible or incompressible, laminar or turbulent, chemically reacting or nonreacting, single- or multi-phase. Of the numerical methods developed to address equations governing fluid-flow problems, *finite-difference methods* (FDMs) and *finite-volume methods* (FVMs) are the most widely used. In this chapter, we first give an introduction to these methods. Afterwards, several popular FD and FV methods are presented for computing compressible and incompressible flows.

15.2 AN OVERVIEW OF FINITE-DIFFERENCE AND FINITE-VOLUME METHODS

Since computers are used to obtain solutions, it is important to understand the constraints that they impose. Of these constraints, four are critical. The first of these is that computers can only perform arithmetic (i.e., $+$, $-$, $\times$, and $\div$) and logic (i.e., true and false) operations. This means non-arithmetic operations, such as derivatives and integrals, must be represented in terms of arithmetic and logic operations. The second constraint is that computers represent numbers by a finite number of digits. This means that there are round-off errors, and that these errors must be controlled. The third constraint is that computers have limited storage memories. This means solutions can only be obtained at a finite number of points in space and time. Finally, computers perform a finite number of operations per unit time. This means solution procedures should minimize the computer time needed to achieve a computational task by fully utilizing all available processors on a computer and minimizing the number of operations.

With these constraints, FD and FV methods generate solutions to PDEs through the following three major steps:

1. **Discretize the domain.** The continuous spatial and temporal domain of the problem must be replaced by a discrete one made up of grid points or cells and time levels. The ideal discretization uses the fewest number of grid points/cells and time levels to obtain solutions of the desired accuracy.

2. **Discretize the PDEs.** The PDEs governing the problem must be replaced by a set of algebraic equations with the grid points/cells and the time levels as their domain. Ideally, the algebraic equations—referred to as finite-difference equations (FDEs) or finite-volume equations (FVEs) depending on whether FD or FV methods are used to construct them—should describe the same physics as those by the governing PDEs.

3. **Specify the algorithm.** The step-by-step procedure by which solutions at each grid points/cells are obtained from the FD or FV equations when advancing from one time level to the next must be described in detail. Ideally, the algorithm should ensure not only accurate solutions, but also efficiency in utilizing the computer.

These steps are illustrated through example problems in Sections 15.3 and 15.4.

15.3 EXAMPLES OF SIMPLE FINITE-DIFFERENCE METHODS

To illustrate FDMs, consider the unsteady, incompressible, laminar flow of a fluid with constant kinematic viscosity ν between two parallel plates separated by a distance H, as shown in Fig. 15.1. Initially, both plates are stationary, and the fluid between them is stagnant. Suddenly at time $t = 0$, the lower plate moves horizontally to the right (in the positive x-direction) at a constant speed of V_0.

The equations governing this flow are the continuity equation and the x- and y-component Navier–Stokes equations. Since the flow is parallel (i.e., $v = 0$) and the pressure is the same everywhere, these equations reduce to the following single linear PDE:

$$\frac{\partial u}{\partial t} = \nu \frac{\partial^2 u}{\partial y^2} \qquad (15.3.1)$$

The initial and boundary conditions for Eq. (15.3.1) are

$$u(y, t = 0) = 0, \qquad u(y = 0, t) = V_0, \qquad u(y = H, t) = 0 \qquad (15.3.2)$$

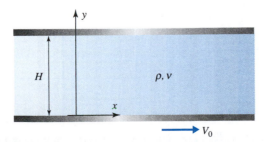

FIGURE 15.1   Flow between stationary and moving parallel plates.

Since Eq. (15.3.1) is linear and the boundary conditions given by Eq. (15.3.2) are homogeneous, separation of variables readily yields the exact analytical solution

$$u/V_0 = 1 - y/H - 2 \sum_{n=1}^{\infty} \frac{1}{n\pi} \sin(n\pi y/H) \exp(-n^2\pi^2 vt/H^2) \quad (15.3.3)$$

This exact solution can be used to evaluate the accuracy of FDMs to be presented.

As mentioned, three steps are involved in generating solutions to PDEs by using a FDM. These three steps are illustrated below, one at a time for Eqs. (15.3.1) and (15.3.2).

15.3.1 Discretization of the Domain

To discretize the domain, we note that the spatial domain is a line segment between 0 and H, and that the temporal domain is a ray emanating from $t = 0$. Though the temporal domain is of infinite extent, the duration of interest is finite, say from $t = 0$ to $t = T$, where T is the time when steady-state is achieved. Here, the spatial domain, $0 \le y \le H$, is discretized by replacing it with JL equally distributed stationary grid points, and the temporal domain, $0 \le t \le T$ is discretized by replacing it with equally incremented time levels (see Fig. 15.2). This discretization of the domain is one of many. For example, the grid points do not have to be uniformly distributed nor do they have to be stationary. A discussion on ways to discretize the domain for computational efficiency and accuracy is given in Section 15.6.

Each point in the discretized domain, shown in Fig. 15.2, has coordinates (y_j, t^n), given by

$$y_j = (j - 1)\,\Delta y, \quad j = 1, 2, 3, \ldots, JL \quad (15.3.4)$$

$$t^n = n\,\Delta t, \quad n = 0, 1, 2, \ldots \quad (15.3.5)$$

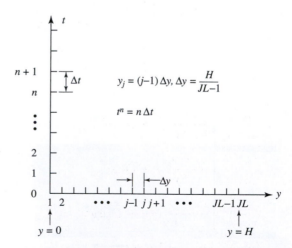

FIGURE 15.2 Grid system and time levels for problem depicted in Figure 15.1.

where $\Delta y = H/(JL - 1)$ is the distance between two adjacent grid points (or the grid spacing), and Δt is the time-step size. The sizes of the grid spacing and the time step depend on the length and time scales that need to be resolved and on the properties of the discretized PDEs used to obtain solutions.

The solution sought—$u(y, t)$ in Eq. (15.3.1)—will be obtained only at the grid points and the time levels. That solution is denoted as

$$u_j^n = u(y_j, t^n) \qquad (15.3.6)$$

where subscripts denote the locations of the grid points and superscripts denote time levels.

15.3.2   Discretization of the Governing Equations

With the domain discretized, the next step is to replace the PDEs governing the problem by a set of algebraic or FDEs that use the grid points and time levels as the domain. With FDMs, PDEs are discretized by replacing derivatives with difference operators. Thus, discretization involves two parts. First, derive *difference operators*. Then, select difference operators.

Derive Difference Operators.   There are many different ways to derive difference operators. For problems with smooth solutions (i.e., solutions without discontinuities such as shock waves), a method based on the following theorem is often used: *If a function, u, and its derivatives are continuous, single-valued, and finite, then the value of that function at any point can be expressed in terms of u and its derivatives at any other point by using a Taylor series expansion, provided the other point is within the radius of convergence of the series.* With this theorem, u_{j+1} and u_{j-1} can be expressed in terms of u_j and its derivatives as

$$u_{j+1} = u_j + \left(\frac{\partial u}{\partial y}\right)_j \Delta y + \left(\frac{\partial^2 u}{\partial y^2}\right)_j \frac{\Delta y^2}{2!} + \left(\frac{\partial^3 u}{\partial y^3}\right)_j \frac{\Delta y^3}{3!} + \cdots \qquad (15.3.7)$$

$$u_{j-1} = u_j - \left(\frac{\partial u}{\partial y}\right)_j \Delta y + \left(\frac{\partial^2 u}{\partial y^2}\right)_j \frac{\Delta y^2}{2!} - \left(\frac{\partial^3 u}{\partial y^3}\right)_j \frac{\Delta y^3}{3!} + \cdots \qquad (15.3.8)$$

From the above two Taylor series, we can readily derive the following four difference operators by solving for $(\partial u/\partial y)_j$ and $(\partial^2 u/\partial y^2)_j$ either directly or by adding or subtracting the two equations:

$$\left(\frac{\partial u}{\partial y}\right)_j = \frac{u_{j+1} - u_j}{\Delta y} + O(\Delta y) \qquad (15.3.9)$$

$$\left(\frac{\partial u}{\partial y}\right)_j = \frac{u_j - u_{j-1}}{\Delta y} + O(\Delta y) \qquad (15.3.10)$$

$$\left(\frac{\partial u}{\partial y}\right)_j = \frac{u_{j+1} - u_{j-1}}{2\Delta y} + O(\Delta y^2) \qquad (15.3.11)$$

$$\left(\frac{\partial^2 u}{\partial y^2}\right)_j = \frac{u_{j+1} - 2u_j + u_{j-1}}{\Delta y^2} + O(\Delta y^2) \qquad (15.3.12)$$

In the above difference operators, $O(\Delta y)$ and $O(\Delta y^2)$ denote the truncation errors (i.e., terms in the Taylor series that have been truncated). In Eqs. (15.3.9) and (15.3.10), the power of Δy in $O(\Delta y)$ is one because the leading term of the truncation errors for these two difference operators is multiplied by Δy raised to the first power. Such difference operators are said to be first-order accurate. In Eqs. (15.3.11) and (15.3.12), that power is two because the leading term in the truncation errors is multiplied by Δy raised to the second power. Such difference operators are said to be second-order accurate. The higher the order of accuracy, the greater is the number of terms retained in the truncated Taylor series. For sufficiently small Δy, a higher-order accurate difference operator is a more accurate representation of a smooth function than a lower-order accurate difference operator. In general, second-order accuracy is adequate.

The grid points used to construct a difference operator form the stencil of that difference operator. If the stencil for a difference operator at y_j involves only grid points with indices greater than or equal to j such as Eq. (15.3.9), then that difference operator is said to be a forward-difference operator. If grid points all have indices less than or equal to j such as Eq. (15.3.10), then that difference operator is said to be a backward-difference operator. If the number of grid points behind and ahead of j are exactly the same such as Eqs. (15.3.11) and (15.3.12), then that difference operator is said to be a central-difference operator. If the number of grid points ahead of and after j are not the same, then that difference operator is said to be either biased backward or biased forward.

Taylor series can be used to derive difference operators for any derivative to any order of accuracy and using any stencil. To illustrate, consider the derivation of a difference operator for a first-order derivative, $(\partial u/\partial y)_j$, with the following specifications: third-order accurate, $O(\Delta y^3)$, with a stencil that includes only one downstream grid point. The derivation of this difference operator involves the following four steps:

Step 1: Determine the number of terms to be kept in each Taylor series. That number, denoted as N, is equal to the order of the derivative for which a difference operator is sought plus the order of accuracy desired. For this case, the order of the derivative is 1, and the order of accuracy desired is 3. Thus, N is 4.

Step 2: Decide on a stencil based on N points for the difference operator at point j. For this case with $N = 4$ and $u > 0$, the 4 points of the stencil are $j - 2, j - 1, j, j + 1$. Other possible stencils include: $(j - 3, j - 2, j - 1, j)$ or completely backward, $(j - 1, j, j + 1, j + 2)$ or biased forward, and $(j, j + 1, j + 2, j + 3)$ or completely forward.

Step 3: Construct $N - 1$ truncated Taylor series about u_j from u at all points of the stencil except point j. Performing this operation gives

$$u_{j+1} = u_j + \left(\frac{\partial u}{\partial y}\right)_j \Delta y + \left(\frac{\partial^2 u}{\partial y^2}\right)_j \frac{\Delta y^2}{2!} + \left(\frac{\partial^3 u}{\partial y^3}\right)_j \frac{\Delta y^3}{3!} + O(\Delta y^4) \qquad (15.3.13a)$$

$$u_{j-1} = u_j - \left(\frac{\partial u}{\partial y}\right)_j \Delta y + \left(\frac{\partial^2 u}{\partial y^2}\right)_j \frac{\Delta y^2}{2!} - \left(\frac{\partial^3 u}{\partial y^3}\right)_j \frac{\Delta y^3}{3!} + O(\Delta y^4) \qquad (15.3.13b)$$

$$u_{j-2} = u_j - \left(\frac{\partial u}{\partial y}\right)_j 2\Delta y + \left(\frac{\partial^2 u}{\partial y^2}\right)_j \frac{(2\Delta y)^2}{2!} - \left(\frac{\partial^3 u}{\partial y^3}\right)_j \frac{(2\Delta y)^3}{3!} + O(\Delta y^4) \quad (15.3.13c)$$

In the above three equations, the unknowns are the first, second, and third derivatives, and they must be expressed in terms of u at points in the stencil.

Step 4:   Solve $N - 1$ coupled linear equations for the difference operator sought. For the first derivative, solving the above three equations gives

$$\left(\frac{\partial u}{\partial y}\right)_j = \frac{2u_{j+1} + 3u_j - 6u_{j-1} + u_{j-2}}{6\Delta y} + O(\Delta y^3) \qquad (15.3.14)$$

Difference operators for the second and third derivatives are also derived in the process of solving for the above difference operator. The orders of accuracy of these difference operators are second order for the second derivative and first order for the third derivative.

In Table 15.1, some commonly used difference operators are summarized. Note that if j and Δy are replaced by n and Δt, then the difference operators can also be used for time derivatives.

TABLE 15.1   Summary of Commonly Used Difference Operators

Description	Finite-Difference	Finite-Volume $(u_{j+1/2} - u_{j-1/2})/\Delta y$	Stencil
First Derivative: $(\partial u/\partial y)_j$			
central, $O(\Delta y^2)$	$(u_{j+1} - u_{j-1})/2\Delta y$	$u_{j\pm 1/2} = (u_j + u_{j\pm 1})/2$	
backward, $O(\Delta y)$	$(u_j - u_{j-1})/\Delta y$	$u_{j+1/2} = u_j, \quad u_{j-1/2} = u_{j-1}$	
forward, $O(\Delta y)$	$(u_{j+1} - u_j)/\Delta y$	$u_{j+1/2} = u_{j+1}, \quad u_{j-1/2} = u_j$	
backward, $O(\Delta y^2)$	$(3u_j - 4u_{j-1} + u_{j-2})/2\Delta y$	$u_{j+1/2} = \frac{3}{2}u_j - \frac{1}{2}u_{j-1}$ $u_{j-1/2} = \frac{3}{2}u_{j-1} - \frac{1}{2}u_{j-2}$	
forward, $O(\Delta y^2)$	$(-3u_j + 4u_{j+1} - u_{j+2})/2\Delta y$	$u_{j+1/2} = \frac{3}{2}u_{j+1} - \frac{1}{2}u_{j+2}$ $u_{j-1/2} = \frac{3}{2}u_j - \frac{1}{2}u_{j+1}$	
Second Derivative: $(\partial^2 f/\partial y^2) = (\partial u/\partial y)_j, u = \partial f/\partial y$			
central, $O(\Delta y^2)$	$(f_{j+1} - 2f_j + f_{j-1})/\Delta y^2$	$u_{j+1/2} = (f_{j+1} - f_j)/\Delta y$ $u_{j-1/2} = (f_j - f_{j-1})/\Delta y$	
backward, $O(\Delta y)$	$(f_j - 2f_{j-1} + f_{j-2})/\Delta y^2$	$u_{j+1/2} = (f_j - f_{j-1})/\Delta y$ $u_{j-1/2} = (f_{j-1} - f_{j-2})/\Delta y$	
forward, $O(\Delta y)$	$(f_j - 2f_{j+1} + f_{j+2})/\Delta y^2$	$u_{j+1/2} = (f_{j+2} - f_{j+1})/\Delta y$ $u_{j-1/2} = (f_{j+1} - f_j)/\Delta y$	

Select Difference Operators. To select difference operators, we apply the governing PDE to be discretized (Eq. (15.3.1)) at an arbitrary interior grid point (a grid point not on the boundary) and at a time between t^n and t^{n+1}:

$$\left(\frac{\partial u}{\partial t} = \nu \frac{\partial^2 u}{\partial y^2}\right)_j^{n'} \tag{15.3.15}$$

where n' is n, $n + 1$, or some value between n and $n + 1$. In this equation, it is assumed that the solution at time t^n is known, and that the solution at t^{n+1} is sought. This assumption is generally applicable because the solution is always known at the previous time level. For example, at the zeroth time level ($n = 0$), the solution is the initial condition.

An Explicit Method. If $n' = n$ in Eq. (15.3.15), then all spatial derivatives are evaluated at time t^n, the previous time level where the solution is known. If we choose the forward-difference operator given by Eq. (15.3.9) for $(\partial u/\partial t)_j^n$ (except replace y and j with t and n) and the central-difference operator given by Eq. (15.3.12) for $(\partial^2 u/\partial y^2)_j^n$, then Eq. (15.3.15) becomes

$$\frac{u_j^{n+1} - u_j^n}{\Delta t} = \nu \frac{u_{j+1}^n - 2u_j^n + u_{j-1}^n}{\Delta y^2} + O(\Delta t, \Delta y^2) \tag{15.3.16a}$$

The above FDE is first-order accurate in time and second-order accurate in space as indicated by $O(\Delta t, \Delta y^2)$. In this equation, u_j^{n+1} is the unknown sought, and solving for it gives

$$u_j^{n+1} = \beta u_{j-1}^n + (1 - 2\beta)u_j^n + \beta u_{j+1}^n, \quad \beta = \nu \, \Delta t/\Delta y^2 \tag{15.3.16b}$$

The FDE in the form of Eq. (15.3.16a) or (15.3.16b) can be applied at any interior grid point ($j = 2, 3, \ldots, JL - 1$). The FDEs at the boundary grid points ($j = 1$ and JL) are obtained by using *boundary conditions* (BCs). For this simple problem, the BCs, given by Eq. (15.3.2), readily yield the following FDEs:

$$u_1^n = u_1^{n+1} = V_0 \tag{15.3.17a}$$

$$u_{JL}^n = u_{JL}^{n+1} = 0 \tag{15.3.17b}$$

Note that each of the FDEs given by Eqs. (15.3.16) and (15.3.17) has only one unknown in it, namely, u_j^{n+1}, a consequence of setting $n' = n$ in Eq. (15.3.15). A method made exclusively of such FDEs is said to be explicit. This particular explicit method with a first-order accurate forward-time differencing is known as the *Euler explicit scheme*.

An Implicit Method. If $n' = n + 1$ in Eq. (15.3.15), then all spatial derivatives are evaluated at t^{n+1}, the new time level where the solution is unknown. If we choose the backward-difference operator given by Eq. (15.3.10) for $(\partial u/\partial t)_j^{n+1}$ to couple t^{n+1} to t^n and the central-difference operator given by Eq. (15.3.11) for $(\partial^2 u/\partial y^2)_j^{n+1}$, then Eq. (15.3.15) becomes

$$\frac{u_j^{n+1} - u_j^n}{\Delta t} = \nu \frac{u_{j+1}^{n+1} - 2u_j^{n+1} + u_{j-1}^{n+1}}{\Delta y^2} + O(\Delta t, \Delta y^2) \qquad (15.3.18)$$

which can be re-written as

$$-\beta u_{j-1}^{n+1} + (1 + 2\beta)u_j^{n+1} - \beta u_{j+1}^{n+1} = u_j^n, \quad \beta = \nu \, \Delta t/\Delta y^2 \qquad (15.3.19)$$

Similar to Eq. (15.3.16), the above equation is first-order accurate in time and second-order accurate in space. But, it differs in that there are three unknowns, u_{j-1}^{n+1}, u_j^{n+1}, and u_{j+1}^{n+1}, instead of one in the FDE. This difference is a consequence of evaluating spatial derivatives at t^{n+1} instead of t^n. By applying Eq. (15.3.19) at every interior grid point and by using Eq. (15.3.17) for the boundary grid points, we obtain the following system of equations:

$$\mathbf{Ax = b} \qquad (15.3.20a)$$

where

$$\mathbf{A} = \begin{bmatrix} 1+2\beta & -\beta & & & & \\ -\beta & 1+2\beta & -\beta & & & \\ & -\beta & 1+2\beta & -\beta & & \\ & & \cdot & \cdot & \cdot & \\ & & & \cdot & \cdot & \cdot \\ & & & -\beta & 1+2\beta & -\beta \\ & & & & -\beta & 1+2\beta \end{bmatrix}, \; \mathbf{x} = \begin{bmatrix} u_2^{n+1} \\ u_3^{n+1} \\ u_4^{n+1} \\ \vdots \\ \vdots \\ u_{JL-2}^{n+1} \\ u_{JL-1}^{n+1} \end{bmatrix}, \; \mathbf{b} = \begin{bmatrix} u_2^n + \beta V_0 \\ u_3^n \\ u_4^n \\ \vdots \\ \vdots \\ u_{JL-2}^n \\ u_{JL-1}^n \end{bmatrix}$$

$$(15.3.20b)$$

In the above system of equations, the FDEs at $j = 1$ and 2 and at $j = JL - 1$ and JL have been combined. A method in which each FDE contains more than one unknown—so that solutions of simultaneous equations are needed—is said to be implicit. This particular implicit method with a first-order accurate backward-time differencing is known as the Euler implicit scheme.

A Generalized Method. If n' is a value between n and $n + 1$, then Eq. (15.3.15) can be written as

$$\frac{u_j^{n+1} - u_j^n}{\Delta t} = \theta \left(\frac{\partial u}{\partial t}\right)_j^{n+1} + (1 - \theta)\left(\frac{\partial u}{\partial t}\right)_j^n, \; \theta = \text{constant} \in (\text{i.e., } [0, 1]) \quad (15.3.21)$$

where the time derivatives are set equal to the spatial derivatives according to the PDE being solved (e.g., $\partial u/\partial t = \nu \partial^2 u/\partial y^2$). If we choose the central-difference operator given by Eq. (15.3.11) for $(\partial^2 u/\partial y^2)$, then the above equation becomes

$$\frac{u_j^{n+1} - u_j^n}{\Delta t} = \theta \nu \left(\frac{u_{j+1}^{n+1} - 2u_j^{n+1} + u_{j-1}^{n+1}}{\Delta y^2}\right) + (1 - \theta)\nu \left(\frac{u_{j+1}^n - 2u_j^n + u_{j-1}^n}{\Delta y^2}\right)$$

$$(15.3.22)$$

It can readily be seen that the above equation reduces to the explicit scheme given by Eq. (15.3.16) and the implicit scheme given by Eq. (15.3.19) when $\theta = 0$ and 1, respectively. When $\theta = \frac{1}{2}$, the above equation is second-order accurate in space and in time. The resulting time-differencing formula is known as the *trapezoidal* or *Crank-Nicolson* method.

15.3.3   Define Solution Algorithm

With the domain and the governing equations discretized, the final step is to specify the solution algorithm (i.e., the step-by-step process of obtaining a solution). Since three sets of FDEs have been derived for Eqs. (15.3.1) and (15.3.2), there are three different solution algorithms.

For the explicit method given by Eqs. (15.3.16) and (15.3.17), the solution algorithm is as follows:

1.   Specify the problem by inputting values for H, V_0, and ν.
2.   Specify the number of grid points desired by inputting JL. As noted, the more the number of grid points, the higher the accuracy.
3.   Specify the time-step size Δt and the duration of interest T. Later, we will find that specification of Δt depends not only on the temporal accuracy sought, but also on Δy.
4.   Calculate the grid spacing with $\Delta y = H/(JL - 1)$.
5.   Calculate constants: $\beta = \nu \, \Delta t / \Delta y^2$ and $N = T/\Delta t$.
6.   Set the time level counter n to 0.
7.   Specify the solution at every grid point at time level n by using the initial condition given by Eq. (15.3.2) ($u_j^n = 0$ for all j).
8.   Calculate u_j^{n+1} at every interior grid point ($j = 2, 3, \ldots, JL - 1$) by using Eq. (15.3.16b).
9.   Calculate u_j^{n+1} at every boundary grid point ($j = 1$ and JL) by using Eq. (15.3.17).
10.   Write solution at time level n to disk for data analysis if desired.
11.   Increment the time level counter by 1; i.e., set n to $n + 1$.
12.   If $n < N$, repeat Steps 7 to 12.

For the implicit method given by Eqs. (15.3.19) and (15.3.17), the solution algorithm is identical to the one described above for the explicit method except that Step 8 is replaced by the following:

8.   Calculate u_j^{n+1} at every interior grid point by solving the linear system of equations given by Eq. (15.3.20).

For the FDEs given by Eqs. (15.3.22) and (15.3.17), the solution algorithm depends on the value of θ. If $\theta = 0$ or 1, then the solution algorithms are identical to those just summarized. If θ greater than 0 but less than 1, then the solution algorithm is identical to one given for the implicit method except that the elements in the matrix **A** and the vector *b* in Eq. (15.3.20) must be replaced by those corresponding to Eq. (15.3.22) instead of (15.3.19).

The exact solution given by Eq. (15.3.3) can be used to evaluate the accuracy of the Euler explicit, the Euler implicit, and the Crank-Nicolson methods as well as give guidelines on grid spacing and time-step size needed to obtain solutions of the desired accuracy.

Before leaving this section, we present the *Thomas algorithm*, which is a highly efficient method for solving linear systems of equations with coefficient matrices that are tridiagonal such as Eq. (15.3.20). Such systems of equations are often encountered in CFD. To illustrate the Thomas algorithm, consider the following system of N independent linear equations:

$$\mathbf{A}\, x = b \tag{15.3.23a}$$

where

$$
\mathbf{A} =
\begin{bmatrix}
A_1 & B_1 \\
C_2 & A_2 & B_2 \\
 & C_3 & A_3 & B_3 \\
 & & & \ddots & \ddots & \ddots \\
 & & & & C_{N-1} & A_{N-1} & B_{N-1} \\
 & & & & & C_N & A_N
\end{bmatrix}
, \quad
x =
\begin{bmatrix}
x_1 \\ x_2 \\ x_3 \\ \vdots \\ x_{N-1} \\ x_N
\end{bmatrix}
, \quad
b =
\begin{bmatrix}
b_1 \\ b_2 \\ b_3 \\ \vdots \\ b_{N-1} \\ b_N
\end{bmatrix}
\tag{15.3.23b}
$$

In the above equations, the elements of $\mathbf{A}$ and b are known, and the elements of x are sought. The first step in the Thomas algorithm is to factor the coefficient matrix $\mathbf{A}$ into two bidiagonal matrices $\mathbf{L}$ and $\mathbf{U}$:

$$
\mathbf{A} = \mathbf{LU} =
\begin{bmatrix}
L_1 \\
P_2 & L_2 \\
 & P_3 & L_3 \\
 & & & \ddots & \ddots \\
 & & & & P_{NA-1} & L_{N-1} \\
 & & & & & P_N & L_N
\end{bmatrix}
\begin{bmatrix}
1 & Q_1 \\
 & 1 & Q_2 \\
 & & 1 & Q_3 \\
 & & & \ddots & \ddots \\
 & & & & 1 & Q_{N-1} \\
 & & & & & 1
\end{bmatrix}
\tag{15.3.24}
$$

$$
=
\begin{bmatrix}
L_1 & L_1 Q_1 \\
P_2 & P_2 Q_1 + L_2 & L_2 Q_2 \\
 & P_3 & P_3 Q_2 + L_3 & L_3 Q_3 \\
 & & & \ddots & \ddots & \ddots \\
 & & & & P_{N-1} & P_{N-1} Q_{N-2} + L_{N-1} & L_{N-1} Q_{N-1} \\
 & & & & & P_N & P_N Q_{N-1} + L_N
\end{bmatrix}
\tag{15.3.25}
$$

From the above, it can be seen that the product of the two bidiagonal matrices is a tridiagonal matrix. The next step is to determine the elements of $\mathbf{L}$ and $\mathbf{U}$. This can be accomplished by equating Eq. (15.3.25) to $\mathbf{A}$ of Eq. (15.3.23b) and comparing term by term. This gives the following recursive formula for computing the elements in $\mathbf{L}$ and $\mathbf{U}$, provided $L_i \neq 0$ for $i = 1, 2, \ldots, N$:

$$P_i = C_i, \ i = 2, 3, \ldots, N; \ L_1 = A_1; \ Q_1 = B_1/L_1$$
$$L_i = A_i - P_i Q_{i-1} \text{ and } Q_i = B_i/L_i, \ i = 2, 3, \ldots, N \tag{15.3.26}$$

The third step is to substitute the **LU** factorization of **A** into Eq. (15.3.23a):

$$\mathbf{A}\,x = \mathbf{L}\,\mathbf{U}\,x = b \tag{15.3.27}$$

and then split as

$$\mathbf{L}\,z = b \tag{15.3.28}$$
$$\mathbf{U}\,x = z \tag{15.3.29}$$

The fourth and final step is to solve for z in Eq. (15.3.28) by forward substitution:

$$z_1 = b_1/L_1; \ z_i = (b_i - P_i z_{i-1})/L_i, \ i = 2, 3, \ldots, N \tag{15.3.30}$$

and to solve for **x** in Eq. (15.3.29) by backward substitution; i.e.,

$$x_N = z_N; \ x_i = z_i - Q_i x_{i+1}, \ i = N - 1, N - 2, \ldots, 1 \tag{15.3.31}$$

This completes the Thomas algorithm, which requires only $5N - 4$ arithmetic operations to obtain a solution. If **A** given by Eq. (15.3.23a) is diagonally dominant (i.e., $|A_i| > |B_i| + |C_i|$ for $i = 1, 2, \ldots, N$), then it can be shown that round-off errors will not grow, which implies that N can be very large (up to 100,000 or more) and still generate very accurate solutions.

15.3.4 Comments on Choice of Difference Operators

From Section 15.3.2, it is clear that the FDE for a given PDE is not unique because a variety of difference operators can be used for each derivative. If the highest-order time derivative is one (which is the case for the continuity, Navier–Stokes, and energy equations), then second-order spatial derivatives represent diffusion, and first-order spatial derivatives represent convection or advection. Since diffusion spreads a disturbance in all directions, second-order spatial derivatives are replaced by central-difference operators. Since convection is directional, upwind and biased-upwind operators are preferred for the first-order spatial derivatives. Upwind and biased-upwind operators are operators whose stencils have more grid points on one side than the other. The side with more grid points is the one where the flow is coming from. For example, if $u > 0$, then a backward or biased-backward difference operator should be used. If $u < 0$, then a forward or biased-forward difference operator should be used. See Hirsch (1991) and Tannehill, et al. (1997) for further discussion.

The difference operator used to replace the time derivative is very important because it determines whether the FDEs for a PDE at different grid points will be coupled to each other or not (i.e., implicit or explicit). In Section 15.3.2, a generalized time-difference operator involving two time levels is given by Eq. (15.3.21). A generalized time-difference operator involving three time levels is as follows (Beam & Warming (1978)):

$$\frac{u_j^{n+1} - u_j^n}{\Delta t} = \frac{\theta}{1 + \gamma} \left(\frac{\partial u}{\partial t}\right)_j^{n+1} + \frac{(1 - \theta)}{1 + \gamma} \left(\frac{\partial u}{\partial t}\right)_j^{n} + \frac{\gamma}{1 + \gamma} \left(\frac{u_j^n - u_j^{n-1}}{\Delta t}\right) \quad (15.3.32)$$

where θ and γ are constants specified by the user. The above formula includes Eq. (15.3.21) as a special case ($\gamma = 0$). Similar to Eq. (15.3.21), Eq. (15.3.32) is explicit when θ equals 0 and implicit otherwise. Also, it is second-order accurate in time when $\theta = \gamma + \frac{1}{2}$ and first-order accurate otherwise. Some commonly used time-difference operators that can be obtained from Eq. (15.3.32) by choosing appropriate values of θ and γ are as follows: Euler explicit ($\theta = \gamma = 0$), Euler implicit ($\theta = 1$ and $\gamma = 0$), Crank-Nicolson ($\theta = 1/2$ and $\gamma = 0$), and three-points backward ($\theta = 1$ and $\gamma = 1/2$).

The time-difference operators in Eqs. (15.3.21) and (15.3.32) belong to the class of "single-stage" operators used to integrate *ordinary differential equations* (ODEs) that are *initial-value problems* (IVPs). In fact, any method for integrating ODEs that are IVPs, whether single-stage or multi-stage, can be used or generalized for use as difference operators for the time derivatives in PDEs. Thus, one could use any of the single-stage ODE integration operators given by the *Adams-Bashforth* or the *Adams-Moulton formulas*. The Adams-Bashforth formulas are explicit and can be written as

$$\frac{u_j^{n+1} - u_j^n}{\Delta t} = \sum_{k=0}^{m} \alpha_{mk} \left(\frac{\partial u}{\partial t}\right)_j^{n-k} + O(\Delta t^{m+1}) \quad (15.3.33)$$

The coefficients α_{mk} in the above equation for up to fourth-order accuracy are as follows: For first-order accuracy, $m = 0$ and $\alpha_{00} = 1$, which gives the Euler explicit formula. For second-order accuracy, $m = 1$, $\alpha_{10} = 3/2$, and $\alpha_{11} = -1/2$. For third-order accuracy, $m = 2$, $\alpha_{20} = 23/12$, $\alpha_{21} = -16/12$, and $\alpha_{22} = 5/12$. For fourth-order accuracy, $m = 3$, $\alpha_{30} = 55/24$, $\alpha_{31} = -59/24$, $\alpha_{32} = 37/24$, and $\alpha_{33} = -9/24$.

The Adams-Moulton formulas are implicit and can be written as

$$\frac{u_j^{n+1} - u_j^n}{\Delta t} = \sum_{k=0}^{m} \beta_{mk} \left(\frac{\partial u}{\partial t}\right)_j^{(n+1)-k} + O(\Delta t^{m+1}) \quad (15.3.34)$$

The coefficients β_{mk} in the above equation for up to fourth-order accuracy are as follows: For first-order accuracy, $m = 0$ and $\beta_{00} = 1$, which gives the Euler implicit formula. For second-order accuracy, $m = 1$, $\beta_{10} = 1/2$, and $\beta_{11} = 1/2$, which is the *Crank-Nicholson formula*. For third-order accuracy, $m = 2$, $\beta_{20} = 5/12$, $\beta_{21} = 8/12$, and $\beta_{22} = -1/12$. For fourth-order accuracy, $m = 3$, $\beta_{30} = 9/24$, $\beta_{31} = 19/24$, $\beta_{32} = -5/24$, and $\beta_{33} = 1/24$.

When using the Adams-Bashforth and Adams-Moulton formulas, note that the higher-order formulas are not "self-starting" so that lower-order formulas must be used to start the computations. Also, note that the higher the order of accuracy, the greater is the number of time levels involved, which increases the amount of computer memory required since solutions at each of those time levels must be stored. Typically, we want to store solutions at no more than two to three time levels.

As an alternative to the single-stage operators, one could use "multi-stage" IVP, ODE integration operators such as Runge-Kutta. The *Runge-Kutta formulas* are constructed by interpolating and extrapolating previously computed derivatives. All Runge-Kutta methods are explicit and self-starting, regardless of the order of accuracy. One second-order Runge-Kutta operator is as follows:

$$\frac{u_j^{n+1/2} - u_j^n}{\Delta t/2} = \left(\frac{\partial u}{\partial t}\right)_j^n \tag{15.3.35a}$$

$$\frac{u_j^{n+1} - u_j^n}{\Delta t} = \left(\frac{\partial u}{\partial t}\right)_j^{n+12} \tag{15.3.35b}$$

Note that the $\partial u/\partial t$ terms in Eqs. (15.3.21) and (15.3.32) to (15.3.35) are to be replaced by spatial derivatives according to the PDE as illustrated by Eqs. (15.3.21) and (15.3.22).

In summary, a wide variety of single- and multi-stage operators can be used to approximate the time derivatives in PDEs. Which one to use depends on whether steady or unsteady solutions are sought. When only steady-state solutions are of interest, temporal accuracy is meaningless, and the time-difference operator is chosen to accelerate convergence to steady state. Typically, the lowest-order accurate time implicit operator is used, which is the Euler implicit operator. When unsteady solutions are of interest, either explicit or implicit operators could be used. The order of accuracy in time is typically two for single-stage difference operators to avoid self-starting problems and increased memory requirements. For multi-stage operators such as Runge-Kutta, order of accuracy can be higher than two without increasing memory requirements though requirements in CPU time do increase somewhat.

15.4 EXAMPLES OF SIMPLE FINITE-VOLUME METHODS

The main difference between FD and FV methods is in the interpretation of the solutions at the grid points. In FDM, the solution u_j^n is viewed as a point function i.e., it is the solution at (y_j, t^n). This implies that u at any point can be interpolated from u at grid points and time levels. In FVMs, the solution u_j^n is viewed as the average value of u in a cell. There are two choices in the selection of the cells. One choice, referred to as *vertex-centered approach*, chooses the cell boundaries to be the bisector between a line joining adjacent grid points. The other choice, referred to as *cell-centered approach*, chooses the cell boundaries to be grid points in one-dimensions (1D), the lines connecting the grid points in two-dimensions (2D), and the planes formed by the lines connecting the grid points in three-dimensions (3D). Whether vertex- or cell-centered, the important thing to note is that abutting cells share a common cell boundary so that there are no voids and overlaps between cells. Figure 15.3 illustrates the difference between vertex-centered and cell-centered approaches, and Fig. 15.4 illustrates the difference between FD and FV interpretations of the solution at a grid point or cell.

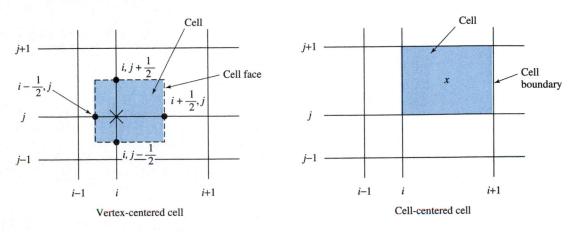

FIGURE 15.3   Vertex-centered and cell-centered cells.

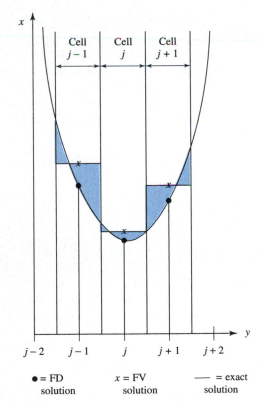

● = FD solution   *x* = FV solution   —— = exact solution

FIGURE 15.4   Finite-difference (FD) and finite-volume (FV) interpretations of the solution at a grid point or cell.

Because of the difference in the interpretation of solutions at grid points or cells, FD and FV methods differ in how PDEs are discretized. Below, the essence of FVMs is given for the problem shown in Fig. 15.1 and governed by Eqs. (15.3.1) and (15.3.2). For this 1D problem, the cells are line segments—either $y_{j-1/2} \leq y \leq y_{j+1/2}, j = 2, 3, \ldots, JL - 1$ for vertex centered or $y_j \leq y \leq y_{j+1}$, $j = 1, 2, \ldots, JL - 1$ for cell centered.

In FVMs, discretization of the PDE proceeds in two steps. The first step is to integrate the PDE over a cell, say from $y_{j-1/2}$ to $y_{j+1/2}$, and from t^n to t^{n+1}. For Eq. (15.3.1), doing this yields

$$\frac{u_j^{n+1} - u_j^n}{\Delta t} = \frac{1}{\Delta y}\left(\tau_{j+\frac{1}{2}} - \tau_{j-\frac{1}{2}}\right) \tag{15.4.1}$$

In the above equation, u_j^n is the average of u over the cell about y_j at time t^n, and $\tau_{j-1/2}$ and $\tau_{j+1/2}$ are the average fluxes that enter/leave the cell boundaries at $y_{j-1/2}$ and $y_{j+1/2}$ over the time period between t^n and t^{n+1}, respectively. This is stated mathematically as

$$u_j^{n/n+1} = \frac{1}{\Delta y}\int_{y_{j-1/2}}^{y_{j+1/2}} u^{n/n+1}\, dy, \quad \tau_{j\pm1/2} = \frac{1}{\Delta t}\int_{t^n}^{t^{n+1}} \mu\left(\frac{\partial u}{\partial y}\right)_{j\pm1/2} dt \tag{15.4.2}$$

The second and final step is to approximate the fluxes at the cell boundaries or faces in terms of u that represent the cell averages. One simple approach is to interpolate from neighboring cells at some time between t^n and t^{n+1}. If linear interpolation is used, then the following approximations are possible:

$$\tau_{j-\frac{1}{2}} = \tau_{j-1/2}^n = \nu\frac{u_j^n - u_{j-1}^n}{\Delta y}, \quad \tau_{j+\frac{1}{2}} = \tau_{j+1/2}^n = \nu\frac{u_{j+1}^n - u_j^n}{\Delta y} \tag{15.4.3}$$

$$\tau_{j-\frac{1}{2}} = \tau_{j-1/2}^{n+1} = \nu\frac{u_j^{n+1} - u_{j-1}^{n+1}}{\Delta y}, \quad \tau_{j+\frac{1}{2}} = \tau_{j+1/2}^{n+1} = \nu\frac{u_{j+1}^{n+1} - u_j^{n+1}}{\Delta y} \tag{15.4.4}$$

$$\tau_{j-\frac{1}{2}} = \frac{1}{2}\left(\tau_{j-1/2}^n + \tau_{j-1/2}^{n+1}\right), \quad \tau_{j+\frac{1}{2}} = \frac{1}{2}\left(\tau_{j+1/2}^n + \tau_{j+1/2}^{n+1}\right) \tag{15.4.5}$$

Substitution of Eqs. (15.4.3), (15.4.4), and (15.4.5) into Eq. (15.4.1) yields Eqs. (15.3.16), (15.3.19), and (15.3.22) with $\theta = \frac{1}{2}$, respectively. The solution algorithms for these FV equations are the same as those described in Section 15.3.3.

Thus, FD and FV methods can give rise to identical algebraic equations. But, FVMs do have a number of advantages. The first is that the interpretation of the solution and the method for discretizing the PDE ensure the correct integral value of the solution over each cell and over the entire spatial domain (i.e., FVEs maintain conservation, an important topic that will be addressed in Section 15.5.4). The second is that the shape of the cell need not be rectangular. In fact, it can be any shape, and that shape can vary from cell to cell as long as there are no overlaps or voids between cells.

15.5 OTHER CONSIDERATIONS

In the previous section, a number of FDEs/FVEs were obtained for the PDE given by Eq. (15.3.1). The question now is how can we judge whether a FDE or a FVE is an algebraic analog of a PDE? Also, how can we be sure that round-off errors do not grow when the solution is advanced from time level to time level? Finally, how can we assess the errors in the computed solutions?

The answers to these questions can be found by examining consistency, numerical stability, convergence, and numerical errors, which are explained in the next four subsections. To facilitate the presentation of these concepts, we define the following:

u_j^n = exact solution of PDE at grid point y_j and time level t^n

U_j^n = exact solution of the FDE or FVE at grid point y_j and time level t^n (i.e., round-off errors do not exist)

N_j^n = numerical solution of the FDE or FVE at grid point y_j and time level t^n (i.e., round-off errors exist)

Based on the above definitions, the total error E_j^n of a solution at grid point y_j and time level t^n is given by

$$E_j^n = u_j^n - N_j^n = (u_j^n - U_j^n) + (U_j^n - N_j^n) \qquad (15.5.1)$$

From the above equation, it can be seen that the total error is made up of two parts. The first part, $u_j^n - U_j^n$, is the error from discretizing the domain and the PDE. Hence, it is referred to as the discretization error:

$$u_j^n - U_j^n = \text{discretization error} \qquad (15.5.2)$$

The second part, $U_j^n - N_j^n$, is concerned with the propagation of round-off errors. This error is referred to as the stability error:

$$U_j^n - N_j^n = \text{stability error} \qquad (15.5.3)$$

Note that the stability error depends only on the FDE or FVE, and is independent of the PDE that it is supposed to represent.

The truncation error, TE, of a FDE or FVE is defined as

$$\text{TE} = (\text{FDE or FVE}) - \text{PDE} \qquad (15.5.4)$$

15.5.1 Consistency

A FDE or FVE is said to be consistent if for every j and n, the following is true:

$$\lim_{\Delta y, \Delta t \to 0} \text{TE}_j^n = 0 \quad \text{or} \quad \lim_{\Delta y, \Delta t \to 0} (\text{FDE or FVE})_j^n = \text{PDE}_j^n \qquad (15.5.5)$$

Thus, consistency measures how well a FDE or a FVE approximates a PDE in the limit of zero grid spacing and time-step size.

To illustrate how consistency is analyzed, consider the FDE given by Eq. (15.3.16a) for the PDE given by Eq. (15.3.1). For convenience, that FDE is repeated:

$$\frac{u_j^{n+1} - u_j^n}{\Delta t} = \nu \frac{u_{j+1}^n - 2u_j^n + u_{j-1}^n}{\Delta y^2} \qquad (15.5.6)$$

Since the FDE given above is algebraic, it is necessary to convert it into a PDE before it can be substituted into Eq. (15.5.5). This can be accomplished by expanding all terms in the FDE about a common grid point and time level by using Taylor series. For Eq. (15.5.6), we choose to expand u_j^{n+1}, u_{j+1}^n, and u_{j-1}^n about u_j^n:

$$u_j^{n+1} = u_j^n + \left(\frac{\partial u}{\partial t}\right)_j^n \Delta t + \left(\frac{\partial^2 u}{\partial t^2}\right)_j^n \frac{\Delta t^2}{2!} + \left(\frac{\partial^3 u}{\partial t^3}\right)_j^n \frac{\Delta t^3}{3!} + \cdots \qquad (15.5.7)$$

$$u_{j+1}^n = u_j^n + \left(\frac{\partial u}{\partial y}\right)_j^n \Delta y + \left(\frac{\partial^2 u}{\partial y^2}\right)_j^n \frac{\Delta y^2}{2!} + \left(\frac{\partial^3 u}{\partial y^3}\right)_j^n \frac{\Delta y^3}{3!} + \cdots \qquad (15.5.8)$$

$$u_{j-1}^n = u_j^n - \left(\frac{\partial u}{\partial y}\right)_j^n \Delta y + \left(\frac{\partial^2 u}{\partial y^2}\right)_j^n \frac{\Delta y^2}{2!} - \left(\frac{\partial^3 u}{\partial y^3}\right)_j^n \frac{\Delta y^3}{3!} + \cdots \qquad (15.5.9)$$

Substituting the above three equations into Eq. (15.5.6) yields

$$\left(\frac{\partial u}{\partial t}\right)_j^n + \left(\frac{\partial^2 u}{\partial t^2}\right)_j^n \frac{\Delta t}{2!} + \left(\frac{\partial^3 u}{\partial t^3}\right)_j^n \frac{\Delta t^2}{3!} + \cdots = \nu \left[\left(\frac{\partial^2 u}{\partial y^2}\right)_j^n + 2\left(\frac{\partial^4 u}{\partial y^4}\right)_j^n \frac{\Delta y^2}{4!} + \cdots\right] \qquad (15.5.10)$$

which becomes in the limit as Δt and Δy approach zero

$$\left(\frac{\partial u}{\partial t} = \nu \frac{\partial^2 u}{\partial y^2}\right)_j^n \qquad (15.5.11)$$

Since the above equation is identical to the PDE given by Eq. (15.3.1) at (y_j, t^n), the FDE given by Eq. (15.5.6) is said to be consistent.

On consistency, it is important to differentiate between unconditional and conditional consistency. Unconditional consistency implies that the FDE approaches the PDE regardless of how Δt and Δy approach zero, which is the case for Eq. (15.5.10). Conditional consistency refers to FDEs that will approach the PDE only if Δt and Δy approach zero in certain prescribed fashion (e.g., Δt must approach zero first before Δy can approach zero). A conditionally consistent FDE may approach a PDE that is fundamentally different from the PDE that it is intended to represent as the grid spacing and time-step size are reduced (see Problem 15.10 for an example). Thus, for a conditionally consistent FD/FVE, reducing the grid spacing or the time-step size could increase instead of decrease errors because the FD/FVE is approaching a different PDE. Also, note that any term multiplied by Δx^p or Δt^p

(e.g., $\Delta y^4 u_i^n$, $\Delta t^2 u_i^n$) can be added to any FD/FVE without violating consistency. Finally, since Δy and Δt are never zero, a consistent FD/FVE is very different from the original PDE (contrast Eqs. (15.5.10) and (15.5.11)). Thus, consistency, though important, is clearly inadequate in ensuring accuracy.

15.5.2 Numerical Stability

A FD/FVE is said to be stable if the stability error given by Eq. (15.5.3) approaches zero or is bounded as n approaches infinity. That is, the round-off error either decays or does not grow as the solution is advanced from time level to time level. Note that this definition is only valid for PDEs with a time-like coordinate. Also, note that stability depends only on the FD/FVE and is independent of the PDE that the FD/FVE is intended to represent.

It turns out that the stability of a FD/FVE depends on whether some parameter ϕ relating grid spacing and time-step size is satisfied or not:

$$\phi \le \phi_{cr} \tag{15.5.12}$$

As an example, for the explicit Euler method given by Eq. (15.3.16), stability is ensured if Δy and Δt satisfy the following criteria:

$$\frac{\nu \Delta t}{\Delta y^2} \le \frac{1}{2} \tag{15.5.13}$$

Thus, $\phi = \nu \Delta t / \Delta y^2$ and $\phi_{cr} = \frac{1}{2}$. For this equation, if the grid spacing is Δy_1, then the largest time-step size permitted is $\Delta t_{max} = (\Delta y_1)^2 / 2\nu$. If Δt used is greater than Δt_{max}, then the solution will oscillate wildly from grid point to grid point with the amplitude of the oscillations increasing from time level to time level so that "blow-up" will eventually take place (the numbers get so large that the computer can no longer represent them). Such solutions in which the round-off errors keep growing are said to be unstable. If Δt used is less than or equal to Δt_{max}, then the solution obtained will either have round-off errors that approach zero or stay bounded. Such solutions are said to be stable. Note, however, though stable solutions are free of round-off errors, they may still be inaccurate from discretization errors (see Eqs. (15.5.1) and (15.5.2)), a topic to be addressed in Section 15.5.4.

For the stability criterion given by Eq. (15.5.13), $\phi_{cr} = 1/2$, which is finite and greater than zero. FD/FVEs, whose stability criteria have finite and nonzero ϕ_{cr}, are said to be conditionally stable. FD/FVEs of most explicit methods are conditionally stable. If ϕ_{cr} is infinite, then the FD/FVE is said to be unconditionally stable. For such FD/FVEs, there are no restrictions on the time-step size as far as stability is concerned. For accuracy, however, the time-step size should be small enough to resolve the temporal physics. FD/FVEs of most implicit methods are unconditionally stable (this can only be proven if the PDE is linear). If ϕ_{cr} equals zero, then the FD/FVE is said to be unconditionally unstable since for a given grid spacing, there is no time-step size, other than zero, that can be used. Clearly, unconditionally unstable FD/FVEs are useless. Problem 15.14 shows a method that appears reasonable but is unconditionally unstable.

There are many ways to analyze the numerical stability of FD/FVEs. One popular and versatile method is the *Fourier method*. The Fourier method analyzes numerical stability via the following three steps. First, introduce an arbitrary disturbance at every grid point at some arbitrary time level t^n. This disturbance represents a random round-off error. Next, expand that disturbance into a Fourier series. Finally, follow the evolution of each Fourier component separately. If any Fourier component grows, then the FD/FVE is said to be unstable. If none grow, then the FD/FVE is said to be stable.

To illustrate the Fourier method for analyzing stability, consider the FD/FVE given by Eq. (15.3.16), which is repeated for convenience:

$$\frac{u_j^{n+1} - u_j^n}{\Delta t} = \nu \frac{u_{j+1}^n - 2u_j^n + u_{j-1}^n}{\Delta y^2} \tag{15.5.14}$$

As mentioned, the first step is to introduce an arbitrary disturbance into the solution at every grid point at t^n. Doing so yields

$$\frac{(u + \varepsilon)_j^{n+1} - (u + \varepsilon)_j^n}{\Delta t} = \nu \frac{(u + \varepsilon)_{j+1}^n - 2(u + \varepsilon)_j^n + (u + \varepsilon)_{j-1}^n}{\Delta y^2} \tag{15.5.15}$$

Subtracting Eq. (15.5.14) from Eq. (15.5.15) gives an evolution equation for the disturbance, ε:

$$\frac{\varepsilon_j^{n+1} - \varepsilon_j^n}{\Delta t} = \nu \frac{\varepsilon_{j+1}^n - 2\varepsilon_j^n + \varepsilon_{j-1}^n}{\Delta y^2} \tag{15.5.16}$$

The next step is to represent the disturbance in a Fourier series. Since the domain has a finite number of equally distributed grid points (see Fig. 15.2), the most general Fourier series is finite and is given by

$$\varepsilon_j^n = \sum_{m=-M}^{M} A_m^n \exp\left(I k_m x_j\right) \tag{15.5.17}$$

where $I = \sqrt{-1}$ is the imaginary number; $M = (IL - 1)/2$ is the maximum number of Fourier components that can be resolved on a grid with IL equally spaced grid points; and $k_m = \pi m/(M \Delta y)$ is the wave number, which is related to the wavelength by $\lambda_m = 2\pi/k_m$. By substituting Eq. (15.5.17) into Eq. (15.5.16) and by noting that $y_{j\pm 1/2} = y_j \pm \Delta y$ and $\exp(y_{j\pm 1/2}) = \exp(y_j)\exp(\pm \Delta y)$, we obtain

$$\sum_{m=-M}^{M} \left[\frac{(A_m^{n+1} - A_m^n)e^{I k_m y_j}}{\Delta t} - \nu \frac{(A_m^n e^{I k_m \Delta y} - 2A_m^n + A_m^n e^{-I k_m \Delta y})e^{I k_m y_j}}{\Delta y^2} \right] = 0 \tag{15.5.18}$$

Since there are no interactions between the Fourier components (i.e., no multiplication of two different Fourier components), Eq. (15.5.18) requires every Fourier component to be zero. This coupled with $\exp(\pm I\phi) = \cos \phi \pm I \sin \phi$ gives

$$A_m^{n+1} = A_m^n + \frac{\nu\Delta t}{\Delta y^2} A_m^n (e^{Ik_m\Delta y} - 2 + e^{-Ik_m\Delta y})$$

$$= A_m^n \left[1 - 2\frac{\nu\Delta t}{\Delta y^2}(1 - \cos(k_m\Delta y)) \right]$$

$$= A_m^n \left[1 - 4\frac{\nu\Delta t}{\Delta y^2} \sin^2 \left(\frac{k_m\Delta y}{2} \right) \right] \tag{15.5.19}$$

Now, define the amplification factor as

$$G_m = A_m^{n+1}/A_m^n \tag{15.5.20}$$

The third and final step is to follow the evolution of the amplification factor as n approaches infinity. For numerical stability, it is clear that the following condition must be satisfied for every Fourier component:

$$|G_m| \le 1 \tag{15.5.21}$$

From Eqs. (15.5.19) and (15.5.20), the above equation implies the following for every m:

$$-1 \le 1 - 4\frac{\nu\Delta t}{\Delta y^2} \sin^2 \left(\frac{k_m\Delta y}{2} \right) \le 1, \tag{15.5.22}$$

This inequality becomes critical when $\sin^2 (k_m\Delta y/2)$ reaches a maximum of unity. At this critical point, we get the correct stability criteria, which is (see Eq. (15.5.13))

$$\frac{\nu\Delta t}{\Delta y^2} \le \frac{1}{2} \tag{15.5.23}$$

From the above example, we make a number of observations. First, the most unstable Fourier component is $m = M$, which corresponds to the largest wave number, but shortest wavelength. This turns out to be generally true for most FD/FVEs. Second, this analysis assumes periodic boundary conditions because of the use of Fourier series. Thus, it cannot account for FD/FVEs on the boundaries. If FD/FVEs at boundary grid points can impose a more stringent stability criteria, then the matrix stability method can be used. Third, this analysis can only be applied to linear FD/FVEs since only then can each Fourier component be analyzed separately. Nonlinear FD/FVEs must be linearized before they can be analyzed. Finally, A_m^n in the Fourier series expansion of ε_m^n in Eq. (15.5.17) can be replaced by $(G_m)^{(n)}A_m^0$ because $A_m^{n+1} = G_mA_m^n = G_mG_mA_m^{n-1} = \cdots = (G_m)^{(n+1)}A_m^0$, where (n) and $(n + 1)$ denote to the power of (see Eq. (15.5.20)). For methods involving more than two time levels, this will be more convenient. For FD/FVEs that involve two dimensions, replace Eq. (15.5.17) by

$$\varepsilon_{ij}^n = \sum_{m_x=-M_x}^{M_x} \sum_{m_y=-M_y}^{M_y} A_{m_xm_y}^n \exp \left[I(k_{m_x}x_i + k_{m_y}y_j) \right] \tag{15.5.24}$$

The three-dimensional form of the above equation is a straightforward extension.

15.5.3 Convergence

A FD/FVE is said to be convergent if its solutions approach the exact solution of the PDE in the limit of infinitesimal Δy and Δt:

$$\lim_{\Delta y, \Delta t \to 0} N_j^n = u_j^n \tag{15.5.25}$$

Equation (15.5.25) indicates that for a convergent FD/FVE, one can get a numerical solution to any desired accuracy by simply reducing the grid spacing and time-step size. For FD/FVEs of linear PDEs, convergence can always be analyzed analytically though the process can be tedious. Fortunately, there is a general theory known as the *Lax equivalence theorem*, which states that *a FD/FVE of a well-posed, linear PDE is convergent, if that FD/FVE is consistent and stable*. With this theorem, a FD/FVE of a well-posed, linear PDE is guaranteed to be convergent if we can show that it is consistent and stable as explained in Sections 15.5.1 and 15.5.2. Since consistency and stability are much easier to analyze than convergence, the Lax equivalence theorem is quite useful.

For FD/FVEs of PDEs that are not linear (e.g., quasi-linear or nonlinear PDEs), we do not know how to prove convergence. But, this inability to prove convergence does not imply that these FD/FVEs are not convergent. For these FD/FVEs, the Lax equivalence theorem is still useful, except it becomes a necessary, but not a sufficient condition for convergence. For these FD/FVEs, a comparison between numerical solutions and experimental measured values may be needed.

15.5.4 Numerical Errors

From the discussions on consistency, numerical stability, and convergence given in the previous three sub-sections, we note the following. First, if a FD/FVE is stable, then the total error given by Eq. (15.5.1) reduces to

$$E_j^n = u_j^n - N_j^n = (u_j^n - U_j^n) + (U_j^n - N_j^n) \approx u_j^n - U_j^n \tag{15.5.26}$$

That is, discretization error dominates with stability error being negligible. Second, when grid spacing and time-step size are finite, which is always the case, then consistency analyses show the FD/FVEs to differ considerably from the original PDEs that they are suppose to represent (compare Eqs. (15.5.10) and (15.5.11)). It turns out that the FD/FVEs are, in general, much more complex. They contain spurious modes (solutions not contained in the original PDEs). Thus, even if a FD/FVE is convergent, we still need to ask the following questions to ensure that physically meaningful solutions are obtained:

- What are some of the most important properties that a FD/FVE must possess when grid spacing and time-step sizes are finite?

- How does one choose the grid spacing and the time-step size to ensure accurate solutions?

The answers to some of these questions are briefly described below.

Conservative Property.   One of the most important properties of PDEs governing fluid-flow problems is the conservation principle. For example, mass cannot be created or destroyed; momentum must be balanced; and total energy is conserved. Thus, it is important for the FD/FVEs to represent this principle correctly. Fortunately, this is easy to do. Just make sure that at every common face between two abutting cells, what leaves one cell must enter the other. But, to do this requires the PDE to be written in what is referred to as the conservative form. A PDE is written in conservative form if all coefficients outside of the outermost derivatives are constants. If any coefficient is a variable, then that PDE is written in non-conservative form. Thus, for example, $\partial u/\partial t + \frac{1}{2}\partial u^2/\partial x = 0$ is written in conservative form, but $\partial u/\partial t + u\partial u/\partial x = 0$ is not. Problems 15.16 and 5.17 illustrate difficulties associated with using PDEs cast in non-conservative form.

Transportive Property.   For a fluid, a disturbance at any location in the flow field can spread to other locations by convection due to fluid motion, by diffusion due to random molecular motion, and by pressure waves. Convection can transport a disturbance only in the direction of fluid velocity. Diffusion and pressure waves can spread a disturbance in every direction. FD/FVEs should possess the same transportive properties as the PDEs that they are intended to represent. Otherwise, there is transportive error. It turns out that this is not easy to do correctly. As noted in Section 15.3.4, derivatives representing convection are upwind or biased-upwind differenced, and derivatives representing diffusion and pressure are centrally differenced. For compressible flows, the convective and pressure terms should be treated collectively, and differencing is based on characteristic theory (see Hirsch (1991)). Unfortunately, the theory on upwind differencing is rigorous only for one-dimensional flows.

Dissipative Property.   One major error in many FD/FVEs is excessive artificial or numerical diffusion (i.e., the "effective" diffusion coefficients such as viscosity or conductivity in the FD/FVEs are much higher than the ones in the PDEs). This can cause shear layers, viscous forces, surface heat transfer, and conversion of mechanical to thermal energy to be computed incorrectly. There are two main culprits of high numerical diffusion. The first is how convective terms are differenced. If upwind differencing is used, then first-order formulas can produce excessive numerical diffusion. If second-order upwind differencing formulas are used, then reasonable results can be obtained. High-resolution differencing schemes such as flux-difference splitting with limiters are aimed at getting the transportive properties correct with minimal artificial diffusion (see Hirsch (1991)). The second culprit is having cells with high aspect ratios and not aligning the grid lines with the flow directions.

Grid-Independent Solution.   Even for a convergent set of FD/FVEs, many different solutions can be obtained, some totally wrong, some qualitatively correct but quantitatively wrong, and some sufficiently accurate. Which solution is obtained depends on the grid spacing and time-step size used. Thus, once generating a solution on a grid, it is important to generate another solution on a finer grid and correspondingly smaller time-step size. Ideally, this process should be repeated until the relative difference between successive solutions obtained on finer and finer grids approach the level of the round-off errors. Unfortunately, for complicated engineering problems, it is often not feasible to obtain grid-independent solutions because of resource constraints. If the solution generated is not grid-independent, then its interpretation must be handled with extreme care.

15.6 GRID GENERATION

As noted in Sections 15.2 and 15.3.1, the domain of the problem must be replaced by a system of grid points or cells. This process is referred to as grid or mesh generation. Grid systems or meshes can be classified as structured, unstructured, or hybrid (Fig. 15.5). *Structured grids* have grid lines that form a curvilinear coordinate system. As a result, the cells are rectangular in 2D and hexahedral in 3D. *Unstructured grids* have grid lines that do not form curvilinear coordinate systems. Their cell shapes can be triangular/tetrahedral or a mixture of various shapes. *Hybrid grids* involve a combination of structured and unstructured grids. For example, the grid could be structured near walls and unstructured elsewhere. The state-of-the-art on this subject is summarized in Thompson, et al. (1998).

Whether one uses structured, unstructured, or hybrid grids, the ideal grid system is one that resolves all relevant physics in the flow, minimizes errors that grids can induce, and uses as few grid points or cells as possible for computational efficiency. Unfortunately, the ideal and even the not-so-ideal grid system can be extremely difficult to generate for problems with complicated flow features and geometries, especially in 3D. At present, grid generation is the most tedious and time-consuming part of a CFD analysis. In addition, it requires considerable expertise, not just in understanding CFD and grid generation, but also in understanding the physics of the problem being studied.

In this section, we illustrate a simple but versatile method for generating structured grid systems, known as the *two-boundary method*. This method involves generating a boundary-conforming, curvilinear coordinate system. We present this method in three steps. First, describe the relationship between structured grids and boundary-conforming coordinate systems. Then, present the

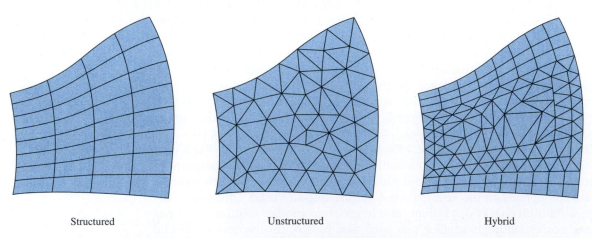

| Structured | Unstructured | Hybrid |

FIGURE 15.5   Structured, unstructured, and hybrid grid systems.

two-boundary method for generating such systems. Finally, illustrate how this method is used to generate solutions to PDEs in complicated geometries.

15.6.1 Boundary-Fitted Coordinate Systems

All structured grids have a corresponding boundary-conforming coordinate system in which coordinate lines in the boundary-fitted coordinate system correspond to grid lines in the physical domain. To illustrate, consider the structured grid shown in Fig. 15.6 in which x-y is the coordinate system of the physical domain, and ξ-η is the boundary-conforming coordinate system of the transformed domain.

From that figure, we note six important points. First, the mapping between the physical and the transformed domains must be one-to-one:

$$(x, y) \Leftrightarrow (\xi, \eta)$$

or

$$x = x(\xi, \eta) \; \& \; y = y(\xi, \eta) \quad \text{and} \quad \xi = \xi(x, y) \; \& \; \eta = \eta(x, y) \quad (15.6.1)$$

where the Jacobian of the coordinate transformation, $J = |\partial(x, y)/\partial(\xi, \eta)|$, must be greater than zero for right-handed coordinate systems. Second, coordinate lines $\eta = 0$ and $\eta = 1$ correspond respectively to the lower (surface 1) and upper

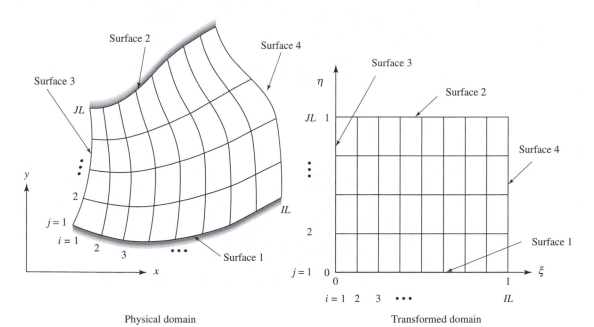

Physical domain   Transformed domain

FIGURE 15.6   Boundary-conforming coordinate system in the physical and transformed domains.

(surface 2) boundaries of the channel in the physical domain. The position vector of the channel walls are then

$$\mathbf{r}_1 = x(\xi, 0)\,\mathbf{i} + y(\xi, 0)\,\mathbf{j} = X_1(\xi)\,\mathbf{i} + Y_1(\xi)\,\mathbf{j} \tag{15.6.2}$$

$$\mathbf{r}_2 = x(\xi, 1)\,\mathbf{i} + y(\xi, 1)\,\mathbf{j} = X_2(\xi)\,\mathbf{i} + Y_2(\xi)\,\mathbf{j} \tag{15.6.3}$$

Third, coordinate lines $\xi = 0$ and $\xi = 1$ correspond to the boundaries where flow enters (surface 3) and exits (surface 4) the channel, referred to as inflow and outflow boundaries. The position vectors for these boundaries are:

$$\mathbf{r}_3 = x(0, \eta)\,\mathbf{i} + y(0, \eta)\,\mathbf{j} = X_3(\eta)\,\mathbf{i} + Y_3(\eta)\,\mathbf{j} \tag{15.6.4}$$

$$\mathbf{r}_4 = x(1, \eta)\,\mathbf{i} + y(1, \eta)\,\mathbf{j} = X_4(\eta)\,\mathbf{i} + Y_4(\eta)\,\mathbf{j} \tag{15.6.5}$$

Fourth, coordinate lines between $\eta = 0$ and $\eta = 1$ correspond to grid lines between the lower and the upper boundaries of the channel, and coordinate lines between $\xi = 0$ and $\xi = 1$ correspond to grid lines between the inflow and the outflow boundaries. Fifth, the number of coordinate lines selected to be grid lines is finite, IL in ξ direction, and JL in the η direction. IL does not need to equal JL. Sixth, the spacing between grid lines in the physical domain can be variable, but they are always uniform in the transformed domain.

Since all grid points are equally spaced in the transformed domain, their locations are known and given by (ξ_i, η_j), where

$$\xi_i = (i - 1)\,\Delta\xi \qquad\qquad \eta_j = (j - 1)\,\Delta\eta \tag{15.6.6}$$

$$\Delta\xi = 1/(IL - 1) \qquad\qquad \Delta\eta = 1/(JL - 1) \tag{15.6.7}$$

Once the boundary-conforming coordinate system has been generated that satisfies the statements given above, the location of each grid point in the physical domain is readily determined by Eq. (15.6.1): $x_{i,j} = x(\xi_i, \eta_j)$ and $y_{i,j} = y(\xi_i, \eta_j)$, where ξ_i, and η_j are given by Eqs. (15.6.6) and (15.6.7).

15.6.2   An Algebraic Method for Structured Grids

There are many methods that can be used to generate boundary-conforming coordinate systems in connection with structured grids (see Thompson, et al. (1998)). The two-boundary method belongs to a class of algebraic grid generation methods based on transfinite interpolation (see, e.g., Shih et al. (1991) and Steinthorsson, et al. (1992)). This method can ensure two boundaries of a spatial domain to be mapped to coordinate lines or surfaces. There are many problems that require only two boundaries of the spatial domain to be mapped correctly. Examples include flow between channels and flow over airfoils. If more than two boundaries must be mapped correctly, then the four- and the six-boundary methods can be used, which are straightforward extensions of the two-boundary method (see Shih, et al. (1991)).

To illustrate the two-boundary method, consider the converging-diverging nozzle shown in Fig. 15.7. Six steps are involved. Each of these steps is described below.

Step 1:   Define the coordinate transformation. For a 2D problem involving grid points that do not move in time, the coordinate transformation is given by Eq. (15.6.1), where x-y is the coordinate system of the physical domain, and ξ-η is the boundary-conforming coordinate system.

Step 2:   Select the two boundaries of the spatial domain that must be mapped correctly, and decide which boundary corresponds to which coordinate line in the boundary-conforming coordinate system. These two boundaries must not intersect each other at any point. For the problem shown in Fig. 15.7, we choose the lower (curve 1) and upper (curve 2) walls of the nozzle to be the ones that must be mapped correctly. And, we choose curve 1 to correspond to $\eta = 0$ and curve 2 to correspond to $\eta = 1$. This implies that the inflow (curve 3) and the outflow (curve 4) boundaries correspond to $\xi = 0$ and $\xi = 1$, respectively. These selections are described by Eqs. (15.6.2) to (15.6.5).

Step 3:   Describe the boundaries selected in parametric form. The expressions $\mathbf{r}_i = X_i(\xi)\,\mathbf{i} + Y_i(\xi)\,\mathbf{j}$ with $i = 1, 2$ imply that each curve must be described in parametric form in terms of the parameter ξ. Suppose the y-coordinates of the lower and upper walls of the nozzle are described by

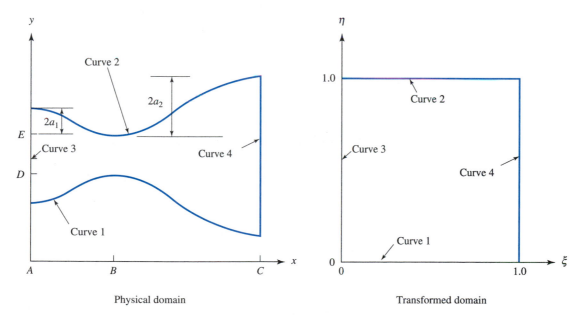

FIGURE 15.7   Geometry of converging-diverging nozzle in physical and transformed domain.

$$y_1 = (D - a_1) - a_1 \cos [\pi(x - A)/(B - A)], \text{ when } x \in [A, B]$$
$$y_1 = (D - a_2) - a_2 \cos [\pi(x - B)/(C - B)], \text{ when } x \in [B, C] \qquad (15.6.8)$$

$$y_2 = (E + a_1) - a_1 \cos [\pi(x - A)/(B - A)], \text{ when } x \in [A, B]$$
$$y_2 = (E + a_2) - a_2 \cos [\pi(x - B)/(C - B)], \text{ when } x \in [B, C] \qquad (15.6.9)$$

where A, B, C, D, E, a_1, and a_2 are constants. The above explicit representations of curves can readily be converted to parametric representations by letting

$$X_1 = X_2 = A + (C - A) \xi \qquad (15.6.10)$$

$$Y_1 = (D - a_1) - a_1 \cos [\pi(C - A) \xi/(B - A)], \text{ when } \xi \in \left[0, \frac{B - A}{C - A} \right]$$

$$Y_1 = (D - a_2) - a_2 \cos \{\pi[A + (C - A) \xi]/(C - B)\}, \text{ when } \xi \in \left[\frac{B - A}{C - A}, 1 \right]$$
$$(15.6.11)$$

$$Y_2 = (E + a_1) - a_1 \cos [\pi(C - A) \xi/(B - A)], \text{ when } \xi \in \left[0, \frac{B - A}{C - A} \right]$$

$$Y_1 = (E + a_2) - a_2 \cos \{\pi[A + (C - A) \xi]/(C - B)\}, \text{ when } \xi \in \left[\frac{B - A}{C - A}, 1 \right]$$
$$(15.6.12)$$

Step 4: Define curves that connect the two selected boundaries by using transfinite interpolation. If *Lagrange interpolation* is employed to connect points on the lower and upper walls (curves 1 and 2) that have the same ξ value, then

$$x(\xi, \eta) = X_1 (\xi) L_1(\eta) + X_2 (\xi) L_2(\eta) \qquad (15.6.13)$$
$$y(\xi, \eta) = Y_1 (\xi) L_1(\eta) + Y_2 (\xi) L_2(\eta) \qquad (15.6.14)$$

In the above equations, X_1, X_2, Y_1, and Y_2 are given by Eqs. (15.6.10) to (15.6.12). The Lagrange polynomials, L_1 and L_2, are given by

$$L_1(\eta) = 1 - \eta \qquad (15.6.15a)$$
$$L_2(\eta) = \eta \qquad (15.6.15b)$$

Equations (15.6.13) to (15.6.15) constitute a boundary-conforming coordinate system in which every point in the ξ-η coordinate system corresponds to a point in the x-y coordinate system, and one set of coordinate lines in the ξ-η coordinate system correspond to the problem boundaries in the x-y coordinate system.

Boundary-conforming coordinate systems formed by Lagrange interpolation involve one set of grid lines that are all straight (e.g., lines of constant ξ in Eqs. (15.6.13) to (15.6.15) will all be straight in the x-y coordinate system). To enable grid lines that can curve in space and be perpendicular to boundary surfaces, Hermite interpolation can be used. If *Hermite interpolation* is employed to connect points on the lower and upper walls (curves 1 and 2) that have the same ξ value, then

$$x\,(\xi, \eta) = X_1(\xi)\,H_1(\eta) + X_2\,(\xi)\,H_2(\eta) + \frac{\partial x(\xi, 0)}{\partial \eta}\,H_3(\eta) + \frac{\partial x(\xi, 1)}{\partial \eta}\,H_4(\eta)$$
$$(15.6.16)$$

$$y\,(\xi, \eta) = Y_1\,(\xi)\,H_1(\eta) + Y_2\,(\xi)\,H_2(\eta) + \frac{\partial\, y(\xi, 0)}{\partial \eta}\,H_3(\eta) + \frac{\partial\, y(\xi, 1)}{\partial \eta}\,H_4(\eta)$$
$$(15.6.17)$$

where the Hermite polynomials are given by

$$H_1 = 2\eta^3 - 3\eta^2 + 1 \qquad (15.6.18a)$$
$$H_2 = -\,2\eta^3 + 3\eta^2 \qquad (15.6.18b)$$
$$H_3 = \eta^3 - 2\eta^2 + \eta \qquad (15.6.18c)$$
$$H_4 = \eta^3 - \eta^2 \qquad (15.6.18d)$$

The derivatives in Eqs. (15.6.16) and (15.6.17) are chosen so that constant ξ coordinate lines intersect the curves 1 and 2 perpendicularly:

$$\frac{\partial x(\xi, 0)}{\partial \eta} = -K_1(\xi)\frac{\partial Y_1}{\partial \xi} \qquad (15.6.19a)$$

$$\frac{\partial y(\xi, 0)}{\partial \eta} = K_1(\xi)\frac{\partial X_1}{\partial \xi} \qquad (15.6.19b)$$

$$\frac{\partial x(\xi, 1)}{\partial \eta} = -K_2(\xi)\frac{\partial Y_2}{\partial \xi} \qquad (15.6.20a)$$

$$\frac{\partial y(\xi, 1)}{\partial \eta} = K_2(\xi)\frac{\partial X_2}{\partial \xi} \qquad (15.6.20b)$$

where $K_1\,(\xi)$ and $K_2\,(\xi)$ are functions of the geometry. The higher the value of these parameters, the more the grid lines push orthogonally away from the boundary (see Fig. 15.8). For the geometry described by Eqs. (15.6.8) and (15.6.9) and shown in Fig. 15.7, these parameters were chosen to be

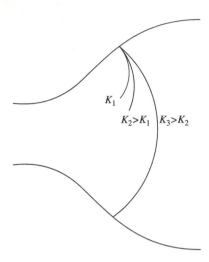

FIGURE 15.8 Effects of increasing the value of K (Eqs. (15.6.19) and (15.6.20)).

$$K_1 (\xi) = K_2 (\xi) = F [Y_2(\xi) - Y_1(\xi)] \qquad (15.6.21)$$

where F is a constant. Equations (15.6.16) to (15.6.21) constitute another boundary-conforming coordinate system.

Step 5: Discretize the transformed domain to generate a structured grid system. Once the boundary-conforming coordinate system has been generated (say, by applying Eqs. (15.6.16) to (15.6.21)), a structured grid system can readily be generated in two steps. First, discretize the transformed domain according to Eqs. (15.6.6) and (15.6.7). Here, it is only necessary to decide on the number of grid lines in the ξ and η directions; that is, IL and JL. Second, determine the location of the grid points in the physical domain by substituting Eqs. (15.6.6) and (15.6.7) into the boundary-conforming coordinate system. If the boundary-conforming coordinate system is the one given by Eqs. (15.6.16) to (15.6.21), then one obtains the grid shown in Fig. 15.9. For that grid, the following values were used for the geometry: $A = 0, B = 2, C = 6, D = 2, E = 3, a_1 = 0.35$, and $a_2 = 0.75$. The following values were chosen for the Hermite interpolant and discretization: $F = 0.08, IL = 21$, and $JL = 11$. Note that the grid system generated has grid lines that intersect nozzle walls orthogonally.

Step 6: Redistribute grid points by clustering them where most needed. The grid system shown in Fig. 15.9 has grid points that are nearly uniformly distributed. Since the total number of grid points used should be kept to the minimum needed to resolve the flow physics, it is efficient to place them where most needed by clustering. Clustering can be accomplished by the use of stretching functions. For flow through a nozzle, grid points should be clustered near the

solid walls because the no-slip condition ($\mathbf{V} = 0$) on the walls creates sharper gradients. Grid points can be clustered to curves 1 and
2 in Fig. 15.7 by replacing the η in Eqs. (15.6.13) to (15.6.15) or
Eqs. (15.6.16) to (15.6.21) by the following stretching function:

$$\frac{(\beta + 1)[(\beta + 1)/(\beta - 1)]^{2\eta-1} - \beta + 1}{2\{1 + [(\beta + 1)/(\beta - 1)]^{2\eta-1}\}} \qquad (15.6.22)$$

where β is a constant greater than unity. More clustering takes
place near $\eta = 0$ and $\eta = 1$ as β approaches unity.

15.6.3   Solving PDEs on Boundary-Conforming Grid Systems

To illustrate how PDEs are solved on boundary-conforming grid systems such as
the one shown in Fig. 15.9, consider the advection of a trace contaminant in an
incompressible flow through a converging-diverging nozzle. If the flow field (i.e.,
u and v) is known, then the governing equation is

$$\frac{\partial Y}{\partial t} + u\frac{\partial Y}{\partial x} + v\frac{\partial Y}{\partial y} = \frac{\partial}{\partial x}\left(D\frac{\partial Y}{\partial x}\right) + \frac{\partial}{\partial y}\left(D\frac{\partial Y}{\partial y}\right) \qquad (15.6.23)$$

where Y is the mass fraction of the contaminant, and D is the mass-diffusion coefficient. The FD/FVE for this PDE on the grid system shown in Fig. 15.9 can be
obtained using six steps. In the following, each of these six steps is described in
detail.

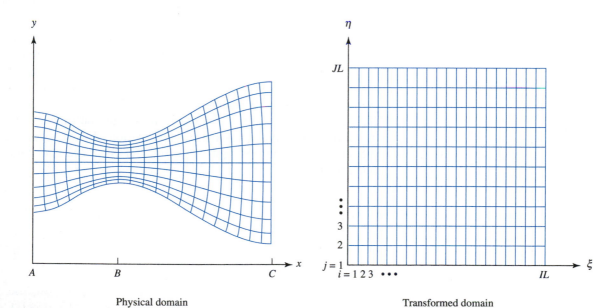

Physical domain   Transformed domain

FIGURE 15.9   Grid system in the physical and transformed domains.

Step 1: Cast the PDE in conservative form. It is imperative for the PDE to be cast in conservative form before replacing derivatives by difference operators, otherwise the resulting FD/FVE equation will not maintain the conservative property of the PDE. By using the continuity equation, $\partial u/\partial x + \partial v/\partial y = 0$, Eq. (15.6.23) can be written as

$$\frac{\partial Y}{\partial t} + \frac{\partial uY}{\partial x} + \frac{\partial vY}{\partial y} = \frac{\partial}{\partial x}\left(D\frac{\partial Y}{\partial x}\right) + \frac{\partial}{\partial y}\left(D\frac{\partial Y}{\partial y}\right) \qquad (15.6.24)$$

which is the conservative form sought.

Step 2: Transform the PDE from the x-y coordinate system to the boundary-fitted coordinate system. Since $x = x(\xi, \eta)$ and $y = y(\xi, \eta)$, the chain-rule gives

$$\frac{\partial}{\partial x} = \frac{\partial \xi}{\partial x}\frac{\partial}{\partial \xi} + \frac{\partial \eta}{\partial x}\frac{\partial}{\partial \eta} = \xi_x \frac{\partial}{\partial \xi} + \eta_x \frac{\partial}{\partial \eta} \qquad (15.6.25a)$$

$$\frac{\partial}{\partial y} = \frac{\partial \xi}{\partial y}\frac{\partial}{\partial \xi} + \frac{\partial \eta}{\partial y}\frac{\partial}{\partial \eta} = \xi_y \frac{\partial}{\partial \xi} + \eta_y \frac{\partial}{\partial \eta} \qquad (15.6.25b)$$

where ξ_x, ξ_y, η_x, and η_y are called metric coefficients. By replacing the x- and y-derivatives in Eq. (15.6.24) by Eqs. (15.6.25a) and (15.6.25b), we obtain

$$\frac{\partial Y}{\partial t} + \left(\xi_x \frac{\partial}{\partial \xi} + \eta_x \frac{\partial}{\partial \eta}\right)uY + \left(\xi_y \frac{\partial}{\partial \xi} + \eta_y \frac{\partial}{\partial \eta}\right)vY$$

$$= \left(\xi_x \frac{\partial}{\partial \xi} + \eta_x \frac{\partial}{\partial \eta}\right)\left[D\left(\xi_x \frac{\partial}{\partial \xi} + \eta_x \frac{\partial}{\partial \eta}\right)Y\right] + \left(\xi_y \frac{\partial}{\partial \xi} + \eta_y \frac{\partial}{\partial \eta}\right)\left[D\left(\xi_y \frac{\partial}{\partial \xi} + \eta_y \frac{\partial}{\partial \eta}\right)Y\right]$$

$$(15.6.26)$$

Since the metric coefficients are variables, the above equation is again in non-conservative form. To convert the above equation into conservative form, multiply the above equation by the Jacobian of the coordinate transformation, $J = |\partial(x, y)/\partial(\xi, \eta)|$, and then use chain-rule differentiation to move the metric coefficients inside the derivatives. For example, application of this procedure to the second term on the right-hand-side (RHS) of the above equation gives

$$J\left(\xi_x \frac{\partial}{\partial \xi} + \eta_x \frac{\partial}{\partial \eta}\right)uY = \frac{\partial}{\partial \xi}(\xi_x J)uY + \frac{\partial}{\partial \eta}(\eta_x J)uY - uY\left(\frac{\partial}{\partial \xi}\xi_x J + \frac{\partial}{\partial \eta}\eta_x J\right)$$

$$= \frac{\partial}{\partial \xi}(\xi_x J)uY + \frac{\partial}{\partial \eta}(\eta_x J)uY \qquad (15.6.27)$$

The third term on the RHS of Eq. (15.6.27) vanishes because of geometric conservation laws such as (see Tannehill, et al. (1997) for derivation)

$$\frac{\partial}{\partial \xi}\hat{\xi}_x + \frac{\partial}{\partial \eta}\hat{\eta}_x = 0 \qquad (15.6.28a)$$

$$\frac{\partial}{\partial \xi}\hat{\xi}_y + \frac{\partial}{\partial \eta}\hat{\eta}_y = 0 \qquad (15.6.28b)$$

where

$$\hat{\xi}_x = \xi_x J, \ \hat{\xi}_y = \xi_y J, \ \hat{\eta}_x = \eta_x J, \ \hat{\eta}_y = \eta_y J \qquad (15.6.28c)$$

By repeating the above procedure for every term in Eq. (15.6.26) and making use of Eq. (15.6.28), we obtain

$$\frac{\partial Y}{\partial t} + \frac{\partial}{\partial \xi}(\hat{\xi}_x u + \hat{\xi}_y v)Y + \frac{\partial}{\partial \eta}(\hat{\eta}_x u + \hat{\eta}_y v)Y$$

$$= \frac{\partial}{\partial \xi}\left(\hat{\xi}_x D\frac{\partial Y}{\partial x} + \hat{\xi}_y D\frac{\partial Y}{\partial y}\right) + \frac{\partial}{\partial \eta}\left(\hat{\eta}_x D\frac{\partial Y}{\partial x} + \hat{\eta}_y D\frac{\partial Y}{\partial y}\right) \quad (15.6.29a)$$

where

$$\frac{\partial Y}{\partial x} = \left(\xi_x \frac{\partial}{\partial \xi} + \eta_x \frac{\partial}{\partial \eta}\right)Y, \quad \frac{\partial Y}{\partial y} = \left(\xi_y \frac{\partial}{\partial \xi} + \eta_y \frac{\partial}{\partial \eta}\right)Y \quad (15.6.29b)$$

Equation (15.6.29) is the desired PDE in boundary-fitted coordinates, which is in conservative form.

Step 3: Replace the time derivative by a time-difference operator. Any one of the time-differencing formulas presented in Section 15.3.4 can be used. Here, we choose the following second-order Runge-Kutta method (see Eq. (15.3.35)):

$$\frac{Y_{ij}^{n+1/2} - Y_{ij}^n}{\Delta t/2} = \left(\frac{\partial Y}{\partial t}\right)_{ij}^n \qquad (15.6.30a)$$

$$\frac{Y_{ij}^{n+1} - Y_{ij}^n}{\Delta t} = \left(\frac{\partial Y}{\partial t}\right)_{ij}^{n+/12} \qquad (15.6.30b)$$

where

$$\frac{\partial Y}{\partial t} = -\frac{\partial}{\partial \xi}(\hat{\xi}_x u + \hat{\xi}_y v)Y - \frac{\partial}{\partial \eta}(\hat{\eta}_x u + \hat{\eta}_y v)Y$$

$$+ \frac{\partial}{\partial \xi}\left(\hat{\xi}_x D\frac{\partial Y}{\partial x} + \hat{\xi}_y D\frac{\partial Y}{\partial y}\right) + \frac{\partial}{\partial \eta}\left(\hat{\eta}_x D\frac{\partial Y}{\partial x} + \hat{\eta}_y D\frac{\partial Y}{\partial y}\right) \quad (15.6.30c)$$

Step 4: Approximate the spatial derivatives by algebraic equations. For FDMs, this involves replacing spatial derivatives by difference operators. For FVMs, this involves integration of Eq. (15.6.30) over a cell and then modeling the fluxes that cross the cell boundaries.

Finite-Difference. We first illustrate how FDMs approximate the spatial derivatives in Eq. (15.6.30). Two approaches are possible. One approach is to replace all spatial derivatives by difference operators to the order of accuracy desired (see Table 15.1 for the choices). Typically, upwind difference operators are used for convective terms, and central-difference operators are used for diffusive terms. This approach works fine in Cartesian coordinate systems, but not in generalized coordinate systems because geometric conservation laws given by Eq. (15.6.28) will not be satisfied in general. Thus, the second approach described below is recommended, which turns out to be equivalent to FVMs to be described later.

With this approach, the first step is to replace all spatial derivatives by central-difference operators that extend from cell boundary to cell boundary along the boundary-fitted coordinates. Then, decide how to evaluate the variables at the cell interfaces. For the first derivatives, the first step gives

$$\left[\frac{\partial}{\partial \xi}(\hat{\xi}_x u + \hat{\xi}_y v)Y\right]_{i,j} = \frac{[(\hat{\xi}_x u + \hat{\xi}_y v)Y]_{i+1/2,j} - [(\hat{\xi}_x u + \hat{\xi}_y v)Y]_{i-1/2,j}}{\Delta \xi} \qquad (15.6.31a)$$

$$\left[\frac{\partial}{\partial \eta}(\hat{\eta}_x u + \hat{\eta}_y v)Y\right]_{i,j} = \frac{[(\hat{\eta}_x u + \hat{\eta}_y v)Y]_{i,j+1/2} - [(\hat{\eta}_x u + \hat{\eta}_y v)Y]_{i,j-1/2}}{\Delta \eta} \qquad (15.6.31b)$$

The next step is to decide how to evaluate $u, v,$ and Y at the cell boundaries (the evaluation of $\hat{\xi}_x, \hat{\xi}_y, \hat{\eta}_x,$ and $\hat{\eta}_y$ at the cell boundaries are given in Step 5). If second-order central differencing is used, then

$$\phi_{1\pm 1/2,j} = \frac{1}{2}(\phi_{i,j} + \phi_{i\pm 1,j}), \quad \phi_{i,j\pm 1/2} = \frac{1}{2}(\phi_{i,j} + \phi_{i,j\pm 1}) \qquad (15.6.32)$$

where $\phi = uY$ or vY. Since central differencing for convective terms can create numerical stability problems, it is not recommended unless numerical dissipation is added. If upwind-difference operators are used, then one does not need to add additional artificial dissipation to maintain numerical stability, though "limiters" may still be necessary in some cases to remove non-physical oscillations in the solutions (see Hirsch (1991)). In order to use upwind differencing, one must know which way the "wind" blows. Thus, we define

$$\phi = \phi^+ + \phi^-, \quad \phi^+ = \frac{1}{2}(|\phi| + \phi), \quad \phi^- = \frac{1}{2}(|\phi| - \phi) \qquad (15.6.33)$$

The above definitions ensure that ϕ^+ is always positive, and ϕ^- is always negative regardless of the values of u and v (note that Y is always positive). As a result, terms involving ϕ^+ can be backward differenced, and terms involving ϕ^- can be forward differenced. From Table 15.1, we see that first-order upwind differencing implies

$$\phi_{i\pm1/2,j}^{+} = \phi_{(i\pm1/2)-1/2],j}^{+}, \quad \phi_{i,j\pm1/2}^{+} = \phi_{i,[(j\pm1/2)-1/2]}^{+} \qquad (15.6.34a)$$

$$\phi_{i\pm1/2,j}^{-} = \phi_{[(i\pm1/2)+1/2],j}^{-}, \quad \phi_{i,j\pm1/2}^{-} = \phi_{i,[(j\pm1/2)+1/2]}^{-} \qquad (15.6.34b)$$

For second-order upwind differencing, we have

$$\phi_{i\pm1/2,j}^{+} = \frac{3}{2}\,\phi_{[(i\pm1/2)-1/2],j}^{+} - \frac{1}{2}\,\phi_{[(i\pm1/2)-3/2],j}^{+}$$

$$\phi_{i,j\pm1/2}^{+} = \frac{3}{2}\phi_{i,[(j\pm1/2)-1/2]}^{+} - \frac{1}{2}\,\phi_{i,[j\pm1/2)-3/2]}^{+}$$

$$\hspace{9cm} (15.6.35)$$

$$\phi_{i\pm1/2,j}^{-} = \frac{3}{2}\,\phi_{[(i\pm1/2)+1/2],j}^{-} - \frac{1}{2}\,\phi_{[(i\pm1/2)+3/2],j}^{-}$$

$$\phi_{i,j\pm1/2}^{-} = \frac{3}{2}\,\phi_{i,[(j\pm1/2)+1/2]}^{-} - \frac{1}{2}\,\phi_{i,[(j\pm1/2)+3/2]}^{-}$$

The first derivatives are the most difficult ones to approximate on a discrete domain. Typically, first-order upwind differencing will lead to inordinately high numerical diffusion (i.e., the diffusion in the FDEs will be much larger than the binary diffusion coefficient, D). With second-order upwind differencing, more accurate results can be obtained.

The remaining derivatives in Eq. (15.6.30) are the second- and the cross-derivative terms. Since these terms represent diffusion, the following second-order central difference operators shown below are adequate:

$$\left[\frac{\partial}{\partial\xi}\left(\hat{\xi}_x\xi_x D\frac{\partial Y}{\partial\xi}\right)\right]_{ij} = \frac{\left(\hat{\xi}_x\xi_x D\frac{\partial Y}{\partial\xi}\right)_{i+1/2,j} - \left(\hat{\xi}_x\xi_x D\frac{\partial Y}{\partial\xi}\right)_{i-1/2,j}}{\Delta\xi}$$

$$= \frac{(\hat{\xi}_x\xi_x D)_{i+1/2,j}\,(Y_{i+1,j} - Y_{i,j}) - (\hat{\xi}_x\xi_x D)_{i-1/2,j}\,(Y_{i,j} - Y_{i-1,j})}{\Delta\xi^2} \qquad (15.6.36)$$

$$\left[\frac{\partial}{\partial\xi}\left(\hat{\xi}_x\eta_x D\frac{\partial Y}{\partial\eta}\right)\right]_{ij} = \frac{(\hat{\xi}_x\eta_x D)_{i+1/2,j}\,(Y_{i+1/2,j+1/2} - Y_{i+1/2,j-1/2}) - (\hat{\xi}_x\eta_x D)_{i-1/2,j}\,(Y_{i-1/2,j+1/2} - Y_{i-1/2,j-1/2})}{\Delta\xi\Delta\eta}$$

$$\hspace{9cm} (15.6.37)$$

Finite-Volume.   Now, we show how FVMs approximate the spatial derivatives in Eq. (15.6.30). Note that in this example, we are using FDM for the temporal derivative and FVM for the spatial derivatives. This decoupling of the temporal derivative from the spatial derivatives is the usual practice because coupling space and time correctly is nontrivial. To facilitate presentation, the diffusion terms in Eq. (15.6.30) are deleted in this discussion.

Integrating Eq. (15.6.30) over a cell from $\xi_{i-1/2,j}$ to $\xi_{i+1/2,j}$ and from $\eta_{i,j-1/2}$ to $\eta_{i,j+1/2}$ gives

$$\frac{Y_{ij}^{n+1/2} - Y_{ij}^n}{\Delta t/2} = \iint_{\text{cell}} \left[-\frac{\partial}{\partial \xi} (\hat{\xi}_x u + \hat{\xi}_y v) Y - \frac{\partial}{\partial \eta} (\hat{\eta}_x u + \hat{\eta}_y v) Y \right]^n d\xi \, d\eta$$

$$= \iint_{\text{cell}} \left[-\frac{\partial u Y}{\partial x} - \frac{\partial v Y}{\partial y} \right]^n dx \, dy$$

$$= \{[(\mathbf{V} \cdot \mathbf{A}_\xi) Y]_{n+1/2,j} - [(\mathbf{V} \cdot \mathbf{A}_\xi) Y]_{n-1/2,j}\}$$

$$+ \{[(\mathbf{V} \cdot \mathbf{A}_\eta) Y]_{n,j+1/2} - [(\mathbf{V} \cdot \mathbf{A}_\eta) Y]_{n,j-1/2}\} \qquad (15.6.38a)$$

where

$$\mathbf{V} = u \, \mathbf{i} + v \, \mathbf{j} \qquad (15.6.38b)$$

$$\mathbf{A}_\xi = |(\mathbf{A}_\xi)_y| \, \mathbf{i} + |(\mathbf{A}_\xi)_x| \, \mathbf{j} = \hat{\xi}_x \, \mathbf{i} + \hat{\xi}_y \, \mathbf{j} \qquad (15.6.38c)$$

$$\mathbf{A}_\eta = |(\mathbf{A}_\eta)_y| \, \mathbf{i} + (\mathbf{A}_\eta)_x| \, \mathbf{j} = \hat{\eta}_x \, \mathbf{i} + \hat{\eta}_y \, \mathbf{j} \qquad (15.6.38d)$$

In the above equations, $\mathbf{V}$ is the velocity vector, and $\mathbf{A}_\xi$ and $\mathbf{A}_\eta$ are the length/area vectors of the cell boundaries that point along the ξ and η direction, respectively (or more precisely, perpendicular to the η and ξ directions, respectively). Unlike the traditional definition of the area vector, which is always an outward normal, this vector always points along the positive ξ and η directions because the signs of the dot products with $\mathbf{V}$ have already been accounted for in the last line of Eq. (15.6.38a).

Similar to the FDEs given by Eq. (15.6.31), we need to decide how to evaluate u, v, and Y at the cell boundaries. The procedure for Eq. (15.6.38) is identical to that for Eq. (15.6.31) through Eqs. (15.6.32) to (15.6.35), and so will not be repeated. It turns out that the FVE thus derived is identical to the FDE given by Eq. (15.6.31) to (15.6.35) minus the diffusion terms, if $\Delta \xi = \Delta \eta = 1$ in Eqs. (15.6.31), (15.6.36), and (15.6.37).

Step 5: Evaluate the metric coefficients and Jacobians. Expressions for the metric coefficients and Jacobians can easily be derived (see, e.g., Tannehill, et al. (1997)). Though these expressions are exact, they should not be used because they are inexact on the discretized domain. The beauty of the FVM is in giving physical meaning to these mathematical terms so that they can be evaluated more accurately in ensuring no overlap or voids between abutting cells (cells that share a common boundary). Though not shown, it turns out that for 3D geometries, the Jacobian is the volume of the cell (area if 2D), and the metric coefficients multiplied by the Jacobian is the area of the cell face (length if 2D). This interpretation of the cell area can be inferred from Eqs. (15.6.38c) and (15.6.38d).

For the 2D geometry and grid system shown in Fig. 15.9, we define the cell surrounding the grid point with indices i and j to be the area enclosed by straight lines connecting the following four points (see Fig. 15.10):

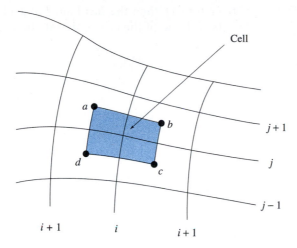

FIGURE 15.10 Definition of corners forming the cell.

$$x_a = (x_{i-1,j} + x_{i-1,j+1} + x_{i,j} + x_{i,j+1})/4$$
$$x_b = (x_{i,j} + x_{i,j+1} + x_{i+1,j} + x_{i+1,j+1})/4$$
$$x_c = (x_{i,j} + x_{i-1,j-1} + x_{i+1,j} + x_{i+1,j-1})/4 \qquad (15.6.39)$$
$$x_d = (x_{i-1,j-1} + x_{i-1,j} + x_{i,j-1} + x_{i,j})/4$$

The y-coordinates of points $a, b, c,$ and d are obtained in a similar manner. The volume of this cell (actually area since 2D) is the magnitude of the vector product given by

$$V_{abcd} = |[(\mathbf{r}_b - \mathbf{r}_c) \times (\mathbf{r}_a - \mathbf{r}_d)]|/2 \qquad (15.6.40a)$$
$$\mathbf{r}_m = x_m \mathbf{i} + y_m \mathbf{j}, m = a, b, c, d \qquad (15.6.40b)$$

where $\mathbf{r}_m$ is the position vector of point m with m being $a, b, c,$ or d. The area vector of the cell boundary or face (actually length since 2D) is given by

$$\mathbf{A}_{mn} = |(y_m - y_n)| \mathbf{i} + |(x_m - x_n)| \mathbf{j} \qquad (15.6.41)$$

where m and n are the two end points forming the cell face (a and b or a and c). The absolute value was used because Eq. (15.6.38a) already accounted for the sign in the dot product.

With the cell volume (area in 2D) and cell-face area (length in 2D) computed in the manner described above, we can now show how the metric coefficients and Jacobians are evaluated. By setting $\Delta\xi = \Delta\eta = 1$ (so that ξ and η coordinates now extend from 1 to IL and JL, the number of grid points in their respective directions instead of from 0 to 1), then the Jacobian J at grid point i,j can be set equal to the volume of the cell centered about grid point i,j:

$$J_{i,j} = V_{abcd} \tag{15.6.42}$$

where V_{abcd} is given by Eqs. (15.6.39) and (15.6.40). The metric coefficient multiplied by the Jacobian is equal to the area of the cell face as indicated by Eqs. (15.6.38c) and (15.6.38d). Thus, for example (see Fig. 15.10),

$$(\hat{\xi}_x)_{i-1/2,j} = |(\mathbf{A}_\xi)_y|_{i-1/2,j} = |(y_a - y_c)|, \quad (\hat{\xi}_y)_{i-1/2,j} = |(\mathbf{A}_\xi)_x|_{i-1/2,j} = |(x_a - x_c)| \tag{15.6.43a}$$

$$(\hat{\eta}_x)_{i,j+1/2} = |(\mathbf{A}_\eta)_y|_{i,j+1/2} = |(y_a - y_b)|, \quad (\hat{\eta}_y)_{i,j+1/2} = |(\mathbf{A}_\eta)_x|_{i,j+1/2} = |(x_a - x_b)| \tag{15.6.43b}$$

It turns out that metric coefficients at cell faces are needed only by diffusion terms such as Eqs. (15.6.36) and (15.6.37). To ensure conservative property of the resulting FD/FVEs, metric coefficients can be evaluated as follows:

$$(\xi_x)_{i\pm1/2,j} = 2(\hat{\xi}_x)_{i\pm1/2,j}/(J_{i,j} + J_{i\pm1,j}), \quad (\xi_x)_{i,j\pm1/2} = 2(\hat{\xi}_x)_{i,j\pm1/2}/(J_{i,j} + J_{i,j\pm1}) \tag{15.6.44}$$

Other metric coefficients can be evaluated in a similar manner.

Step 6: Derive FD/FVEs at the boundary grid points/cells. The FD/FVEs at the boundary grid points are derived by applying boundary conditions (BCs). One simple approach often used in FDMs is to simply apply the BCs at the grid point. Thus, if the mass fraction on the nozzle wall is fixed at say Y_w, then the FDEs at the nozzle walls located at $j = 1$ and $j = JL$ are (see Fig. 15.9)

$$Y_{i,1} = Y_{i,JL} = Y_w, \quad i = 1, 2, \ldots, IL \tag{15.6.45}$$

If the nozzle walls are impermeable to mass diffusion, then the BCs are

$$(\partial Y/\partial n)_{i,1} = (\partial Y/\partial n)_{i,JL} = 0, \quad i = 1, 2, \ldots, IL \tag{15.6.46}$$

where n is direction normal to the wall. Replacing the above spatial derivatives by one-sided difference operators (forward for $j = 1$ and backward for $j = JL$) gives

$$Y_{i,1} = Y_{i,2}, \quad Y_{i,JL} = Y_{i,JL-1}, i = 1, 2, \ldots, IL \tag{15.6.47}$$

which is first-order accurate or

$$Y_{i,1} = (4\,Y_{i,2} - Y_{i,3})/3, \quad Y_{i,JL} = (4\,Y_{i,JL-1} - Y_{i,JL-2})/3 \qquad (15.6.48)$$

which is second-order accurate.

The above approach in deriving FDEs on boundary grid points is often adequate, but is not conservative. A more accurate way is to first integrate the governing PDE over the half or quarter cell surrounding each boundary grid point, and then apply appropriate BCs. Since this procedure is tedious for vertex-centered cells, cell-center cells are often preferred (see Fig. 15.3). If cell-centered cells are used, then all cells are "full" cells including those that abut the boundaries, and application of BCs at cell faces becomes straightforward.

This completes the derivation of the FD/FVEs for this problem. At this point, note that if an implicit method was selected instead of the explicit Runge-Kutta method, then the FD/FVEs at every grid point or cell will be coupled to each other. For 2D and 3D problems, the resulting system of equations can be quite large and inefficient to solve. One highly efficient method for handling such systems of equations is known as approximate factorization. Refer to Beam & Warming (1978) and Steinthorsson & Shih (1993) for further discussions.

15.7 METHODS FOR COMPRESSIBLE NAVIER–STOKES EQUATIONS

The equations governing unsteady, 2D compressible flow of a thermally and calorically perfect gas can be written as (see Tannehill, et al (1997))

$$\frac{\partial \rho}{\partial t} + \frac{\partial \rho u}{\partial x} + \frac{\partial \rho v}{\partial y} = 0 \qquad (15.7.1)$$

$$\frac{\partial \rho u}{\partial t} + \frac{\partial \rho u u}{\partial x} + \frac{\partial \rho u v}{\partial y} = -\frac{\partial p}{\partial x} + \frac{\partial \tau_{xx}}{\partial x} + \frac{\partial \tau_{yx}}{\partial y} \qquad (15.7.2)$$

$$\frac{\partial \rho v}{\partial t} + \frac{\partial \rho u v}{\partial x} + \frac{\partial \rho v v}{\partial y} = -\frac{\partial p}{\partial y} + \frac{\partial \tau_{xy}}{\partial x} + \frac{\partial \tau_{yy}}{\partial y} \qquad (15.7.3)$$

$$\frac{\partial e}{\partial t} + \frac{\partial (e+p)u}{\partial x} + \frac{\partial (e+p)v}{\partial y} = \frac{\partial}{\partial x}(u\tau_{xx} + v\tau_{yx} + q_x) + \frac{\partial}{\partial y}(u\tau_{xy} + v\tau_{yy} + q_y) \qquad (15.7.4)$$

where e is total energy (mechanical and thermal) and

$$\tau_{xx} = \frac{2}{3}\mu\left(2\frac{\partial u}{\partial x} - \frac{\partial v}{\partial y}\right), \quad \tau_{yy} = \frac{2}{3}\mu\left(2\frac{\partial v}{\partial y} - \frac{\partial u}{\partial x}\right), \quad \tau_{xy} = \tau_{yx} = \mu\left(\frac{\partial u}{\partial y} + \frac{\partial v}{\partial x}\right) \qquad (15.7.5)$$

$$q_x = -k\frac{\partial T}{\partial x}, \quad q_y = -k\frac{\partial T}{\partial y} \qquad (15.7.6)$$

$$p = (\gamma - 1)\left(e - \rho\frac{u^2 + v^2}{2}\right), \quad \gamma = \frac{c_p}{c_v} \qquad (15.7.7)$$

For convenience, these equations can be written in the following more compact form:

$$\frac{\partial U}{\partial t} + \frac{\partial F}{\partial x} + \frac{\partial G}{\partial y} = \frac{\partial F_v}{\partial x} + \frac{\partial G_v}{\partial y} \tag{15.7.8}$$

where

$$U = \begin{bmatrix} \rho \\ \rho u \\ \rho v \\ e \end{bmatrix}, F = \begin{bmatrix} \rho u \\ \rho u^2 + p \\ \rho uv \\ (e + p)u \end{bmatrix}, G = \begin{bmatrix} \rho v \\ \rho uv \\ \rho v^2 + p \\ (e + p)v \end{bmatrix}, F_v = \begin{bmatrix} 0 \\ \tau_{xx} \\ \tau_{xy} \\ u\tau_{xx} + v\tau_{xy} - q_x \end{bmatrix}, G_v = \begin{bmatrix} 0 \\ \tau_{yx} \\ \tau_{yy} \\ u\tau_{yx} + v\tau_{yy} - q_y \end{bmatrix}$$

$$\tag{15.7.9}$$

In the above equations, U is the vector that contains the unknown variables.

Suppose that the flow domain is a rectangle of length L and height H with two inlets and one outlet. Also, suppose that the grid points are given by

$$x_i = (i - 1)\,\Delta x, \quad \Delta x = L/(IL - 1), \quad i = 1, 2, \ldots, IL \tag{15.7.10a}$$

$$y_j = (j - 1)\,\Delta y, \quad \Delta y = H/(JL - 1), \quad j = 1, 2, \ldots, JL \tag{15.7.10b}$$

The FD/FVEs for this coupled system of PDEs can be derived in a manner very similar to that described for Eq. (15.6.24). Here, only the essence is described. The first step is to replace the temporal derivatives by a time-differencing formula. Any of the time-differencing formulas in Section 15.3.4 can be used. For simplicity, we choose the Euler explicit method, and this gives

$$\frac{U_{i,j}^{n+1} - U_{ij}^{n}}{\Delta t} = \left(\frac{\partial U}{\partial t}\right)_{i,j}^{n} = \left(-\frac{\partial F}{\partial x} - \frac{\partial G}{\partial y} + \frac{\partial F_v}{\partial x} + \frac{\partial G_v}{\partial y}\right)_{i,j}^{n} \tag{15.7.11}$$

Implied in Eq. (15.7.11) is that the dependent variables (the solution variables sought) are no longer $\rho, u, v,$ and e, but rather the conserved variables ρ, m (ρu), n (ρv), and e, and that every element in $F, G, F_v,$ and G_v in Eqs. (15.7.8) and (15.7.9) including temperature and pressure must be expressed in terms of $\rho, m, n,$ and e. Note that once $\rho, m, n,$ and e are determined, u and v can be obtained by m/ρ and n/ρ.

The next step is to replace the spatial derivatives. The diffusion terms can be replaced by central difference formulas in a manner similar to that described by Eqs. (15.6.36) and (15.6.37), and will not be repeated. The convective terms and pressure are more complex because the direction of upwinding does not depend just on the direction of the velocity vector. To illustrate, consider the derivative involving F. In terms of $\rho, m, n,$ and e, the flux-vector F is homogeneous to degree one so that $F = A\,U$, where $A = \partial F/\partial U$ is the Jacobian matrix. The eigenvalues of A are $u, u, u + c,$ and $u - c$, where c is the speed of sound ($c = \sqrt{\gamma p/\rho}$). According to theory of characteristics, if the flow is subsonic ($u < c$), then even though $u > 0$, only information along characteristics $dx/dt = u$ and $dx/dt = u + c$ can be backward differenced; the information along $dx/dt = u - c$ must be forward differenced (see Hirsch (1991)). Thus, it is the sign of the eigenvalues in **A** that determine the direction of the upwinding.

There are many different ways to separate the positive and negative eigenvalues in the Jacobian matrices, $\mathbf{A} = \partial F/\partial U$ and $\mathbf{B} = \partial G/\partial U$, so that F and G can be split as follows:

$$F = \mathbf{A} U = (\mathbf{A}^+ + \mathbf{A}^-)U = \mathbf{A}^+ U + \mathbf{A}^- U = F^+ + F^- \quad (15.7.12)$$

$$G = \mathbf{B} U = (\mathbf{B}^+ + \mathbf{B}^-)U = \mathbf{B}^+ U + \mathbf{B}^- U = G^+ + G^- \quad (15.7.13)$$

where $\mathbf{A}^+$ and $\mathbf{B}^+$ contain only non-negative eigenvalues, and $\mathbf{A}^-$ and $\mathbf{B}^-$ contain only non-positive eigenvalues. Examples include flux-vector splitting and flux-difference splitting with and without limiters (see Hirsch (1991)). One of the simplest way to separate the positive and negatives eigenvalues is to recognize that if λ_i ($i = 1, 2, 3, 4$) are the eigenvalues of ϕ, then $\lambda_i \pm a$ are the eigenvalues of $\phi^* = \phi \pm a\mathbf{I}$, where $\mathbf{I}$ is an identity matrix. Thus, by choosing $a = \max\{|\lambda_i|\} = |\lambda_{\max}|$ (i.e., the spectral radius), then **A** and **B** can readily be split as follows:

$$\mathbf{A}^+ = \frac{1}{2}(\mathbf{A} + |\lambda_{\max,A}|\mathbf{I}), \quad \mathbf{A}^- = \frac{1}{2}(\mathbf{A} - |\lambda_{\max,A}|\mathbf{I}) \quad (15.7.14)$$

$$\mathbf{B}^+ = \frac{1}{2}(\mathbf{B} + |\lambda_{\max,B}|\mathbf{I}), \quad \mathbf{B}^- = \frac{1}{2}(\mathbf{B} - |\lambda_{\max,B}|\mathbf{I}) \quad (15.7.15)$$

where $\mathbf{I}$ is an identity matrix and

$$|\lambda_{\max,A}| = |u| + c, \quad |\lambda_{\max,B}| = |v| + c, \quad c = \sqrt{\gamma p/\rho} \quad (15.7.16)$$

Substituting Eqs. (15.7.12) to (15.7.16) into Eq. (15.7.11) yields

$$\frac{U_{i,j}^{n+1} - U_{i,j}^n}{\Delta t} = \left(-\frac{\partial F^+}{\partial x} - \frac{\partial F^-}{\partial x} - \frac{\partial G^+}{\partial y} - \frac{\partial G^-}{\partial y} + \frac{\partial F_v}{\partial x} + \frac{\partial G_v}{\partial y} \right)_{i,j}^n \quad (15.7.17)$$

In FDMs, terms with superscript $+$ are replaced by backward-difference operators, and terms with superscript $-$ are replaced by forward-difference operators (see Table 15.1). If first-order upwind formulas are used, then

$$\left(\frac{\partial F}{\partial x} \right)_{i,j} = \left(\frac{\partial F^+}{\partial x} \right)_{i,j} + \left(\frac{\partial F^-}{\partial x} \right)_{i,j} = \frac{F_{i,j}^+ - F_{i-1,j}^+}{\Delta x} + \frac{F_{i+1,j}^- - F_{i,j}^-}{\Delta x} \quad (15.7.18)$$

$$\left(\frac{\partial G}{\partial y} \right)_{i,j} = \left(\frac{\partial G^+}{\partial y} \right)_{i,j} + \left(\frac{\partial G^-}{\partial y} \right)_{i,j} = \frac{G_{i,j}^+ - G_{i,j-1}^+}{\Delta y} + \frac{G_{i,j+1}^- - G_{i,j}^-}{\Delta y} \quad (15.7.19)$$

Though the above approximations of the convective terms are stable, they are highly diffusive. The FDEs given by Eqs. (15.7.18) and (15.7.19) can readily be shown to be equivalent to

$$\left(\frac{\partial \boldsymbol{F}}{\partial x}\right)_{i,j} = \frac{\boldsymbol{F}_{i+1,j} - \boldsymbol{F}_{i-1,j}}{2\Delta x} - \frac{\Delta x}{2}\left(\frac{|\lambda_{\max,A}|_{i+1,j} - 2|\lambda_{\max,A}|_{i,j} + |\lambda_{\max,A}|_{i-1,j}}{\Delta x^2}\right)$$

$$(15.7.20)$$

$$\left(\frac{\partial \boldsymbol{G}}{\partial y}\right)_{i,j} = \frac{\boldsymbol{G}_{i,j+1} - \boldsymbol{G}_{i,j-1}}{2\Delta y} - \frac{\Delta y}{2}\left(\frac{|\lambda_{\max,B}|_{i,j+1} - 2|\lambda_{\max,B}|_{i,j} + |\lambda_{\max,N}|_{i,j-1}}{\Delta y^2}\right)$$

$$(15.7.21)$$

From the above two equations, it can be seen that first-order upwind differencing is equivalent to central differencing of $\boldsymbol{F}$ and adding the following artificial diffusion terms:

$$\mu_A \frac{\partial^2 |\lambda_{\max,A}|}{\partial x^2} + \mu_B \frac{\partial^2 |\lambda_{\max,B}|}{\partial y^2} \qquad (15.7.22)$$

where the artificial viscosities μ_A and μ_B are $\Delta x/2$ and $\Delta y/2$, respectively, which can be much higher than the physical viscosity given by μ in Eq. (15.7.5). If second-order upwind differencing is used, then the artificial diffusion is much less.

Previously, we showed how FDMs approximate the convective terms. If FVMs or a FV—like FDMs were used, then we first integrate the derivative across the cell:

$$\left(\frac{\partial \boldsymbol{F}}{\partial x}\right)_{i,j} = \left(\frac{\partial \boldsymbol{F}^+}{\partial x}\right)_{i,j} + \left(\frac{\partial \boldsymbol{F}^-}{\partial x}\right)_{i,j} = \frac{\boldsymbol{F}^+_{i+1/2,j} - \boldsymbol{F}^+_{i-1/2,j}}{\Delta x} + \frac{\boldsymbol{F}^-_{i+1/2,j} - \boldsymbol{F}^-_{i-1/2,j}}{\Delta x}$$

$$(15.7.23)$$

$$\left(\frac{\partial \boldsymbol{G}}{\partial y}\right)_{i,j} = \left(\frac{\partial \boldsymbol{G}^+}{\partial y}\right)_{i,j} + \left(\frac{\partial \boldsymbol{G}^-}{\partial y}\right)_{i,j} = \frac{\boldsymbol{G}^+_{i,j+1/2} - \boldsymbol{G}^+_{i,j-1/2}}{\Delta y} + \frac{\boldsymbol{G}^-_{i,j+1/2} - \boldsymbol{G}^-_{i,j-1/2}}{\Delta y}$$

$$(15.7.23)$$

Now, two choices are possible in evaluating $\boldsymbol{F}^\pm$ and $\boldsymbol{G}^\pm$ at the cell faces located at $(i \pm 1/2, j)$ and $(i, j \pm 1/2)$. One choice is to interpolate $\boldsymbol{F}^\pm$ and $\boldsymbol{G}^\pm$. If this is done, then the results are identical to that obtained by FDMs. The second and the more accurate choice is to interpolate $\boldsymbol{U}$ to the cell faces, and then evaluate $\boldsymbol{F}^\pm$ and $\boldsymbol{G}^\pm$ as a function of those values.

15.8 METHODS FOR THE INCOMPRESSIBLE NAVIER–STOKES EQUATIONS

For incompressible flows of a Newtonian fluid with constant viscosity, the continuity and momentum equations are decoupled from the energy equation. In 2D, the continuity and momentum equations reduce to

$$\frac{\partial u}{\partial x} + \frac{\partial v}{\partial y} = 0 \qquad (15.8.1)$$

$$\frac{\partial u}{\partial t} + \frac{\partial uu}{\partial x} + \frac{\partial uv}{\partial y} = -\frac{1}{\rho}\frac{\partial p}{\partial x} + \frac{\nu}{\rho}\left(\frac{\partial^2 u}{\partial x^2} + \frac{\partial^2 u}{\partial y^2}\right) \tag{15.8.2}$$

$$\frac{\partial v}{\partial t} + \frac{\partial vu}{\partial x} + \frac{\partial vv}{\partial y} = -\frac{1}{\rho}\frac{\partial p}{\partial y} + \frac{\nu}{\rho}\left(\frac{\partial^2 v}{\partial x^2} + \frac{\partial^2 v}{\partial y^2}\right) \tag{15.8.3}$$

Since the continuity equation does not have a time derivative, the FD/FVMs presented in previous sections cannot be applied. Two well-known methods for solving such systems of PDEs are the artificial compressibility method (Chorin (1967)) and a pressure-base method such as SIMPLE (see Pantankar (1980)). Here, only the artificial compressibility method is presented because this method will enable methods presented in previous sections to be applied to incompressible flows.

With the artificial compressibility method, we replace the continuity equation by

$$\frac{1}{\beta}\frac{\partial p}{\partial t} + \frac{\partial u}{\partial x} + \frac{\partial v}{\partial y} = 0 \tag{15.8.4}$$

where β is a constant, chosen between a lower and an upper limit. These limits are determined from physical and numerical considerations. Since the speed of sound is infinite in an incompressible flow but finite with artificial compressibility (the higher the value of β, the higher is the speed of sound), β should be chosen as large as possible in order to mimic an infinite sound speed. The lower limit for β is that the speed of sound should be much higher than the diffusive and convective speeds. Though it is desirable to have as high a speed of sound as possible from physical considerations, numerical considerations impose an upper limit on β. This is because the ratio of the largest to the smallest eigenvalues of a system of conservation equations (inviscid part only) affect the convergence rate and stiffness of any algorithm used to analyze it. Specifically, the higher the ratio, the slower is the convergence rate. With the artificial compressibility formulation, the higher the value of β, the higher is the ratio of the largest to the smallest eigenvalues, and hence slows the convergence rate.

Once the continuity equation given by Eq. (15.8.1) is replaced by Eq. (15.8.4), the method presented in Sections 15.6 and 15.7 can readily be applied. Note that when the solution reaches steady state, Eq. (15.8.4) reduces to Eq. (15.8.1).

15.9 FINAL REMARKS

CFD is a very powerful tool for obtaining solutions to fluid flow problems if used and interpreted correctly. Its greatest danger is that it always gives an answer, and that answer may be right, somewhat right, a little wrong, or very wrong. Which solution is obtained and how one interprets that solution depends on the user. Thus, it is imperative that CFD not be used as a "black box." Today, with the proliferation of readily available commercial CFD codes that are so user friendly, using CFD as a "black box" is an easy trap.

How does one avoid the trap? One must think hard about the physics of the problem being solved and ask tough questions. Do the governing equations describe the physics adequately? For example, how is turbulence modeled? How is combustion modeled? Are all of the boundary conditions known to the accuracy needed? If the governing equations and boundary conditions have incorrect information, then there is no chance in obtaining a correct solution.

Next, one needs to think about the flow features and the time and length scales that need to be resolved. Based on this understanding, construct a grid system that has the resolution needed at the right places and is smooth, nearly orthogonal, and aligned with the flow. Then, the type of differencing for time and space derivatives must be selected. If only "rough" solutions are needed, then low-order accurate methods may be adequate, even though numerical diffusion is high, which tends to "smear" details. If more accurate solutions are needed, then higher-order accurate methods should be selected. Once a solution has been generated, it needs to be examined for grid independence and reasonableness. Finally, when interpreting results, consider the inadequacies in the governing equations and the numerical method of solution.

Thus, the key to success in using CFD is to understand the sources of errors, approaches to minimize them, and ways to interpret the results despite practical limitations in the governing equations and the numerical method of solution. Remember, even a validated code can give wrong answers just like a calibrated measurement probe can if used or interpreted incorrectly.

REFERENCES

Beam, R.M. and Warming, R.F., "An Implicit Factored Scheme for Compressible Navier–Stokes Equations," *AIAA Journal*, Vol. 16, 1978, pp. 393–402.

Chorin, A.J., "A Numerical Method for Solving Incompressible Viscous Flow Problems," *Journal of Computational Physics*, Vol. 2, August 1967, pp. 12–26.

Hirsch, C., *Numerical Computation of Internal and External Flows—Vol 2: Computational Methods for Inviscid and Viscous Flows*, John Wiley & Sons, Chichester, 1991.

Patankar, S.V., *Numerical Heat Transfer and Fluid Flow*, Hemisphere, Washington, D.C., 1980.

Schlichting, H. and Gersten, K., *Boundary Layer Theory*, 8th Revised and Enlarged Edition, Springer, Berlin, 2000.

Shih, T.I-P., Bailey, R.T., Nguyen, H.L., and Roelke, R.J., "Algebraic Grid Generation for Complex Geometries," *International Journal for Numerical Methods in Fluids*, Vol. 13, 1991, pp. 1–31.

Steinthorsson, E., Shih, T.I-P., and Roelke, R.J., "Enhancing Control of Grid Distribution in Algebraic Grid Generation," *International Journal for Numerical Methods in Fluids*, Vol. 15, 1992, pp. 297–311.

Steinthorsson, E. and Shih, T.I-P., "Methods for Reducing Approximate-Factorization Errors in Two- and Three-Factored Schemes," *SIAM Journal of Scientific Computing*, Vol. 14, No. 5, 1993, pp. 1214–1236.

Tannehill, J.C., Anderson, A.A., and Pletcher, R.H., *Computational Fluid Mechanics and Heat Transfer*, 2nd Edition, Taylor & Francis, Washington, D.C., 1997.

Thompson, J.F., Soni, B., and Weatherill, N., Editors, *Handbook on Grid Generation*, CRC Press, 1998.

Yih, C.-S., *Fluid Mechanics, A Concise Introduction*, McGraw-Hill Book Company, New York, 1969.

PROBLEMS

15.1 Expand $y = \exp(x)$ into a Taylor series about $x = 0$. What is the radius of convergence of this series? Suppose that only the first three terms of the series are added. Compare that sum against the exact solution from a calculator evaluation of $\exp(x)$ by computing the relative error for $x = 0.1$, 1, and 2. Explain why the error increases as x increases. If you want a 1% relative error in the evaluation of $\exp(x)$, where $x = 1$, how many terms in the series must be included?

15.2 Write a computer program to evaluate $\exp(x)$, using single precision (use seven digits to represent a number). Afterwards, use the program to evaluate $\exp(+20)$ and $\exp(-20)$. Explain why results for $\exp(+20)$ is correct and the results for $\exp(-20)$ is incorrect, when compared to the corresponding solution from the calculator. Note that programming is not trivial. Hint: If n is large, x^{n+1} and $(n+1)!$ are huge numbers, but $x^{n+1}/(n+1)!$ may be small. Thus: $x^{n+1}/(n+1)! = [x/(n+1)][x/n!]$ or $[x/(n+1)] \times$ [previous term].

15.3 For $(\partial u/\partial y)$, derive a forward-difference and a backward-difference operator that is second-order accurate in space.

15.4 Program the Thomas algorithm given by Eqs. (15.3.23) to (15.3.31). Use the program compute $\mathbf{A}\,x = b$, where $\mathbf{A}$ and b are

$$\mathbf{A} = \begin{bmatrix} 20 & 5 & \\ 8 & 15 & 4 \\ & 12 & 14 \end{bmatrix} \quad b = \begin{bmatrix} 1 \\ 2 \\ 3 \end{bmatrix}$$

15.5 Derive $\mathbf{A}\,x = b$ for Eq. (15.3.22) applied at $j = 2, 3, \dots, JL - 1$ along with Eq. (15.3.17).

15.6 Derive the FDEs for Eqs. (15.3.1) and (15.3.2) with a second-order Runge-Kutta time differencing formula.

15.7 Program the solution algorithm for the explicit method in Section 15.3.3. Generate solutions with $H = 0.1$ m, $V_0 = 0.5$ m/s, $\nu = 10^{-3}$ m^2/s. Try several different values of IL. In all cases, set $\Delta t = \Delta y^2/2\nu$. Compare the solutions generated with Eq. (15.3.3).

15.8 Modify FDEs and program develop in Problem 15.7 by making the bottom plate oscillate with time; i.e., $u(y = 0, t) = V_0 \sin(2\pi\sigma t)$. Obtain solutions with different values of σ. Examine the limiting cases of high and low σ.

15.9 Analyze the consistency of Eqs. (15.3.18).

15.10 Analyze the consistency of the following FDE:

$$\frac{u_j^{n+1} - u_j^{n-1}}{2\Delta t} = \nu \frac{u_{j+1}^n - (u_j^{n+1} + u_j^{n-1}) + u_{j-1}^n}{\Delta y^2}$$
$$+ O\left(\Delta t^2, \Delta y^2, \frac{\Delta t^4}{\Delta y^2}\right)$$

Show that it can converge to the following PDE as grid spacing and time-step size approach zero:

$$\frac{\partial u}{\partial t} + \alpha v \frac{\partial^2 u}{\partial t^2} = v \frac{\partial^2 u}{\partial y^2}$$

15.11 For the program developed for Problem 15.7, generate solutions for different time-step sizes to examine stability and accuracy (i.e., remove the requirement that $\Delta t = \Delta y^2 / 2v$).

15.12 Program the solution algorithm for the implicit method in Section 15.3.3. Generate solutions with $H = 0.1$ m, $V_0 = 0.5$ m/s, $v = 10^{-3}$ m²/s. Try several different values of IL and Δt. Compare solution generated with Eq. (15.3.3). If only the steady-state solution is of interest, then what is the optimum Δt (the time-step size that will enable steady-state to be achieved with the fewest time steps)?

15.13 Analyze the numerical stability of the FDEs given by Eqs. (15.3.18).

15.14 Analyze the numerical stability of the following FDE:

$$\frac{u_j^{n+1} - u_j^{n-1}}{2\Delta t} = v \frac{u_{j+1}^n - 2u_j^n + u_{j-1}^n}{\Delta y^2} + O(\Delta t^2, \Delta y^2)$$

$$\frac{u_j^{n+1} - u_j^{n-1}}{2\Delta t} = v \frac{u_{j+1}^n - (u_j^{n+1} + u_j^{n-1}) + u_{j-1}^n}{\Delta y^2}$$

$$+ O\left(\Delta t^2, \Delta y^2, \frac{\Delta t^4}{\Delta y^2}\right)$$

15.15 For the FDE given by Eq. (15.3.22), show that it is unconditionally stable if $\theta \in [0.5, 1]$, and conditionally stable if $\theta \in [0, 0.5)$.

15.16 Though the PDEs $\partial u / \partial t + \frac{1}{2} \partial u^2 / \partial x = 0$ and $\partial u / \partial t + u \partial u / \partial x = 0$ are identical analytically, they differ when finite-differenced. If Euler explicit is used for the time derivative, and first-order backward differencing is used for the spatial derivative, then which FDE will maintain the conservative property?

15.17 Program the FDEs derived in Problem 15.16. Obtain solutions for the following initial conditions: (a) $u = 0$ when $x < -1$ and > 1; $u = 1$, $u = x + 1$ when x is between -1 and 0; $u = -x + 1$ when x is between 0 and 1. (b) $u + 0$, when $x < -1$ and > 1; $u = 1$, when x is between -1 and 1.

15.18 Program the grid generation method given by Eqs. (15.6.13) to (15.6.15) for the geometry given by Eqs. (15.6.8) to (15.6.12) for a variety of values for A, B, C, D, a_1, a_2, IL, and JL.

15.19 Program the grid generation method given by Eqs. (15.6.16) to (15.6.21) for the geometry given by Eqs. (15.6.8) to (15.6.12) for a variety of values for $A, B, C, D, a_1, a_2, IL, JL$, and F.

15.20 Incorporate the stretching function given by Eq. (15.6.22) into the computer program developed under Problem 15.19.

15.21 Apply Two-Boundary Method to generate grid system in an U-shaped channel of width L with a 90-degree bend of radius R. Develop equations, program, and generate grids for several different L, R, IL, and JL.

15.23 Develop a set of FDEs for the advection equation given by Eq. (15.6.23) by using second-order Runge-Kutta for the time derivative, second-order upwind for the advection terms, and second-order central differencing for the diffusion terms.

15.24 Program an algorithm based on the FDEs developed in Problem 15.23.

15.25 For the one-dimensional form of Eq. (15.7.8), show that $F = A\ U$.

15.26 Show that if second-order upwind is to approximate $(\partial F / \partial x)$ in Eq. (15.7.8), then the artificial diffusion is fourth-order.

15.27 Derive FDEs for the unsteady compressible Navier–Stokes equations in one-dimensions.

15.28 Program the algorithm developed in Problem 15.27, and apply it to study the shock tube problem.

15.29 Derive FDEs for the incompressible Navier–Stokes equations in one dimension by using the artificial compressibility method.

15.30 Derive FDEs and a solution algorithm for solving the equations governing potential flow in a channel with one flat wall and one wall with a forward-facing step. The governing equation is $\nabla^2 \phi = 0$ or $\nabla^2 \psi = 0$, where ϕ is the velocity potential and ψ is the stream function. Similar to the artificial compressibility method, $\nabla^2 \phi = \nabla^2 \psi = 0$ can be modified to $K(\partial \psi / \partial t) = \nabla^2 \psi$ and then seek the steady-state solution.

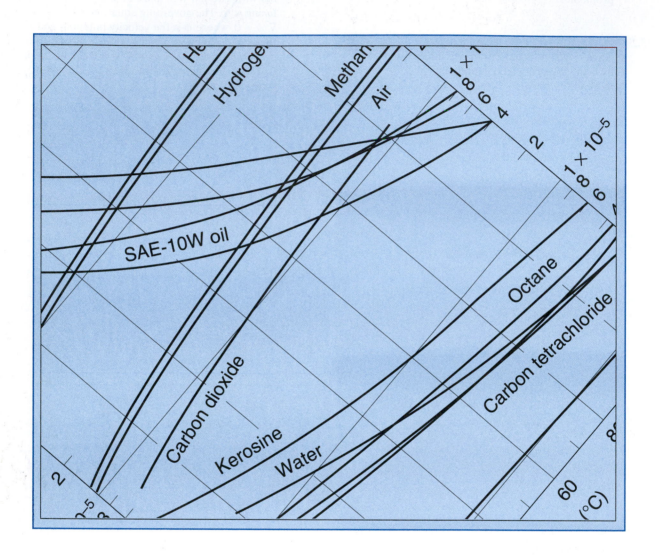

Appendix

TABLE A.1 SI Units of different physical quantities

Quantity	International system[a] SI
Length	millimeter
	meter
	kilometer
Area	square centimeter
	square meter
Volume	cubic centimeter
	cubic meter
Mass	kilogram
Density	kilogram/cubic meter
Force	newton
Work or torque	newton-meter
Pressure	newton/square meter (pascal)
Temperature	degree Celsius
	kelvin
Energy	joule
Power	watt
Velocity	meter/second
Acceleration	meter/second squared
Frequency	hertz
Viscosity	newton-second/square meter

[a]The reversed initials in this abbreviation come from the French form of the name: Systéme International.

TABLE A.2 Vector Relationships

$\mathbf{A} \cdot \mathbf{B} = A_x B_x + A_y B_y + A_z B_z$

$\mathbf{A} \times \mathbf{B} = (A_y B_z - A_z B_y)\hat{\mathbf{i}} + (A_z B_x - A_x B_z)\hat{\mathbf{j}} + (A_x B_y - A_y B_x)\hat{\mathbf{k}}$

If $\mathbf{A} \perp \mathbf{B}, \mathbf{A} \cdot \mathbf{B} = 0$

If $\mathbf{A} \parallel \mathbf{B}, \mathbf{A} \times \mathbf{B} = 0$

gradient operator $= \nabla = \dfrac{\partial}{\partial x}\hat{\mathbf{i}} + \dfrac{\partial}{\partial y}\hat{\mathbf{j}} + \dfrac{\partial}{\partial z}\hat{\mathbf{k}}$

divergence of $\mathbf{V} = \nabla \cdot \mathbf{V} = \dfrac{\partial u}{\partial x} + \dfrac{\partial v}{\partial y} + \dfrac{\partial w}{\partial z}$

curl of $\mathbf{V} = \nabla \times \mathbf{V} = \left(\dfrac{\partial w}{\partial y} - \dfrac{\partial v}{\partial z}\right)\hat{\mathbf{i}} + \left(\dfrac{\partial u}{\partial z} - \dfrac{\partial w}{\partial x}\right)\hat{\mathbf{j}} + \left(\dfrac{\partial v}{\partial x} - \dfrac{\partial u}{\partial y}\right)\hat{\mathbf{k}}$

Laplace's equation $= \nabla^2 \phi = 0$

Irrotational (conservative) vector field: $\nabla \times \mathbf{V} = 0$

Stokes' theorem: $\displaystyle\oint_C \mathbf{V} \cdot d\mathbf{l} = \iint_A (\nabla \times \mathbf{V}) \cdot \hat{\mathbf{n}}\, dA$

Gauss' Theorem: $\displaystyle\oiint_A \mathbf{V} \cdot \hat{\mathbf{n}}\, dA = \iiint_V \nabla \cdot \mathbf{V}\, d\forall$

B. FLUID PROPERTIES

TABLE B.1 Properties of Water

Temperature T (°C)	Density ρ (kg/m³)	Viscosity μ (N · s/m²)	Kinematic viscosity ν (m²/s)	Surface tension σ (N/m)	Vapor pressure p_v (kPa)	Bulk modulus B (Pa)
0	999.9	1.792×10^{-3}	1.792×10^{-6}	0.0762	0.610	204×10^7
5	1000.0	1.519	1.519	0.0754	0.872	206
10	999.7	1.308	1.308	0.0748	1.13	211
15	999.1	1.140	1.141	0.0741	1.60	214
20	998.2	1.005	1.007	0.0736	2.34	220
30	995.7	0.801	0.804	0.0718	4.24	223
40	992.2	0.656	0.661	0.0701	3.38	227
50	988.1	0.549	0.556	0.0682	12.3	230
60	983.2	0.469	0.477	0.0668	19.9	228
70	977.8	0.406	0.415	0.0650	31.2	225
80	971.8	0.357	0.367	0.0630	47.3	221
90	965.3	0.317	0.328	0.0612	70.1	216
100	958.4	0.284×10^{-3}	0.296×10^{-6}	0.0594	101.3	207×10^7

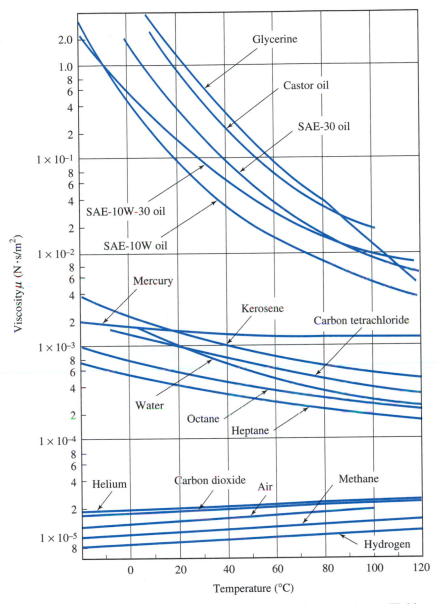

FIGURE B.1   Viscosity as a function of temperature. (Introduction to Fluid Mechanics, 2nd ed., R. W. Fox and T. A. McDonald, © 1978 John Wiley & Sons, Inc., New York. Reproduced with permission of John Wiley & Sons, Inc.)

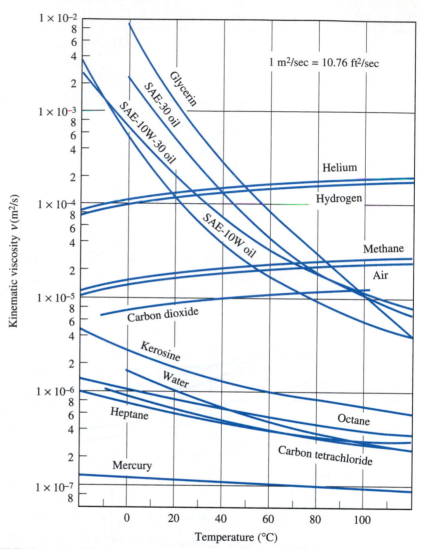

FIGURE B.2 Kinematic viscosity (at atmospheric pressure) as a function of temperature. (Introduction to Fluid Mechanics, 2nd ed., R. W. Fox and T. A. McDonald, © 1978 John Wiley & Sons, Inc., New York. Reproduced with permission of John Wiley & Sons, Inc.)

TABLE B.2 Properties of Air at Atmospheric Pressure

Temperature T (°C)	Density ρ (kg/m^3)	Viscosity μ (N · s/m^2)	Kinematic viscosity ν (m^2/s)	Velocity of sound c (m/s)
−50	1.582	1.46×10^{-5}	0.921×10^{-5}	299
−30	1.452	1.56	1.08×10^{-5}	312
−20	1.394	1.61	1.16	319
−10	1.342	1.67	1.24	325
0	1.292	1.72	1.33	331
10	1.247	1.76	1.42	337
20	1.204	1.81	1.51	343
30	1.164	1.86	1.60	349
40	1.127	1.91	1.69	355
50	1.092	1.95	1.79	360
60	1.060	2.00	1.89	366
70	1.030	2.05	1.99	371
80	1.000	2.09	2.09	377
90	0.973	2.13	2.19	382
100	0.946	2.17	2.30	387
200	0.746	2.57	3.45	436
300	0.616	2.93×10^{-5}	4.75×10^{-5}	480

TABLE B.3 Properties of the Standard Atmosphere

Altitude (m)	Temperature (K)	Pressure (kPa)	Density (kg/m^3)	Velocity of sound (m/s)
0	288.2	101.3	1.225	340
500	284.9	95.43	1.167	338
1 000	281.7	89.85	1.112	336
2 000	275.2	79.48	1.007	333
4 000	262.2	61.64	0.8194	325
6 000	249.2	47.21	0.6602	316
8 000	236.2	35.65	0.5258	308
10 000	223.3	26.49	0.4136	300
12 000	216.7	19.40	0.3119	295
14 000	216.7	14.17	0.2278	295
16 000	216.7	10.35	0.1665	295
18 000	216.7	7.563	0.1216	295
20 000	216.7	5.528	0.0889	295
30 000	226.5	1.196	0.0184	302
40 000	250.4	0.287	4.00×10^{-3}	317
50 000	270.7	0.0798	1.03×10^{-3}	330
60 000	255.8	0.0225	3.06×10^{-4}	321
70 000	219.7	0.00551	8.75×10^{-5}	297
80 000	180.7	0.00103	2.00×10^{-5}	269

TABLE B.4 Properties of Ideal Gases at 300 K

Gas	Chemical formula	Molar mass	R (kJ/kg·K)	c_p (kJ/kg·K)	k
Air	—	28.97	0.287	1.004	1.40
Argon	Ar	39.94	0.2081	0.5203	1.667
Carbon dioxide	CO_2	44.01	0.1889	0.8418	1.287
Carbon monoxide	CO	28.01	0.2968	1.041	1.40
Ethane	C_2H_6	30.07	0.2765	1.766	1.184
Helium	He	4.003	2.077	5.193	1.667
Hydrogen	H_2	2.016	4.124	14.21	1.40
Methane	CH_4	16.04	0.5184	2.254	1.30
Nitrogen	N_2	28.02	0.2968	1.042	1.40
Oxygen	O_2	32.00	0.2598	0.9216	1.394
Propane	C_3H_8	44.10	0.1886	1.679	1.12
Steam	H_2O	18.02	0.4615	1.872	1.33

$c_v = c_p - R$, $k = c_p/c_v$

TABLE B.5 Properties of Common Liquids at Atmospheric Pressure and Approximately 16 to 21°C

Liquid	Specific weight γ (N/m³)	Density ρ (kg/m³)	Surface tension[a] σ (N/m)	Vapor pressure p_v (kPa)
Alcohol, ethyl	7 744	789	0.022	—
Benzene	8 828	902	0.029	10.3
Carbon tetrachloride	15 629	1 593	0.026	86.2
Gasoline	6 660	680	—	—
Glycerin	12 346	1 258	0.063	1.4×10^{-5}
Kerosene	7 933	809	0.025	—
Mercury	132 800	13 550	0.467	1.59×10
SAE 10 oil	9 016	917	0.036	—
SAE 30 oil	9 016	917	0.035	—
Turpentine	8 529	871	0.026	5.31×10^{-2}
Water	9 810	1000	0.073	2.34

[a]In contact with air.

C. PROPERTIES OF AREAS AND VOLUMES

TABLE C.1 Areas

	Sketch	Area	Centroid	Second moment
Rectangle		bh	$\bar{y} = h/2$	$\bar{I} = bh^3/12$ $\bar{I}_{xy} = 0$
Triangle		$bh/2$	$\bar{y} = h/3$	$\bar{I} = bh^3/36$ $\bar{I}_{xy} = (b - 2d)bh^3/72$
Circle		$\pi D^2/4$	$\bar{y} = r$	$\bar{I} = \pi D^4/64$
Semicircle		$\pi D^2/8$	$\bar{y} = 4r/3\pi$	$I_x = \pi D^4/128$
Ellipse		πab	$\bar{y} = b$	$\bar{I} = \pi ab^3/4$
Semiellipse		$\pi ab/2$	$\bar{y} = 4b/3\pi$	$I_x = \pi ab^3/8$

TABLE C.2 Volumes

	Sketch	Surface area	Volume	Centroid
Cylinder		$\pi Dh + \pi D^2/2$	$\pi D^2 h/4$	$\bar{y} = h/2$
Sphere		πD^2	$\pi D^3/6$	$\bar{y} = r$
Cone		$\pi(r^2 + r\sqrt{r^2 + h^2})$	$\pi D^2 h/12$	$\bar{y} = h/4$
Hemisphere		$3\pi D^2/4$	$\pi D^2/12$	$\bar{y} = 3r/8$

D. COMPRESSIBLE-FLOW TABLES FOR AIR

TABLE D.1 Isentropic Flow

M	p/p_0	T/T_0	A/A^*	M	p/p_0	T/T_0	A/A^*
0	1.0000	1.0000	∞	.96	.5532	.8444	1.0014
.02	.9997	.9999	28.9421	.98	.5407	.8389	1.0003
.04	.9989	.9997	14.4815	1.00	.5283	.8333	1.000
.06	.9975	.9993	9.6659	1.02	.5160	.8278	1.000
.08	.9955	.9987	7.2616	1.04	.5039	.8222	1.001
.10	.9930	.9980	5.8218	1.06	.4919	.8165	1.003
.12	.9900	.9971	4.8643	1.08	.4800	.8108	1.005
.14	.9864	.9961	4.1824	1.10	.4684	.8052	1.008
.16	.9823	.9949	3.6727	1.12	.4568	.7994	1.011
.18	.9776	.9936	3.2779	1.14	.4455	.7937	1.015
.20	.9725	.9921	2.9635	1.16	.4343	.7879	1.020
.22	.9668	.9904	2.7076	1.18	.4232	.7822	1.025
.24	.9607	.9886	2.4956	1.20	.4124	.7764	1.030
.26	.9541	.9867	2.3173	1.22	.4017	.7706	1.037
.28	.9470	.9846	2.1656	1.24	.3912	.7648	1.043
.30	.9395	.9823	2.0351	1.26	.3809	.7590	1.050
.32	.9315	.9799	1.9219	1.28	.3708	.7532	1.058
.34	.9231	.9774	1.8229	1.30	.3609	.7474	1.066
.36	.9143	.9747	1.7358	1.32	.3512	.7416	1.075
.38	.9052	.9719	1.6587	1.34	.3417	.7358	1.084
.40	.8956	.9690	1.5901	1.36	.3323	.7300	1.094
.42	.8857	.9659	1.5289	1.38	.3232	.7242	1.104
.44	.8755	.9627	1.4740	1.40	.3142	.7184	1.115
.46	.8650	.9594	1.4246	1.42	.3055	.7126	1.126
.48	.8541	.9560	1.3801	1.44	.2969	.7069	1.138
.50	.8430	.9524	1.3398	1.46	.2886	.7011	1.150
.52	.8317	.9487	1.3034	1.48	.2804	.6954	1.163
.54	.8201	.9449	1.2703	1.50	.2724	.6897	1.176
.56	.8082	.9410	1.2403	1.52	.2646	.6840	1.190
.58	.7962	.9370	1.2130	1.54	.2570	.6783	1.204
.60	.7840	.9328	1.1882	1.56	.2496	.6726	1.219
.62	.7716	.9286	1.1657	1.58	.2423	.6670	1.234
.64	.7591	.9243	1.1452	1.60	.2353	.6614	1.250
.66	.7465	.9199	1.1265	1.62	.2284	.6558	1.267
.68	.7338	.9153	1.1097	1.64	.2217	.6502	1.284
.70	.7209	.9107	1.0944	1.66	.2151	.6447	1.301
.72	.7080	.9061	1.0806	1.68	.2088	.6392	1.319
.74	.6951	.9013	1.0681	1.70	.2026	.6337	1.338
.76	.6821	.8964	1.0570	1.72	.1966	.6283	1.357
.78	.6691	.8915	1.0471	1.74	.1907	.6229	1.376
.80	.6560	.8865	1.0382	1.76	.1850	.6175	1.397
.82	.6430	.8815	1.0305	1.78	.1794	.6121	1.418
.84	.6300	.8763	1.0237	1.80	.1740	.6068	1.439
.86	.6170	.8711	1.0179	1.82	.1688	.6015	1.461
.88	.6041	.8659	1.0129	1.84	.1637	.5963	1.484
.90	.5913	.8606	1.0089	1.86	.1587	.5910	1.507
.92	.5785	.8552	1.0056	1.88	.1539	.5859	1.531
.94	.5658	.8498	1.0031	1.90	.1492	.5807	1.555

TABLE D.1   Isentropic Flow (*continued*)

M	p/p_0	T/T_0	A/A^*	M	p/p_0	T/T_0	A/A^*
1.92	.1447	.5756	1.580	2.92	$.3071^{-1}$	.3696	3.924
1.94	.1403	.5705	1.606	2.94	$.2980^{-1}$	.3665	3.999
1.96	.1360	.5655	1.633	2.96	$.2891^{-1}$	.3633	4.076
1.98	.1318	.5605	1.660	2.98	$.2805^{-1}$	.3602	4.155
2.00	.1278	.5556	1.688	3.00	$.2722^{-1}$	.3571	4.235
2.02	.1239	.5506	1.716	3.02	$.2642^{-1}$	.3541	4.316
2.04	.1201	.5458	1.745	3.04	$.2564^{-1}$	.3511	4.399
2.06	.1164	.5409	1.775	3.06	$.2489^{-1}$	.3481	4.483
2.08	.1128	.5361	1.806	3.08	$.2416^{-1}$	.3452	4.570
2.10	.1094	.5313	1.837	3.10	$.2345^{-1}$	.3422	4.657
2.12	.1060	.5266	1.869	3.12	$.2276^{-1}$	.3393	4.747
2.14	.1027	.5219	1.902	3.14	$.2210^{-1}$	.3365	4.838
2.16	$.9956^{-1}$	.5173	1.935	3.16	$.2146^{-1}$	.3337	4.930
2.18	$.9649^{-1}$	.5127	1.970	3.18	$.2083^{-1}$	.3309	5.025
2.20	$.9352^{-1}$	.5081	2.005	3.20	$.2023^{-1}$	.3281	5.121
2.22	$.9064^{-1}$	.5036	2.041	3.22	$.1964^{-1}$	.3253	5.219
2.24	$.8785^{-1}$	.4991	2.078	3.24	$.1908^{-1}$	.3226	5.319
2.26	$.8514^{-1}$	.4947	2.115	3.26	$.1853^{-1}$	.3199	5.420
2.28	$.8251^{-1}$	.4903	2.154	3.28	$.1799^{-1}$	.3173	5.523
2.30	$.7997^{-1}$	.4859	2.193	3.30	$.1748^{-1}$	.3147	5.629
2.32	$.7751^{-1}$	.4816	2.233	3.32	$.1698^{-1}$	.3121	5.736
2.34	$.7512^{-1}$	.4773	2.274	3.34	$.1649^{-1}$	.3095	5.845
2.36	$.7281^{-1}$	.4731	2.316	3.36	$.1602^{-1}$	.3069	5.956
2.38	$.7057^{-1}$	.4688	2.359	3.38	$.1557^{-1}$	.3044	6.069
2.40	$.6840^{-1}$	.4647	2.403	3.40	$.1512^{-1}$	.3019	6.184
2.42	$.6630^{-1}$	.4606	2.448	3.42	$.1470^{-1}$	.2995	6.301
2.44	$.6426^{-1}$	.4565	2.494	3.44	$.1428^{-1}$	.2970	6.420
2.46	$.6229^{-1}$	.4524	2.540	3.46	$.1388^{-1}$	.2946	6.541
2.48	$.6038^{-1}$	.4484	2.588	3.48	$.1349^{-1}$	.2922	6.664
2.50	$.5853^{-1}$	.4444	2.637	3.50	$.1311^{-1}$	.2899	6.790
2.52	$.5674^{-1}$	.4405	2.686	3.52	$.1274^{-1}$	.2875	6.917
2.54	$.5500^{-1}$	.4366	2.737	3.54	$.1239^{-1}$	.2852	7.047
2.56	$.5332^{-1}$	.4328	2.789	3.56	$.1204^{-1}$	.2829	7.179
2.58	$.5169^{-1}$	.4289	2.842	3.58	$.1171^{-1}$	.2806	7.313
2.60	$.5012^{-1}$	.4252	2.896	3.60	$.1138^{-1}$	.2784	7.450
2.62	$.4859^{-1}$	.4214	2.951	3.62	$.1107^{-1}$	.2762	7.589
2.64	$.4711^{-1}$	.4177	3.007	3.64	$.1076^{-1}$	.2740	7.730
2.66	$.4568^{-1}$	.4141	3.065	3.66	$.1047^{-1}$	.2718	7.874
2.68	$.4429^{-1}$	.4104	3.123	3.68	$.1018^{-1}$	.2697	8.020
2.70	$.4295^{-1}$	.4068	3.183	3.70	$.9903^{-2}$	.2675	8.169
2.72	$.4165^{-1}$	.4033	3.244	3.72	$.9633^{-2}$	.2654	8.320
2.74	$.4039^{-1}$	.3998	3.306	3.74	$.9370^{-2}$	.2633	8.474
2.76	$.3917^{-1}$	.3963	3.370	3.76	$.9116^{-2}$	.2613	8.630
2.78	$.3799^{-1}$	.3928	3.434	3.78	$.8869^{-2}$	.2592	8.789
2.80	$.3685^{-1}$	.3894	3.500	3.80	$.8629^{-2}$	.2572	8.951
2.82	$.3574^{-1}$	.3860	3.567	3.82	$.8396^{-2}$	.2552	9.115
2.84	$.3467^{-1}$	.3827	3.636	3.84	$.8171^{-2}$	.2532	9.282
2.86	$.3363^{-1}$	.3794	3.706	3.86	$.7951^{-2}$	.2513	9.451
2.88	$.3263^{-1}$	.3761	3.777	3.88	$.7739^{-2}$	.2493	9.624
2.90	$.3165^{-1}$	.3729	3.850	3.90	$.7532^{-2}$	.2474	9.799

TABLE D.1 Isentropic Flow (*continued*)

M	p/p_0	T/T_0	A/A^*	M	p/p_0	T/T_0	A/A^*
3.92	$.7332^{-2}$	.2455	9.977	4.54	$.3288^{-2}$	.1952	17.13
3.94	$.7137^{-2}$	.2436	10.16	4.56	$.3207^{-2}$	.1938	17.42
3.96	$.6948^{-2}$	.2418	10.34	4.58	$.3129^{-2}$	.1925	17.72
3.98	$.6764^{-2}$	.2399	10.53	4.60	$.3053^{-2}$	.1911	18.02
4.00	$.6586^{-2}$	.2381	10.72	4.62	$.2978^{-2}$	.1898	18.32
4.02	$.6413^{-2}$	.2363	10.91	4.64	$.2906^{-2}$	.1885	18.63
4.04	$.6245^{-2}$	.2345	11.11	4.66	$.2836^{-2}$	.1872	18.94
4.06	$.6082^{-2}$	.2327	11.31	4.68	$.2768^{-2}$	.1859	19.26
4.08	$.5923^{-2}$	.2310	11.51	4.70	$.2701^{-2}$	.1846	19.58
4.10	$.5769^{-2}$	.2293	11.71	4.72	$.2637^{-2}$	.1833	19.91
4.12	$.5619^{-2}$	.2275	11.92	4.74	$.2573^{-2}$	.1820	20.24
4.14	$.5474^{-2}$	.2258	12.14	4.76	$.2512^{-2}$	.1808	20.58
4.16	$.5333^{-2}$	.2242	12.35	4.78	$.2452^{-2}$	.1795	20.92
4.18	$.5195^{-2}$	.2225	12.57	4.80	$.2394^{-2}$	.1783	21.26
4.20	$.5062^{-2}$	.2208	12.79	4.82	$.2338^{-2}$	.1771	21.61
4.22	$.4932^{-2}$	.2192	13.02	4.84	$.2283^{-2}$	.1759	21.97
4.24	$.4806^{-2}$	.2176	13.25	4.86	$.2229^{-2}$	.1747	22.33
4.26	$.4684^{-2}$	.2160	13.48	4.88	$.2177^{-2}$	.1735	22.70
4.28	$.4565^{-2}$	.2144	13.72	4.90	$.2126^{-2}$	.1724	23.07
4.30	$.4449^{-2}$	.2129	13.95	4.92	$.2076^{-2}$	.1712	23.44
4.32	$.4337^{-2}$	.2113	14.20	4.94	$.2028^{-2}$	.1700	23.82
4.34	$.4228^{-2}$	.2098	14.45	4.96	$.1981^{-2}$	.1689	24.21
4.36	$.4121^{-2}$	.2083	14.70	4.98	$.1935^{-2}$	.1678	24.60
4.38	$.4018^{-2}$	.2067	14.95	5.00	$.1890^{-2}$	.1667	25.00
4.40	$.3918^{-2}$	.2053	15.21	6.00	$.0633^{-2}$	.1219	53.19
4.42	$.3820^{-2}$	.2038	15.47	7.00	$.0242^{-2}$	.0926	104.14
4.44	$.3725^{-2}$	.2023	15.74	8.00	$.0102^{-2}$	.0725	109.11
4.46	$.3633^{-2}$	.2009	16.01	9.00	$.0474^{-3}$	.0582	327.19
4.48	$.3543^{-2}$	.1994	16.28	10.00	$.0236^{-3}$	.0476	535.94
4.50	$.3455^{-2}$	.1980	16.56	∞	0	0	∞
4.52	$.3370^{-2}$	.1966	16.84				

TABLE D.2 Normal-Shock Flow

M_1	M_2	p_2/p_1	T_2/T_1	p_{02}/p_{01}
1.00	1.000	1.000	1.000	1.000
1.02	.9805	1.047	1.013	1.000
1.04	.9620	1.095	1.026	.9999
1.06	.9444	1.144	1.039	.9997
1.08	.9277	1.194	1.052	.9994
1.10	.9118	1.245	1.065	.9989
1.12	.8966	1.297	1.078	.9982
1.14	.8820	1.350	1.090	.9973
1.16	.8682	1.403	1.103	.9961
1.18	.8549	1.458	1.115	.9946
1.20	.8422	1.513	1.128	.9928
1.22	.8300	1.570	1.141	.9907
1.24	.8183	1.627	1.153	.9884
1.26	.8071	1.686	1.166	.9857
1.28	.7963	1.745	1.178	.9827

TABLE D.2   Normal-Shock Flow (*continued*)

M_1	M_2	p_2/p_1	T_2/T_1	p_{02}/p_{01}
1.30	.7860	1.805	1.191	.9794
1.32	.7760	1.866	1.204	.9758
1.34	.7664	1.928	1.216	.9718
1.36	.7572	1.991	1.229	.9676
1.38	.7483	2.055	1.242	.9630
1.40	.7397	2.120	1.255	.9582
1.42	.7314	2.186	1.268	.9531
1.44	.7235	2.253	1.281	.9476
1.46	.7157	2.320	1.294	.9420
1.48	.7083	2.389	1.307	.9360
1.50	.7011	2.458	1.320	.9298
1.52	.6941	2.529	1.334	.9233
1.54	.6874	2.600	1.347	.9166
1.56	.6809	2.673	1.361	.9097
1.58	.6746	2.746	1.374	.9026
1.60	.6684	2.820	1.388	.8952
1.62	.6625	2.895	1.402	.8877
1.64	.6568	2.971	1.416	.8799
1.66	.6512	3.048	1.430	.8720
1.68	.6458	3.126	1.444	.8640
1.70	.6405	3.205	1.458	.8557
1.72	.6355	3.285	1.473	.8474
1.74	.6305	3.366	1.487	.8389
1.76	.6257	3.447	1.502	.8302
1.78	.6210	3.530	1.517	.8215
1.80	.6165	3.613	1.532	.8127
1.82	.6121	3.698	1.547	.8038
1.84	.6078	3.783	1.562	.7948
1.86	.6036	3.870	1.577	.7857
1.88	.5996	3.957	1.592	.7765
1.90	.5956	4.045	1.608	.7674
1.92	.5918	4.134	1.624	.7581
1.94	.5880	4.224	1.639	.7488
1.96	.5844	4.315	1.655	.7395
1.98	.5808	4.407	1.671	.7302
2.00	.5774	4.500	1.688	.7209
2.02	.5740	4.594	1.704	.7115
2.04	.5707	4.689	1.720	.7022
2.06	.5675	4.784	1.737	.6928
2.08	.5643	4.881	1.754	.6835
2.10	.5613	4.978	1.770	.6742
2.12	.5583	5.077	1.787	.6649
2.14	.5554	5.176	1.805	.6557
2.16	.5525	5.277	1.822	.6464
2.18	.5498	5.378	1.839	.6373
2.20	.5471	5.480	1.857	.6281
2.22	.5444	5.583	1.875	.6191
2.24	.5418	5.687	1.892	.6100
2.26	.5393	5.792	1.910	.6011
2.28	.5368	5.898	1.929	.5921

TABLE D.2 Normal-Shock Flow (*continued*)

M_1	M_2	p_2/p_1	T_2/T_1	p_{02}/p_{01}
2.30	.5344	6.005	1.947	.5833
2.32	.5321	6.113	1.965	.5745
2.34	.5297	6.222	1.984	.5658
2.36	.5275	6.331	2.002	.5572
2.38	.5253	6.442	2.021	.5486
2.40	.5231	6.553	2.040	.5401
2.42	.5210	6.666	2.059	.5317
2.44	.5189	6.779	2.079	.5234
2.46	.5169	6.894	2.098	.5152
2.48	.5149	7.009	2.118	.5071
2.50	.5130	7.125	2.138	.4990
2.52	.5111	7.242	2.157	.4991
2.54	.5092	7.360	2.177	.4832
2.56	.5074	7.479	2.198	.4754
2.58	.5056	7.599	2.218	.4677
2.60	.5039	7.720	2.238	.4601
2.62	.5022	7.842	2.259	.4526
2.64	.5005	7.965	2.280	.4452
2.66	.4988	8.088	2.301	.4379
2.68	.4972	8.213	2.322	.4307
2.70	.4956	8.338	2.343	.4236
2.72	.4941	8.465	2.364	.4166
2.74	.4926	8.592	2.386	.4097
2.76	.4911	8.721	2.407	.4028
2.78	.4896	8.850	2.429	.3961
2.80	.4882	8.980	2.451	.3895
2.82	.4868	9.111	2.473	.3829
2.84	.4854	9.243	2.496	.3765
2.86	.4840	9.376	2.518	.3701
2.88	.4827	9.510	2.540	.3639
2.90	.4814	9.645	2.563	.3577
2.92	.4801	9.781	2.586	.3517
2.94	.4788	9.918	2.609	.3457
2.96	.4776	10.06	2.632	.3398
2.98	.4764	10.19	2.656	.3340
3.00	.4752	10.33	2.679	.3283
3.02	.4740	10.47	2.703	.3327
3.04	.4729	10.62	2.726	.3172
3.06	.4717	10.76	2.750	.3118
3.08	.4706	10.90	2.774	.3065
3.10	.4695	11.05	2.799	.3012
3.12	.4685	11.19	2.823	.2960
3.14	.4674	11.34	2.848	.2910
3.16	.4664	11.48	2.872	.2860
3.18	.4654	11.63	2.897	.2811
3.20	.4643	11.78	2.922	.2762
3.22	.4634	11.93	2.947	.2715
3.24	.4624	12.08	2.972	.2668
3.26	.4614	12.23	2.998	.2622
3.28	.4605	12.38	3.023	.2577

TABLE D.2   Normal-Shock Flow (*continued*)

M_1	M_2	p_2/p_1	T_2/T_1	p_{02}/p_{01}
3.30	.4596	12.54	3.049	.2533
3.32	.4587	12.69	3.075	.2489
3.34	.4578	12.85	3.101	.2446
3.36	.4569	13.00	3.127	.2404
3.38	.4560	13.16	3.154	.2363
3.40	.4552	13.32	3.180	.2322
3.42	.4544	13.48	3.207	.2382
3.44	.4535	13.64	3.234	.2243
3.46	.4527	13.80	3.261	.2205
3.48	.4519	13.96	3.288	.2167
3.50	.4512	14.13	3.315	.2129
3.52	.4504	14.29	3.343	.2093
3.54	.4496	14.45	3.370	.2057
3.56	.4489	14.62	3.398	.2022
3.58	.4481	14.79	3.426	.1987
3.60	.4474	14.95	3.454	.1953
3.62	.4467	15.12	3.482	.1920
3.64	.4460	15.29	3.510	.1887
3.66	.4453	15.46	3.539	.1855
3.68	.4446	15.63	3.568	.1823
3.70	.4439	15.81	3.596	.1792
3.72	.4433	15.98	3.625	.1761
3.74	.4426	16.15	3.654	.1731
3.76	.4420	16.33	3.684	.1702
3.78	.4414	16.50	3.713	.1673
3.80	.4407	16.68	3.743	.1645
3.82	.4401	16.86	3.772	.1617
3.84	.4395	17.04	3.802	.1589
3.86	.4389	17.22	3.832	.1563
3.88	.4383	17.40	3.863	.1536
3.90	.4377	17.58	3.893	.1510
3.92	.4372	17.76	3.923	.1485
3.94	.4366	17.94	3.954	.1460
3.96	.4360	18.13	3.985	.1435
3.98	.4355	18.31	4.016	.1411
4.00	.4350	18.50	4.047	.1388
4.02	.4344	18.69	4.078	.1364
4.04	.4339	18.88	4.110	.1342
4.06	.4334	19.06	4.141	.1319
4.08	.4329	19.25	4.173	.1297
4.10	.4324	19.45	4.205	.1276
4.12	.4319	19.64	4.237	.1254
4.14	.4314	19.83	4.269	.1234
4.16	.4309	20.02	4.301	.1213
4.18	.4304	20.22	4.334	.1193
4.20	.4299	20.41	4.367	.1173
4.22	.4295	20.61	4.399	.1154
4.24	.4290	20.81	4.432	.1135
4.26	.4286	21.01	4.466	.1116
4.28	.4281	21.20	4.499	.1098

TABLE D.2 Normal-Shock Flow (*continued*)

M_1	M_2	p_2/p_1	T_2/T_1	p_{02}/p_{01}
4.30	.4277	21.41	4.532	.1080
4.32	.4272	21.61	4.566	.1062
4.34	.4268	21.81	4.600	.1045
4.36	.4264	22.01	4.633	.1028
4.38	.4260	22.22	4.668	.1011
4.40	.4255	22.42	4.702	$.9948^{-1}$
4.42	.4251	22.63	4.736	$.9787^{-1}$
4.44	.4247	22.83	4.771	$.9628^{-1}$
4.46	.4243	23.04	4.805	$.9473^{-1}$
4.48	.4239	23.25	4.840	$.9320^{-1}$
4.50	.4236	23.46	4.875	$.9170^{-1}$
4.52	.4232	23.67	4.910	$.9022^{-1}$
4.54	.4228	23.88	4.946	$.8878^{-1}$
4.56	.4224	24.09	4.981	$.8735^{-1}$
4.58	.4220	24.31	5.017	$.8596^{-1}$
4.60	.4217	24.52	5.052	$.8459^{-1}$
4.62	.4213	24.74	5.088	$.8324^{-1}$
4.64	.4210	24.95	5.124	$.8192^{-1}$
4.66	.4206	25.17	5.160	$.8062^{-1}$
4.68	.4203	25.39	5.197	$.7934^{-1}$
4.70	.4199	25.61	5.233	$.7809^{-1}$
4.72	.4196	25.82	5.270	$.7685^{-1}$
4.74	.4192	26.05	5.307	$.7564^{-1}$
4.76	.4189	26.27	5.344	$.7445^{-1}$
4.78	.4186	26.49	5.381	$.7329^{-1}$
4.80	.4183	26.71	5.418	$.7214^{-1}$
4.82	.4179	26.94	5.456	$.7101^{-1}$
4.84	.4176	27.16	5.494	$.6991^{-1}$
4.86	.4173	27.39	5.531	$.6882^{-1}$
4.88	.4170	27.62	5.569	$.6775^{-1}$
4.90	.4167	27.85	5.607	$.6670^{-1}$
4.92	.4164	28.07	5.646	$.6567^{-1}$
4.94	.4161	28.30	5.684	$.6465^{-1}$
4.96	.4158	28.54	5.723	$.6366^{-1}$
4.98	.4155	28.77	5.761	$.6268^{-1}$
5.00	.4152	29.00	5.800	$.6172^{-1}$
6.00	.4042	41.83	7.941	$.2965^{-1}$
7.00	.3974	57.00	10.469	$.1535^{-1}$
8.00	.3929	74.50	13.387	$.0849^{-1}$
9.00	.3898	94.33	16.693	$.0496^{-1}$
10.00	.3875	116.50	20.388	$.0304^{-1}$
∞	.3780	∞	∞	0

TABLE D.3 Prandtl-Meyer Function

M	θ	μ	M	θ	μ
1.00	0	90.00	2.02	26.929	29.67
1.02	.1257	78.64	2.04	27.476	29.35
1.04	.3510	74.06	2.06	28.020	29.04
1.06	.6367	70.63	2.08	28.560	28.74
1.08	.9680	67.81	2.10	29.097	28.44
1.10	1.336	65.38	2.12	29.631	28.14
1.12	1.735	63.23	2.14	30.161	27.86
1.14	2.160	61.31	2.16	30.689	27.58
1.16	2.607	59.55	2.18	31.212	27.30
1.18	3.074	57.94	2.20	31.732	27.04
1.20	3.558	56.44	2.22	32.250	26.77
1.22	4.057	55.05	2.24	32.763	26.51
1.24	4.569	53.75	2.26	33.273	26.26
1.26	5.093	52.53	2.28	33.780	26.01
1.28	5.627	51.38	2.30	34.283	25.77
1.30	6.170	50.28	2.32	34.783	25.53
1.32	6.721	49.25	2.34	35.279	25.30
1.34	7.280	48.27	2.36	35.771	25.07
1.36	7.844	47.33	2.38	36.261	24.85
1.38	8.413	46.44	2.40	36.746	24.62
1.40	8.987	45.58	2.42	37.229	24.41
1.42	9.565	44.77	2.44	37.708	24.19
1.44	10.146	43.98	2.46	38.183	23.99
1.46	10.731	43.23	2.48	38.655	23.78
1.48	11.317	42.51	2.50	39.124	23.58
1.50	11.905	41.81	2.52	39.589	23.38
1.52	12.495	41.14	2.54	40.050	23.18
1.54	13.086	40.49	2.56	40.509	22.99
1.56	13.677	39.87	2.58	40.963	22.81
1.58	14.269	39.27	2.60	41.415	22.62
1.60	14.861	38.68	2.62	41.863	22.44
1.62	15.452	38.12	2.64	42.307	22.26
1.64	16.043	37.57	2.66	42.749	22.08
1.66	16.633	37.04	2.68	43.187	21.91
1.68	17.222	36.53	2.70	43.621	21.74
1.70	17.810	36.03	2.72	44.053	21.57
1.72	18.397	35.55	2.74	44.481	21.41
1.74	18.981	35.08	2.76	44.906	21.24
1.76	19.565	34.62	2.78	45.327	21.08
1.78	20.146	34.18	2.80	45.746	20.92
1.80	20.725	33.75	2.82	46.161	20.77
1.82	21.302	33.33	2.84	46.573	20.62
1.84	21.877	32.92	2.86	46.982	20.47
1.86	22.449	32.52	2.88	47.388	20.32
1.88	23.019	32.13	2.90	47.790	20.17
1.90	23.586	31.76	2.92	48.190	20.03
1.92	24.151	31.39	2.94	48.586	19.89
1.94	24.712	31.03	2.96	48.980	19.75
1.96	25.271	30.68	2.98	49.370	19.61
1.98	25.827	30.33	3.00	49.757	19.47
2.00	26.380	30.00	3.02	50.142	19.34

TABLE D.3　Prandtl-Meyer Function (*continued*)

M	θ	μ	M	θ	μ
3.04	50.523	19.20	4.04	66.309	14.33
3.06	50.902	19.07	4.06	66.569	14.26
3.08	51.277	18.95	4.08	66.826	14.19
3.10	51.560	18.82	4.10	67.082	14.12
3.12	52.020	18.69	4.12	67.336	14.05
3.14	52.386	18.57	4.14	67.588	13.98
3.16	52.751	18.45	4.16	67.838	13.91
3.18	53.112	18.33	4.18	68.087	13.84
3.20	53.470	18.21	4.20	68.333	13.77
3.22	53.826	18.09	4.22	68.578	13.71
3.24	54.179	17.98	4.24	68.821	13.64
3.26	54.529	17.86	4.26	69.053	13.58
3.28	54.877	17.75	4.28	69.302	13.51
3.30	55.222	17.64	4.30	69.541	13.45
3.32	55.564	17.53	4.32	69.777	13.38
3.34	55.904	17.42	4.34	70.012	13.32
3.36	56.241	17.31	4.36	70.245	13.26
3.38	56.576	17.21	4.38	70.476	13.20
3.40	56.907	17.10	4.40	70.706	13.14
3.42	57.237	17.00	4.42	70.934	13.08
3.44	57.564	16.90	4.44	71.161	13.02
3.46	57.888	16.80	4.46	71.386	12.96
3.48	58.210	16.70	4.48	71.610	12.90
3.50	58.530	16.60	4.50	71.832	12.84
3.52	58.847	16.51	4.52	72.052	12.78
3.54	59.162	16.41	4.54	72.271	12.73
3.56	59.474	16.31	4.56	72.489	12.67
3.58	59.784	16.22	4.58	72.705	12.61
3.60	60.091	16.13	4.60	72.919	12.56
3.62	60.397	16.04	4.62	73.132	12.50
3.64	60.700	15.95	4.64	73.344	12.45
3.66	61.000	15.86	4.66	73.554	12.39
3.68	61.299	15.77	4.68	73.763	12.34
3.70	61.595	15.68	4.70	73.970	12.28
3.72	61.899	15.59	4.72	74.176	12.23
3.74	62.181	15.51	4.74	74.381	12.18
3.76	62.471	15.42	4.76	74.584	12.13
3.78	62.758	15.34	4.78	74.786	12.08
3.80	63.044	15.26	4.80	74.986	12.03
3.82	63.327	15.18	4.82	75.186	11.97
3.84	63.608	15.10	4.84	75.383	11.92
3.86	63.887	15.02	4.86	75.580	11.87
3.88	64.164	14.94	4.88	75.775	11.83
3.90	64.440	14.86	4.90	75.969	11.78
3.92	64.713	14.78	4.92	76.162	11.73
3.94	64.983	14.70	4.94	76.353	11.68
3.96	65.253	14.63	4.96	76.544	11.63
3.98	65.520	14.55	4.98	76.732	11.58
4.00	65.785	14.48	5.00	76.920	11.54
4.02	66.048	14.40			

E. NUMERICAL SOLUTIONS FOR CHAPTER 10

```
clear;
%Input diameter, Manning coefficient, and channel slope
d=5;
n=0.013;
So=0.0005;
c1=1.0;

%Set range and increments of y
y=0.01:.01:d;

%Define geometric functions
alpha=acos(1-2*y/d);
A=d^2/4*(alpha-sin(alpha).*cos(alpha));
P=alpha*d;
R=A./P;

%Define discharge function, i.e., Manning's equation
Q=c1/n*A.*R.^(2/3)*sqrt(So);

%Plot y verses Q
plot(Q,y);
xlabel('Q(y)');
ylabel('y');
```

(a) Algorithms

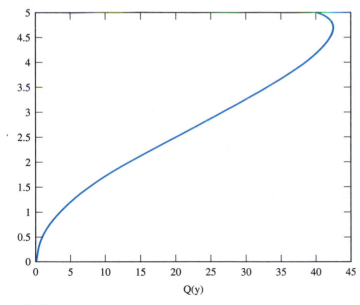

(b) Solutions

FIGURE E.1   Example 10.2 using MATLAB®

Input side slope, bottom width, discharge, critical depth, upstream depth, and gravitational constant:

$$m := 2.5 \qquad b := 0 \qquad Q := 20$$

$$y_c := 1.67 \qquad y_1 := 0.75 \qquad g := 9.81$$

Define the momentum function:

$$M(y) := \frac{y^2}{6} \cdot (2 \cdot m \cdot y + 3 \cdot b) + \frac{Q^2}{g \cdot \left(b \cdot y + m \cdot y^2\right)}$$

Find the root of M(y) within the limits $y_c < y < 3y_c$:

$$y_2 := \text{root}\left(M(y) - M(y_1), y, y_c, 3 \cdot y_c\right)$$

Hence the solution is $\qquad y_2 = 3.218$

FIGURE E.2 Example 10.11 Mathcad solution

```
clear;
%Input side slope, bottom width, discharge, critical depth,
%upstream depth, and gravitational constant
m=2.5;
b=0;
Q=20;
yc=1.67;
y1=0.75;
g=9.81;

%Set appropriate upper and lower limits
y=yc:.001:3*yc;

%Reducing Eq.10.5.16 with F=0 yields
f=y.^3+19.58./y.^2-35.23;

%Find value of y where f crosses zero
[s,t]=min(abs(f));
y2=y(t)
```

FIGURE E.3 Example 10.11 MATLAB® algorithm

(a) Solution

	A	B	C	D	E	F	G	H	I
1									
2			Depth [m]	Residual					
3		Critical	0.865	1.08665E-06					
4									
5		Normal	1.292	1.81248E-06					
6									
7									
8	Station	y [m]	A [m^2]	V [m/s]	E [m]	y$_m$ [m]	S(y$_m$)	Δx [m]	x [m]
9	1	0.865	8.358	2.632	1.218				2000
10	2	0.950	9.381	2.345	1.230	0.908	2.165E-03	-8	1992
11	3	1.050	10.631	2.069	1.268	1.000	1.527E-03	-41	1951
12	4	1.150	11.931	1.844	1.323	1.100	1.081E-03	-114	1837
13	5	1.250	13.281	1.656	1.390	1.200	7.866E-04	-357	1480
14	6	1.270	13.557	1.623	1.404	1.260	6.574E-04	-250	1230
15									

	B	C	D
2		Depth [m]	Residual
3	Critical	0.86479527789122	=22^2*(7.5+2*2.5*C3)/(9.81*(7.5*C3+2.5*C3^2)^3)-1
4			
5	Normal	1.29156940192658	=22*0.015*(7.5+2*C5*SQRT(1+2.5^2))^0.6667/((7.5*C5+2.5*C5^2)^1.6667*SQRT(0.0006))-1

	A	B	C	D	E	F
8	Station	y [m]	A [m^2]	V [m/s]	E [m]	y$_m$ [m]
9	1	0.865	=7.5*B9+B9^2*2.5	=22/C9	=B9+D9^2/(2*9.81)	
10	2	0.95	=7.5*B10+B10^2*2.5	=22/C10	=B10+D10^2/(2*9.81)	=0.5*(B9+B10)
11	3	1.05	=7.5*B11+B11^2*2.5	=22/C11	=B11+D11^2/(2*9.81)	=0.5*(B10+B11)
12	4	1.15	=7.5*B12+B12^2*2.5	=22/C12	=B12+D12^2/(2*9.81)	=0.5*(B11+B12)
13	5	1.25	=7.5*B13+B13^2*2.5	=22/C13	=B13+D13^2/(2*9.81)	=0.5*(B12+B13)
14	6	1.27	=7.5*B14+B14^2*2.5	=22/C14	=B14+D14^2/(2*9.81)	=0.5*(B13+B14)

	G	H	I
8	S(y$_m$)	Δx [m]	x [m]
9			2000
10	=22^2*0.015^2*(7.5*F10+F10^2*2.5)^-3.3333*((7.5+2*F10*SQRT(1+2.5^2))^1.3333)	=(E10-E9)/(0.0006-G10)	=I9+H10
11	=22^2*0.015^2*(7.5*F11+F11^2*2.5)^-3.3333*((7.5+2*F11*SQRT(1+2.5^2))^1.3333)	=(E11-E10)/(0.0006-G11)	=I10+H11
12	=22^2*0.015^2*(7.5*F12+F12^2*2.5)^-3.3333*((7.5+2*F12*SQRT(1+2.5^2))^1.3333)	=(E12-E11)/(0.0006-G12)	=I11+H12
13	=22^2*0.015^2*(7.5*F13+F13^2*2.5)^-3.3333*((7.5+2*F13*SQRT(1+2.5^2))^1.3333)	=(E13-E12)/(0.0006-G13)	=I12+H13
14	=22^2*0.015^2*(7.5*F14+F14^2*2.5)^-3.3333*((7.5+2*F14*SQRT(1+2.5^2))^1.3333)	=(E14-E13)/(0.0006-G14)	=I13+H14

(b) Spreadsheet formulas

FIGURE E.4   Example 10.15 using Excel®

```
clear;
%Input data
Q=22;
M=2.5;
b=7.5;
L=2000;
So=0.0006;
n=0.015;
g=9.81;

%Set range and increments of y
y=0.01:.001:5;

%Define Functions
A=b*y+M*y.^2;
B=b+2*M*y;
P=b+2*y*sqrt(1+M^2);
S=Q^2*n^2./(A.^3.3333.*P.^-1.3333);
Fr=sqrt(Q^2*B./(g*A.^3));

%Find values of y where (S-So) and (Fr-1) cross zero
[s,t]=min(abs(S-So));
yN=y(t)
[s,t]=min(abs(Fr-1));
yc=y(t)

%Reset range and increments of y to go from critical
%to normal depths
clear y;
dy=0.1;
y=yc:dy:yN

%Calculate specific energy at new y values
A=b*y+M*y.^2;
E=y+Q^2./(2*g*A.^2)

%Calculate x locations according to Eq.10.7.6
ym=(y(2:5)+y(1:4))/2;
A_ym=b*ym+M*ym.^2;
P_ym=b+2*ym*sqrt(1+M^2);
S_ym=Q^2*n^2./(A_ym.^(10/3).*P_ym.^(-4/3));

x(1)=L;
for i=2:5;
   x(i)=x(i-1)+(E(i)-E(i-1))/(So-S_ym(i-1));
end;
```

FIGURE E.5 Example 10.15 using MATLAB® (a) Algorithm

yN =

   1.2920

yc =

   0.8650

y =

   0.8650   0.9650   1.0650   1.1650   1.2650

E =

   1.2181   1.2346   1.2756   1.3326   1.4006

x =

  1.0e+003 *

   2.0000.   1.9890   1.9406   1.8076   1.3591

(b) Solution

FIGURE E.5   Example 10.15 using MATLAB® (b) Solution

TABLE E.1 Varied Flow Function[a], Equation 10.7.11

u	N 2½	3	3⅓	u	N 2½	3	3⅓
0.00	0.000	0.000	0.000	0.82	1.057	0.993	0.963
0.02	0.020	0.020	0.020	0.83	1.083	1.016	0.985
0.04	0.040	0.040	0.040	0.84	1.110	1.040	1.007
0.06	0.060	0.060	0.060	0.85	1.139	1.065	1.030
0.08	0.080	0.080	0.080	0.86	1.171	1.092	1.055
0.10	0.100	0.100	0.100	0.87	1.205	1.120	1.081
0.12	0.120	0.120	0.120	0.88	1.241	1.151	1.109
0.14	0.140	0.140	0.140	0.89	1.279	1.183	1.139
0.16	0.161	0.160	0.160	0.90	1.319	1.218	1.172
0.18	0.181	0.180	0.180	0.91	1.362	1.257	1.206
0.20	0.201	0.200	0.200	0.92	1.400	1.300	1.246
0.22	0.222	0.221	0.220	0.93	1.455	1.348	1.290
0.24	0.243	0.241	0.240	0.94	1.520	1.403	1.340
0.26	0.263	0.261	0.261	0.950	1.605	1.467	1.398
0.28	0.284	0.282	0.281	0.960	1.703	1.545	1.468
0.30	0.305	0.302	0.301	0.970	1.823	1.644	1.559
0.32	0.326	0.323	0.322	0.975	1.899	1.707	1.615
0.34	0.347	0.343	0.342	0.980	1.996	1.783	1.684
0.36	0.368	0.364	0.363	0.985	2.111	1.880	1.772
0.38	0.391	0.385	0.383	0.990	2.273	2.017	1.895
0.40	0.413	0.407	0.404	0.995	2.550	2.250	2.106
0.42	0.435	0.428	0.425	0.999	3.195	2.788	2.590
0.44	0.458	0.450	0.447	1.000	∞	∞	∞
0.46	0.481	0.472	0.469	1.001	2.786	2.184	1.907
0.48	0.504	0.494	0.490	1.005	2.144	1.649	1.425
0.50	0.528	0.517	0.512	1.010	1.867	1.419	1.218
0.52	0.553	0.540	0.535	1.015	1.705	1.286	1.099
0.54	0.578	0.563	0.557	1.020	1.602	1.191	1.014
0.56	0.604	0.587	0.580	1.03	1.436	1.060	0.896
0.58	0.631	0.612	0.604	1.04	1.321	0.967	0.813
0.60	0.658	0.637	0.628	1.05	1.242	0.896	0.749
0.61	0.673	0.650	0.641	1.06	1.166	0.838	0.697
0.62	0.686	0.663	0.653	1.07	1.111	0.790	0.651
0.63	0.700	0.676	0.666	1.08	1.059	0.749	0.618
0.64	0.716	0.690	0.679	1.09	1.012	0.713	0.586
0.65	0.731	0.703	0.692	0.10	0.973	0.681	0.558
0.66	0.746	0.717	0.705	1.11	0.939	0.652	0.532
0.67	0.762	0.731	0.718	1.12	0.907	0.626	0.509
0.68	0.777	0.746	0.732	1.13	0.878	0.602	0.488
0.69	0.795	0.761	0.746	1.14	0.851	0.581	0.479
0.70	0.811	0.776	0.760	1.15	0.824	0.561	0.452
0.71	0.828	0.791	0.775	1.16	0.802	0.542	0.436
0.72	0.845	0.807	0.790	1.17	0.782	0.525	0.421
0.73	0.863	0.823	0.805	1.18	0.760	0.509	0.406
0.74	0.881	0.840	0.821	1.19	0.740	0.494	0.393
0.75	0.900	0.857	0.837	1.20	0.723	0.480	0.381
0.76	0.919	0.874	0.853	1.22	0.692	0.454	0.358
0.77	0.940	0.892	0.870	1.24	0.662	0.431	0.338
0.78	0.962	0.911	0.887	1.26	0.633	0.410	0.320
0.79	0.985	0.930	0.905	1.28	0.609	0.391	0.303
0.80	1.008	0.950	0.924	1.30	0.587	0.373	0.289
0.81	1.032	0.971	0.943	1.32	0.568	0.357	0.275

TABLE E.1   Varied Flow Function[a], Equation 10.7.11 (*continued*)

u	N 2½	3	3⅓	u	N 2½	3	3⅓
1.34	0.549	0.342	0.262	2.20	0.220	0.107	0.071
1.36	0.531	0.329	0.251	2.3	0.204	0.098	0.064
1.38	0.513	0.316	0.239	2.4	0.190	0.089	0.057
1.40	0.496	0.304	0.229	2.5	0.179	0.082	0.052
1.42	0.481	0.293	0.220	2.6	0.169	0.076	0.048
1.44	0.467	0.282	0.211	2.7	0.160	0.070	0.043
1.46	0.455	0.272	0.203	2.8	0.150	0.065	0.040
1.48	0.444	0.263	0.196	2.9	0.142	0.060	0.037
1.50	0.432	0.255	0.188	3.0	0.135	0.056	0.034
1.55	0.405	0.235	0.172	3.5	0.106	0.041	0.024
1.60	0.380	0.218	0.158	4.0	0.087	0.031	0.017
1.65	0.359	0.203	0.145	4.5	0.072	0.025	0.013
1.70	0.340	0.189	0.135	5.0	0.062	0.019	0.010
1.75	0.322	0.177	0.125	6.0	0.048	0.014	0.007
1.80	0.308	0.166	0.116	7.0	0.038	0.010	0.005
1.85	0.293	0.156	0.108	8.0	0.031	0.008	0.004
1.90	0.279	0.147	0.102	9.0	0.027	0.006	0.003
1.95	0.268	0.139	0.095	10.0	0.022	0.005	0.002
2.00	0.257	0.132	0.089	20.0	0.015	0.002	0.001
2.10	0.238	0.119	0.079				

[a] $F(u, N) = \int_0^u du/(1 - u^N)$ with constant of integration adjusted so that $F(0, N) = 0$ and $F(\infty, N) = 0$.

Source: Henderson, Open Channel Flow, 1st,©1966. Electronically reproduced by permission of Pearson Education, Inc., Upper Saddle River, New Jersey.

F. NUMERICAL SOLUTIONS FOR CHAPTER 11

Given data:

$$L := \begin{pmatrix} 100 \\ 150 \\ 200 \end{pmatrix} \quad D := \begin{pmatrix} 0.05 \\ 0.075 \\ 0.085 \end{pmatrix} \quad e := \begin{pmatrix} 0.0001 \\ 0.0002 \\ 0.0001 \end{pmatrix} \quad K := \begin{pmatrix} 10 \\ 3 \\ 2 \end{pmatrix}$$

$$Q_T := 0.02 \quad g := 9.81 \quad v := 1 \cdot 10^{-6}$$

Define function for resistance coeffcient:

$$R(e,D,Q,L,K,C) := \frac{1.07 \cdot L}{g \cdot D^5} \cdot \left[\ln \left[0.27 \cdot \frac{e}{D} + C \cdot 5.74 \cdot \left(\frac{\pi \cdot v \cdot D}{4 \cdot Q} \right)^{0.9} \right] \right]^{-2} + \frac{K}{2 \cdot g \cdot \left(\frac{\pi \cdot D^2}{4} \right)^2}$$

Estimate initial values of unknowns:

$$W := Q_T^{\,2} \cdot \left(\sum_{i=0}^{2} \frac{1}{\sqrt{R(e_i,D_i,Q_T,L_i,K_i,0)}} \right)^{-2}$$

$$i := 0..2 \quad Q_i := \sqrt{\frac{W}{R(e_i,D_i,Q_T,L_i,K_i,1)}}$$

Solve system of equations:

$$\text{Given} \quad W = Q_T^{\,2} \cdot \left(\sum_{i=0}^{2} \frac{1}{\sqrt{R(e_i,D_i,Q_i,L_i,K_i,1)}} \right)^{-2} \quad Q_T = \sum_{i=0}^{2} Q_i$$

$$\text{Find}(W,Q_0,Q_1,Q_2) = \begin{pmatrix} 7.858 \\ 3.122 \times 10^{-3} \\ 7.274 \times 10^{-3} \\ 9.604 \times 10^{-3} \end{pmatrix}$$

FIGURE F.1   Example 11.3 Mathcad solution

$$L := \begin{pmatrix} 500 \\ 750 \\ 1000 \end{pmatrix} \qquad D := \begin{pmatrix} 0.10 \\ 0.15 \\ 0.13 \end{pmatrix} \qquad f := \begin{pmatrix} 0.025 \\ 0.020 \\ 0.018 \end{pmatrix}$$

$$H := \begin{pmatrix} 5 \\ 20 \\ 13 \end{pmatrix} \qquad K := \begin{pmatrix} 3 \\ 2 \\ 7 \end{pmatrix} \qquad g := 9.81$$

Resistance coefficients:

$$i := 0..2 \qquad R_i := \frac{8 \cdot (f_i \cdot L_i + D_i \cdot K_i)}{g \cdot \pi^2 \cdot (D_i)^5}$$

Estimate initial values of unknowns:

$$Q := \begin{pmatrix} 0.01 \\ 0.01 \\ 0.01 \end{pmatrix} \qquad H_J := 14$$

Solve for unknowns:

Given

$$H_J - H_0 = R_0 \cdot Q_0 \cdot |Q_0| \qquad\qquad H_J - H_2 = R_2 \cdot Q_2 \cdot |Q_2|$$

$$H_1 - H_J = R_1 \cdot Q_1 \cdot |Q_1| \qquad\qquad Q_0 - Q_1 + Q_2 = 0$$

$$\text{Find}(H_J, Q_0, Q_1, Q_2) = \begin{pmatrix} 15.182 \\ 9.812 \times 10^{-3} \\ 1.701 \times 10^{-2} \\ 7.201 \times 10^{-3} \end{pmatrix}$$

FIGURE F.2   Example 11.4 Mathcad solution

(a)

```
clear;
%Given Data
global H R;
L=[500 750 1000];
D=[0.10 0.15 0.13];
f=[0.025 0.020 0.018];
K=[3 2 7];
H=[5 20 13];
g=9.81;

%Evaluate equivalent lengths and resistence
coefficients
Le=D.*K./f;
R=8*f.*(L+Le)./(g*pi^2*D.^5);

%Initial estimates of unknowns x0=[HB Q1 Q2 Q3]
x0=[14 0.01 0.01 0.01];

%Call function file f.m and solve for unknowns
options=optimset('Precondbandwidth',Inf);
[x,fval] = fsolve('f',x0,options);
x
```

(b)

```
%Define each function as f(x)=0

function F = f(x);
global H R;

F=[x(1) - H(1) - R(1)*x(2)*abs(x(2));
   H(2) - x(1) - R(2)*x(3)*abs(x(3));
   x(1) - H(3) - R(3)*x(4)*abs(x(4));
   x(2) - x(3) + x(4)];
```

(c)

```
>> ex11_4
Optimization terminated successfully:
 Relative function value changing by less than
OPTIONS.TolFun

x =

   15.1818    0.0098    0.0170    0.0072
```

FIGURE F.3 Example 11.4 MATLAB solution: (a) main algorithm, (b) function subroutine, (c) output. Note that the Optimization Toolbox is required.

Given data:

$$L_1 := 50 \qquad D_1 := 0.15 \qquad f_1 := 0.02 \qquad K_1 := 2 \qquad H_1 := 10$$

$$L_2 := 100 \qquad D_2 := 0.10 \qquad f_2 := 0.015 \qquad K_2 := 1 \qquad H_2 := 30$$

$$L_3 := 300 \qquad D_3 := 0.10 \qquad f_3 := 0.025 \qquad K_3 := 1 \qquad H_3 := 15$$

$$g := 9.81 \qquad P := 20 \cdot 10^3$$

Equivalent lengths:

$$Leq_2 := \frac{D_2 \cdot K_2}{f_2} \qquad Leq_1 := \frac{D_1 \cdot K_1}{f_1} \qquad Leq_3 := \frac{D_3 \cdot K_3}{f_3}$$

Resistance coefficients:

$$R_1 := \frac{8 \cdot f_1 \cdot (L_1 + Leq_1)}{g \cdot \pi^2 \cdot D_1^5} \qquad R_2 := \frac{8 \cdot f_2 \cdot (L_2 + Leq_2)}{g \cdot \pi^2 \cdot D_2^5} \qquad R_3 := \frac{8 \cdot f_3 \cdot (L_3 + Leq_3)}{g \cdot \pi^2 \cdot D_3^5}$$

Initial estimates of unknowns:

$$Q_1 := 0.05 \qquad Q_2 := 0.05 \qquad Q_3 := 0.05 \qquad H_B := 20$$

Solve for unknowns:

Given

$$H_1 + \frac{P}{9800 \cdot Q_1} - H_B = R_1 \cdot Q_1 \cdot |Q_1| \qquad H_B - H_3 = R_3 \cdot Q_3 \cdot |Q_3|$$

$$H_B - H_2 = R_2 \cdot Q_2 \cdot |Q_2| \qquad Q_1 - Q_2 - Q_3 = 0$$

$$\mathrm{Find}(H_B, Q_1, Q_2, Q_3) = \begin{pmatrix} 43.845 \\ 0.054 \\ 0.032 \\ 0.021 \end{pmatrix}$$

FIGURE F.4 Example 11.5 Mathcad solution

(a)
```
clear;
%Given Data
global H R P;
L=[50 100 300];
D=[0.15 0.10 0.10];
f=[0.020 0.015 0.025];
K=[2 1 1];
H=[10 30 15];
g=9.81;
P=20000;

%Evaluate equivalent lengths and resistence
coefficients
Le=D.*K./f;
R=8*f.*(L+Le)./(g*pi^2*D.^5);

%Initial estimates of unknowns x0=[HB Q1 Q2 Q3]
x0=[20 0.05 0.05 0.05];

%Call function file g.m and solve for unknowns
options=optimset('Precondbandwidth',Inf);
[x,fval] = fsolve('g',x0,options);
x
```

(b)
```
%Define each function as g(x)=0

function G = g(x);
global H R P;

G=[H(1) + P/(9800*x(2)) - x(1) - R(1)*x(2)*abs(x(2));
   x(1) - H(2) - R(2)*x(3)*abs(x(3));
   x(1) - H(3) - R(3)*x(4)*abs(x(4));
   x(2) - x(3) - x(4)];
```

(c)
```
>> ex11_5
Optimization terminated successfully:
 Relative function value changing by less than
OPTIONS.TolFun

x =

    43.8449    0.0538    0.0324    0.0214
```

FIGURE F.5 Example 11.5 MATLAB® solution: (a) main algorithm, (b) function subroutine, (c) output. Note that the Optimization Toolbox is required.

	A	B	C	D	E	F	G	H	I	J	K	L	M	N	O	P
1					Iteration 1			Iteration 2			Iteration 3			Iteration 4		
2																
3			R	Q	RQ|Q|	2R|Q|	Q	RQ|Q|	2R|Q|	Q	RQ|Q|	2R|Q|	Q	RQ|Q|	2R|Q|	Q
4																
5		ΔH			20.000			20.000			20.000			20.000		
6		Pipe 4	100	-0.020	-0.040	4.000	-0.022	-0.051	4.495	-0.019	-0.038	3.899	-0.020	-0.039	3.968	-0.019
7	Loop 1	Pipe 3	200	-0.060	-0.720	24.000	-0.064	-0.809	25.440	-0.062	-0.770	24.817	-0.063	-0.787	25.098	-0.062
8		Pipe 2	500	-0.130	-8.450	130.000	-0.137	-9.433	137.352	-0.133	-8.907	133.466	-0.135	-9.150	135.276	-0.134
9		Pipe 1	100	-0.320	-10.240	64.000	-0.322	-10.399	64.495	-0.319	-10.208	63.899	-0.320	-10.230	63.968	-0.319
10																
11					0.550	222.000		-0.691	231.783		0.078	226.081		-0.206	228.309	
12																
13					ΔQ =	-2.48E-03		ΔQ =	2.98E-03		ΔQ =	-3.44E-04		ΔQ =	9.03E-04	
14																
15																
16		Pipe 2	500	0.130	8.450	130.000	0.137	9.433	137.352	0.133	8.907	133.466	0.135	9.150	135.276	0.134
17	Loop 2	Pipe 8	300	0.070	1.470	42.000	0.074	1.632	44.251	0.071	1.530	42.853	0.073	1.578	43.519	0.072
18		Pipe 7	400	-0.040	-0.640	32.000	-0.035	-0.494	28.101	-0.036	-0.519	28.823	-0.035	-0.478	27.650	-0.035
19		Pipe 6	300	-0.190	-10.830	114.000	-0.185	-10.281	111.075	-0.186	-10.382	111.617	-0.185	-10.219	110.738	-0.185
20																
21					-1.550	318.000		0.290	320.779		-0.464	316.759		0.031	317.183	
22																
23					ΔQ =	4.87E-03		ΔQ =	-9.03E-04		ΔQ =	1.47E-03		ΔQ =	-9.83E-05	
24																
25																
26		Pipe 3	200	0.060	0.720	24.000	0.064	0.809	25.440	0.062	0.770	24.817	0.063	0.787	25.098	0.062
27	Loop 3	Pipe 5	400	0.040	0.640	32.000	0.041	0.676	32.898	0.043	0.724	34.039	0.043	0.736	34.325	0.043
28		Pipe 8	300	-0.070	-1.470	42.000	-0.074	-1.632	44.251	-0.071	-1.530	42.853	-0.073	-1.578	43.519	-0.072
29																
30					-0.110	98.000		-0.146	102.589		-0.036	101.710		-0.054	102.941	
31																
32					ΔQ =	1.12E-03		ΔQ =	1.43E-03		ΔQ =	3.57E-04		ΔQ =	5.29E-04	

FIGURE F.6 Example 11.6 using Excel® (a) Solution

	A	B	C	D	E	F	G
1							
2					Iteration 1		
3			R	Q	RQIQI	2RIQI	Q
4	ΔH				20		
5							
6		Pipe 4	100	-0.02	=C6*D6*ABS(D6)	=2*C6*ABS(D6)	=D6+F13
7	Loop 1	Pipe 3	200	-0.06	=C7*D7*ABS(D7)	=2*C7*ABS(D7)	=D7+F13-F32
8		Pipe 2	500	-0.13	=C8*D8*ABS(D8)	=2*C8*ABS(D8)	=D8+F13-F23
9		Pipe 1	100	-0.32	=C9*D9*ABS(D9)	=2*C9*ABS(D9)	=D9+F13
10							
11					=SUM(E5:E9)	=SUM(F6:F9)	
12							
13					ΔQ = = -E11/F11		
14							
15							
16		Pipe 2	500	0.13	=C16*D16*ABS(D16)	=2*C16*ABS(D16)	=D16+F23-F13
17	Loop 2	Pipe 8	300	0.07	=C17*D17*ABS(D17)	=2*C17*ABS(D17)	=D17+F23-F32
18		Pipe 7	400	-0.04	=C18*D18*ABS(D18)	=2*C18*ABS(D18)	=D18+F23
19		Pipe 6	300	-0.19	=C19*D19*ABS(D19)	=2*C19*ABS(D19)	=D19+F23
20							
21					=SUM(E16:E19)	=SUM(F15:F19)	
22							
23					ΔQ = = -E21/F21		
24							
25							
26		Pipe 3	200	0.06	=C26*D26*ABS(D26)	=2*C26*ABS(D26)	=D26+F32-F13
27	Loop 3	Pipe 5	400	0.04	=C27*D27*ABS(D27)	=2*C27*ABS(D27)	=D27+F32
28		Pipe 8	300	-0.07	=C28*D28*ABS(D28)	=2*C28*ABS(D28)	=D28+F32-F23
29							
30					=SUM(E26:E28)	=SUM(F26:F28)	
31							
32					ΔQ = = -E30/F30		

FIGURE F.6 (cont'd) Example 11.6 using Excel® (b) Spreadsheet formulas

Given data:

$R_1 := 100$ $R_3 := 200$ $R_5 := 400$ $R_7 := 400$ $H_A := 50$ $Q_F := 0.15$

$R_2 := 500$ $R_4 := 100$ $R_6 := 300$ $R_8 := 300$ $H_B := 30$ $Q_G := 0.15$

Initial estimates of unknowns:

$Q_1 := 0.50$ $Q_4 := 0.50$ $Q_7 := 0.50$ $H_C := 35$ $H_F := 35$

$Q_2 := 0.50$ $Q_5 := 0.50$ $Q_8 := 0.50$ $H_D := 35$ $H_G := 35$

$Q_3 := 0.50$ $Q_6 := 0.50$ $H_E := 35$

Solve for unknowns:

Given

$$H_A - H_C = R_1 \cdot Q_1 \cdot |Q_1| \qquad H_E - H_G = R_5 \cdot Q_5 \cdot |Q_5| \qquad Q_1 = Q_2 + Q_6$$

$$H_C - H_D = R_2 \cdot Q_2 \cdot |Q_2| \qquad H_C - H_F = R_6 \cdot Q_6 \cdot |Q_6| \qquad Q_2 = Q_3 + Q_8$$

$$H_D - H_E = R_3 \cdot Q_3 \cdot |Q_3| \qquad H_F - H_G = R_7 \cdot Q_7 \cdot |Q_7| \qquad Q_3 = Q_4 + Q_5$$

$$H_E - H_B = R_4 \cdot Q_4 \cdot |Q_4| \qquad H_D - H_G = R_8 \cdot Q_8 \cdot |Q_8| \qquad Q_6 = Q_7 + Q_F$$

$$Q_7 + Q_8 + Q_5 = Q_G$$

$$\text{Find}(H_C, H_D, H_E, H_F, H_G, Q_1, Q_2, Q_3, Q_4, Q_5, Q_6, Q_7, Q_8) = \begin{pmatrix} 39.862 \\ 30.814 \\ 30.034 \\ 29.717 \\ 29.257 \\ 0.318 \\ 0.135 \\ 0.062 \\ 0.018 \\ 0.044 \\ 0.184 \\ 0.034 \\ 0.072 \end{pmatrix}$$

FIGURE F.7   Example 11.6 Mathcad solution

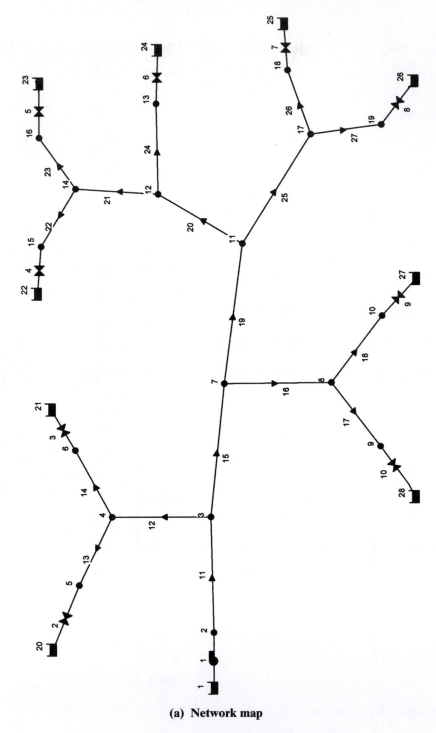

(a) Network map

FIGURE F.8 Figure 11.7a EPANET 2 solution

Link - Node Table:

Link ID	Start Node	End Node	Length m	Diameter mm	
11	2	3	195	100	
12	3	4	158	50	
13	4	5	115	50	
14	4	6	155	50	
15	3	7	188	100	
16	7	8	117	100	
17	8	9	59	50	
18	8	10	82	50	
19	7	11	130	100	
20	11	12	102	100	
21	12	14	78	100	
22	14	15	55	50	
23	14	16	60	50	
24	12	13	114	50	
25	11	17	165	100	
26	17	18	64	50	
27	17	19	127	50	
1	1	2	#N/A	#N/A	Pump
2	5	20	#N/A	50	Valve
3	6	21	#N/A	50	Valve
4	15	22	#N/A	50	Valve
5	16	23	#N/A	50	Valve
6	13	24	#N/A	50	Valve
7	18	25	#N/A	50	Valve
8	19	26	#N/A	50	Valve
9	10	27	#N/A	50	Valve
10	9	28	#N/A	50	Valve

(b) Computed results

```
Node Results:
-----------------------------------------------------------------------
Node               Demand      Head   Pressure   Quality
ID                    LPS         m          m
-----------------------------------------------------------------------
2                    0.00    135.85     105.85      0.00
3                    0.00     89.79      57.79      0.00
4                    0.00     47.59      12.59      0.00
5                    0.00     40.02       3.02      0.00
6                    0.00     36.36       3.36      0.00
7                    0.00     58.08      28.08      0.00
8                    0.00     53.37      26.37      0.00
9                    0.00     37.67      13.67      0.00
10                   0.00     36.41      10.41      0.00
11                   0.00     49.22      22.22      0.00
12                   0.00     47.26      14.26      0.00
13                   0.00     38.55       3.55      0.00
14                   0.00     46.56       6.56      0.00
15                   0.00     42.98       2.98      0.00
16                   0.00     41.74       3.74      0.00
17                   0.00     46.01      20.01      0.00
18                   0.00     34.75       8.75      0.00
19                   0.00     30.07       6.07      0.00
1                  -38.91     30.00       0.00      0.00 Reservoir
20                   3.15     37.00       0.00      0.00 Reservoir
21                   3.32     33.00       0.00      0.00 Reservoir
22                   3.13     40.00       0.00      0.00 Reservoir
23                   3.51     38.00       0.00      0.00 Reservoir
24                   3.42     35.00       0.00      0.00 Reservoir
25                   5.36     26.00       0.00      0.00 Reservoir
26                   4.47     24.00       0.00      0.00 Reservoir
27                   5.85     26.00       0.00      0.00 Reservoir
28                   6.70     24.00       0.00      0.00 Reservoir
```

FIGURE F.8 (b) Computed results (*continued*)

Link Results:
```
-----------------------------------------------------------------
Link                Flow  Velocity  Headloss   Status
ID                   LPS      m/s      m/km
-----------------------------------------------------------------
11                 38.91     4.96    236.24     Open
12                  6.47     3.30    267.05     Open
13                  3.15     1.61     65.81     Open
14                  3.32     1.69     72.49     Open
15                 32.44     4.13    168.65     Open
16                 12.55     1.60     40.21     Open
17                  6.70     3.42    266.15     Open
18                  5.85     2.98    206.84     Open
19                 19.89     2.53     68.13     Open
20                 10.06     1.28     19.27     Open
21                  6.64     0.85      8.94     Open
22                  3.13     1.60     65.00     Open
23                  3.51     1.79     80.23     Open
24                  3.42     1.74     76.36     Open
25                  9.83     1.25     19.45     Open
26                  5.36     2.73    176.01     Open
27                  4.47     2.28    125.52     Open
1                  38.91     0.00   -105.85     Open Pump
2                   3.15     1.61      3.02     Active Valve
3                   3.32     1.69      3.36     Active Valve
4                   3.13     1.60      2.98     Active Valve
5                   3.51     1.79      3.74     Active Valve
6                   3.42     1.74      3.55     Active Valve
7                   5.36     2.73      8.75     Active Valve
8                   4.47     2.28      6.07     Active Valve
9                   5.85     2.98     10.41     Active Valve
10                  6.70     3.42     13.67     Active Valve
```

(b) Computed results (*continued*)

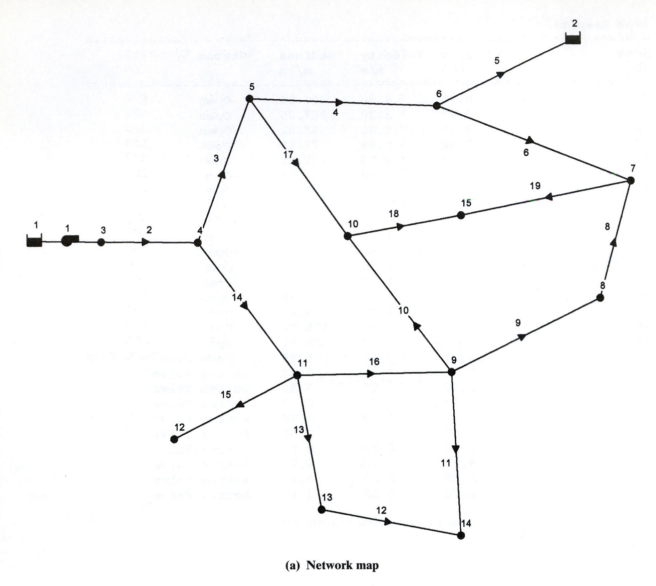

(a) Network map

FIGURE F.9   Figure 11.7b EPANET 2 solution

Link - Node Table:

Link ID	Start Node	End Node	Length m	Diameter mm	
2	3	4	3000	600	
3	4	5	1520	450	
4	5	6	1520	400	
5	6	2	305	150	
6	6	7	1680	350	
8	7	8	1070	300	
9	8	9	1680	350	
10	9	10	1680	300	
11	9	14	1380	300	
12	14	13	760	150	
13	13	11	1100	300	
14	11	4	2000	450	
15	11	12	1200	400	
16	9	11	670	380	
17	10	5	1520	350	
18	10	15	900	350	
19	15	7	1200	300	
1	1	3	#N/A	#N/A	Pump

(b) Computed results

Node Results:

Node ID	Demand LPS	Head m	Pressure m	Quality	
3	0.00	192.15	177.15	0.00	
4	0.00	155.49	109.49	0.00	
5	140.00	132.55	83.55	0.00	
6	0.00	126.03	76.03	0.00	
7	100.00	122.64	73.64	0.00	
8	100.00	123.04	77.04	0.00	
9	0.00	130.00	87.00	0.00	
10	140.00	124.13	80.13	0.00	
11	0.00	136.56	92.56	0.00	
12	55.00	136.02	96.02	0.00	
13	55.00	133.12	92.12	0.00	
14	55.00	128.61	88.61	0.00	
15	85.00	122.59	76.59	0.00	
1	-823.11	15.00	0.00	0.00	Reservoir
2	93.11	61.00	0.00	0.00	Reservoir

Link Results:

Link ID	Flow LPS	Velocity m/s	Headloss m/km	Status	
2	823.11	2.91	12.22	Open	
3	460.26	2.90	15.10	Open	
4	177.63	1.41	4.29	Open	
5	93.11	5.27	213.20	Open	
6	84.52	0.88	2.01	Open	
8	-22.87	0.32	0.37	Open	
9	-122.87	1.28	4.15	Open	
10	74.98	1.06	3.50	Open	
11	39.19	0.55	1.01	Open	
12	-15.81	0.89	5.94	Open	
13	-70.81	1.00	3.13	Open	
14	-362.85	2.28	9.46	Open	
15	55.00	0.44	0.45	Open	
16	-237.04	2.09	9.80	Open	
17	-142.63	1.48	5.54	Open	
18	77.61	0.81	1.71	Open	
19	-7.39	0.10	0.05	Open	
1	823.11	0.00	-177.15	Open	Pump

FIGURE F.9 (b) Computed results (*continued*)

Bibliography

REFERENCES

ADRIAN, R. J., "Multi-point Optical Measurements of Simultaneous Vectors in Unsteady Flow — A Review," *Int. J. Heat Fluid Flow*, Vol. 7, No. 2, June 1986, pp. 127–145.

BAKHMETEFF, B. A., *Hydraulics of Open Channels*, McGraw-Hill Book Company, New York, 1932.

BEAN, H. S., ed., *Fluid Meters: Their Theory and Application*, 6th ed., American Society of Mechanical Engineers, New York, 1971.

BENEDICT, R. P., *Fundamentals of Pipe Flow*, John Wiley & Sons, Inc., New York, 1980.

CHAPRA, S. C., and CANALE, R. P., *Numerical Methods for Engineers*, 3rd ed., McGraw-Hill Book Company, New York, 1998.

CHOW, V. T., *Open Channel Hydraulics*, McGraw-Hill Book Company, New York, 1959.

COLEMAN, HUGH W., and STEELE, W. GLENN, JR, *Experimentation and Uncertainty Analysis for Engineers*, 2nd ed. John Wiley & Sons, New York, 1998.

CROSS, H., "Analysis of Flow in Networks of Conduits or Conductors," *University of Illinois Bulletin 286*, November 1936.

GOLDSTEIN, R. J., ed., *Fluid Mechanics Measurements*, 2nd ed., Taylor and Francis, Washington, DC, 1996.

HEC-RAS River Analysis System, Hydrologic Engineering Center, U.S. Army Corps of Engineers, Davis, Calif., July 1995.

HENDERSON, F. M., *Open Channel Flow*, Macmillan Publishing Company, New York, 1966.

KARASSIK, I. J., MESSINA, J. P., COOPER, P., HEALD, C. C., eds., *Pump Handbook*, 3rd ed., McGraw-Hill Professional Publishing, 2000.

KING, H. W., and BRATER, E. F., *Handbook of Hydraulics*, 7th ed., McGraw-Hill Book Company, New York, 1996.

KLINE, S. J., ed., Proceedings of the Symposium on Uncertainty Analysis. *J. Fluids Eng.*, Vol. 107, No. 2, June 1985, pp. 153–178.

MCBEAN, E. A., and PERKINS, F. E., "Convergence Schemes in Water Profile Computation," *J. Hydraulics Div., ASCE*, Vol. 101, No. HY10, October 1975, pp. 1380–1384.

MARTIN, C. S., "Experimental Investigation of Column Separation with Rapid Valve Closure," *Proceedings, 4th International Conference on Pressure Surges*, BHRA Fluid Engineering, Cranfield, England, 1983, pp. 77–88.

ORMSBEE, L. E., and WOOD, D. J., "Hydraulic Design Algorithms for Pipe Networks," *J. Hydraulic Eng., ASCE*, Vol. 112, No. 12, December 1986, pp. 1195–1207.

ROBERSON, J. A., CASSIDY, J. J., and CHAUDHRY, M. H., *Hydraulic Engineering*, Houghton Mifflin Company, Boston, Mass., 1988.

ROBERSON, J. A., and CROWE, C. T., *Engineering Fluid Mechanics*, 6th ed., John Wiley & Sons, 1997.

ROSSMAN, L. A., *EPANETZ Users Manual*, United States Environmental Protection Agency, June 2000.

STEPANOFF, A. J., *Centrifugal and Axial Flow Pumps*, 2nd ed., John Wiley & Sons, Inc., New York, 1957.

SWAMEE, P. K., and JAIN, A. K., "Explicit Equations for Pipe-Flow Problems," *J. Hydraulics Div., ASCE*, Vol. 102, No. HY5, May 1976, pp. 657–664.

TAYLOR, G. I., "Dispersion of Soluble Matter in Solvent Flowing Slowly Through a Tube*," Proc. Royal Soc. London*, Vol. 219A, 1953, pp. 186–203.

U. S. DEPARTMENT OF INTERIOR, BUREAU OF RECLAMATION, *Design of Small Dams*, U.S. Government Printing Office, Washington D.C., 1974.

VAN DYKE, M., *An Album of Fluid Motion*, Parabolic Press, Stanford, Calif., 1982.

WARNICK, C. C., *Hydropower Engineering*, Prentice-Hall, Inc., Englewood Cliffs, N.J., 1984.

WOOD, D. J., "Algorithms for Pipe Network Analysis and Their Reliability," *Research Report 127*, Water Resources Research Institute, University of Kentucky, Lexington, Ky., 1981.

WYLIE, E. B., and STREETER, V. L., *Fluid Transients in Systems*, Prentice Hall, Englewood Cliffs, N.J., 1993.

GENERAL INTEREST

DAILY, J. W., and HARLEMAN, D. R. F., *Fluid Dynamics*, Addison-Wesley Publishing Company, Inc., Reading, Mass., 1968.

ESKINAZI, S., *Principles of Fluid Mechanics*, 2nd ed., Allyn and Bacon, Inc., Needham Heights, Mass., 1968.

FOX, R. W., and MCDONALD, A. T., *Introduction to Fluid Mechanics*, 4th ed., John Wiley & Sons, Inc., New York, 1992.

FUNG, Y. C., *Continuum Mechanics*, 2nd ed., Prentice-Hall, Inc., Englewood Cliffs, N.J., 1977.

HANSON, A. G., *Fluid Mechanics*, John Wiley & Sons, Inc., New York, 1967.

JOHN, E. A. J., *Gas Dynamics*, Allyn and Bacon, Inc., Needham Heights, Mass., 1969.

MUNSON, B. R., YOUNG, D. F., and OKIISHI, T. H., *Fundamentals of Fluid Mechanics*, 2nd ed., John Wiley & Sons, Inc., New York, 1994.

OWCZAREK, J. A., *Introduction to Fluid Mechanics*, International Textbook Company, Scranton, Pa., 1968.

POTTER, M. C., and FOSS, J. F., *Fluid Mechanics*, Great Lakes Press, Okemos, Mich., 1979.

ROBERSON, J. A., and CROWE, C. T., *Engineering Fluid Mechanics*, 6th ed., Houghton-Mifflin Company, Boston, 1996.

SABERSKY, R. H., ACOSTA, A. J., and HAUPTMAN, E. G., *Fluid Flow*, 5th ed., Macmillan Publishing Company, New York, 2001.

SCHLICHTING, H., *Boundary Layer Theory*, 8th ed., McGraw-Hill Book Company, New York, 2000.

SHAMES., I., *Mechanics of Fluids*, 3rd ed., McGraw-Hill Book Company, New York, 1992.

SHAPIRO, A. H., *The Dynamics and Thermodynamics of Compressible Fluid Flow*, Ronald Press, New York, 1953.

STREETER, V. L., and WYLIE, E. B., *Fluid Mechanics*, 8th ed., McGraw-Hill Book Company, New York, 1985.

VAN WYLEN, G. J., and SONNTAG, R. E., *Fundamentals of Classical Thermodynamics*, 2nd ed., John Wiley & Sons, Inc., New York, 1976.

WHITE, F. M. *Fluid Mechanics*, 3rd ed., McGraw-Hill Book Company, New York, 1994.

YIH, C. S., *Fluid Mechanics*, West River Press, 1988.

YUAN, S. W., *Foundations of Fluid Mechanics*, Prentice-Hall, Inc., Englewood Cliffs, N.J., 1967.

Answers to Selected Problems

Chapter 1

1.2	**(b)** M/LT^2	**(d)** ML^2/T^2	
1.4	C		
1.5	B		
1.6	**(b)** L/T^2		
1.8	**(b)** $kg \cdot m^2/s^2$	**(d)** $kg/m \cdot s$	
1.10	**(b)** 572 GPa	**(d)** 17.6 cm^3	
	(f) 76 mm^3		
1.12	A		
1.14	**(b)** 209.4 rad/s	**(d)** 0.0472 m^3/s	
	(f) 0.0083 kg/s	**(h)** 1.8×10^9 J	
1.16	C		
1.18	C		
1.20	**(b)** 142.2 kPa	**(d)** 78.8 kPa	
1.22	62.8 kPa, 1.95%		
1.24	$-46.5°C$		
1.25	B		
1.26	1296 MPa		
1.28	1000 kg/m^3, 9810 N/m^3		
1.30	-0.88%		
1.32	0.125 m^3		
1.33	D		
1.34	0.36 N/m^2		
1.36	0.69 hp		
1.38	133.2×10^{-4} N·m		
1.40	A		
1.42	$E(e^{-Cy} - 1)$		

1.46	899 m
1.48	**(b)** 1500 m/s
1.50	59.3 kPa
1.52	0.13 m
1.53	D
1.54	-0.385 mm
1.56	$\sqrt{8\sigma/\pi\rho g}$
1.58	$2\sigma\pi D$
1.60	50°C
1.62	7.4 kPa
1.64	16.83 km
1.65	C
1.66	12.03 N/m^3
1.68	27 kg
1.69	C
1.70	9333 N
1.72	25.2 m
1.74	**(b)** 15.27 m/s
1.76	69.2°C
1.78	B
1.80	**(b)** 999 kJ
1.82	2970 m, $-121.2°C$
1.84	534 kPa, -129 kJ/kg
1.86	D
1.88	**(b)** 2854 m

Chapter 2

2.2	**(b)** 78.5 kPa	**(d)** 156.0 kPa	
2.4	C		
2.6	**(b)** 10.2 m		
2.8	-36.2 kPa		
2.9	D		
2.10	1.84 Pa		
2.12	$-\rho(\vec{a} + \vec{g})$		
2.14	-0.237 Pa		
2.16	**(b)** 30.8 kPa		
2.18	70886 m		

2.20	**(b)** 3.373 m, 0.0%
2.22	**(b)** 39.9 m/s
2.23	C
2.24	30.965 kPa
2.26	17.43 cm
2.28	15.62 kPa
2.30	14.0 kPa
2.32	17.89 kPa
2.33	A
2.34	11.12 kPa

2.36 20 cm
2.38 3.4 mm
2.40 **(b)** 21.4 cm
2.42 **(b)** 98.0 kN   **(d)** 90.5 kN
2.44 167.4 kN
2.46 **(b)** (0.65, 0.9) m
2.47 B
2.48 **(b)** (0.849, −0.167) m
        (d) (2.4, 0.343) m
2.50 523 kN
2.52 $H/3$
2.54 3350 N
2.56 **(b)** 0.667 m
2.57 A
2.58 **(b)** 1.732 m
2.60 **(b)** will tip
2.62 616 kN
2.64 D
2.66 70.07 kN
2.68 4580 N/m^3

2.70 **(b)** −23.4 kN
2.72 A
2.74 7645 cm^3, 13 080 N/m^3
2.76 0.535 m
2.78 77.4°C
2.80 832.5 kg/m^3, 8167 N/m^3, 0.8325
2.82 **(b)** 0.959
2.84 **(b)** not float
2.86 unstable if $0.211 < S < 0.789$
2.88 **(b)** 11.3°
2.90 stable
2.91 A
2.92 **(b)** 59 620 Pa
2.94 **(b)** 4.8 m/s^2
2.96 **(b)** 1163 kN
2.98 **(b)** −9000 Pa, −3114 Pa, 5890 Pa
2.100 **(b)** −18 000 Pa, −14 080 Pa, 3924 Pa
2.102 **(b)** 10 780 Pa   **(d)** 57 900 Pa
2.104 **(b)** 7210 N   **(d)** 26 400 N

Chapter 3

3.6 $udy − vdx$
3.8 **(b)** 0, 8.246 m/s
3.9 D
3.10 **(b)** 104°, $(−4\hat{\mathbf{i}} + \hat{\mathbf{j}})/\sqrt{17}$
3.12 C
3.14 **(b)** 0   **(d)** $−2\hat{\mathbf{i}} + 3\hat{\mathbf{k}}$
3.16 **(b)** $\begin{bmatrix} 2 & 0 & 0 \\ 0 & 2 & 0 \\ 0 & 0 & 0 \end{bmatrix}$

        (d) $\begin{bmatrix} 1 & 3 & 0 \\ 3 & -12 & 2 \\ 0 & 2 & 2 \end{bmatrix}$

3.18 **(b)** 0, 0, 0
3.22 −0.3693°C/s
3.24 −2500 kg/m^3 · s
3.26 D
3.30 $−51.4 \times 10^{-5}\hat{\mathbf{j}} + 0.0224\,\hat{\mathbf{i}}$ m/s^2
3.32 steady: a, c, e, f, h
3.36 **(b)** inviscid   **(d)** viscous inside b.l.
        (f) viscous
3.39 C

3.40 39 000, turbulent
3.42 14 100, turbulent
3.44 3 cm
3.46 0.325, compressible
3.48 incompressible
3.49 B
3.50 104 m/s
3.52 57 m/s
3.53 C
3.54 B
3.56 **(b)** $\rho U_\infty^2/2$ **(d)** $−3\rho U_\infty^2/2$
3.58 **(b)** 50ρ **(d)** 450ρ
3.60 9.39 m/s
3.62 12.76 m/s
3.64 **(b)** 36.1 m/s   **(d)** 133.6 m/s
3.66 A
3.68 **(b)** 3.516 m/s, 193 600 Pa
3.70 **(b)** $\sqrt{2g(H − h)}$
3.72 51.0 m
3.74 −36.9 Pa
3.75 D

Chapter 4

4.3	B	**4.86**	0.903
4.6	**(b)** conservation of mass	**4.88**	71.8 hp
	(d) energy equation	**4.90**	D
4.12	1559 cm^3	**4.92**	480 hp
4.18	$\dot{m} - \rho Q$	**4.94**	A
4.20	D	**4.96**	1.85 m or 2.22 m
4.22	1.736 m/s, 19.63 kg/s, 0.0196 m^3/s	**4.98**	0.815 m
4.24	114.8 m/s, 1456.3 m/s	**4.100**	**(b)** 1.54
4.26	−83°C	**4.104**	7.37×10^{-5} m^3/s
4.28	A	**4.106**	26.7 kW
4.30	**(b)** 6.667 m/s, 427 kg/s, 0.427 m^3/s	**4.108**	**(b)** 679 N
4.32	5.08 kg/s	**4.110**	**(b)** 1479 N **(d)** 2900 N
4.34	0.82 kg/s	**4.111**	C
4.36	2.28 kg/s, 6.4 kg/s	**4.112**	6283 N
4.38	0.565 m	**4.114**	373 kN
4.40	1.44×10^{-2} kg/m^3/s	**4.116**	1986 N
4.42	0.244 m/s	**4.118**	**(b)** 3.23 m
4.44	27.3 m/s	**4.120**	1.90 m, 1.58 m/s
4.48	3.99×10^{-4} kg/s	**4.122**	135 kPa
4.50	**(b)** 8.84 mm/s	**4.124**	103 N, 759 N
4.52	1847 W	**4.126**	**(b)** 28.8 m/s
4.54	0.836°C	**4.128**	D
4.55	D	**4.130**	147.3 kW
4.56	0.01656 m^3/s	**4.132**	986 kW
4.58	1.273 m/s	**4.133**	A
4.60	2.47 m, 0.646 m	**4.134**	**(b)** 481.4 hp
4.62	7.41 m/s	**4.136**	**(b)** 41.4°, 48.2°, 188 kW
4.64	403.95 kPa	**4.138**	432 kW
4.66	32.1×10^6 Pa	**4.140**	A
4.68	5.01 m^3/s	**4.142**	647 hp
4.70	**(b)** 0.485 m^3/s	**4.144**	10.3 m/s, 2.88 m/s
4.72	C	**4.146**	2890 Pa, 818 hp
4.74	**(b)** 0.0503 m^3/s	**4.148**	2000 N, 26.8 hp
4.75	B	**4.150**	**(b)** 1.021
4.76	0.663 m	**4.152**	2.11 N
4.78	60.4 m	**4.154**	**(b)** 2.1 N
4.80	0.131 m	**4.156**	0.562
4.82	24.3 hp	**4.160**	8313 N·m/s
4.84	C	**4.162**	1.16 m^3/s

Chapter 5

5.4	$\rho\, du/dx + u\, d\rho/dx = 0$	**5.16**	$(10 - 80\, r^3)\cos\theta$
5.6	$u\, \partial\rho/\partial x + w\, \partial\rho/\partial z = 0,\ \partial u/\partial x + \partial w/\partial z = 0$	**5.18**	0.541 m/s
5.8	**(b)** $v_r = C/r^2$	**5.20**	2 m, 0.296 m/s
5.10	$v = $ const	**5.22**	**(b)** 2772 m/s^2
5.12	$-Ay$	**5.38**	$\rho\, \partial u/\partial t = \mu\, \partial^2 u/\partial y^2$
5.14	$(10 - 0.4/r^2)\sin\theta$	**5.44**	$\rho\, D\tilde{u}/Dt = k\, \nabla^2 T$

Chapter 6

6.2 **(b)** F/L^2 **(d)** FT/L^2 **(f)** FL/T

6.3 A

6.5 A

6.6 $\rho\, Vl/\mu = C$

6.8 $\sqrt{gH/C}$

6.10 $\dfrac{F_D}{\rho l^2 V^2} = f\left(\dfrac{d}{l}, \dfrac{\mu}{\rho l V}\right)$

6.12 $\dfrac{h}{d} = f\left(\dfrac{\sigma}{\gamma d^2}, \beta\right)$

6.14 $\sigma = C\, My/I$

6.16 $V = C\sqrt{gH}$

6.18 $\dfrac{\Delta p}{\rho V^2} = f\left(\dfrac{\nu}{Vd}, \dfrac{L}{d}, \dfrac{e}{d}\right)$

6.20 $\dfrac{Q}{\sqrt{gR^5}} = f\left(\dfrac{A}{R^2}, \dfrac{e}{R}, S\right)$

6.22 $\dfrac{F_D}{\rho V^2 d^2} = f\left(\dfrac{\mu}{Vd\rho}, \dfrac{e}{d}, I\right)$

6.24 $\dfrac{F_D}{\rho V^2 d^2} = f\left(\dfrac{\mu}{Vd\rho}, \dfrac{e}{d}, \dfrac{r}{d}, cd^2\right)$

6.26 $\dfrac{F_L}{\rho V^2 l_c^2} = f\left(\dfrac{c}{V}, \dfrac{t}{l_c}, \alpha\right)$

6.28 $\dfrac{F_D}{\rho V^2 d^2} = f\left(\dfrac{Vd\rho}{\mu}, \dfrac{d}{L}, \dfrac{\rho}{\rho_c}, \dfrac{V}{\omega d}\right)$

6.30 $\dfrac{T}{\rho\omega^2 d^5} = f_1\left(\dfrac{f}{\omega}, \dfrac{H}{d}, \dfrac{l}{d}, N, \dfrac{h}{d}\right)$

6.34 $\dfrac{\mu}{\rho VD} = f\left(\dfrac{H}{D}, \dfrac{l}{D}, \dfrac{gD}{V^2}\right)$

6.36 $\dfrac{y_2}{y_1} = f\left(\dfrac{gy_1}{V_1^2}\right)$

6.39 A

6.40 A

6.42 **(b)** 112 kg/s, 7390 kPa

6.44 1000 km/hr

6.48 wind tunnel

6.50 C

6.51 A

6.52 1.29 m/s, 2.16×10^6 N

6.54 18.97 m/s, 4000 N

6.56 278

6.58 $\geq$ 37.5 m/s

6.60 $\geq$ 72 m/s

6.62 276 m/s, 262 m/s, 34.6 kPa abs, 5°

6.64 750 N · m, 12.5 rpm

6.66 fl/V

6.68 gl/U^2

6.72 1/Pr Re

Chapter 7

7.2 **(b)** 0.0015 m/s

7.4 7×10^5

7.6 **(b)** 16.4 m   **(d)** 45.1 m

7.8 7.2 m, developed

7.10 3.7 m, 0.72 m

7.12 4.8 m, 0.4 m

7.17 D

7.18 74.6 m $\times$ 0.032 N/m^2, 0.04

7.20 **(b)** 510 Pa

7.22 0.211 m, 68.2 N/m^2

7.24 0.704 mm

7.26 2210, 1.1 Pa

7.28 $R/\sqrt{2}$

7.30 $r_0/2$

7.32 5.95×10^{-3} m, 0.073 Pa, 65 m

7.34 0.462 m/s, 0.0054 Pa

7.37 A

7.38 **(b)** 14.5 L/s

7.40 **(b)** 8 m/s

7.42 0.049 m^3/s, 0.015 Pa, 0.05, 490

7.44	0.0107		
7.46	**(b)** 13.6 Pa/m	**(d)** 9.05 Pa/m	
7.48	14.7 N		
7.50	1.26 N · m, 1240		
7.52	0.694 N · m		
7.54	$\omega R^2/r$, 9.38×10^{-4} N · m		
7.56	0.0134 N · s/m^2, 1070		
7.58	0.66%		
7.59	D		
7.64	16.2 m/s, -1.6 m/s, 51.2 m^2/s^2		
7.66	0.012 m^2/s		
7.68	**(b)** rough		
7.70	**(b)** 0.262 m/s		
7.72	7.23 m/s		
7.74	-40.8 kPa/m		
7.76	24.2 m/s		
7.78	**(b)** 0.0491 m^3/s		
7.80	D		
7.81	A		
7.82	**(b)** 0.0146		
7.84	**(b)** 0.04	**(d)** 0.033	
7.85	B		
7.86	D		
7.88	**(b)** 19.14 m	**(d)** 11.89 m	
7.90	**(b)** 29.4 kPa	**(d)** 17 kPa	
7.92	661×10^3 Pa		
7.94	147 kPa		
7.96	B		

7.98	**(b)** 0.0157 m^3/s	**(d)** 0.0979 m^3/s
7.100	**(b)** 13.46 m^3/s	
7.102	**(b)** 0.96 m	
7.104	A	
7.106	1.43×10^{-4} m^3/s	
7.108	**(b)** 0.257 m^3/s	
7.109	B	
7.110	$(A_2/A_1 - 1)^2$	
7.112	**(b)** 50.7 kPa	
7.114	**(b)** 29	
7.116	**(b)** 0.869	
7.117	A	
7.118	0.0044 m^3/s	
7.120	**(b)** 0.011 m^3/s	
7.122	1.3 min	
7.124	46.1 m/s	
7.126	oscillatory flow	
7.128	13.0 m	
7.130	204 hp, 12 m	
7.132	190 kW	
7.134	**(b)** -81 kPa	
7.136	1.05 MW	
7.137	C	
7.138	8 Pa	
7.140	**(b)** 1.63 m^3/s	
7.142	1.01 m/s, 0.86 m^3/s	
7.144	3.91 m	
7.146	0.66 m	

Chapter 8

8.5	C
8.6	C
8.7	B
8.8	**(b)** 9.7×10^{-5} m/s
8.12	0.25, 0.0179
8.14	5.4 N
8.15	B
8.16	D
8.18	**(b)** 50.9 m/s
8.20	81.9 kN, 2.46 MN · m
8.22	39.3 m/s, 3.76 cm
8.24	**(b)** 4.72 m/s
8.26	140 fps
8.28	9.5 m/s
8.30	**(b)** 4.29 m/s
8.32	38.8 hp
8.34	269 N

8.36	54.2 m/s
8.38	48.2 kmph
8.40	C
8.42	5.7 mm $< D < 2.28 \times 10^{-5}$ m
8.44	0.0191 m/s
8.46	C
8.48	0.822 N, 70.8%
8.50	1.04 hp, 0.26 hp
8.52	20.6 m/s
8.54	5400 N, no cavitation
8.56	0.69 m
8.58	16.7% error
8.59	B
8.60	13.8 hp
8.62	38 m/s
8.64	39.9 m/s
8.66	**(b)** -10.3%

8.70 $D\omega_3/Dt = \nu\nabla^2\omega_z$
8.74   **(b)** incompressible   **(d)** 5.43 m
8.78 $(3 - 10y)y^2/6, 6.67 \times 10^{-3}$ m²/s, no ϕ
8.80   2.098 m, 0.286 m, -280 Pa
8.82 $1 \cdot 2\hat{\mathbf{i}} + 0.4\hat{\mathbf{j}}$ m/s
8.84   **(b)** 32 kPa   **(d)** 128 kPa
8.86   **(b)** -6.96 m/s, -9.28 m/s
8.88   49.8 N/m², -530 N/m²
8.90 -106 Pa, -4.04 Pa
8.94   **(b)** 7.4 cm
8.96   **(b)** 10 cm   **(d)** 3 cm
8.98 $20 - 200 \sin^2(x/2)$ kPa
8.100 $dp/dx = 256/(2.67 - x)^3$

8.104 **(b)** $0.328\mu\, U_\infty \sqrt{U_\infty/\nu x}$
8.106 $0.451\, \rho U_\infty^2\, \mathrm{Re}_x^{-1/2}$, 33%
8.108 2.1 cm, 1 mm
8.110 **(b)** 6.4%, 13.5%
8.112 **(b)** 0.00498 Pa   **(d)** 0.00391 m/s
8.114 **(b)** 286 Pa
8.116 **(b)** 1.41 N
8.118 **(b)** 0.151 Pa
8.120 **(b)** 12.99 N/m²   **(d)** 0.0685 m
8.122 163 kN, 0.89 m
8.128 **(b)** 0.0122 m   **(d)** 0.04 m³/s/m
8.130 **(b)** 24.8 mm
8.132 3.85 mm, 0.00127 m/s, 0.011 Pa

Chapter 9

9.2 $Rk/(k - 1)$
9.8   1453 m/s
9.10   0.0069 sec
9.12   309 m
9.14   3.78 s
9.16 -0.018 N/m², 0.006°C
9.18   **(b)** 111 m/s
9.20   **(b)** 0.1393 kg/s
9.22   **(b)** 0.1423 kg/s
9.24   7.29 kg/s
9.26   191.2 kPa, 1.78 kg/s, 3.56 kg/s
9.28   97.45 kPa, 199.4 kPa abs
9.30   1.3 kg/s
9.32   494, 4.29 kPa abs
9.34   10 cm, 515 m/s, 0.013 m
9.36   8.16 cm
9.38   0.00221 m², 772°C, 3670 kPa

9.40   1260 m/s
9.42   412 kN
9.44   **(b)** 2.97, 0.477, 810 kPa, 475°C,
      3.771 kg/m³
9.48   908 m/s, 1600 kPa abs, 8.33 kg/m³
9.50   391 kPa abs, 448°C
9.52   102.5 kPa abs, 0.471 kg/s, 0.302 kg/s
9.54   341.7 kPa, 180.5 kPa, 310.5 m/s
      10.18 kPa, 605.6 m/s, 101 kPa, 159.3 m/s
9.56   6 cm, 9.2 cm
9.58   5.8 cm
9.60   **(b)** 0.631, 230 kPa abs, 303 m/s
9.62   780 m/s
9.64   555 kPa, 413 kPa
9.66   867 m/s, -156°C
9.68   **(b)** 2.56, 2.58   **(d)** 0.0122
9.70   0.0854, 0.010

Chapter 10

10.2   2.86 m
10.6   **(a)** 2.15 m, 23.5 m², 18.1 m
      (c) 1.71 m, 5.08 m², 5.93 m
10.8   **(a)** 0.012 m³/s
10.12 **(a)** -0.17 m   **(c)** -0.43 m
10.14 **(a)** 3.46 m
10.16 **(a)** 0.30 m
10.18 **(a)** 5.75 m   **(b)** 3.89 m
10.20 **(a)** 1.00 m   **(c)** 1.81 m
10.22 **(a)** 1.75 m
10.28 **(a)** 0.443 m³/s

10.30 **(a)** height 1.03 m, width 28.8 m
10.34 **(a)** 3.87 m, 363 N
10.36 7.03 m³/s
10.38 **(a)** 1.77 m, 0.339 m/s, 3.33 m/s
10.40 **(a)** 3.43 m³/s   **(c)** 52 kW
10.42 2.55 m, 338 kW
10.44 11.51 m³/s, 0.346 m
10.46 hydraulic jump and S_1 upstream of
      transition, S_3 downstream
10.48 M_3 followed by hydraulic jump to M_2
10.50 M_3 followed by hydraulic jump to M_1

10.56 M_3 followed by hydraulic jump to M_2
10.58 **(a)** 20.2 m
10.60 15.5 m^3/s, S_3
10.62 M_1 upstream, M_3 downstream
10.64 S_3 followed by hydraulic jump to S_1

10.66 **(a)** 0.87 m   **(b)** 48 m
10.68 **(a)** 720 kW   **(b)** 60 m
10.70 at a depth of 2 m, the distance is approximately 55 km
10.72 3.2 m

Chapter 11

11.2 **(a)** 750 W
11.4 **(a)** 420 mm
11.6 **(a)** 0.265 m^3/s
11.8 **(a)** 0.32 m^3/s
11.10 1.78 m^3/s, 0.285 m^3/s, 0.935 m^3/s, 1.07 MW
11.14 16 L/s, 7 L/s, 5 L/s, 4 L/s
11.16 **(c)** 28.2 L/s, 10.7 L/s, 8.6 L/s, 9.1 L/s
11.18 **(a)** 143 L/min, 290 L/min, 166 L/min
11.20 1.05 m^3/s, 90.8 m, 73.6 m, 810 kW
11.22 **(a)** 2.27 m^3/s, 0.811 m^3/s, 0.580 m^3/s, 0.880 m^3/s

11.24 **(a)** 0.77 m   **(c)** 22.7 m^3/h
11.26 **(a)** 36 m^3/h
11.28 2.30 flow units, 0.43 flow units, 2.73 flow units
11.32 21.6 L/s, 6.6 L/s, 28.4 L/s
11.34 2.77 m^3/s
11.36 9.3 L/s, 1.6 L/s, 7.7 L/s, 0.6 L/s, 7.1 L/s
11.38 **(d)** 2.06 m^3/s
11.44 69 s
11.46 1.57 L/s, 8.03 s
11.48 **(a)** 1090 m/s   **(b)** 590 kPa

Chapter 12

12.2 42.1 L/s, 7.22 m, 35.6 N · m, 2980 W
12.4 14.46 m, 1.22 hp
12.6 1.23 m
12.8 17.6 m, 172 kW
12.12 computed specific speed is 1.30, use a mixed flow pump
12.14 **(a)** 5.38 m, 0.69 m, 111.5 hp, 6.88 m
12.16 computed specific speed is 0.751, use a radial flow pump
12.18 the speed and diameter are 1.2 times greater
12.20 mixed flow pump, 0.251 m, 282 rad/s
12.22 2.75 m^3/min, 12.6 m, 7.2 kW
12.24 2.44 m^3/min, 10.0 m, 5.1 kW

12.26 computed specific speed is 5.2
12.28 **(a)** 280 m^3/h, 64 kW, 8.3 m
12.30 three pumps, 155 hp
12.32 **(a)** 5.67 m
12.34 **(a)** radial flow pump, 4680 rev/min, 0.09 m
12.36 6.1°, 2.83 × 10^7 N · m, 243 m, 360 MW
12.38 0.253 m, 380 rev/min
12.40 190 kW, Francis, 2900 rev/min
12.42 4.78 m^3/s, 1.95 m, 244 mm, 3 jets, 0.204
12.44 6.62 m, Francis or pump/turbine
12.46 **(a)** 1.45 m
12.48 two units, 4.6 MW
12.50 **(a)** 25 kW
   (c) 0.30 m, 680 rev/min, 26 kW

Chapter 13

13.2 **(a)** 46.8
13.4 3.33 × 10^{-6} m^3/s, 0.678 m/s
13.6 0.0592 m^3/s, 7.54 m/s

13.8 **(a)** 0.064 m^3/s
13.10 **(a)** 0.0974 m^3/s **(b)** 0.035 m^3/s
13.14 0.00396, 0.490

Chapter 14

14.2 **(a)** temperature changes in x-direction
    (c) 4.3×10^3 W/m^2
14.4 **(a)** first-order reaction
14.6 **(a)** 0.62 mg/L
14.8 **(a)** concentration after one year is
    0.16 mg/L

14.10 **(a)** 92.9 W   **(c)** 5.7×10^{-7} m
14.12 -2.5×10^2 W/m^2
14.14 **(a)** 5 mg/L
14.16 **(d)** approximately 8.5 km
14.18 **(a)** 0.013 cm^2/s at 15 m
14.20 **(a)** 3.6×10^{-2} cm^2/s

Index

Credits

Chapter 1 2: CO photo, U.S. National Aeronautics and Space Administration; CO photo, California Department of Water Resource; CO photo, U.S. National Aeronautics and Space Administration. 22: Figure 1.11, Food Pix, Inc. 23: Figure 1.12, S. A. Kinnas.

Chapter 2 38: CO photo, U.S. Bureau of Reclamation.

Chapter 3 86: CO photo, Ottawa River Rafting Co. 90: Figure 3.1, Wallet and Ruellan/M. C. Vasseur. 91: Figure 3.2, Sadatashi Taneda. 104: Figure 3.13, Burkhalter and Koschmieder. 104: Figure 3.14, Parabolic Press. 111: Figure 3.21, Japan Society of Mechanical Engineers/Pergamon Press.

Chapter 4 126: CO photo, Corbis, Inc.

Chapter 5 202: CO photo, Corbis, Inc. 221: Figure 5.4, Pullin & Perry/The Parabolic Press.

Chapter 6 236: CO photo, Calspan Corporation. 239: Figure 6.1, Fluid Mechanics/Diffusion Laboratory, Colorado State University. 252: Figure 6.5, U.S. Army Corps of Engineers Waterways Experiment Station.

Chapter 7 270: CO photo, Alaska Division of Tourism. 275: Figure 7.3, PhD Thesis of Dr. Jack Backus, Michigan State University. 308: Figure 7.13, L. F. Moody, Trans. ASME, Vol. 66, 1944. 315: Figure 7.15, A. H. Gibson, *Engineering*, Vol. 93, 1912.

Chapter 8 344: CO photo, Goodyear Inc. 344: Figure 8.1, Parabolic Press. 353: Figure 8.8, U.S. Navy photographs. 359: Figure 8.10, NACA Rep. 1191, by A. Roshko, 1954. 360: Figure 8.11, Thomas Corke & Hassan Nagib; Sadatoshi Taneda. 386: Figure 8.23, R. E. Falco. 407: Figure 8.27, R. E. Falco.

Chapter 9 425: CO photo, U.S. National Aeronautics and Space Administration. 443: Figure 9.8, A. C. Charters.

Chapter 10 474: CO photo, U.S. Bureau of Reclamation.

Chapter 11 544: CO photo, Southwestern Refining Company. 582: Figure 11.10, After Rust, 1979. 583: Figure 11.12, After Martin, 1983.

Chapter 12 600: CO photo, Voith Hydro, Inc. 612: Figure 12.6, Sulzer Pumps Ltd. 614: Figure 12.8, Sulzer Pumps Ltd. 616: Figure 12.9, Sulzer Pumps Ltd. 617: Figure 12.10, Sulzer Pumps Ltd. 629: Figure 12.15, Karassik. 636: Figure 12.19, Sulzer Pumps Ltd. 638: Figure 12.21, Voith Siemens Hydro Power Generation, Inc. 640: Figure 12.23, Gilbert Gilkes and Gordon, Ltd. 641: Figure 12.24, Gilbert Gilkes and Gordon, Inc. 642: Figure 12.25, Voith Siemens Hydro Power Generation, Inc. 643: Figure 12.26, Voith Siemens Hydro Power Generation, Inc. 644: Figure 12.27, Voith Siemens Hydro Power Generation, Inc. 646: Figure 12.30, Gilbert Gilkes and Gordon, Ltd. 647: Figure 12.31, Gilbert Gilkes and Gordon, Ltd. 649: Figure 12.32, Voith Siemens Hydro Power Generation, Inc. 650: Figure 12.33, Voith Siemens Hydro Power Generation, Inc.

Chapter 13 660: CO photo, Corbis Inc. 674: Figure 13.10, Roberson and Crowe, 1990. 680: Figure 13.18, M. Koochesfahani. 681: Figure 13.19, R. Falco. 682: Figure 13.20, R. Adrian. 683: Figure 13.21, R. Bouwmeester. 684: Figure 13.22, C. Gendrich/M. Koochesfahani. 685: Figure 13.23, C. MacKinnon/M. Koochesahani. 686: Figure 13.24, R. Goldstein. 686: Figure 13.25, R. Goldstein. 687: Figure 13.26, R. Goldstein.

Chapter 14 702: CO photo, Corbis, Inc.

Chapter 15 752: CO photo, Courtesy of Y.-L., Lin, T.I-P., Shih, M. A., Stephens, and M. K., Chyu, "A Numerical Study of Flow and Heat Transfer in a Smooth and a Ribbed U-Duct with and without Rotation," *ASME J. of Heat Transfer*, Vol. 123, April 2001, pp. 219–232.